Automotive Brake Systems

Classroom Manual

Fourth Edition

Chek-Chart

Jeffrey Rehkopf
Revision Author

James D. Halderman
Series Advisor

Upper Saddle River, New Jersey
Columbus, Ohio

Library of Congress Cataloging-in-Publication Data
Rehkopf, Jeffrey.
   Automotive brake systems. Classroom manual / Jeffrey Rehkopf, revision author ; James D. Halderman, series editor.—4th ed.
     p. cm
   Includes index.
   ISBN 0-13-048203-X
 1. Automobiles—Brakes. I. Halderman, James D., 1943- Automotive brake systems. II. Title.
   TL269.R44 2006
   629.2'46—dc22

                                                                                                                                                                            2005020689

Acquisitions Editor: Tim Peyton
Associate Editor: Jill Jones-Renger
Editorial Assistant: Nancy Kesterson
Production Coordination: Carlisle Publishers Services
Production Editor: Christine Buckendahl
Design Coordinator: Diane Ernsberger
Cover Designer: Jeff Vanik
Cover Art: Corel
Production Manager: Deidra Schwartz
Marketing Manager: Ben Leonard
Senior Marketing Coordinator: Liz Farrell
Marketing Assistant: Les Roberts

This book was set in Times by Carlisle Communications, Ltd. It was printed and bound by Bind-Rite Graphics. The cover was printed by The Lehigh Press, Inc.

Portions of materials contained herein have been reprinted with permission of General Motors Corporation, Service and Parts Operations License Agreement #0510862.

Copyright © 2006 by Pearson Education, Inc., Upper Saddle River, New Jersey 07458. Pearson Prentice Hall. All rights reserved. Printed in the United States of America. This publication is protected by Copyright and permission should be obtained from the publisher prior to any prohibited reproduction, storage in a retrieval system, or transmission in any form or by any means, electronic, mechanical, photocopying, recording, or likewise. For information regarding permission(s), write to: Rights and Permissions Department.

Pearson Prentice Hall™ is a trademark of Pearson Education, Inc.
Pearson® is a registered trademark of Pearson plc
Prentice Hall® is a registered trademark of Pearson Education, Inc.

Pearson Education Ltd.
Pearson Education Singapore Pte. Ltd.
Pearson Education Canada, Ltd.
Pearson Education—Japan

Pearson Education Australia Pty. Limited
Pearson Education North Asia Ltd.
Pearson Educación de Mexico, S.A. de C.V.
Pearson Education Malaysia Pte. Ltd.

10 9 8 7 6 5 4 3 2
0-13-048203-X

Introduction

Automotive Brake Systems is part of the Chek-Chart automotive series. The entire series is job-oriented and is designed especially for students who intend to work in the automotive service profession. The package for each course consists of two volumes, a *Classroom Manual* and a *Shop Manual*.

This fourth edition of *Automotive Brake Systems* has been revised to include in-depth coverage of the latest developments in automotive braking systems. Students will be able to use the knowledge gained from these books and from the instructor to diagnose and repair automotive brake systems used on today's automobiles.

This package retains the traditional thoroughness and readability of the Chek-Chart automotive series. Furthermore, both the *Classroom Manual* and the *Shop Manual,* as well as the *Instructor's Manual,* have been greatly enhanced.

CLASSROOM MANUAL

New features in the *Classroom Manual* include:

- More than 60 new figures and photographs
- Updated information on Federal regulations
- Updated information on brake fluids
- Expanded coverage of antilock brake systems (ABS)
- Introduction to advanced antilock brake functions, including traction control, stability control, and braking assist systems
- Introduction to new vehicle communication systems related to ABS

SHOP MANUAL

The chapters of the revised *Shop Manual* correlate closely with the *Classroom Manual*. There are over 130 new or revised illustrations. New features in the *Shop Manual* include:

- Addition of shop safety practices in Chapter 1
- Consolidate repair procedures for disc and drum brakes
- New photo sequences showing brake pad replacement on three common types of brake systems
- New photo sequence on the use of the on-vehicle-lathe
- Update on the latest antilock brake systems and components

INSTRUCTOR'S MANUAL

The *Instructor's Manual* includes task sheets that cover many of the NATEF tasks for *Automotive Brake Systems*. Instructors may reproduce these task sheets for use by the students in the lab or during an internship. The *Instructor's Manual* also includes a test bank and answers to end-of-chapter questions in the *Classroom Manual.*

The *Instructor's Resource* CD that accompanies the *Instructor's Manual* includes Microsoft® PowerPoint® presentations and photographs that appear in the *Classroom Manual* and *Shop Manual*. These high-resolution photographs are suitable for projection or reproduction.

Because of the comprehensive material, hundreds of high-quality illustrations, and inclusion of the latest automotive technology, these books will keep their value over the years. In fact, *Automotive Brake Systems* will form the core of the master technician's professional library.

How to Use This Book

WHY ARE THERE TWO MANUALS?

This two-volume text—*Automotive Brake Systems*—is unlike most other textbooks. It is actually two books, a *Classroom Manual* and a *Shop Manual* that should be used together. The *Classroom Manual* teaches you what you need to know about brake system theory and the braking systems on cars. The *Shop Manual* will show you how to repair and adjust complete systems, as well as individual components.

WHAT IS IN THESE MANUALS?

These key features of the *Classroom Manual* make it easier to learn and remember the material:

- Each chapter is based on detailed learning objectives, which are listed in the beginning of each chapter.
- Each chapter is divided into self-contained sections for easier understanding and review. This organization clearly shows which parts make up which systems, and how various parts or systems that perform the same task differ or are the same.
- Most parts and processes are fully illustrated with drawings and photographs.
- A list of Key Terms is located at the beginning of each chapter. These are printed in **boldface type** in the text and are defined in a glossary at the end of the manual. Use these words to build the vocabulary needed to understand the text.
- Review Questions follow each chapter. Use them to test your knowledge of the material covered.
- A brief summary at the end of each chapter helps you review for exams.

The *Shop Manual* has detailed instructions on the test, service, and overhaul of automotive brake systems and their components. These are easy to understand and often include step-by-step explanations of the procedure. Key features of the *Shop Manual* include:

- Each chapter is based upon ASE/NATEF tasks, which are listed in the beginning of each chapter.
- Helpful information on the use and maintenance of shop tools and test equipment.
- Detailed safety precautions.
- Clear illustrations and diagrams to help you locate trouble spots while learning to read the service literature.
- Test procedures and troubleshooting hints that help you work better and faster.
- Repair tips used by professionals, presented clearly and accurately.

WHERE SHOULD I BEGIN?

If you already know something about automotive brake systems and know how to repair them, you will find that this book is a helpful review. If you are just starting in car repair, then the book will give you a solid foundation on which to develop professional-level skills.

Your instructor will design a course to take advantage of what you already know, and what facilities and equipment are available to work with. You may be asked to read certain chapters of this manual out of order. That is fine; the important thing is to fully understand each subject before you move on to the next. Study the vocabulary words, and use the review questions to help you comprehend the material.

While reading the *Classroom Manual,* refer to your *Shop Manual* and relate the descriptive text to the service procedures. When working on actual car brake systems, look back to the *Classroom Manual* to keep basic information fresh in your mind. Working on such a complicated modern brake system isn't always easy. Take advantage of the information in the *Classroom Manual,* the procedures in the *Shop Manual,* and the knowledge of your instructor to help you.

How to Use This Book

Remember that the *Shop Manual* is a good book for work, not just a good workbook. Keep it on hand while you're working on a brake system. For ease of use, the *Shop Manual* will fold flat on the workbench or under the car, and it can withstand quite a bit of rough handling.

When you perform actual test and repair procedures, you need a complete and accurate source of manufacturer specifications and procedures for the specific vehicle. As the source for these specifications, most automotive repair shops have the annual service information (on paper, CD, or Internet formats) from the vehicle manufacturer or an independent guide.

Acknowledgments

In producing this series of textbooks for automotive technicians, Chek-Chart has drawn extensively on the technical and editorial knowledge of the vehicle manufacturers and their suppliers. Automotive design is a technical, fast-changing field, and Chek-Chart gratefully acknowledges the help of the following companies and organizations that provided help and information allowing us to present the most up-to-date text and illustrations possible. These companies and organizations are not responsible for any errors or omissions in the instructions or illustrations, or for changes in procedures or specifications made by the manufacturers or suppliers, contained in this book or any other Chek-Chart product:

Alfred Teves Technologies, Inc.
Ammco Tools, Inc.
Audi of America, Inc.
Bendix Aftermarket Brake Division, Allied Automotive
Benwil Industries, Inc.
DaimlerChrysler Motors Corporation
DMC Technical Products/Diemolding Corporation
Earl's Performance Products
Easco/K-D Tools
EIS Division, Parker Hannifin Corporation
Failure Analysis Associates
Ford Motor Company
FMC Corporation, Automotive Service Equipment Division
General Motors Corporation
    AC-Delco Division
    Buick Motor Division
    Cadillac Motor Car Division
    Chevrolet Motor Division
    Oldsmobile Division
    Pontiac-GMC Division

Goodyear Tire & Rubber Company
Hako Minuteman
Honda Motor Company, Inc.
Kelsey-Hayes Group
Kent-Moore Tool Group, Sealed Power Corporation
Kwik-Way Manufacturing Company
MAC Tools, Inc.
Mazda Motor Corporation
Mercedes-Benz USA, Inc.
Nissan Motors
OTC Division, Sealed Power Corporation
Porsche Cars North America
Robert Bosch Corporation
Snap-On Tools Corporation
Stainless Steel Brakes Corporation
Subaru of America, Inc.
Sumitomo Corporation
Toyota Motor Corporation
Vim Tools, Durston Manufacturing
The White Lung Association of New York

The comments, suggestions, and assistance of the following reviewers and technical consultants were invaluable:

Jerry Ciraolo, *Skyline College, San Bruno, CA*
Steve Martin, *Snap-on Diagnostics, San Jose, CA*
Lance J. David, *College of Lake County*
James D. Halderman

The authors have made every effort to ensure that the material in this book is as accurate and up-to-date as possible. However, Chek-Chart, Prentice Hall, or any related companies are not responsible for mistakes or omissions, or for changes in procedures or specifications by the manufacturers or suppliers.

Contents

Chapter 1 — Brake System Overview 1
    Objectives 1
    Key Terms 1
    Introduction 1
    Early Brake Designs 2
    Automotive Brake Designs 3
    Brake System Operation 6
    Service Brakes 6
    Wheel Friction Assemblies 8
    Parking Brakes 10
    Methods of Brake Actuation 11
    Antilock Brakes 17
    Summary 20
    Review Questions 21

Chapter 2 — Brake Legal and Health Issues 23
    Objectives 23
    Key Terms 23
    Brakes and the Law 23
    Federal Brake Standards 24
    Brake Repair and the Law 26
    Brakes and Health 26
    Asbestos Exposure 27
    Asbestos Precautions 29
    Asbestos Waste Disposal 32
    Chemical Poisoning 32
    Chemical Precautions 33
    Health Care Rights 33
    Summary 34
    Review Questions 35

Chapter 3 — Principles of Brake Operation 37
    Objectives 37
    Key Terms 37
    Introduction 37
    Energy Principles 38
    Kinetic Energy 38
    Inertia 40
    Mechanical Principles 41
    Hydraulic Principles 42
    Friction Principles 48

    Friction and Heat 51
    Brake Fade 52
    Summary 54
    Review Questions 55

Chapter 4 — Brake Fluid and Lines 57
    Objectives 57
    Key Terms 57
    Introduction 57
    Brake Fluid 58
    Brake Fluid Specifications 58
    Brake Fluid Types 61
    Brake Fluid Storage and Handling 63
    Brake Bleeding 64
    Brake Fluid Changes 66
    Brake Lines 67
    Brake Line Fittings 70
    Brake Line Routing 73
    Summary 73
    Review Questions 75

Chapter 5 — Pedal Assemblies and Master Cylinders 77
    Objectives 77
    Key Terms 77
    Introduction 77
    Pedal Assemblies 78
    Master Cylinder Construction 82
    Dual-Circuit Brake Systems 87
    Master Cylinder Operation 89
    Summary 95
    Review Questions 97

Chapter 6 — Hydraulic Valves and Switches 99
    Objectives 99
    Key Terms 99
    Introduction 99
    Hydraulic Valves 100
    Residual Pressure Check Valve 100
    Metering Valve 101
    Proportioning Valve 104
    Combination Valves 108

ix

Brake System Switches 108
Pressure Differential Switch 109
Fluid Level Switch 111
Stoplight Switch 112
Summary 115
Review Questions 116

Chapter 7 — Wheel Cylinders and Brake Caliper Hydraulics 119
Objectives 119
Key Terms 119
Introduction 119
Wheel Cylinders 119
Wheel Cylinder Designs 121
Wheel Cylinder Operation 122
Brake Calipers 122
Brake Caliper Body 123
Brake Caliper Pistons 126
Caliper Piston Seals 128
Brake Caliper Dust Boots 131
Brake Caliper Operation 132
Summary 133
Review Questions 134

Chapter 8 — Drum Brake Friction Assemblies 137
Objectives 137
Key Terms 137
Drum Brake Advantages 137
Drum Brake Disadvantages 138
Drum Brake Construction 140
Drum Brake Design 144
Non-Servo Brakes 144
Dual-Servo Brakes 147
Uni-Servo Brake 150
Brake Adjusters 150
Automatic Brake Adjusters 150
Manual Brake Adjusters 156
Summary 157
Review Questions 159

Chapter 9 — Disc Brake Friction Assemblies 161
Objectives 161
Key Terms 161
Introduction 161
Disc Brake Advantages 162
Disc Brake Disadvantages 165
Disc Brake Construction 166
Disc Brake Design 168
Rear Disc Brakes 173
Inboard Disc Brakes 174
Summary 174
Review Questions 176

Chapter 10 — Brake Shoes and Pads 179
Objectives 179
Key Terms 179
Introduction 179
Drum Brake Shoe Construction 180
Disc Brake Pad Construction 182
Shoe and Pad Friction Materials 184
Lining Material Coefficient of Friction 188
Lining Assembly Methods 190
Brake Shoe-to-Drum Fit 192
Summary 194
Review Questions 196

Chapter 11 — Brake Drums and Rotors 199
Objectives 199
Key Terms 199
Introduction 199
Brake Drum Construction 200
Brake Rotor Construction 202
Drum and Rotor Mounting Methods 205
Brake Drum and Rotor Wear 207
Brake Drum and Rotor Damage 208
Brake Drum and Rotor Distortion 210
Drum and Rotor Refinishing 214
Drum and Rotor Metal Removal Limits 216
Special Drum Refinishing Considerations 217
Special Rotor Refinishing Considerations 217
Summary 219
Review Questions 220

Chapter 12 — Parking Brakes 223
Objectives 223
Key Terms 223
Introduction 223
Pedals, Levers, and Handles 224
Warning Lights and Switches 226
Parking Brake Linkages 226
Drum Parking Brakes 231
Caliper-Actuated Disc Parking Brakes 234
Driveline Auxiliary Parking Brakes 237
Summary 239
Review Questions 240

Chapter 13 — Power Brakes 243
Objectives 243
Key Terms 243
Introduction 243
The Need for Power Brakes 244
Ways to Increase Braking Power 244
Air Pressure—High and Low 245
Vacuum Booster Theory 248
Integral Vacuum Boosters 250
Multiplier Vacuum Boosters 256
Hydraulic Boosters 256

Contents

Mechanical-Hydraulic Boosters 256
Electro-Hydraulic Boosters 261
The Powermaster Brake Booster 261
Powermaster Operation 262
Dual-Power Brake Systems 265
Summary 265
Review Questions 267

Chapter 14 — Antilock Brake Basics 269
Objectives 269
Key Terms 269
Introduction 269
ABS Characteristics 269
ABS Operation 272
System Configurations 273
ABS Components 276
Advanced ABS Functions 286
Summary 289
Review Questions 291

Chapter 15 — Antilock Brake Systems 293
Objectives 293
Key Term 293
Introduction 293
Bendix ABS 293
Bosch ABS 304
Delphi Chassis (Delco Moraine) ABS 319
Kelsey-Hayes ABS 333

Nippondenso ABS 344
Sumitomo ABS 345
Teves ABS 347
Toyota ABS 360
Review Questions 368

Chapter 16 — The Brake System and Vehicle Suspension 371
Objectives 371
Key Terms 371
Introduction 371
Tires and Braking 372
Tire Construction 372
Tire Size 373
Tire Tread Design 375
Tire Inflation 376
Tire Runout 376
Wheels and Braking 377
Wheel Runout 379
Brake Shrouding 379
Suspension and Braking 380
Wheel Bearings and Braking 380
Summary 383
Review Questions 384

Glossary 387

Index 395

1

Brake System Overview

OBJECTIVES

Upon completion and review of this chapter, you will be able to:

- Explain the importance of brakes in terms of safety.
- Identify the major parts of the brake system.
- Explain the purpose of the service brakes.
- Explain the purpose of the parking brakes.
- Describe the operation of drum brakes in general terms.
- Describe the operation of disc brakes in general terms.
- Explain the purpose of the power booster.
- Describe the operation of the master cylinder in general terms.
- Explain the purpose of hydraulic lines.
- Explain the purpose of hydraulic valves and switches.
- List and explain the four basic methods of brake actuation.

KEY TERMS

actuating system	drum brake
air chambers	dual master cylinder
air-hydraulic system	emergency brakes
annular electric brake	foundation brakes
backing plate	master cylinder
base brakes	parking brake
brake band	power booster
brake caliper	rotor
brake drum	service brakes
brake lines	spot magnetic electric brake
brake lining	
brake pad	straight-air system
brake shoe	wheel cylinder
controller	wheel friction assemblies
disc brake	

INTRODUCTION

The brake system is the single most important safety feature of every vehicle on the road. Every time someone's foot pushes down on a brake pedal, the health and welfare of that person, those riding with him or her, and those in surrounding vehicles depend on the brakes working properly and slowing or stopping the car. No matter how well a vehicle performs in any other area, it remains unsafe if it cannot be brought to a fast, straight, and controlled stop whenever necessary.

This chapter covers the evolution of automotive brakes, and describes the typical brake system found on most late-model cars and trucks. It also looks at the four methods commonly used to actuate vehicle brake systems, and discusses the advantages and disadvantages of each. This overview provides the foundation for the detailed examinations of brake system components that follow in later chapters.

EARLY BRAKE DESIGNS

Like every other part of the modern automobile, brake systems evolved from relatively simple beginnings. Many other types of vehicles had brake systems long before the automobile was invented, and some of the braking technology used on these earlier vehicles was carried over to the first cars. Although the actual brake components on new cars are far more advanced than those on older vehicles, the same basic principles of brake operation apply equally to old and new designs.

Shortly after the invention of the wheel, a new challenge emerged: carts and wagons with wheels were reluctant to slow down and stop once set in motion. This fact led to one of the first known vehicular brakes, the locked wheel, figure 1-1. This type of brake consisted of a hole drilled in the side of the wooden wheel disc into which a stick was manually inserted to jam against the wagon chassis and stop the wheel's rotation. Although this type of brake was not practical at much more than a walking pace, it could prevent a wagon from rolling down a hill if the animal pulling it tired, and it also worked well as a parking brake.

By the time of the Roman Empire, civilization had advanced and so had braking technology. It is believed that the Romans used a chariot brake, figure 1-2, that employed lengths of chain or rope wrapped around each hub and attached to a cross-lever at the front of the chariot. To slow the chariot, the driver pulled the chain on either side, which would then tighten the chain around the hubs. The major advance of this brake was that it allowed the amount of braking force to vary—the harder the chain was pulled the greater the braking action. It also provided a form of self-energizing brake action as the rotation of the wheel helped wind the chain more tightly around the hub.

The next advance in vehicle brakes saw widespread use by the 1800s. The typical farm wagon of the period used a wooden **brake shoe** that was pressed against the wheel rim by a lever-operated linkage, figure 1-3. Applying the brake at the wheel rim increased the amount of stopping power available at the axle, but the wooden brake shoes wore quickly and provided only a moderate amount of friction, especially when wet. Although later designs used different types of **brake lining** materials

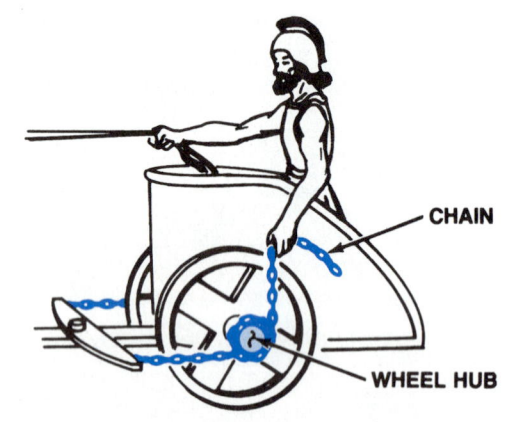

Figure 1-2. The chariot brake allowed braking force to be varied and provided a self-energizing effect.

Figure 1-1. The locked wheel brake was one of the earliest types of vehicle brakes.

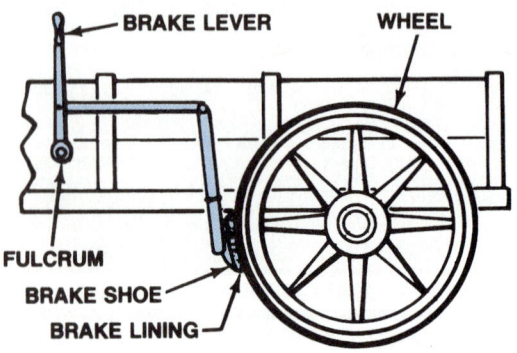

Figure 1-3. The rim-contact farm-wagon brake used levers to increase the brake application force.

Brake System Overview

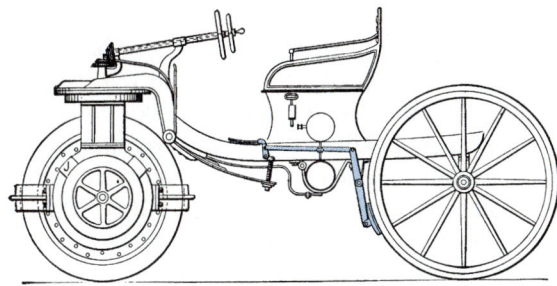

Figure 1-4. This vehicle from about 1900 used a wagon-type brake system on the rear wheels. (Courtesy of Lindsay Publications, Inc.)

Figure 1-5. An early advertisement for automobile tires. (Courtesy of Lindsay Publications, Inc.)

on the faces of the brake shoes to try to solve these problems, farm-wagon brakes, like locked-wheel brakes, were most effective as parking brakes or for holding a stopped wagon in position on a hill.

AUTOMOTIVE BRAKE DESIGNS

The very first automobiles were little more than farm wagons or carriages fitted with engines and drivetrains. As a result, early cars had brake systems much like those of farm wagons, including lever-operated brake shoes that contacted the iron rims of spoked wooden wheels, figure 1-4. With the invention of rubber tires, figure 1-5, the rim-contact farm-wagon brake became impractical. A brake shoe pressing against the new tires would rapidly wear them out or tear them from their rims. To solve this problem, a metal **brake drum,** figure 1-6, was attached to the inside of the wheel to provide a rubbing surface for the brake shoe. The practice of using a separate rubbing surface for the vehicle brakes is still followed today.

External Contracting-Band Brakes

Automotive applications of the farm-wagon brake worked well enough for a short time, but the pace of automotive progress was such that the speeds attained by newer models quickly exceeded the ability of farm-wagon brakes to stop them safely and reliably. The small brake shoes would overheat or wear out, and braking power would be lost.

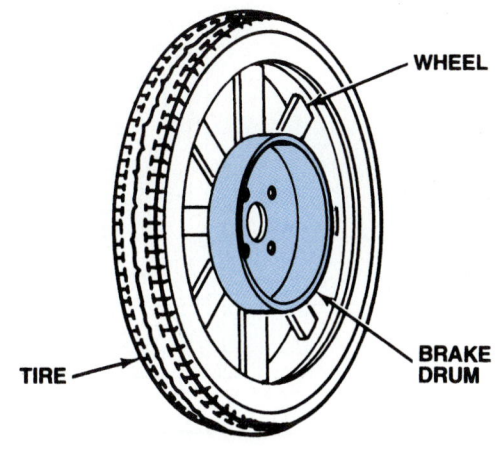

Figure 1-6. The brake drum was introduced when rubber tires made rim-contact brakes impractical.

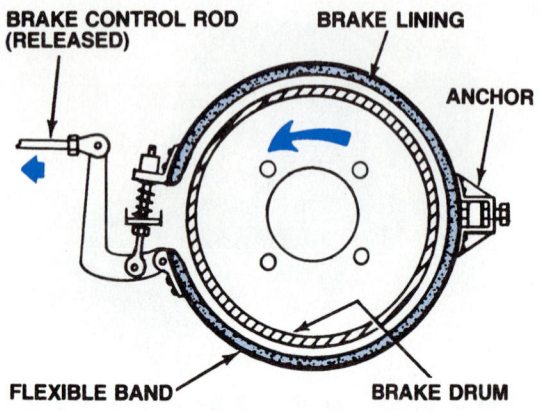

Figure 1-7. The external contracting-band brake provided additional friction area for better stopping power.

However, like other brakes that had gone before, the external contracting-band design had some problems. The brake lining material was exposed to the elements which resulted in poor braking in wet weather, and dirt and grit thrown up from unpaved roads caused the unprotected brake lining to wear very rapidly. In addition, under hard use the metal brake drums would heat up and expand inside the bands, causing the brakes to drag. This created even more heat and expansion, which eventually locked the brakes up solid unless the driver slowed or stopped to allow them to cool.

The initial solution to this problem was to increase the surface area of the lining material. One of the first brake designs to do this was the external contracting-band brake, figure 1-7. In this system, the brake shoe was replaced with a flexible metal **brake band** lined with a suitable friction material. The band was wrapped around the outside of the brake drum and anchored to the vehicle chassis. When the brakes were applied, a lever mechanism pulled the band tight around the drum to slow the car, figure 1-8.

The external contracting-band brake was used for quite some time, and the basic design is still used today in the parking brakes of some trucks.

Internal Expanding-Band Brakes

To combat the problems of the external contracting-band design, car manufacturers introduced the internal expanding-band brake, figure 1-9. This design moved the band to the inside of the brake drum where it was actuated by a rotating cam. Placing the band inside the drum shielded it from the elements and eliminated the possibility of brake lockup from drum expansion, but it also created some new problems. If internal expanding-band brakes were used hard enough for heat to expand the drums, braking power was lost because the band was unable to extend out far enough to maintain solid drum contact. In addition, the enclosed band was no longer exposed to a direct flow of cooling air, so the lining material could more

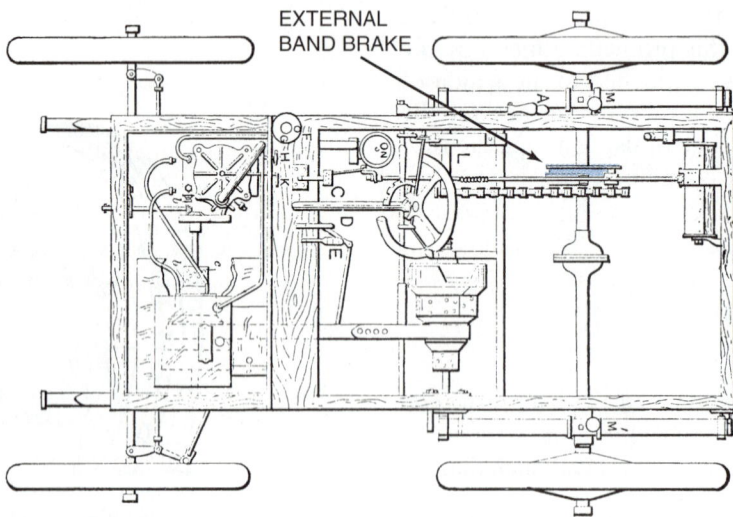

Figure 1-8. This automobile used an external band brake on the rear axle, operated by pedal "C". (Courtesy of Lindsay Publications, Inc.)

Brake System Overview

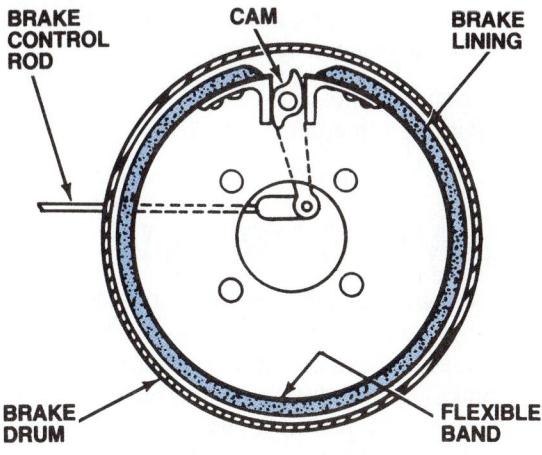

Figure 1-9. The internal expanding-band brake saw service primarily as a parking brake.

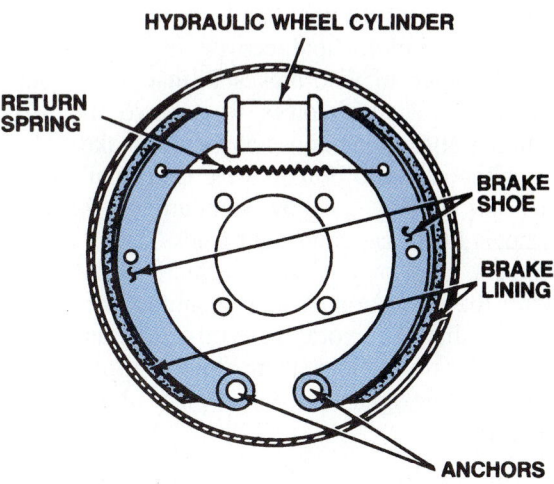

Figure 1-10. Internal expanding-shoe brakes have been the dominant friction assembly design for most of the automobile's life.

easily become overheated and damaged, which reduced braking efficiency even further.

Because of these inherent problems, most internal expanding-band brakes were used as parking brakes in systems that had farm-wagon or contracting-band wheel brakes. A few cars did use internal expanding-band brakes as their primary system, but the drawbacks described previously made the design less than ideal.

Internal Expanding-Shoe Brakes

The cure for the problems of internal expanding-band brakes was the internal expanding-shoe brake, figure 1-10, which replaced the flexible brake band with two or more rigid brake shoes. In early systems, the shoes were expanded by a rotating cam just as in the expanding-band brake. Many truck air brakes still use this method of actuation. Later systems, including all modern automobile drum brakes, use a hydraulic cylinder to force the shoes out against the brake drum.

The rigid brake shoes eliminated the breakage problems that sometimes occurred with flexible bands. And because the brake lining material no longer had to flex on application, it could be made of stronger, more rigid materials that had better braking properties. Eventually, as lining materials and brake drum designs improved, the internal expanding-shoe brake became highly reliable under all but the most extreme conditions. This design remained the standard of the braking industry well into the 1960s.

Disc Brakes

As highway speeds, traffic congestion, and the size and weight of cars continued to increase, it became apparent that even the largest practical drum brakes were unable to adequately dissipate the heat generated during repeated hard braking. After a number of stops, the brake linings would overheat, the drums would expand away from the brake shoes, and just as with earlier internal expanding-band brakes, braking performance would decrease. In order to safely stop a two-ton-plus automobile from the 70- and 80-mile-per-hour (mph) highway speeds of the time, the disc brake was adopted.

Disc brakes operate by pressing two **brake pads** against a spinning disc, or **rotor,** figure 1-11. The major advantage of disc brakes is that all of the friction components are out in the airstream where they

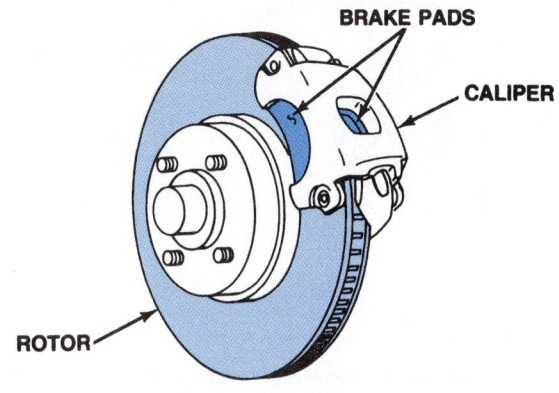

Figure 1-11. A disc brake friction assembly.

are more easily cooled; this leads to longer brake pad life and more rapid recovery after hard use. A disadvantage of disc brakes is that they require higher application forces to achieve the same amount of stopping power as a drum brake.

Although disc brakes were first available as an option, they are now standard on the front wheels of all cars and light trucks. They are the current state of the art in vehicle brakes, and, with the exception of new manufacturing techniques that use exotic materials, such as ceramics and carbon fiber, there do not appear to be any new designs that will replace them in the near future.

BRAKE SYSTEM OPERATION

An automobile's brake system, figure 1-12, must do several things: It must be able to slow or stop the vehicle when it is in motion, and it must be able to hold the car in position when it is stopped on an incline. In addition, the brake system should be able to mechanically lock at least two of the individual wheel brakes when the car is parked.

To perform these tasks, modern automotive brakes have two interrelated systems:

- Service brakes
- Parking (emergency) brakes.

SERVICE BRAKES

When technicians talk about a vehicle's brake system, they are usually referring to the **service brakes.** Manufacturers may also refer to the service brakes as **foundation brakes** or **base brakes.** The term *service brakes* refers to the primary system used to slow and stop the vehicle during everyday driving and does not include the antilock or traction control system. Service brakes are applied by pushing on the brake pedal, which pressurizes the brake hydraulic system and applies the friction assemblies at all four wheels to stop the vehicle.

The time interval between force applied at the brake pedal and braking action at the wheels is only a fraction of a second, but these events are linked by a long chain of mechanical and hydraulic actions that combine to provide the braking power needed to stop the car. Each link in this chain is carefully engineered to perform a specific function

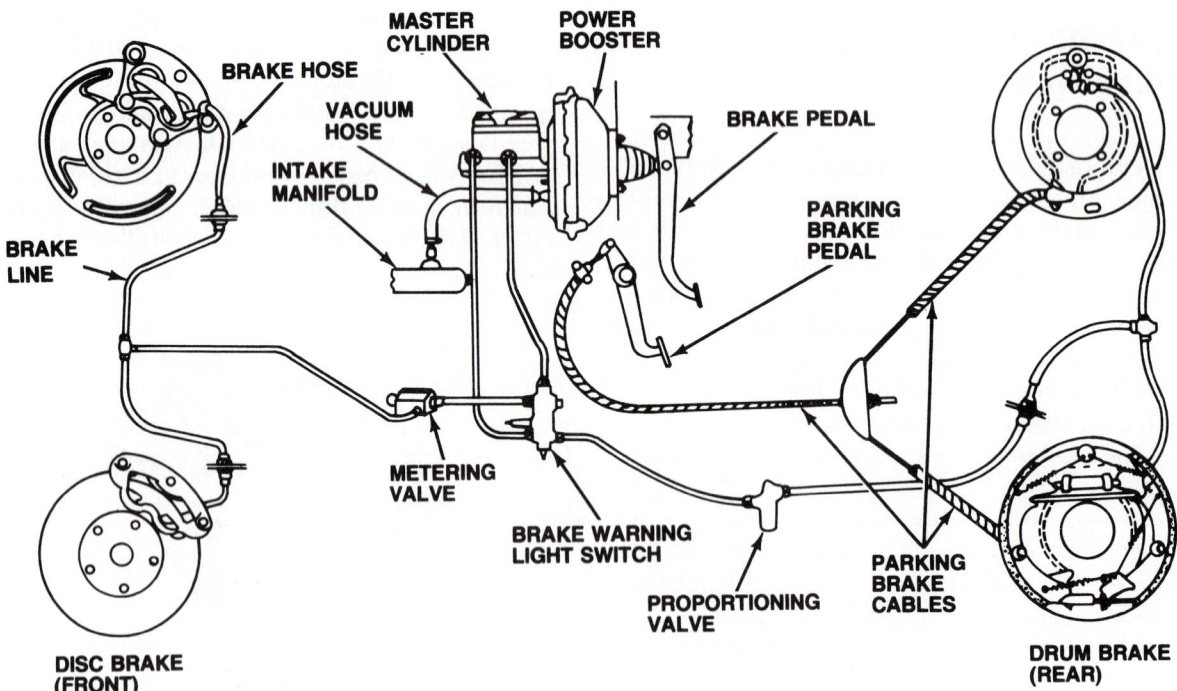

Figure 1-12. A typical brake system with front disc and rear drum friction assemblies.

Brake System Overview

in the overall braking process, and a weakness or failure at any point can result in greatly reduced stopping power or total brake failure. The following sections trace the flow of braking force through a typical modern brake system and provide basic descriptions of the most common components.

Pedal Assembly

When the driver steps on the brake pedal, figure 1-13, it moves a pushrod attached near the top of the pedal arm. Because of the mechanical advantage designed into the pedal assembly, the pedal pushrod transmits a force several times greater than that applied to the pedal pad.

Power Booster

On most late-model brake systems, the pedal pushrod enters a vacuum-, hydraulic-, or electro-hydraulic-assisted **power booster**, figure 1-14, where it actuates the booster control valve. The booster further increases the braking force which passes out of the booster through the booster pushrod.

Master Cylinder

The booster pushrod (or the pedal pushrod on systems without a power booster) enters the brake **master cylinder**, figure 1-15. The master cylinder converts the mechanical force of the pushrod into hydraulic pressure that is distributed to the wheels.

The master cylinder is similar to a hollow tube. One end is closed except for a fluid line exiting to the wheels; the other end is sealed off by a sliding piston that is moved by the pedal or booster pushrod. A brake fluid reservoir keeps the cylinder full of fluid at all times. When the pushrod moves the piston toward the closed end of the cylinder, fluid trapped ahead of the piston is forced out

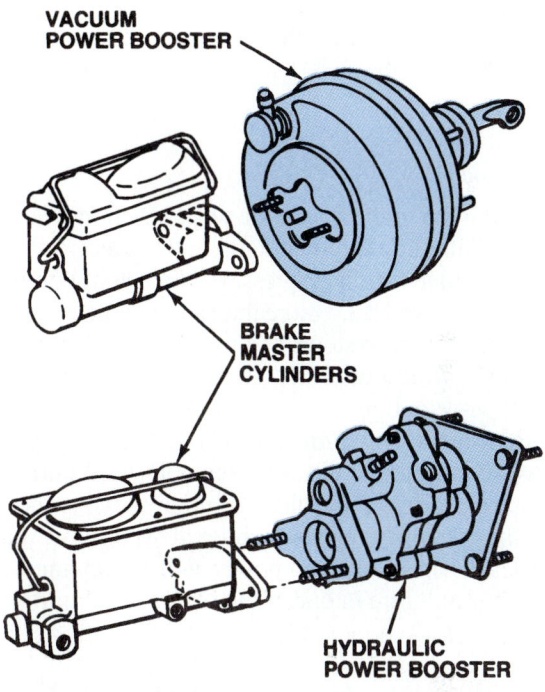

Figure 1-14. Typical vacuum- and hydraulic-powered brake boosters.

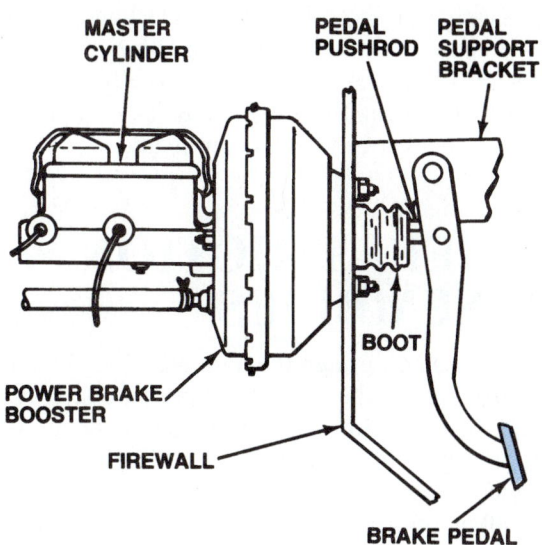

Figure 1-13. The brake pedal is the driver's control mechanism for the brake system.

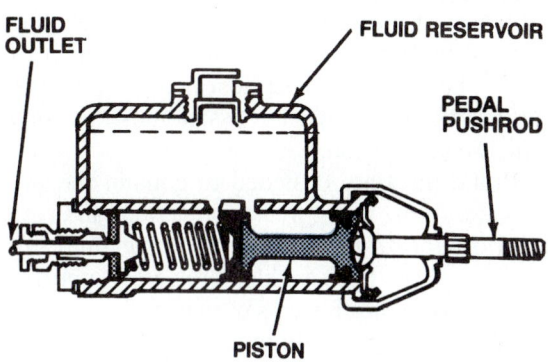

Figure 1-15. A brake master cylinder converts mechanical brake pedal force into hydraulic pressure.

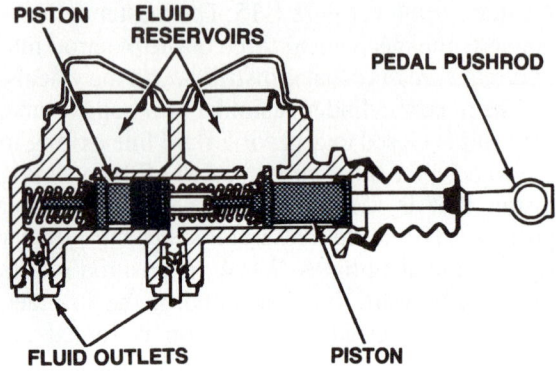

Figure 1-16. Since 1967, all cars sold in the United States have been equipped with a dual-circuit master cylinder for added safety.

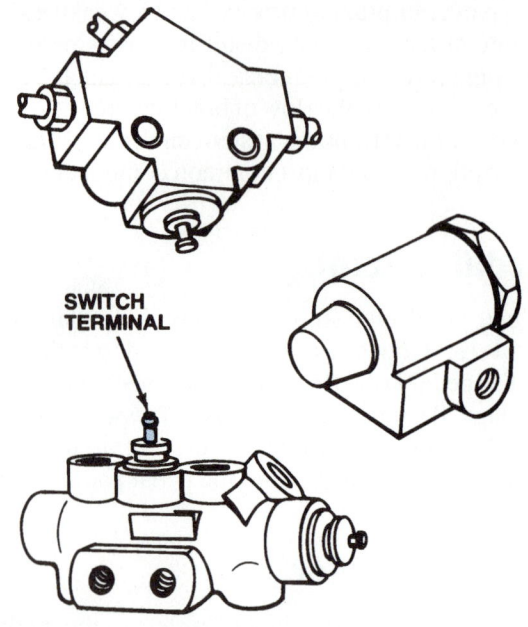

Figure 1-17. Hydraulic switches and control valves are part of most brake systems.

through the brake line to the wheel friction assemblies. Although the fluid leaving the master cylinder is under very high pressure, the actual volume of fluid moved is no more than a few ounces.

Early brake systems had master cylinders with a single fluid outlet that supplied pressure to all four wheels, figure 1-15. All modern systems use a **dual master cylinder,** figure 1-16, that contains two hydraulic circuits with separate fluid outlets. Each of the hydraulic circuits supplies two or more of the wheel friction assemblies. This ensures that at least partial braking power will be retained if there is a failure in one of the hydraulic circuits.

Hydraulic Lines and Hoses

As pressurized fluid leaves the master cylinder, it is routed to the wheel friction assemblies through **brake lines** made of steel tubing or reinforced rubber hose. Steel tubing that is solidly attached to the vehicle chassis makes up most of the brake lines. The flexible rubber hoses connect the steel lines to the wheel friction assemblies that move with the suspension.

Brake lines are designed to contain the high pressure of the brake hydraulic system and resist deterioration from exposure to brake fluid. The outsides of the lines are also protected to prevent damage from the elements or materials thrown up by the tires. Special fittings are required on all brake lines to provide fluid-tight connections, and where a single line must be split into two, steel junction blocks are used.

Hydraulic Switches and Valves

Once the fluid leaves the master cylinder, it does not necessarily go directly to the wheels; it usually passes through one or more hydraulic valves or switches along the way, figure 1-17. The valves and switches serve a number of purposes that are explained in Chapter 6, but for the most part, they either sense variations in hydraulic pressure as an indication of fluid leakage, or they modulate hydraulic pressure in some manner to ensure better braking performance. These valves and switches can be individual units, or they may be put together in various combinations.

WHEEL FRICTION ASSEMBLIES

After passing through the hydraulic switches and control valves, the brake fluid pressure arrives at the **wheel friction assemblies** that generate the friction needed to stop the car. As already discussed, there are two basic designs of friction assemblies on today's cars: drum brakes and disc brakes. On modern automobiles, drum brakes are used only as rear-wheel brakes, whereas disc brakes are used at the front wheels because of their

Brake System Overview

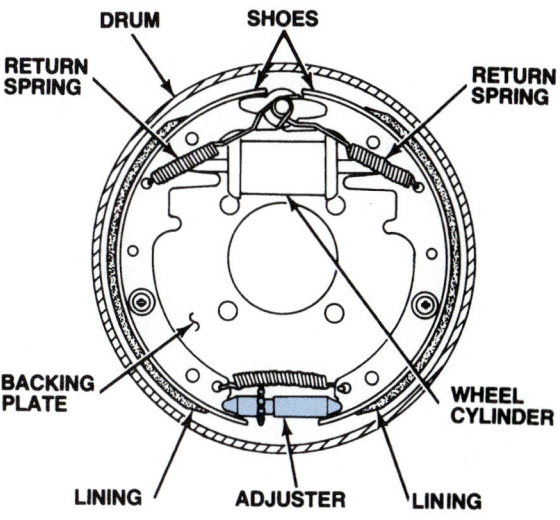

Figure 1-18. A basic drum brake equipped with a manual adjuster.

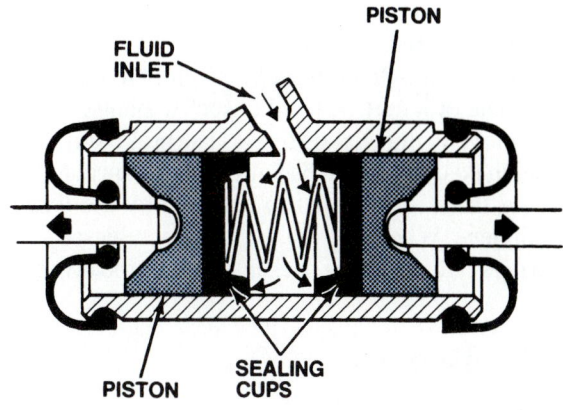

Figure 1-19. A wheel cylinder converts hydraulic pressure into mechanical force.

superior performance. Sports cars, heavyweight luxury cars, and models with high-performance packages use disc brakes at all four wheels for maximum stopping power.

Drum Brakes

In a **drum brake** friction assembly, figure 1-18, the brake system hydraulic pressure is routed into a slave cylinder. The slave cylinder in an automotive drum brake is commonly called a **wheel cylinder,** and it operates in reverse of the master cylinder—it converts hydraulic pressure back into mechanical force, which is used to actuate the wheel friction assembly. The majority of today's drum brakes use a single two-piston wheel cylinder, figure 1-19, at each wheel. However, some systems use a pair of single-piston cylinders instead.

In a two-piston wheel cylinder, hydraulic pressure enters the cylinder between the two pistons. The pressure forces the pistons outward toward the open ends of the cylinder where they act directly, or through short pushrods, on the brake shoes. As the brake shoes are pushed outward by the wheel cylinder pistons, their lining material contacts the brake drum, which is attached to the rotating hub or axle of the vehicle. The friction created between the brake linings and the drum slows the rotation of the drum, the wheel and tire attached to it, and thus the entire car.

■ The Dangerous Safety Feature

As logical as a one-brake-per-wheel system sounds today, it was not until the 1920s that brakes became a common sight on front wheels. Early experiments with mechanically actuated front-wheel brakes had been disastrous. Hard braking easily locked the front wheels, which impaired steering and caused accidents. Many manufacturers actually believed front brakes made a car uncontrollable, and that a car might flip over on its nose if front brakes were applied with too much force!

In 1909, Isotta-Fraschini (I-F), a flamboyant yet conservative Italian carmaker known for its exotic, exclusive, and well-engineered touring cars, became the first automaker to offer four-wheel brakes. Ironically, I-F was still using the nearly obsolete chain drive (although elegantly enclosed) on the same machine, which introduced the first practical front-wheel brakes.

Another expensive make, the French Hispano-Suiza, followed shortly after I-F with a well-designed four-wheel brake system. However, the proper engineering of front-wheel brakes eluded most contemporary manufacturers, and some newly introduced four-wheel brake systems had to be hastily withdrawn. Many cautious carmakers stayed with rear-wheel brakes only well into the 1920s when traffic congestion made four-wheel brakes an increasing necessity for safe operation. Fortunately, by that time even less expensive cars had suspension and steering systems that could cope with powerful front brakes.

The wheel cylinder and brake shoes mount to a **backing plate** that is attached to the vehicle's suspension. The cylinder may bolt solidly to the backing plate or it may be fixed in such a manner that it can move within certain limits. The brake shoes are held in position on the backing plate by springs that allow them to move when actuated by the wheel cylinders. When the brake pedal is released, the springs pull the shoes back away from the drum.

Disc Brakes

A **disc brake** friction assembly, figure 1-20, performs the same job as a drum brake; it creates friction to stop the car. However, it does so in a somewhat different manner. In place of a wheel cylinder, disc brakes use a **brake caliper** to convert hydraulic system pressure back into mechanical force. The caliper assembly is mounted to the vehicle suspension and straddles the brake rotor, which is the disc brake equivalent of a brake drum. The rotor is made of cast iron or steel and is keyed to the rotating hub or axle.

When hydraulic pressure enters the brake caliper, figure 1-21, it pushes on the back of one or more pistons and forces them outward in their bores. As the pistons move outward, they push the brake pads against the brake rotor. Brake pads are the disc brake equivalent of drum brake shoes, and they create the friction that slows the rotation of the brake rotor, the wheel and tire attached to it, and thus the car. Compared to brake shoes, disc pads are usually rather simple, consisting of only a flat metal backing plate with a block of friction material attached to it.

Brake calipers can contain one, two, three, or four hydraulic pistons. In a multipiston caliper, the pistons are located so they clamp down on the rotor with equal force from either side. In a single-piston caliper, the portion of the caliper containing the piston extends over and around the rotor and is free to slide or "float" on the portion of the caliper that is solidly mounted to the suspension. When the piston is forced toward the rotor by hydraulic pressure, the caliper housing moves with equal force in the opposite direction and applies the brake pad on the far side of the rotor.

PARKING BRAKES

The **parking brake** system, as its name implies, is a secondary braking system used to hold a parked car in position. The parking brakes operate independently of the service brakes, although on most cars they share some of the same friction assembly components. Because less braking power is required to hold a stopped car in place than to slow a moving vehicle, the parking brakes apply with less force than the service brakes and operate on only two of a car's four wheels.

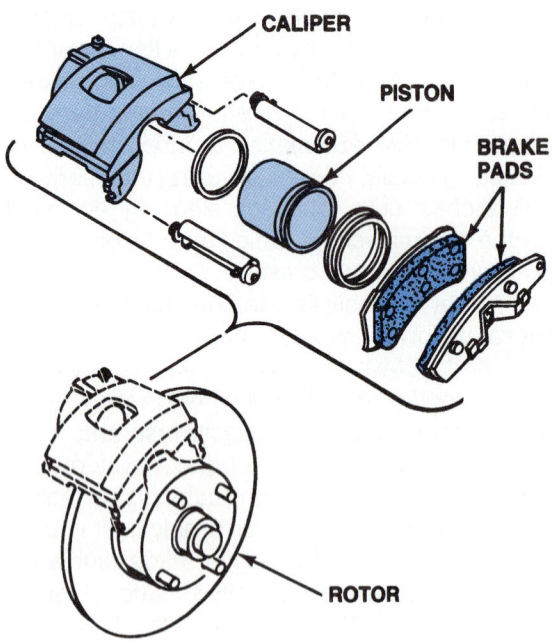

Figure 1-20. The disc brake is found on the front wheels of all modern automobiles. This is a single piston, sliding caliper.

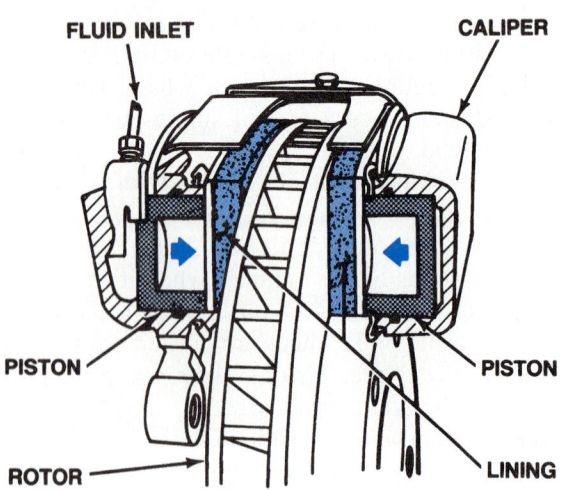

Figure 1-21. The brake caliper assembly forces the brake pads against the rotor.

Brake System Overview

Parking Brake Actuation

Unlike the service brakes, the parking brakes, figure 1-22, are 100 percent mechanically actuated—hydraulic pressure plays no direct part in their application. For practical reasons, this is done because hydraulic systems will leak if required to maintain a constant pressure for a long time. Federal Department of Transportation (DOT) regulations also specify that the parking brake be mechanically operated (see Chapter 2). Elements of the hydraulic system may be involved in some parking brake designs, but if so, they are applied mechanically.

Parking brakes are applied by pushing on a floor-mounted pedal or pulling on a hand-operated lever. A ratchet-type locking device holds the pedal or lever in the applied position until released. The pedal or lever transfers the force of its movement to the wheels through a system of rods, cables, and levers. These pieces are designed to provide a mechanical advantage that will apply the brakes hard enough to hold the car in place on inclines.

Because they are mechanically operated, parking brakes can also be used to stop the car in the event of a total failure of the service brake hydraulic system. For this reason, parking brakes are sometimes called **emergency brakes**. However, the amount of stopping power available from the parking brakes is substantially less than that provided by the service brakes. Not only do the parking brakes operate on only two wheels, but the amount of braking force that can be applied manually is far less than that supplied by the service brake hydraulic system and power booster.

Parking Brake Friction Assemblies

The parking brake system usually employs two of the service brake friction assemblies. The rear wheel drum brakes are used in most cases, but a few systems do apply the front disc brakes. Drum brakes make better parking brakes than disc brakes because they require less application force to achieve the same braking power. Cars equipped with four-wheel disc brakes use two different types of parking brakes. Some systems apply the disc brake pads, others incorporate small drum brakes in the hubs of the rear rotors that function solely as parking brakes.

Drum-type parking brakes are usually applied through a lever attached to a connecting link between the two brake shoes, figure 1-22: As the lever is pulled into the applied position, it effectively lengthens the connecting link, which forces the shoes out against the brake drum. A similar system is used in the small drum-type parking brake assemblies fitted to some cars with four-wheel disc brakes, figure 1-23.

A disc-type parking brake, figure 1-24, operates quite differently. The parking brake cable pulls on a lever which rotates a rod that enters the brake caliper. The rotating rod applies the caliper piston by unscrewing a threaded portion of the piston, which then extends outward to apply the brake. Another design uses the motion of the rod to rotate an intermediate member that drives a number of ball bearings up tapered ramps, which in turn forces the piston outward to apply the brake. Both systems incorporate automatic adjusting mechanisms that compensate for brake pad wear.

METHODS OF BRAKE ACTUATION

Brake actuation is the process of applying the wheel friction assemblies; the **actuating system** includes all of the components that help transmit braking force from the brake pedal to the wheels. The combination of mechanical and hydraulic systems described previously is the most common form of automotive brake actuation, but compressed air is used on many heavy-truck brakes, and some trailer brakes are actuated electrically.

In addition to transmitting braking force to the wheels, the actuating system must also help increase the force so it is sufficient to stop the car. If the pressure applied to the brake pedal were applied

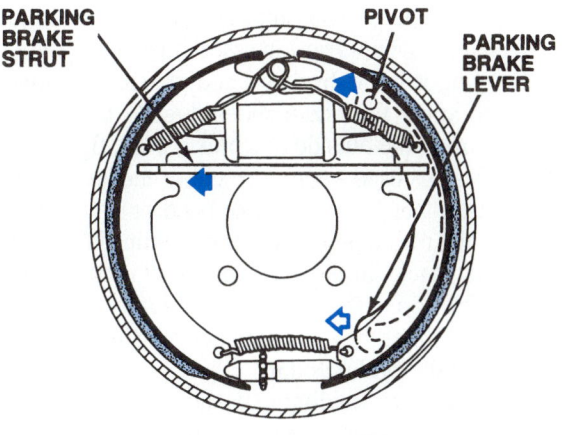

Figure 1-22. Most drum-type parking brakes use some variation of lever-and-strut actuation.

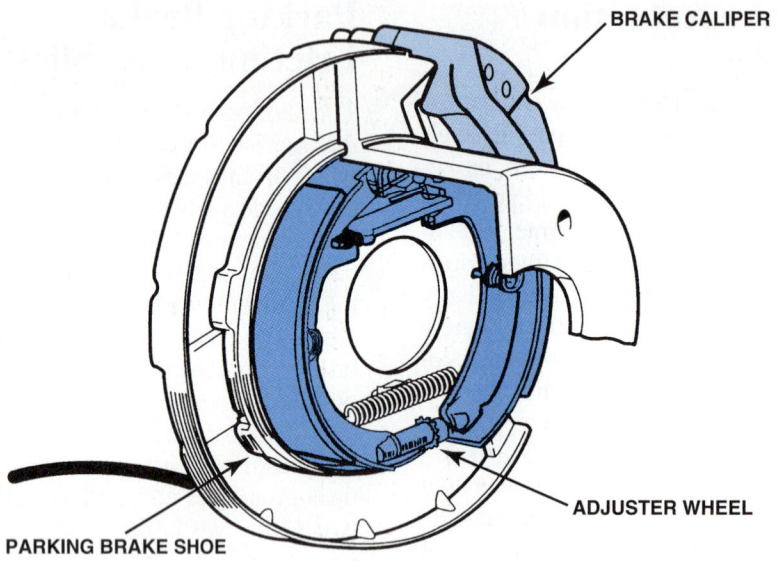

Figure 1-23. This parking brake assembly uses a pair of brake shoes and the hub of the brake disc as a small brake drum. (Courtesy of General Motors Corporation, Service and Parts Operations)

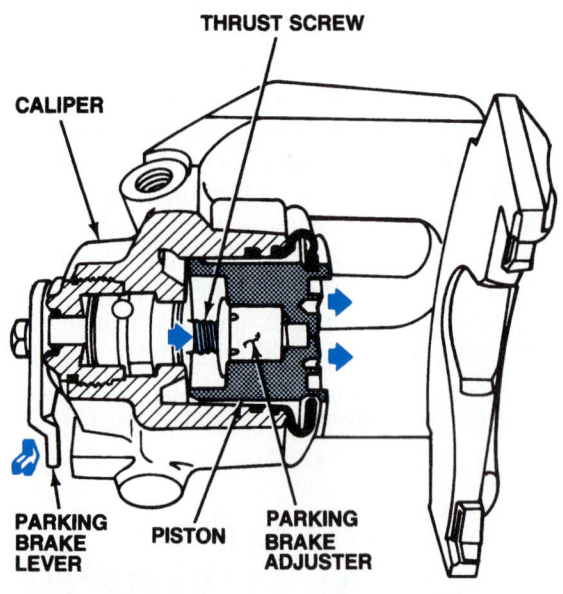

Figure 1-24. Some rear disc parking brakes use this thrust-screw design to apply the brake pads.

directly to the road, the amount of braking force would be negligible; but when this force is assisted by the actuating system, a 100-pound driver can stop a vehicle weighing several tons. The physical "laws" and principles that allow actuating systems to increase braking force are discussed in Chapter 3.

All service brake actuating systems begin with mechanical foot pressure on the brake pedal and end with mechanical application pressure on the brake shoes or pads. However, between these two points, the actuating system can take many forms as it transmits braking force to the wheels. The remainder of this chapter will examine the four major types of brake actuating systems:

- Mechanical
- Hydraulic
- Pneumatic
- Electric.

Mechanical Actuation

The earliest vehicle brakes were actuated entirely by mechanical means. In a typical mechanical actuating system, pressure on the brake pedal operated a linkage of rods and/or cables that moved levers at the wheel friction assemblies, figure 1-25. Farm-wagon-type brakes used the lever movement to apply the brake shoes directly against the wheel rim or the outside of a brake drum, figure 1-3. Later designs used the movement of the lever to rotate a cam that forced a brake band or brake shoes against the inside of a brake drum, figure 1-9. Many modern truck brake systems still actuate the wheel friction assemblies mechanically, although they no longer have a mechanical linkage connected to the brake pedal.

Completely mechanical actuating systems were the best technology available in their day, but they had several drawbacks. A primary prob-

Brake System Overview

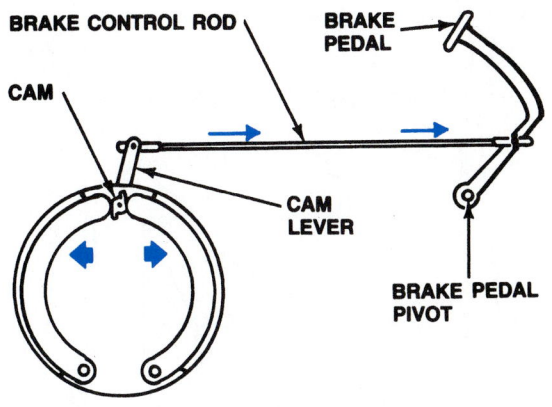

Figure 1-25. A simple mechanically actuated brake system.

lem was designing a mechanical linkage that would operate the brakes on the steered front wheels. For a long time, mechnically actuated brakes were limited to rear wheel use only. However, by using complex lever and cable linkages, four-wheel mechanical actuating systems were eventually developed.

Unfortunately, both two- and four-wheel mechanical brakes suffered from another problem, an inability to supply the proper amount of braking power to each wheel under all conditions. Through proper engineering of the leverage ratios of the various connecting links, mechanical actuating systems could achieve fairly good front-to-rear balance. However, changing vehicle loads, suspension movement, and wear within the brake system created additional difficulties.

On the earliest cars, the mechanical brake linkage and the axles supporting the wheel friction assemblies were all solidly mounted to the vehicle chassis. However, as cars evolved, suspensions were developed that allowed the axles and brakes to move while the chassis and brake pedal remained stable. When suspension-mounted brakes were mechanically attached to a chassis-mounted brake pedal, the braking force would vary as the suspension moved through its range of travel. Additional freeplay also had to be included in the system to prevent suspension movement from actually applying the brakes. This freeplay took up a portion of the brake pedal travel, and slowed brake application time.

The most significant problems with mechanical actuating systems occurred because the brake shoes wore at different rates at each wheel. When this happened, the brake application force varied from wheel to wheel, and the result was a pull to one side and premature locking of individual friction assemblies. Wear also affected overall braking efficiency. Cam-operated brakes, for example, required that the cam lever be adjusted at a right angle to the operating linkage for best performance. As the shoes wore and the cam had to rotate farther before braking began, the effective length of the lever was changed and the amount of braking power reduced.

Later mechanical actuating systems had methods of compensating for wear, but they were often complicated or worked poorly. They also added to another problem of mechanical actuation systems, linkage wear. Not only did mechanically actuated brake systems have to be adjusted regularly but the many rotating and sliding surfaces also had to be lubricated periodically to prevent rust and wear from causing sticking brakes or excessive freeplay.

The ultimate demise of mechanically actuated brakes came about because of changes in the nature of the automobile. Cars became less like wagons with drivetrains, and more like the vehicles of today. They got larger and heavier until mechanical linkages alone could no longer supply enough braking force to stop them in a reasonable distance. And the lower chassis and enclosed bodies of the newer cars did not lend themselves well to the routing of mechanical rods, cables, levers, and linkages. Cars were ready for an advance in brake actuation, and hydraulics were the answer to the problem.

Hydraulic Actuation

All modern cars and light trucks are equipped with hydraulically actuated service brake systems. In a hydraulic brake system, the mechanical force applied at the brake pedal is converted into hydraulic (fluid) pressure in the master cylinder then routed through various lines and hoses to the wheel friction assemblies where it is reconverted into mechanical force by wheel cylinders (drum brakes) or brake calipers (disc brakes), figure 1-26. The use of hydraulic pressure to transmit braking force has several advantages over the rods and cables of mechanical brakes: It is more consistent and reliable, it allows the amount of braking force delivered to each wheel to be precisely regulated, and it is able to supply the higher application forces required to stop heavier cars.

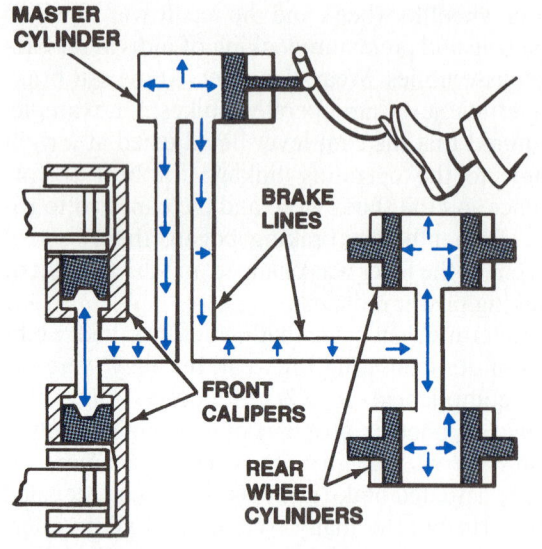

Figure 1-26. A simple hydraulically actuated brake system.

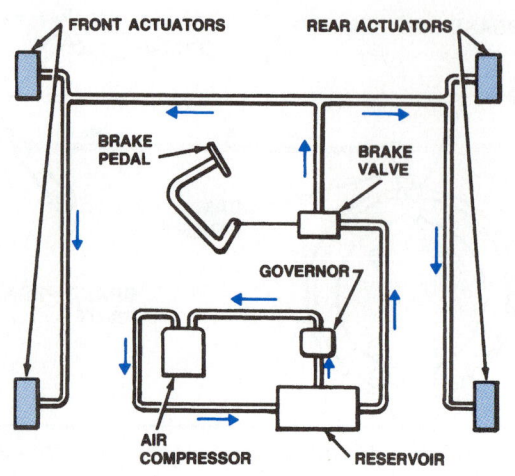

Figure 1-27. A simple pneumatically actuated brake system.

■ Duesenbrakes from the Jazz Age

The Duesenburg brothers, Fred and Augie, introduced a new super car in 1921, the Model A Duesenburg. This model bristled with advanced features: a straight-eight overhead-cam engine, light-alloy pistons, and a molybdenum-steel chassis. Definitely not to be confused with Ford's product of the same name, the Model A was "built to outclass, outrun and outlast any car on the road." One might also paraphrase the builders and state that it was built to outstop any car on the road as well. The car sported hydraulically actuated, four-wheel brakes with huge, finned drums. The brake system hydraulic fluid was a mixture of glycerin and water.

The Model A's hydraulic brakes were a world first for a production automobile and a giant step forward. Although the specific layout was a Duesenburg design, it drew its inspiration from the original hydraulic brake patented in 1918 by Malcolm Lockheed. The Duesenburg did not lord its superior brake system over other cars for long, however. By 1924 hydraulic brakes were being fitted to a mass-produced automobile, the Chrysler Light Six.

Compared to mechanical brake linkages, hydraulic actuating systems are relatively maintenance free. They do not require periodic adjustment or lubrication, and wear of the hydraulic components is minimal unless the system becomes contaminated in some way. Generally, a hydraulic actuating system does not require maintenance, other than simple topping up of the fluid level, until the wheel friction assemblies also need service.

The real advantage of hydraulic actuation is that it supplies high application pressures that can be precisely regulated to deliver the desired braking force to each wheel. How and why this is possible is explained in Chapter 3, but once a hydraulic actuating system has been designed and installed on a car, it will continue to supply the proper application pressures regardless of mechanical wear within the wheel friction assemblies. Also, because the chassis-mounted hydraulic components are connected to the suspension-mounted friction assemblies with flexible rubber hoses, routing application force to the steered wheels poses no problem, and suspension movement has no effect on the braking force applied at the wheels.

Pneumatic Actuation

Pneumatic, or air-actuated, brakes are used primarily on heavy trucks, buses, and off-highway vehicles. Although they are not covered in depth in this text, air-actuated brakes operate on many of the same principles as other brakes, and their wheel friction assemblies are usually somewhat similar.

A simple air brake system, figure 1-27, consists of an air compressor, a governor, an air storage reservoir, a pedal-controlled brake valve, and one or more air chambers (actuators). During operation, the compressor builds air pressure in the storage reservoir to a level determined by the governor. When the brake pedal is depressed, the brake valve releases a

Brake System Overview

portion of the compressed air to the **air chambers**, figure 1-28, which then apply the brakes.

Straight-air Systems

There are two different designs of air actuation systems presently being used. The first type uses the air chambers to directly operate the cam levers of mechanical drum brakes, figure 1-29. This is an air-mechanical combination and is commonly called a **straight-air system**. Disc brakes also use straight-air systems. On an air-actuated disc brake, the air chamber operates a lever that turns a helical cam, figure 1-30. The *rotating* cam transfers its force through a thrust bearing to a helical *sliding* cam that forces a thrust spindle outward to apply the brake pad. This design is similar in concept to the parking brake mechanism of some hydraulic disc brake calipers.

The primary advantage of straight-air brakes is that they are particularly safe. A total loss of braking power cannot result from small air leaks because reserve pressure is continually supplied by the compressor and stored in the reservoir. In fact, federal regulations require that the air reservoir volume of the system be 12 times as great as the combined volume of all the air chambers. Air brakes also simplify the problem of coupling and uncoupling tractor-trailer rigs; hydraulic brake systems must be air free to operate properly, but air brakes use air itself as the operating medium.

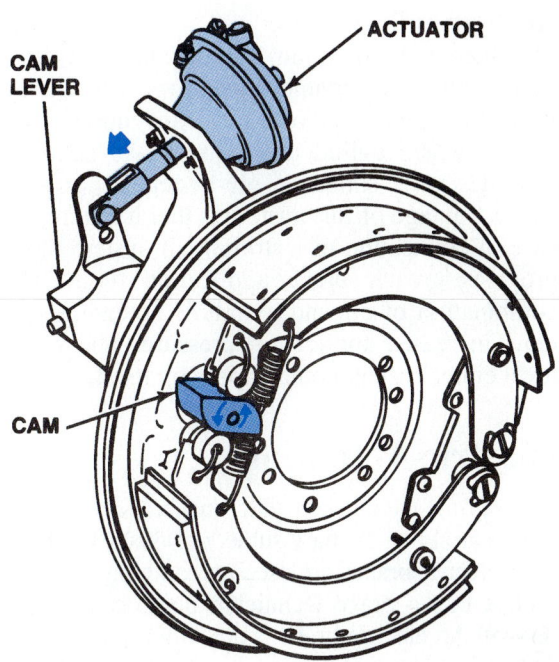

Figure 1-29. A straight-air system uses air pressure to actuate the wheel friction assembly through a mechanical lever and cam.

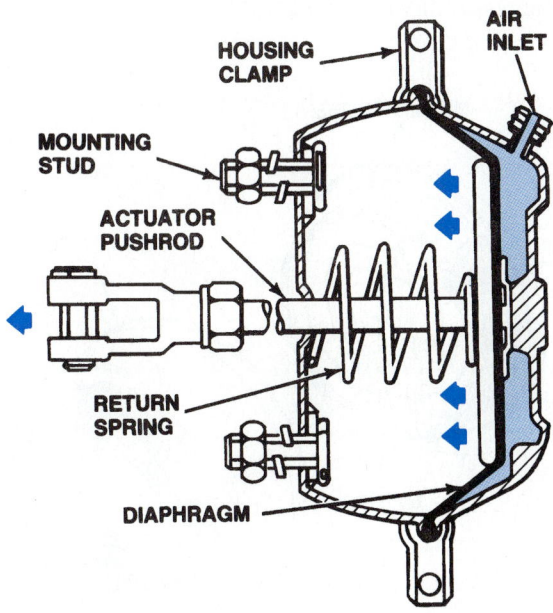

Figure 1-28. An air chamber (shown here in the released position) converts air pressure into mechanical force.

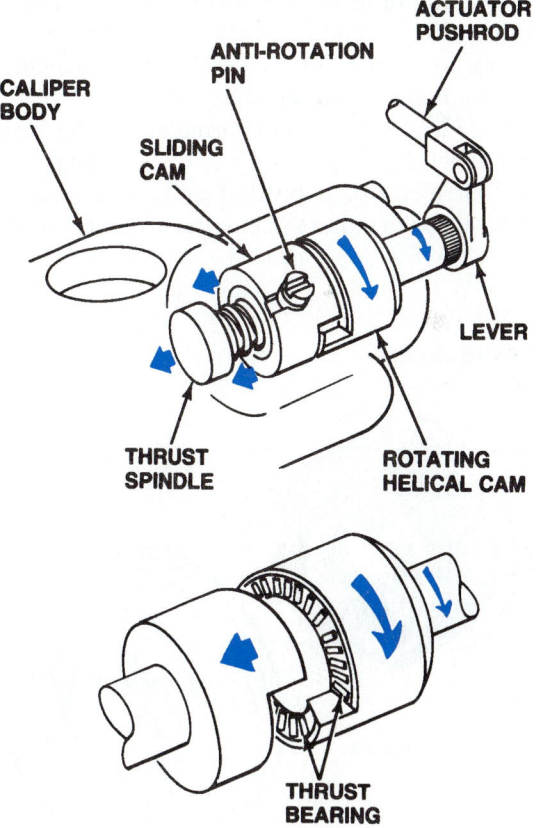

Figure 1-30. An air-actuated disc brake uses air pressure to actuate the brake caliper through a lever and two helical cams.

Air-hydraulic Systems

The second kind of air-actuated brake system uses a single air chamber to power a hydraulic master cylinder that operates conventional hydraulic wheel cylinders or disc brake calipers. This design is called an **air-hydraulic system**. The advantage of this system is that it retains the ease of connection of a straight-air system but it provides greater brake application force. The combination of air and hydraulic actuating systems increases application pressures above the level either system could supply by itself.

Electric Actuation

Like air-actuated truck and bus brakes, electric brakes are not a primary subject of this text. However, electric brakes are used on camping and boat trailers that a brake technician may occasionally service. An electric brake is actually nothing more than a drum brake friction assembly applied by an electromagnet. There are two main types of electric brakes, annular and spot magnetic. Both share two distinguishing parts: an armature and a magnet.

In an **annular electric brake**, figure 1-31, the armature is built into the brake drum and the magnet forms a circle surrounded by the brake shoes. The **spot magnetic electric brake** design, figure 1-32, also houses the armature in the brake drum, but the magnet is a small disc mounted on the outer end of a pivoted arm. When current flows through either system, the magnet is pulled toward the armature and the resulting movement expands a linkage that forces the brake shoes against the drum.

The stopping power of an electric brake is determined, to a large degree, by the amount of electrical current supplied to the electromagnet. A variable resistance device called a **controller** is used to regulate the current flow, figure 1-33. Recreational vehicle electric brakes are usually operated by a dash-mounted hand controller or a foot controller mounted on the brake pedal pad. Commercial-duty electric brake systems use a controller keyed to work from the hydraulic or air pressure of the vehicle's main braking system.

Electric brakes are not suitable for use as a vehicle's primary braking system because they do not generate as much stopping power as air or hydraulic systems, and an electrical failure could result in an immediate and total loss of braking power. However, electric brakes are well suited for use on trailers because they do not require any additional plumbing, they use only a small amount of electrical energy, they apply and release very rapidly, and the connection between the trailer and tow vehicle can easily be made or broken. On multiaxle trailers, resistors can be put into the brake wiring to alter current flow and balance the braking performance of each axle. Finally, electric brakes allow the trailer brakes to be applied independently of the tow vehicle's brakes. This stabilizes the trailer during mountain driving and helps prevent the possibility of a "jackknife."

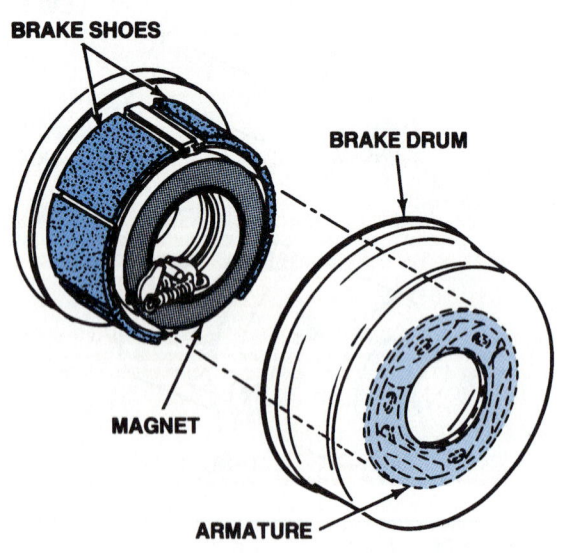

Figure 1-31. An annular electric brake.

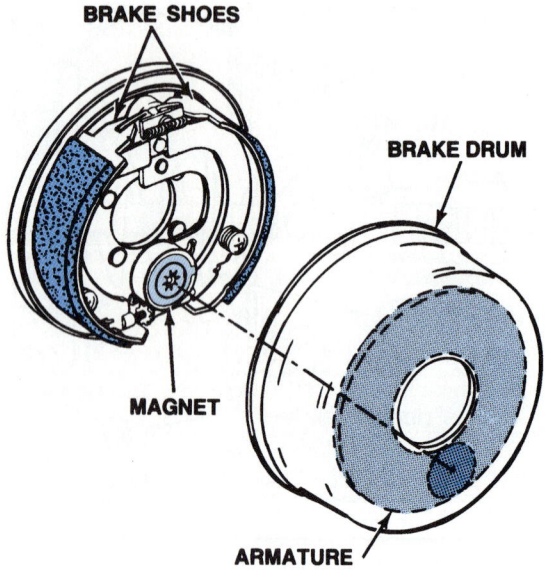

Figure 1-32. A spot magnetic electric brake.

Brake System Overview

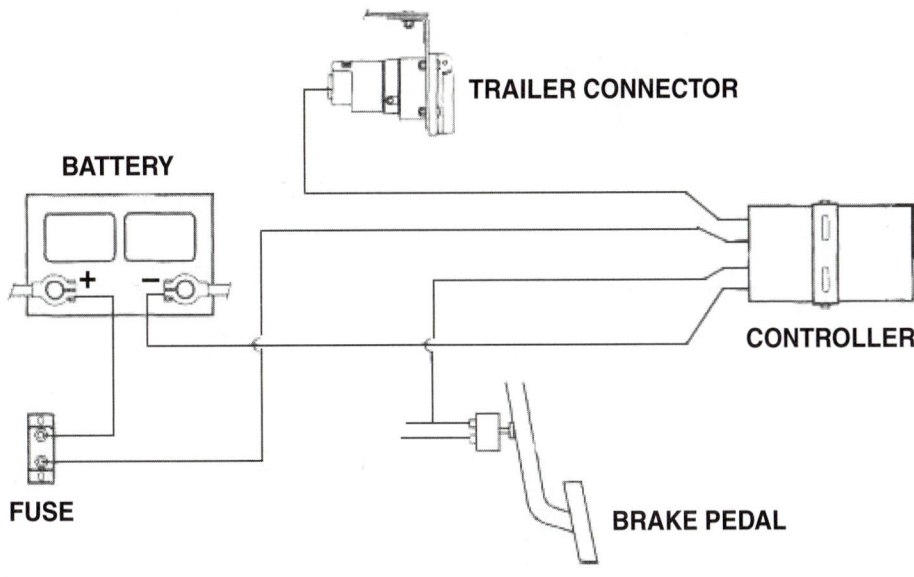

Figure 1-33. The application force of this electric brake system is regulated by the brake switch and an electronic controller.

ANTILOCK BRAKES

The components discussed to this point comprise what is known as the base brakes (foundation brakes), that is, the brake components that all vehicles have, regardless of whether a vehicle is equipped with antilock brakes. To review, the base brakes consist of:

- Master cylinder and booster (if equipped)
- Brake calipers and wheel cylinders
- Brake pads and shoes
- Brake rotors and drums
- Brake lines
- Parking brake
- Proportioning and other control valves.

In addition to the base brakes, many vehicles are equipped with antilock brakes, either as standard or optional equipment. These systems are covered in detail in later chapters.

The purpose of the antilock brake system (ABS) is to maintain vehicle stability and directional control during braking events that normally would cause the wheels to lock up and the car to skid. The systems are designed to help the driver:

- Maintain steering control (directional control)
- Stop straight (directional stability)
- Stop in the shortest distance.

Braking performance depends on the friction between the road and the tire. Zero percent slippage is a free rolling wheel and 100 percent slippage is a locked wheel sliding on the road; testing shows that optimum braking occurs at a slippage of from 10 to 30 percent. Using electronic and hydraulic controls, the goal of the ABS is to meet this 10 to 30 percent slip rate.

ABS Components

ABS is combined together with the base brake system, located between the master cylinder and the wheel cylinders. The system may be combined with the master cylinder as a single unit (integral ABS), figure 1-34, or mounted separately from the master cylinder (nonintegral ABS), figure 1-35. Referring to the manner in which the wheels are controlled, the systems are also classified as one-, three-, or four-channel systems.

- One channel controls both rear wheels together
- Three channel controls each front wheel individually and both rear wheels together
- Four channel controls each of the four wheels individually.

ABS adds a hydraulic control unit, speed sensors, and an electronic control unit to the base brakes, figure 1-36. Depending on the vehicle and manufacturer, the components are mounted in a variety of ways, figure 1-37.

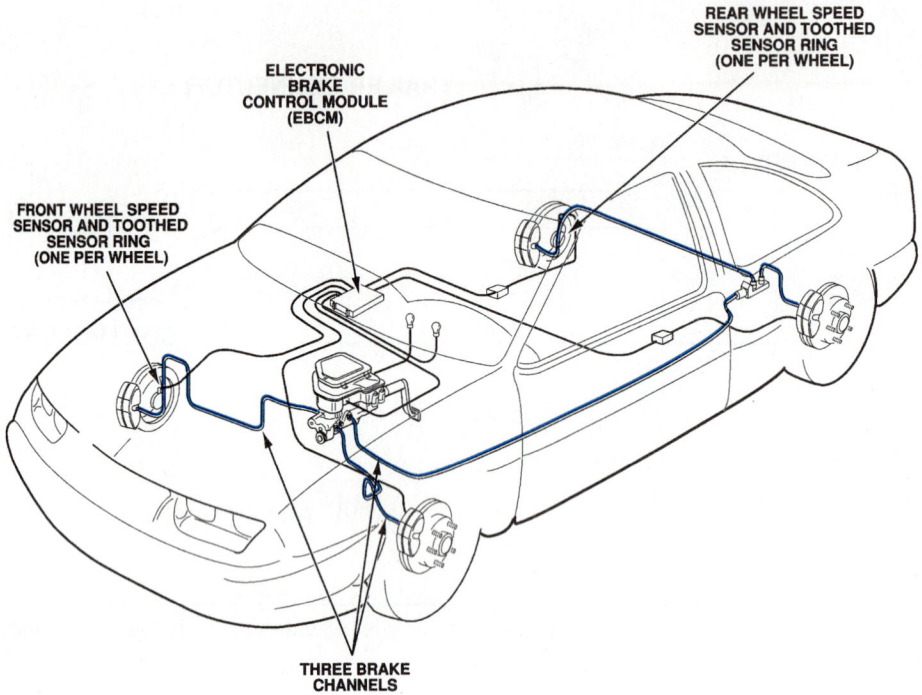

Figure 1-34. This GM ABS is an integral, three-channel system. (Courtesy of General Motors Corporation, Service and Parts Operations)

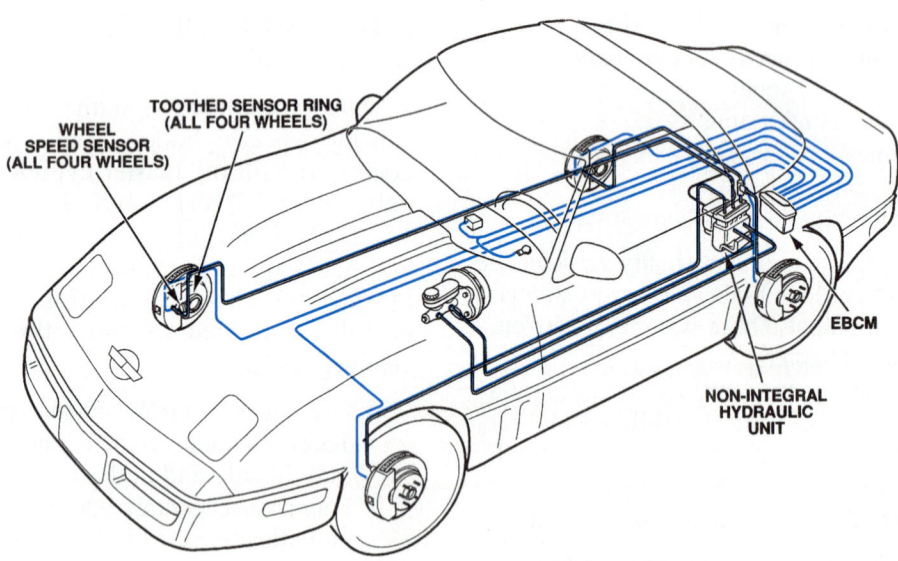

Figure 1-35. The Corvette uses a nonintegral, four-channel ABS. (Courtesy of General Motors Corporation, Service and Parts Operations)

18

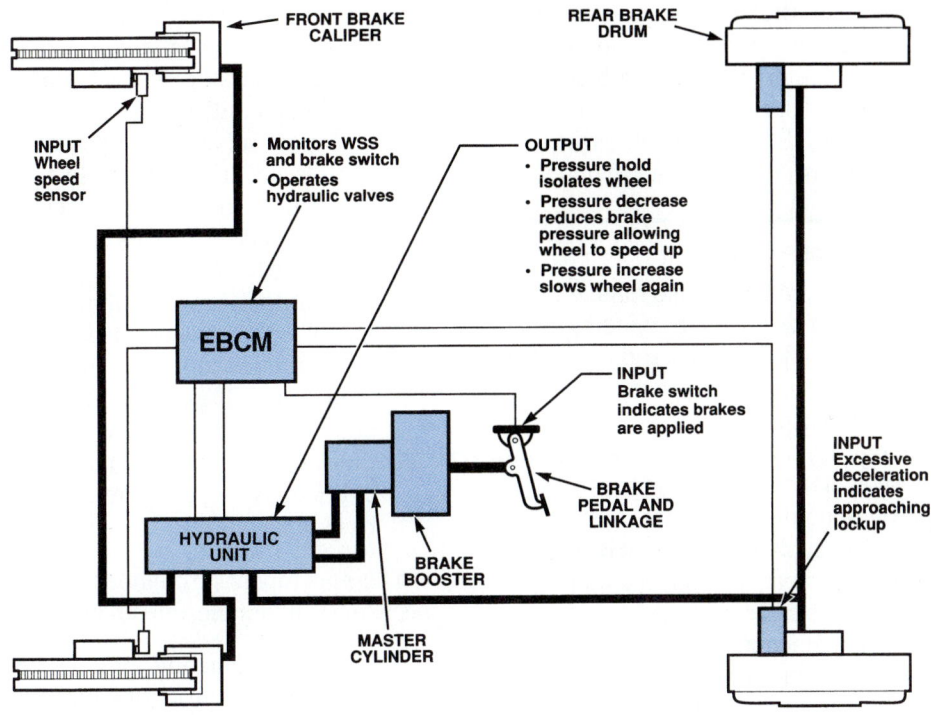

Figure 1-36. This diagram shows the various parts, inputs, and outputs of a typical antilock brake system. (Courtesy of General Motors Corporation, Service and Parts Operations)

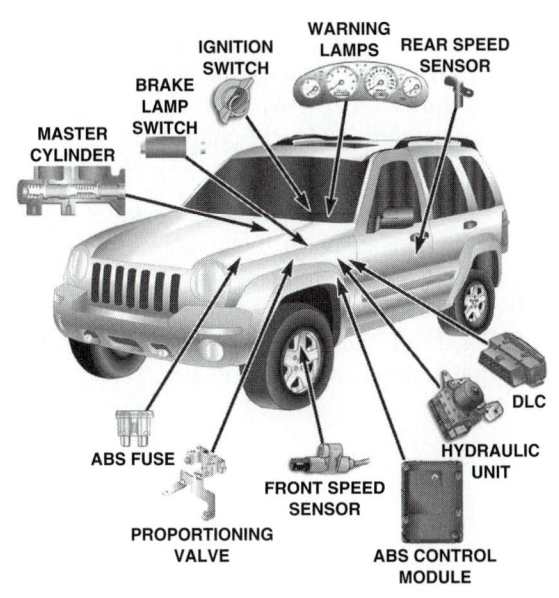

Figure 1-37. Location of the various ABS components on the Jeep Liberty. (Courtesy of DaimlerChrysler Corporation. Used by permission)

Other ABS Functions

Developments in vehicle technology have expanded the use of the ABS control system and allowed the development of other systems that use the ABS components in their operation. Some of these are:

- **Traction control:** In addition to using engine torque control, traction control systems may use the ABS controller to control wheel spin at the drive wheels during acceleration and cornering. Some systems use the ABS in place of a limited-slip differential unit.
- **Stability control:** With the addition of other sensors, stability control uses the ABS to maintain vehicle control during cornering and high-speed maneuvering.
- **Dynamic proportioning:** Dynamic proportioning allows the elimination of the hydraulic proportioning valve(s) and uses the ABS controller to maintain the proper front-to-rear brake balance during normal braking.

19

- **Brake-by-wire:** Brake-by-wire, the latest development, replaces the hydraulic master cylinder and booster assembly with an electronic brake pedal assembly to control the brakes using the ABS components as the total brake system. There is no master cylinder.

■ Electronic Brakes

Although driving a car with no brake master cylinder seems unsafe, brake-by-wire systems must meet the same DOT requirements and more. The "by-wire" technology has been proven safe and reliable, having been used on throttle-by-wire systems for a number of years. The expected change from 12- to 14-volt electrical systems to 36- to 42-volt systems will bring more "by-wire" systems to future vehicles. One such development is a totally electronic braking system with no hydraulic system of any kind. In this system, the wheel calipers are electrically actuated under the control of a computer using, among other sensors, an electronic brake pedal assembly. The higher 42-volt system provides the power that is needed for quick, powerful braking without the need for brake fluid, hydraulic cylinders, hoses, lines, and the maintenance that these systems require.

SUMMARY

The brake system is the most important safety device on any vehicle. Since the invention of the wheel, most vehicles have used some sort of brakes to bring them to a stop. Early cars had brake systems adapted from farm-wagon brakes; as vehicle weights and speeds increased, existing brake designs were improved and new systems were developed to maintain adequate stopping power. Modern cars and trucks use both drum and disc brake designs.

Vehicle brake systems are broken down into three subsystems: service brakes, parking brakes, and the antilock brake system.

- The service brakes, also called foundation brakes or base brakes, operate on all four wheels and stop the vehicle under normal driving conditions.
- The parking brakes operate on only two wheels and are used to hold a parked vehicle in position. In the event of service brake failure, the parking brakes can provide a limited amount of emergency stopping power.
- The antilock brake system (ABS) uses a system of wheel speed sensors, a computer, and a hydraulic controller to help the driver maintain vehicle stability and directional control during emergency and slippery conditions. ABS may be built into the master cylinder (integral system) or added to the base brake system.

There are four basic methods of actuating brakes: mechanical, hydraulic, pneumatic (air), and electric.

- All brake actuating systems begin and end with mechanical actions, but total mechanical actuation is now used only for parking brakes.
- Hydraulic actuation is the preferred choice for modern car and light truck service brakes because of the high, yet controllable, application pressures it can develop.
- Large trucks and off-highway vehicles commonly use pneumatic (air) actuation, which offers a large reserve braking capacity that provides a great margin of safety.
- Electric brake actuation is most often found on trailer brakes, where it offers additional control under adverse conditions.

All automotive service and parking brakes operate basically as outlined in this chapter. Of course, modern brake systems are much more complex than described, and there are many variations among the systems now in use. Later chapters cover these variations in detail and highlight the specific components that make each system unique.

Brake System Overview

Review Questions

Choose the letter that represents the best possible answer to the following questions:

1. The first brake design to offer variable braking force was the _____ brake.
   a. Locked-wheel
   b. Farm-wagon
   c. Chariot
   d. External contracting-band

2. A brake band may be used on:
   a. Internal expanding brakes
   b. External contracting brakes
   c. Truck parking brakes
   d. All of the above

3. Internal expanding-shoe brakes are preferred over earlier designs because they:
   a. Expose the friction elements to a cooling airflow
   b. Allow the use of thinner linings for better heat dissipation
   c. Do not suffer from lining breakage problems
   d. Offer twice the friction area

4. A brake _____ is the disc brake equivalent of a brake drum.
   a. Pad
   b. Rotor
   c. Caliper
   d. Backing plate

5. Which of the following is *not* a job of the service brake system?
   a. Mechanically lock at least two wheels
   b. Slow a vehicle that is in motion
   c. Hold a car in position on an incline
   d. Provide braking power to all four wheels

6. The service brake system is applied using:
   a. Mechanical force
   b. Hydraulic pressure
   c. Both a and b
   d. Neither a nor b.

7. The brake booster is located:
   a. On the brake pedal
   b. Inside the vehicle, near the instrument cluster
   c. Between the firewall and the master cylinder
   d. On the power steering pump

8. A brake master cylinder:
   a. Often contains two hydraulic circuits
   b. Moves only a small volume of brake fluid
   c. Creates high hydraulic pressures
   d. All of the above

9. Hydraulic valves and switches in a brake system:
   a. Control the amount of fluid leakage
   b. Create additional hydraulic pressure to ensure safe braking
   c. Both a and b
   d. Neither a nor b

10. A drum brake wheel cylinder:
    a. Converts hydraulic pressure into mechanical pressure
    b. May contain one to four pistons
    c. Performs the same job as the master cylinder
    d. None of the above

11. The wheel cylinder and brake shoes mount on the:
    a. Backing plate
    b. Brake drum
    c. Steering knuckle
    d. Suspension

12. The disc brake caliper is the equivalent of the drum brake:
    a. Master cylinder
    b. Hydraulic valves and switches
    c. Lining
    d. Slave cylinder

13. The amount of force required to hold a parked car in position is _____ that required to stop a slow-moving vehicle.
    a. Exactly the same as
    b. Less than
    c. More than
    d. Approximately the same as

14. Parking brakes:
    a. Are hydraulically operated
    b. Are completely separate from the service brakes
    c. Sometimes operate on the front wheels
    d. Have a tendency to leak if used for long periods

15. On cars with four-wheel disc brakes, the parking brake may be applied with a:
    a. Thrust screw
    b. Ball-and-ramp mechanism
    c. Lever and strut
    d. All of the above

16. The most common form of brake actuation system is:
    a. Mechanical-hydraulic
    b. Static-electric
    c. Air-hydraulic
    d. None of the above

17. The job of the actuating system is to:
    a. Transmit braking force to the wheels
    b. Increase the available braking force
    c. Both a and b
    d. Neither a nor b

18. Which of the following was *not* a problem with mechanically actuated brakes:
    a. Linkage wear
    b. More modern car designs
    c. Front-to-rear brake balance
    d. Uneven brake lining wear

19. The primary advantage of hydraulic actuation is that it:
    a. Provides high-application pressures
    b. Is virtually maintenance free
    c. Eliminates braking force variations
    d. None of the above

20. One of the reasons pneumatic and electric brakes were adopted was because they both:
    a. Are extremely safe
    b. Require little energy to operate
    c. Can be used with disc brakes
    d. Are easy to connect and disconnect

2

Brake Legal and Health Issues

OBJECTIVES

Upon completion and review of this chapter, you will be able to:

- Relate what is expected of a modern brake system by the FMVSS 135 test.
- State who is liable for damages or injuries resulting from poor brake repair.
- List the health hazards of asbestos exposure.
- List the sources of asbestos exposure.
- Describe the methods for safe removal of asbestos from brake assemblies.
- Describe the methods for safe handling of asbestos.
- List the dangers inherent in the chemicals used in brake repair.
- Describe the methods to avoid chemical poisoning.
- Describe the health care rights of technicians.

KEY TERMS

arcing
asbestos
asbestosis
brake dust
Economic Commission for Europe (ECE)
European Union (EU)
Ferodo
forsterite
gross vehicle weight rating (GVWR)
lightly loaded vehicle weight (LLVW)
Occupational Safety and Health Administration (OSHA)
regenerative braking system (RBS)

BRAKES AND THE LAW

Like many parts of the modern automobile, the brake system is regulated by federal laws. These laws specify minimum performance levels for motor vehicle braking systems, and were enacted to ensure that auto manufacturers produce vehicles with adequate stopping power. However, modern brake systems are not designed simply to meet minimum legal requirements, they are also built with the performance capability of the entire vehicle in mind. A high-performance car has much more powerful brakes than a compact economy car, but both have brake systems that conform to federal regulations *and* provide adequate stopping power for the car's intended uses.

FEDERAL BRAKE STANDARDS

The statutes pertaining to automotive brake systems are part of the Federal Motor Vehicle Safety Standards (FMVSS) established by the U.S. Department of Transportation (DOT, www.dot.gov). These Standards cover many areas of car design, figure 2-1, but the Standards directly affecting brake systems are found in Part 571. Several Standards apply to specific components within the brake system, and these are covered in the appropriate chapters later in the text. The overall service and parking brake systems are dealt with in Standard 105 for vehicles manufactured before September 1, 2000, and Section 135 for vehicles manufactured after that date. The current federal requirement established for the **European Union (EU)** by the **Economic Commission for Europe (ECE)** is ECE R13H, mandated May 11, 1998, by the European community. Because the requirements are similar, only FMVSS 135 is considered for the purposes of this text.

FMVSS 135 was first mandated on September 1, 2000, for passenger cars, and September 1, 2002, for multipurpose vehicles, trucks, and buses with gross vehicle weight ratings (GVWR$) of 7,716 pounds (3,500 kilograms) or less. This brake requirement replaces the previous governing regulation, FMVSS 105. In its own words, the scope of FMVSS 135 is to specify "requirements for service brake and associated parking brake systems." Its purpose is to "insure safe braking performance under normal and emergency conditions." FMVSS 135 applies "to passenger cars, multipurpose passenger vehicles, trucks, and buses."

FMVSS 135 deals with brake system safety by establishing specific brake *performance* requirements. It does not dictate the *design* of the system, although some requirements may make older technologies impractical or obsolete. Only four parts of the brake system are specifically regulated: the fluid reservoir and labeling; the dashboard warning lights; a method of automatic adjustment; and a mechanically engaging, friction-type parking brake system. However, room is allowed for variations in final design.

The bulk of FMVSS 135 consists of a comprehensive test procedure designed to reveal any weaknesses in a vehicle's braking system. The test is used by manufacturers to certify the braking performance of new models and is never performed on a vehicle in service.

FMVSS 135 Brake Test

The overall FMVSS 135 brake test procedure consists of up to 24 steps, depending on the vehicle's configuration and braking system. The actual performance tests are made with the vehicle loaded to both the manufacturer's specified **gross vehicle weight rating (GVWR)** and the **lightly loaded vehicle weight (LLVW),** with certain applied brake forces, figure 2-2. There are precise instructions for every step of the test, including the number of times the tests must be repeated, the sequence of the testing, and the allowable stopping distance for the particular type of vehicle. These instructions must be followed exactly for the test to be valid. The procedure is summarized as follows. Before testing can be performed, the vehicle must be properly instrumented ac-

FMVSS	TITLE
105	Hydraulic Brake Systems (Sept 1, 2000 and earlier)
106	Brake Hoses
108	Lines, Reflective Devices, and Associated Equipment
116	Motor Vehicle Brake Fluids
121	Air Brake Systems
122	Motorcycle Brake Systems
135	Passenger Car Brake Systems
211	Wheel Nuts, Wheel Discs, and Hub Caps

Figure 2-1. Several Standards in Part 571 of the FMVSS deal with vehicle brakes and related systems.

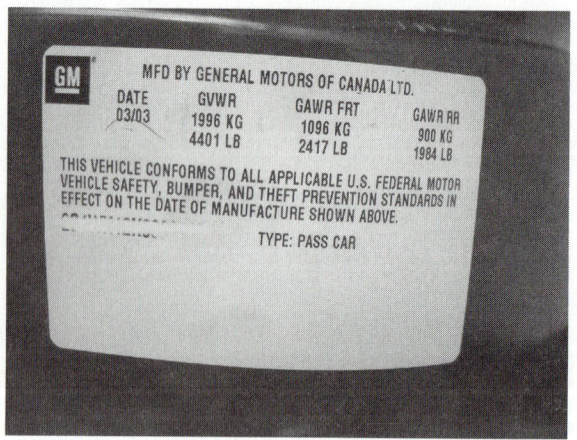

Figure 2-2. The vehicle GVWR is found on this label, located on the door edge or jamb.

Brake Legal and Health Issues

cording to the FMVSS 135 instructions. The test instrumentation is mounted on a vehicle with new brakes and checked for proper operation.

1. **Burnish procedure.** The brakes are burnished by making 200 stops from 50 mph (80 km/h) at a fixed rate of deceleration with a controlled cool-down period after each stop. Afterward, the brakes are adjusted according to the vehicle manufacturer's recommendations.
2. **Wheel lockup sequence.** For non–ABS-equipped vehicles, a wheel lockup sequence test is performed to ensure that the lockup of both front wheels occur simultaneously or at a rate of deceleration lower than the lockup of both rear wheels. This is to ensure that the vehicle will remain stable during an emergency stop.
3. **ABS performance (reserved).** ABS performance testing is currently under development by the Department of Transportation (DOT) members.
4. **Adhesion utilization (torque wheel method).** For vehicles not equipped with ABS, the adhesion utilization test is performed at LLVW and GVWR to determine whether the brake system will make adequate use of the road surface in stopping the vehicle.
5. **Cold effectiveness.** This cold effectiveness test is performed at both GVWR and LLVW to determine whether the vehicle will have sufficient stopping power when the brake lining materials are not preheated by previous stops.
6. **High-speed effectiveness.** The high-speed effectiveness test is performed only on vehicles capable of exceeding 78 mph (125 km/h) to determine whether the brake system will provide adequate stopping power for all loading conditions. The allowable stopping distance is calculated from the maximum speed the vehicle can attain.
7. **Stops with engine off.** The test for stopping with the engine off is for vehicles equipped with brake power units or power assist units. The test is also for electric vehicles. The vehicle, loaded to GVWR, must stop within 230 ft (70 m) from a speed of 62 mph (100 km/h). For an electric vehicle (EV) the test is conducted with the EV's propulsion motor disabled but with the vehicles' **regenerative braking system (RBS)** still functioning. This test must be repeated six times.
8. **Antilock functional failure.** The antilock functional failure test ensures that service brakes will function correctly in the event of an antilock functional failure and that the brake system warning indicator is activated when an ABS electrical function failure occurs.
9. **Variable brake proportioning system.** The variable brake proportioning system test is performed on vehicles equipped with either a mechanical or an electrical variable proportioning system. It ensures that, in the event of a failure, the vehicle can still come to a stop in an acceptable distance. In addition, if the vehicle uses an electrically operated variable brake proportioning system, the brake warning system must immediately alert the driver of any electrical functional failure.
10. **Hydraulic circuit failure.** The hydraulic circuit failure test is performed to ensure that the driver will be alerted via the brake warning system indicator that a failure has occurred and that the vehicle can still be stopped in an acceptable distance.
11. **Brake power assist unit inoperative.** The test for the brake power assist unit makes sure the service brake can stop the vehicle in an acceptable distance with the brake power assist unit in an inoperative state. It is performed on vehicles with brake power assist units turned off or rendered inoperative. For a vehicle in which the brake signal is transmitted electrically between the brake pedal and some or all of the foundation brakes, this may be any single failure in any circuit that electrically transmits the brake signal. For an EV with RBS that is part of the service brake system, this may be any single failure in the RBS.
12. **Parking brake.** The parking brake alone will hold the vehicle stationary in either forward or reverse direction on a 20 percent grade for a period of at least five minutes.
13. **Heating snubs.** The heating snubs procedure heats the brake system by making a series of 15 stops from a high speed. The vehicle is loaded to GVWR, with rapid acceleration between each stop to minimize cooling the brakes.
14. **Hot performance.** After the brake system has been heated by a series of heating snubs,

the hot performance test is immediately performed. The vehicle is loaded to GVWR and two stops are made. The stopping distance must be within acceptable limits as specified in the test. This test ensures that the brake system on the vehicle will not fade following a series of high speed stops at GVWR.

15. **Brake cooling stops.** During the brake cooling test, a series of three cycles of driving at 31 mph (50 km/h) for .91 mi (1.5 km) and then stopping is followed by one cycle of driving at 62 mph (100 km/h) for .91 mi (1.5 km), which is then followed by the next test. The cooling stops are conducted immediately after completion of the hot performance test.
16. **Recovery performance.** The recovery performance test is performed immediately after the cooling test to determine whether the brake system will recover braking performance after the hot performance test and after four brake cooling stops are made.
17. **Final inspection.** Following all of these steps, the brake system is inspected for any physical damage or hydraulic fluid leakage; none is permitted. Also, the brake fluid reservoir and the system warning light are checked to ensure that they comply with specific requirements set forth in the FMVSS Standard.

Although these tests may seem extreme, remember that they are only a minimum standard of performance. Any brake repair work performed on a vehicle should leave the brake system capable of meeting FMVSS 135.

BRAKE REPAIR AND THE LAW

Once automobiles leave the factory, the responsibility for maintaining the designed-in level of braking performance falls on car owners. Those owners look to trained automotive technicians to service their brake systems. To help ensure safe repairs, many states have laws that regulate brake work. These laws vary from one area to another, but they may require special licensing for brake technicians or special business practices when selling brake work. In some cases, the laws provide the consumer with specific warranties and the right to outside arbitration in cases of defective or substandard repairs.

■ The War on the Automobile

Auto manufacturers and enthusiasts today complain about the attacks environmentalists and legislators make on the automobile, and of the maze of rules and regulations that have resulted. Yet, compared to the early days of the automobile, today's conflicts seem minor. At the turn of the century, drivers speeding about American cities were perceived as rich and arrogant, and they might be subjected to a shower of rocks on the streets of poorer neighborhoods. This was nothing, however, compared to the problems encountered in the countryside.

Farmers resented the noisy automobile intruding on their quiet lifestyles and considered it a menace to their livestock. Some went on the offensive and planted broken bottles, jagged edge up, on paths that autos were likely to take. Others held careless motorists at gunpoint and made them pay the exorbitant price of one dollar for every chicken they had run down. Some of the worst opposition came from a group called the Farmers' Anti-Automobile Society. They demanded laws requiring any motorist whose car frightened a horse to immediately disassemble the offending vehicle and hide the pieces in the bushes.

The farmer and the automobile finally made peace when Henry Ford brought out his Model T. This affordable car, and its pickup truck offshoots, seduced farmers with speed, mobility, and carrying capacity that horse-drawn wagons could never hope to match.

Regardless of whether there are specific laws governing brake repair, a technician is always liable for damage or injuries resulting from repairs performed in an unprofessional manner. Considering the lives and property that depend on good brakes, there is only one acceptable goal when making brake system repairs: to restore the system and its component parts so they perform to original specifications. To do any less could result in an accident, injury, or death and may leave you and your business open to legal action.

BRAKES AND HEALTH

In the normal day-to-day course of events, automotive repair exposes technicians to many opportunities for injury. Most of these hazards are

Brake Legal and Health Issues

self-evident, however, and with experience a technician learns work habits that help prevent injuries. The hazards involved in brake repair are different. In addition to the normal safety concerns involved in any mechanical repair, brake work exposes the technician to other dangers that are almost invisible yet considered deadly. These hazards occur in two specific areas:

- Asbestos exposure
- Chemical poisoning

ASBESTOS EXPOSURE

Asbestos is the general name for a group of silicate minerals that are resistant to heat and corrosive chemicals. Asbestos has more than 3,000 commercial uses, but in the automotive field its primary applications are clutch disc facings and brake friction materials. All types of asbestos pose equally serious health threats.

Many technicians believe that the use of asbestos in clutches and brakes has been banned and the precautions discussed here are no longer necessary. Environmental Protection Agency (EPA) rules written in 1986 and 1987 banned the use of asbestos in automotive friction materials but the rules were overturned in 1989 and never took effect. Most vehicle manufacturers have voluntarily discontinued the use of asbestos, but it is still used in some vehicles and in replacement and aftermarket parts. Because the brake technician cannot tell for sure, it is safer to assume that all brake dust may contain some asbestos.

Asbestos is made up of millions of separate and bundled fibers, figure 2-3, that give it both strength and flexibility. To create commercially useful products, the fibers are woven or bonded together in a variety of ways. For automotive brakes, asbestos fibers are combined with other elements and molded into friction materials for brake shoes and pads. Typical asbestos-based brake friction materials are composed of 25 to 75 percent asbestos fibers.

The health risks from asbestos occur when individual asbestos fibers are freed from woven or molded materials and become suspended in the air where they can be inhaled or ingested. In automotive brakes, the fibers are normally released as the molded friction material is manufactured, machined, or worn into dust during use.

Figure 2-3. Asbestos is a very useful material, but its fibers are a source of several significant health hazards.

Asbestos Exposure Standards

The evolution of the current **Occupational Safety and Health Administration (OSHA)** asbestos exposure standard (1910.1001) indicates the minute sizes and quantities of fibers that can cause health problems. The OSHA standard defines an asbestos fiber as any strand longer than 5 microns (0.00020"). OSHA standards state that no employee may be exposed to more than 0.1 fiber per cubic centimeter of air averaged over an eight-hour period.

In addition to the maximum exposure limit, OSHA has established three "action levels" of exposure that require different responses. If air quality testing shows the exposure level is less than 0.1 fiber per cubic centimeter, no action is required. Most brake repair facilities will fall into this category, provided special cleaning tools are used to remove and collect the brake dust from the cars being serviced.

If testing shows the exposure level is between 0.1 and 0.2 fiber per cubic centimeter of air, regular air quality tests are required, the employer must conduct employee training to warn individuals about the dangers of asbestos, and a medical record-keeping system must be established to track the health of all employees exposed to asbestos. Unless proper safety procedures are implemented, it is possible some brake repair outlets may fall into this category.

Finally, if testing shows that the exposure level is above the legal limit of 0.2 fiber per cubic centimeter of air, a number of extreme steps, up to and including building redesign, are required to bring the workplace into compliance with regulations. This exposure level is likely to occur only in a brake friction material manufacturing plant; it is doubtful a brake repair facility would fall into this category unless it did a very large volume of business and safety measures were ignored altogether.

■ The Long and Short of Brake Theory

What are the practical results of all the laws, principles, and equations of brake theory? Just how effective is the conversion of kinetic energy into heat energy? If driver reaction time is not included, tests have shown that the average car can stop from 60 mph in about 135 feet—less than half the length of a football field. This distance represents the time it takes to convert the vehicle's kinetic energy into heat energy.

Average stopping distance is a generalization, however, and there are many variables from one vehicle to the next. Full-size trucks and vans tested took as long as 182 feet to stop from 60 mph; extra weight (more inertia) and unfavorable weight bias are the factors that lengthened their stopping distances. At the other end of the spectrum, high-performance cars stopped in as little as 107 feet. Wide sticky tires, low weight, and good weight balance are factors in their excellent performance.

To answer the first question, brake theory directly relates to vehicle stopping distances, and thus safety. The conversion of kinetic energy into heat energy is not yet as efficient as it might be, but the brakes on modern cars are far better than those of the past.

Sources of Asbestos Exposure

The most potentially hazardous asbestos exposure conditions in the brake industry exist at factories where shredded raw asbestos fibers are blended with other ingredients to form new friction materials. Most of these fibers are longer than 5 microns (0.00020") and pose a serious danger. Workers and plant owners take extraordinary precautions in these circumstances to avoid exposure to the fibers.

Friction material manufacture and machining are significant sources of asbestos exposure. The most common source for the brake technician is the dust created as brake shoes and pads wear. However, although the original friction material may contain up to 75 percent asbestos, the fiber content of **brake dust** is usually much lower and contains fewer fibers larger than 5 microns (0.00020"). The quantity and size of the asbestos fibers are reduced by the friction and heat of braking. Friction tears apart the fibers and shortens them, whereas heat causes a chemical breakdown of a portion of the asbestos and turns it into **forsterite**.

Effects of Asbestos Exposure

Diseases resulting from exposure to asbestos fibers can reach the incurable stage before they are detected and can cause severe disability or death. The symptoms of the diseases are delayed and may occur 20 to 30 or more years after first exposure. Disease is more likely to occur as a result of repeated, long-term inhalation or ingestion of fibers, rather than from occasional exposure to higher concentrations.

Asbestosis

One major health hazard related to asbestos exposure is **asbestosis.** Asbestosis is a progressive and disabling lung disease caused by inhaling asbestos fibers. The fibers become lodged in the lungs and cause inflammation of the air sacs and tubes. As the inflammation heals, it leaves scar tissue that thickens the linings of the air sacs making it difficult for oxygen to get into the bloodstream, figure 2-4. Strained breathing caused by a lack of oxygen can lead to heart failure. Asbestosis is irreversible once scarring has begun, and in advanced cases it can continue to destroy lung tissue even though the victim is no longer exposed to asbestos.

The symptoms of asbestosis are shortness of breath, pain in the upper chest or back, and a dry sound (rales) during inhalation. As the ability to breathe becomes more limited, the fingers and toes become "clubbed"—rounded with flattened nails. A bluish discoloration of the skin and lining of the mouth and tongue may also appear.

Brake Legal and Health Issues

■ Brake Shoe Arcing

For best braking performance, the curvature of the lining on a brake shoe should closely match the contour of the brake drum. It was not uncommon in the past for the replacement shoes to be made oversized. Then, after turning the drum, the shoe was machined to fit the drum. This was achieved by grinding the lining to the proper curvature in a process called **arcing**. The lining material ground away is converted into dust that contains about the same size and percentage of asbestos fibers as the friction material.

Most commercial brake shops no longer need machine linings for passenger cars and light trucks. However, for large trucks and buses, where asbestos is still a typical friction material, arcing is not uncommon. In these cases, the arcing equipment must have an approved dust collection system.

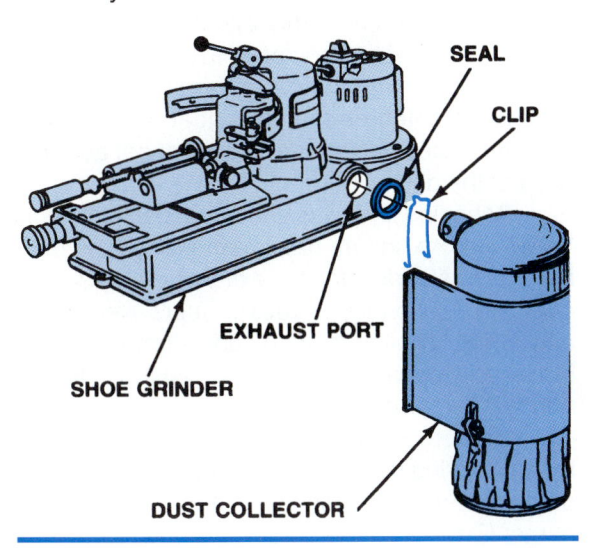

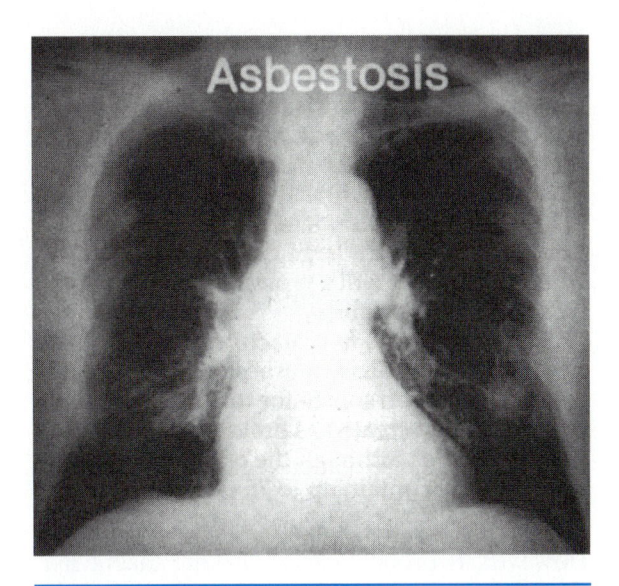

Figure 2-4. Asbestos fibers irritate and scar lung tissue, which slows the transfer of oxygen into the bloodstream.

Cancer

Lung cancer has the highest mortality rate of any asbestos-related disease and has caused up to 25 percent of all deaths in some groups of individuals heavily exposed to asbestos. Among the general population, only 4 to 5 percent die from lung cancer. In cases of cancer, the action of asbestos fibers alone is multiplied by the presence of other cancer-causing substances, such as cigarette smoke.

The symptoms of lung cancer vary and depend to a great extent on where the cancer first occurs. If it starts in the air passages (bronchi), it will cause partial obstruction and irritation. The symptom for such bronchial cancers will probably be a cough, and the sputum may contain blood. However, lung cancer can also start in any other part of the lung, and these cancers are not suspected until they show up on X-rays, or until late in their development when pain and a shortness of breath appear.

Mesothelioma is an extremely rare form of cancer in the general population. However, mesotheliomas involving cancer of the lining of the lungs or abdominal cavity are frequent in persons exposed to high concentrations of asbestos. Mesothelioma is incurable and can cause death in six months to two years. The symptoms are a shortness of breath and chest or abdominal pain.

Heavy exposure to asbestos can also lead to cancers of the digestive system—the esophagus, stomach, colon, and rectum. All of these cancers are thought to be caused by ingesting asbestos fibers.

ASBESTOS PRECAUTIONS

As the descriptions in the previous section make clear, the dangers posed by exposure to asbestos fibers are serious. The only way to reduce the risk of disease is to reduce exposure to the lowest possible level. This is done by adopting proper shop practices, using special brake cleaning equipment, paying close attention to personal hygiene, and making sure all co-workers do the same. One person who violates asbestos exposure safety rules can endanger everyone in the shop.

The first step in reducing exposure is to set aside a special location for brake repairs. Isolate brake work in a part of the shop away from highly traveled areas. The brake repair area should never be swept with a dry broom that would raise dust. The floor can be washed clean or vacuumed with a brake vacuum equipped with a special filter (discussed later). The OSHA regulations require that entrances to areas where levels of airborne asbestos exceed the standards must be posted with a warning sign, figure 2-5. However, if proper precautions are taken, it is doubtful that unsafe exposure levels will occur in an auto repair shop.

The disassembly of worn friction assemblies is the major source of asbestos exposure for brake technicians. This operation can create serious hazards unless certain rules are followed.

- Never use compressed air or a dry brush to clean accumulated dust from a brake assembly; these methods put hundreds of thousands of asbestos fibers into the air.
- Another method of cleaning brake assemblies that has been recommended in the past was to cover the brake with a wet cloth and carefully wipe away the dust; this technique should also be discontinued because it will release levels of fibers above legal limits.
- Do not remove the brake drum without first using one of the containment procedures that follow.
- Do not bang the drum on the floor to "knock the dust out."

Danger
Asbestos
Cancer And Lung Disease Hazard
Authorized Personnel Only
Respirator And Protective Clothing
Are Required In This Area

Figure 2-5. This sign must be posted in areas where airborne asbestos levels exceed OSHA regulations.

■ **A Brief History of the Use of Asbestos in Automobile Brake Linings**

Before the introduction of brake lining materials as we know them today, horse drawn carriages used leather and wood as friction materials. In fact—old boots were often used for this purpose. Because vehicle speeds were slow, the friction materials needed were not critical. Herbert Frood developed the first friction materials in 1897 in England. The first friction linings were composed of hair and cotton belting. The belting was then charged with coal dust to give it higher friction and greater stopping power. Frood's invention resulted in superior stopping power and was so successful that it was used by both early motorized vehicles and horse drawn carriages.

The problem with the woven cotton belting used in brake linings was that it would char and lose its frictional characteristics at about 300°F (149°C). To overcome this effect, in 1908 Frood introduced woven Chrysotile asbestos fibers to replace the woven cotton and hair belting. This provided greater resistance to heat generated by the ever-increasing speed and weight of automobiles of the day.

Asbestos fiber was chosen because it could be woven like cotton, has great strength, is flexible, and stands up to extreme temperature changes. Generally, asbestos fibers were spun into string, similar to yarn, and combined with brass and copper wires for increased strength and temperature resistance.

Herbert Frood started a well-known friction manufacturing company, **Ferodo,** by combining a combination of his last name, and his wife's first name, Elizabeth.

The use of asbestos fibers became prevalent in the brake lining industry until the EPA issued the restraints on its use. Because of concerns over the use of asbestos and health-related illnesses, asbestos has been almost completely eliminated from original equipment manufacturing use—but it is still prevalent in aftermarket brake linings.

Today, asbestos is being replaced in brake linings by high-tech materials such as Kevlar. However, the long-term health effects of such materials are still not known.

Brake Legal and Health Issues

OSHA has mandated that, as of April 10, 1995, brake repair shops must use one of these methods (or another equally effective method) when cleaning brake assemblies. The vendor that supplies the shop with washer equipment may also be able to supply the OSHA-required brake cleaning equipment.

- An enclosure system that uses an industrial vacuum equipped with an approved High Efficiency Particulate Air (HEPA) filter, figure 2-6. Do not attempt to use a conventional shop vacuum; asbestos fibers are small enough to easily pass through a normal dust filter.
- An alternative to a vacuum with an HEPA filter is a low-pressure brake washer, figure 2-7. These devices use water or a special cleaning solution to flush the dust from the friction assembly into a holding tank.

Unlike old worn friction materials, new brake shoes or pads pose relatively little threat of airborne asbestos contamination. Most of the fibers are locked into the molded friction material, and residual dust in the packaging can be vacuumed out or wiped away with a damp cloth. The majority of replacement brake shoes today are preground to an arc designed to fit a brake drum that has been machined oversize. However, if further arcing is needed, it should only be performed on equipment fitted with an approved dust collection device.

In addition to good shop practices, brake technicians need to take extra personal care; good hygiene is a must. Wash your hands thoroughly before eating, and definitely avoid smoking in the work area. If exposure levels are high enough, wear special work clothing on the job, change it regularly, and have it cleaned often. Always wear an approved respirator when arcing brake shoes.

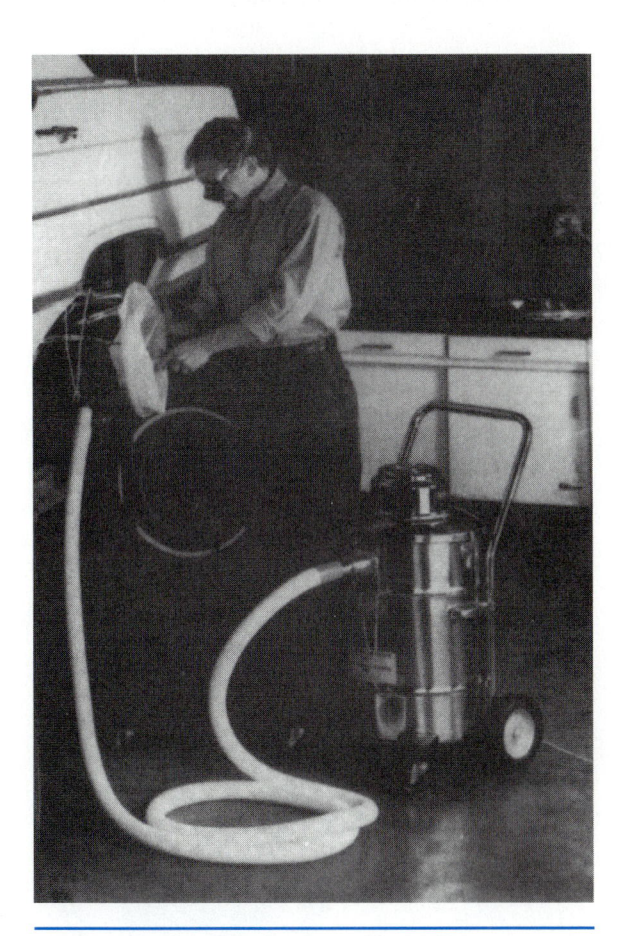

Figure 2-6. A vacuum with an HEPA filter is an excellent means of cleaning dust from brake assemblies.

Figure 2-7. A low-pressure liquid brake washer is another method of brake cleaning that helps minimize exposure to asbestos fibers.

ASBESTOS WASTE DISPOSAL

EPA regulations require that all waste materials containing more than 1 percent asbestos must be disposed of in a manner that will not endanger the public health. Because brake dust exceeds this limit, vacuum cleaner bags and filters, cloths used for wiping up brake dust, and the dust removed from brake washers must be sealed in double plastic bags or some other form of nonpermeable container. The bags must then be marked with an asbestos exposure warning label, figure 2-8, and disposed of properly.

Because the hazards created by asbestos are primarily a result of airborne fibers, it is generally acceptable to turn the bagged asbestos waste over to a local trash collection agency for burial in a sanitary landfill. Local hazardous waste disposal regulations differ, however, and may require special disposal procedures for asbestos-contaminated materials. Check with the local authorities or EPA office if you have any questions.

CHEMICAL POISONING

Chemical poisoning in brake service is not as severe a problem as asbestos exposure, but it still demands concern on the part of technicians. The major sources of chemical danger are liquid and aerosol brake cleaning fluids that contain chlorinated hydrocarbon solvents. The most common such solvents are 1,1,1-trichlorethane, trichlorethylene, and tetrachlorethylene, which is also known as perchlorethylene or "perk" for short.

These solvents are all members of the same chemical family and share the same basic characteristics. They are colorless liquids with an odor of chloroform or ether. They are narcotic, and in large enough quantities can dull the senses, induce sleep, or cause a stupor. Very high levels of exposure over even a short period of time can be fatal. And if these solvents are exposed to high heat or an open flame, they decompose into deadly gases, such as hydrogen chloride, phosgene, and carbon monoxide.

Because 1,1,1-trichlorethane and trichlorethylene are known to be ozone depleters, their manufacture has been prohibited since January 1, 1996 by the EPA. Perks do not deplete the ozone and their use will continue. Several other chemicals that do not deplete the ozone, such as heptane, hexane, and xylene, are now being used in nonchlorinated brake cleaning solvents. Some manufacturers are also producing solvents they describe as environmentally responsible, which are biodegradable and noncarcinogenic, figure 2-9.

Sources of Chemical Poisoning

The health hazards presented by brake cleaning solvents occur from three different forms of exposure: ingestion, inhalation, or physical contact. It should be obvious that swallowing brake cleaning solvent is harmful, and such occurrences are not a common problem. Still, brake cleaning sol-

**Danger
Contains Asbestos Fibers
Avoid Creating Dust
Cancer And Lung Disease Hazard**

Figure 2-8. All containers holding asbestos wastes must be identified with this label before disposal.

Figure 2-9. Biodegradable brake cleaning sprays and liquids make service safer, but they must still be used with proper precautions.

Brake Legal and Health Issues

vents should always be handled and stored properly and kept out of reach of children.

The dangers of inhalation are perhaps the most serious problem with these chemicals; even very low levels of solvent vapors are hazardous. For example, the current OSHA standard (1910.1000) for airborne trichlorethylene is 100 parts per million (ppm) in the ambient air averaged during an eight-hour work shift. The ceiling level for exposure is 200 ppm, and there is a maximum acceptable peak level of 300 ppm for five minutes in any two-hour period. The limits for other chlorinated hydrocarbon solvents, and for other chemicals now replacing the chlorinated ones, are similar. These alternative chemicals are being used because they do not deplete the ozone layer, not because they are necessarily any safer to breathe, ingest, or contact.

Ingestion and inhalation are common forms of poisoning from many hazardous substances, but allowing brake cleaning solvents to come in contact with the skin presents a danger unknown to many people. Not only do these solvents strip natural oils from the skin and cause irritation of the tissues but they also have the ability to be absorbed through the skin directly into the bloodstream. The transfer begins immediately on contact and continues until the liquid is wiped or washed away.

There is no specific standard for physical contact with chlorinated hydrocarbon solvents or the chemicals replacing them; all contact should be avoided whenever possible. The law requires an employer to provide appropriate protective equipment and ensure proper work practices by employees handling these chemicals.

Effects of Chemical Poisoning

The effects of exposure to chlorinated hydrocarbon and other types of solvents can take many forms. Short-term exposure at low levels can cause headache, nausea, drowsiness, dizziness, uncoordination, or unconsciousness. It may also cause irritation of the eyes, nose, and throat, and flushing of the face and neck. Short-term exposure to high concentrations can cause liver damage with symptoms such as yellow jaundice or dark urine. Liver damage may not become evident until several weeks after the exposure.

In addition to the symptoms already mentioned, long-term or repeated exposure to perk may cause irritation or burning of the skin; it also increases the risk of damage to the liver or kidneys. If you experience any of these symptoms, and suspect that they may be a result of exposure to brake cleaning solvents, seek medical treatment immediately.

CHEMICAL PRECAUTIONS

Unlike many industrial applications of chlorinated hydrocarbon solvents, automotive brake cleaning sprays and liquids present relatively limited opportunity for exposure to dangerous levels of contamination. The possibility still exists, however, and just as with asbestos there are safety precautions that should be followed to minimize the risk.

Always use any brake cleaning solvent in an open, well-ventilated area, and avoid breathing the vapors. Take precautions to prevent physical contact with the liquid solvent, and clean any spills onto the skin promptly using soap and water. Wear protective clothing and immediately remove any piece of clothing that becomes wetted with solvent; do not wear the item again until it has been cleaned. Wear safety goggles or other eye protection when spraying brake cleaning solvents. Finally, observe good personal hygiene; wash your hands before eating, smoking, or using the toilet.

HEALTH CARE RIGHTS

The OSHA regulations concerning on-the-job safety place certain responsibilities on the employer and give employees specific rights. Any person who believes there might be unsafe conditions where he or she works, whether asbestos exposure, chemical poisoning, or any other problem, should discuss the issue with fellow workers, the union representative (where applicable), and the supervisor or employer. If no action is taken and there is reason to believe the employer is not complying with OSHA standards, a complaint can be filed with OSHA and it will investigate.

The law forbids employers from taking action against employees who file complaints concerning health or safety hazards. However, if workers fear reprisal as the result of a complaint, they may request that OSHA withhold their names from the employer.

SUMMARY

The performance of vehicle braking systems is regulated by federal law. Every new brake system must pass a comprehensive test before it is approved for market. When performing brake repair work, the objective should be to restore the system to like-new performance and condition.

Just as brake system performance is federally legislated, brake system repair may be controlled by state or local laws. Brake repair outlets must operate within local statutes, but even where there are no specific regulations, brake repairs are covered under general laws pertaining to the responsibility of those in the service trades to perform their work in a professional manner.

Brake work exposes technicians to several hazards that other automotive repairs might not. Among these are asbestos and chemical poisoning. The manufacture and arcing of brake shoes release raw asbestos fibers, and brake dust created as friction materials wear also contains asbestos fibers, although in lower concentrations.

Asbestos fibers can cause a variety of ills, including asbestosis and lung cancer. The only way to prevent such diseases is to reduce exposure to the lowest practical level. This is done through proper shop practices, the use of special brake cleaning equipment, and good personal hygiene.

Chemical poisoning in the brake repair industry occurs primarily from brake cleaning solvents that can be inhaled, ingested, or absorbed through the skin. Short-term effects of exposure can be as simple as a headache or dizziness, whereas long-term effects may involve liver and kidney damage. The danger of chemical poisoning from brake cleaning solvents is minimal, but steps taken to avoid breathing the vapors or coming into physical contact with the solvents can eliminate the danger almost entirely.

Limits for both asbestos and chemical exposure are set by OSHA to protect persons in the workplace. If a hazard exists and an employer is unwilling to correct the situation, employees may file a complaint with OSHA, which will investigate and require corrective action.

Because of the known dangers to human health and the environment, asbestos and chlorinated solvents are being phased out and replaced with less hazardous substances. The technician must still observe precautions, but both he or she and the environment benefit.

Brake Legal and Health Issues

Review Questions

Choose the letter that represents the best possible answer to the following questions:

1. Modern vehicle brake systems are designed:
   a. To meet federal vehicle safety standards
   b. To match the vehicle's performance capability
   c. To provide safe stopping under most conditions
   d. All of the above

2. FMVSS 135 regulates brake systems primarily by requiring that they meet certain _____ standards.
   a. Design
   b. Performance
   c. Safety
   d. All of the above

3. Which of the following is not a part of the FMVSS 135 brake test?
   a. Spike stop test
   b. Fluid reservoir inspection
   c. Stopping stability test
   d. Water recovery test

4. Legal liability for brake system repairs rests with the:
   a. Vehicle manufacturer
   b. Technician
   c. Vehicle owner
   d. All of the above

5. The greatest danger posed by asbestos comes from:
   a. Particles
   b. Brake dust
   c. Fibers
   d. Vapors

6. Asbestos-based friction materials for automotive brakes normally contain _____ percent asbestos.
   a. 15 to 50
   b. 25 to 75
   c. 50 to 100
   d. 75 to 100

7. A brake technician is most commonly exposed to asbestos from:
   a. Brake shoe arcing
   b. Brake dust
   c. Friction material manufacture
   d. None of the above

8. Which of the following is not true of brake dust?
   a. It contains less asbestos than the brake lining material
   b. It contains forsterite
   c. It is reduced by the actions of braking
   d. It is a significant health hazard

9. Which of the following may be caused by exposure to asbestos?
   a. Stomach cancer
   b. Asbestosis
   c. Mesothelioma
   d. All of the above

10. Exposure to asbestos can be limited by:
    a. Sweeping brake repair areas regularly
    b. Using a soft bristle brush to clean the brake assemblies
    c. Changing the filter in the shop ventilation system regularly
    d. Wiping friction assemblies clean with a wet rag

11. Asbestos waste materials from automobile brake work must:
    a. Be marked with an identifying label
    b. Disposed of at a hazardous waste disposal site
    c. Both a and b
    d. Neither a nor b

12. Which of the following is not a trait of perk?
    a. It causes liver damage
    b. It is a narcotic
    c. It acts as a stimulant
    d. It emits hazardous vapors

13. The current OSHA standard forbids skin contact with solutions containing more than _____ ppm of chlorinated hydrocarbon solvents.
    a. 111
    b. 200
    c. 300
    d. None of the above

14. Brake cleaning sprays and liquids:
    a. Can be cleaned from the skin with soap and water
    b. Present extremely serious danger of chemical poisoning
    c. Both a and b
    d. Neither a nor b

15. If a brake technician believes the shop environment is unsafe because of asbestos or chemical exposure, the best course of action is to:
    a. File a complaint with the DOT
    b. Discuss the matter with a supervisor
    c. Look for other employment
    d. Contact a lawyer

16. The EPA ban on the use of asbestos in brakes:
    a. Is in force today
    b. Means technicians can ignore most precautions for asbestos
    c. Requires new cars to use asbestos-free brake linings
    d. Was overturned in 1989 and is not in force today

17. The 1,1,1-triclorethane and trichlorethylene-type brake cleaning solvents were not manufactured after January 1, 1996, because _____
    a. They presented safety hazards to the technicians using them
    b. They depleted the ozone layer
    c. Both a and b
    d. Neither a nor b

18. The part of the braking system not specifically regulated by the DOT is:
    a. The fluid reservoir
    b. The automatic transmission park lock mechanism
    c. The parking brake mechanism
    d. The dashboard warning lights

19. Which chemical is not a new replacement for the ozone-depleting chlorinated brake cleaning solvents?
    a. Hexane
    b. Xylene
    c. Perchlorethylene
    d. Heptane

3
Principles of Brake Operation

OBJECTIVES

Upon completion and review of this chapter, you will be able to:

- Define the terms "energy" and "work."
- Explain the relationship between speed, weight, and kinetic energy.
- Explain the concepts of inertia and weight transfer.
- Explain the concepts of leverage and mechanical advantage.
- Explain the significance of the concept "non-compressibility of liquids."
- Explain the concept "constancy of hydraulic pressure."
- List the factors that affect the coefficient of friction.
- Define the terms "static" and "kinetic" friction.
- Explain the relationship between friction and heat.
- Define and explain brake fade.

KEY TERMS

brake fade
coefficient of friction
energy
first law of
  thermodynamics
friction
gas fade
inertia
kinetic energy
kinetic friction
leverage
lining fade
mass
mechanical advantage
mechanical fade
pressure
static friction
thermodynamics
weight bias
weight transfer
work

INTRODUCTION

All brake systems, regardless of design or actuation, work in accordance with physical "laws" or principles that describe the relationships between elements of the physical world. Some of the elements involved in brake systems are the weight and speed of the vehicle, the hydraulic pressure of the actuating system, the mechanical force used to apply the brakes, and the heat created by the wheel friction assemblies. Weight and speed involve matters of energy and inertia, hydraulic pressure is based on specific hydraulic principles, application force is determined by both hydraulic and mechanical principles, and heat is explained by physical laws concerning friction. The physical laws of energy, inertia, hydraulics, mechanics, and friction explain the relationships that enable a brake system to work.

ENERGY PRINCIPLES

Energy is the ability to do work. There are many forms of energy, figure 3-1, but chemical, mechanical, and electrical energy are the most familiar kinds involved in the operation of an automobile. For example, when the ignition key is turned to the start position, chemical energy in the battery is converted into electrical energy to operate the starter motor. The starter motor then converts the electrical energy into mechanical energy that is used to crank the engine.

In the preceding example, energy is being used to perform work. **Work** is the transfer of energy from one physical system to another—especially the transfer of energy to an object through the application of force. This is precisely what occurs when a vehicle's brakes are applied: The *force* of the actuating system *transfers* the energy of the vehicle's motion to the brake drums or rotors where friction *converts* it into heat energy and stops the car. To understand this process clearly, let's take a close look at energy in the brake system.

KINETIC ENERGY

Kinetic energy is a fundamental form of mechanical energy; it is the energy of mass in motion. Every moving object possesses kinetic energy, and the amount of that energy is determined by the object's mass and speed. The greater the mass of an object and the faster it moves, the more kinetic energy it possesses. Even at low speeds, a moving automobile has enough kinetic energy to cause serious injury and damage, figure 3-2. The job of the brake system is to dispose of that energy in a safe and controlled manner.

It is important to note that kinetic energy is based on speed and **mass**, not weight. The difference is subtle, but one way to describe it is to say that mass remains constant whereas weight can vary. For example, an astronaut and a space shuttle resting on the launching pad have a *weight* difference of several hundred tons, but once in orbit they weigh the same, nothing. Their respective *masses*, however, remain very different. If a spacewalking astronaut grasps a wing and tries to move the shuttle, the only thing that will move is the astronaut's body. Despite the fact that both are weightless, the mass of the shuttle remains thousands of times

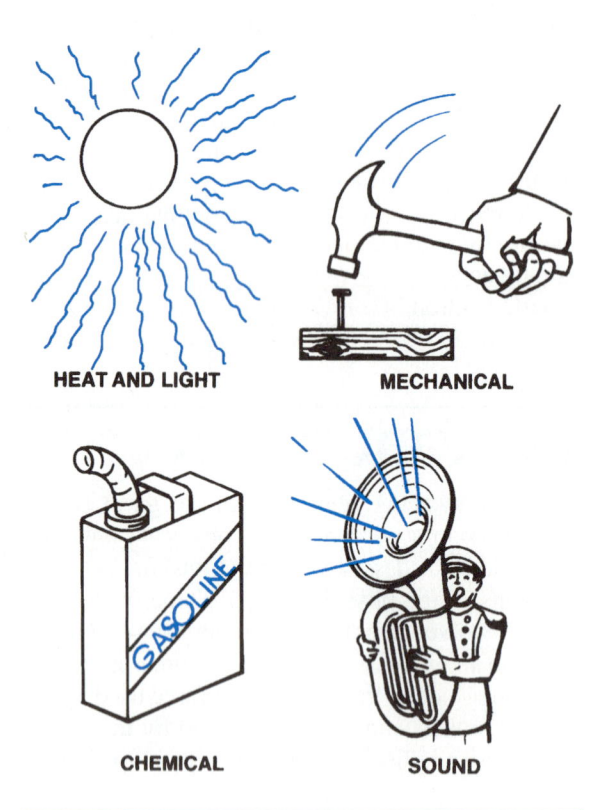

Figure 3-1. Energy, the ability to perform work, exists in many forms.

Figure 3-2. This vehicle shows the result if a car's kinetic energy is not disposed of in a controlled manner.

Principles of Brake Operation

greater than that of the astronaut. Technically, weight is the mass of an object acted on by the force of gravity. Because the force of gravity is relatively constant on Earth, we will use the terms weight and mass interchangeably in this text.

Weight and Speed Effects

Although both weight and speed contribute to kinetic energy, they do not affect it to the same degree; speed has a much greater effect. Consider two balls, a lightweight foam baseball and a regulation hardball that weighs nine times as much. If both are thrown by a child at 10 miles per hour (mph), neither is likely to cause injury if it hits someone. Both balls travel at the same *speed*, but despite the relatively large difference in *weight*, there is not a significant difference in kinetic energy.

Now bring in a major-league pitcher who can throw the hardball at 90 mph, nine times faster than the child. Whereas a baseball thrown at 10 mph is nearly harmless, the fastball can break bones or cause a concussion. The baseball has the same *weight* in both cases, but the ninefold increase in *speed* results in much greater kinetic energy than did the ninefold weight increase of the previous example.

The relative effects of speed and weight on kinetic energy hold true for all physical objects including automobiles. To explain why this occurs, it is necessary to look at kinetic energy from a mathematical viewpoint. Engineers calculate kinetic energy using the formula:

$$\frac{mv^2}{29.9} = E_k$$

Where:
m = mass or weight of the car in pounds
v = velocity of the car in miles per hour
E_k = kinetic energy in foot-pounds (ft-lb)

Another way to express this equation is:

$$\frac{\text{weight} \times \text{speed}^2}{29.9} = \text{kinetic energy}$$

Although the preceding baseball examples indicate the effect of weight on kinetic energy is less than that of speed, weight does have a major effect. The equation for computing kinetic energy can show exactly what that effect is. If a 3,000-pound car traveling at 30 mph is compared to a 6,000-pound car also traveling at 30 mph, figure 3-3, the equa for computing their respective kinetic energies look like this:

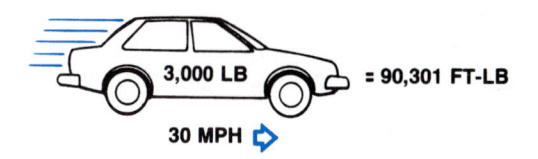

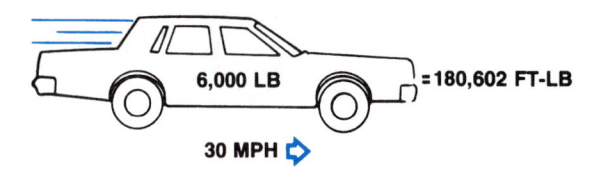

Figure 3-3. Kinetic energy increases in direct proportion to vehicle weight.

$$\frac{3{,}000 \text{ lb} \times 30^2 \text{ mph}}{29.9} = 90{,}301 \text{ ft-lb}$$

$$\frac{6{,}000 \text{ lb} \times 30^2 \text{ mph}}{29.9} = 180{,}602 \text{ ft-lb}$$

The results show that when the weight of a car is doubled from 3,000 to 6,000 pounds, its kinetic energy is also doubled from 90,301 foot-pounds to 180,602 foot-pounds. In mathematical terms, kinetic energy increases *proportionally* as weight increases. In other words, if the weight of a moving object doubles, its kinetic energy also doubles; if the weight quadruples, the kinetic energy becomes four times greater.

The baseball examples also indicated that speed has a much greater effect on kinetic energy than does weight. The equation for computing kinetic energy can again be used to explain why this occurs. If a 3,000-pound car traveling at 30 mph is compared to the same car traveling at 60 mph, figure 3-4, the equations for computing their respective kinetic energies look like this:

$$\frac{3{,}000 \text{ lb} \times 30^2 \text{ mph}}{29.9} = 90{,}301 \text{ ft-lb}$$

$$\frac{3{,}000 \text{ lb} \times 60^2 \text{ mph}}{29.9} = 361{,}204 \text{ ft-lb}$$

The results show that the car traveling 30 mph has more than 90,000 foot-pounds of kinetic energy, but at 60 mph the figure increases to more than 350,000 foot-pounds. In fact, at twice the speed,

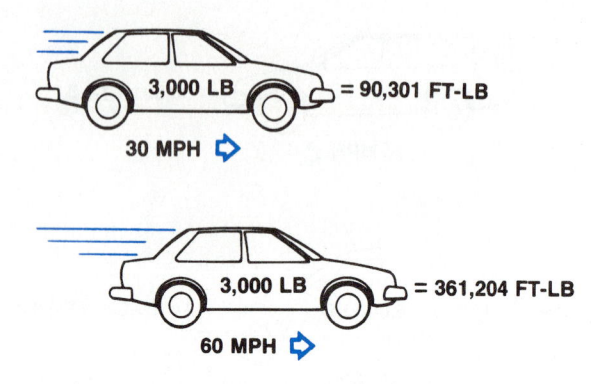

Figure 3-4. Kinetic energy increases as the square of any increase in vehicle speed.

the car has exactly four times more kinetic energy. If the speed were doubled again to 120 mph, the amount of kinetic energy would grow to almost 1,500,000 foot-pounds! In mathematical terms, kinetic energy increases as the *square of its speed*. In other words, if the speed of a moving object doubles (2), the kinetic energy becomes four times as great ($2^2 = 4$). And if the speed quadruples (4), say from 15 to 60 mph, the kinetic energy becomes 16 times as great ($4^2 = 16$). This is the reason speed has such an impact on kinetic energy.

Kinetic Energy and Brake Design

The relationships between weight, speed, and kinetic energy have significant practical consequences for the brake system engineer. If car A weighs twice as much as car B, it needs a brake system that is twice as powerful. But if car C has twice the speed potential of car D, it needs brakes that are, not twice, but four times more powerful. In the 1950s when horsepower, vehicle weights, and highway speeds all increased dramatically, many large, heavy cars had marginal brakes. Since the introduction of disc brakes, however, most cars have had brake systems with a good margin of safety.

INERTIA

Although brake engineers take both weight and speed capability into account when designing a brake system, these are not the only factors involved. Another physical property, inertia, also affects the braking process and the selection of brake components. **Inertia** is defined by Isaac Newton's first law of motion which states that a body at rest tends to remain at rest, and a body in motion tends to remain in motion in a straight line unless acted on by an outside force.

The space shuttle described earlier is at rest on the launch pad and remains so until acted on by an outside force, in this case its rocket engines. As the shuttle reaches space its engines are shut off and, following Newton's law, it attempts to continue in a straight line away from the Earth. However, the outside force of gravity acts on the shuttle and bends its course into a curved path called an orbit.

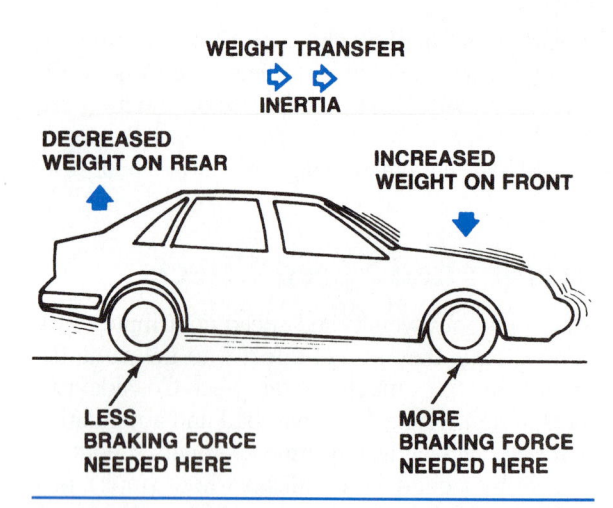

Figure 3-5. Inertia creates weight transfer that requires the front brakes to provide most of a car's braking power.

Weight Transfer and Bias

Inertia, in the form of **weight transfer,** plays a big part in a vehicle's braking performance. Newton's first law of motion dictates that a moving car will remain in motion unless acted on by an outside force. The vehicle brakes provide that outside force, but when the brakes are applied at the wheel friction assemblies, only the wheels and tires begin to slow immediately. The rest of the car, all of the weight carried by the suspension, attempts to remain in forward motion. The result is that the front suspension compresses, the rear suspension extends, and weight is transferred toward the front of the car, figure 3-5. The total weight of the car does not change, only the amount supported by each axle.

To compound the problem of weight transfer, most cars also have a forward **weight bias,** even when stopped, more than 50 percent of their weight is supported by the front wheels. This occurs because the engine, transmission, and most other heavy parts are located toward the front of

Principles of Brake Operation

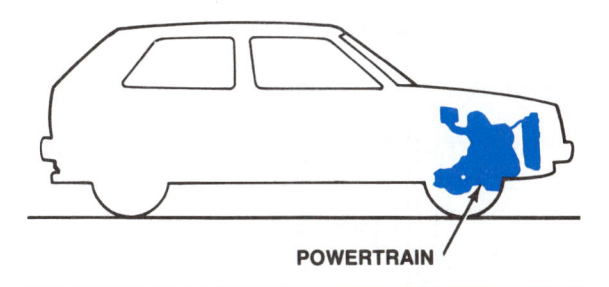

Figure 3-6. This FWD car has a forward weight bias that places more than 60 percent of its weight over the front wheels.

the vehicle, figure 3-6. Front-wheel-drive (FWD) cars, in particular, have a forward weight bias.

Whenever the brakes are applied, weight transfer and weight bias greatly increase the load on the front wheels, whereas the load on the rear wheels is substantially reduced. This requires the front brakes to provide 60 to 80 percent of the total braking force. To deal with the extra load, the front brakes are much more powerful than the rear brakes; they are able to convert more kinetic energy into heat energy.

■ Braker Beware!

The effects of weight and speed on kinetic energy can have serious consequences in the real world. Professional brake technicians need to be on the lookout for vehicle owners who overload their cars' brake systems. For instance, when Robert Recreation asks if he can pull a 2,500-lb boat and trailer with his subcompact economy car, you had better recommend against it unless the vehicle is equipped with a towing package that includes upgraded brakes to handle the increase in weight. And when Peter Performance asks you to help swap a Corvette engine into his Corolla you had better make sure the sports car's brakes are also bolted onto the resulting rocketship.

The standard brakes in the previously mentioned cars would probably work just fine during normal driving, but the first time Robert tows his boat and trailer down a mountain, his brakes could go up in smoke. And when Peter lets his V-8 Corolla unwind on that long straight stretch, he might find out to his dismay that the car has a lot more go than whoa. Most people take their car's brake system for granted, but ignorance of the principles involved can easily lead to dangerous brake failure.

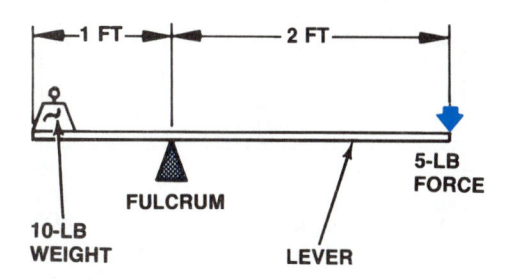

Figure 3-7. A first-class lever increases force and changes the direction of the force.

MECHANICAL PRINCIPLES

The physical principles of kinetic energy and inertia describe the forces the brake system must convert into heat through friction. For the wheel friction assemblies to develop that heat, they must be applied with great force—in fact, the force required is so great that leverage and hydraulics must be employed to allow a human being to apply it. This section details the mechanical principles used by brake actuating systems to create the necessary application force. Hydraulic actuating principles are covered in the next section.

The primary mechanical principle used to increase application force in every brake system is **leverage.** In the science of mechanics, a lever is a simple machine that consists of a rigid object, typically a metal bar, that pivots about a fixed point called a fulcrum. There are three basic types of levers, but the job of all three is to change a quantity of energy into a more useful form. The type of lever chosen for a particular job is normally determined by the situation and the results desired. The following paragraphs examine the effects the three kinds of levers have on the job of moving a 10-pound weight.

A first-class lever, figure 3-7, increases the force applied to it and also changes the direction of the force. With a first-class lever, the weight is placed at one end while the lifting force is applied at the other; the fulcrum is positioned at some point in between. If the fulcrum is placed twice as far from the long end of the lever as from the short end, a 10-pound weight on the short end can be lifted by only a 5-pound force at long end. However, the short end of the lever will travel only half as far as the long end. Moving the fulcrum closer to the weight will further reduce the force required to lift it, but it will also decrease the distance the weight is moved.

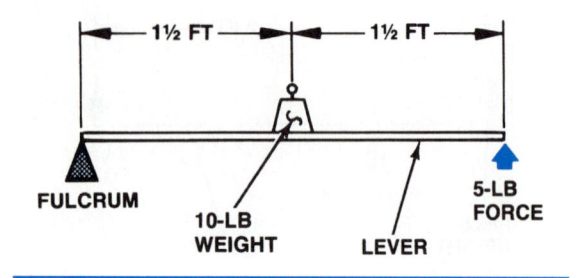

Figure 3-8. A second-class lever increases force in the same direction it is applied.

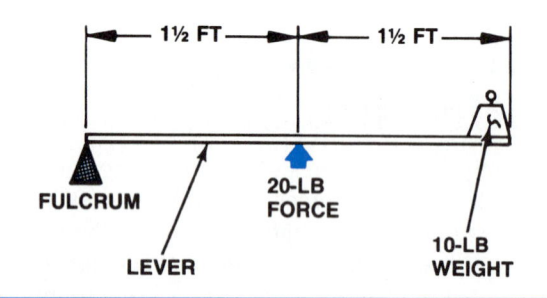

Figure 3-9. A third-class lever reduces force but increases the speed and travel of the resulting work.

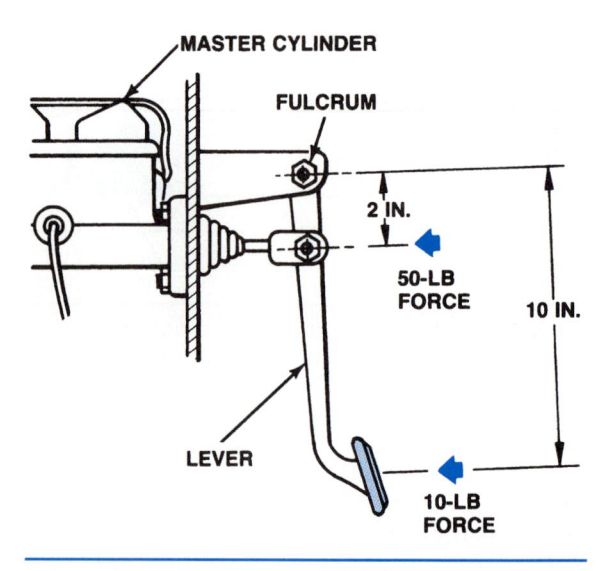

Figure 3-10. This brake pedal assembly is a second-class lever that provides a 5 to 1 mechanical advantage.

A second-class lever, figure 3-8, increases the force applied to it and passes it along in the same direction. With a second-class lever, the fulcrum is located at one end while the lifting force is applied at the other; the weight is positioned at some point in between. If a 10-pound weight is placed at the center of the lever, it can be lifted by only a 5-pound force at the end of the lever. However, the weight will only travel half the distance the end of the lever does. As the weight is moved closer to the fulcrum, the force required to lift it, and the distance it travels, are both reduced.

A third-class lever, figure 3-9, actually reduces the force applied to it, but the resulting force moves farther and faster. With a third-class lever, the fulcrum is located at one end and the weight is placed at the other; the lifting force is applied at some point in between. If a 10-pound weight is placed at the end of the lever, it can be lifted by a 20-pound force applied at the middle of the lever. Although the force required to move the weight has doubled, the weight is moved twice as far and twice as fast as the point on the lever where the force was applied. The closer to the fulcrum the lifting force is applied, the greater the force required but the farther and faster the weight will move.

The levers in brake systems are used to increase force, so they are either first- or second-class levers. Second-class levers are the most common, and the service brake pedal is a good example. In a typical suspended brake pedal, figure 3-10, the pedal arm is the lever, the pivot point is the fulcrum, and the force is applied at the foot pedal pad. The force applied to the master cylinder by the pedal pushrod attached to the pivot is much greater than the force applied at the pedal pad, but the pushrod does not travel nearly as far.

Leverage creates a **mechanical advantage** that, at the brake pedal, is called the pedal ratio. For example, a pedal ratio of 5 to 1 is common for manual brakes, which means that a pressure of 10 pounds at the brake pedal will result in a pressure of 50 pounds at the pedal pushrod. In practice, leverage is used at many points in both the service and parking brake systems to increase braking force while making it easier for the driver to control the amount of force applied.

HYDRAULIC PRINCIPLES

In addition to the mechanical advantage provided by leverage, all modern cars, and many trucks, use hydraulic pressure to help increase brake application force. They do so because hydraulic actuating systems are governed by physical laws that make them very efficient at transmitting motion and force. In addition, hydraulic principles make it easier and more convenient to achieve larger in-

Principles of Brake Operation

creases in application force than if the same gains were obtained by mechanical methods, such as leverage. The laws of hydraulics explain how this is done.

■ The Father of Mechanics

The Greek philosopher Archimedes, who lived in the second century B.C., is justly credited with being the father of the science of mechanics. Archimedes brought to light the principle of leverage, the first important mechanical law. Next to the wheel, leverage is the earliest discovery of technology that eventually allowed the creation of the automobile.

Archimedes was puzzled by the fact that a small weight could balance a larger one when they were placed on opposite ends of a pivoting rod. From experimentation, he learned that the distance of each weight from the pivot was the crucial factor. If one of the weights was halved, its distance from the pivot had to be doubled to maintain even balance. From this, Archimedes deduced that a three-foot rod (lever) with one foot extending past the pivot point (fulcrum) would allow a man to lift a weight on the short end by applying only half the force at the long end.

Archimedes bragged to his friend King Hieron of Syracuse, "give me a place to stand and I will move the earth." He was implying that he would do so using a fulcrum and a very long, very stout lever. The levers used in automobile brakes do not have to lift the world, but without leverage brakes would be nearly impossible to operate effectively.

Noncompressibility of Liquids

Hydraulic systems use liquids to transmit motion. This is possible because, for all practical purposes, a liquid cannot be compressed. No matter how much pressure or force is placed on a quantity of liquid, its volume will remain the same. This trait enables liquids in a closed system to transmit motion. Figure 3-11 shows a simple hydraulic system. If piston A is moved a distance of 1 inch, the liquid will be displaced ahead of it and piston B will move 1 inch as well.

Whereas liquids cannot be compressed, the same is not true of gasses. A gas such as air *will* compress, and hydraulic systems must be free of air to work properly. The simple hydraulic system shown in figure 3-12 has been contaminated with air. Even though piston A is moved a distance of 1

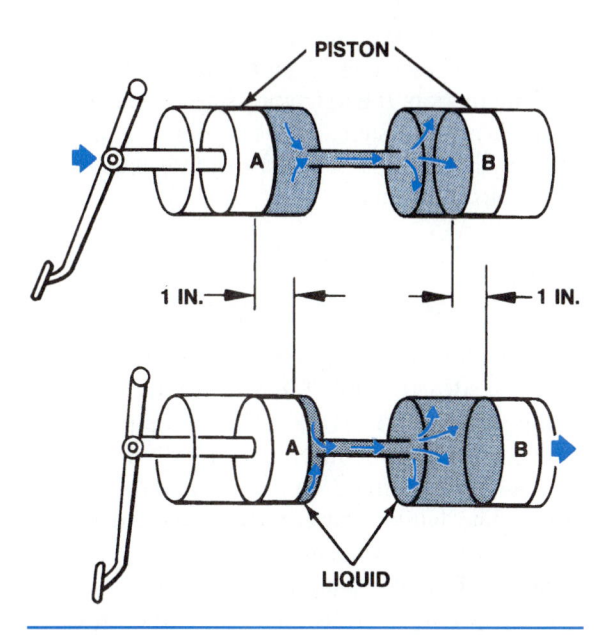

Figure 3-11. Because liquids cannot be compressed, they are able to transmit motion in a closed system.

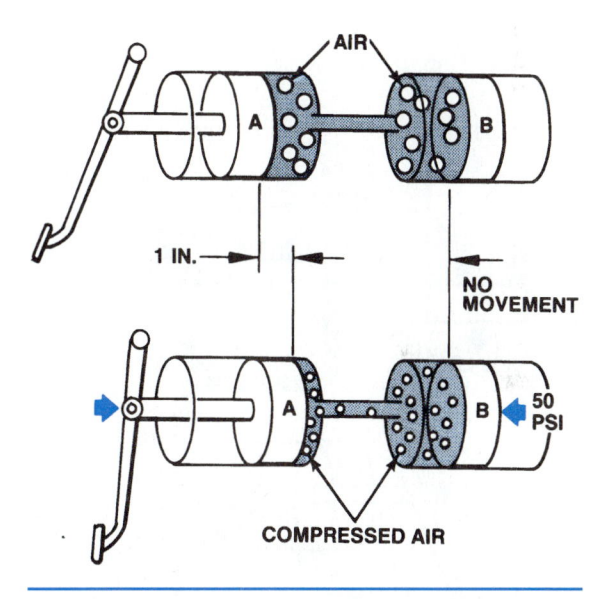

Figure 3-12. Hydraulic systems must be air free to operate properly.

inch, piston B will not move if the load on it is greater than the pressure of the air in the system. For example, if the load on piston B is 50 pounds per square inch, the movement of piston A must compress the air in the system to that same pressure before piston B will begin to move.

Unfortunately, air requires a great deal of work to compress to high pressures. The amount of piston travel in a hydraulic brake system is insufficient to compress even a small amount of air to any appreciable degree. This is not to say that air cannot transmit motion. If enough pressure is available, as in an air brake system, air works quite well. But a brake hydraulic system must be air free or there will be serious problems.

Constancy of Pressure

Brake hydraulic systems not only transmit motion but they also transmit force in the form of hydraulic pressure. **Pressure** is the amount of *force* applied to a specific *area*. It is usually measured in pounds per square inch (psi) or kilo-Pascals (kPa). The latter term comes, in part, from the name of Blaise Pascal, who in 1650 discovered an important law governing pressurized liquids. Pascal found that pressure on a confined liquid is transmitted equally in all directions and acts with equal force on equal areas, figure 3-13.

In a brake hydraulic system, Pascal's law dictates that if 100 psi is produced by the master cylinder, 100 psi will exist at every point throughout the system. In practice, however, this is not always true because engineers install different kinds of control valves in the system to modify pressures for more balanced braking.

■ Computing Piston Area

For clarity of explanation, all of the hydraulic component examples in this chapter use pistons with surface areas that are easy to divide and multiply. In real brake systems, piston sizes are chosen for other reasons, and their areas are usually not simple numbers. The surface area of any hydraulic piston can be calculated with the formula:

$$\pi R^2 = A$$

Where:
π = 3.142
R = the radius of the piston diameter in inches
A = the piston surface area in square inches

Another way to express this equation is:

$$3.142 \times radius^2 = area$$

For example, if a caliper piston has a 2-inch diameter (1-inch radius), the equation reads:

$$3.142 \times 1^2 = 3.142 \text{ sq. in.}$$

Likewise, if a wheel cylinder piston has a 0.5- or 1/2-inch diameter (0.25- or 1/4-inch radius), the equation reads:

$$3.142 \times 0.25^2 = 0.196 \text{ sq. in.}$$

As the results show, these common piston sizes result in uneven surface area totals. Choosing the piston sizes and determining their surface areas so that the vehicle will have proper braking balance is one of the many jobs of the brake system engineer.

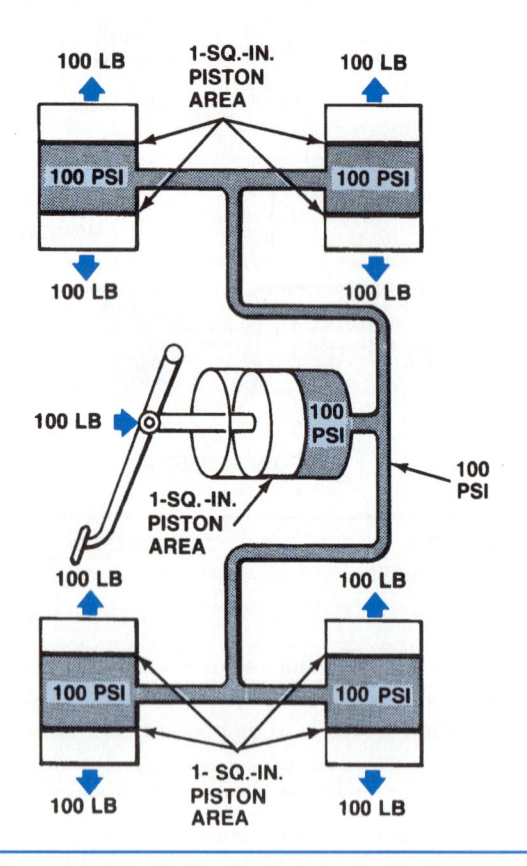

Figure 3-13. Hydraulic pressure is the same throughout a closed system, and acts with equal force on equal areas.

Hydraulic Pressure and Piston Size

Brake hydraulic systems are designed to operate within a certain range of pressures. The amount of pressure at any given moment is determined by two factors: the force applied to the brake pedal multiplied by the mechanical advantage of the pedal ratio, and the surface area of the master cylinder piston. The manner in which brake pedal pressure and pedal ratio result in

Principles of Brake Operation

mechanical force was described earlier. To understand how changes in piston area affect hydraulic pressure, it is necessary to once again take a mathematical approach. The formula used by engineers to compute pressures within a brake system is:

$$\frac{F}{A} = p$$

Where:
F = mechanical force applied to the piston
A = piston area in square inches (in.²)
p = pressure in psi

Another way to express this equation is:

$$\frac{\text{force}}{\text{piston area}} = \text{pounds per square inch}$$

It is important when using this equation to realize that it is the surface *area* of the piston, not its diameter, that affects the pressure.

Consider the examples shown in figure 3-14. If a mechanical force of 100 pounds is exerted by the brake pedal pushrod onto a master cylinder piston with 1 square inch of surface area, the equation reads:

$$\frac{100 \text{ lb}}{1 \text{ in.}^2} = 100 \text{ psi}$$

The result in this case is 100 psi of brake system hydraulic pressure. However, if the same 100-pound force is applied to a master cylinder piston with twice the area (2 square inches) the equation will read:

$$\frac{100 \text{ lb}}{2 \text{ in.}^2} = 50 \text{ psi}$$

Doubling the area of the master cylinder piston cuts the hydraulic system pressure in half. Conversely, if the same 100-pound force is applied to a master cylinder piston with only half the area (0.5 or H square inch) the equation will show that the system pressure is doubled:

$$\frac{100 \text{ lb}}{0.5 \text{ in.}^2} = 200 \text{ psi}$$

Application Force and Piston Size

Although the size of the master cylinder piston affects the hydraulic pressure of the entire brake system, weight shift and bias require that the heavily loaded front brakes receive much higher application forces than the lightly loaded rear brakes. These differences in force are obtained by using different sized pistons in the wheel cylinders and brake calipers. Remember, Pascal's law states that a pressurized liquid in a confined space acts with equal pressure on equal *areas;* as long as all the pistons in a hydraulic system have the same area, as in figure 3-13, 100 psi from the master cylinder will result in 100 psi of friction assembly application force. However, when equal pressure acts on *unequal* areas (i.e., different sized pistons), the brake application force will differ as well.

The mathematical equation in the previous section described how mechanical *force* at the brake pedal pushrod is applied to the master cylinder piston *area* and converted into brake system hydraulic *pressure*. Brake calipers and wheel cylinders perform exactly the opposite; hydraulic *pressure* applied to the wheel cylinder or brake caliper piston *area* is converted back into mechanical *force* that is used to apply the wheel friction assemblies. Because the variables are identical, the same equation can be rewritten to explain how changes in piston

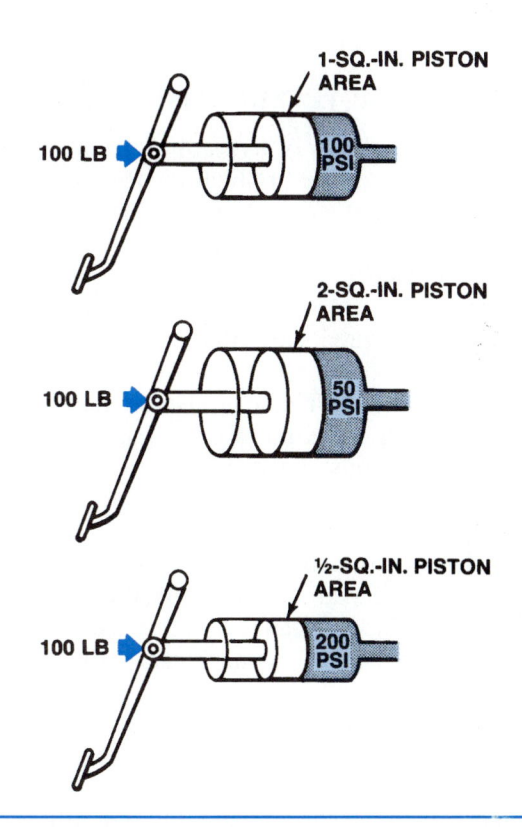

Figure 3-14. Mechanical force and master cylinder piston area determine hydraulic pressure within the brake system.

size affect brake application force. When the equation is rewritten to solve for mechanical force instead of hydraulic pressure, it reads:

$$p \times A = F$$

Where:
p = system hydraulic pressure in psi
A = piston area in square inches
F = application force in pounds

Another way to express this equation is:

$$psi \times area = force$$

Once again, remember that it is piston surface *area*, not diameter, that affects force. Some examples will demonstrate how this equation works.

In the simple brake system shown in figure 3-15, the pedal and linkage apply a 100-pound force on a master cylinder piston with an area of 1 square inch. This results in a pressure of 100 psi throughout the hydraulic system. At the front wheels, the 100 psi is applied to a brake caliper piston that has an area of 4 square inches. The equation for this example is:

$$100 \text{ psi} \times 4 \text{ in.}^2 = 400 \text{ pounds}$$

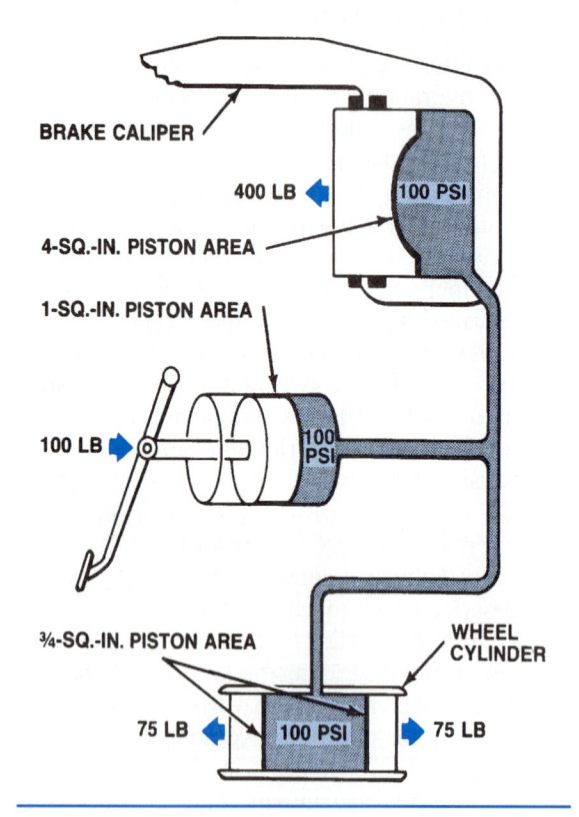

Figure 3-15. Differences in brake caliper and wheel cylinder piston area have a significant effect on brake application force.

In this case the difference in piston areas (1 square inch compared to 4 square inches) results in the 100-psi brake pedal pushrod force being increased to 400 pounds of application force at the wheel friction assembly. Note, however, that the hydraulic pressure is still 100 psi at all points within the system—the increase in application pressure is solely the result of 100 psi acting on a 4-square-inch piston; the 400 pounds is a mechanical force, not hydraulic pressure.

The drum brakes at the rear wheels of the same brake system, figure 3-15, use wheel cylinders whose pistons have three quarters of an inch (¾ or 0.75) of surface area. If the hydraulic system pressure remains 100 psi, the equation for this example is:

$$100 \text{ psi} \times 0.75 \text{ in.}^2 = 75 \text{ pounds}$$

Just as larger pistons increase application force, this example shows that smaller pistons decrease it. Once again the system hydraulic pressure remains 100 psi at all points, but the smaller piston is unable to transmit all of the available pressure. As a result, the mechanical application force is reduced to only 75 pounds.

Piston Size versus Piston Travel

Although the ability of hydraulic systems to increase and decrease application forces would seem to make it easy to build very powerful brakes, there is another side to the process that must be considered. The first law of thermodynamics, which is discussed in greater detail later in the chapter, states that energy cannot be destroyed, it can only be changed from one form into another. This can also be states such that whenever one kind of energy is increased, another kind must be decreased. Or, in even simpler terms, you don't get something for nothing.

In the previous disc brake example, the mechanical force available to apply the brakes is four times greater because of the size difference between the master cylinder and caliper pistons. Some of the hydraulic energy is converted into *increased* mechanical force. The tradeoff is that the larger caliper piston with the greater force will not move as far as the smaller master cylinder piston. The amount of hydraulic energy converted into mechanical motion is *decreased*. The relative movement of pistons

Principles of Brake Operation

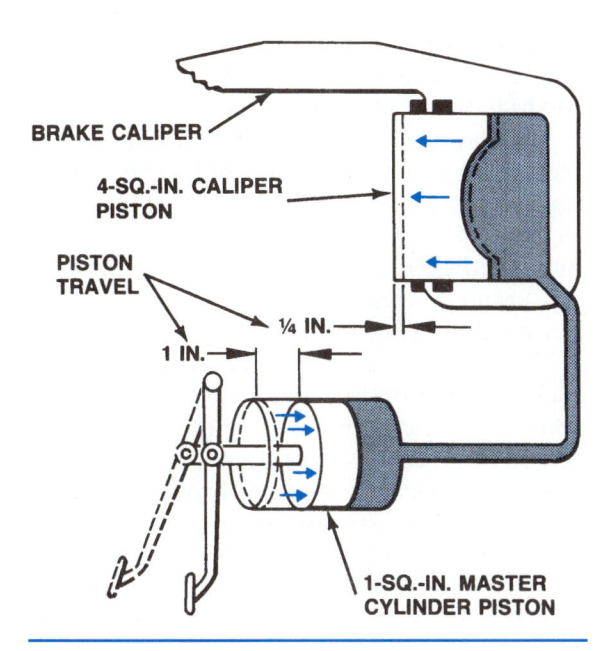

Figure 3-16. The increase in application force created by a large brake caliper piston is offset by a decrease in piston travel.

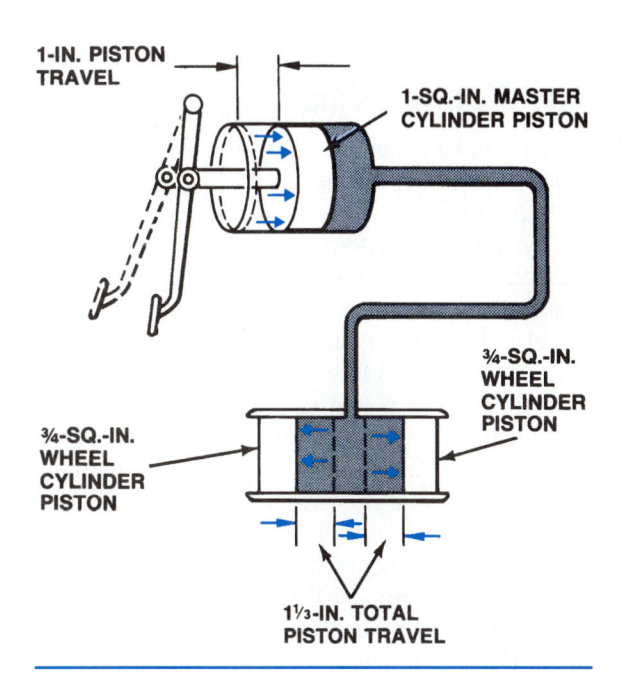

Figure 3-17. The decrease in application force created by a small wheel cylinder piston is offset by an increase in piston travel.

within the brake system can be calculated with the equation:

$$\frac{A_1}{A_2} \times S = M$$

Where:

A_1 = the area of the master cylinder piston
A_2 = the area of the wheel cylinder or caliper piston
S = master cylinder piston stroke length
M = wheel cylinder or caliper piston movement

Another way to express this equation is:

$$\frac{\text{area}_1}{\text{area}_2} \times \text{stroke} = \text{movement}$$

In the case of the previous disc brake example, the equation would read:

$$\frac{1 \text{ in.}^2}{4 \text{ in.}^2} \times 1 \text{ in.} = \frac{1}{4} \text{ in.}$$

The results show that in this example if the master cylinder piston stroke is 1 inch, the caliper piston will move only ¼ inch, figure 3-16. If the caliper piston area were reduced to only 2 square inches, the application force would increase to only 200 pounds, but the caliper piston would travel ½ inch for a 1-inch master cylinder stroke.

The equation for computing the difference in piston movement works for wheel cylinders as well. In the previous drum brake example, the amount of force transmitted by the wheel cylinder is less than the 100 psi that exists within the hydraulic system. If energy cannot be destroyed, the extra 25 psi of pressure must be converted into another form. The equation for this problem reveals where the energy goes:

$$\frac{1 \text{ in.}^2}{0.75 \text{ in.}^2} \times 1 \text{ in.} = 1.333 \text{ in.}$$

The answer shows that if the master cylinder again travels 1 inch, the wheel cylinder piston will travel 1⅓ inches. With a dual-piston wheel cylinder like that shown in figure 3-17, the total travel is divided between the two pistons. If the wheel cylinder piston area were reduced to only 0.5 or ½ inch, the application force would be further reduced to only 50 pounds, but the wheel cylinder piston would travel 2 inches for a 1-inch master cylinder stroke.

In mathematical terms, piston travel changes as the inverse of the change in the application force. In other words, if the application force is doubled (2 = ²⁄₁), the piston travel is halved (½). But if the application force is halved (½), the piston travel is doubled (²⁄₁ = 2).

Hydraulic Principles and Brake Design

When a brake system is designed, the hydraulic relationships discussed previously play a major part in determining the sizes of the many pistons within the system. The piston sizes selected must move enough fluid to operate the wheel cylinder and brake caliper pistons through a wide range of travel, while at the same time they must create enough application force to lock the wheel friction assemblies. The piston sizes chosen should also provide the driver with good brake pedal "feel" so the brakes are easy to apply in a controlled manner.

For example, a very small master cylinder piston can provide a lot of hydraulic pressure with light pedal effort, but it will not move enough fluid to operate brake calipers with large pistons. In addition, a small piston will give the brake pedal a very "touchy" feel that makes modulation difficult and leads to premature brake lockup. A large piston, however, provides less pressure and requires higher pedal effort, but it provides plenty of fluid volume and results in a less sensitive pedal feel that makes the brakes easier to control. Most cars with disc brakes have large master cylinder pistons to move the required volume of fluid and a power booster to reduce the required brake pedal force.

Brake caliper and wheel cylinder piston sizes must also be selected to provide the proper force, travel, and feel for balanced braking. In practice, caliper pistons cannot be too large because, although they would provide great force, excessive pedal travel and fluid volume would be needed to move them the required distance. Wheel cylinders are found only on the rear brakes of modern cars, and obtaining sufficient fluid volume and application force is not a problem. The main concern in choosing between different size wheel cylinder pistons is to obtain proper front-to-rear brake balance under a wide range of stopping conditions.

FRICTION PRINCIPLES

The opening section of this chapter explained that a moving automobile has a great deal of kinetic energy. The previous two sections discussed the mechanical and hydraulic principles used by actuating systems to increase brake application force. This final section examines the physical laws and principles that affect how the wheel friction assemblies use the application force to convert kinetic energy into heat energy and stop the car.

The way the brake system disposes of kinetic energy is explained by the laws of **thermodynamics,** which is the study of the relationship between mechanical and heat energy. The **first law of thermodynamics,** mentioned earlier, states that energy cannot be created or destroyed; it can only be converted from one form into another.

The wheel friction assemblies, as their name implies, use **friction** to convert kinetic energy into heat energy. Friction is the resistance to movement between two surfaces in contact with one another. Brake performance is improved by increasing friction (at least to a point), and brakes that apply enough friction to use all the grip the tires have to offer will always have the potential to stop a car faster than brakes with less ability to apply friction.

Coefficient of Friction

The amount of friction between two objects or surfaces is commonly expressed as a value called the **coefficient of friction.** The coefficient of friction, also referred to as the friction coefficient, is determined by dividing tensile force by weight force. The tensile force is the pulling force required to slide one of the surfaces across the other. The weight force is the force pushing down on the object being pulled. The equation for calculating the coefficient of friction is:

$$\frac{F_t}{G} = \mu$$

Where:
F_t = tensile force in pounds
G = weight force in pounds
μ = coefficient of friction

Another way to express the equation is:

$$\frac{\text{tensile force}}{\text{weight force}} = \text{coefficient of friction}$$

This equation can be used to show the effect different variables have on the coefficient of friction. At any given weight (application) force there are three factors that affect the friction coefficient of vehicle brakes:

- Surface finish
- Friction material
- Heat.

Principles of Brake Operation

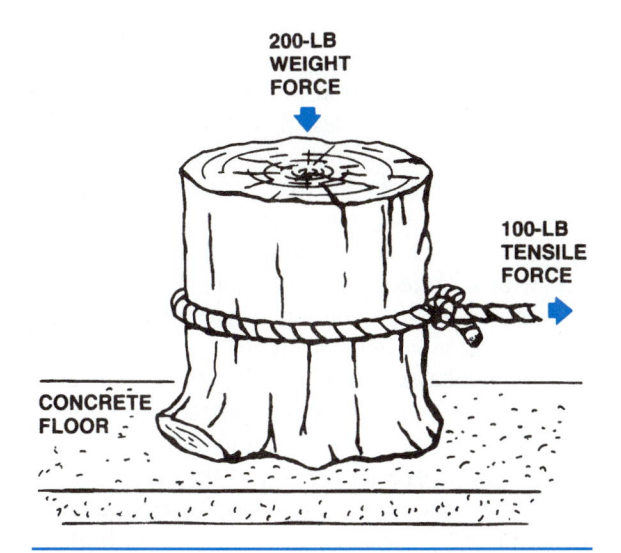

Figure 3-18. In this example, the coefficient of friction between the wood block and concrete floor is 0.5.

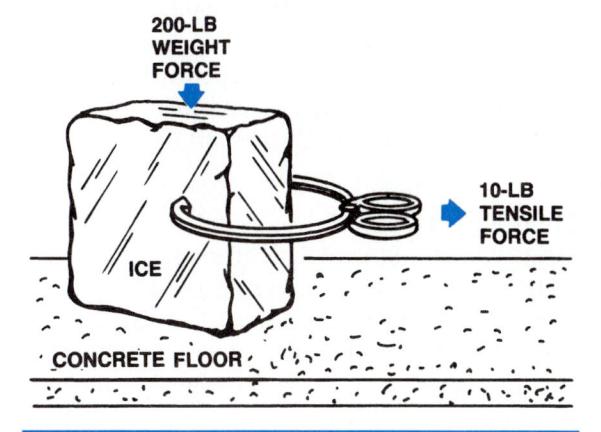

Figure 3-19. The types of friction materials affect the friction coefficient; the value in this example is only 0.05.

For reasons that will be explained later, the friction coefficient of the wheel friction assemblies of vehicle brake systems is always less than one.

Surface Finish Effects

The effect of surface finish on the friction coefficient can be seen in figure 3-18. In this case, 100 pounds of tensile force is required to pull a 200-pound block of wood across a concrete floor. The equation for computing the coefficient of friction is:

$$\frac{100 \text{ lb}}{200 \text{ lb}} = 0.5$$

The friction coefficient in this instance is 0.5. Now take the same example, except assume that the block of wood has been sanded smooth, which improves its surface finish and reduces the force required to move it to only 50 pounds. In this case the equation reads:

$$\frac{50 \text{ lb}}{200 \text{ lb}} = 0.25$$

The friction coefficient drops by half, and it would decrease even further if the surface finish of the floor were changed from rough concrete to smooth marble.

It is obvious that the *surface finish* of two contacting surfaces has a major effect on their coefficient of friction. However, on automotive brakes, the surface finish of the drums, rotors, and linings is predetermined by the fact that they must be smooth enough for good wear. The brake engineer cannot alter their smooth surface finishes to change the friction coefficient of the brakes because greatly increased wear would result. For example, 60-grit sandpaper on a soft pine brake rotor would provide an excellent coefficient of friction, but only for a few stops!

Friction Material Effects

Taking the preceding example one step further, consider the effect if a 200-pound block of ice, a totally different type of material, is substituted for the wood block. In this case, figure 3-19, it requires only a 10-pound force to pull the block across the concrete, so the equation reads:

$$\frac{10 \text{ lb}}{200 \text{ lb}} = 0.05$$

The coefficient of friction in this example decreases dramatically to only 0.05, and once again, even further reductions would be seen if the floor surface were changed to polished marble or some other similar smooth surface.

It is obvious that the *type* of materials being rubbed together has a very significant effect on the coefficient of friction. But, just as with surface finish, the choice of materials for brake drums and rotors is limited. Iron and steel are used most often because they are relatively inexpensive and can stand up under the extreme friction brake drums and rotors must endure.

The brake lining material, however, can be replaced relatively quickly and inexpensively, and, therefore, does not need to have as long a service life. Brake shoe and pad friction materials play a major part in determining coefficient of friction, and brake engineers use special care in selecting them. There are several fundamentally different materials to choose from, and each has its own unique friction coefficient and performance characteristics. These materials and their effects on braking are detailed in Chapter 10.

Heat Effects

Heat is the third factor that affects the coefficient of friction, but its effects are the most difficult to generalize about because heat has varied influences at different times and on different types of friction material. A little heat actually improves the friction coefficient of most automotive brakes, figure 3-20; they work best, just as an engine does, when warmed to operating temperature. However, if temperatures rise much beyond a certain point, the coefficient of friction begins to drop and braking efficiency is reduced. The effects of heat on brakes are discussed further in the *Brake Fade* section later in this chapter and also in Chapter 10.

Friction Contact Area

It might seem logical that the amount of contact area between the brake lining and drum or disc would be a fourth factor that affects the coefficient of friction. However, for *sliding* surfaces, such as those in wheel friction assemblies, the amount of contact area has no affect on the amount of friction generated. This fact is related to the earlier statement that brake friction materials always have a friction coefficient of less than 1.0. To have a friction coefficient of 1.0 or more, material must be *transferred* between the two friction surfaces.

Tires are an example where contact area makes a difference. All other things being equal, a wide tire with a large contact area on the road has a higher coefficient of friction than a narrow tire with less contact area. This occurs because the tire and road *do not* have a sliding relationship. A tire conforms to and engages the road surface, and during a hard stop, a portion of the braking force comes from shearing or tearing away of the tire tread rubber. The rubber's tensile strength, its internal resistance to being pulled apart, adds to the braking efforts of friction. A racing tire making a hard stop on dry pavement, for example, has a friction coefficient of 1.0 or better; the transfer of material between the two friction surfaces can be seen as skid marks on the pavement.

The fact that brake friction materials have a friction coefficient of less than 1.0 should not be seen as a deficiency. If they behaved like tires and transferred material to the brake drum or disc, they would wear far too quickly, be much too "grabby" in operation, and would not be able to withstand the heat generated in the wheel friction assemblies. And although the amount of contact area does not affect the coefficient of friction, it does have significant effects on lining life and the dissipation of heat that can lead to brake fade.

Static and Kinetic Friction

There are actually two measurements of the coefficient of friction, the **static friction** coefficient and the **kinetic friction** coefficient. The static value is the coefficient of friction with the two friction surfaces at rest. The kinetic value is the coefficient of friction while the two surfaces are sliding against one another.

The coefficient of static friction is always higher than that of kinetic friction, which explains why it is harder to *start* an object moving than to *keep* it moving. In the example in figure 3-21, it takes 100 pounds of tensile force to start the wooden block sliding, but once it is in motion, it takes only 50 pounds to keep the block sliding. The relatively high static friction is harder to overcome than the somewhat lower kinetic friction. The static and kinetic friction coefficients for several combinations of materials are shown in figure 3-22.

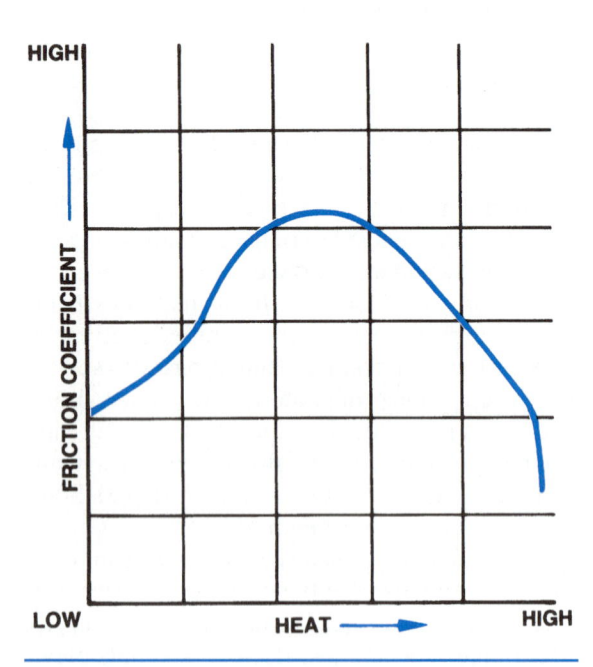

Figure 3-20. Some heat increases the coefficient of friction, but too much heat can cause it to drop off sharply.

Principles of Brake Operation

the stationary vehicle has no kinetic energy, and the brake lining and drum or disc are not moving when they are applied. To start the car moving, enough force would have to be applied to overcome the relatively high static friction of the parking brakes. The service brakes, however, have a much more difficult job. The moving car has a great deal of kinetic energy, and the fact that the brake friction surfaces are in relative motion means that kinetic friction makes them less efficient.

FRICTION AND HEAT

As already stated, the function of the brake system is to convert kinetic energy into heat energy through friction. But just how much heat is created by this conversion process? Once again, a mathematical equation can answer the question. Although there are too many variables to obtain the exact temperature increase of any specific component, the *average* temperature rise of the brakes during a single stop can be computed as follows:

$$\frac{K_c}{77.8 \, W_b} = T_r$$

Where:
- K_c = kinetic energy change in ft-lb
- W_b = weight of all the rotors and drums in pounds
- T_r = temperature rise in Fahrenheit degrees

Another way to express this equation is:

$$\frac{\text{energy change}}{77.8 \times \text{drum/rotor weight}} = \text{temperature rise}$$

To see how this works, consider a 3,000-pound car with a combined brake drum and rotor weight of 20 pounds that is brought to a complete stop from 30 mph, figure 3-23. In the first section of this chapter we calculated that this vehicle has 90,301 foot-pounds of kinetic energy, and because the car is coming to a full stop, the change in kinetic energy during the stop will equal the entire 90,301 foot-pounds. Based on this information, the equation for computing the rise in brake temperature reads:

$$\frac{90,301 \text{ ft-lb}}{77.8 \times 20 \text{ lb}} = 58°F \, (32°C)$$

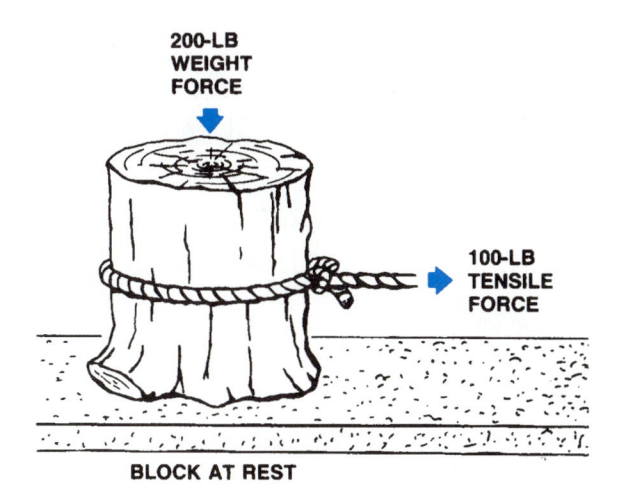

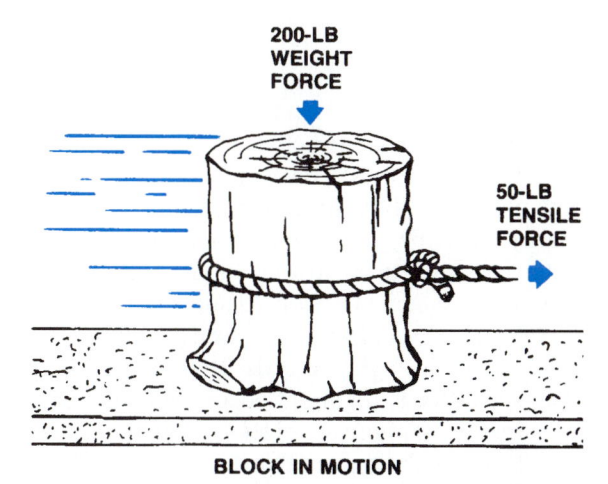

Figure 3-21. The static friction coefficient of an object at rest is higher than its kinetic friction coefficient once in motion.

Contacting Surfaces	Coefficient of Friction	
	Static	Kinetic
Steel on steel (dry)	0.6	0.4
Steel on steel (greasy)	0.1	0.05
Teflon on steel	0.04	0.04
Brass on steel (dry)	0.5	0.4
Brake lining on cast iron	0.4	0.3
Rubber tires on smooth pavement (dry)	0.9	0.8
Metal on ice	—	0.02

Figure 3-22. Every combination of materials has different static and kinetic friction coefficients.

The difference between static and kinetic friction explains why parking brakes, although much less powerful than service brakes, are still able to hold a car in position on a hill. The job of the parking brakes is relatively easy because

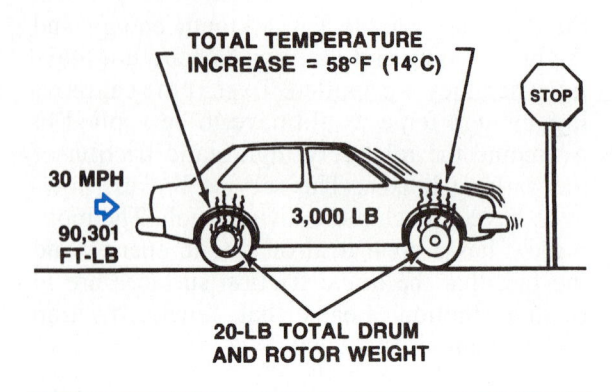

Figure 3-23. Brake temperature increase is determined primarily by vehicle weight, drum and rotor weight, and the change in kinetic energy.

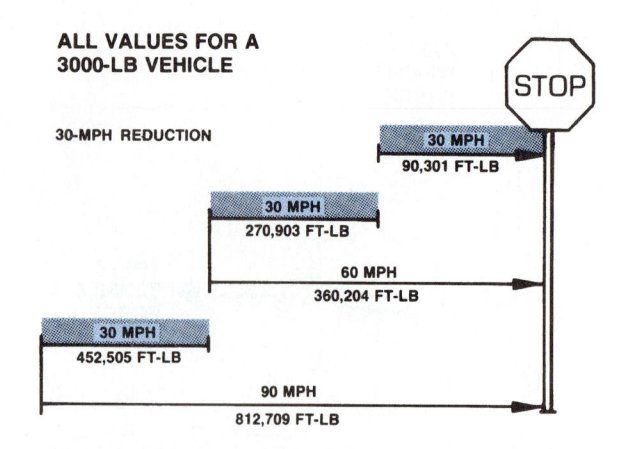

Figure 3-24. The change in kinetic energy required for a 30-mph speed reduction increases dramatically at higher speeds.

The total brake temperature increase in this case is 58°F (32°C). This increase is relatively small, but the weight and speed of the vehicle are also rather low. Keep in mind it is the *change* in kinetic energy that determines the amount of temperature increase, and as we learned in the first part of this chapter, kinetic energy increases proportionately with increases in weight, and as the square of any increase in speed. As a result, the rate of temperature increase will follow these same patterns. For instance, if the weight of the car is doubled to 6,000 pounds, the change in kinetic energy required to bring it to a full stop will be 180,602 foot-pounds. In this case the equation will read:

$$\frac{108,602 \text{ ft-lb}}{77.8 \times 20 \text{ lb}} = 116°F \ (64°C)$$

Note that just as doubling the weight doubles the amount of kinetic energy, it also doubles the amount of temperature rise.

If the weight of the car remains 3,000 pounds but it is brought to a stop from 60 mph where it has four times the kinetic energy, the equation reads:

$$\frac{361,204 \text{ ft-lb}}{77.8 \times 20 \text{ lb}} = 232°F \ (129°C)$$

The temperature increase in this example again matches the increase in kinetic energy which is four times greater as a result of doubling the speed.

All of the preceding examples are for vehicles brought to a complete stop, but because kinetic energy increases at a much greater rate as speeds go up, a full stop from 30 mph creates far less temperature increase than slowing a car from 60 to 30 mph. For a 3,000-pound car, the *change* in kinetic energy between 30 and 0 mph is only 90,301 foot-pounds, but the change between 60 and 30 mph is 270,903 foot-pounds! The car is slowed 30 mph in both cases, but three times the amount of kinetic energy must be converted into heat when the faster moving car is slowed. Figure 3-24 shows how a speed reduction of 30 mph requires the brake system to deal with increased amounts of kinetic energy at higher speeds. Repeated braking from such speeds, as might occur while descending a mountain highway, places much greater demands on a brake system than stop-and-go traffic around town.

Remember that the temperature increase computed with this equation is the average of all the friction-generating components. Some of the heat is absorbed by the brake drums and rotors, some goes into the shoes and pads, and some is conducted into the wheel cylinders, calipers, and brake fluid. In addition, keep in mind that the front brakes provide 60 to 80 percent of the total braking force. Because of this, they receive a similar percentage of the average temperature increase. The increase at each axle is divided evenly between the two wheel friction assemblies unless there is unequal traction from one side to the other or there is a problem within the brake system itself.

BRAKE FADE

It is a fact of brake life that drums and rotors are forced to absorb the heat of braking much faster than they can dissipate it into the surrounding air.

Principles of Brake Operation

The temperature of a brake drum or rotor may rise more than 100°F (55°C) in only seconds during a hard stop, but it could take 30 seconds or more for the rotor to cool to the temperature that existed before the stop. If repeated hard stops are demanded of a brake system, it can overheat and lose effectiveness, or possibly fail altogether. This loss of braking power is called **brake fade**.

The point at which brakes overheat and fade is determined by a number of factors, including the brake design, its cooling ability, and the type of friction material being used. There are three primary types of brake fade caused by heat:

- Mechanical fade
- Lining fade
- Gas fade.

■ The Long and Short of Brake Theory

What are the practical results of all the laws, principles, and equations of brake theory? Just how effective is the conversion of kinetic energy into heat energy? If driver reaction time is not included, tests have shown that the average car can stop from 60 mph in 154 feet—about half the length of a football field. This distance represents the time it takes to convert the vehicle's kinetic energy into heat energy.

Average stopping distance is a generalization, however, and there are many variables from one vehicle to the next. Station wagons and vans tested took as long as 182 feet to stop from 60 mph; extra weight (more inertia) and unfavorable weight bias are probably the factors that lengthened their stopping distances. At the other end of the spectrum, sports cars stopped in as little as 136 feet. Wide sticky tires, low weight, and good weight balance were likely factors in their excellent performance.

To answer the first question, brake theory directly relates to vehicle stopping distances, and thus safety. The conversion of kinetic energy into heat energy is not yet as efficient as it might be, but the brakes on modern cars are far better than those of the past.

Mechanical Fade

Mechanical fade, figure 3-25, occurs when a brake drum overheats and expands away from the brake lining. To maintain braking power, the brake shoes must move farther outward, which requires additional brake pedal travel. When the drum expands to a point where there is not enough

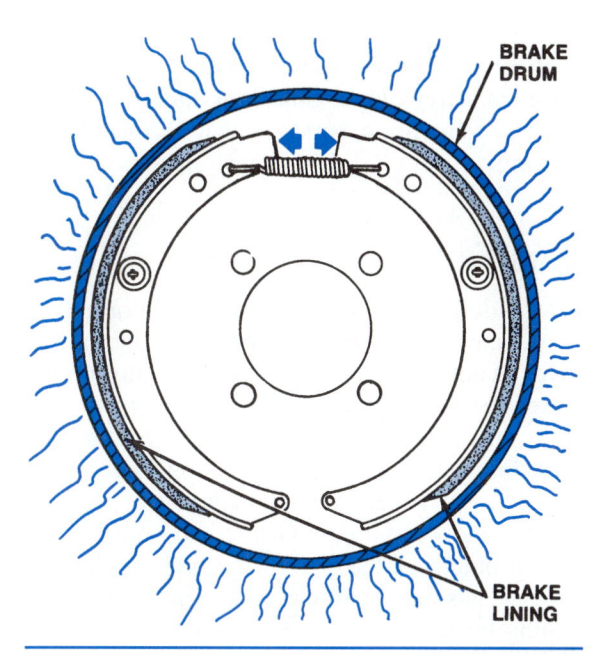

Figure 3-25. Mechanical fade occurs when the brake drums become so hot they expand away from the brake lining.

pedal travel to keep the lining in contact with the drum, brake fade occurs. Sometimes, partial braking power can be restored by rapidly pumping the brake pedal to move the brake shoes farther outward and back into contact with the drums. Manufacturers combat mechanical fade in drum brakes by using larger or heavier drums that can absorb more heat before they expand too far. They also fin the drums or make them partially of aluminum to help speed heat transfer to the passing air. Mechanical fade is not a problem with disc brakes because as a brake rotor heats up it expands *toward* the brake linings (the brake pads) rather than away from them.

Lining Fade

Lining fade affects both drum and disc brakes, and occurs when the friction material overheats to the point where its coefficient of friction drops off, figure 3-20. There are several possible reasons for this drop off, and they are discussed in detail in Chapter 10. When lining fade occurs on drum brakes, partial braking power can sometimes be restored by increasing pressure on the brake pedal, although this may only make matters worse because the extra pressure increases the amount of heat and thus fade. With disc brakes, lining fade is possible, but less of a problem because of disc brakes' superior ability to dissipate heat. The rotor friction surfaces are exposed to the passing air, and rotors in heavy-duty applications commonly have internal ventilation passages that further aid in cooling.

Gas Fade

Gas fade is a relatively rare type of brake fade that occurs under very hard braking when a thin layer of hot gases and dust particles builds up between the brake drum or rotor and linings; the gas layer acts as a lubricant and reduces friction. As with lining fade, greater application force at the brake pedal is required to maintain a constant level of stopping power. Gas fade becomes more of a problem as the size of the brake lining increases; gases and particles have a harder time escaping from under a drum brake shoe than a disc brake pad. Some high-performance brake shoes and pads have slotted linings to provide paths for gas and particles to escape.

In most cases brake fade is a temporary condition; the brakes will return to normal once they have been allowed to cool. This is true in all except extreme situations where the heat has been so great it has damaged the friction material or melted rubber seals within the hydraulic system.

SUMMARY

Brake systems work in accordance with unchanging physical "laws" or principles. The principles involved with vehicle brake systems involve energy, mechanics, hydraulics, friction, and heat.

Moving automobiles possess kinetic energy. The amount of that energy is affected by a car's weight and speed, but speed has a much greater affect. The job of the brake system is to convert a car's kinetic energy into heat energy. During a stop, inertia transfers the majority of a car's weight onto the front wheels. This requires the front brakes to provide 60 to 80 percent of the total stopping power. As a result, front brakes are generally much more powerful than rear brakes.

A great deal of force is required to convert kinetic energy into heat energy at the wheel friction assemblies. This force is provided by the actuating system, which uses mechanical and hydraulic means to increase the force applied at the brake pedal. The lever is the primary mechanical device used. The brake pedal assembly is a good example of a lever that provides a mechanical advantage as high as 5 to 1 to increase application force.

The hydraulic system uses liquids to transmit the pedal assembly force to the wheels. Because liquids cannot be compressed, they are able to transmit movement and force within a closed system. Air, however, can be compressed, and brake hydraulic systems must be air free to operate properly. Hydraulic system pressure is created by a combination of the mechanical force at the brake pedal pushrod and the size of the master cylinder piston. The resulting pressure is distributed equally throughout the system.

To obtain changes in application force, different sized pistons are used in the wheel cylinders and brake calipers. A piston larger than that in the master cylinder will increase application force, whereas a smaller piston will decrease it. The various piston sizes in the system are engineered to move enough fluid to operate the brake calipers and wheel cylinders, supply enough pressure to lock the wheels, and give the driver good pedal feel for controlled braking.

The wheel friction assemblies perform the actual work of converting kinetic energy into heat energy. They do this by rubbing two materials together to create friction. The amount of resistance the materials have when rubbed together is called their coefficient of friction, and this value is determined by the types of materials, their surface finishes, and their temperatures.

When two materials are forced together at rest, their resistance to movement is called the static coefficient of friction. When the materials are forced together in motion, their resistance to movement is called the kinetic coefficient of friction. The static friction coefficent is always higher than the kinetic coefficient, which allows parking brakes used to hold a stopped car in position to be much smaller than service brake, which must slow a vehicle in motion.

The wheel friction assemblies are often required to absorb more heat than they can immediately dissipate into the surrounding air. If too much heat is forced on them in too short a time, brake fade will result. Drum brakes may suffer from mechanical fade where the brake drum expands away from the brake lining, but both drum and disc brakes can suffer from lining fade where the lining overheats and its coefficient of friction is reduced.

Principles of Brake Operation

Review Questions

Choose the letter that represents the best possible answer to the following questions:

1. Work is the transfer of _____ from one physical system to another.
   a. Force
   b. Energy
   c. Mechanics
   d. All of the above

2. Which of the following does *not* have an effect on kinetic energy?
   a. Weight
   b. Speed
   c. Temperature
   d. Motion

3. If the weight of a car is reduced by one-half, its kinetic energy becomes _____ as great.
   a. $1/16$
   b. $1/8$
   c. $1/4$
   d. $1/2$

4. If the speed of a car is doubled, its kinetic energy becomes _____ times as great.
   a. 2
   b. 4
   c. 6
   d. 8

5. Inertia causes a body _____ to remain so unless acted on by an outside force.
   a. In motion
   b. At rest
   c. Both a and b
   d. Neither a nor b

6. Inertia requires the front brakes to provide most of the stopping power because of weight:
   a. Bias
   b. Force
   c. Effect
   d. Transfer

7. A _____ lever provides a mechanical advantage that increases force.
   a. First-class
   b. Second-class
   c. Both a and b
   d. Neither a nor b

8. When a lever is used to increase force, the _____ of the resulting work is decreased.
   a. Speed
   b. Travel
   c. Both a and b
   d. Neither a nor b

9. Liquids are used in hydraulic systems because:
   a. They can be compressed
   b. They absorb heat
   c. Their volume can be varied
   d. None of the above

10. Brake system hydraulic pressure is determined in part by:
    a. Master cylinder piston area
    b. Pascal's law
    c. Brake pedal area
    d. All of the above

11. If the brake system hydraulic pressure is 100 psi at the master cylinder, the hydraulic pressure in a brake caliper with a 4-square-inch piston area will be:
    a. 25 psi
    b. 50 psi
    c. 100 psi
    d. 400 psi

12. If 50 psi of hydraulic pressure is applied to a 0.5- or ½-square-inch wheel cylinder piston, the brake application force will be:
    a. 25 psi
    b. 50 psi
    c. 100 psi
    d. None of the above

13. When the wheel cylinder piston travel is less than that of the master cylinder piston, which of the pistons is smaller?
    a. The master cylinder piston
    b. The wheel cylinder piston
    c. They are the same size
    d. Any of the above, depending on pressure

14. There must be sufficient pressure in the brake hydraulic system to:
    a. Lock the wheel friction assemblies
    b. Prevent brake slippage
    c. Provide light pedal effort
    d. None of the above

15. A large brake master cylinder piston:
    a. Is often used with disc brakes
    b. Moves large fluid volumes
    c. Reduces pedal travel
    d. All of the above

16. The first law of thermodynamics states that energy:
    a. Cannot be created or destroyed
    b. Is always converted from one form into another when work is performed
    c. Both a and b
    d. Neither a nor b

17. The _____ is the lower measure of friction between two materials.
    a. Static coefficient of friction
    b. Kinetic friction coefficient
    c. Minimum drag coefficient
    d. Coefficient of grip

18. Engineers alter brake friction by changing the _____ of the wheel friction assemblies.
    a. Surface finish
    b. Friction materials
    c. Contact area
    d. All of the above

19. The amount of heat generated when stopping a car from 60 mph will be:
    a. Greater at the front brakes
    b. Greater at the rear brakes
    c. The same at the front and rear brakes
    d. Less on the driver's side

20. Brake fade can result from:
    a. Brake drum contraction
    b. Brake rotor expansion
    c. Lining wear
    d. Temporary brake cooling problems

4

Brake Fluid and Lines

OBJECTIVES

Upon completion and review of this chapter, you will be able to:

- Describe the SAE and DOT specifications required of brake fluids and lines.
- List the types of brake fluids and explain their characteristics.
- Describe the correct methods for storage and handling of brake fluids.
- List and describe the methods for bleeding hydraulic brake systems.
- Explain the reasons for brake fluid changes and describe the procedure.
- Explain the use and construction of brake tubing.
- Explain the use and construction of brake hoses.
- Describe the construction of brake line fittings.
- Describe the construction of ISO flares and SAE double flares.

KEY TERMS

banjo fitting
bleeder screw
bleeder valve
brake lines
corrosion
Equilibrium Reflux
 Boiling Point (ERBP)
homogeneous
hygroscopic
inert
oxidation
particulates
solvent
specific gravity
swaged
vapor lock
viscosity

INTRODUCTION

Hydraulic brake actuating systems are used on virtually all modern automobiles and light trucks. For these systems to work properly, they must be filled with a liquid that will not damage mechanical components or break down under the extreme conditions that often exist within the brake system. In automotive brake hydraulic systems this liquid is commonly called brake fluid. The first part of this chapter examines the properties required of brake fluid and the types of fluid now in use.

In addition to having the proper fluid, a hydraulic system must also be free of air to do its job properly. Any time the hydraulic integrity of a brake system is disturbed, the system must be flushed free of air. This process is called bleeding the brakes, and the second part of the chapter explains how it is done.

Finally, once a brake fluid has been chosen, a means must be found to transport it between the various components in the brake system. The only methods used today are steel tubing and flexible high-pressure hose, collectively called the brake lines. The final part of this chapter looks at how brake lines are constructed and used.

BRAKE FLUID

Brake fluid, figure 4-1, is the lifeblood of the brake hydraulic system. If inferior quality fluid or fluid of the wrong type is used, brake failure can result. To ensure a safe level of performance, and compatibility among different brake fluids, both the Society of Automotive Engineers (SAE) and the Department of Transportation (DOT) have set standards for automotive brake fluids.

The latest SAE Standard, J1703 of June 2003, serves primarily as a guideline for engineers involved with brake system design. Brake fluid manufacturers, however, are required by law to meet the specifications set forth in the Federal Motor Vehicle Safety Standard (FMVSS) 116. Fluids classified according to FMVSS 116 are assigned DOT numbers. There are currently three grades of approved fluid and one variant. These are DOT 3, DOT 4, DOT 5, and DOT 5.1. The higher the DOT number, the stricter the specifications the fluid must meet.

The DOT *grades* of brake fluid should not necessarily be identified with different *types* of brake fluid, although there is some correlation. The DOT gradings indicate compliance with certain minimum performance specifications, discussed later, and it is entirely possible for different types of fluids to meet the same specifications. There are also many additives that can be blended into brake fluid to give it specific performance capabilities, and several automakers specify requirements over and above DOT regulations when they buy brake fluid for factory fill. When adding or changing brake fluid, it is important to make sure the fluid used meets the vehicle manufacturer's specifications as well as those of the government.

BRAKE FLUID SPECIFICATIONS

Both the SAE and DOT Standards establish many requirements for brake fluid. In general, brake fluid must not boil or otherwise be affected by the highest temperature a brake system is designed to reach. However, brake fluid must remain free flowing at all temperatures, and not thicken or freeze when exposed to very low temperatures. Brake fluid needs to provide lubrication for the internal parts of the hydraulic system, but it must not attack any of the metal or rubber parts. Finally, brake fluid must remain chemically stable throughout the time it remains in the system, and fluids from different manufacturers must be compatible with other fluids of the same type.

Boiling Point

FMVSS 116 deals with many aspects of a brake fluid's performance, but boiling point is among the most important. As pointed out in the previous chapter, the friction of braking creates a great deal of heat. Most of this heat is absorbed by the brake drums and rotors and then radiated into the surrounding air, but some of the heat is absorbed by the wheel cylinders and brake calipers and transmitted to the brake fluid inside, figure 4-2. If the fluid absorbs too much heat, it will boil into a compressible gas. This causes an increase in brake pedal travel because the vapor must be compressed before the fluid can transmit force. If enough fluid boils, the brake pedal will go all the way to the floor and braking power will be lost. This condition is called **vapor lock.**

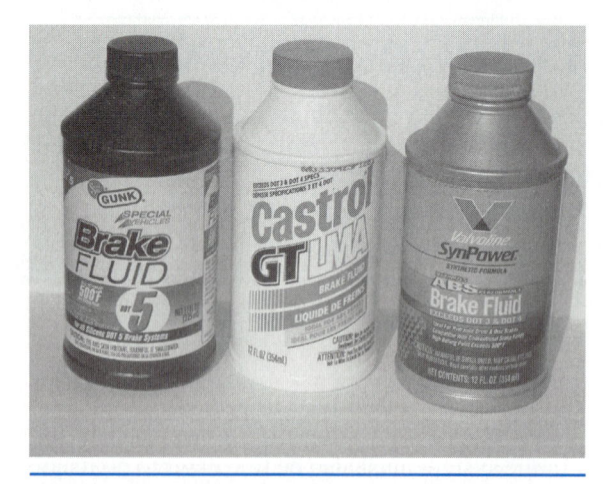

Figure 4-1. Brake fluid makes operation of the hydraulic system possible.

Brake Fluid and Lines

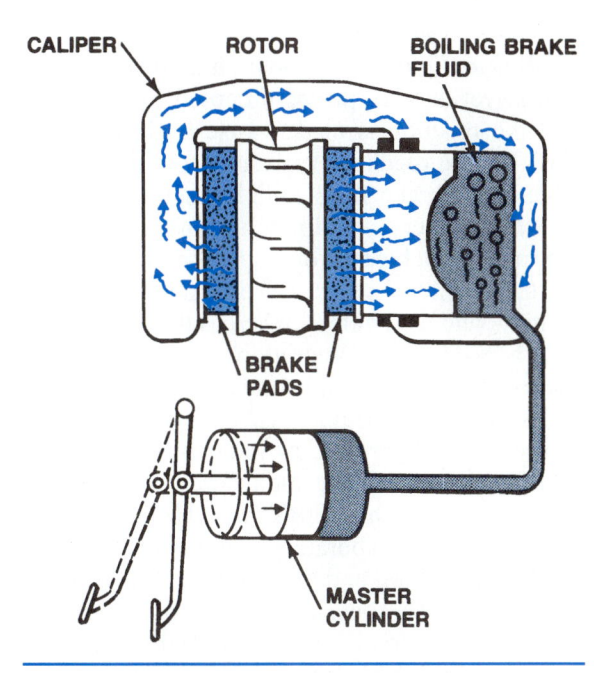

Figure 4-2. If sufficient heat is transferred to the brake fluid, it will boil and cause vapor lock.

The increased brake pedal travel caused by vapor lock should not be confused with the extra pedal travel that results from mechanical fade of a drum brake, although both problems are caused by excessive heat and may occur at the same time. When a wheel friction assembly overheats to the point of fade, it may get hot enough to boil the brake fluid in the wheel cylinder or brake caliper and cause vapor lock as well. It is when both problems occur at the same time that the most dangerous condition occurs: sudden brake loss. As a result of repeated hard braking, the calipers and fluid may exceed the boiling temperature of the fluid, but because the fluid is under pressure from the master cylinder, the fluid does not boil and the pedal remains functional. As soon as the driver releases the pedal, thus releasing the pressure in the caliper, the fluid boils rapidly, and the next pedal application will find the pedal going to the floor, with no braking occurring. This sudden brake loss is difficult to diagnose because as soon as the fluid returns to normal temperature the pedal returns to normal operation and there appears to be no fault in the brake system.

Dry and Wet Boiling Points

To help prevent vapor lock, FMVSS 116 specifies both a dry and a wet **Equilibrium Reflux Boiling Point (ERBP)** for each of the three grades of brake fluid. The term *equilibrium reflux* describes the method used to determine the point at which the fluid boils.

The dry ERBP is the minimum boiling point of new, uncontaminated brake fluid. The boiling point of new brake fluid is of concern to fluid manufacturers, however, the boiling point of brake fluid that has been in service for any length of time will be substantially lower. This change in boiling point occurs because most of the brake fluid used today is **hygroscopic**—it absorbs water. Because water boils at only 212°F (100°C), the boiling point of brake fluid contaminated with water is greatly reduced. The wet ERBP test required by FMVSS 116 determines a fluid's boiling point after it has absorbed a specified amount of water; this provides a better indication of how the fluid will perform in normal service.

Federal law requires that the wet ERBP of all brake fluid be printed on the label of the container. Following are the standards, as listed in FMVSS 571.116:

- **S5.1.1 ERBP:** When brake fluid is tested according to S6.1, the ERBP shall not be less than the following value for the grade indicated:
  a. DOT 3: 401°F. (205°C.).
  b. DOT 4: 446°F. (230°C.).
  c. DOT 5: 500°F. (260°C.).
- **S5.1.2 Wet ERBP:** When brake fluid is tested according to S6.2, the wet ERBP shall not be less than the following value for the grade indicated:
  a. DOT 3: 284°F. (140°C.).
  b. DOT 4: 311°F. (155°C.).
  c. DOT 5: 356°F. (180°C.).

Brake fluid absorbs moisture from many sources. Most of the moisture in brake hydraulic systems is absorbed from the air. If the master cylinder cap is missing or loose, water will quickly enter; and even if the cap is tightly installed, water will still be absorbed if the seal is torn. In fact, brake fluid's affinity for water is so great that even in a perfectly maintained brake hydraulic system water is continually drawn in through the walls of the rubber brake hoses and seals!

Temperature Compatibility

In addition to a high boiling point, brake fluids are tested for several other properties. A number of these tests make sure the fluid remains stable throughout the wide range of temperatures in which a brake system must operate.

Heat is one of the adverse conditions brake fluid has to endure, and in addition to the boiling point tests described previously, brake fluid is also tested for high-temperature stability—its ability to maintain a high boiling point after prolonged exposure to heat. Of course, a fluid that can maintain a high boiling point is not very useful if it evaporates so quickly that fluid must constantly be added to the system. For this reason, FMVSS 116 also includes a test to ensure that the brake fluid is not overly prone to evaporation and that the residue left when it does evaporate is not abrasive.

Until now, most of the points discussed deal with the behavior of brake fluid at high temperatures, but low-temperature performance is important as well. Brake fluid must remain liquid at low temperatures; if the fluid thickens, it moves very slowly through the system and brake application is delayed. In the worst possible case, moisture in the brake fluid can freeze and braking power will be lost.

Like motor oil, brake fluid's resistance to flow is called its **viscosity.** FMVSS 116 requires a viscosity test at both high and low temperatures. For brake fluid in service, the problem of high viscosity at low temperatures is made worse by water absorption. Because water freezes at only 32°F (0°C), water-contaminated brake fluid will thicken or freeze at a higher temperature than uncontaminated fluid.

Mechanical Compatibility

In addition to remaining stable across a wide range of temperatures, brake fluid must also be compatible with the metals and rubbers used in the brake system. To ensure that a fluid will not damage brake components, FMVSS 116 specifies tests for pH value, corrosion, oxidation resistance, effects on the rubber cups, and lubrication or stroking properties.

If a brake fluid is too acidic, it will etch the metals in the hydraulic system. The pH test determines the overall acidity of the brake fluid. In general, brake fluid is more alkaline than acidic. To get a better idea of the effect of the fluid on actual brake system parts, FMVSS 116 requires a **corrosion** test in which various metals and wheel cylinder cups are submerged in the brake fluid for a specified time. Afterward, the metals are visually inspected and weighed to make sure the fluid has not eaten them away, and the cups are checked for signs of disintegration.

Brake system parts are also susceptible to **oxidation,** figure 4-3. Oxidation in the brake system is yet another problem caused by moisture, and the reaction is made worse by heat. In the FMVSS 116 oxidation resistance test, moisture is added to a brake fluid sample, metal strips and rubber brake cups are immersed in the fluid, and the mixture is heated for a specified time. Afterward, the metal strips are checked for pitting, etching, and weight loss, whereas the rubber cups are checked for disintegration, swelling, and any increase or decrease in hardness.

Finally, brake fluid must have the ability to lubricate the hydraulic cylinder pistons as they slide back and forth in their bores during brake operation. In the FMVSS 116 lubrication (stroking) test, a laboratory brake system is filled with the test fluid and operated under controlled conditions for a specified number of cycles. Afterward the hydraulic cylinders are disassembled and checked for pitting, etching, and wear.

Fluid Compatibility

Although the performance specifications for the three grades of DOT-approved brake fluid are different, FMVSS 116 requires the three grades to be compatible. This means that any DOT-approved brake fluid can be mixed with any other approved fluid. For example, the DOT 4 fluid of one manufacturer can be poured into a master cylinder that contains DOT 3 fluid made by another manufacturer and no damaging chemical reaction will take place. Keep in mind, however, that if a lower DOT grade of fluid is used in a system that re-

Figure 4-3. Moisture in the brake fluid combines with the heat of braking to oxidize hydraulic system parts.

Brake Fluid and Lines

quires a higher grade, the overall boiling point of the fluid in the system will be reduced.

Although DOT fluids are compatible, they are not necessarily **homogeneous**; they do not always blend together into a single solution. As explained in the next section, only fluids of the same *type* mix together completely. Different types remain separated within the hydraulic system, and certain non-DOT fluids will cause severe damage if mixed. The best practice, even with compatible brake fluids, is to stick with a single high-quality brand of the correct type and DOT grade fluid recommended for the particular system.

BRAKE FLUID TYPES

As hydraulic brake systems evolved, many different liquids were used as brake fluid; each was selected to match the design of a particular system. Castor oil, mineral oil, alcohol, polyglycol, and silicone are only a few of the fluids that have been used. As brake hydraulic systems became more similar in design and construction, brake fluids also came to have more in common until, today, only three types of brake fluid are used in automotive brake hydraulic systems:

- Polyglycol
- Silicone
- Hydraulic System Mineral Oil (HSMO).

Polyglycol fluid is by far the most common type now in use, but both silicone fluid and HSMO have advantages in some applications. To help prevent intermixing of the three types of fluids, federal law requires that each type be a specific color: polyglycol fluids are clear to amber, silicone fluids are purple, and hydraulic mineral oils are green.

Polyglycol Brake Fluid

The most common type of brake fluid is a polyalkylene-glycol-ether mixture, called polyglycol for short. Ployglycol brake fluid has been used since the 1940s, and it works predictably over a fairly wide range of temperatures. Because of its widespread use, polyglycol fluid has the advantages of being inexpensive and readily available. It also causes rubber parts of the hydraulic system to swell slightly, which improves sealing and helps prevent leaks. Early polyglycol fluids had problems with mechanical compatibility, but additive improvements over the years have made the latest fluids less corrosive and more chemically stable.

Although it basically performs quite well, polyglycol brake fluid does have two drawbacks. One is that polyglycol is a strong **solvent**; if spilled on a car's finish it will almost immediately dissolve the paint, figure 4-4. Polyglycol's major drawback is its hygroscopicity, or water-attracting nature. This trait leads to a reduced boiling point, higher viscosity at low temperatures, and increased rust and corrosion within the hydraulic system. Polyglycol is the only one of the three fluid types to have this trait.

DOT 3, DOT 4, and the recently designated DOT 5.1 are all polyglycol fluids and they all will blend together into a homogeneous mixture. DOT 3 brake fluid is specified by most automakers, but some manufacturers have begun to require DOT 4 fluid. DOT 5.1 meets the DOT 5 standards but is a glycol-based fluid rather than a silicone-based one; some suppliers may also call it "synthetic" brake fluid. The Ford Motor Company uses a special heavy duty DOT 3 fluid that has a dry ERBP that exceeds the DOT 5 specification by a wide margin, although the fluid conforms to DOT 3 requirements in others areas.

Silicone Brake Fluid

Silicone brake fluid has been on the market since the late 1970s, and it offers some advantages over

Figure 4-4. Polyglycol brake fluid is a strong solvent that will damage paint in seconds.

polyglycol fluid. Silicone fluid has high dry and wet boiling points and was the only fluid to meet the DOT 5 standards when it was first marketed. Glycol-based fluids have since been developed that also meet the DOT 5 standards, although at a high cost. Silicone fluid retains an advantage in the wet boiling test. One reason for this is that silicone brake fluid is not hygroscopic; it does not attract or absorb moisture. Because silicone fluid lacks an affinity for water, rust and corrosion problems are almost eliminated when it is used. In addition, silicone fluid is basically **inert,** not a solvent, and will not harm painted surfaces.

Silicone brake fluid is a relatively new and unknown product (compared to glycol fluids), and there have been reports that it can cause a variety of problems. Some concerns are:

- Silicone fluid does not work well at low temperatures. It is true that some early non–DOT-approved silicone fluids did have high viscosity at lower temperatures; however, the latest formulations meet DOT 5 standards, including the low temperature standard at −40°F. Silicone brake fluid does have a higher viscosity at room temperature than polyglycol fluid, but this does not reflect its cold weather performance.
- Silicone fluids are poor lubricants of the mechanical parts in the brake system. This is only partially true. Silicone fluid is an excellent lubricant of rubber parts, but it is not quite as good as polyglycol at lubricating the metal pistons. Testing has shown, however, that piston wear when using silicone fluid is not a significant problem in the real world.
- Silicone fluid will compress slightly and cause a spongy brake pedal. This does happen, but is most noticeable at higher temperatures. These conditions are likely to occur only in racing or extreme duty use.

Silicone brake fluid does have certain other disadvantages that have all but eliminated its use in passenger cars.

- It is expensive—three to four times the cost of polyglycol fluid.
- Silicone fluid does not have enough conductivity to operate the electrical brake fluid level sensors used in some cars.
- Silicone fluid has a greater affinity for air than polyglycol fluid, which can make it more difficult to bleed air from the system.
- Because silicone fluid is inert, it may not provide the small amount of rubber swell

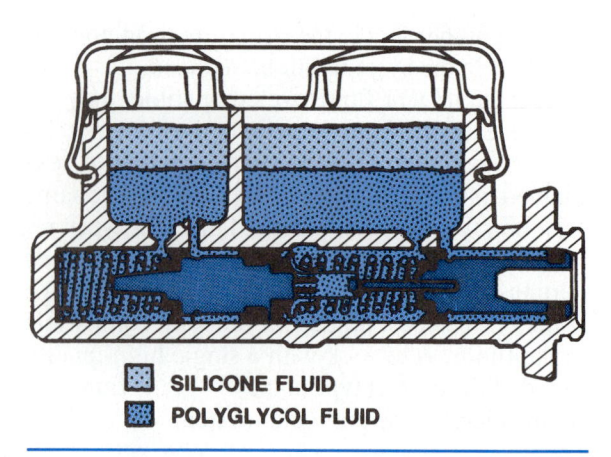

Figure 4-5. Silicone and polyglycol brake fluids do not form a homogeneous mixture when combined in a brake system.

that polyglycol does to improve sealing and help prevent leaks.
- If moisture gets into the system, rather than being dispersed into the fluid as it is with glycol fluids, the moisture will fall to the lowest point in the system where it can freeze, boil, or cause corrosion.
- Silicone fluid is compatible with DOT 3 and 4 fluids, but it does not blend with them; it remains separate in the hydraulic system. This can create problems because silicone brake fluid has a lower **specific gravity** than polyglycol fluid; if equal quantities of the two fluids are compared, the silicone fluid will weigh less. If the two are mixed together, figure 4-5, the polyglycol fluid will sink to the bottom of the system (i.e., the wheel cylinders and brake calipers), where its hygroscopic properties will cause rust and corrosion problems.

Manufacturers that do use silicone brake fluids have, in most cases, designed the whole brake system to use silicone fluid, thus minimizing or eliminating these concerns. In vehicles that may see intermittent use or long periods of storage, such as military vehicles and show cars, silicone fluids greatly reduce required maintenance.

Hydraulic System Mineral Oil (HSMO)

HSMO, figure 4-6, is made up of a mineral oil base and several additives. It is the least common type of brake fluid, but in many ways it is also the best: It has an extremely high boiling point, it is not hygroscopic, and it does not contribute to rust

Brake Fluid and Lines

Figure 4-6. HSMO is a petroleum-based brake fluid that should only be used in systems specially engineered for it.

Figure 4-7. Brake fluid containers are sealed at the factory to prevent contamination.

or corrosion in any way. In fact, HSMO will guard against such damage because it is petroleum based and, therefore, a natural rust inhibitor.

HSMO has been used mainly by only three automakers. Citroen has used HSMO since 1966 in some models in a central hydraulic system that powers the brakes, steering, and suspension. Rolls-Royce went to HSMO in mid-1980 to operate the brakes and leveling system. Audi uses HSMO in the hydraulic brake booster of some models, although the brake actuating system remains separate and uses DOT 4 polyglycol brake fluid.

HSMO does not fall under any DOT grade classification, and its primary drawback is that it is not compatible with polyglycol and silicone brake fluids. Because HSMO is petroleum based, the seals in systems designed to use it must be made of special rubber. Polyglycol fluids attack this rubber, and HSMO attacks the seals in brake systems that use polyglycol fluid. HSMO and polyglycol fluids must be used only in systems designed for them and must never be mixed in the same system.

BRAKE FLUID STORAGE AND HANDLING

Storage and handling precautions for brake fluid depend, to a large degree, on the type of fluid being considered. Polyglycol fluid, for example, has very limited storage life. Once a can of polyglycol fluid has been opened, its entire contents should be used as soon as possible because it immediately begins to absorb moisture that degrades its performance. In contrast, silicone brake fluid and HSMO can be stored almost indefinitely. They are not hygroscopic and there are no known limits to the length of time they retain their original properties.

Certain precautions apply to all brake fluids. They should be stored in a clean, dry location, and only in the original container. Federal regulations require that brake fluid packaging must be designed so that it will not react with the fluid. The container must also have an airtight seal, figure 4-7, that has to be destroyed the first time fluid is dispensed. Metal plugs and foil barriers are the most common seals. Never use brake fluid from a new container if the seal is broken. Partially used cans of polyglycol fluid must be kept tightly capped to help reduce moisture absorption, and if the fluid is not used within a short time, it should be discarded.

When handling brake fluid, take care to avoid contaminating it with petroleum products that attack rubber or **particulates**, such as dirt or rust, that increase wear in the hydraulic system. Always cap a fluid container tightly when not dispensing fluid to prevent moisture and contaminants from entering. Never reuse brake fluid drained from the hydraulic system because it is always contaminated to some degree. Finally, when filling a system with polyglycol brake fluid, avoid spilling any on the car's finish.

Although new brake fluid may or may not be considered a hazardous waste according to local or state regulations, used brake fluid is almost always considered hazardous and must be disposed of properly. Check with local authorities to determine the proper storage and disposal methods. In many cases, there are private companies that specialize in disposal and recycling of waste oil; they may also take care of used brake fluid and antifreeze.

BRAKE BLEEDING

Brake bleeding is the process in which brake fluid is forced through the hydraulic system until all air is purged from the system. The brakes must be bled when the hydraulic system is first filled, every time the system is opened for service, and any time air has entered the system creating a spongy brake pedal. Naturally, in the latter case, the technician should inspect the system, find the leak that is allowing air to enter, and repair it before bleeding the brakes.

Brake bleeding is covered in depth in the *Shop Manual* volume of this text. In general, there are a number of ways to bleed a hydraulic brake system. Some methods are:

- Manual bleeding
- Gravity bleeding
- Pressure bleeding
- Vacuum bleeding.

All-three methods work in the same manner: A **bleeder valve**, sealed by a **bleeder screw**, figure 4-8, is opened at the wheel friction assembly, then pressure is created in the system to force brake fluid out of the bleeder along with any air. This procedure is repeated at all four wheels until the air is removed from the system.

Manual Bleeding

When manually bleeding brakes, figure 4-9, one person opens the bleeder valve at the wheel cylinder or brake caliper. Another person then presses the brake pedal slowly to the floor, which creates pressure within the system and forces brake fluid and air bubbles out of the bleeder valve. The first person then closes the bleeder valve, and the brake pedal is released allowing fresh fluid to be drawn into the master cylinder bore.

■ **Mountaineering for Mechanics**

Why would a mechanic want to drive his customer's car to the top of a mountain? It might sound crazy, but it could make bleeding the brakes easier. Small air bubbles have an unfortunate tendency to cling to the internal surfaces of brake hydraulic components. The smaller they are the greater they resist efforts to flush them out by bleeding. If there was a way to make the bubbles larger, they would be easier to bleed from the system. This is where the mountain comes in.

Atmospheric pressure is approximately 14.7 psi at sea level, but at 5,000 feet it is only 11.9 psi, a drop of approximately 20 percent (⅘ the original pressure remains). Because air *volume* is inversely proportional to *pressure,* the air bubbles in a brake system first bled at sea level will be 25 percent larger (5/4 = 1.25) at 5,000 feet, providing the temperature remains the same. If the brakes are bled again at the higher altitude, the change in bubble size will make the air easier to remove from the system.

This effect works in reverse as well. A car with a spongy brake pedal in Denver will have a much firmer pedal in Death Valley! Of course, the sponginess will return if the car is driven back to the higher altitude.

Manual bleeding is the most inexpensive method of bleeding brakes because it does not waste fluid or require special tools. It does, however, normally require two people: one to pump

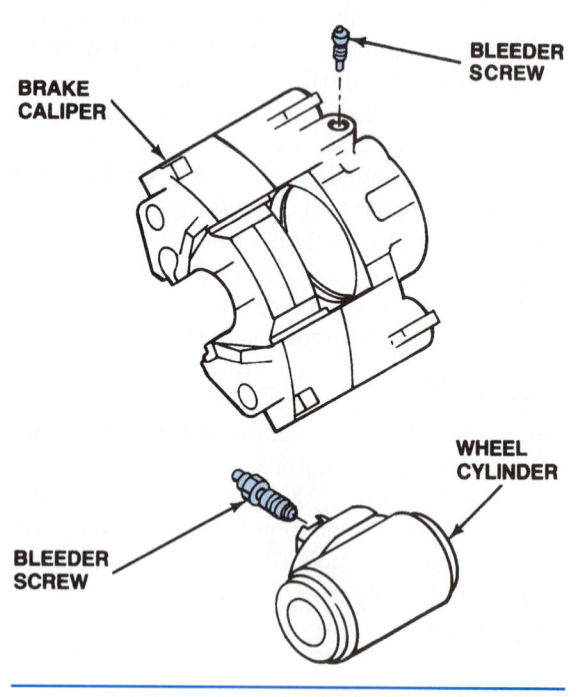

Figure 4-8. Brake bleeder screws are used when flushing air from the hydraulic system.

Brake Fluid and Lines

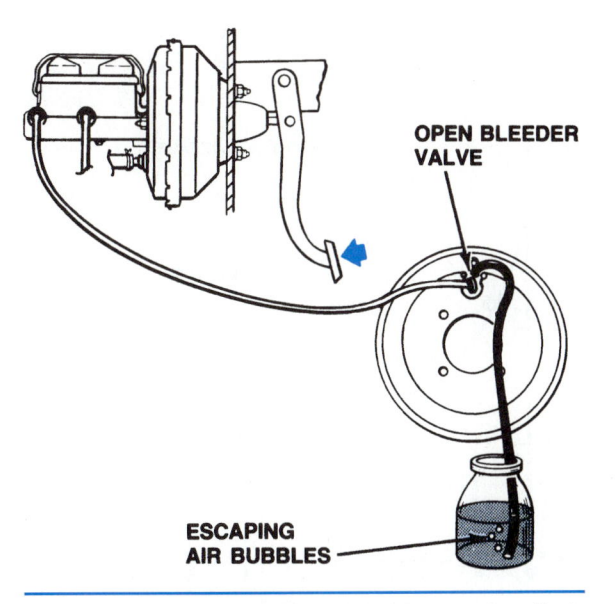

Figure 4-9. Manual bleeding can be used to flush the air from any brake system.

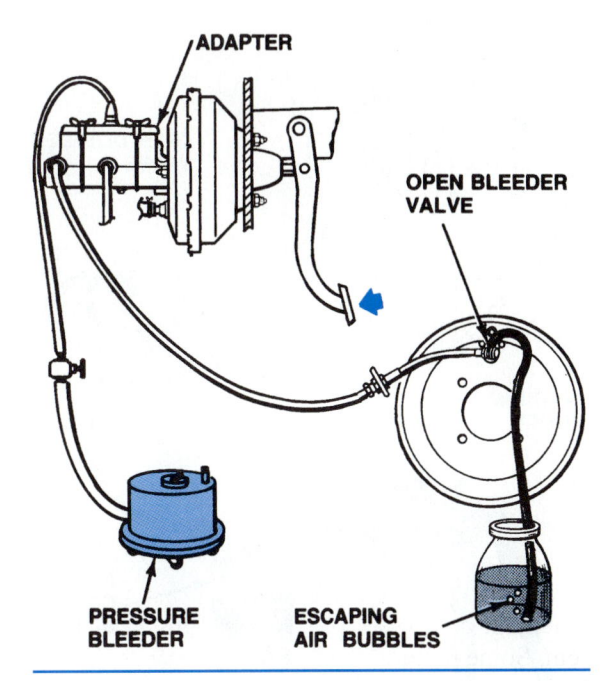

Figure 4-10. A pressure bleeder uses compressed air to force brake fluid through the hydraulic system.

the brake pedal, and the other to open and close the bleeder valves and periodically refill the master cylinder reservoir with fluid.

Gravity Bleeding

Gravity bleeding uses the force of gravity to pull the brake fluid through the hydraulic system. In this process, the master cylinder is filled with fluid, the bleeder screw at each wheel in turn is opened, and the system is allowed to drain naturally until the fluid coming from the bleeder is free of air.

Gravity bleeding is a slow process. In addition, this procedure cannot be used on older vehicles using residual check valves in the master cylinder because the valves restrict the fluid flow. An advantage of gravity bleeding is that it can be done by a single technician, but the technique does not work with all systems.

Pressure Bleeding

With the pressure bleeding method, a special tool called a pressure bleeder, figure 4-10, is attached to the master cylinder. The bleeder contains brake fluid under pressure, and forces the fluid through the system as each bleeder valve is opened.

Pressure bleeding is a fast and effective way to bleed brakes, and only one person is needed to perform the operation. The drawbacks to pressure bleeding are that the equipment is expensive, it wastes more fluid than the other methods, and if the pressure bleeder is not used regularly, the fluid in its reservoir may become contaminated with moisture.

Vacuum Bleeding

Vacuum bleeding is a relatively recent development. This method uses an electric or hand-operated vacuum pump to create low pressure at the bleeder valve, figure 4-11. When the valve is opened, atmospheric pressure forces fluid and air bubbles through the system.

Like pressure bleeding, vacuum bleeding requires only one person. However, it is less expensive because the equipment costs less and the process does not waste as much fluid. There is one significant drawback to vacuum bleeding; vacuum can pull air past wheel cylinder cup seals and into the hydraulic system unless the seals are equipped with cup expanders. Wheel cylinders and cup expanders are detailed in Chapter 7.

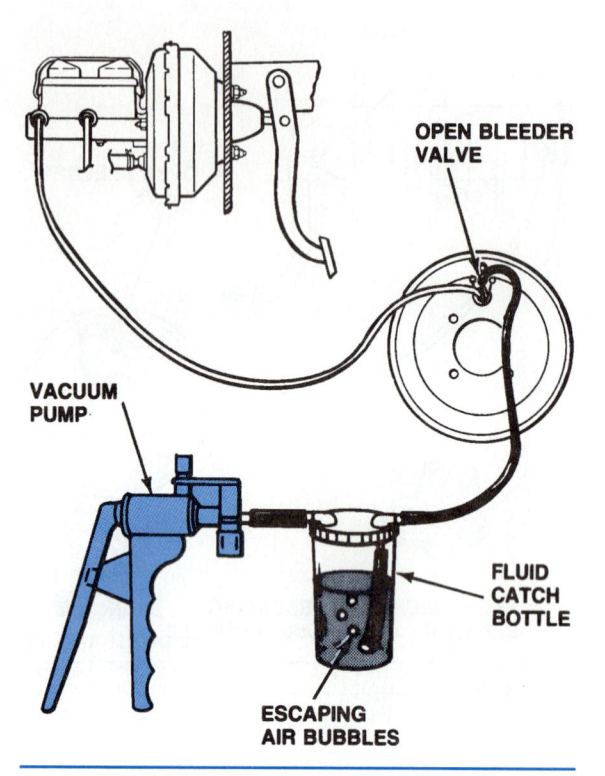

Figure 4-11. Vacuum bleeding uses atmospheric pressure to force brake fluid through the hydraulic system.

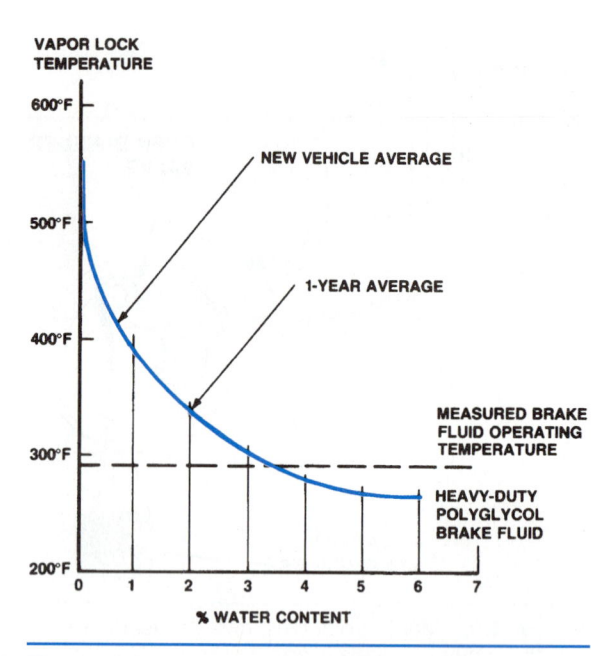

Figure 4-12. Moisture contamination has a significant effect on brake fluid boiling point.

BRAKE FLUID CHANGES

Brake bleeding is also the process by which old contaminated brake fluid is flushed from the system and replaced with fresh fluid. Three forms of contamination that affect all types of brake fluid are brake dust, rubber dust, and metallic particles from wear that turn the brake fluid dark brown, gray, or black. These particles are abrasive and can increase wear, so fluid that is no longer transparent should be changed. In the case of polyglycol brake fluid, moisture contamination is also a serious problem, but only in extreme cases will the fluid be turned cloudy. Determining the need for a fluid change can be based on time, mileage, or on the testing of fluid for moisture content.

The time required for polyglycol brake fluid to absorb enough water to create a problem depends on factors such as system design, vehicle maintenance, and local weather conditions. Information from Dow-Corning fleet tests and the SAE R11 Field Evaluation Program, figure 4-12 shows that the average 1-year-old vehicle has about 2 percent water content in the brake fluid. This small amount of moisture is enough to reduce the boiling point of a DOT 3 fluid from over 400°F (200°C) to less than 320°F (160°C), which is within 35°F (20°C) of normal operating temperatures. To maintain a high brake fluid boiling point and a good margin of safety, some automakers recommend that polyglycol brake fluid be changed every 1 or 2 years.

Actual testing of the fluid for moisture content may be a more accurate method of determining the need for fresh fluid. There are a number of ways to do this. One method uses test strips that are dipped in the fluid and then checked against a color chart. Another method uses a tester that boils the fluid sample and then reports the actual boiling temperature. Any fluid that tests at less than the DOT required wet boiling point should be replaced. These tests are detailed in the *Shop Manual*.

> **NOTE:** Studies have shown that the moisture content can vary widely throughout the brake system. These tests only test the fluid in the master cylinder, which may have significantly less water than the wheel cylinders or calipers. For this reason most vehicle manufacturers recommend, and some European countries mandate, flushing the brake system every 2 years and replacing the fluid with fresh brake fluid.

Silicone Brake Fluid Changes

Silicone brake fluid is not hygroscopic, and because it suffers no drop in boiling point, it does not need to be changed annually. However, silicone fluid should be changed if it becomes contaminated with particulates that change its color. There are also some precautions to consider if a brake system is being converted from polyglycol to silicone fluid.

Technically, the best time to outfit a vehicle with silicone fluid is on the assembly line when all the parts in the system are new. If a system in the field is flushed of polyglycol fluid and filled with new silicone fluid, traces of the old fluid will remain to attract water. Although the boiling point of the fluid in the system will be increased and stabilized, the full anticorrosion benefits of silicone fluid will not be realized. The best time to convert a brake system to silicone fluid is following an overhaul in which all of the components have been replaced or rebuilt and the brake lines have been flushed of all traces of polyglycol fluid.

BRAKE LINES

Brake lines carry brake fluid from the master cylinder to the wheel cylinders and brake calipers. The brake lines contain and direct the pressure of the brake hydraulic system. To do this they must be strong enough to contain pressure, flexible enough to prevent fracture from vibration, and tough enough to resist rust and corrosion. Because brake lines are an important safety-related item, special caution and care must be taken in their manufacture, selection, and installation.

Brake Tubing

Most of the total length of the brake lines consists of rigid tubing. Single-wall copper tubing was used for early brake lines, but this type of tubing has a number of disadvantages. The single-wall construction combined with the softness of copper results in a relatively low burst strength. In addition, vibration causes copper to a harden and become brittle, which leads to cracking and leaks. Finally, copper is very susceptible to corrosion.

For maximum strength and durability, all modern brake systems use double-wall brake tubing made from copper-plated steel sheet. There are two basic types of double-wall tubing: seamless and multiple ply, figure 4-13. To make seamless

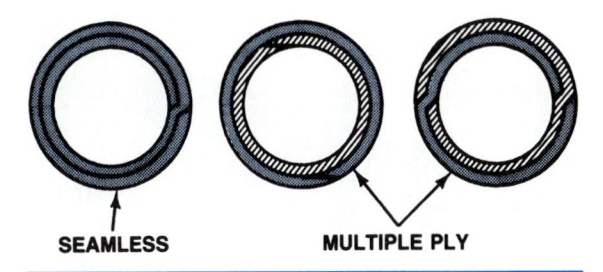

Figure 4-13. Modern steel brake tubing is double-walled for strength, and plated for corrosion resistance.

tubing, the steel sheet is rolled a minimum of two times and then run through a furnace where the copper plating melts and brazes the tubing into a single piece. Multiple-ply double-wall brake tubing is built like two single-wall tubes one inside the other. The inner and outer tubes are individually seamed, and the seams must be positioned at least 120 degrees apart. Then, just as with seamless construction, the two-ply tubing is run through a furnace where it is brazed into a single piece. All double-walled brake tubing is plated with tin, zinc, or a similar substance for protection against rust and corrosion.

Just as with brake fluid, the SAE has guidelines for brake tubing. SAE standard J1047 specifies that an 18-inch section of brake tubing should be able to withstand an internal pressure of 8,000 psi. The tubing should also be capable of being bent 360 degrees around a mandrel five times its diameter without kinking, cracking, or developing other flaws. Additional tests check the tubing's resistance to fatigue, heat, impacts, rust, and corrosion.

Brake tubing is available in a variety of diameters. Many U.S. built vehicles are fitted with 3/16-inch (4.75-mm) outside diameter tubing as are some import vehicles. Imported vehicles also may use 5-mm tubing as a standard size; the technician must take care to use the proper size because the 4.75-mm tubing and the 5-mm tubing cannot be easily identified by eye. Other brake tubing sizes that may be used in automotive brake systems are 1/8 inch, 1/4 inch, 5/16 inch, and 3/8 inch; metric sizes include 4 mm, 5 mm, 6 mm, and 8 mm. Special "armored" brake tubing, figure 4-14 is sometimes used where the tubes are exposed under the car. Armored tubing has a hardened steel spring coiled around it to provide extra protection against impacts and abrasion.

Brake tubing can be purchased in various lengths with fittings already attached, or it can be custom made from tube stock by the technician.

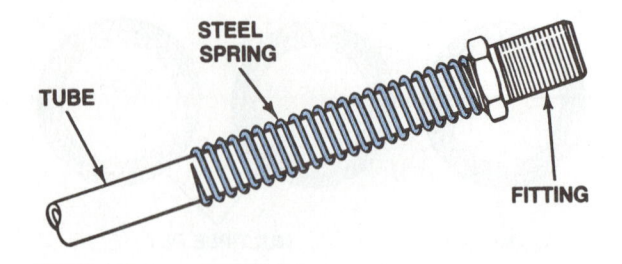

Figure 4-14. Armored brake tubing is protected by a coiled steel spring in areas that may be exposed to damage from impacts.

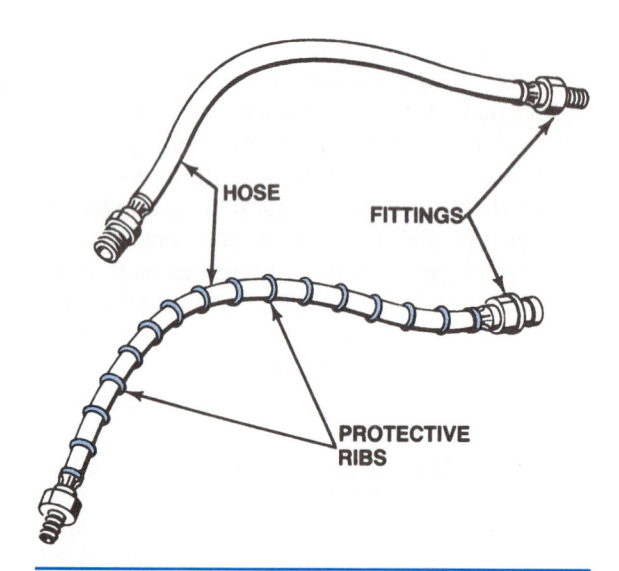

Figure 4-15. Brake hoses span the gap between fixed brake tubing and moving wheel friction assemblies.

Replacement brake tubing is sold only in straight sections so any time a tube is replaced the new part must be bent to fit. Special care must be taken when bending brake lines to prevent kinks that restrict fluid flow and lead to cracks. A few special types of brake tubing can be bent by hand, but most tubing requires the use of a special sleeve or tubing bender to prevent kinks.

Brake Hoses

Whereas brake tubing is solidly mounted to the vehicle chassis, the wheel cylinders and brake calipers are mounted on suspension parts that move. Flexible brake hose, figure 4-15, is used to carry brake fluid between the fixed and moving portions of the hydraulic system. The hoses attach to the tubing at brackets on the chassis, figure 4-16, that position the hose away from moving parts that might cause damage. The outboard end of the hose often attaches directly to the wheel cylinder or brake caliper, but on

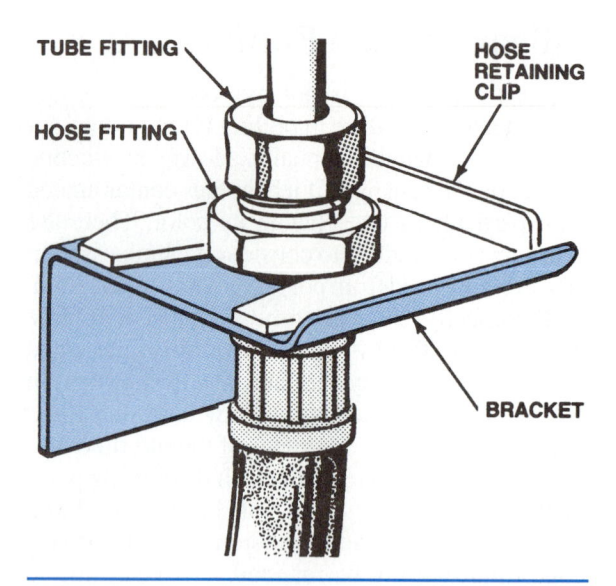

Figure 4-16. Special brackets are used where brake hoses attach to brake tubing.

some front-wheel brakes, the hose is routed to a second bracket where it connects to a short section of steel tubing that supplies the fluid to the caliper. The extra steel line helps isolate the rubber hose from heat and allows easier routing of the hose.

Like brake fluid and tubing, brake hose is a safety-related part subject to a number of performance requirements. FMVSS 106 specifies that brake hoses must have a burst strength of not less than 5,000 psi and be able to withstand 4,000 psi of pressure for two minutes without rupturing. To meet these requirements, brake hoses are made from several layers of fabric impregnated with synthetic rubber, figure 4-17. Cotton fabric was used up until the mid-1960s, but rayon has been used almost exclusively since 1968.

■ Early Brake Lines

Today's flexible brake hoses are a tour de force of design simplicity compared to the methods used to transmit pressure on the first hydraulic brake systems. High-pressure hose was simply not available in the 1920s, so brake fluid was routed through hollow suspension and steering parts instead. These parts were fitted with seals that slid and rotated as the suspension moved or the wheels were turned. Exposed to the elements and made of the relatively crude materials of the day, these seals were prone to leakage that could quickly lead to a loss of stopping power. It's no wonder Henry Ford and others stayed with mechanical brakes until almost 1940.

Brake Fluid and Lines

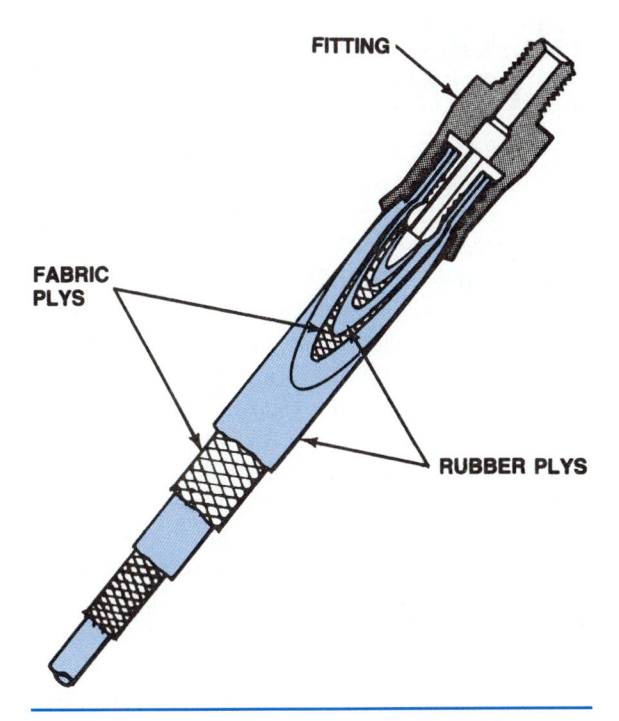

Figure 4-17. Brake hose gets its strength from several layers of fabric and synthetic rubber.

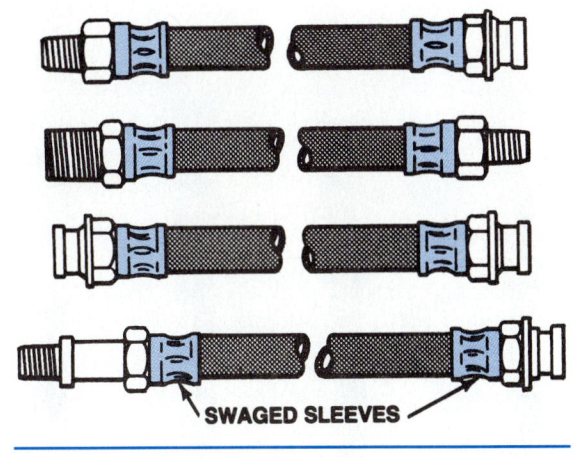

Figure 4-18. These typical hose fittings are all swaged into place for a permanent, leak-free seal.

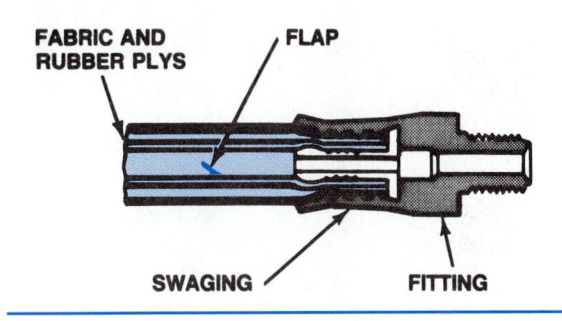

Figure 4-19. A rubber flap torn loose inside a brake hose can create braking problems.

To ensure that the end fittings on the brake hose will not leak or come loose, FMVSS 106 requires that they be permanently attached. Normally, the fittings are crimped or **swaged** onto the hose at very high pressures, figure 4-18. Special equipment is required for this operation, so unlike brake tubing, brake hose cannot easily be custom made by a technician in the field.

FMVSS 106 also mandates that replacement brake hoses be marked with raised ribs or two 1/16-inch-wide stripes on opposite sides of the hose along its length. These markings help the technician install the hose without twisting it. A twisted brake hose is under much greater stress and may rupture under pressure. Other tests require the hose to meet minimum standards for expansion, tensile strength, water absorption, and brake fluid compatibility. The resistance of the hose to whipping action, low temperatures, ozone, and end fitting corrosion is examined as well.

In many ways, the rubber hoses are the weakest links in the brake system. They are porous and allow polyglycol brake fluid to absorb moisture through their walls from out of the air. Ozone in the atmosphere attacks the rubber causing it to age and crack. And the rough environment under the car usually means that brake hoses have shorter service lives than other parts of the brake system.

External wear can also result if a hose rubs against the frame or suspension, particularly at the front suspension when the wheels are turned to full lock. Hoses are carefully selected and positioned to avoid such contact, and some hoses have raised rubber ribs around them to provide protection from any abrasion. For this reason, a brake hose should only be replaced with the proper original equipment part or its equivalent.

Brake hoses are also subject to internal failures that can be difficult to diagnose. A bulge or bubble on the outside of a hose is a sign of an internal leak, but a hose problem impossible to see is a rubber flap torn loose inside the hose that acts as a one-way check valve, figure 4-19. If the flap is positioned toward the master cylinder it will delay or prevent hydraulic pressure reaching the wheel cylinder or brake caliper. If the flap faces the friction assembly the brake will not release or will release slowly and drag. In both cases, the car may pull to one side. Because of the possibility of internal damage, brake calipers or other parts should never be hung by the brake hoses during repairs.

Yet another problem with rubber brake hoses is that they expand slightly when holding pressure. This expansion can be felt as sponginess at the

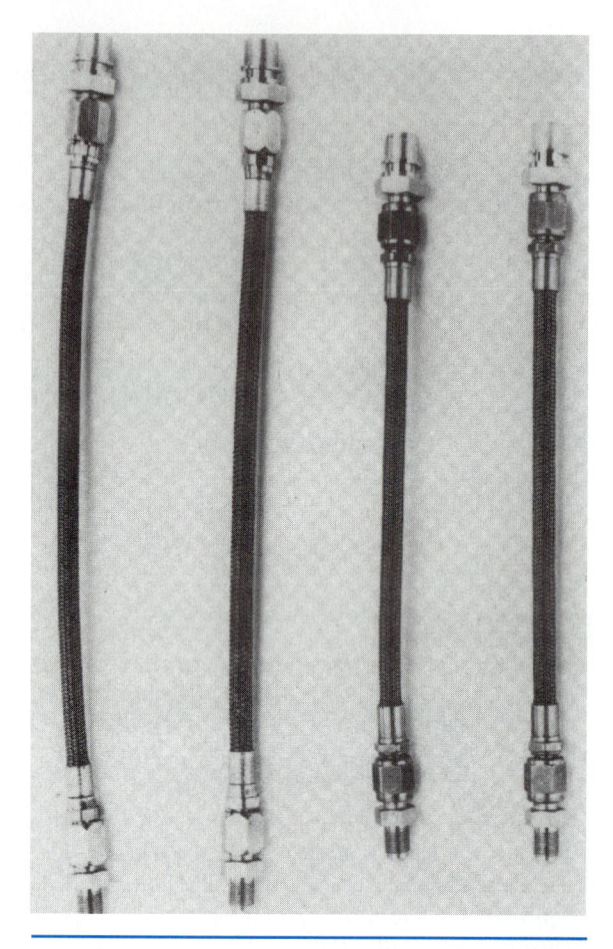

Figure 4-20. Steel braid-covered teflon brake hose is an expensive solution to the problems caused by rubber hose.

brake pedal. FMVSS 106 specifies that brake hose must be labeled with the letters "HR" which indicate the hose has regular expansion characteristics, or "HL" indicating that the hose has low expansion characteristics.

One answer to the problems of water absorption and hose expansion is steel braid-covered teflon brake hose, figure 4-20. Originally developed for aircraft use, this type of hose is made of a strong, flexible teflon tube covered with braided stainless steel wire. The teflon tubing does not expand nearly as much as the fabric and rubber of conventional brake hose, and the steel braid cover helps control any expansion that does take place. In addition, the cover is so chafe resistant it takes a hacksaw to cut through it. At the present time, steel braid-covered teflon brake lines are used only on racing cars, although aftermarket replacement hose kits are available for high-performance street cars.

BRAKE LINE FITTINGS

Because of the high pressures within the brake system, brake lines use special fittings that ensure strong connections and prevent leaks. Brake tubing always has male fittings on both ends, whereas brake hose ends may have either male or female fittings. The fittings on master cylinders, wheel cylinders, and brake calipers are invariably female, and female fittings are also used on junction blocks, such as tees, elbows, and unions.

The threads of fittings for domestic cars are standard SAE fine sizes, such as ⅜ × 24 or ½ × 20. Imported cars use metric fittings, with 10 × 1.0 being the most common thread size.

Tapered pipe threads are not used in brake hydraulic systems because they leak under high pressures, and the female portion of the fitting can split if overtightened.

There are several types of fittings used in brake systems. Some are based on SAE specifications, whereas others are metric designs regulated by the International Organization for Standardization (ISO). All of the fittings, however, fall into two basic groups:

- Compression fittings
- Flare fittings.

Compression Fittings

Compression fittings, as their name implies, make a connection by compressing a sealing washer between parallel surfaces that are tightened against one another. In brake systems, the sealing washers are usually made of copper. The compression fittings in brake systems are either straight compression fittings or banjo fittings.

Straight compression fittings, figure 4-21, are sometimes used on the end of a brake hose that attaches to a wheel cylinder or brake caliper. The sealing washer is compressed between the fitting and the cylinder or caliper body. Straight compression fittings are usually attached to the end of a brake hose and are not free to rotate. For this reason they must be installed and tightened before the other end of the hose is connected to the brake tubing.

The **banjo fitting,** figure 4-22, gets its name from its shape, which resembles the instrument of the same name. A banjo fitting allows the brake hose to exit the wheel cylinder or brake caliper at a right angle, which can be a major benefit where

Brake Fluid and Lines

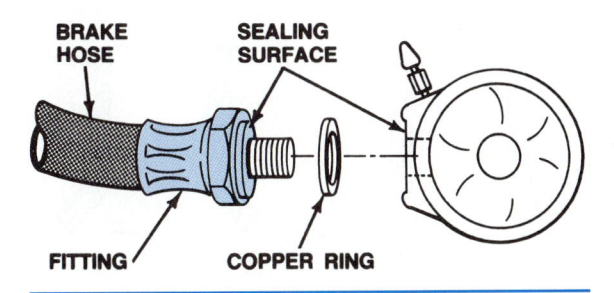

Figure 4-21. Straight compression fittings are sometimes used where hoses connect to wheel cylinders or brake calipers.

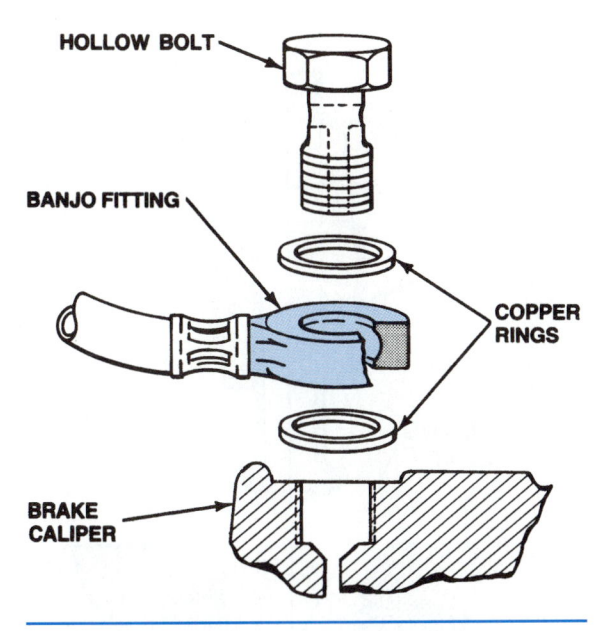

Figure 4-22. The banjo fitting enables a brake line to make a 90-degree turn.

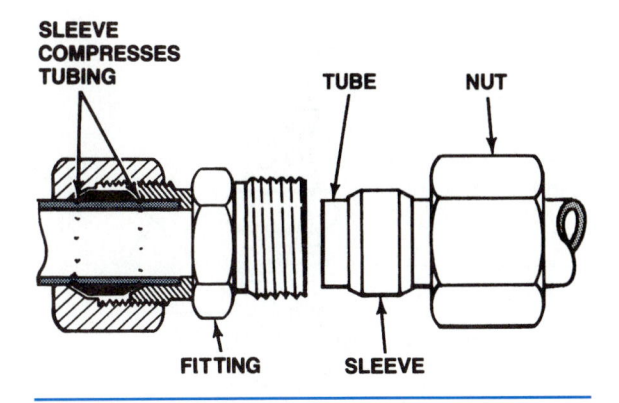

Figure 4-23. Spherical-sleeve compression fittings should never be used to repair steel brake lines.

clearance is limited. Brake fluid passes from the hose into the banjo fitting, then passes out through the hollow bolt into the cylinder or caliper. The connection is sealed with two washers installed on the bolt on either side of the banjo. Banjo fittings are subject to leakage from distortion if overtightened.

The spherical-sleeve compression fitting shown in figure 4-23 is used to join sections of tubing. This fitting works by compressing two brass rings, or ferrules, that clamp down on the tubing and create a seal. Although this type of fitting is acceptable in low-pressure applications, it is unreliable and unsafe for use in high-pressure steel brake lines. Spherical-sleeve compression fittings should *never* be used in brake hydraulic systems.

Flare Fittings

Flare fittings are used for most of the connections in modern brake systems. Although flare fittings do use compression to effect a seal, they have tapered seats and do not employ a separate sealing washer. Both SAE inch and metric flare fittings are used in brake systems, however, the two are not interchangeable.

SAE Flare Fittings

All SAE flare fittings used in automotive brake applications have a 45-degree taper on the male part of the fitting and a 42-degree taper on tubing seat in the female part of the fitting. The 3-degree difference in angles creates an interference fit that provides a better seal and strengthens the connection.

SAE flare fittings come in two varieties: standard flare and inverted flare. Brake hose ends sometimes use a standard flare fitting, figure 4-24, where they attach to a fixed component. A standard flare fitting does not use any type of separate sealing surface; when the fitting is tightened, the tapered faces of the male and female parts of the fitting are forced into contact to create the seal.

Far more common is the inverted flare fitting, figure 4-25, that is used on all SAE brake tubing connections. In an inverted flare fitting, a male tubing nut compresses a flared section on the end of the tube against the tapered face of the tubing seat in the female part of the fitting. The brake tubing actually serves as the sealing surface in this design.

There are two types of tubing flares that can be used with SAE fittings; single flares and double flares, figure 4-26. The single flare is the simplest type, but single flares are never used on brake tubing because they have a tendency to

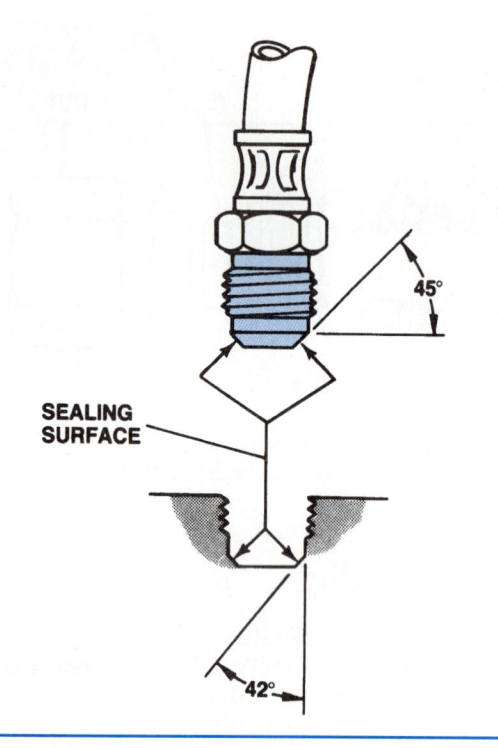

Figure 4-24. Standard SAE flare fittings are used for brake line connections in rare instances.

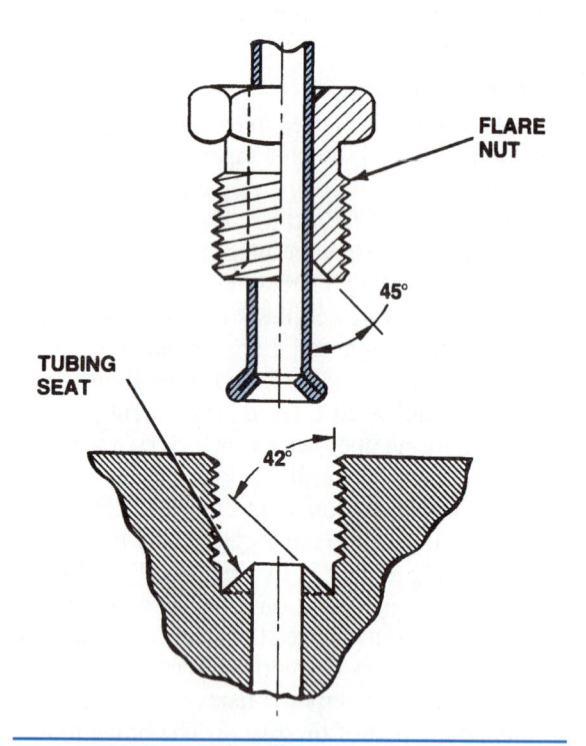

Figure 4-25. The SAE inverted flare fitting is the most common type used in domestic brake systems.

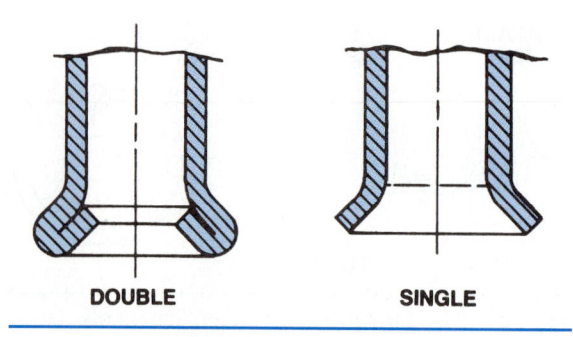

Figure 4-26. Both single and double flares will work in an SAE flare fitting, but only the double flare is strong enough to be safe in brake lines.

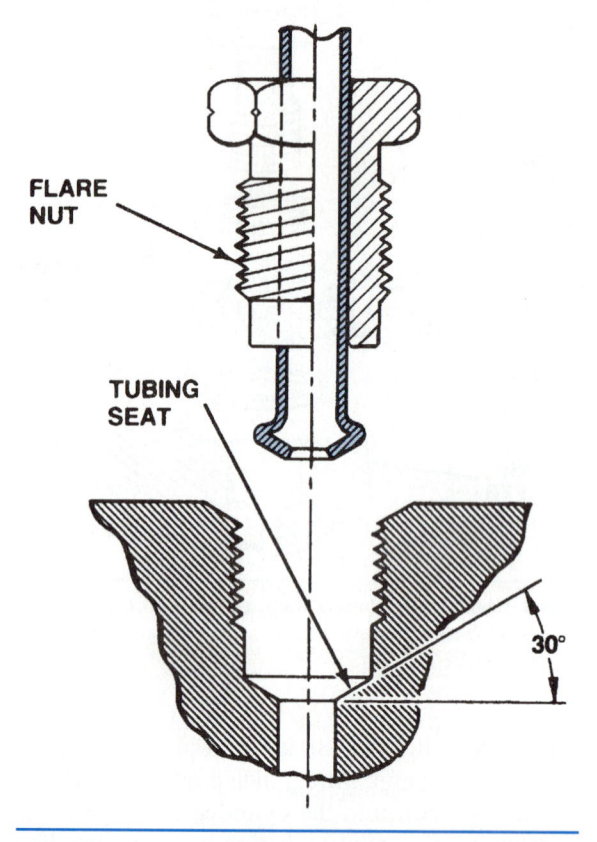

Figure 4-27. The ISO flare is used in most European, and many domestic, brake systems.

crack and leak. The double flare is much stronger and is used on all SAE brake tubing fittings. Replacement brake tubing is sold with a preformed double flare on each end. When a section of tubing is cut to length in the field, a special flaring tool is used to form a new double flare.

ISO Flare Fittings

The ISO flare fitting, figure 4-27, is a metric design first used on the brake tubing of many imported cars. Because it offers several advantages, domestic manufacturers are also adopting the ISO flare fitting on many newer models.

Brake Fluid and Lines

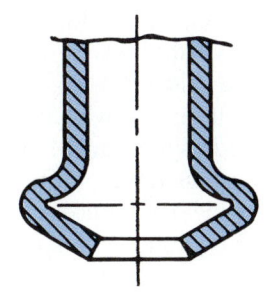

Figure 4-28. The ISO tubing flare has a unique shape that is not compatible with SAE fittings.

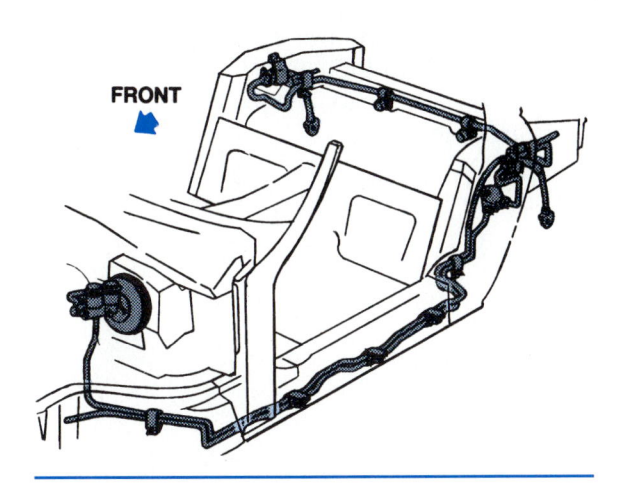

Figure 4-29. The brake line routing is carefully engineered into the design of the overall system.

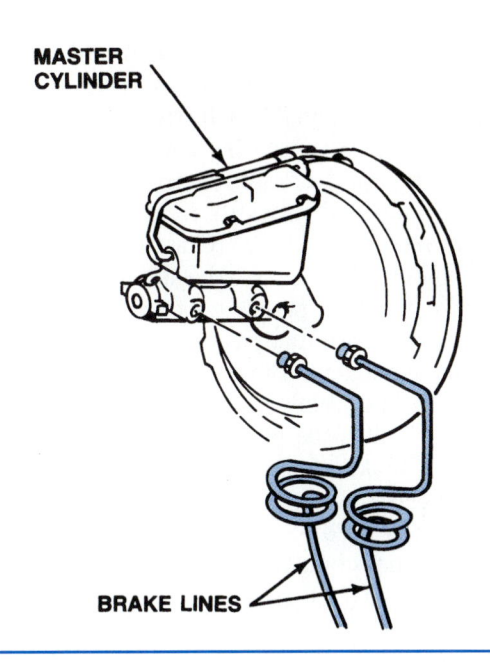

Figure 4-30. The coils found in some brake lines help prevent cracks caused by vibration.

The shape of the ISO flare is different from that of the SAE double flare. The ISO design, figure 4-28, is open rather than folded back on itself, and the resulting shape is sometimes referred to as a "bubble flare." As the tubing nut is tightened, the flare is compressed and forced outward to create the seal. The interference angle designed into the taper seats actually pushes the flare more tightly into the seal rather than squeezing it out as in the SAE flare fitting. This makes for a stronger, more leak-free connection that is also less subject to overtightening.

BRAKE LINE ROUTING

Brake system engineers take special care in routing brake lines from the master cylinder to the wheel friction assemblies. Brake lines on the underside of the vehicle are generally routed along the inside of the frame rails, or inside the unit body, where they are protected from impacts and the corrosion caused by water and salt, figure 4-29. Brake lines should never pass over or under an exhaust pipe, muffler, or catalytic converter where excessive heat might contribute to vapor lock.

Brake tubing is usually rubber mounted in strong brackets that prevent and absorb vibration that could cause metal fatigue and cracking. In many cases, the tubing leaving the master cylinder is coiled to further ensure that it is unaffected by vibration, figure 4-30. The coils allow the brake line to flex without breaking.

SUMMARY

Brake fluids must meet performance standards established by the federal government. These specifications deal with a fluid's boiling point, how it deals with temperature variations, its effect on mechanical parts in the brake system, and its compatibility with other brake fluids. Fluids that meet the requirements are classified as DOT 3, DOT 4, DOT 5, or DOT 5.1 grade fluids.

There are three types of brake fluids in use today. Polyglycol fluid is by far the most common, but its extreme hygroscopicity creates a number of problems in the brake system. Silicone brake fluid is a newer type that does not attract water, but it is not currently used by any major vehicle manufacturer. Hydraulic system mineral oil (HSMO) is the rarest type of brake fluid, and it is used in only a few specialized systems. Polyglycol and silicone brake fluids are compatible, although they do not blend together; HSMO is not compatible with either of the other brake fluid types.

All brake fluids must be stored and handled with care to keep them clean and moisture free. Polyglycol fluid should not be stored for extended periods because once the container has been opened the fluid absorbs water that lowers its boiling point and raises its freezing point. Silicone brake fluid and HSMO are relatively unaffected by moisture and can be stored indefinitely.

Brake bleeding purges air from the hydraulic system. There are four basic ways to bleed brake systems: manual bleeding, gravity bleeding, pressure bleeding, and vacuum bleeding. Brake bleeding is also the process used to change the fluid in a brake system. Polyglycol fluid should be changed once a year. Silicone and HSMO brake fluids need to be changed only when the fluid becomes contaminated with particulates.

Brake lines transmit hydraulic pressure between components in the brake system. They contain extreme pressure and work in a harsh environment. Only proper double-walled steel brake tubing or reinforced, synthetic rubber brake hose should be used. Brake tubing is often fabricated in the field, but brake hose cannot be easily made up because special equipment is required to attach the fittings.

Brake line fittings are of either the compression or flare design. Straight compression and banjo fittings seal with separate copper washers. Flare fittings have tapered seats and commonly use a flared portion of brake tubing to form the sealing surface. Both SAE and ISO flared fittings are used on newer cars, but the two types are not interchangeable.

Brake Fluid and Lines

Review Questions

Choose the letter that represents the best possible answer to the following questions:

1. Which of the following is not related to brake fluid quality standards?
   a. FMVSS
   b. DOT
   c. HSMO
   d. SAE

2. DOT numbers refer specifically to _____ of brake fluid.
   a. Types
   b. Grades
   c. Both a and b
   d. Neither a nor b

3. When brake fluid is overheated, _____ may result.
   a. Brake fade
   b. Vapor lock
   c. Increased pedal pressure
   d. All of the above

4. A brake fluid's wet ERBP:
   a. Reflects its performance in service
   b. Is printed on the fluid container
   c. Simulates the effects of hygroscopicity
   d. All of the above

5. Most of the moisture in a brake system enters through the:
   a. Rubber parts
   b. Master cylinder
   c. Wheel friction assemblies
   d. None of the above

6. Which of the following is *not* an FMVSS test requirement for brake fluid?
   a. Low-temperature viscosity
   b. Evaporation residue
   c. Contamination level
   d. Oxidation resistance

7. Which of these brake fluids is *not* compatible with the other types?
   a. Polyglycol
   b. DOT 4
   c. Silicone
   d. None of the above

8. Polyglycol brake fluid is:
   a. Not inert
   b. Hydroscopic
   c. Required to be amber colored by law
   d. All of the above

9. One of the problems with silicone brake fluid is that it:
   a. May not operate a brake fluid level sensor
   b. Has higher room temperature viscosity than polyglycol fluid
   c. May attack rubber parts
   d. Becomes compressible at low temperatures

10. If polyglycol and silicone brake fluids are mixed, problems may result from:
    a. Chemical reactions
    b. Differences in specific gravity
    c. Both a and b
    d. Neither a nor b

11. HSMO can be considered:
    a. Compatible with DOT 5 brake fluid
    b. A marginal brake fluid
    c. An inert material
    d. A natural rust preventive

12. Brake fluid containers are designed:
    a. With an airtight seal
    b. Not to react with the fluid they contain
    c. To meet federal regulations
    d. All of the above

13. Brake bleeding is required:
    a. When the brake fluid is changed
    b. Before the car leaves the factory
    c. Both a and b
    d. Neither a nor b

14. Which of the following is not a method of bleeding brakes?
    a. Pressure
    b. Manual
    c. Static
    d. Vacuum

15. Pressure bleeding is preferred by many professional repair shops because it:
    a. Saves time
    b. Is inexpensive
    c. Does not require special tools
    d. All of the above

16. The boiling point of DOT 3 polyglycol brake fluid can be reduced to a marginal level by only _____ percent water contamination.
    a. 1
    b. 2
    c. 5
    d. 10

17. Silicone brake fluid should be changed:
    a. Every one or two years
    b. When it is particulate contaminated
    c. Both a and b
    d. Neither a nor b

18. Modern brake tubing:
    a. Makes up 50 percent of the brake lines
    b. Is made from a mixture of tin and steel
    c. Has a burst strength of more than 5,000 psi
    d. None of the above

19. Brake tubing may be:
    a. Bent by hand
    b. Coiled to prevent cracking
    c. Either seamed or seamless
    d. All of the above

20. Which of the following is *not* true of replacement brake hoses.
    a. They are made of natural rubber
    b. They must have permanently attached ends
    c. They must pass an ozone resistance test
    d. They must be striped

21. A damaged brake hose can prevent a wheel friction assembly from:
    a. Releasing
    b. Applying
    c. Both a and b
    d. Neither a nor b

22. A _____ brake line fitting seals with two copper washers.
    a. Straight compression
    b. Taper
    c. Double flare
    d. Banjo

23. A _____ fitting should never be used in a high-pressure brake line.
    a. Single-flare
    b. Spherical-sleeve compression
    c. Both a and b
    d. Neither a nor b

24. A 30-degree seat is used only on a(n) _____ fitting.
    a. SAE
    b. Inverted flare
    c. Male
    d. ISO

25. Brake lines are *not* routed:
    a. Through rubber mounts
    b. Over exhaust system components
    c. On the frame rails
    d. Inside the car body

5
Pedal Assemblies and Master Cylinders

OBJECTIVES

Upon completion and review of this chapter, you will be able to:

- Describe the construction of the pedal assembly.
- Explain the term "pedal ratio."
- Explain brake pedal freeplay and describe freeplay adjustment.
- Describe master cylinder construction.
- Explain single-piston and dual-piston master cylinder operation.
- Explain the need for and the function of dual-circuit brake systems.
- Explain the function of the quick-take-up master cylinder.
- Explain what a portless master cylinder is and why it is used.
- Explain the purpose and operation of dual-circuit brake systems.

KEY TERMS

anodized
clevis
compensating port
cup seals
diaphragm
dual-circuit brake systems
dual master cylinder
eccentric
effective pedal travel
freeplay
grommets
pedal ratio
quick-take-up master cylinder
quick-take-up valve
replenishing port

INTRODUCTION

Brake pedal assemblies have not changed much over the years, and the changes that have occurred affect pedal assembly construction much more than basic function. The braking process still begins at the brake pedal, and the force that ultimately stops the car is controlled by the much smaller force applied to the pedal pad. The first part of this chapter examines the pedal assemblies used in brake systems.

In contrast to the pedal assembly, brake master cylinders have undergone a great deal of modification and development. Disc brakes, antilock brakes, safety considerations, and fuel economy concerns have all caused signficant changes in master cylinder design. The second part of the chapter covers the evolution, construction, and operation of brake master cylinders. It also examines the different ways hydraulic pressure from the master cylinder is distributed to the wheel friction assemblies.

PEDAL ASSEMBLIES

Most modern brake pedal assemblies share a common design, figure 5-1. A steel bar called the pedal arm pivots on a bracket that is solidly attached to the vehicle chassis. The end of the pedal arm opposite the bracket is fitted with a metal foot pedal, and at some point between the pivot and the foot pedal, a pushrod is attached to the pedal arm. The pedal pushrod transmits force from the brake pedal assembly to the power booster or master cylinder.

The foot pedal is usually covered with a rubber pad that helps prevent the driver's shoe from slipping off the pedal if either is wet or covered with mud or snow. Some cars with power brakes and automatic transmissions have an especially wide brake pedal that allows for two-foot braking if the power booster fails, figure 5-2. The wide pedal also allows for either left or right foot braking. Cars with a manual transmission, and therefore a clutch pedal, generally do not have enough room in the footwell to install an extrawide brake pedal.

Although the basic elements of all pedal assemblies are the same, assemblies do differ in the way the pedal is supported. There are two basic designs:

- Frame-mounted pedals
- Suspended brake pedals.

Frame-Mounted Brake Pedals

The frame-mounted brake pedal, figure 5-3, is the earlier of the two pedal designs. Frame-mounted brake pedals attach directly to the vehicle chassis, and pivot at the bottom or center of the pedal arm. In the case of unit body cars that do not have a separate chassis, the pedal bracket attaches to the floorpan inside the car, figure 5-4. This design is often called a floor-mounted pedal. Frame-mounted brake pedals were widely used on cars into the 1950s, and on light trucks through the 1960s.

Frame-mounted brake pedals worked well with the mechanically actuated brakes of early automobiles; it was easy to connect mechanical rods and linkages to a pedal arm located under the car. In addition, mounting the brake pedal on the frame made it possible to build and test a complete rolling chassis before the body was fitted. Many of the earliest cars were sold as a running chassis only; the bodywork was added later by an independent coachbuilder.

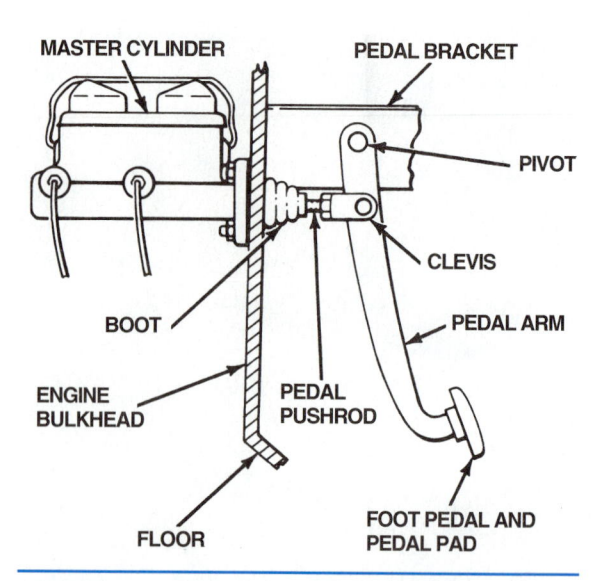

Figure 5-1. All brake pedal assemblies have certain parts in common.

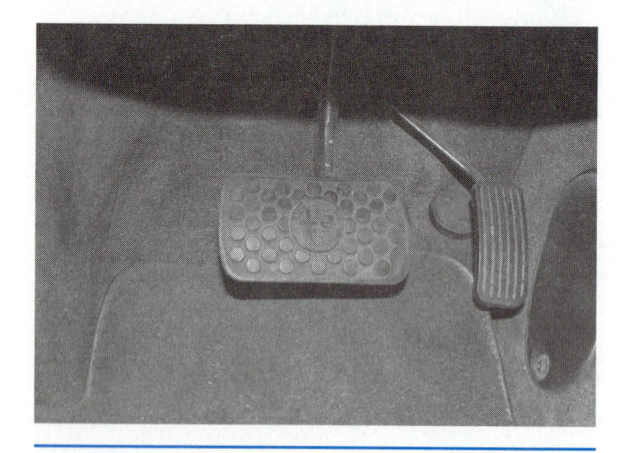

Figure 5-2. Many vehicles with power brakes and automatic transmission have an extrawide pedal.

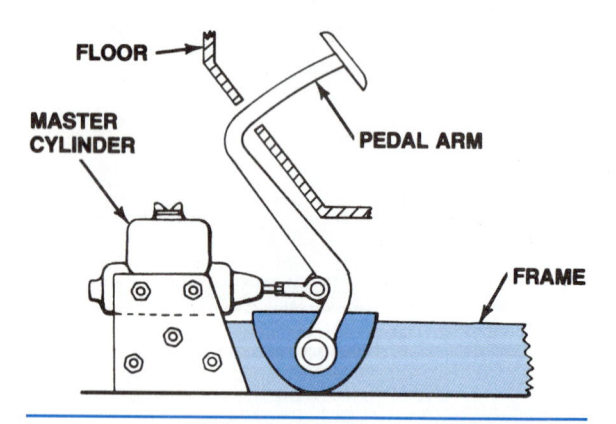

Figure 5-3. Frame-mounted brake pedals were widely used through the 1950s.

Pedal Assemblies and Master Cylinders

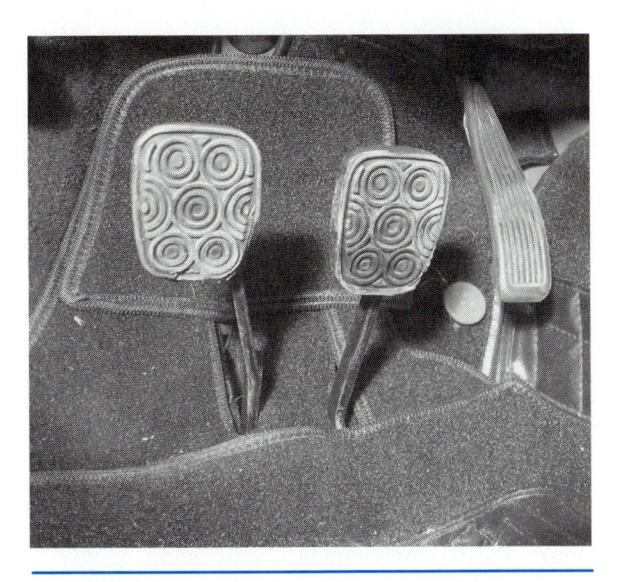

Figure 5-4. Floor-mounted pedals were last used on some imports and may still be found on hot rods and custom built vehicles.

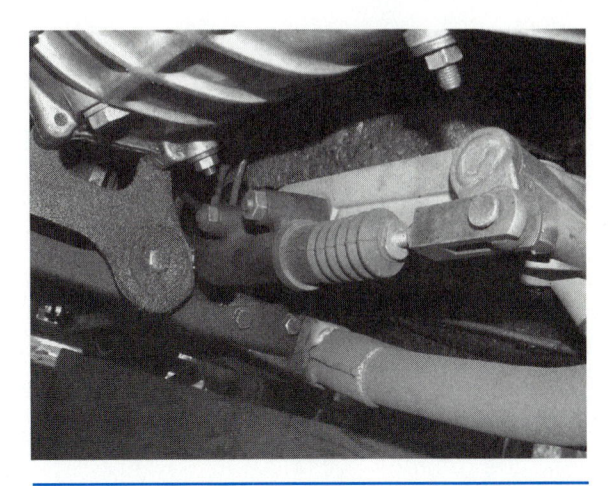

Figure 5-5. This frame-mounted master cylinder is located under the car, between the transmission and the exhaust system. The rubber boot is critical in keeping it operational.

Frame-mounted pedal designs do have a number of drawbacks. They are susceptible to wear and corrosion because the pedal assembly is exposed below the vehicle. And even floor-mounted pedals locate the pivot near or on the floor where dirt and other grit collect and cause wear. Frame-mounted brake pedals also require an opening in the floor for the pedal arm to pass through, and it can be difficult to seal this opening against moisture, dirt, and noise. Finally, frame-mounted pedal assemblies position the master cylinder under the car where it is difficult to service and exposed to road splash, figure 5-5.

Suspended Brake Pedals

As the production of car bodies shifted to the vehicle manufacturers, and hydraulic brakes and unit body construction were introduced, the disadvantages of frame-mounted brake pedals soon outweighed any advantages they once had. In their place, almost all modern cars and light trucks use a suspended brake pedal like that shown in figure 5-1. Suspended pedals mount high on the engine bulkhead in the interior of the vehicle behind the instrument panel. In some cars, the pedal bracket helps save weight by doubling as a steering column support.

Suspended brake pedals have none of the disadvantages of frame-mounted pedals, and they offer several benefits. They can be made fairly long to create greater leverage, and they are located off the floor in a clean environment where wear and corrosion are less of a problem. An opening in the bulkhead is still required for the pedal pushrod, but the master cylinder or power booster being actuated usually bolts directly to the bulkhead in the engine compartment and completely seals the opening.

Perhaps the most significant advantage of a suspended brake pedal assembly is that it positions the master cylinder relatively high in the engine compartment. In all but the most unusual circumstances, this location places the cylinder out of reach of road splash and other moisture that can contaminate the brake fluid. It also makes the cylinder easier to reach for service, and even contributes to a small amount of self-bleeding action because air bubbles in the system will tend to rise to the top of the master cylinder fluid reservoir.

Adjustable Pedals

Beginning in 1999, some manufacturers began offering adjustable pedal assemblies. These systems use electric actuators that allow the driver to move the brake and accelerator pedals forward or backward, for a total travel of about 3 inches, figure 5-6. The pedals are adjusted with a dash-mounted switch, figure 5-7, and may also be combined with the body control system so that the pedals automatically adjust to a memorized position, depending on the driver. Having the ability to adjust the pedals helps drivers keep their seat at a safe distance from the steering wheel and at a height that provides comfort and good visibility.

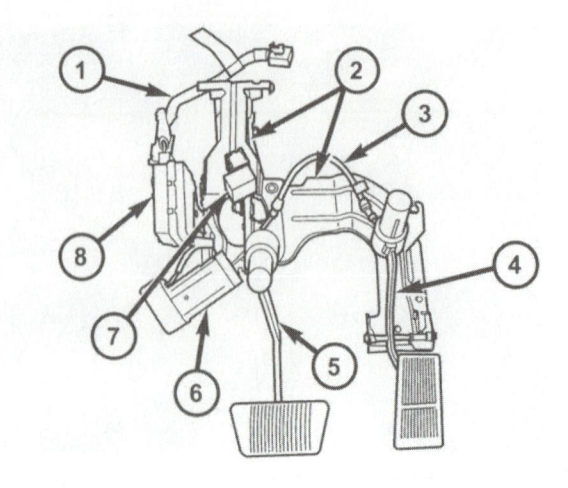

1 - HARNESS
2 - ADJUSTABLE PEDAL BRACKET
3 - CABLE
4 - ACCELERATOR PEDAL
5 - BRAKE PEDAL
6 - ADJUSTABLE PEDAL MOTOR
7 - BRAKE LIGHT SWITCH
8 - ADJUSTABLE PEDALS MODULE

Figure 5-6. This assembly allows the pedals to adjust electrically up to 3 inches closer to the driver. (Courtesy of DaimlerChrysler Corporation. Used with permission)

Figure 5-7. The adjustable pedal assembly can be controlled by a switch on the dashboard.

Brake Pedal Ratio

The brake pedal assembly is basically a lever. The length of the pedal arm combined with the support bracket pivot point creates a mechanical advantage. The mechanical advantage increases the force applied at the foot pedal pad and applies it through the pedal pushrod to the brake booster or master cylinder piston.

As discussed in Chapter 3, the amount of mechanical advantage provided by the brake pedal assembly is determined by the **pedal ratio.** Different pedal ratios are used on different cars, and the ratio chosen depends to a large degree on whether the brake system is equipped with a power booster. If a booster is fitted, the pedal assembly does not need to supply as great an increase in force. Manual brakes use a pedal ratio of about 5 to 1, but the pedal ratio in a power brake system is generally closer to 3 to 1, figure 5-8.

Brake system engineers often choose to use a numerically lower pedal ratio *with* a power booster rather than a high pedal ratio *without* a booster, although the latter design would work just as well in normal driving. One reason they do this is to provide a greater margin of safety in the event of mechanical brake fade, fluid vapor lock, or a slow fluid leak. All other factors being equal, a low pedal ratio provides more master cylinder piston movement for a given amount of pedal travel. This means more **effective pedal travel** is available to move fluid through the hydraulic system to compensate for problems.

■ Le Champignon

Since 1955, the French car maker Citroen has used a brake pedal unique in both appearance and operation. On the floor of most Citroens, in the location normally reserved for a brake pedal, is a small, round, rubber bulb that looks like a toadstool—hence its affectionate French nickname, le champignon, the mushroom. Citroens that use the mushroom have a high-pressure central hydraulic system that powers the suspension, steering, and brakes. The mushroom operates a spool valve that controls the flow of pressurized hydraulic fluid to the wheel friction assemblies.

Because high pressure always exists in the hydraulic system, long brake pedal travel or a separate power booster are not needed to actuate the brakes. The mushroom moves only a very short distance when foot pressure is applied. The mushroom is also more pressure sensitive and less travel sensitive than a conventional brake pedal. This makes it easier to modulate the brakes because stopping power is more proportional to the pressure applied at the mushroom.

Pedal Assemblies and Master Cylinders

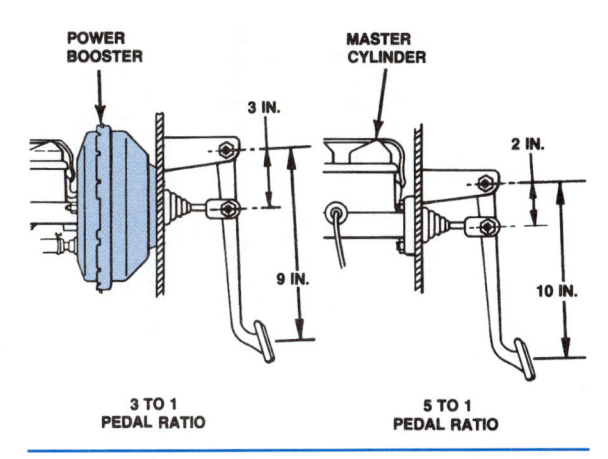

Figure 5-8. Brake pedal ratio often varies based on whether the brake system is equipped with a power booster.

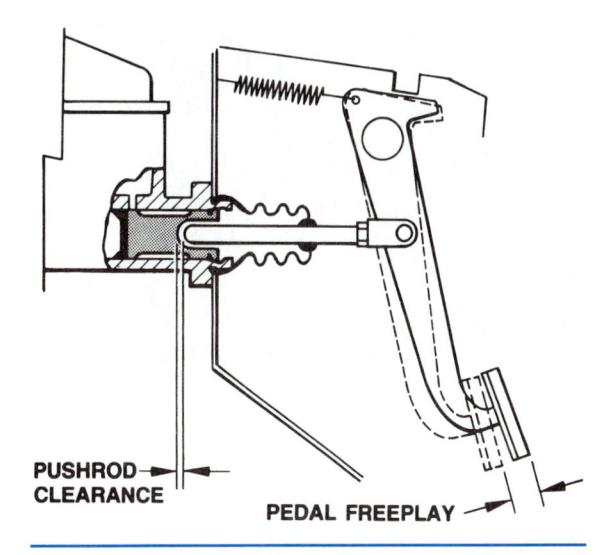

Figure 5-9. The clearance between the pedal pushrod and master cylinder piston is measured as pedal freeplay.

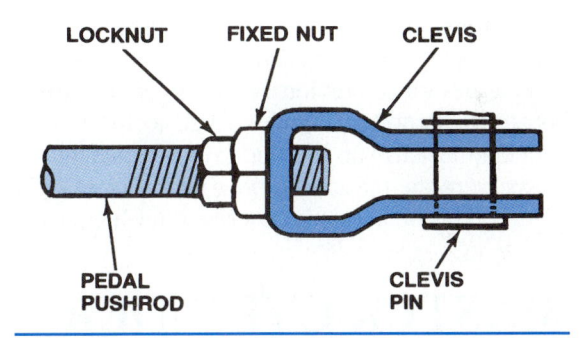

Figure 5-10. Most adjustable pedal pushrods thread into the clevis that attaches to the brake pedal arm.

Brake Pedal Freeplay

The design of brake system master cylinders requires a small amount of clearance, or **freeplay**, before movement of the brake pedal brings the brake pedal pushrod into contact with the master cylinder piston. This freeplay allows the piston to fully retract when the brake pedal is released, and this in turn ensures that a compensating port inside the master cylinder is opened. The function of the compensating port is explained later in the chapter.

Depending on the system, the amount of freeplay at the brake pedal may be as little as $\frac{1}{16}$ inch or as great as 1 inch, figure 5-9. It is important to realize that the pedal freeplay is actually the clearance between the brake pedal pushrod and master cylinder piston magnified by the pedal ratio. If the pedal ratio is 5 to 1 and there is $\frac{1}{16}$ inch of clearance between the pushrod and the piston, the freeplay at the pedal will be $\frac{5}{16}$ inch—plus any mechanical play that exists in the pedal assembly itself.

Generally, the clearance between the brake pedal pushrod and master cylinder piston is kept to a minimum. All that is really needed is enough to ensure that the piston can fully retract. By keeping the clearance as small as possible, pedal freeplay is reduced, which allows the brakes to be applied faster. Reduced freeplay also increases the effective pedal travel in the event of mechanical brake fade, fluid vapor look, or a slow fluid leak.

Freeplay Adjustment

Vehicle manufacturers usually specify a range of pedal freeplay for their brake systems. On many cars there is no provision for adjustment because manufacturing tolerances are specified that ensure any combination of pedal assembly and master cylinder will result in acceptable freeplay. On other cars, the length of the brake pedal pushrod can be varied to adjust the freeplay. Adjustable pushrods are preset at the factory and normally do not require adjustment unless the master cylinder is rebuilt or a new cylinder is installed.

Most adjustable pushrods thread into a **clevis** that mounts to the pedal arm, figure 5-10. The pushrod is threaded in or out of the clevis to establish the specified clearance and then fixed in position with a locknut that tightens against the clevis. There are several minor variations of the thread-adjustable pushrod, but they all function in essentially the same manner.

Another, less common, type of adjustment uses a pushrod end that attaches to the pedal arm

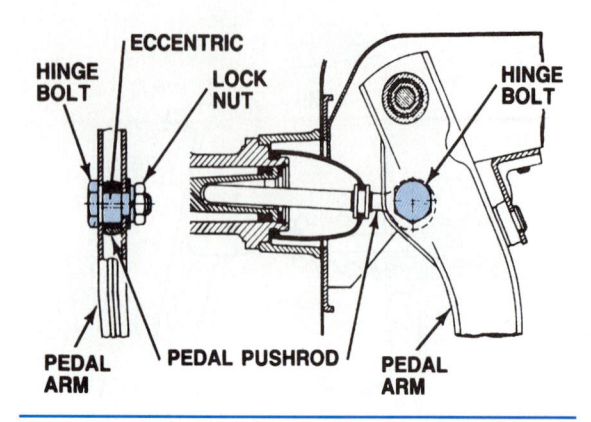

Figure 5-11. An eccentric hinge bolt is used to adjust the pedal freeplay on some cars.

with an **eccentric** hinge bolt, figure 5-11. The eccentric bolt is rotated until the proper clearance is obtained, then the bolt is fixed in position with a locknut that tightens against the pedal arm.

Vehicles with brake power boosters also have either fixed or adjustable pedal pushrods. The pedal freeplay on systems with adjustable pushrods is set as described previously. However, the power booster pushrod that applies force to the master cylinder must also be checked to ensure that it does not prevent the master cylinder piston from fully seating. This operation is described in Chapter 13.

MASTER CYLINDER CONSTRUCTION

The master cylinder converts the mechanical force provided by leg power, leverage, and in some cases a power booster, into hydraulic pressure. This pressure is then distributed to the wheel friction assemblies where it is converted back into mechanical force that applies the brakes. Although there are several different designs of master cylinders, they all have certain basic parts in common. Some of the parts, such as pistons and seals, are also used in wheel cylinders and brake calipers. The basic elements that make up a master cylinder are described in the following sections.

Brake Fluid Reservoir

To operate properly, a hydraulic brake system must be completely filled with air-free brake fluid. However, the amount of fluid required to fill the system will vary. Temperature changes cause brake fluid to expand and contract; brake

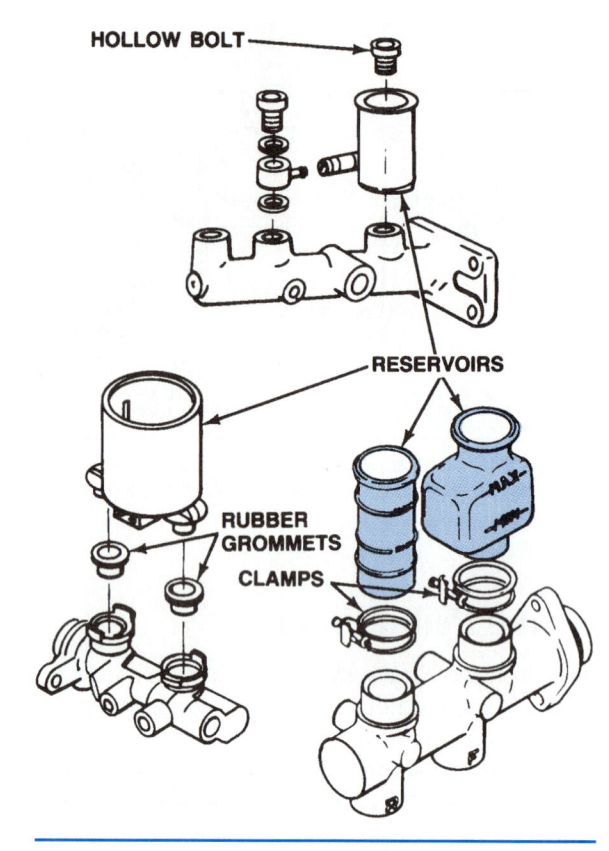

Figure 5-12. Brake fluid reservoirs come in many shapes and sizes.

application forces fluid out to the wheel friction assemblies, which increases total hydraulic system volume; and brake lining wear leaves the wheel cylinder and brake caliper pistons farther out in their bores, which requires additional fluid to fill the resulting space. The brake fluid reservoir stores a supply of brake fluid to compensate for these changes.

Brake fluid reservoirs come in many forms; some are cast in metal as part of the master cylinder body, whereas others are separate parts made of plastic or nylon, figure 5-12. Separate reservoirs either attach to the master cylinder with rubber **grommets**, or they clamp or bolt directly to the cylinder body. The fluid reservoir may include a fluid level sensor, figure 5-13. Where the master cylinder is mounted in a difficult-to-reach location, a remote-mounted reservoir is sometimes used, figure 5-14. Remote reservoirs can be made of metal, plastic, or nylon, and they are connected to the master cylinder with steel tubing or rubber hose.

All cars built after 1966 are equipped with a dual master cylinder that divides the brake system into two separate hydraulic circuits. The two circuits are served by two independent fluid reser-

Pedal Assemblies and Master Cylinders

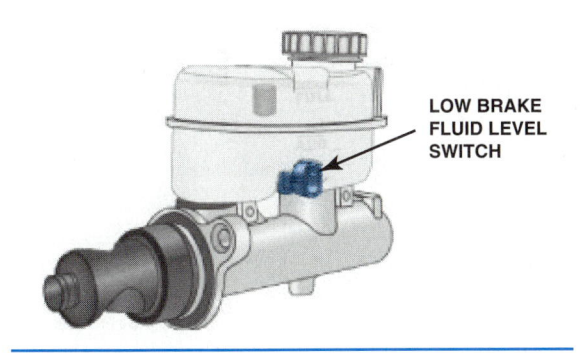

Figure 5-13. The brake fluid level sensor is located in the fluid reservoir. (Courtesy of DaimlerChrysler Corporation. Used with permission)

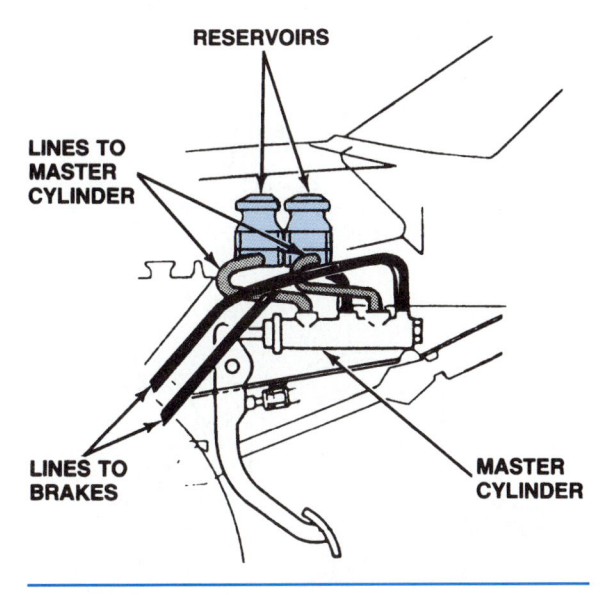

Figure 5-14. Remote-mounted fluid reservoirs are used where it would be difficult to check and fill a reservoir mounted on the master cylinder.

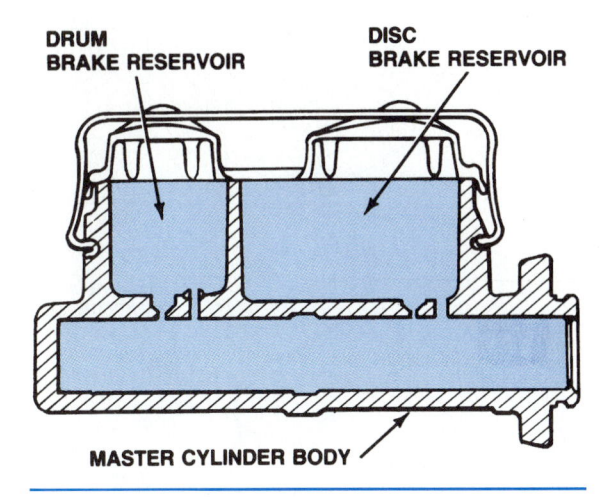

Figure 5-15. When the two sides of a dual reservoir have unequal fluid volumes, the larger chamber serves the front disc brakes.

voirs, although in most cases they are combined into a single housing, figure 5-15. When the hydraulic system is split so the front disc brakes are applied by one circuit, and rear drum brakes are applied by the other, the reservoir for the front brakes is bigger. This ensures there will be enough fluid to keep the brake calipers filled as the brake pads wear and the relatively large caliper pistons remain farther out in their bores. When the hydraulic system is split so a front disc brake and a rear drum brake are applied by each circuit, the reservoirs for both circuits are the same size.

Reservoir Covers

Brake fluid reservoirs are fitted with covers to prevent dirt and other particulates from contaminating the fluid. When the reservoir is cast into the master cylinder body, the cover is often a metal stamping held in place by a wire bail, figure 5-16. Nylon and plastic reservoirs have caps made of plastic, nylon, or rubber that thread or snap into place, figure 5-16. The reservoir covers on many cars have fluid level sensors built into them. These sensors activate a warning light on the instrument panel if the fluid level falls below a safe level, figure 5-17. Fluid level sensors are covered in detail in the next chapter.

All brake fluid reservoir covers are vented to equalize pressure in the air space above the fluid as the level rises and falls. If the cover is sealed airtight, a partial vacuum would be formed whenever the fluid level fell. The vacuum would prevent fluid from entering the hydraulic system, and if strong enough, could pull air into the system past the wheel cylinder seals.

Venting the brake fluid reservoir creates a problem in systems that use polyglycol brake fluid. If polyglycol fluid is not kept out of contact with the air, it will quickly absorb enough moisture to reduce its boiling point to a dangerous level. To isolate the fluid from the air while still allowing pressure to equalize, a rubber **diaphragm** may be installed between the reservoir and the cover. The diaphragm seals tightly around the edge of the reservoir cover to isolate the fluid from the air; at the same time it is able to flex up and down as the fluid level changes.

The fluid reservoirs on the master cylinders of some import cars have plastic floats that serve much the same purpose as a rubber diaphragm. The floats ride on top of the fluid and limit air exposure to only a small ring around their outer edge.

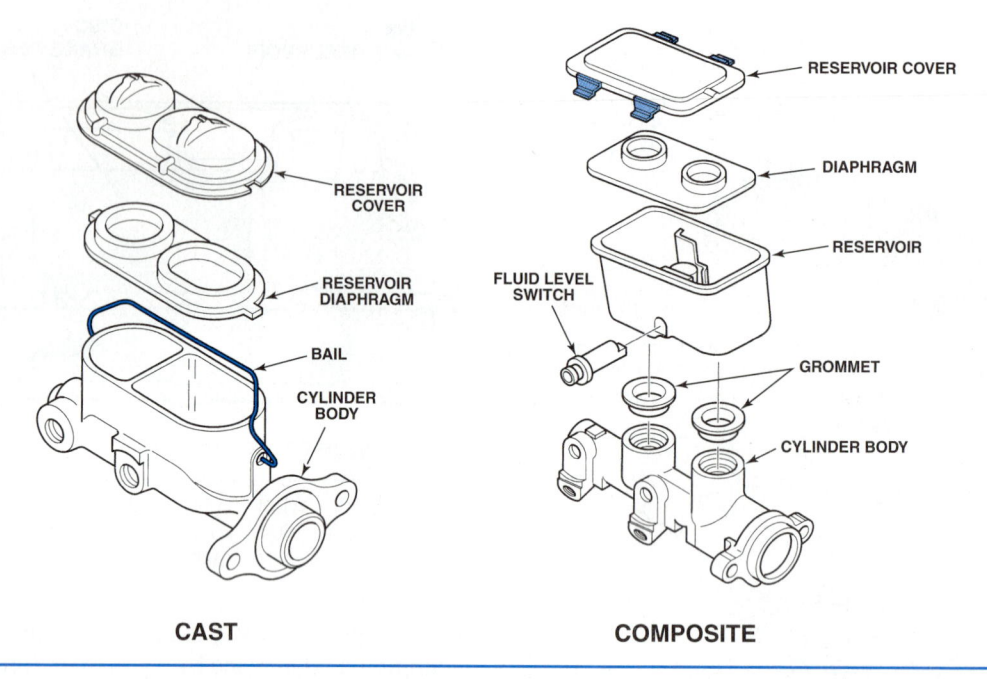

Figure 5-16. The reservoir cover may be held in place by a wire bail, left, or a snap fit, right. (Courtesy of General Motors Corporation, Service and Parts Operations)

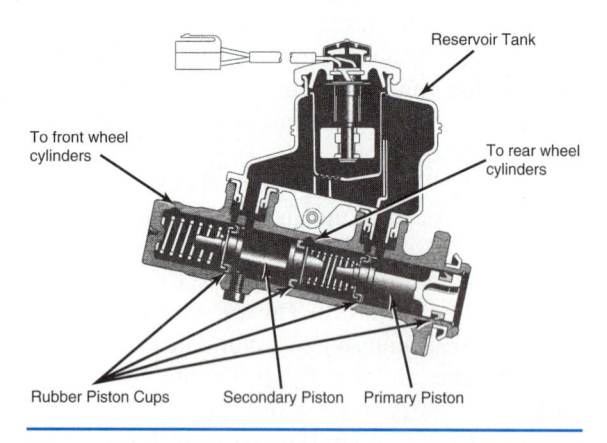

Figure 5-17. This master cylinder uses a fluid level sensor built into the cap. A float and magnetic switch operate a warning light on the instrument panel. (Courtesy of Toyota Motor Sales U.S.A., Inc.)

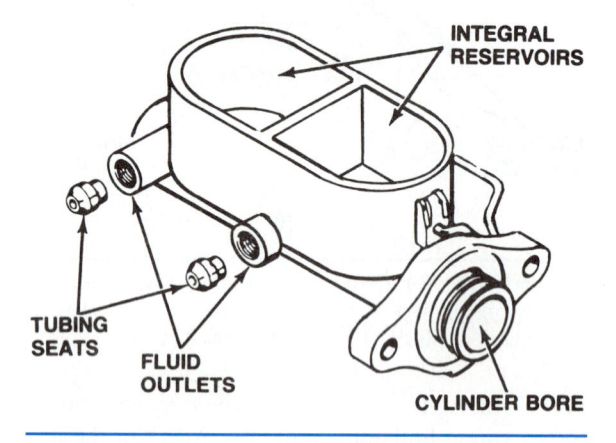

Figure 5-18. The master cylinder body provides the foundation for the brake hydraulic system.

Master Cylinder Body

The cylinder body, figure 5-18, is the basic building block of the entire master cylinder assembly. The brake fluid reservoir attaches to the outside of the cylinder body, as do the brake lines that carry the hydraulic pressure to the wheel friction assemblies. Inside the cylinder is the bore that contains the pistons and seals that create the hydraulic pressure that operates the brake system.

For each master cylinder piston there are two small openings between the fluid reservoir and the cylinder bore, figure 5-19. The function of these holes, called compensating and replenishing ports, is explained later in the chapter.

All early master cylinders and many built today are made of cast iron. Cast-iron master cylinders are sturdy, although heavy, and the fluid reservoir is usually cast as part of the cylinder body. To provide a good sealing surface, the bore

Pedal Assemblies and Master Cylinders

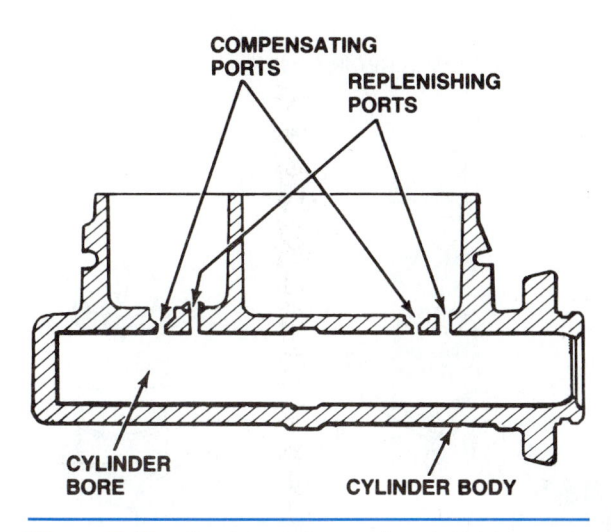

Figure 5-19. Compensating and replenishing ports allow brake fluid to pass between the reservoir and the cylinder bore.

of a cast-iron master cylinder body is drilled slightly undersize and heated red hot in a furnace. A steel ball bearing of the appropriate diameter is then forced into the cylinder bore to expand the opening to the finished size. This compresses the metal that forms the bore and produces a "bearingized" surface that is exceptionally smooth and durable.

Most newer brake systems have aluminum master cylinders that use a separate plastic or nylon fluid reservoir. The main reason for the change is a reduction in weight that contributes to better fuel economy, although aluminum master cylinders are also less costly to manufacture. Aluminum is a much softer metal than cast iron, so the bore of an aluminum master cylinder must be **anodized** to protect it from corrosion and provide a suitable surface for good sealing and a long service life.

Master Cylinder Pistons

Simple master cylinders, figure 5-20, contain a single piston that slides in the bore with only a few thousandths of an inch clearance. The piston is larger at its ends where the seals fit, and narrower in the center so a small chamber of fluid is created between the two ends of the piston. When force is applied to the piston by the brake pedal or power booster pushrod, it moves forward in the master cylinder bore and creates the hydraulic pressure that actuates the wheel friction assemblies. When the brake pedal is released, a spring in front of the piston helps return it to its retracted position.

Dual master cylinders, figure 5-21, contain a pair of pistons, one for each of the hydraulic circuits. The piston closest to the mouth of the cylinder, the one that contacts the brake pedal or power booster pushrod, is called the primary piston. The piston located toward the front of the cylinder is called the secondary piston. Each piston has its own return spring, and the spring on the primary piston may be attached to the piston with a retainer and screw.

Master cylinder pistons can be made of either steel or aluminum, although aluminum pistons are used almost exclusively in modern brake systems because they are lighter and less expensive to manufacture. The softness of aluminum pistons also provides increased cylinder body life when they are used in a cast-iron master cylinder. Despite the lubricating effects of brake fluid there is still friction as the pistons slide back and forth. An aluminum piston, which is softer than the cast-iron cylinder, will absorb most of the wear from this friction.

Piston Seals

Piston seals prevent brake fluid from escaping through the small gap between the pistons and the master cylinder bore. Without these seals, fluid would bypass the pistons and the master cylinder would be unable to produce and hold the hydraulic pressure necessary to operate the brakes. Some seals have a hole through their center where they mount on the piston; others have solid centers with a flat back that butts against a parallel surface on the end of the piston.

At least two seals are used on every master cylinder piston, figure 5-20: a primary seal and a secondary seal. The primary seal is located at the front of the piston; its job is to create hydraulic pressure. The secondary seal is located at the rear of the piston; its job is to prevent fluid from escaping the master cylinder.

The secondary piston in a dual master cylinder may have one or two secondary seals. When two secondary seals are used, figure 5-21, one secondary seal faces the primary piston and contains pressure created by the primary piston. The other secondary seal faces the front of the master cylinder and prevents fluid from the secondary circuit from leaking into the primary circuit. Some master cylinders

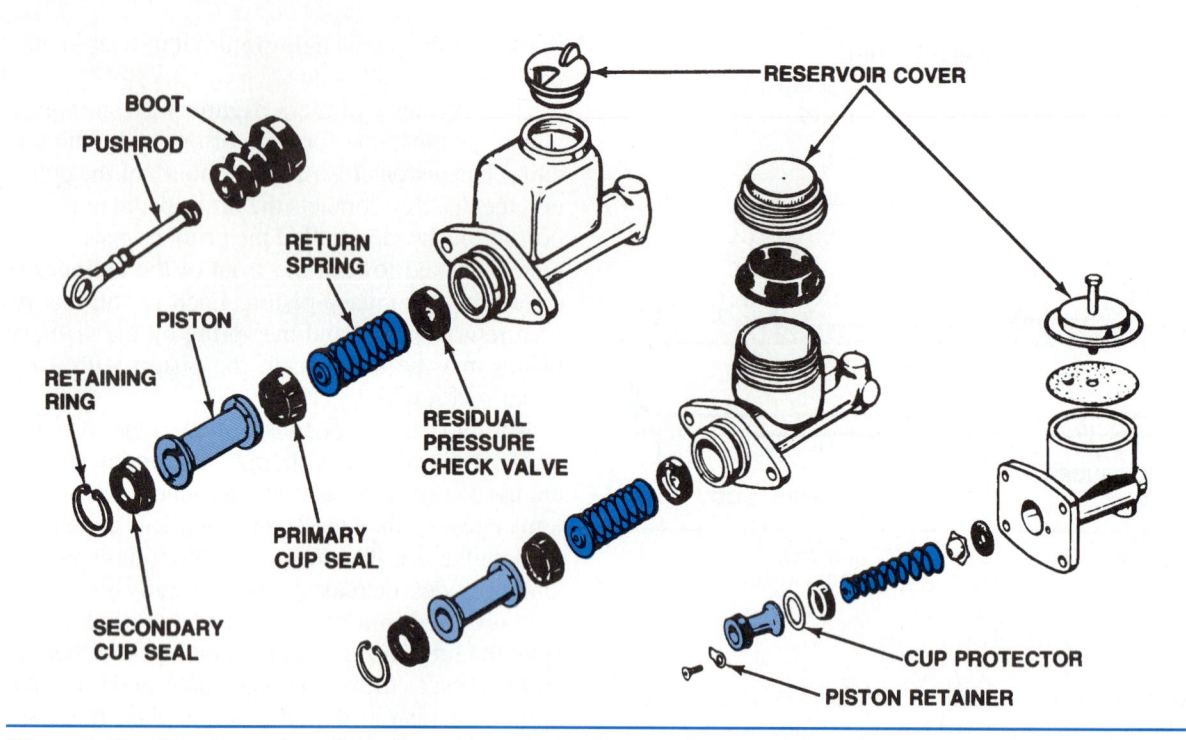

Figure 5-20. Simple master cylinders contain a single piston and return spring.

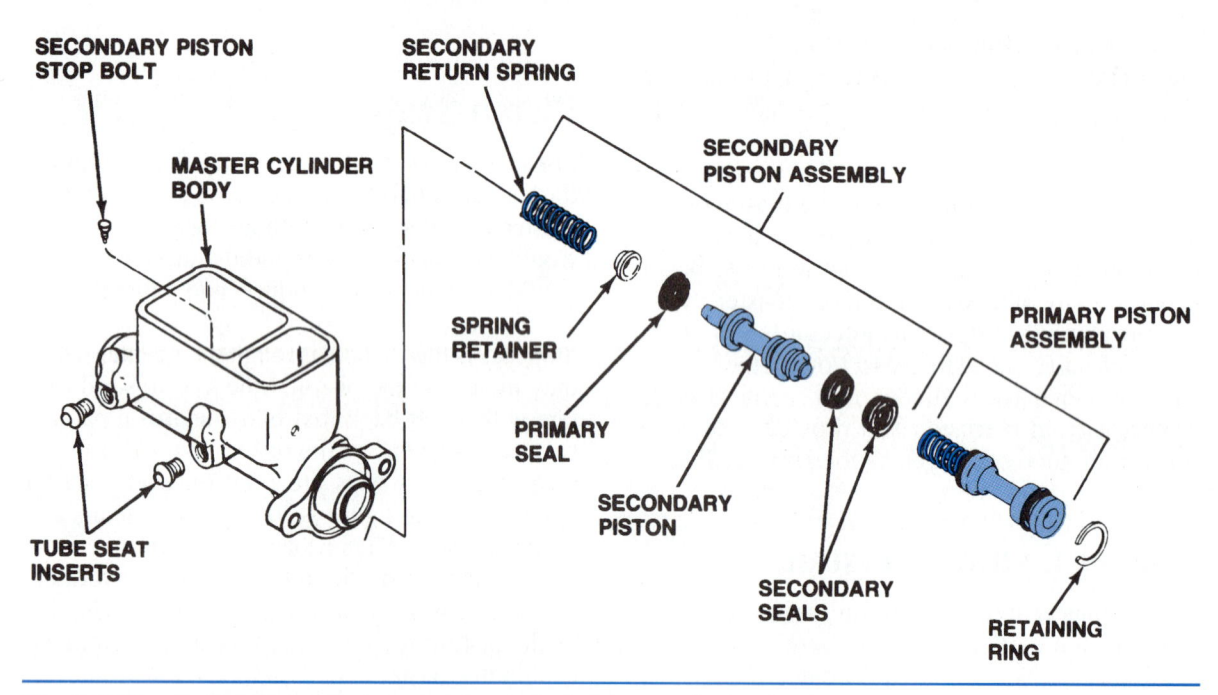

Figure 5-21. Dual master cylinders contain a pair of pistons and return springs.

are designed in such way so as to eliminate the need for the second secondary seal, figure 5-22.

Because of their shape, the seals used on master cylinder pistons are called **cup seals**. With a cup seal, the fluid seal is created by an angled lip on the outer edge of the cup that contacts the cylinder bore, figure 5-23. The outside diameter of the lip is a little larger than that of the cylinder bore, and this causes the lip to be compressed slightly when the seal is installed.

When the brakes *are not* applied, the small amount of lip compression provides all of the

Pedal Assemblies and Master Cylinders

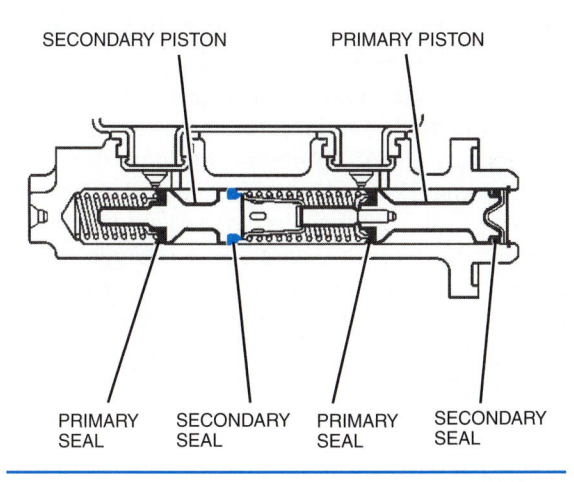

Figure 5-22. This master cylinder uses only one secondary seal on the secondary piston. (Courtesy of General Motors Corporation, Service and Parts Operations)

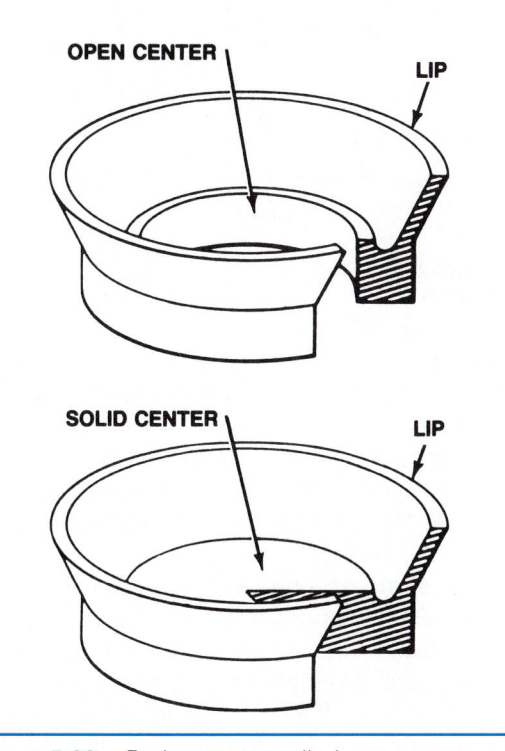

Figure 5-23. Brake master cylinders use cup seals to prevent fluid leakage between the pistons and cylinder walls.

sealing force. When the brakes *are* applied, pressure created in front of the cup forces the lip tightly against the cylinder bore; this additional sealing force enables cup seals to contain very high hydraulic pressures. With a cup seal, the sealing lip always faces the pressure to be contained.

Another important trait of cup seals is that they seal only in one direction. If the pressure on the back side of the seal becomes greater than the pressure on the cup side, the lip will collapse and fluid will bypass the seal. In effect, a cup seal acts like a one-way check valve. As explained later, the one-way sealing ability of cup seals plays an essential part in master cylinder operation.

DUAL-CIRCUIT BRAKE SYSTEMS

A single-piston master cylinder provides only one brake hydraulic circuit, figure 5-24. Because all the brakes are operated by a single pressure source, a leak at any point in the system will cause complete brake failure. To provide an extra margin of safety, **dual-circuit brake systems** were developed. A dual-circuit brake system uses a master cylinder with two piston assemblies, commonly called a **dual master cylinder,** that creates pressure for two separate hydraulic circuits. The cylinder is designed so a leak in one of the hydraulic circuits will not affect the other circuit. This ensures that in the event of a failure at least partial braking power will be retained. Most rear-wheel-drive (RWD) cars use a dual-circuit brake system with a front/rear split, figure 5-25, half of the master cylinder actuates the front brakes while the other half actuates the rear brakes.

A dual-circuit braking system with a front/rear split is satisfactory for a RWD car, but such a system could cause serious problems if used on a front-wheel-drive (FWD) vehicle. FWD cars have a strong forward weight bias that is made even more pronounced by weight transfer during braking. This means the rear brakes of FWD cars do very little work in stopping the car; in a hard stop they may supply as little as 15 percent of the total braking power. If the dual-circuit brake system on a FWD car were split front to rear, failure of the front brakes would leave the car with almost no stopping ability.

To solve this problem, brake system engineers developed the diagonal-split dual-circuit brake system, figure 5-26. With this design the right front and left rear brakes are operated by one hydraulic circuit, whereas the left front and right rear brakes are operated by the other. In the event of partial system failure, at least one front and one rear brake will continue to work. This means approximately 50 percent of the total braking power is always available with this system.

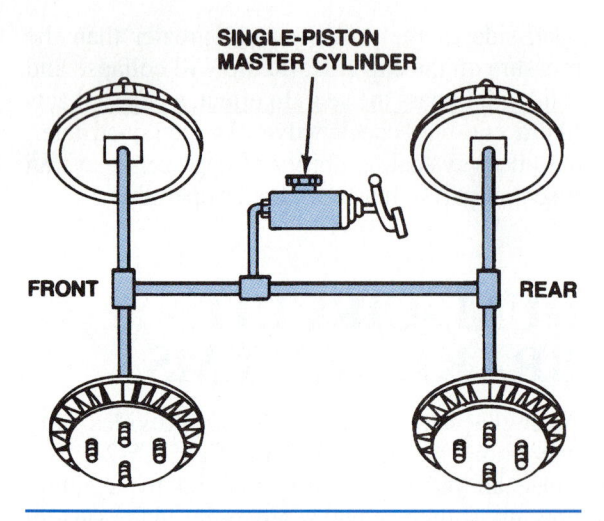

Figure 5-24. Until 1967, most brake systems had only a single hydraulic circuit.

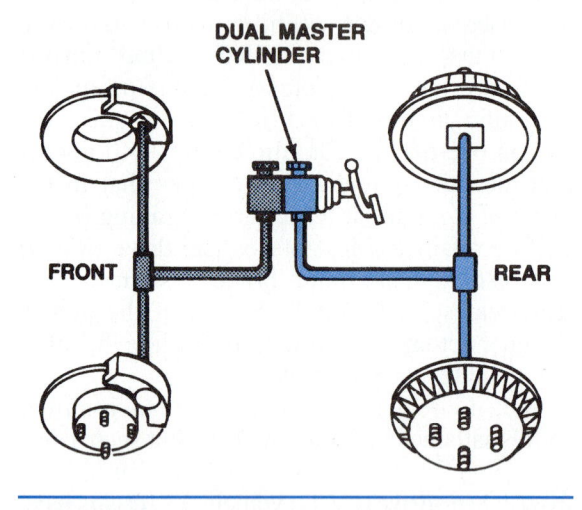

Figure 5-25. The dual-circuit brake system on most RWD cars has a front/rear split.

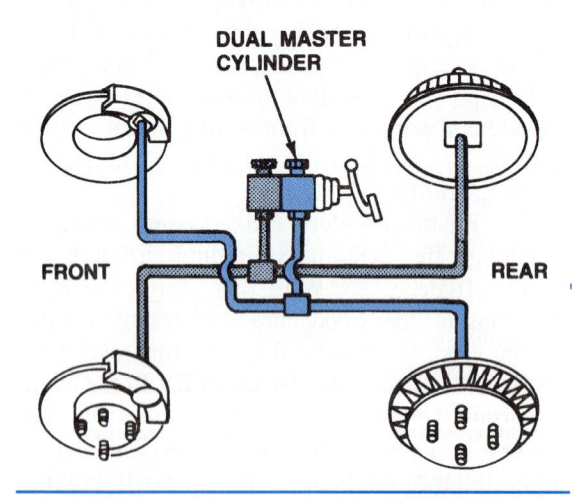

Figure 5-26. A diagonally split dual-circuit brake system is used on most FWD cars.

■ Balancing Act

Race cars use dual-circuit brake systems just like road cars but not for the same reasons. The dual-circuit brake systems on race cars are there primarily to achieve precise and easily adjustable brake balance. In place of a dual master cylinder, many racing cars use two single-piston master cylinders mounted side by side. The brake system is split so that one cylinder actuates the front brakes, whereas the other actuates the rear.

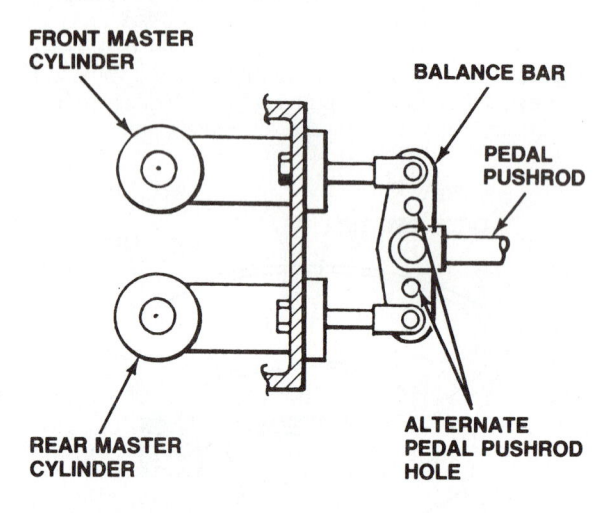

Brake balance is obtained in two ways. First, the bore size of each master cylinder is selected to provide approximately the correct amount of stopping power at each axle. Then, fine-tuning is done with a device called a balance bar. The pushrods that operate the two master cylinders attach to the ends of the balance bar, and the pedal pushrod attaches to the bar at some point in between. If the pedal pushrod is attached midway between the master cylinder pushrods, pedal force is distributed evenly between the two master cylinders. If the pedal pushrod is attached so it is closer to one of the master cylinders, that cylinder will receive a greater proportion of the pedal force.

The sophisticated balance bars used in some race cars can be adjusted by the driver during a race simply by moving a lever in the cockpit. On the position of this small lever riders the balance of the brakes—and perhaps the balance of the driver's life.

Pedal Assemblies and Master Cylinders

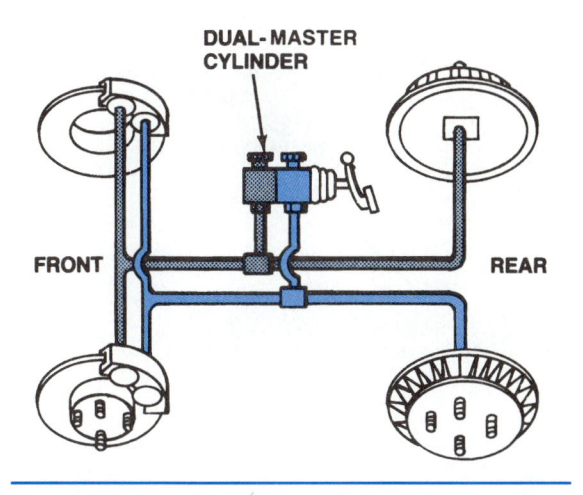

Figure 5-27. A triagonal dual-circuit brake system maintains up to 80 percent stopping power with one of the circuits disabled.

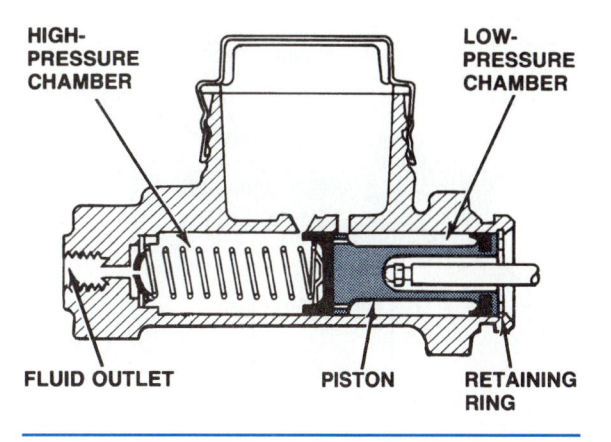

Figure 5-28. A basic single-piston master cylinder in the retracted position.

A variation of the diagonal-split braking system is found on some import cars. Often called a triagonal system, figure 5-27, this design requires the use of a special two- or four-piston brake caliper at each front wheel. These calipers appear similar to conventional parts, but they actually contain two separate hydraulic circuits and operate like two independent calipers in a common housing. Both hydraulic circuits actuate one-half of each front brake caliper, as well as one of the rear wheel brakes. In the event of a partial failure, half of each front brake and one of the rear brakes will always continue to operate. This means that approximately 80 percent of the total braking power is always available with this system.

MASTER CYLINDER OPERATION

All master cylinders operate in basically the same way because they must obey the hydraulic laws and principles described in Chapter 3. A major change in master cylinders occurred in 1967 when there was a change from the single-piston master cylinder to the dual-piston master cylinder. Because all of today's vehicles use a dual-piston master cylinder of some type, the single-piston master cylinder, figure 5-28, will not be described in detail. Its operation is similar to the operation of one piston in the dual-piston master cylinder. There are many types of dual-piston master cylinders. Some of these are:

- Conventional dual-piston master cylinder
- Quick-take-up master cylinder
- Portless master cylinders

Conventional Dual-Piston Master Cylinder

Dual master cylinders have two fluid reservoirs and two pistons fitted end to end in the cylinder bore, figure 5-29. The rear piston, actuated by the master cylinder or power booster pushrod, is called the primary piston. The forward piston, normally actuated by hydraulic pressure, is called the secondary piston.

Each piston has primary and secondary seals. The area *in front* of the primary cup seals is called the high-pressure chamber; this is where the hydraulic pressure that operates the brakes is created. The area *in between* the primary and secondary cup seals is called the low-pressure chamber. Fluid in this area is never under high pressure, but it plays an important part in master cylinder operation.

Brakes Not Applied

When the brakes are not applied, figure 5-29, both pistons of a dual master cylinder are held in the fully retracted position by their respective return springs. The primary piston bottoms against a retaining ring at the mouth of the cylinder. The secondary piston can bottom in one of two ways. On some cylinders, a bolt threaded into the cylinder body provides a stop for the secondary piston. Some stop bolts enter the cylinder bore from inside the fluid reservoir; others enter from outside the cylinder body. A copper washer is fitted under the head of exterior stop bolts to prevent fluid leaks.

On other master cylinders, the primary piston return spring serves as the secondary piston stop. The spring attaches to the primary piston with a retainer and screw, and extends a fixed distance forward to

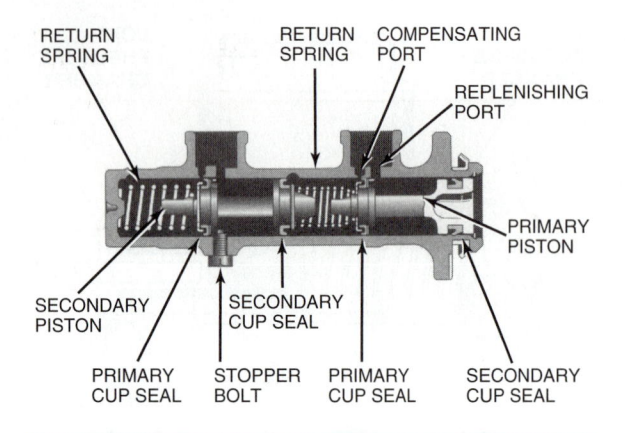

Figure 5-29. A dual master cylinder, brakes not applied. (Courtesy of Toyota Motor Sales U.S.A., Inc.)

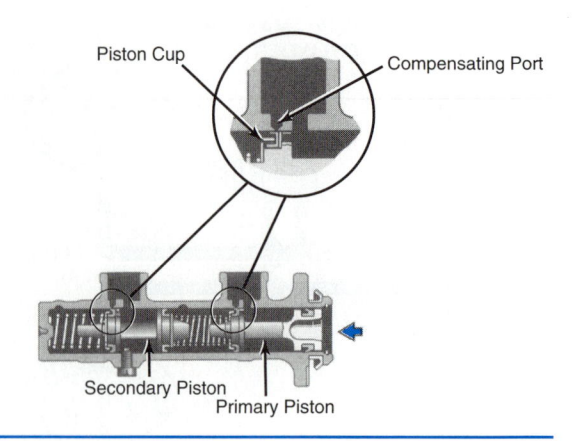

Figure 5-30. A dual master cylinder in the brakes-applied position. (Courtesy of Toyota Motor Sales U.S.A., Inc.)

contact the secondary piston. In these applications, the secondary piston return spring is less powerful than the primary piston return spring so it cannot overpower the primary spring and cause the secondary piston to move too far rearward in the bore.

There is no hydraulic pressure in any part of the master cylinder at this time; all chambers are under atmospheric pressure only. All of the sealing force is provided by the compressed lips of the cup seals.

The **compensating port** in front of each piston primary seal is open when the brakes are not applied. The compensating port provides a very small passage between the brake fluid reservoir and the high-pressure chambers of the master cylinder. The compensating port serves two important functions. First, it allows fluid to flow into the master cylinder when the hydraulic system is initially filled. Second, once the system is in service, it allows fluid to flow between the reservoirs and high-pressure chambers to compensate for changes in hydraulic system fluid volume. These changes are caused by variations in fluid temperature or wear of the brake linings.

The **replenishing port** behind each piston primary cup seal is also open when the brakes are not applied. The replenishing port is larger than the compensating port and provides a passage between the fluid reservoirs and the low-pressure chambers of the master cylinder. The replenishing ports keep the low-pressure chambers filled with brake fluid that will be required when the brakes are released.

Brake Application

As the brake pedal pushrod moves the primary piston forward in the master cylinder bore, a small amount of fluid from the primary high-pressure chamber is forced out of the compensating port into the fluid reservoir. Once the compensating port is closed by the primary cup seal, the primary high-pressure chamber is effectively sealed. This occurs because the lip of the rearmost cup seal on the secondary piston faces the lip of the primary cup seal.

Continued movement of the primary piston forces fluid through the primary circuit to the appropriate wheel friction assemblies. But pressure in the primary high-pressure chamber also moves the secondary piston forward in the master cylinder bore. Like the primary piston, the secondary piston forces a small amount of brake fluid from the secondary high-pressure chamber into the fluid reservoir through the compensating port. Once the secondary piston moves far enough to close the compensating port, the entire master cylinder is sealed and brake fluid is pumped through both hydraulic circuits to the wheel friction assemblies, figure 5-30.

Because the fluid trapped in the high-pressure chambers cannot be compressed, it is forced out of the master cylinder into the brake lines and out to the wheel friction assemblies. Hydraulic pressure is relatively low until all of the clearance is taken up in the brake system, but once the pads and shoes solidly contact the rotors and drums, hydraulic pressure builds up rapidly and braking force is applied. As the pressure builds, it forces the lips of the primary seals tightly against the cylinder bore to prevent any hydraulic pressure from escaping into the low-pressure chamber.

The replenishing ports remain open during brake application to keep the low-pressure chamber filled

Pedal Assemblies and Master Cylinders

with brake fluid. In fact, the replenishing port is always open and the fluid in the low-pressure chamber is never under more than atmospheric pressure unless there is a problem in the master cylinder.

Primary Circuit Failure

The extra margin of safety provided by a dual master cylinder comes into play when there is a leak in either of the hydraulic circuits. If the leak occurs in a part of the brake system served by the primary circuit, hydraulic pressure will not be created in the primary high-pressure chamber and will, therefore, be unavailable to actuate the secondary piston. When this happens, the pedal pushrod moves the primary piston forward until it physically contacts the secondary piston, figure 5-31. This transfers pushrod force directly to the secondary piston, which then creates pressure in the secondary hydraulic circuit.

Secondary Circuit Failure

If the leak occurs in a part of the brake system served by the secondary circuit, the primary high-pressure chamber will remain sealed, but no hydraulic pressure will be developed because the secondary piston will simply shift forward in the bore. In this case, additional brake pedal travel moves the secondary piston until an extension on the front of the piston bottoms on the end of the cylinder body, figure 5-32. This allows any further primary piston movement to create pressure in the primary hydraulic circuit.

Brake Release

When the brake pedal is released, return springs at the brake pedal and in the master cylinder help return the system to the *brakes-not-applied* position described previously. Because the master cylinder pistons are small and light, they return to the retracted position almost immediately. However, the brake fluid pumped into the brake lines and wheel friction assemblies during application has greater inertia and cannot return as quickly. This creates a pressure drop in the high-pressure chambers, and if nothing were done to compensate, the reduced pressure would cause the piston to return very sluggishly.

To avoid this problem, master cylinders take advantage of the one-way sealing ability of the primary cup seals. When the brakes are released and pressure drops below the atmospheric pressure in the low-pressure chambers, the lip of the primary seal collapses and fluid bypasses the seal

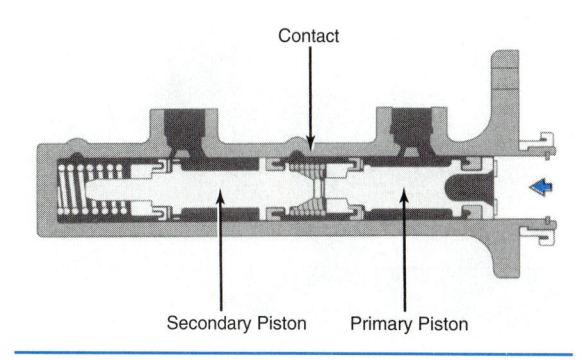

Figure 5-31. Primary circuit failure in a dual master cylinder. (Courtesy of Toyota Motor Sales U.S.A., Inc.)

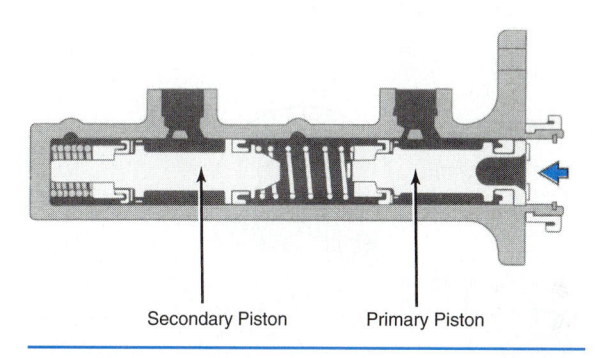

Figure 5-32. Secondary circuit failure in a dual master cylinder. (Courtesy of Toyota Motor Sales U.S.A., Inc.)

until pressure in the chambers is equalized, figure 5-33. Many pistons have small holes drilled behind the seal to let the fluid through.

Once the piston reaches the fully retracted position, the lips of the primary cup seals expand and regain their seal against the cylinder bore. As excess fluid returns from the brake lines and wheel friction assemblies, it is diverted back into the reservoirs through the now open compensating port, figure 5-34.

Brake Pumping

Normally, less than a fluid ounce of brake fluid is moved by the master cylinder to apply the brakes. But when mechanical brake fade or vapor lock occurs, additional fluid is required to keep the brake shoes in contact with the drums and discs or to compress the gas created when the fluid boils. The amount of fluid needed in these situations is usually more than can be delivered by a single stroke of the master cylinder. The one-way sealing characteristics of cup seals combined

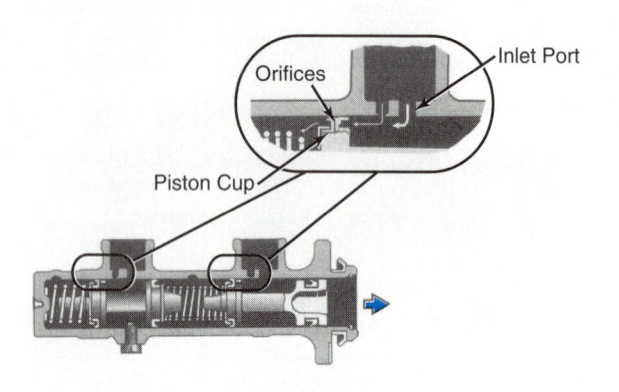

Figure 5-33. The one-way action of the primary cup seal allows fluid to bypass the seal and allows the rapid return of the master cylinder pistons. (Courtesy of Toyota Motor Sales U.S.A., Inc.)

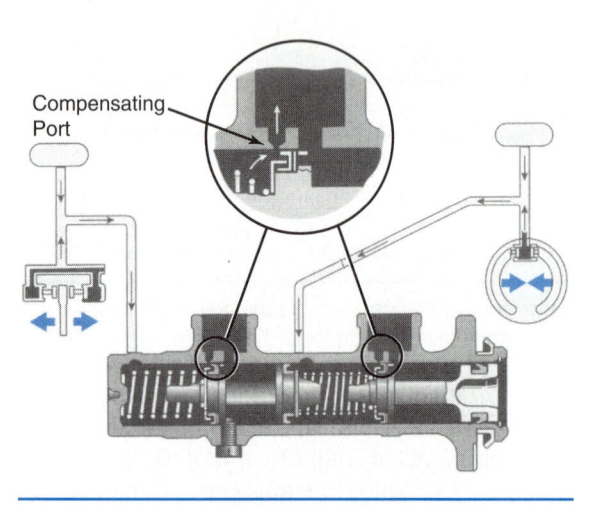

Figure 5-34. Excess fluid returns to the reservoir through the compensating port. (Courtesy of Toyota Motor Sales U.S.A., Inc.)

with the action of the fluid-replenishing port enable the master cylinder to act as a fluid pump that can help restore some of the lost braking power.

As described previously, every time the driver releases the brake pedal, pressure drops in the high-pressure chambers and brake fluid bypasses the primary cup seals to equalize pressure ahead of the piston. Once the piston is retracted, fluid returning from the brake lines passes through the compensating port into the fluid reservoir. However, if the driver pumps the brake pedal rapidly, some of the returning fluid remains in the lines while additional fluid that bypasses the primary seals is pumped into the system. This supplies the extra fluid needed to combat brake fade or vapor lock.

Quick-Take-Up Master Cylinder

The **quick-take-up master cylinder,** also called a fast-fill master cylinder, is a type of dual master cylinder used on cars equipped with low-drag disc brake calipers. When the brakes are not applied, low-drag calipers retract the caliper pistons and brake pads farther from the brake rotors to reduce friction and improve fuel economy. If a standard dual master cylinder was used on a car with low-drag calipers, excessive pedal travel would be required to move enough brake fluid to take up the additional clearance. This would leave little pedal reserve, and braking action would be delayed. To compensate for the increased clearance of low-drag brake calipers, quick-take-up master cylinders provide a large volume of fluid when the brakes are first applied.

There are two main differences between a standard dual master cylinder and a quick-take-up design. The first is that a quick-take-up cylinder has a stepped bore that creates an oversized primary low-pressure chamber, figure 5-35. The fluid stored in this chamber is used to take up the extra clearance in low-drag brake calipers.

The second difference is a **quick-take-up valve,** figure 5-36, located between the fluid reservoir and the cylinder bore. The quick-take-up valves seals the fluid passage between these two areas and allows fluid to pass only under certain conditions. The outside of the valve seals to the cylinder body with a cup seal, whereas an internal fluid passage is kept closed by a spring-loaded check ball. Several holes around the edge of the valve allow fluid access to the back side of the cup seal, and a small bypass groove is cut into the seat of the internal check ball.

The following sections explain how the oversized primary low-pressure chamber and the quick-take-up valve work together to take up the additional clearance in low-drag brake calipers.

Brakes Not Applied

When the brakes are not applied, a quick-take-up master cylinder functions exactly the same as a standard dual master cylinder. Both of the pistons are in their fully retracted position, and all of the compensating and replenishing ports are open. Fluid to the secondary piston ports flows in directly from the reservoir, but all fluid to or from the ports serving the primary piston must travel through the small bypass groove cut in the quick-take-up valve check ball seat.

Pedal Assemblies and Master Cylinders

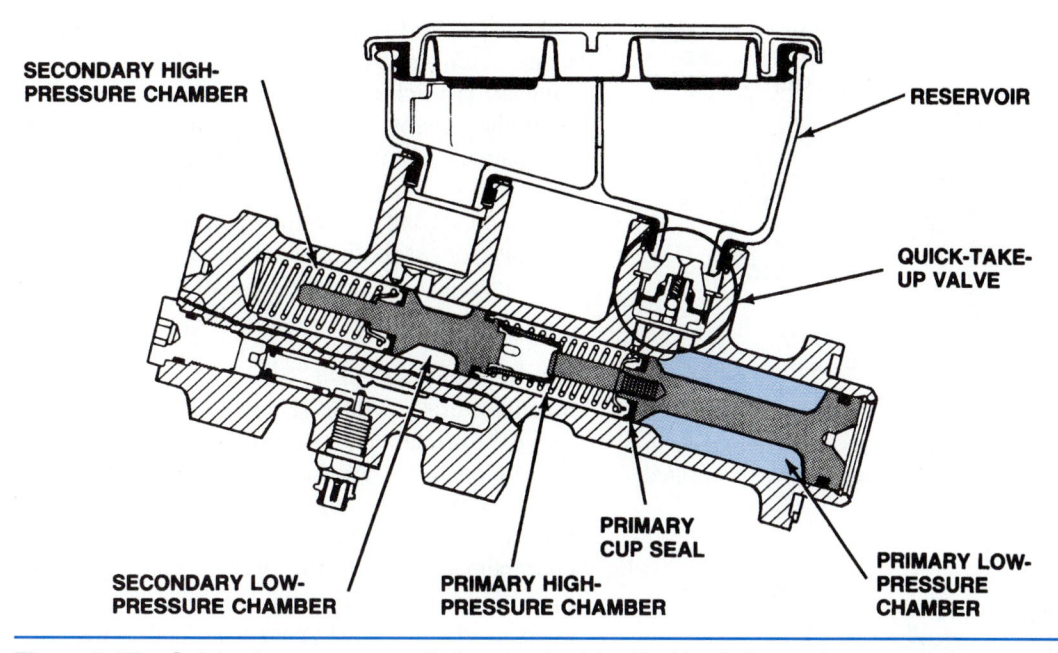

Figure 5-35. Quick-take-up master cylinders can be identified by their oversized primary low-pressure chamber.

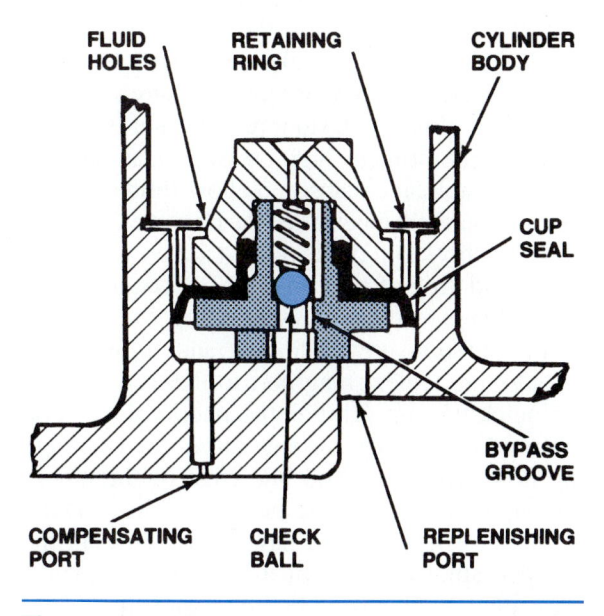

Figure 5-36. The quick-take-up valve controls fluid flow to and from the primary low-pressure chamber.

Brake Application

When the brakes are applied, the primary piston moves forward in the cylinder bore. This reduces the total area of the primary piston low-pressure chamber because the small-diameter area created behind the primary cup seal has less volume than the large-diameter area being reduced by the piston movement. Because the low-pressure chamber is getting smaller and the brake fluid it contains cannot be compressed, a portion of the fluid must exit the chamber in some way.

The quick-take-up valve prevents the fluid from returning to the reservoir; the path around the outside of the valve is blocked by the cup seal, and the hydraulic pressure in the low-pressure chamber is not high enough to unseat the quick-take-up valve check ball at this time. A small amount of fluid does escape through the bypass groove in the check ball seat, but it is not enough to affect the overall operation of the system.

Because the passage to the reservoir is closed, the only other way out for the fluid is past the primary cup seal, figure 5-37. As the volume of the low-pressure chamber shrinks, pressure on the trapped fluid increases until it exceeds that of the fluid in the primary high-pressure chamber. The lip of the primary cup seal then collapses and fluid from the low-pressure chamber bypasses the seal and flows out of the master cylinder to the wheels where it takes up the clearance in the low-drag brake calipers.

Once all of the clearance at the calipers has been eliminated, pressure in both the low-pressure and high-pressure chambers begins to increase. When the pressure reaches 70 to 100 psi, the spring-loaded check ball in the quick-take-up valve opens and allows fluid in the low-pressure chamber to return to the reservoir, figure 5-38A. With the pressure once again greater in the high-pressure chamber than in the low-pressure chamber, the primary cup regains its seal and the cylinder functions

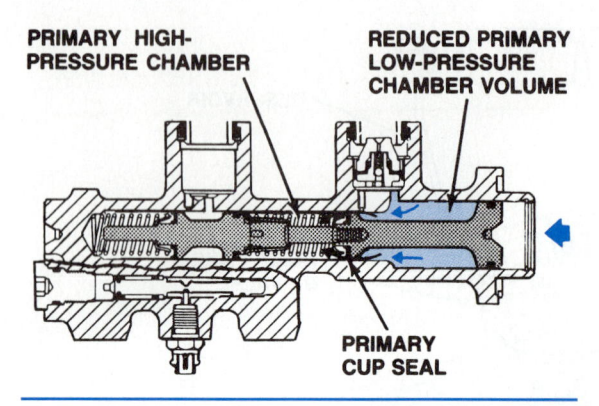

Figure 5-37. As the brakes are applied, reduced low-pressure chamber volume results in a pressure increase that causes fluid to bypass the primary cup seal.

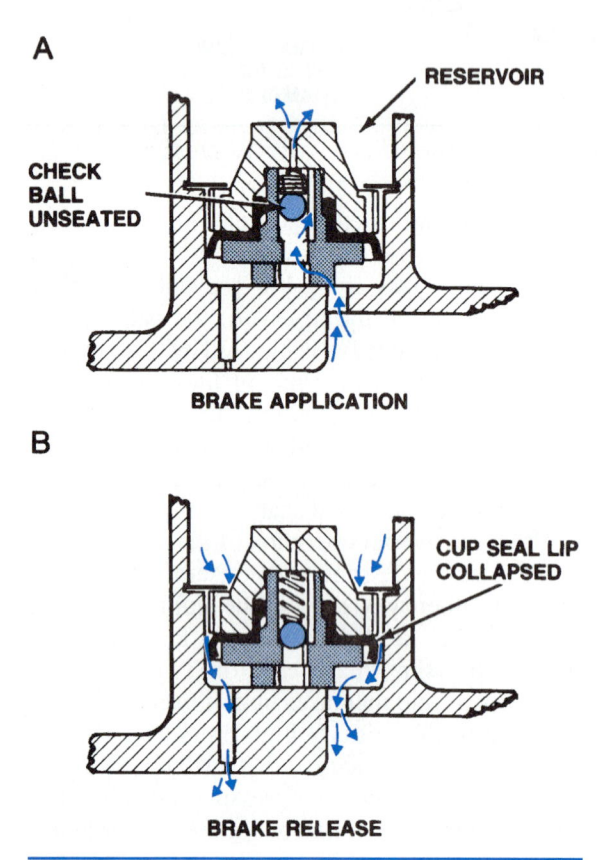

Figure 5-38. The one-way sealing abilities of both a spring-loaded ball check valve and a cup seal are used in the quick-take-up valve.

like a standard dual master cylinder. Because both the primary and secondary pistons and cup seals are now the same diameter, equal pressure is delivered to both hydraulic circuits.

All of the quick-take-up actions occur in the hydraulic circuit served by the primary piston. However, if the brake system is split diagonally, an equal volume of fluid is needed to take up the clearance in the circuit served by the secondary piston. As long as the quick-take-up valve remains *closed,* the secondary piston will move a greater distance than the primary piston. This keeps hydraulic pressure and fluid displacement the same in both the primary and secondary circuits. Once the quick-take-up valve *opens,* both pistons move together at the same rate, just as in a standard dual master cylinder.

Brake Release

When the brake pedal is released, the return springs move the primary and secondary pistons to their retracted positions. As in any master cylinder, the pistons return faster than the fluid in the lines can follow. This reduces pressure in the high-pressure chambers, causing the lips of the primary cup seals to collapse and allowing fluid from the low-pressure chambers to bypass and equalize the pressure. Fluid in the secondary low-pressure. Fluid in the secondary low-pressure chamber is supplied directly from the reservoir through the replenishing port, but the quick-take-up valve affects the flow of fluid to the primary low-pressure chamber.

As the primary piston returns and pressure drops in the primary high-pressure chamber, the primary cup seal collapses to allow fluid from the low-pressure chamber to enter. However, the cup seal and check ball of the quick-take-up valve prevent fluid from freely entering the replenishing port. This extends the pressure drop into the primary low-pressure chamber, and from there through the replenishing port to the underside of the quick-take-up valve. Atmospheric pressure on the fluid in the reservoir then forces fluid through the holes around the outside edge of the quick-take-up valve and past the backside of the cup seal, figure 5-38B. This fluid passes through both the compensating and replenishing ports to equalize pressure in the low- and high-pressure chambers.

Portless Master Cylinders

The master cylinders discussed so far are of the conventional type that use a compensating port that allows for variations in brake fluid volume within the system when the pedal is released. This compensating port is a small hole, figure 5-39, as small as .040 inch in some cases; this allows fluid to expand and return from the wheels when needed. The small hole also minimizes loss of fluid into the reservoir when the pedal is first applied.

Pedal Assemblies and Master Cylinders

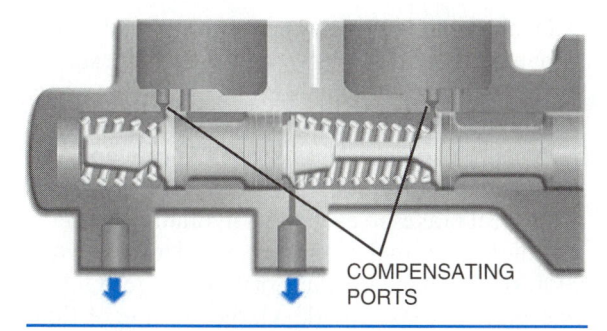

Figure 5-39. The conventional master cylinder uses a small compensating port to allow for fluid return and expansion when the pedal is released. (Courtesy of DaimlerChrysler Corporation. Used with permission)

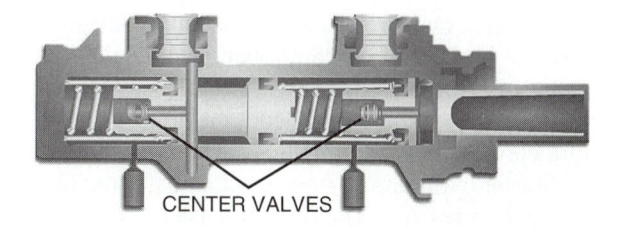

Figure 5-40. The center valve, or portless, master cylinder has no compensating ports and uses two center-mounted valves instead. (Courtesy of DaimlerChrysler Corporation. Used with permission)

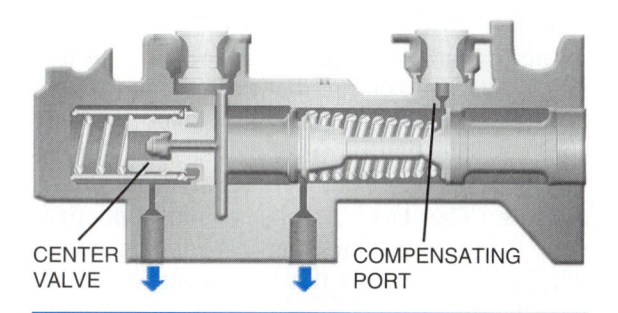

Figure 5-41. The combination master cylinder has both a center valve assembly and a compensating port. (Courtesy of DaimlerChrysler Corporation. Used with permission)

Some vehicles with antilock brakes or traction control systems require a larger volume of fluid flow to and from the reservoir during ABS or traction control events. Modifications to the master cylinder pistons allow this to occur.

Center Port Master Cylinder

Center port master cylinders do not use compensating ports but instead use a fluid port and valve located in the center of the cylinder pistons, figure 5-40. They are also called "portless" master cylinders, referring to the lack of a compensating port.

At rest (pedal released) there is free flow of fluid to the rest of the hydraulic system, both feed and return. For example, during a traction control event, the ABS controller may need to draw fluid from the master cylinder reservoir to apply a brake caliper, and then return the fluid when it is no longer needed. The center port valve allows this to happen. When the brakes are applied, the ports close with the first movement of the pistons to seal off the fluid, and braking occurs in the same manner as with a conventional master cylinder.

Combination Master Cylinder

This variation combines the conventional master cylinder and the portless type so that one piston is of the center valve type and one piston uses the compensating port, figure 5-41.

The type of master cylinder used depends on the brake system options. Typically, the applications are:

- Vehicles without ABS use a conventional master cylinder with compensating ports in the primary and secondary pistons.
- Vehicles with ABS may use a portless (center port) secondary piston and compensating port in the primary.
- Vehicles with ABS and a traction control system may use the portless (center port) type valve on both the secondary and primary pistons.

It is important that the technician is aware of these differences because it is not possible to tell which type a vehicle has without disassembling the cylinder. The external appearance of the cylinders are all the same. When replacing the master cylinder on any vehicle always be sure that the correct replacement part is installed.

SUMMARY

The brake pedal assembly is a type of lever that multiplies pressure applied to the foot pedal and passes it on to the master cylinder or brake power booster. All early brake pedals were either frame- or floor-mounted, but most modern pedals are suspended from the firewall under the dashboard.

Brake pedals are designed to have a small amount of freeplay that ensures the master cylinder piston returns all the way when the brakes are not applied. On many cars the freeplay is designed into the brake system components and cannot be changed. But on other cars the freeplay is adjustable by threading the pedal pushrod in or out of a clevis, or by rotating an eccentric bolt.

Master cylinders convert mechanical force into hydraulic pressure. The typical master cylinder is made up of several parts: a fluid reservoir, a vented reservoir cover, a rubber diaphragm that isolates the fluid from moisture, a cylinder body made of either cast iron or aluminum, and one or two pistons fitted with return springs and rubber cup seals.

Before 1967, most cars used a single-piston master cylinder to supply pressurized fluid to a single hydraulic circuit that operated the brakes at all four wheels. All cars built since 1966 are equipped with a dual master cylinder that divides the brake system into two separate hydraulic circuits. The two circuits operate independently and ensure that partial stopping power will still be available if a leak occurs in one of the circuits.

Dual-circuit braking systems are normally split front to rear on RWD cars, but most FWD cars use a diagonally split system that serves one front wheel and the opposite rear wheel with each circuit. The triagonal brake system on some import cars connects half of each front brake caliper and one of the rear wheel brakes to each of its hydraulic circuits.

Quick-take-up dual master cylinders are a variation of the standard dual-piston design. Quick-take-up cylinders provide a large volume of fluid when the brakes are first applied. This fluid is required to take up the extra clearance designed into special low-drag brake calipers.

Center port (portless) master cylinders have been developed for use on some vehicles with ABS and traction control. These systems require a larger volume of fluid flow to and from the master cylinder reservoir during ABS and/or traction control events. The small compensating port is replaced with a special valve in the center of the master cylinder piston that allows this increased fluid flow to occur.

Pedal Assemblies and Master Cylinders

Review Questions

Choose the letter that represents the best possible answer to the following questions:

1. The _____ transmits force from the brake pedal assembly to the master cylinder piston.
    a. Pedal arm
    b. Pedal pushrod
    c. Pedal bracket
    d. Pedal pivot

2. The most common type of brake pedal assembly is the:
    a. Floor-mounted
    b. Suspended
    c. Frame-mounted
    d. None of the above

3. Which of the following is *not* true about frame-mounted brake pedals.
    a. They require an opening in the floor
    b. They place the master cylinder in a hard-to-service location
    c. Their pivot is located near or at the bottom of the pedal arm
    d. They are relatively immune to wear

4. Suspended brake pedals are used because they:
    a. Can be longer for greater leverage
    b. Make master cylinder service easier
    c. Both a and b
    d. Neither a nor b

5. The brake pedal ratio on a car equipped with a power booster is numerically _____ the ratio on a car with manual brakes.
    a. Lower than
    b. Higher than
    c. The same as
    d. Any of the above, depending on the car

6. Brake pedal freeplay is:
    a. A consequence of pedal pushrod clearance
    b. Absolutely necessary in a brake system
    c. Not adjustable on many cars
    d. All of the above

7. Brake pedal freeplay may be adjusted using:
    a. A threaded clevis
    b. An eccentric hinge bolt
    c. Both a and b
    d. Neither a nor b

8. Brake fluid reservoirs may be:
    a. Made of metal or plastic
    b. Mounted away from the master cylinder
    c. Clamped to the master cylinder body
    d. All of the above

9. Brake fluid reservoir covers are:
    a. Vented to allow air to enter the reservoir
    b. Sometimes sealed with a nylon diaphragm
    c. Both a and b
    d. Neither a nor b

10. Which of the following is *not* a part of the master cylinder body?
    a. Cylinder bore
    b. Wire bail
    c. Replenishing port
    d. Fluid reservoir

11. The interior of an aluminum master cylinder is _____ for long life.
    a. Bearingized
    b. Anodized
    c. Chrome plated
    d. Any of the above

12. Master cylinder pistons:
    a. Fit loosely in the cylinder bore
    b. Return under their own power when the brakes are released
    c. Are usually made of soft aluminum
    d. Always contact the pedal pushrod

13. Master cylinder piston seals:
    a. Can contain pressure from either side
    b. Sometimes have holes around their edges to allow brake fluid to pass
    c. Prevent fluid flow between the low- and high-pressure chambers
    d. None of the above

14. When the brakes are not applied, hydraulic pressure is greatest in the:
    a. Low-pressure chamber
    b. High-pressure chamber
    c. Fluid reservoir
    d. None of the above

15. The compensating port provides a passage between the:
    a. Low- and high-pressure chambers
    b. Reservoir and low-pressure chamber
    c. Reservoir and high-pressure chamber
    d. Low-pressure chamber and bypass seal

16. The small fountain of fluid that appears in the fluid reservoir on brake application is caused by fluid passing through the:
    a. Compensating port
    b. Replenishing port
    c. Both a and b
    d. Neither a nor b

17. When the brakes are released, a pressure drop occurs in the master cylinder because of:
    a. Primary cup seal collapse
    b. Brake pedal freeplay
    c. Piston return
    d. Reservoir inertia

18. As the brake pedal is released, fluid in the master cylinder will bypass the:
    a. Primary cup seal
    b. Tertiary cup seal
    c. Secondary cup seal
    d. Pushrod cup seal

19. The master cylinder is able to pump additional fluid into the hydraulic system when the brakes fade because of:
    a. Brake fluid inertia
    b. Lightweight pistons
    c. The design of cup seals
    d. All of the above

20. A late-model FWD car that retains approximately 80 percent stopping power after a hydraulic failure is probably equipped with a _____ brake system.
    a. Single
    b. Front/rear-split
    c. Diagonal-split
    d. Triagonal-split

21. During normal operation, the secondary piston in a standard dual master cylinder is actuated directly by the:
    a. Pedal pushrod
    b. Primary piston
    c. Either a or b
    d. Neither a nor b

22. A leak in the circuit served by the secondary piston of a dual master cylinder will cause:
    a. The primary piston to contact the secondary piston
    b. High pressure in the fluid reservoir
    c. The secondary piston to bottom out against the end of the cylinder bore
    d. All of the above

23. A quick-take-up master cylinder:
    a. Increases pedal travel
    b. Enables the use of low-drag calipers
    c. Reduces fluid flow
    d. Requires additional pushrod clearance

24. In a quick-take-up master cylinder, fluid will bypass the primary cup seal during:
    a. Brake application
    b. Brake release
    c. Both a and b
    d. Neither a nor b

25. The quick-take-up valve controls:
    a. Fluid flow between the reservoir and the secondary low-pressure chamber
    b. Fluid flow between the primary and secondary high-pressure chambers
    c. Fluid flow between the primary low- and high-pressure chambers
    d. None of the above

6

Hydraulic Valves and Switches

OBJECTIVES

Upon completion and review of this chapter, you will be able to:

- Identify and describe the operation of the residual pressure check valve.
- Identify and describe the operation of the metering valve.
- Identify and describe the operation of the proportioning valve.
- Explain the purpose of the combination valve.
- Describe the operation of the height-sensing proportioning valve.
- Identify and describe the operation of the pressure differential switch.
- Describe the operation of the fluid level switch.
- Describe the operation of the stoplight switch.

KEY TERMS

brake balance
cam
ground
red brake warning light (RBWL)
residual pressure
slope
split point
torque

INTRODUCTION

Pascal's law, discussed in Chapter 3, states that hydraulic pressure will be the same at all points within a hydraulic system. However, after years of development, brake system engineers have determined that vehicle brakes work better if the amount or timing of the hydraulic pressure within the system is altered in certain ways. The hydraulic valves described in this chapter regulate pressure within the brake system to provide faster and more controllable stops.

In addition to hydraulic control valves, most brake systems also have several hydraulic or mechanical switches. The most familiar of these is the stoplight switch, but every late-model vehicle also has one or more switches that trigger warning lights to inform the driver of potential brake system problems.

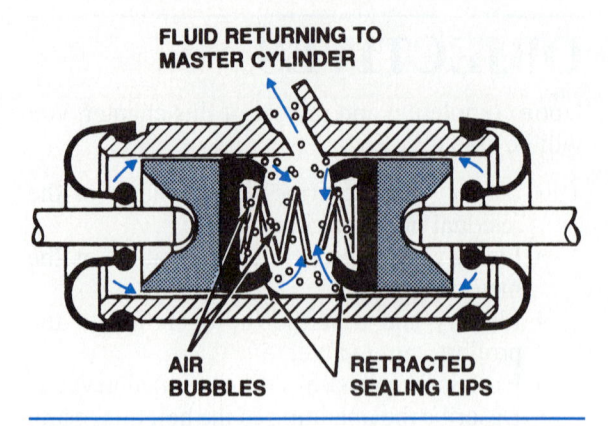

Figure 6-1. The momentary drop in pressure created when the brakes are released can draw air into the hydraulic system.

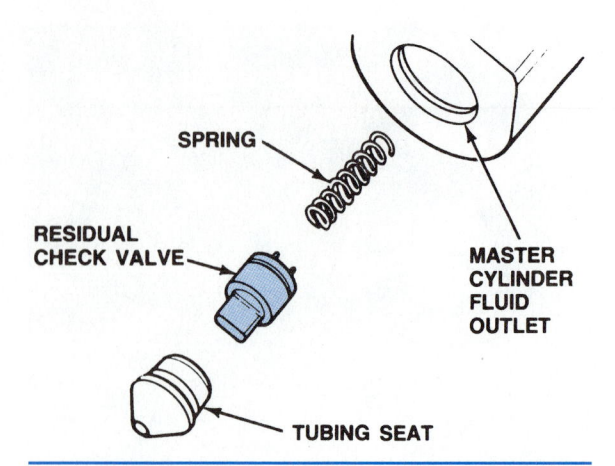

Figure 6-2. Most residual check valves are located under the tubing seats in the master cylinder outlet ports.

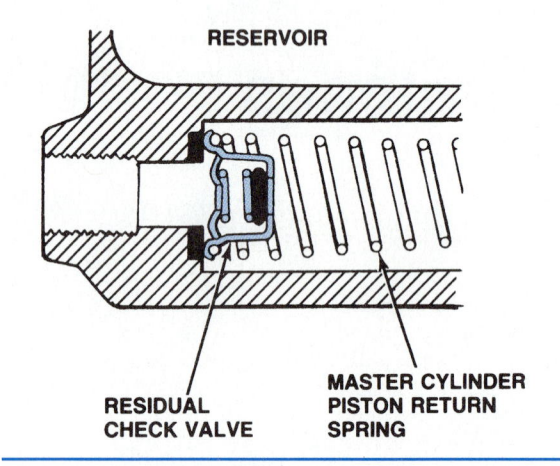

Figure 6-3. Some older brake systems locate the residual check valve at the end of the master cylinder bore.

HYDRAULIC VALVES

Four basic types of hydraulic valves are used in vehicle brake systems:

- Residual pressure check
- Metering
- Proportioning
- Combination.

Residual pressure check valves are used primarily on older vehicles with four-wheel drum brakes. Metering and proportioning valves are found on most newer brake systems, and they are often combined with a hydraulic switch to form a combination valve. The operation of these valves and the reasons they are needed are explained in the following sections.

RESIDUAL PRESSURE CHECK VALVE

Residual pressure check valves, often simply called residual check valves, maintain between 6 and 25 psi of **residual pressure** in the brake lines at all times. Residual check valves are important because the sealing lips of wheel cylinder cup seals tend to relax away from the cylinder walls and allow air to enter the hydraulic system. This is most likely to occur when the brakes are released and the retracting master cylinder piston creates a momentary pressure drop in the wheel cylinders, figure 6-1. The pressure trapped in the brake lines by the residual check valve keeps the sealing lips of the cup seals in firm contact with the wheel cylinder bores to prevent air from entering the system.

Residual check valves are often located under the tube seats in the master cylinder outlet ports that supply fluid to drum brakes, figure 6-2. In some older vehicles, the check valve may be housed in the end of the master cylinder, figure 6-3. Regardless of the type or location of the residual check valve, the brake system is designed so any application force created by residual pressure is less than the force of the brake shoe return springs. If this were not done, residual pressure would cause the brakes to drag.

Residual Check Valve Operation

All residual check valves work in essentially the same manner. When the brakes are applied, the center of the valve opens and allows unrestricted

Hydraulic Valves and Switches

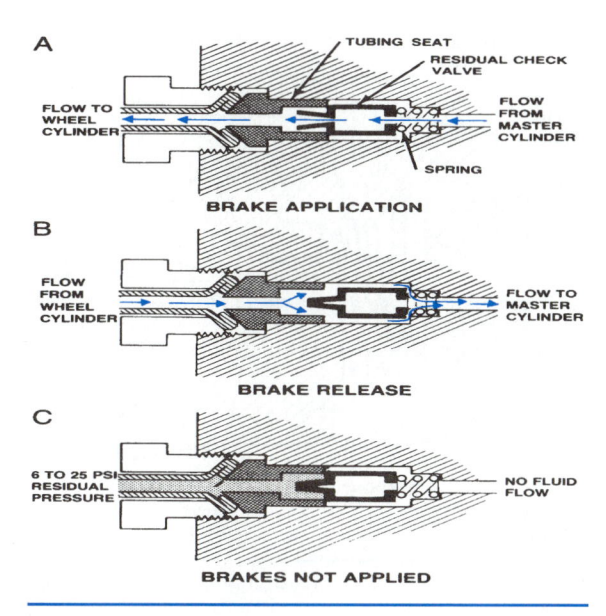

Figure 6-4. Operation of a residual check valve installed under a tubing seat.

fluid flow to the wheel friction assemblies, figure 6-4A. When the brakes are released, returning fluid unseats the check valve against spring pressure, and fluid flows around the outside of the valve back into the master cylinder and fluid reservoir, figure 6-4B. When line pressure drops to the residual value controlled by the spring tension, the check valve seats and pressure is held in the lines, figure 6-4C.

Systems without Residual Check Valves

Although residual check valves were once found in almost every brake system, they have rarely been used since the early 1970s. There are a number of reasons for this. First, mechanical cup expanders, figure 6-5, were developed to keep cup seal lips in contact with the wheel cylinder walls. Compared to residual check valves, cup expanders are simpler, cheaper, less prone to failure, and easier to work on.

The second reason residual check valves are seldom seen today is that they are not used with disc brakes. Early brake caliper pistons with lip seals use cup expanders to prevent air leaks, and the O-ring seals in all late-model calipers maintain an air-tight seal even under a slight negative pressure. In addition, brake calipers do not have return springs; if hydraulic pressure was retained in the brake lines, the brakes would always be

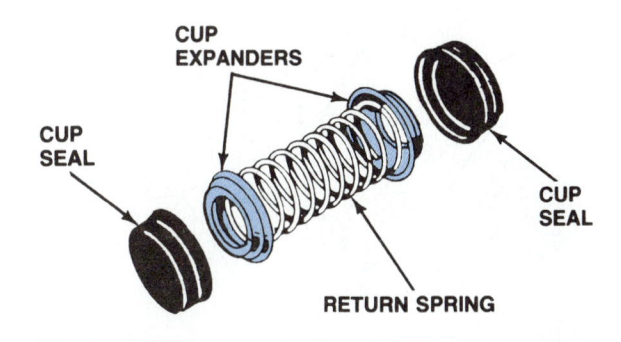

Figure 6-5. Mechanical cup expanders are one factor that led to the elimination of residual check valves.

slightly applied. The resulting drag would reduce fuel economy and cause unnecessary friction that would shorten brake pad life.

■ Residual Fluid Check Valves

The only job of the residual check valve in recent drum brake systems has been to keep the sealing lips of the wheel cylinder cup seals firmly against the cylinder bore. However, on very early cars, the residual *pressure* check valve also served as a residual *fluid* check valve. This was the case on cars with frame-mounted master cylinders that were positioned lower in the car than the wheel cylinders. If a check valve was not installed in these cars, gravity would cause the fluid in the wheel cylinders to slowly drain back into the master cylinder and spill out through the reservoir cover vent onto the pavement. The result was brake failure because half the fluid would be missing! The residual check valve prevented this from happening.

The final reason residual check valves are fast disappearing is the diagonally split dual braking system that operates both a disc and a drum brake off each hydraulic circuit. This type of system would require two residual check valves, one in the line to each rear brake. The extra cost and complication of dual check valves are easily avoided by using cup expanders.

METERING VALVE

Brake systems with front disc and rear drum brakes sometimes use a metering valve, figure 6-6, to withhold hydraulic pressure from the front

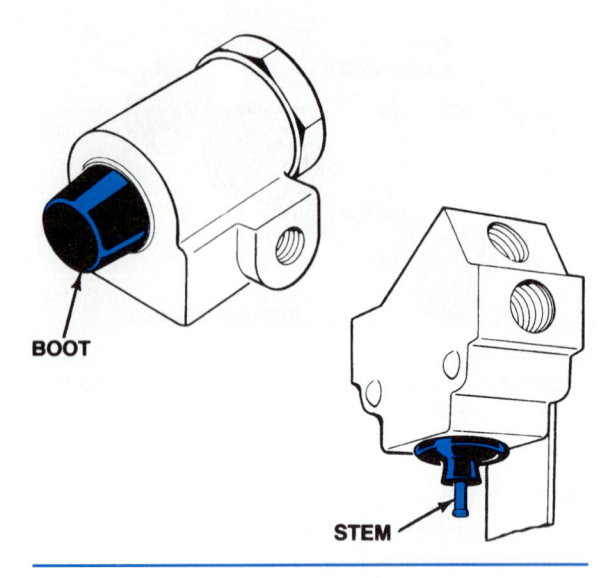

Figure 6-6. Typical metering valves.

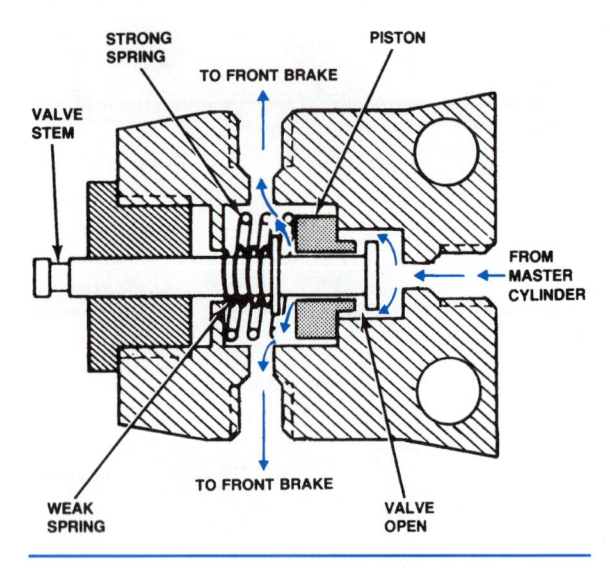

Figure 6-7. A metering valve when the brakes are not applied.

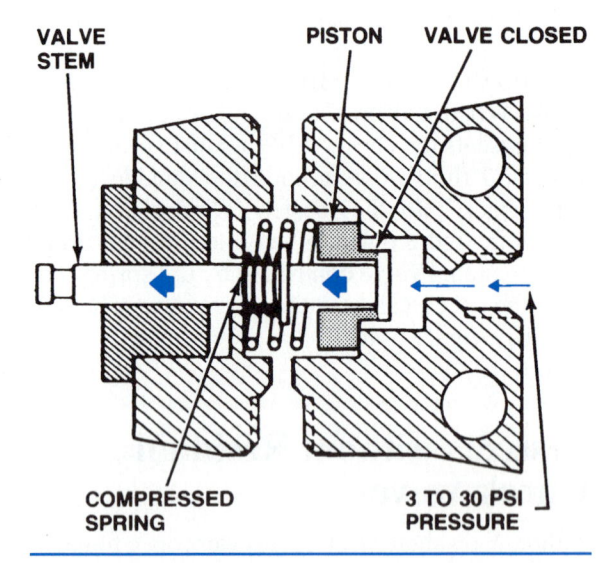

Figure 6-8. A metering valve under light pedal pressure.

brakes until the rear brakes have begun to apply. This is done because disc brakes apply much faster than drum brakes; disc brake pads are always in light contact with the rotors, whereas drum brake shoes are retracted away from the drums. Delaying front brake application until the clearance in the rear brakes has been taken up helps keep the front brakes from momentarily locking when the pedal is lightly applied at slow speeds on slippery pavement. The metering valve also improves **brake balance** during light braking, and helps equalize lining wear at the front and rear axles.

Early metering valves, such as those shown in figure 6-6, were often a separate component installed in the brake line from the master cylinder to the front brakes. Today, most metering valves are part of a combination valve that contains other hydraulic valves and switches.

Metering Valve Operation

A metering valve, figure 6-7, consists of a piston controlled by a strong spring and a valve stem controlled by a weak spring. When the brakes are not applied, the strong spring seats the piston and prevents fluid flow around it. At the same time, the weak spring holds the valve stem to the right and opens a passage through the center of the piston. Brake fluid is free to flow through this passage to compensate for changes in system fluid volume.

When the brakes are applied and pressure in the front brake line reaches 3 to 30 psi, the tension of the weak spring is overcome and the metering valve stem moves to the left, figure 6-8; this closes the passage through the piston and prevents fluid flow to the front brakes. The small amount of pressure applied to the calipers before the metering valve closes is enough to take up any clearance but not enough to generate significant braking force.

While fluid flow to the front calipers is shut off, the rear brake shoes move into contact with

Hydraulic Valves and Switches

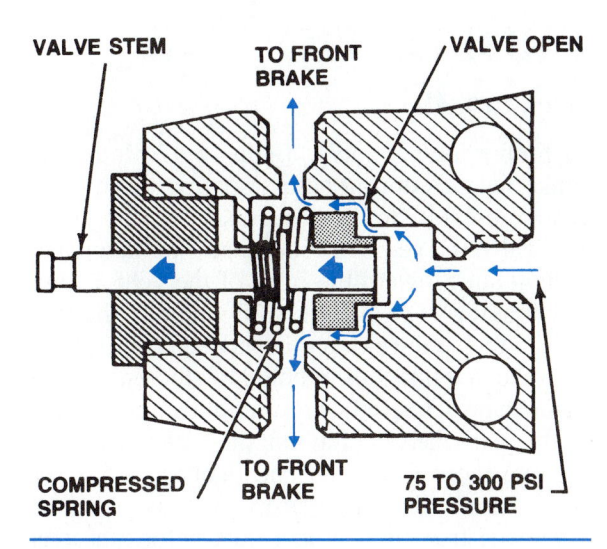

Figure 6-9. A metering valve while normal braking is in progress.

the drums, braking begins, and hydraulic pressure throughout the brake system increases. When the pressure at the metering valve reaches 75 to 300 psi, figure 6-9, the tension of the strong spring is overcome and the valve stem and piston move farther to the left. This opens a passage around the outside of the piston and allows fluid to flow through the valve to the front brake calipers.

When the brakes are released, the strong spring seats the piston and prevents fluid flow around it. At the same time, the weak spring opens the fluid passage through the center of the piston. Excess fluid returns to the master cylinder through this passage and the valve is ready for another brake application.

System without Metering Valves

The metering valve was introduced along with disc brakes in the 1960s. At that time, all domestic cars had RWD and large, powerful engines with extended fast idle times. When engineers did cold weather testing of the first disc brakes on these cars, they found the front brakes often locked for a moment when they were applied on slippery pavement. At the same time, the fast-idling engine would still be driving the rear wheels because clearance in the drum brakes had not been taken up. Metering valves were developed to prevent these problems and provide safer braking. However, automobiles have changed a great deal since the introduction of disc brakes. In recent years the metering valve has been eliminated from the brake systems of most FWD, and some RWD, automobiles because testing has shown it to be unnecessary or undesirable.

There are three reasons FWD cars do not use metering valves. First, they usually have a diagonally split dual braking system that would require a separate metering valve for each hydraulic circuit. This would make the brake system more costly and complicated. Second, FWD cars have a forward weight bias that requires the front brakes to supply up to 80 percent of the total braking power. Because the front brakes do most of the work, it is desirable to apply them as soon as possible when the brake pedal is depressed; a metering valve would create a slight delay. Finally, until all the clearance in the brake system is taken up, there will not be enough pressure in the brake hydraulic system for the front disc brakes to overcome the engine **torque** applied to the driven front wheels. Engine torque and a heavy front weight bias help prevent front wheel lockup from being a problem during light braking or when the brakes are first applied.

Most RWD cars without metering valves are equipped with four-wheel disc brakes. Because the clearance between the pads and rotors is approximately the same at all four wheels, there is no need to delay front brake actuation. Some of these cars also have antilock brake systems that prevent the wheels from locking at *any* time; these systems are described in Chapter 15. Other RWD cars without metering valves have a predominantly forward weight bias, like FWD cars and, therefore, benefit from having the front brakes applied sooner.

Metering Valves and Brake Bleeding

Most metering valves, including those that are part of a combination valve, have a stem or button that protrudes from the valve body. The stem or button is usually covered in whole or part by a rubber boot, figure 6-10. When the brakes are bled with a power bleeder, the valve stem must be pushed in or pulled out, depending on the design, to allow fluid to pass through the valve to the front brakes, figure 6-11. This is necessary because 75 to 300 psi is required to open the metering valve; a power bleeder usually operates at less than 40 psi. If the brakes are

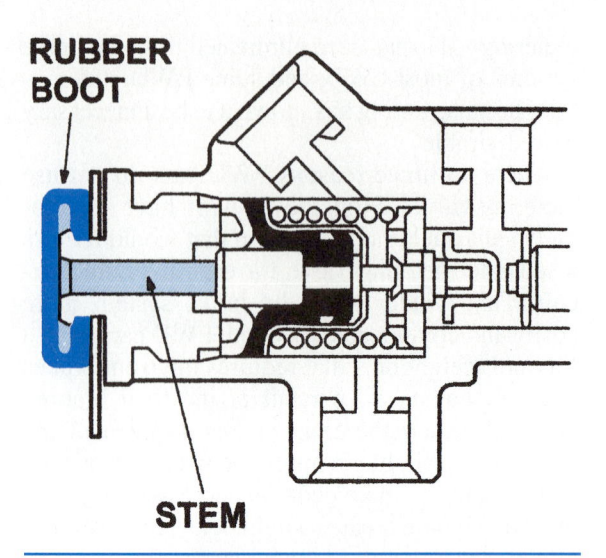

Figure 6-10. This metering valve is part of the combination valve (not shown) and has a rubber boot covering the metering valve stem. (Courtesy of General Motors Corporation, Service and Parts Operations)

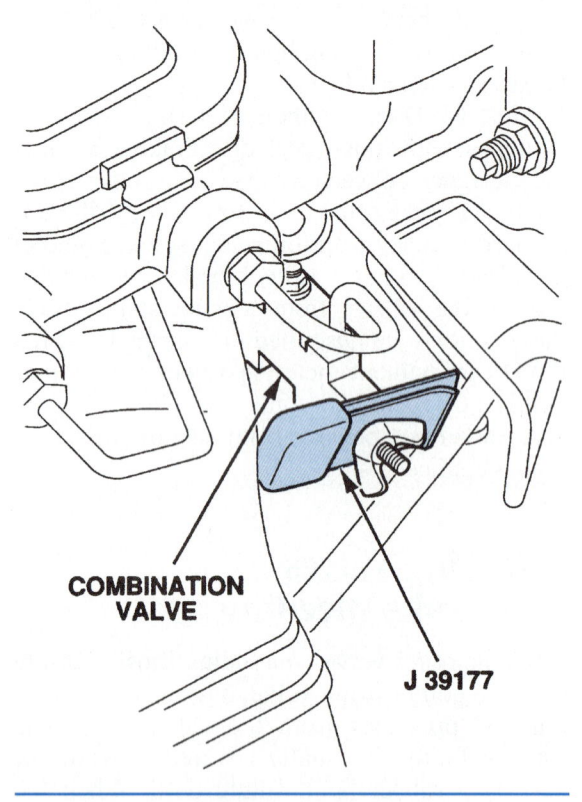

Figure 6-11. This tool is used to hold the metering valve open during pressure bleeding. (Courtesy of General Motors Corporation, Service and Parts Operations)

bled manually, pushing or pulling the metering valve stem is unnecessary because the master cylinder generates enough pressure to open the valve.

PROPORTIONING VALVE

A proportioning valve, figure 6-12, improves brake balance during hard stops by limiting hydraulic pressure to the rear brakes. A proportioning valve is necessary because inertia creates weight shift toward the front of the vehicle during braking. The weight shift unloads the rear axle, which reduces traction between the tires and the road and limits the amount of stopping power that can be delivered. Unless application pressure to the rear wheels is limited, the brakes will lock making the car unstable and likely to spin. The best overall braking performance is achieved when the front brakes lock just before the rear brakes.

Cars with front disc and rear drum brakes require a proportioning valve for additional reasons. First, disc brakes require higher hydraulic pressure for a given stop than do drum brakes; in a disc/drum system, the front brakes always need more pressure than the rear brakes. Second, once braking has begun, drum brakes require less pressure to *maintain* a fixed level of stopping power than they did to *establish* that level; in a disc/drum system, the rear brakes will always need less pressure than the front brakes. The reasons for the differences in disc and drum brake operation are explained in Chapters 8 and 9. A proportioning valve is used to compensate for these differences because it is easier to reduce pressure to the rear brakes than to increase pressure to the front brakes.

The proportioning valve does not work at all times, however. During light or moderate braking, there is insufficient weight transfer to make rear wheel locking a problem. Before proportioning action will begin, brake system hydraulic pressure must reach a minimum level called the **split point.** Below the split point full system pressure is supplied to the rear brakes, figure 6-13. Above the split point, the proportioning valve allows only a portion of the pressure through to the rear brakes.

The proportioning valve gets its name from the fact that it regulates pressure to the rear brakes *in proportion* to the pressure applied to the front brakes. Once system hydraulic pressure exceeds the split point, the rear brakes receive a fixed percentage of any further increase in pressure. Brake engineers refer to the ratio of front to rear brake pressure proportioning as the **slope,** figure 6-13. Full system

Hydraulic Valves and Switches

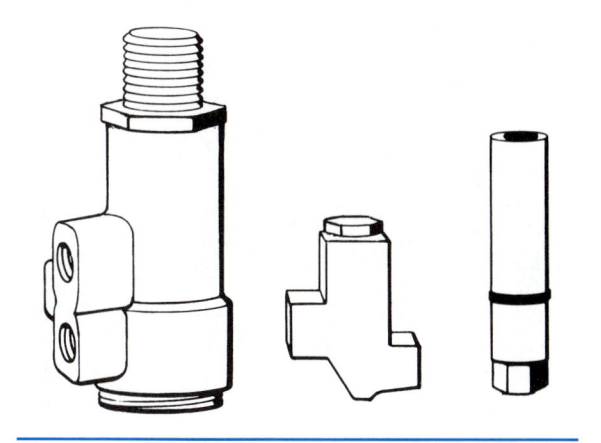

Figure 6-12. Some early proportioning valves look like simple brake line fittings.

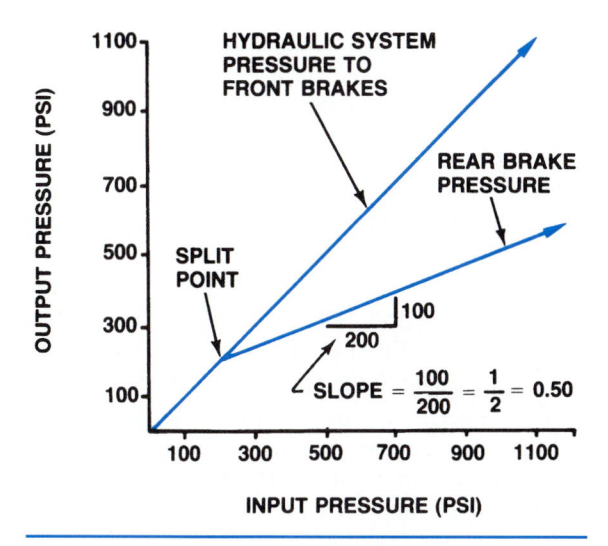

Figure 6-13. The split point and slope control the operation of a proportioning valve.

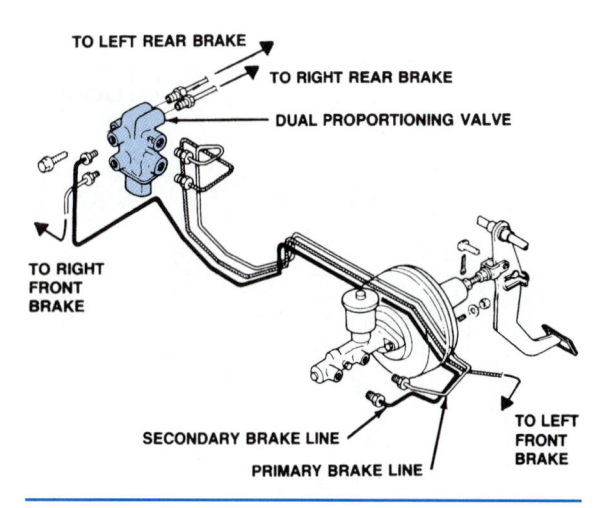

Figure 6-14. A dual proportioning valve is found on some import vehicles.

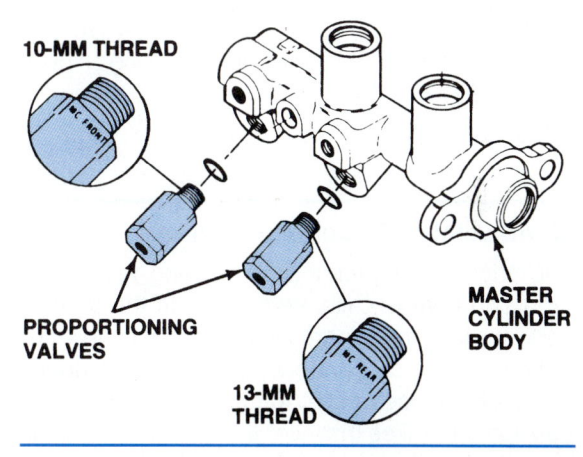

Figure 6-15. Proportioning valves that mount on the master cylinder body are common on many newer brake systems.

pressure to the rear brakes equals a slope of one, but if only half the pressure is allowed to reach the rear brakes, the proportioning valve is said to have a slope of 0.50. The proportioning valves on most cars have a slope between 0.25 and 0.50.

The first proportioning valves, figure 6-12, were located in the brake line between the master cylinder and the rear wheels. Today, many proportioning valves are part of a combination valve. Diagonally split dual braking systems require two proportioning valves, one for each hydraulic circuit. These valves may be combined into a dual proportioning valve, figure 6-14, but many brake systems now use separate valves installed between the master cylinder and the brake lines, figure 6-15. The front and rear valves on the master cylinder are different and must not be interchanged.

Proportioning Valve Operation

A simple proportioning valve consists of a spring-loaded piston that slides in a stepped bore, figure 6-16. The piston is exposed to pressure on both sides: The smaller end of the piston is acted on by pressure from the master cylinder, whereas the larger end reacts to pressure in the rear brake circuit. The actual proportioning valve is located in the center of the piston and is opened or closed depending on the position of the piston in the stepped bore.

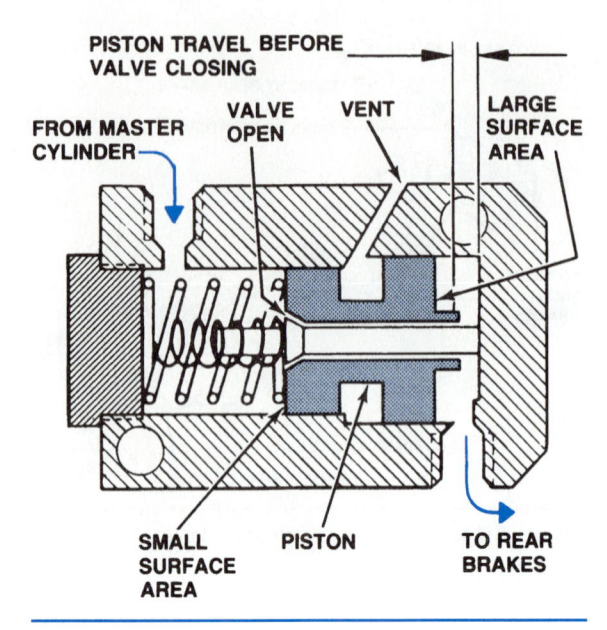

Figure 6-16. The proportioning valve piston can travel within the range shown without shutting off pressure to the rear brakes.

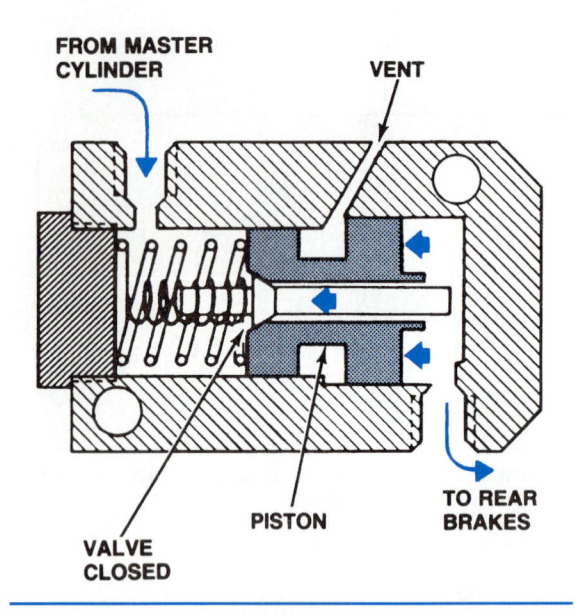

Figure 6-17. At the split point, the proportioning valve piston closes the fluid passage through the valve.

■ Adjustable Proportioning Valves

The original equipment proportioning valve on a production car is calibrated to match the braking performance of that particular car. In fact, the valve will not accurately control brake balance if modifications have been made to the tires, wheels, suspension, or brakes that alter the amount of braking force that can be applied at each axle. Some companies have addressed this problem by equipping their performance cars with adjustable proportioning valves, and similar valves are available from the aftermarket.

An adjustable proportioning valve mounts downstream of the pressure differential switch in the brake line to the rear brakes. Like any proportioning valve, an adjustable valve allows a fixed percentage of pressure to the rear brakes once system pressure has reached the valve's split point. However, the split point of an adjustable proportioning valve can be raised or lowered. Certain Corvettes and Porsches have valves that are adjusted by loosening a locknut and turning a threaded shaft. One common aftermarket valve has a small knob that is turned to set the split point to any pressure between 100 and 1000 psi. On racing cars, a variable proportioning valve is often mounted so the driver can turn the adjusting knob during a race to change the brake bias as fuel consumption and track conditions affect the car's balance.

When the brakes are first applied, hydraulic pressure passes through the proportioning valve to the rear brakes. Hydraulic pressure is the same on both sides of the piston, but because the side facing the rear brakes has more surface area than the side facing the master cylinder, greater force is developed and the piston moves to the left against spring tension. At pressures below the split point, the piston remains within the range of travel shown in figure 6-16; the proportioning valve is open, and pressure to both the front and rear brakes is the same.

As the car is braked harder, increased system pressure forces the piston so far to the left that the proportioning valve is closed, figure 6-17. This seals off the brake line and prevents any additional pressure from reaching the rear brakes. The pressure at the moment the proportioning valve first closes is the split point of the valve. From this point on, the rear brakes receive only a portion of the pressure supplied the front brakes.

As system pressure (the pressure to the front brakes) increases, enough force is developed on the master cylinder side of the piston to overcome the pressure trapped in the rear brake circuit. This forces the piston back to the right and opens the proportioning valve. Some of the higher pressure enters the rear brake circuit, but before pressure in the two circuits can equalize, the force developed on the larger piston area in the rear circuit moves the piston back to the left and closes the valve. The difference in surface area between the two ends of the piston determines the slope of the

Hydraulic Valves and Switches

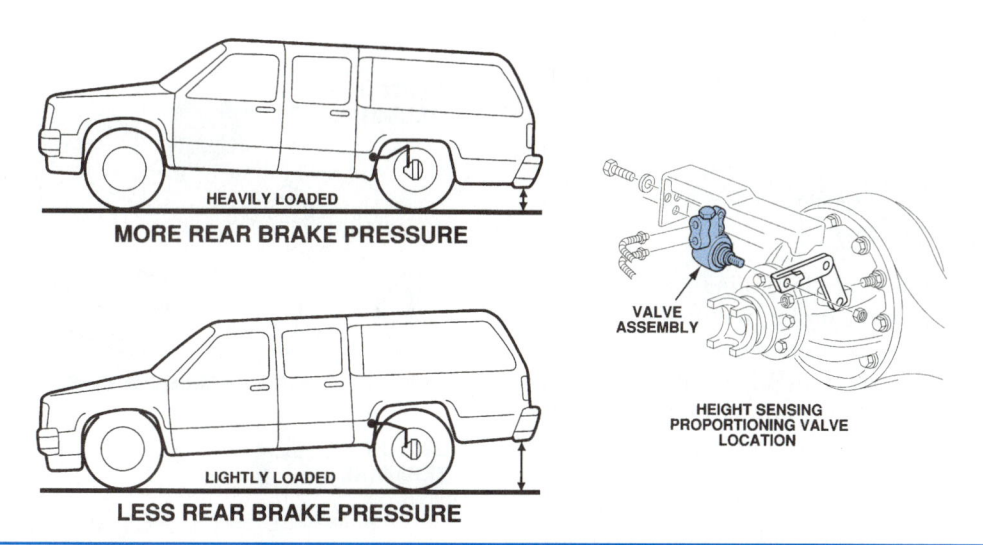

Figure 6-18. A height-sensing proportioning valve provides the vehicle with variable brake balance. (Courtesy of General Motors Corporation, Service and Parts Operations)

valve, and thus the percentage of system pressure allowed to reach the rear brakes.

As long as system pressure continues to increase, the piston will repeatedly cycle back and forth, opening and closing the proportioning valve and maintaining a fixed proportion of full system pressure to the rear brakes. When the brakes are released, the spring returns the piston all the way to the right, which opens the valve and allows fluid to pass in both directions.

Height-Sensing Proportioning Valve

Some light trucks, and a few cars, are fitted with a height-sensing proportioning valve, figure 6-18, which varies the balance between the front and rear brakes based on vehicle loading. This is desirable because trucks undergo a major change in weight bias when they are loaded. A truck with an empty bed can make little use of its rear brakes, but a fully loaded truck is able to apply a good deal of stopping power at the rear axle. The weight bias of a car does not change as dramatically as that of a truck, but a full load of passengers and luggage can still cause a substantial rearward shift, particularly in a FWD sedan.

A height-sensing proportioning valve attaches to the vehicle chassis and is operated by levers or springs attached to the rear axle or a suspension member. As the vehicle is loaded, the distance between the chassis and the axle or suspension member changes, causing the levers to rotate a **cam** inside the proportioning valve, figure 6-19.

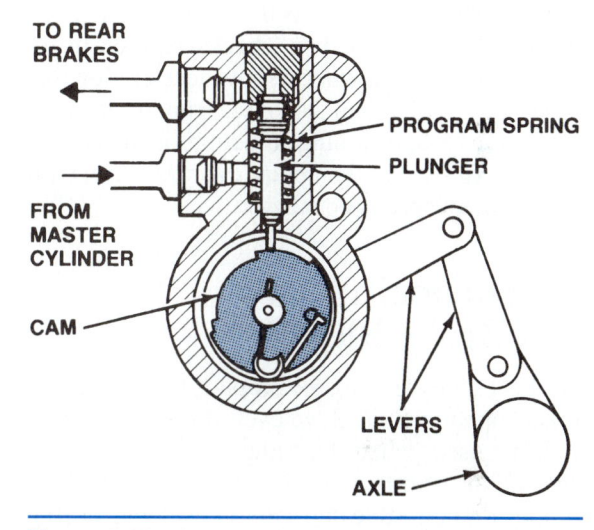

Figure 6-19 A stepped cam is used to alter the split point of this height-sensing proportioning valve.

The cam has two or more steps that adjust the position of the valve in the center of the piston to alter the split point of the valve. With a light load, the split point pressure is low and brake proportioning begins early. With a full load, the split point pressure is raised and full system pressure is applied to the rear brakes for a longer time before proportioning begins.

The distance between the vehicle chassis and the axle or suspension member is carefully engineered into the design of a height-sensing proportioning valve. Any modification to the vehicle that affects the relationship of these parts will adversely affect brake balance and could cause an accident. Lift kits, air bags, air shocks, heavy-duty springs, and coil-over load booster shock absorbers are some of the parts that could cause problems.

Electronic Brake Proportioning

On some vehicles proportioning valves are no longer used to limit brake pressure to the rear brakes. Instead, this becomes a function of the antilock brake system (ABS) controller. Called electronic variable brake proportioning (Daimler Chrysler) or dynamic proportioning (GM), this system eliminates the need for a mechanical proportioning valve. Using the ABS controller software and the rear wheel hydraulic control circuits, optimum front to rear balance is maintained at all times.

The ABS computer compares the rear wheel speed deceleration rate to the front wheel speed deceleration rate based on wheel speed sensor inputs. When the rear wheels show deceleration that is slightly greater than that of the front wheels, the electronic proportioning function is activated. The controller will then pulse the rear wheels until the wheel speeds equalize.

Proportioning Valve and Brake Bleeding

During brake bleeding the proportioning valve remains open to fluid flow, even if high foot pressure is used when bleeding, and requires no special attention. However, the height-sensing proportioning valve must be considered during this process. When the vehicle is lifted by the frame and the wheels are allowed to hang down, the height-sensing proportioning valve may completely cut off any fluid flow to the rear wheels. This results in there being no fluid from the rear wheel cylinders when bleeding. In this case, the technician can either jack the rear wheels upward until the valve is opened or bleed the brakes on the ground or on a drive-on lift.

COMBINATION VALVES

Combination valves that contain several hydraulic valves and switches were introduced in the early 1970s because they were cheaper to manufacture and install than several separate valves. They also

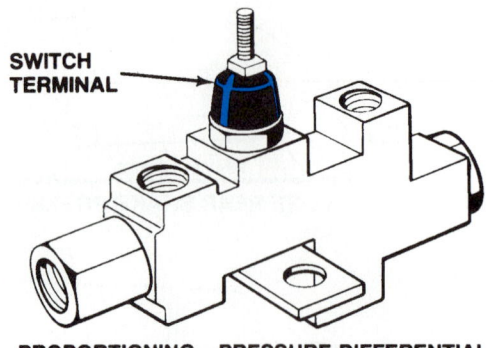

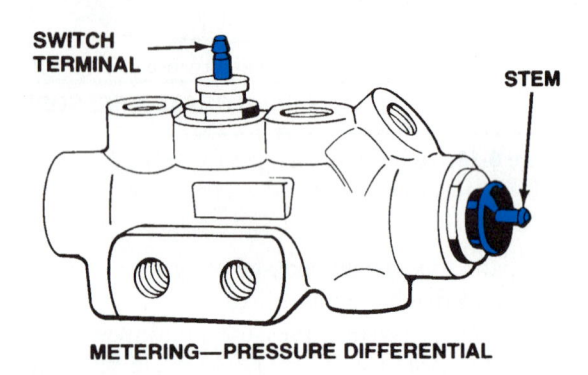

Figure 6-20. Typical two-function combination valves.

made for a neater hydraulic system with fewer fittings where leaks might develop. The various valves and switches in a combination valve function the same as the separate parts described elsewhere in this chapter.

There are several types of combination valves, figure 6-20. Two-function valves combine a pressure differential switch (see the discussion later in the chapter) with *either* a metering valve or a proportioning valve. Three-function valves combine a pressure differential switch with *both* a metering valve and a proportioning valve, figure 6-21.

BRAKE SYSTEM SWITCHES

Most automotive brake systems contain several mechanical and hydraulic switches. Mechanical switches use the physical movement of a brake system component to operate the switch contacts. Hydraulic switches use the pressure within the brake hydraulic system to operate the switch contacts. Sometimes, either a mechanical or a hydraulic switch can be used to perform the same job.

Hydraulic Valves and Switches

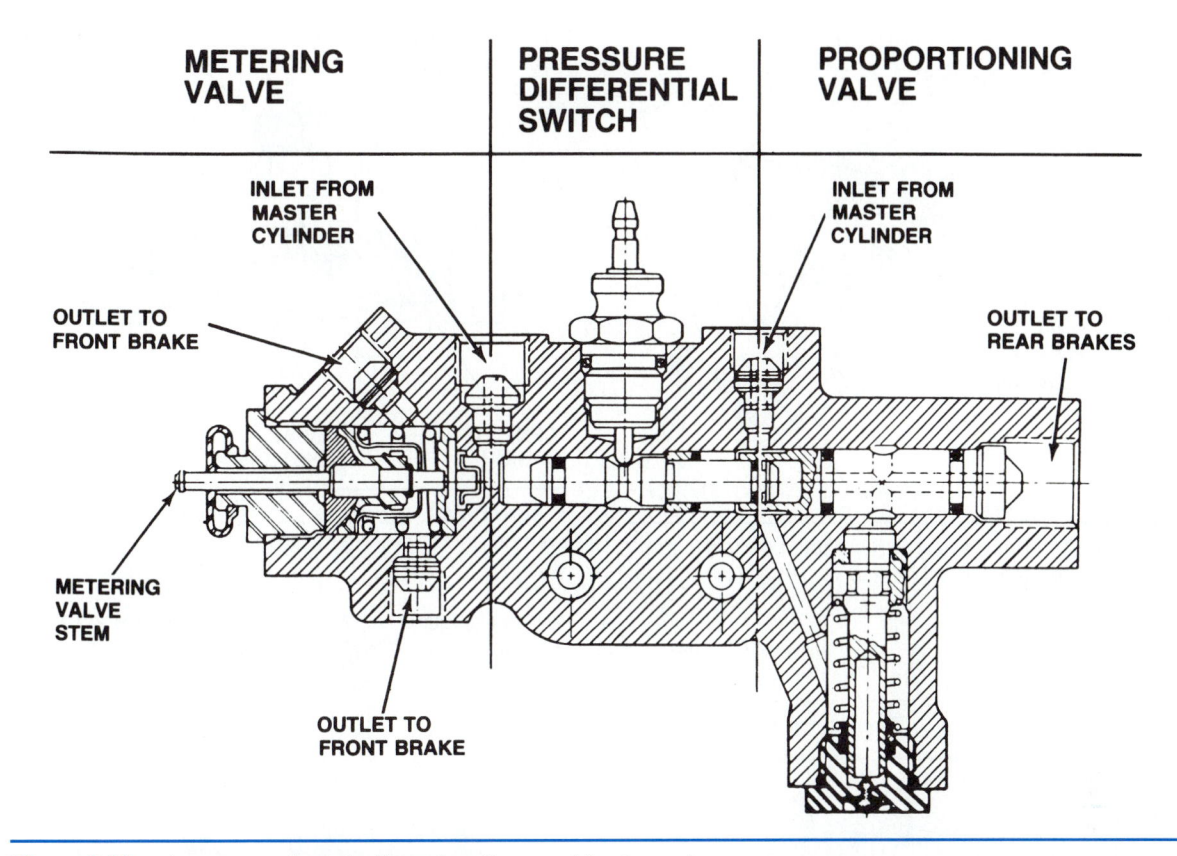

Figure 6-21. A cross section of a three-function combination valve.

Switches in the brake system are commonly used to warn of problems within the hydraulic system and to operate the vehicle brake lights. In addition, the switches may be part of a related system, such as:

- Parking brake system (parking brake "on" indicator)
- Antilock brakes (brake pedal depressed gives input signal)
- Cruise control (disengage when brake pedal is depressed)
- Transmission shift lock (allow/disallow movement of the shifter).

These jobs are done with three basic switches:

- Pressure differential
- Fluid level
- Stoplight.

PRESSURE DIFFERENTIAL SWITCH

Dual-circuit braking systems provide an extra margin of safety by actuating the brakes with two separate hydraulic circuits; if there is a failure in one circuit, the other will supply enough braking power to allow the vehicle to be driven to a shop for repairs. A requirement of DOT regulations is that a warning light indicate when there is a partial system failure of the brakes. One way of doing this is with a pressure differential switch. These switches were common on vehicles manufactured into the 1990s but are becoming less common on late-model vehicles.

Because the brakes still work to a limited degree, some drivers might continue to operate the vehicle unaware of the need for immediate service. The pressure differential switch compares the pressure in the two hydraulic circuits and actuates the **red brake warning light (RBWL)** on the instrument panel if their pressures become unequal as a result of loss of fluid or other fault.

A pressure differential switch may be a separate assembly or part of a combination valve, figure 6-21. In some brake systems the pressure differential switch is built into the master cylinder body, figure 6-22. To prevent false signals, the pressure differential switch is always installed upstream of any hydraulic control valves that alter pressure within the system.

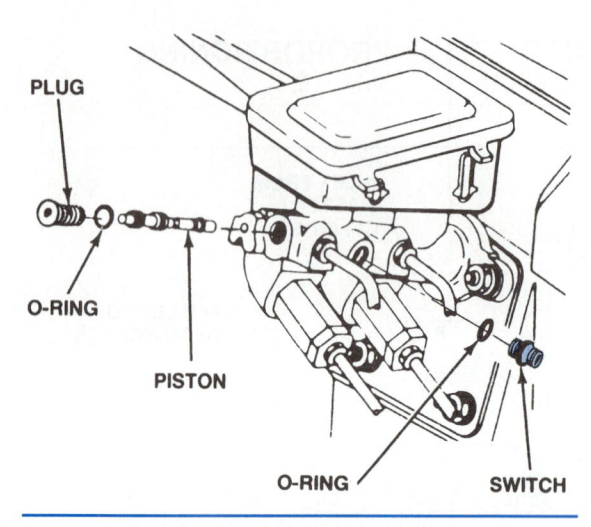

Figure 6-22. This pressure differential switch is integrated into the master cylinder.

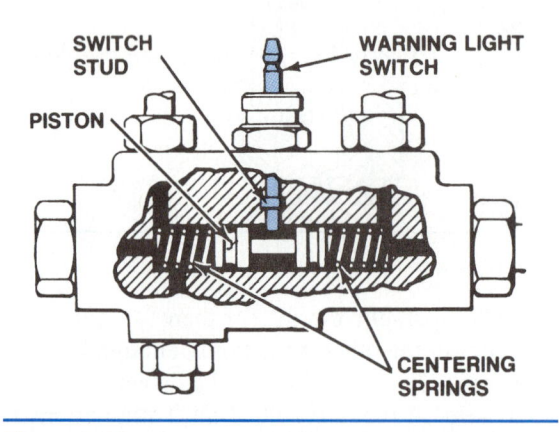

Figure 6-23. A pressure differential switch with centering springs.

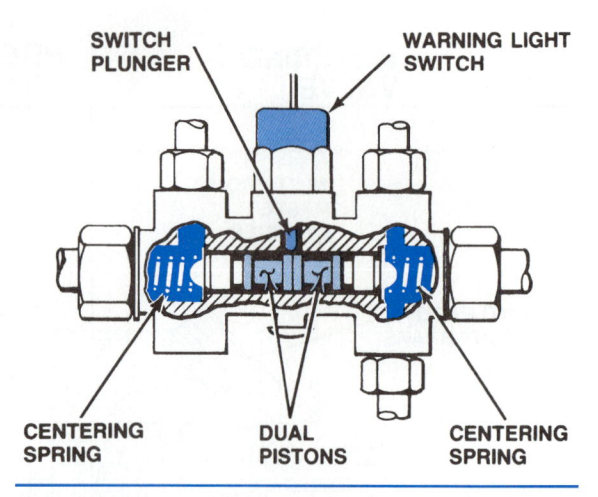

Figure 6-24. A pressure differential switch with dual pistons and centering springs.

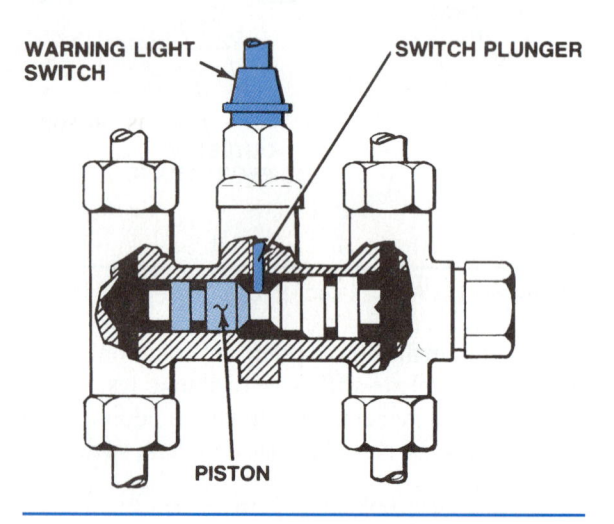

Figure 6-25. A pressure differential switch without centering springs.

Pressure Differential Switch Operation

There are three basic designs of pressure differential switches: those with single pistons and centering springs, figure 6-23; those with two pistons and centering springs, figure 6-24; and those with single pistons but no centering springs, figure 6-25. The basic operation of all three switches is the same, but they do differ in the way they actuate the warning light.

Each end of the piston in the pressure differential switch is exposed to the pressure in one of the dual brake hydraulic circuits. When the brakes are operating properly, pressure in the two circuits is the same, and both ends of the piston are acted on with equal force; this keeps the piston centered in the bore. When a leak occurs in one of the circuits, pressure on that side of the piston drops, and the higher pressure on the opposite end moves the piston toward the weak side, activating the switch. The switch is usually a normally open push-button-type switch, figure 6-26A and 6–26B, or a metal contact, figure 6–26C.

As shown in figure 6-26, the shape of the pressure differential switch piston determines how the warning light circuit is grounded. If the center of the piston is higher than surrounding areas, the switch button drops down to ground the switch, position A. If the center of the piston is lower than surrounding areas, the switch button is pushed upward to ground the switch, position B. And if the center of the piston is open, the piston itself completes the ground when it contacts the metal stud projecting into the bore, position C.

In any dual braking system, pressure between the two circuits will vary slightly until all the clearance in the wheel friction assemblies is taken up.

Hydraulic Valves and Switches

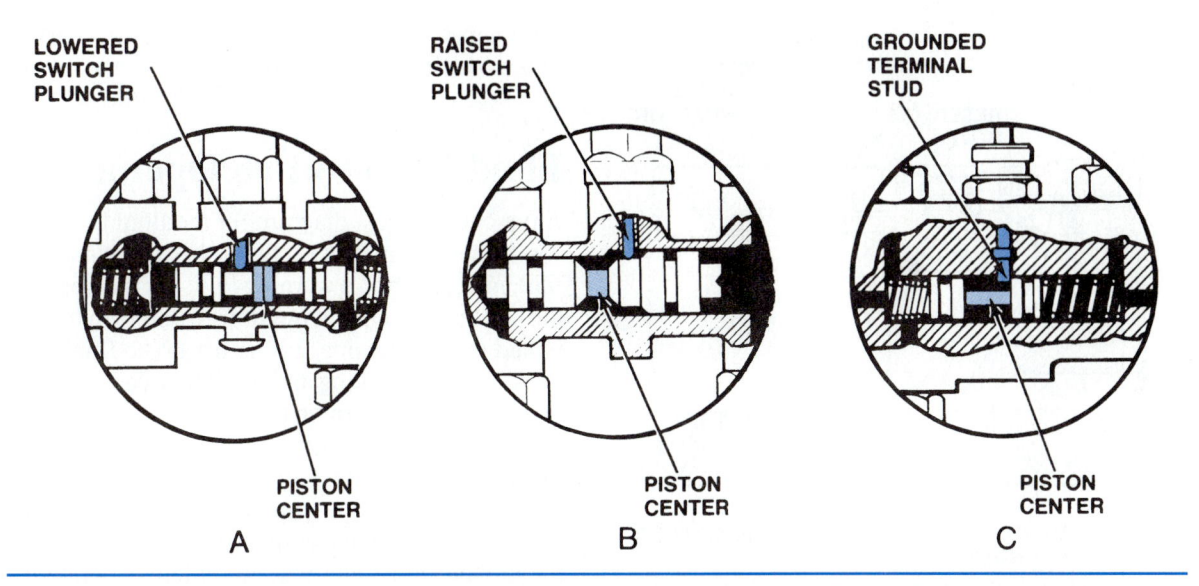

Figure 6-26. The shape of the pressure differential switch piston is used to ground the warning light circuit.

Two methods are used to prevent the warning light from coming on during these variations. In some cases, centering springs prevent the piston from moving far enough to activate the switch unless there is a significant pressure difference between the two circuits. Switches without centering springs use the tapered ramps leading to the raised portions of the piston to accomplish the same job.

Resetting the Pressure Differential Switch

When the pressure differential switch has been activated by a hydraulic failure, it may have to be manually reset after the brake system has been repaired. The technician should also be aware that simply bleeding the brakes can be enough to activate the switch. A final check after a brake repair should always include checking that the brake warning light system is working correctly.

NOTE: In some cases the cylinder bore for the piston will be corroded enough to lock the piston in place once moved. Under normal use the piston never moves into the bore area except with a pressure failure. In this case, the entire valve assembly should be replaced.

Whether a pressure differential switch has centering springs not only affects the way it operates the warning light but also the manner of resetting. Switch types in figure 6-26A and B turn the warning light on the first time a pressure difference occurs. The light then stays on until the problem is repaired and the switch is manually recentered. Switch type in figure 6-26C turns the light on only when an actual pressure difference exists in the switch, in other words, when the brakes are applied. As soon as the brakes are released, pressure is again equal (zero) at both ends of the piston so the springs recenter the piston and turn off the light. See the *Shop Manual* for more details.

FLUID LEVEL SWITCH

Some dual braking systems use a fluid level switch instead of a pressure differential switch. The fluid level switch turns on a warning light on the instrument panel when the amount of brake fluid in the reservoir falls below a safe level. The drop in fluid level can result from either brake wear or leaks within the hydraulic system. One or two fluid level switches may be used, depending on the design of the reservoir.

Fluid level switches are usually built into the reservoir or reservoir cover and consist of two fixed contacts and a rod-mounted contact connected to a float, figure 6-27. When the fluid in the reservoir is at a safe level, the float holds the rod-mounted contact suspended above the fixed contacts. If the fluid level falls below a safe level, the float drops and the rod-mounted contact connects the fixed contacts and completes the circuit that turns on the instrument panel warning light.

A second type of fluid level switch, figure 6-28, uses a magnet mounted in the bottom of the reservoir float. When the fluid level drops low enough,

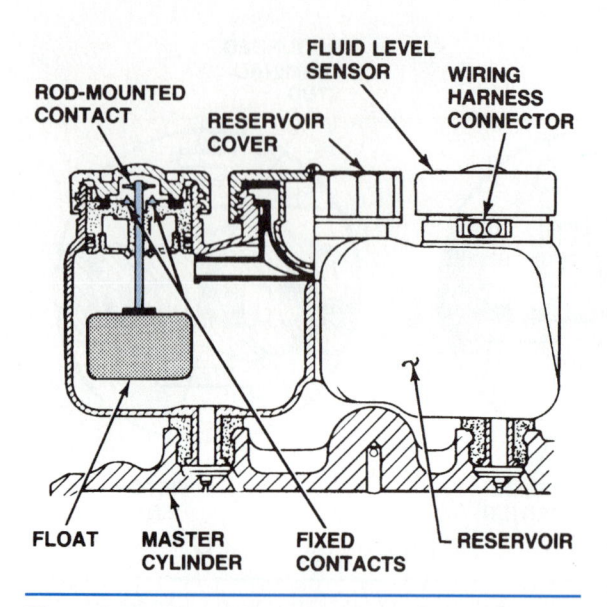

Figure 6-27. A movable-contact brake fluid level switch.

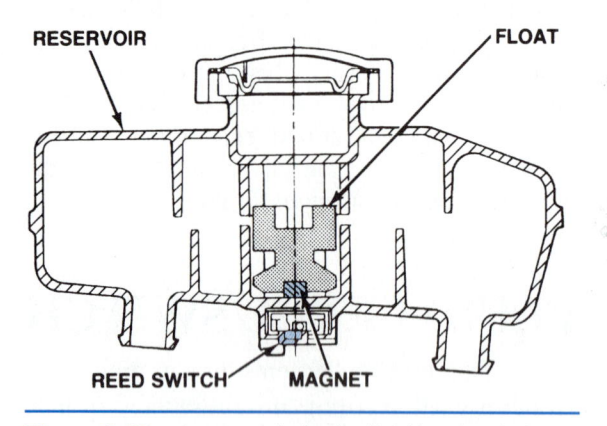

Figure 6-28. A magnetic brake fluid level switch.

the magnet trips a reed switch beneath the reservoir to turn on the warning light.

Fluid Level Switch Operation

A typical electrical diagram for the fluid level sensor is shown in figure 6-29. In this simplified schematic, the brake warning light gets power from the battery through the ignition switch with the key in and the car running. Either the brake fluid level switch or the parking brake switch can supply a **ground** that will turn on the warning light.

On vehicles with antilock brakes or body control modules, the system is a bit more complex, figure 6-30. However, this type of system is much more common than the simple system previously mentioned; most late-model vehicles operate using this system. In this system the brake fluid level switch is an input to the electronic brake control module (EBCM). When the fluid is low, the fluid level switch grounds circuit 333 at the EBCM. The EBCM then communicates with the instrument panel causing the red brake warning light to be illuminated. The instrument cluster then turns on the warning light. See the Chek-Chart book *Automotive Electrical and Electronic Systems* for more details on this type of system.

STOPLIGHT SWITCH

The job of the stoplight switch is to turn on the brake lights at the back of the car when the brakes are applied. A properly adjusted stoplight switch will activate the brake lights as soon as the brake pedal is

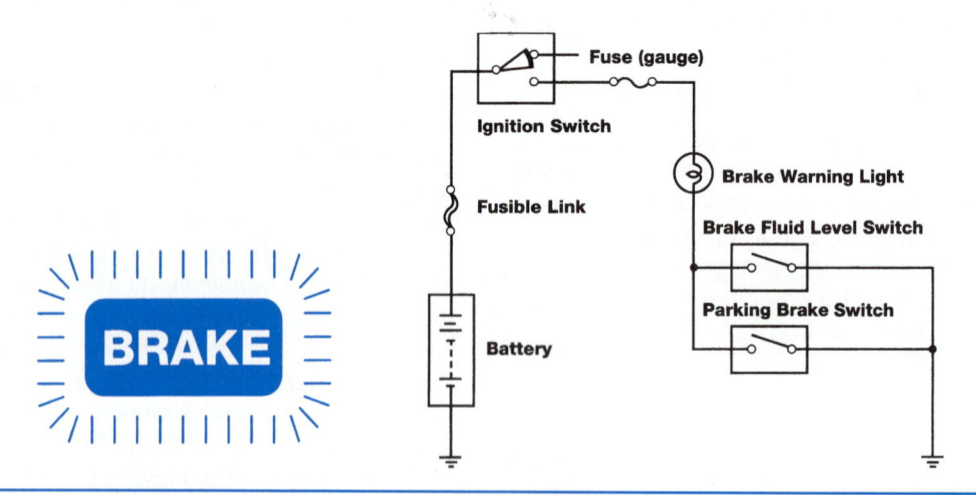

Figure 6-29. This brake warning system allows either the fluid level switch or the parking brake switch to turn on the brake warning light. (Courtesy of Toyota Motor Sales U.S.A., Inc.)

Hydraulic Valves and Switches

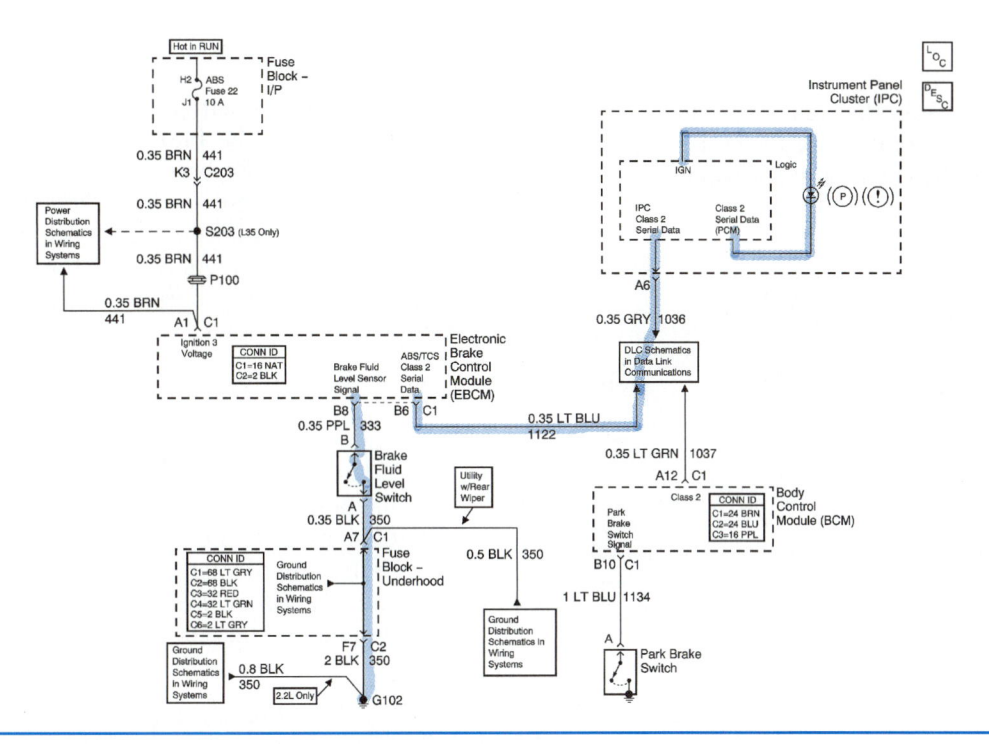

Figure 6-30. This wiring diagram shows how the fluid level switch signals the electronic brake control module (EBCM) that the brake fluid is low. The EBCM then requests the instrument cluster to turn on the red brake warning light (RBWL). (Courtesy of General Motors Corporation, Service and Parts Operations)

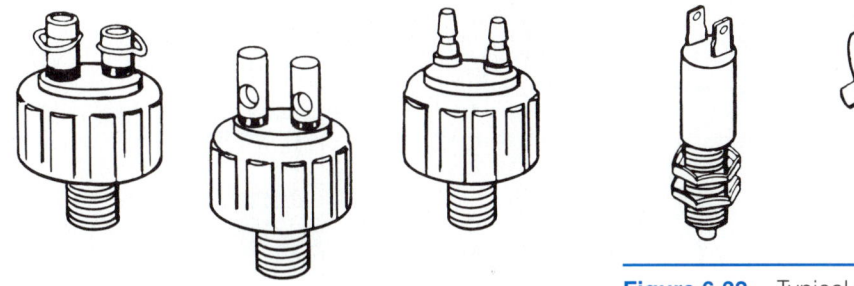

Figure 6-31. Typical hydraulic stoplight switches.

Figure 6-32. Typical mechanical stoplight switches.

applied and before braking action actually begins at the wheels; this allows drivers of following vehicles the maximum amount of time to react and apply their own cars' brakes if necessary.

Many early cars used a hydraulic stoplight switch, figure 6-31, that threaded into the master cylinder. Hydraulic switches worked well enough, but they did not turn on the brake lights until pressure had built up in the hydraulic system and braking had begun. Hydraulic switches also provided another potential source of leaks within the system, and dual braking systems would require two such switches.

Hydraulic switches have been replaced by mechanical switches that operate directly off the brake pedal arm, figure 6-32. These switches are adjusted to turn on the brake lights as soon as the driver's foot moves the pedal off its stop. On some cars, the stoplight switch itself serves as the brake pedal stop and can be used to adjust the pedal freeplay.

Both hydraulic and mechanical stoplight switches are normally open; when the brakes are applied, the switch closes to complete the stoplight circuit. Hydraulic switches are springloaded in the open position; system pressure overcomes the spring tension and closes the switch. Mechanical switches are held open against spring tension, and when the brakes are applied, the spring extends and closes the switch.

Multifunction brake light switches are used on many vehicles, figure 6-33. In addition to the brake light function, the switch may also control the operation of the cruise control or the ABS. The brake light switch will have a set of normally open contacts for the brake lights and a set of normally closed contacts for the cruise control. This eliminates the need for two separate switches.

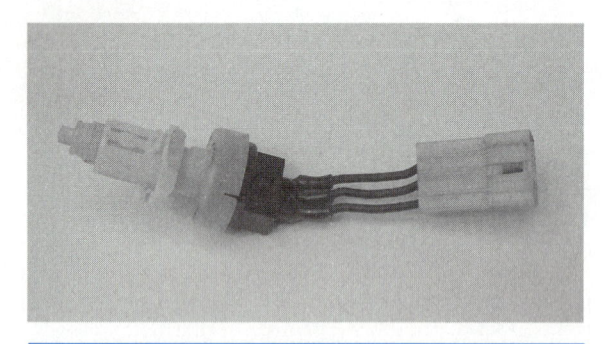

Figure 6-33. The multifunction brake light switch may include a second switch for the cruise control.

Brake Light Switch Operation

The brake lights electrical operation is similar to the operation of the brake fluid level warning system. That is, the brake lights receive power from the brake light switch when the pedal is pressed and turns on the bulbs. As an example, figure 6-34 is a typical wiring schematic for a General Motors pickup truck. The electrical current flow is as follows:

1. Stop/hazard fuse gets voltage from the battery.
2. The voltage goes to the stop lamp switch, orange wire.
3. When the switch is closed, current flows to the turn signal switch.
4. Current flows through the turn signal switch to the stop filament in the stop lamp bulb.
5. Current flows through the bulb to the ground, lighting the stop lamp.

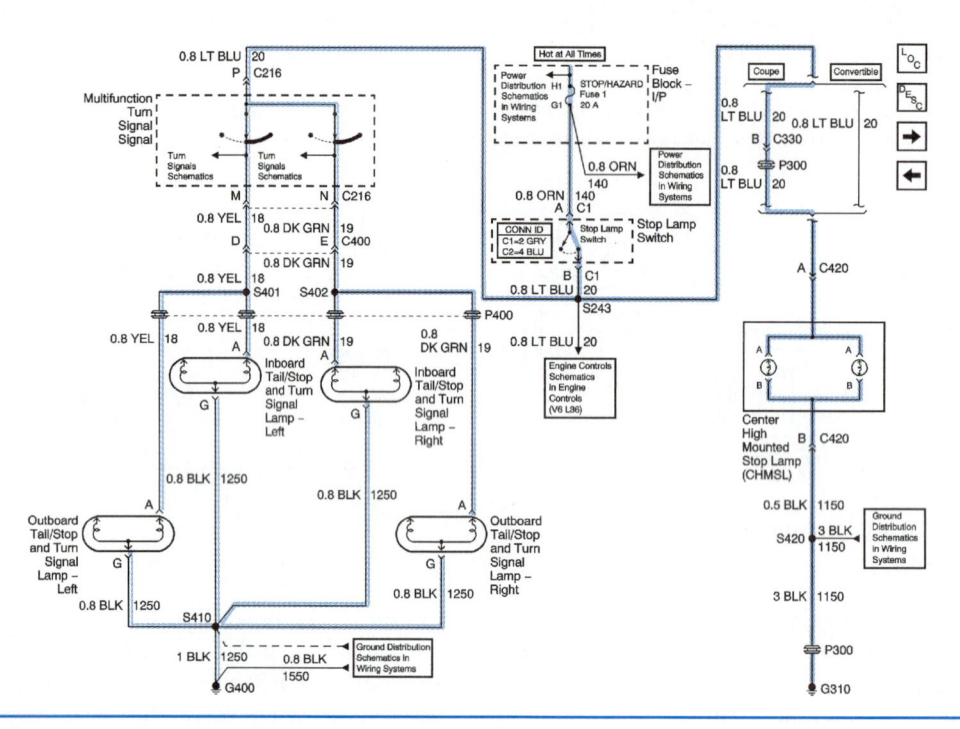

Figure 6-34. This wiring diagram shows the flow of current when the brake lamp switch is closed. (Courtesy of General Motors Corporation, Service and Parts Operations)

Hydraulic Valves and Switches

SUMMARY

Brake systems use a number of different valves to control pressures within the hydraulic system. Residual check, proportioning, and combination valves are the most common.

A residual check valve holds a static pressure of 6 to 25 psi in a brake line leading to a drum brake. This pressure keeps the sealing lips of the cup seals in the wheel cylinder tight against the cylinder walls to prevent air from entering the hydraulic system when the brakes are released.

A metering valve holds off pressure to the front disc brakes until the clearance in the rear drum brakes has been taken up. This improves brake balance and wear, and helps keep the front discs from locking when they are first applied on slippery pavement.

A proportioning valve limits pressure to the rear brakes during hard stops to improve brake balance and prevent rear wheel locking. Once brake hydraulic system pressure reaches a preset split point, only a fixed percentage of any further increase in pressure is allowed to the rear brakes. Some cars and light trucks use a height-sensing proportioning valve that changes brake balance as the vehicle is loaded. In this design, springs or levers rotate a cam to alter the proportioning valve split point.

The combination valve was introduced to simplify the brake hydraulic system and lower manufacturing costs. A combination valve combines a pressure differential switch with a metering valve, a proportioning valve, or both a metering and a proportioning valve.

Brake systems use switches to operate warning lights and the vehicle brake lights. The most common of these are the pressure differential, brake fluid level, and stoplight switches.

The pressure differential switch compares the pressure in the two hydraulic circuits of a dual braking system and turns on a warning light if the pressure becomes significantly lower in one circuit than the other. When a pressure difference occurs, a piston in the switch is moved to one side grounding an electrical contact that completes the warning light electrical circuit.

A fluid level switch detects leaks in a dual-circuit brake system by monitoring the level of brake fluid in the master cylinder reservoir. When a leak occurs, the fluid level drops, and a set of contacts controlled by a float in the reservoir is closed to turn on the warning light.

Two types of stoplight switches are used, hydraulic and mechanical. Hydraulic switches are operated by the pressure in the brake hydraulic system and are rarely used today because they do not activate the brake lights until braking has begun. Most vehicles now use a mechanical stoplight switch operated by the brake pedal arm. This type of switch can be adjusted to turn on the brake lights as soon as the pedal begins to move; this allows the drivers in following cars more time to react. The stoplight switch may include functions for the cruise control and antilock brakes.

Review Questions

Choose the letter that represents the best possible answer to the following questions:

1. Residual check valves hold _____ psi of pressure in the brake lines to the front brakes.
   a. 0 to 15
   b. 6 to 25
   c. 75 to 300
   d. None of the above

2. A residual check valve may be located:
   a. At the end of the master cylinder bore
   b. In the master cylinder outlet port
   c. Both a and b
   d. Neither a nor b

3. A multifunction brake light switch may control any of the following except:
   a. Antilock brakes
   b. Back up lights
   c. Brake lights
   d. Cruise control

4. A vehicle is observed to have *no* proportioning valve. This is because:
   a. The factory forgot to install it
   b. The vehicle is a four-wheel drive
   c. The function is performed by the ABS controller
   d. The vehicle has a 50/50 weight distribution

5. A metering valve does *not:*
   a. Even out brake wear
   b. Reduce pressure to the rear brakes
   c. Improve brake balance
   d. Prevent the front brakes from locking

6. The brake hydraulic system pressure must be _____ psi for fluid to pass through the metering valve.
   a. 75 to 300
   b. Less than 3
   c. Both a and b
   d. Neither a nor b

7. Metering valves are not used on some newer cars because:
   a. They cause a delay in brake application
   b. The car has four-wheel disc brakes
   c. The car has FWD
   d. All of the above

8. When manually bleeding a brake system, the _____ valve must be held open.
   a. Metering
   b. Combination
   c. Both a and b
   d. Neither a nor b

9. Proportioning valves are required because:
   a. Inertia affects a car's stopping ability
   b. Disc brakes require less pressure than drum brakes
   c. Drum brakes require less pressure to begin a stop
   d. All of the above.

10. During a hard stop, the percentage of system hydraulic pressure delivered to the rear brakes is determined by the _____ of the proportioning valve.
    a. Split point
    b. Slope
    c. Both a and b
    d. Neither a nor b

11. A proportioning valve may be mounted:
    a. In the brake line to the rear brakes
    b. In a combination valve
    c. On the master cylinder
    d. All of the above

12. The operation of the proportioning valve is based on:
    a. Hydraulic pressure
    b. Piston surface area
    c. Spring tension
    d. All of the above

13. A height-sensing proportioning valve:
    a. May alter brake proportioning with a cam
    b. Is operated by an electrical switch
    c. Works off of brake pedal position
    d. None of the above

14. Combination valves were developed to:
    a. Reduce manufacturing costs
    b. Help prevent leaks
    c. Simplify the braking system
    d. All of the above

Hydraulic Valves and Switches

15. Which of the following is *not* a type of combination valve?
    a. Metering/pressure differential
    b. Pressure differential/proportioning
    c. Proportioning/residual check/metering
    d. Metering/pressure differential/proportioning

16. A pressure differential switch measures the _____ pressure in the two brake hydraulic circuits.
    a. Total
    b. Difference in
    c. Maximum
    d. Controlled

17. If the brake warning light comes on only when the brakes are applied, the pressure differential switch:
    a. Is equipped with centering springs
    b. Is not equipped with centering springs
    c. Is not equipped with dual pistons
    d. Probably has a loose connection

18. The pressure differential switch compensates for minor pressure variations with:
    a. Centering springs
    b. Tapered ramps
    c. Both a and b
    d. Neither a nor b

19. Brake fluid reservoir level switches are not activated by a:
    a. Reed switch
    b. Hollow float
    c. Hydraulic switch
    d. Movable contact

20. To provide other drivers with maximum warning, the stoplight switch should be:
    a. Hydraulic
    b. Open when the brakes are applied
    c. Closed before braking begins at the wheels
    d. None of the above

7

Wheel Cylinders and Brake Caliper Hydraulics

OBJECTIVES

Upon completion and review of this chapter, you will be able to:

- Explain the operation of the three designs of wheel cylinders.
- Describe the construction of the wheel cylinder pistons, seals, cup expanders, and dust boots.
- Describe the operation of the brake caliper.
- Explain the functional difference between fixed and floating calipers.
- List and explain the function of the four types of brake caliper piston materials.
- Describe the construction and function of fixed, stroking, and low-drag caliper seals.

KEY TERMS

anchor plate
bridge bolts
insulator
knockback
lathe-cut
runout
series
unsprung weight

INTRODUCTION

Chapter 4 described the fluids used in brake hydraulic systems and the lines that carry the fluids between brake components. Chapter 5 explained how the master cylinder uses brake fluid in converting mechanical force into hydraulic pressure. That pressure is then modified by the hydraulic valves described in Chapter 6 and routed through the brake lines to the wheels.

At the wheels, the drum and disc brake friction assemblies contain the final parts in the brake hydraulic system, the wheel cylinders and brake calipers. These parts convert system hydraulic pressure back into the mechanical force needed to stop the car. This chapter explains the hydraulic operation of wheel cylinders and brake calipers. Overall drum and disc brake friction assembly design is covered in Chapters 8 and 9.

WHEEL CYLINDERS

Wheel cylinders actuate drum brake friction assemblies by converting hydraulic pressure from the master cylinder into mechanical force that

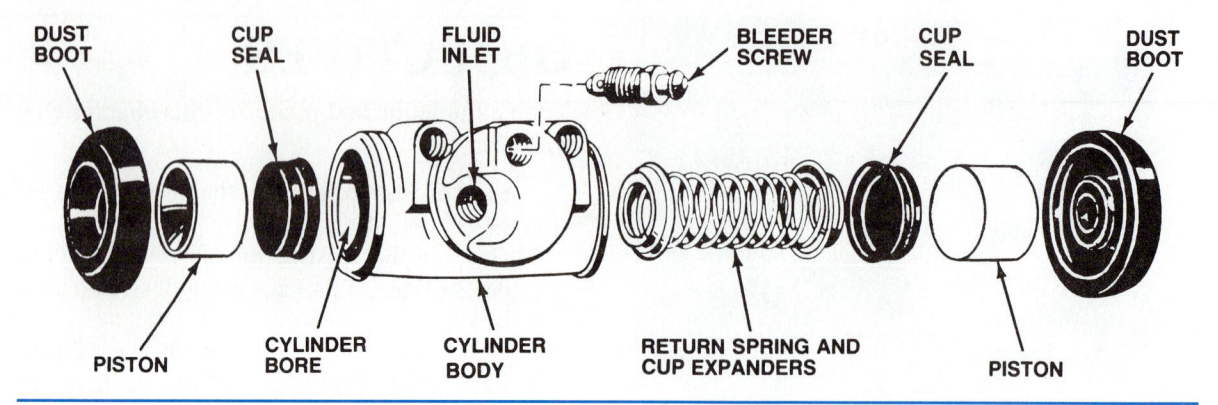

Figure 7-1. A typical drum brake wheel cylinder assembly.

moves the brake shoe linings into contact with the brake drums. Although there are minor differences in wheel cylinder designs, all cylinders operate in essentially the same way and contain certain basic parts:

- Cylinder body
- Pistons
- Seals
- Cup expanders
- Dust boots.

Wheel Cylinder Body

The foundation of any wheel cylinder, figure 7-1, is the cylinder body. The body is made of cast iron or aluminum and contains the bore in which the cylinder pistons operate. As in a master cylinder, the bore is bearingized or anodized (depending on the metal used) to provide a long-wearing and corrosion-resistant surface.

Hydraulic pressure enters the wheel cylinder through a brake line inlet fitting machined into the cylinder body. In a single-piston cylinder, the inlet is located at the closed end of the cylinder; in a dual-piston cylinder, the inlet is at the center of the cylinder between the two pistons.

A bleeder screw is also threaded into the wheel cylinder body. The fluid passage to the bleeder screw is located at the highest point in the cylinder bore when the cylinder is installed on the vehicle. This allows air to be easily purged from the hydraulic system.

In rare cases a wheel cylinder body will have two brake line fittings but no bleeder screw. This type of cylinder is used when the rear brakes are connected in **series,** figure 7-2. In this system, the brake line from the master cylinder connects to one of the rear wheel cylinders. Another brake line routes fluid from that cylinder to the one at the other rear wheel. A single bleeder screw on the second wheel cylinder is used to bleed the entire rear brake circuit.

Wheel Cylinder Pistons and Seals

The bore inside the wheel cylinder body contains one or two pistons that move toward the open ends of the cylinder when hydraulic pressure is applied to them. As in a brake master cylinder, fluid is prevented from escaping through the gaps between the pistons and the bore by rubber seals. Cup seals, described in Chapter 5, are used in virtually all modern wheel cylinders, although O-ring seals have been used in a few applications in the past.

Cup Expanders

When the brakes are applied, fluid pressure flares the sealing lips of the cup seals and holds them tightly against the cylinder bore. However, when the brakes are released, the retracting master cylinder piston creates a momentary pressure drop that can draw air past the cup seals into the hydraulic system. Two methods are used to maintain a good seal under these conditions.

As described in Chapter 6, older vehicles used residual pressure check valves to maintain a small amount of hydraulic pressure in the brake lines and wheel cylinders at all times. Since the early 1970s, however, most cars have used spring-loaded metal cup expanders, figure 7-3, to physically hold the sealing lips of the cup seals against the cylinder bore. Cup expanders consist of small metal cones

Wheel Cylinders and Brake Caliper Hydraulics

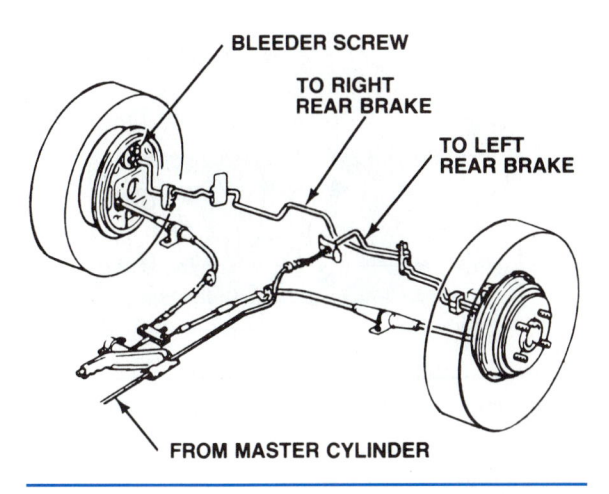

Figure 7-2. Some cars connect the rear wheel cylinders in series and use a single bleeder screw.

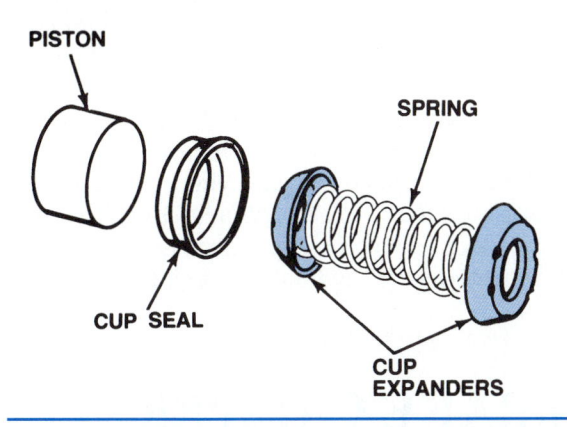

Figure 7-3. Cup expanders mechanically hold the cup sealing lip against the wheel cylinder bore.

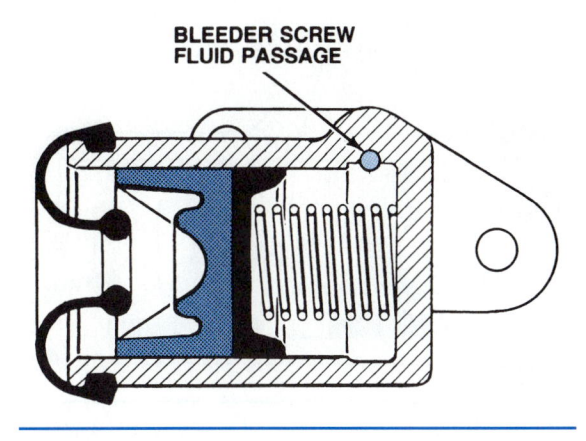

Figure 7-4. A single-piston wheel cylinder.

that are pressed against the inside of the sealing lips by spring tension. Compared to residual check valves, cup expanders are simpler, cheaper, less prone to failure, and easier to work on.

The spring that forces the cup expanders against the seals serves two other purposes as well. First, it takes up any slack between the piston, piston pushrod (if used), and brake shoe. This helps keep brake pedal travel to a minimum. Second, in dual-piston cylinders, the spring centers the pistons in the bore and prevents them from blocking the fluid inlet passage.

Wheel Cylinder Dust Boots

The open ends of the wheel cylinder are sealed with rubber dust boots that keep dirt, brake dust, and moisture out of the cylinder bore. These materials can damage or cause rapid wear of the cup seals and cylinder bore, which result in fluid leaks and shortened service life. The dust boots also prevent minor brake fluid seepage past the cup seals from getting onto the brake linings where it would reduce stopping power and cause the brakes to grab.

WHEEL CYLINDER DESIGNS

There are three basic wheel cylinder designs in use today:

- Single-piston
- Straight-bore dual-piston
- Stepped-bore dual-piston.

All three types of cylinders operate the same, but each is used only in specific kinds of drum brake assemblies. Drum brake design is covered in detail in Chapter 8.

The single-piston wheel cylinder, figure 7-4, is the simplest type. It contains one piston and operates a single brake shoe. A pair of single-piston wheel cylinders is required for a typical drum brake friction assembly.

The straight-bore dual-piston wheel cylinder, figure 7-5, is the most common type. It contains two pistons and operates two brake shoes, one at each end. Most modern drum brakes use one dual-piston wheel cylinder in each drum brake friction assembly.

The stepped-bore dual-piston wheel cylinder, figure 7-6, is the rarest type. This design of cylinder also contains two pistons and operates two brake shoes, but the pistons and their respective

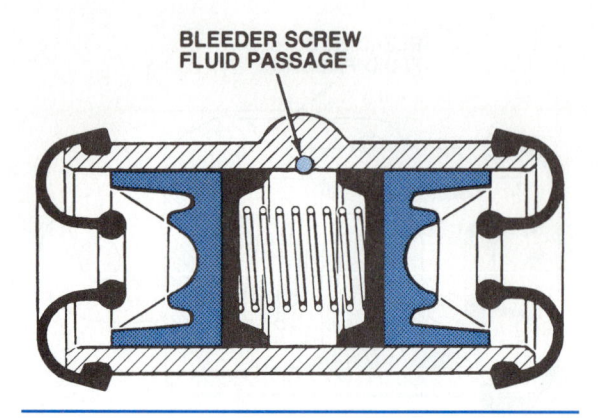

Figure 7-5. A straight-bore dual-piston wheel cylinder.

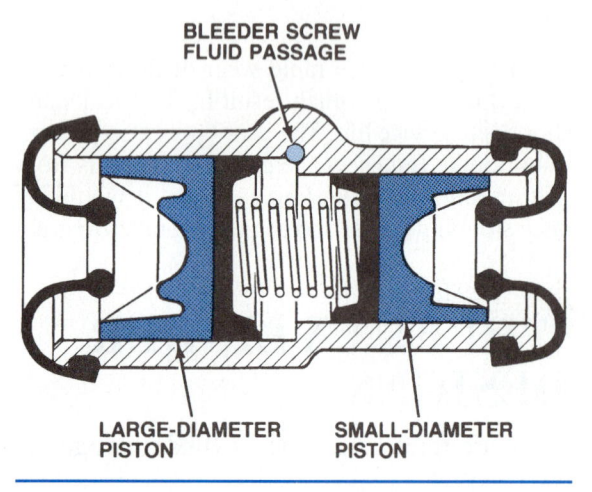

Figure 7-6. Stepped-bore dual-piston wheel cylinders apply the brake shoes with unequal force.

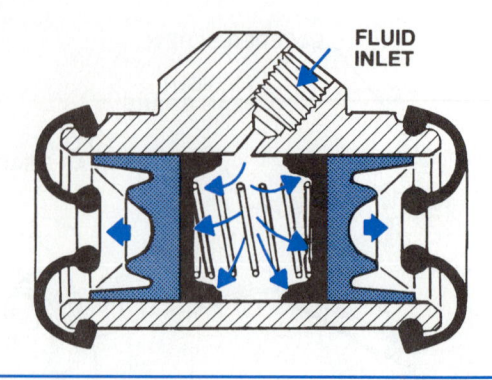

Figure 7-7. When the brakes are applied, hydraulic pressure moves the wheel cylinder pistons outward in their bore.

bores have different diameters. This creates unequal force at the two ends of the wheel cylinder; the brake shoe on the side with the large piston is applied with greater force than the shoe on the side with the smaller piston.

WHEEL CYLINDER OPERATION

All wheel cylinders operate in basically the same manner. When the brakes are applied, pressure is routed from the master cylinder through the brake lines and into the wheel cylinders. The pressure, contained in the cylinders by the cups seals, acts on the back sides of the wheel cylinder pistons forcing them outward in their bores, figure 7-7. The pistons, acting directly or through pushrods, force the brake shoe linings into contact with the brake drum, creating friction to stop the car.

When the brakes are released, the shoes are retracted from the drums by return springs, and fluid returns from the wheel cylinders through the brake lines to the master cylinder. Residual pressure check valves or cup expanders prevent air from entering the hydraulic system at this time.

BRAKE CALIPERS

The wheel cylinder is only one element of a drum brake friction assembly; it serves a single function that is relatively easy to describe. In contrast, the brake caliper is the primary component of a disc brake friction assembly; it serves many functions, not all related to the brake hydraulic system. This chapter covers only the hydraulic operation of brake calipers; overall disc brake friction assembly design is explained in Chapter 9.

Brake calipers actuate disc brake friction assemblies by converting hydraulic pressure from the master cylinder into mechanical force that moves the brake pads into contact with the brake rotors. Despite the design differences referred to earlier, all brake calipers operate essentially the same and contain certain basic parts:

- Caliper body
- Pistons
- Piston seals
- Dust boots.

Wheel Cylinders and Brake Caliper Hydraulics

■ **The "Anticlunk" Wheel Cylinder**

When a car's brakes are released, the low pressure created by the retracting master cylinder piston is not the only factor that affects the return of brake fluid to the reservoir. The relatively strong brake shoe return springs also work to force fluid back through the brake lines. In fact, in the late 1950s and early 1960s, certain Ford models had such strong return springs that the returning brake shoes made an annoying "clunk" as they reached their stops.

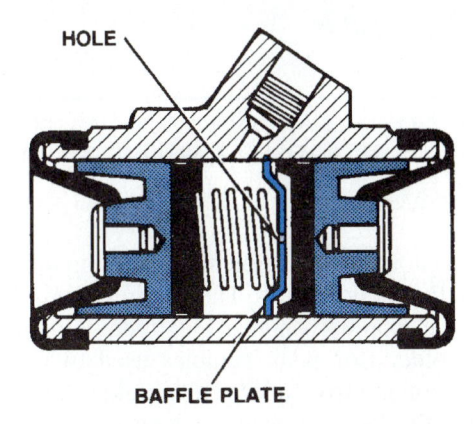

Ford's answer to the noise was to install a baffle in the wheel cylinder that contained a small hole to restrict fluid flow. The restriction had no significant effect on brake application because the high pressures developed in the system easily forced sufficient fluid through the small opening. When the brakes were released, however, fluid return was slowed just enough to prevent the annoying sound. Later brake designs were changed to allow the use of weaker return springs that eliminated noisy shoe return—and the "anticlunk" wheel cylinder.

BRAKE CALIPER BODY

The foundation of any disc brake is the caliper body, a U-shaped casting that wraps around the brake rotor, figure 7-8. Brake caliper bodies are usually made of cast iron, although high-performance cars sometimes have aluminum alloy calipers to reduce weight and dissipate heat better. Single-piston brake caliper bodies are usually made in one piece, but multipiston calipers that position pistons on both sides of the rotor are manufactured in halves that bolt together with high-strength **bridge bolts.** Many

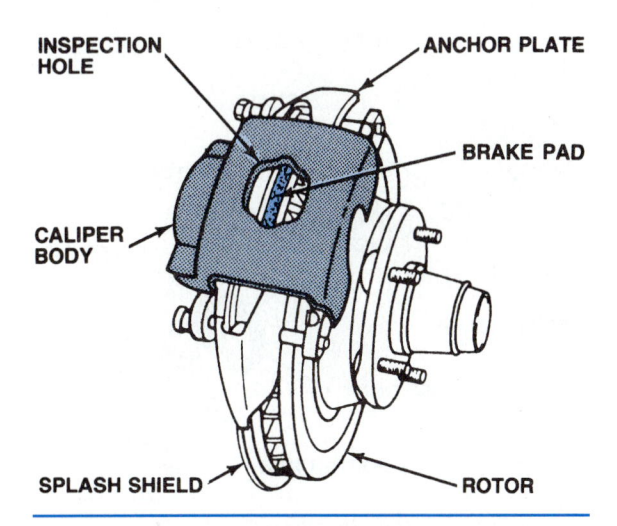

Figure 7-8. The caliper body houses the hydraulic components of a disc brake friction assembly.

caliper bodies incorporate an inspection hole or slot through which a technician can check brake pad wear.

At the front axle, the brake caliper mounts to the spindle or steering knuckle. Rear disc brake calipers mount to the axle flange or a support bracket on the suspension. Calipers that bolt solidly to the suspension are called fixed calipers; the body of a fixed caliper does not move at any time. However, on most late-model cars, the caliper body is free to move within a limited range on an **anchor plate** that is solidly bolted to the suspension. These designs are called sliding or floating calipers. The different caliper designs are explained in detail in Chapter 9.

Piston Bores

The caliper body contains one to four bores that hold the caliper pistons. If one piston is used, the bore is on the inboard side of the caliper, figure 7-9. Single-piston calipers are used on most late-model cars because they are simple, inexpensive, easy to service, and less likely to leak than calipers with more moving parts. They also place the fluid chamber of the caliper closer to the passing airflow for better cooling.

Two-piston calipers are built in two types. One design uses a fixed caliper with a bore on each side of the rotor, figure 7-10. The second type is a floating/sliding caliper with both pistons located on the inboard side of the rotor, figure 7-11. The first type is stronger than a single-piston design

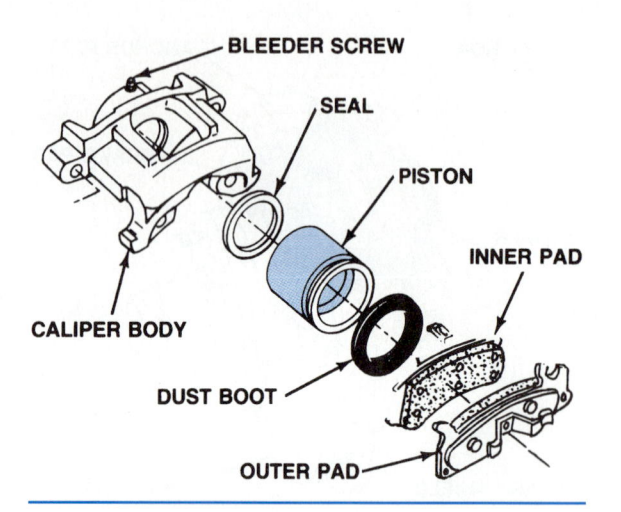

Figure 7-9. A single-piston brake caliper.

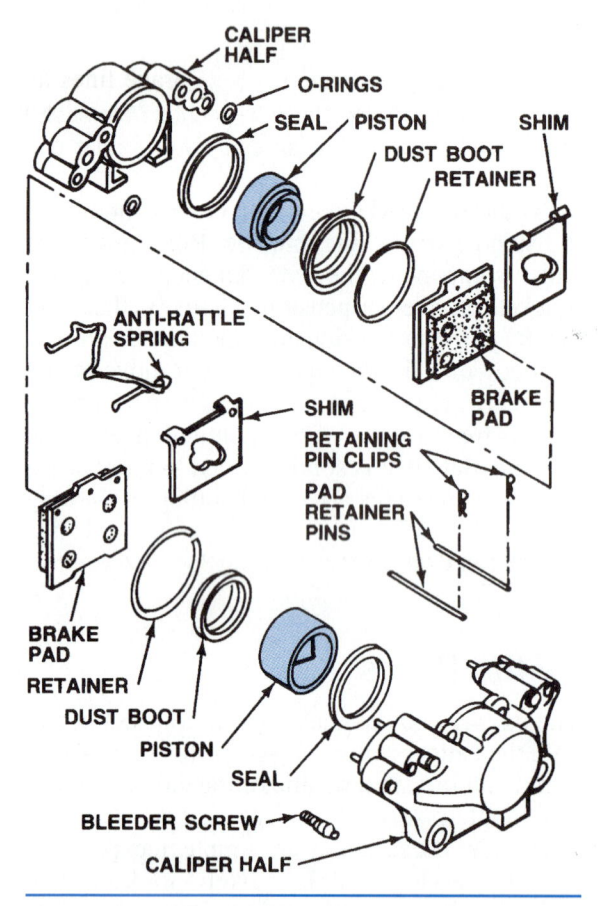

Figure 7-10. Two-piston calipers are common on many European automobiles.

Figure 7-11. This two-piston caliper uses phenolic pistons and an aluminum housing to reduce heat transfer to the fluid.

and both types can provide higher application force because of their larger piston area. In addition, the greater overall mass of two-piston calipers allows them to absorb and dissipate more of the heat from braking without passing that heat on to the brake fluid. The sliding, two-piston caliper is popular on many of today's vehicles. Many older European vehicles used two-piston calipers on the front and/or rear brakes; late models use the two-piston caliper on the rear, with four-piston calipers on the front.

Three-piston calipers are relatively rare and have a single large bore on the inside of the rotor and two smaller bores on the outside, figure 7-12. This style of caliper has the same advantages as a two-piston caliper but also provides additional clearance for the wheel at the outside of the caliper. The pistons are sized so the force created by the two small pistons is equal to that provided by the single large one.

Calipers with four pistons place two bores on both sides of the rotor, figure 7-13. Four-piston calipers are strong and powerful, and provide wheel clearance equal to that of a three-piston caliper. However, compared to single-piston calipers, they are more costly, complex, and difficult to service. Four-piston calipers were used on some early disc brakes but in recent years have been limited to high-performance and luxury applications.

Calipers using more than four pistons are being manufactured for certain high-load and super high-performance applications. Some of the world's fastest cars may use six-piston or even eight-piston calipers in combination with carbon-ceramic brake pads and rotors for an extremely powerful braking system. Special applications may also use carbon-ceramic brake discs and a pair of lightweight four-, six-, or eight-piston calipers, figure 7-14.

Wheel Cylinders and Brake Caliper Hydraulics

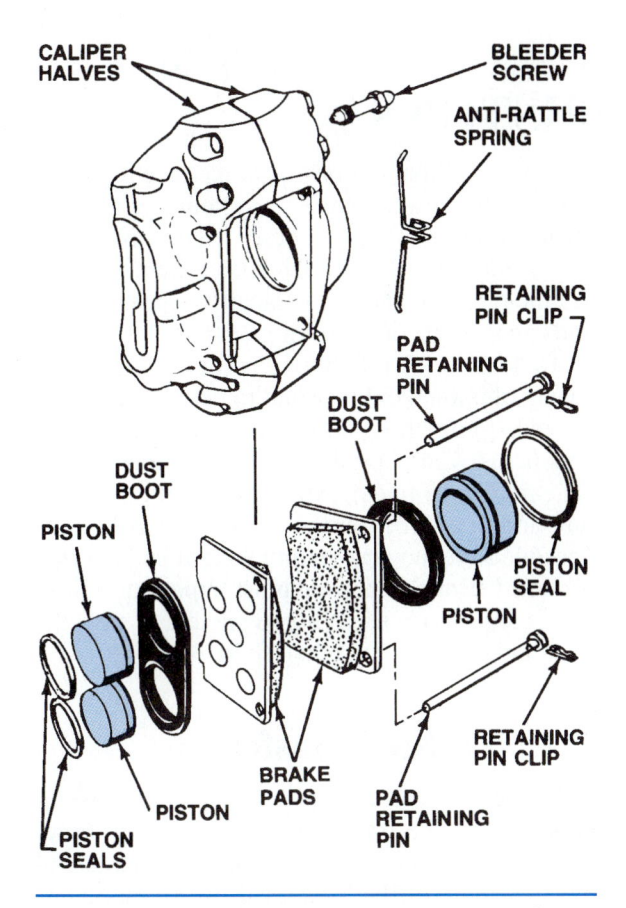

Figure 7-12. Three-piston calipers use two different sizes of pistons.

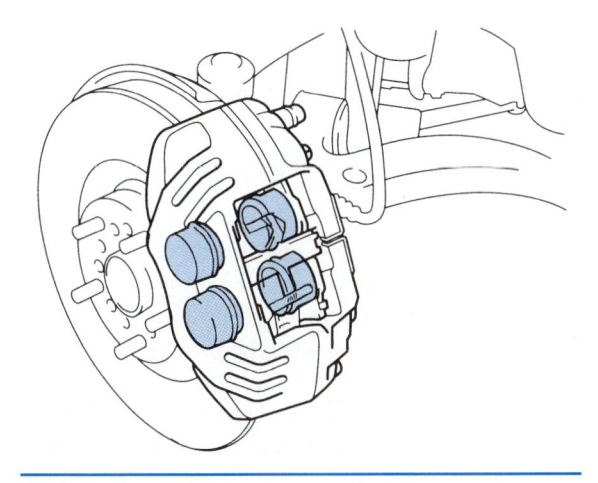

Figure 7-13. The four-piston caliper may be used on some high-performance vehicles. (Courtesy of Toyota Motor Sales U.S.A., Inc.)

Brake Fluid Routing

Brake fluid pressure is supplied to the piston bores through brake line inlet fittings machined into the caliper body. In order to clear the wheels,

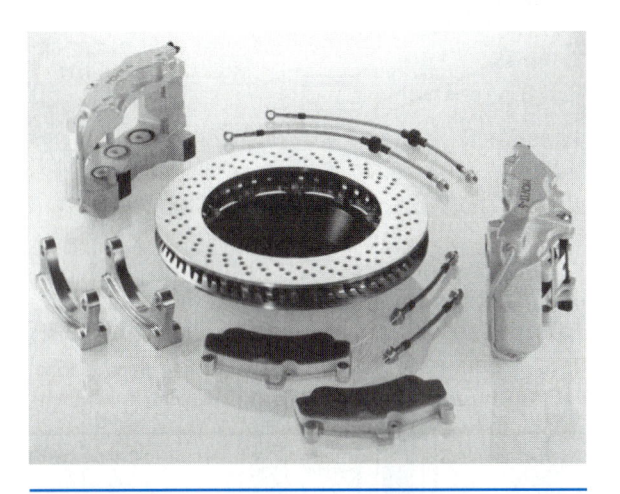

Figure 7-14. These aftermarket brakes feature a pair of six-piston brake calipers. (Courtesy of Mov'It GmbH)

tires, and other suspension parts, brake lines always connect to the inboard side of the caliper body. The majority of calipers have a single brake line inlet fitting. Calipers with two brake line inlet fittings are used in the triagonal dual-circuit brake systems described in Chapter 5.

Calipers manufactured in halves use one of two methods to route fluid to the pistons on the outboard side of the caliper. Most have drilled internal passages that are sealed at the caliper split with rubber O-rings, figure 7-15. The O-rings fit into machined grooves in the caliper halves and are compressed when the caliper is assembled.

Older split calipers, and some newer high-performance designs, use an external crossover tube to provide a fluid path between the halves, figure 7-16. Standard threaded brake line fittings are used at both ends of the tube. External crossover tubes are preferred in heavy-duty calipers because internally drilled passages weaken the caliper body. In addition, extreme application pressures can cause caliper flex that can lead to leaks if O-ring seals are fitted at the caliper split.

One, two, or sometimes three bleeder screws are located on the caliper body to allow trapped air to be bled from the system. As with wheel cylinders, the fluid passages leading to the bleeder screws are located at the highest points of the caliper bores when the caliper is installed on the car. On some cars, brake calipers can be accidently swapped side-to-side and still bolt to the suspension mounting points. The only problem is that the bleeder screws will be at the bottoms of the calipers making it impossible to bleed air from the system!

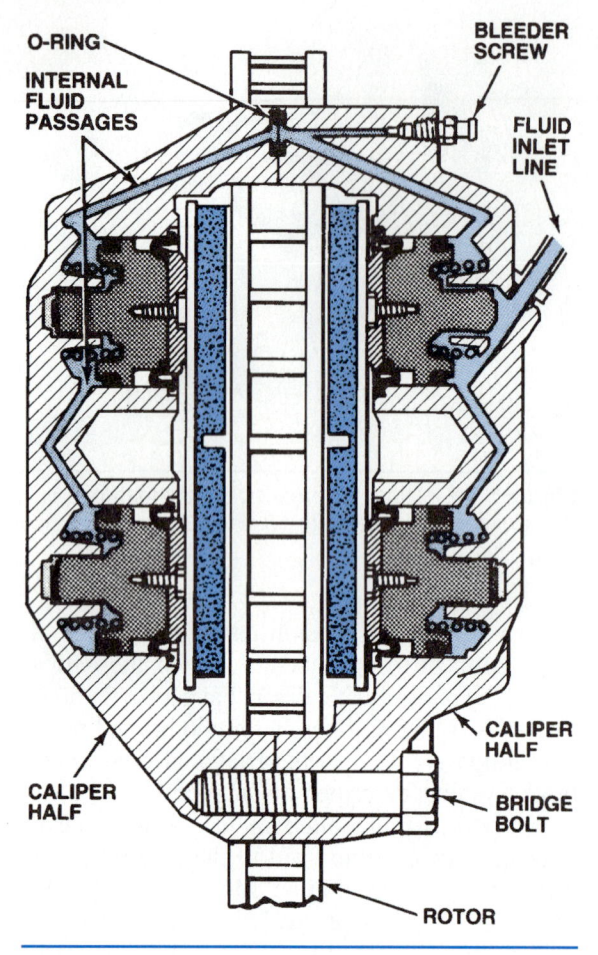

Figure 7-15. Internally drilled fluid passages are common in multipiston calipers.

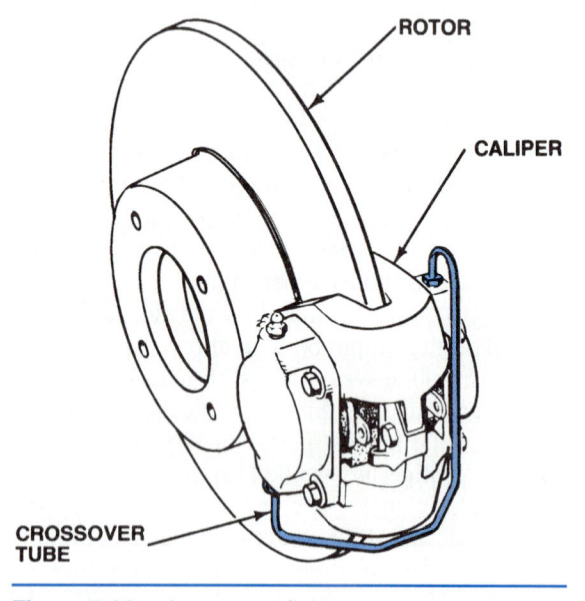

Figure 7-16. An external fluid crossover tube is used on some multipiston calipers.

BRAKE CALIPER PISTONS

The brake caliper pistons form the meeting point between the brake hydraulic system and the mechanical action of the disc brake friction assembly. The pistons move outward in their bores under hydraulic pressure and mechanically force the brake pads against the brake rotor to stop the car.

To do their job reliably, caliper pistons must be strong, durable, and resistant to corrosion caused by moisture, brake fluid, road salt, and other chemicals. Caliper pistons must maintain their size and shape under extremes of both temperature and pressure, and they should be as light as possible to reduce **unsprung weight** and aid fuel economy. But perhaps the most important trait of a caliper piston is its ability to prevent or slow the transfer of heat to the brake fluid.

Piston Heat Transfer

The inner side of a caliper piston is in constant contact with the brake fluid in the caliper bore, whereas the outer side rides against the metal backing plate of the brake pad. Brake pads routinely operate at temperatures above the boiling point of brake fluid, and, during heavy use, brake pad temperatures can exceed the fluid boiling point by several hundred degrees Fahrenheit. Some of this heat is transferred into the caliper piston and from there into the brake fluid. If too much heat is transferred through the piston, the brake fluid will boil and vapor lock will occur.

The design of caliper pistons plays a big part in limiting the heat they transfer into the fluid. The centers of most pistons are hollow to reduce the surface area in contact with the brake pad, and the airspace in the hollow area serves as an additional **insulator.** However, the most important factor affecting heat transfer is the caliper piston material.

At the present time, caliper pistons are made from four materials:

- Aluminum
- Cast iron
- Steel
- Phenolic plastic.

The chart in figure 7-17 shows how pistons made of these materials conduct heat under controlled test conditions. The advantages and disadvan-

Wheel Cylinders and Brake Caliper Hydraulics

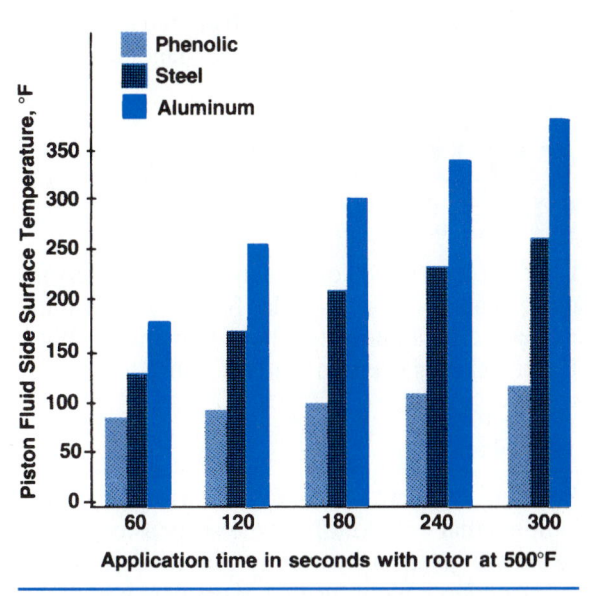

Figure 7-17. Different caliper piston materials have differing rates of heat transfer.

Figure 7-18. Steel and phenolic caliper pistons are the most common types today.

tages of each material are discussed in the following sections.

■ Racing Calipers

Brake engineers rarely specify fixed, multipiston calipers for road cars today; however, such calipers are still widely used on racing cars and high-performance street machines. At high speeds, these cars store incredible amounts of kinetic energy. Their brakes must be able to apply great force to convert this energy into heat and remain flex-free while doing so. It's no coincidence that these are all virtues of fixed, multipiston calipers.

Fixed calipers are large and strong by virtue of their basic design. Their multiple pistons can exert more force than any single piston caliper, and they distribute that force more evenly over larger brake pads that run cooler and provide longer service life. To reduce unsprung weight, most high-performance calipers are made of aluminum alloy that contributes to even better heat dissipation. Fixed, multipiston calipers are still used on the world's fastest cars, and that's no accident!

Aluminum Pistons

Where weight is the main concern, aluminum pistons can be used. They are seldom used on production vehicles because they have a number of disadvantages. When aluminum is heated, it expands at a much greater rate than cast iron. This means that aluminum pistons require fairly large clearances in the caliper bore, which can lead to leaks and other problems. And although aluminum pistons are anodized, they are still more susceptible to corrosion and scuffing than iron or steel parts. But the main problem with aluminum as a material for caliper pistons is that it is an excellent conductor of heat and, therefore, very poor at keeping heat away from the brake fluid.

Aluminum pistons can be used in racing applications with high-temperature synthetic brake fluids, carbon-ceramic rotors, and some type of brake cooling system. Rebuilding the calipers and replacing the fluid every few hours of running is considered normal maintenance for these systems.

Cast-Iron and Steel Pistons

Cast iron and steel have been the primary caliper piston materials for most of the production life of disc brake calipers. Cast-iron pistons were used first, but stronger steel parts are most common today, figure 7-18. Both materials are very strong, and they maintain their size and shape well throughout a wide range of temperatures and pressures. Cast-iron and steel pistons are precision ground to their final size and surface finish, and are often chrome plated for improved corrosion resistance. The main problems with cast-iron and steel pistons are that they are relatively heavy and they conduct heat more readily than is usually desired.

Phenolic Pistons

Phenolic materials are made from phenol (also called carbolic acid) combined with other elements under heat and pressure. Common phenolic parts on today's vehicles include disc brake caliper pistons, power-assist braking components, accessory drive pulleys, water pump housings, solenoids, ashtrays, and transmission components. Brake caliper pistons made of phenolic plastic were first introduced by Chrysler in the mid-1970s.

Different combinations of ingredients produce phenolic materials with different properties. Modern phenolic caliper pistons may contain up to 80 percent glass fiber, carbon fiber, or other materials and 20 percent phenol formaldehyde.

Phenolic caliper pistons, figure 7-18, have a number of distinct advantages over metal pistons; they are inexpensive, lightweight, strong, and highly corrosion resistant. In addition, their outer sealing surface is not as slippery as that of a cast-iron or steel piston. This allows the seal to grip the piston better and provide greater piston retraction (discussed later in the chapter). But most of all, phenolic pistons are very poor conductors of heat; they do an excellent job of insulating the brake fluid in the caliper from the heat of braking.

Although phenolic caliper pistons are used by all of the domestic automakers, they have met with resistance from brake technicians. As with silicone brake fluids, some concerns are justified, whereas others are not. For example, early Chrysler calipers with phenolic pistons suffered from piston sticking. Studies showed that those problems resulted primarily from caliper bore corrosion caused by poor dust boot sealing and moisture absorption from contaminated brake fluid. Newer calipers are designed with improved sealing and improved piston materials that eliminate these concerns.

When phenolic pistons had problems in the past, the common practice was to install steel replacement pistons. Modern brake systems and calipers are engineered to take advantage of the insulating properties of phenolic pistons. Substituting metal parts can cause a substantial rise in brake fluid temperature and an increased possibility of vapor lock. It is not recommended that steel or phenolic pistons be substituted one for the other, except as approved by the manufacturer.

CALIPER PISTON SEALS

Like any hydraulic component, brake calipers require seals to prevent brake fluid from escaping through the gap between the piston and the bore. Brake calipers use two types of seals:

- Stroking
- Fixed.

Stroking seals are used *only* on early fixed-brake calipers that bolt solidly to the suspension. Fixed seals are used on virtually *all* modern brake calipers. Do not confuse fixed calipers with fixed seals; fixed seals are used on all three types of calipers: fixed, sliding, and floating.

The design difference between the two is in the seal and sealing surface. In the stroking seal, the surface of the caliper bore is the critical surface for sealing. In the fixed seal, the sealing surface is on the piston, making the piston surface the most critical part of the sealing system.

Stroking Piston Seals

The stroking seals in disc brake calipers are ring-type lip seals, figure 7-19, similar to those on some master cylinder and wheel cylinder pistons.

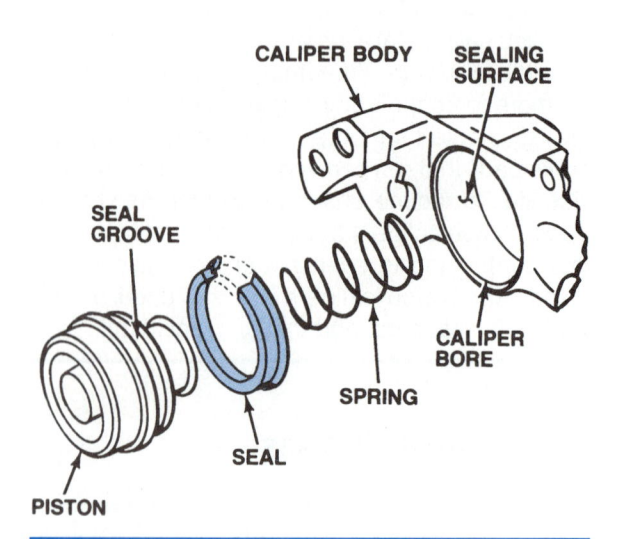

Figure 7-19. A stroking seal mounts on the piston and uses the caliper bore as its sealing surface.

Wheel Cylinders and Brake Caliper Hydraulics

A stroking seal fits into a machined groove at the back of the caliper piston, and moves out and back with the piston as the brakes are applied and released. The bore of the caliper body provides the sealing surface and is finely finished to provide the best possible seal. If the surface of the bore becomes damaged, the caliper must be honed or replaced to prevent leaks. In some cases, the caliper bore can be machined oversize and special sleeves can be installed to restore the sealing surface.

Pistons fitted with stroking seals can be installed relatively loosely in the caliper bore because the seal lip flares out against the caliper bore to prevent leaks. Most manufacturers specify a clearance between .004″ and .010″ (.10 and .25 mm). If the clearance exceeds the maximum value when checked using a new piston, the caliper must be replaced.

Stroking seals are rarely used today for a number of reasons. One reason is the manufacturing cost. A stroking seal caliper requires more parts and machine work to function properly when compared to a fixed-seal caliper.

Because stroking seals exert little force on the caliper bore, any amount of brake rotor **runout** (common in the early days of disc brakes) can force the pistons back away from the rotor when the brakes are released, figure 7-20. The amount of piston **knockback** will vary from one stop to the next, causing inconsistent and excessive brake pedal travel.

To prevent the negative effects of too much knockback, most calipers with stroking seals have springs or some type of self-adjuster behind the pistons to take up any slack between the piston, brake pad, and rotor. This keeps the brake pedal height constant and provides automatic brake adjustment by keeping the pads in light contact with the rotor at all times. However, it also creates a small amount of continuous brake drag that reduces pad life and hurts fuel economy, an additional problem with stroking seals.

Another drawback is that, with the seal installed at the back of the piston, the bore sealing surface outside of the seal is always exposed; this is the area the seal will pass over as the brake pads wear. If the protective dust boot becomes damaged, moisture and grit will attack the bore and seal causing fluid leaks and caliper failure. This can happen very quickly, even on a brand new caliper.

Fixed-Piston Seals

Fixed seals are used in modern calipers because they are superior to stroking seals in almost every way. Fixed seals are **lathe-cut** O-rings that fit into grooves machined at the outer edge of the caliper bore, figure 7-21. The caliper piston slides through the inside of the seal, and the outside of the piston provides the sealing surface. Because the

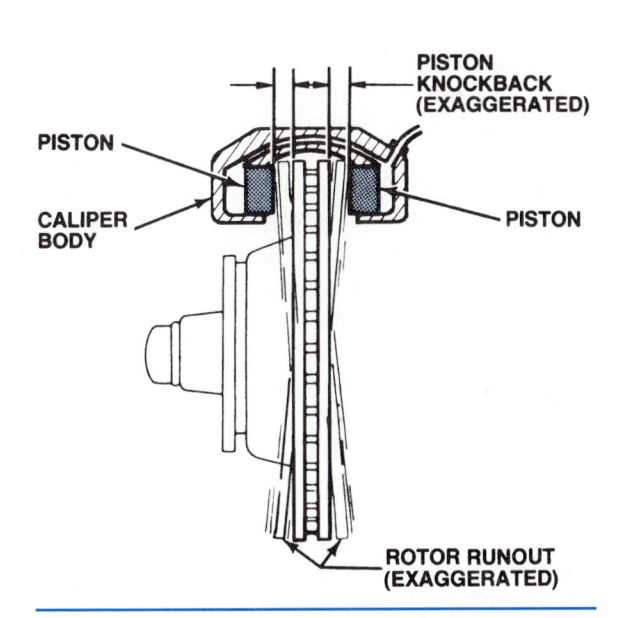

Figure 7-20. Rotor runout can cause excessive piston knockback on calipers fitted with stroking seals.

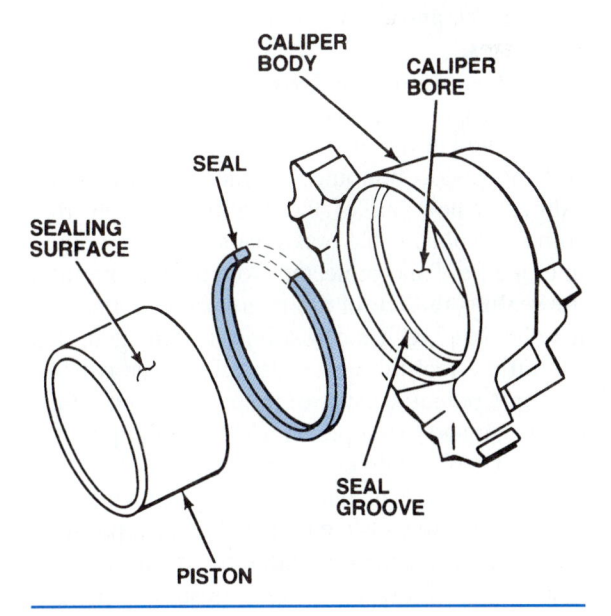

Figure 7-21. A fixed seal mounts in the caliper bore and uses the piston as its sealing surface.

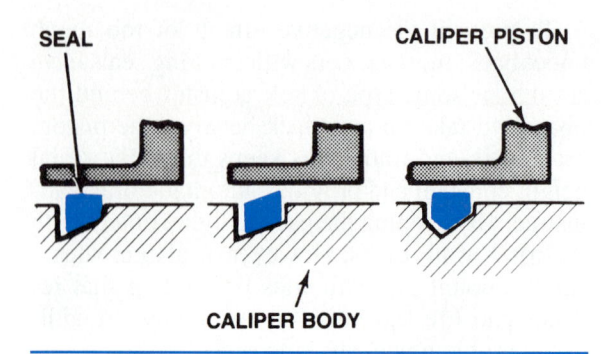

Figure 7-22. Fixed-piston seals are manufactured in a number of different shapes.

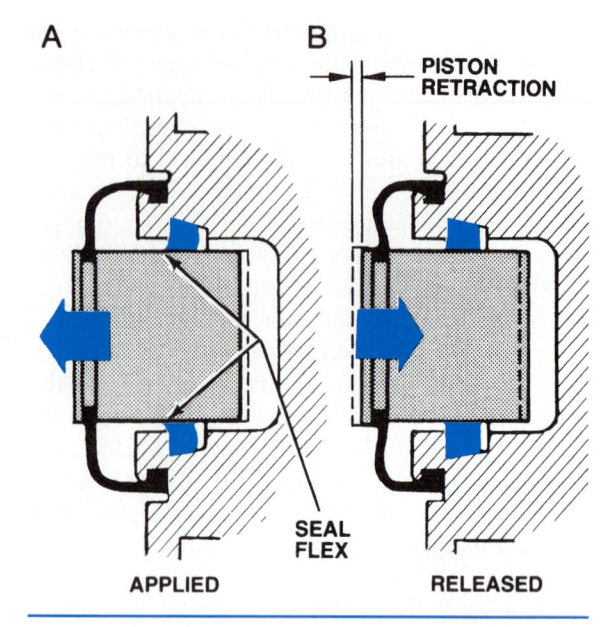

Figure 7-23. The flexing of a fixed seal retracts the caliper piston from the brake rotor.

sealing occurs between the seal and the piston, the surface finish of the caliper bore is not as critical as when a stroking seal is used.

Many caliper piston seals are square-cut like the one shown in figure 7-21. However, seals do come in a variety of cross sections, figure 7-22. The shape of the seal must match the shape of the groove in the caliper bore or a leak will result.

Metal pistons used in calipers with fixed seals are installed with tighter clearances than similar pistons in calipers with stroking seals; most manufacturers recommend between .002″ and .005″ (.06 and .13 mm). However, phenolic pistons used in calipers with fixed seals require clearances similar to those used with stroking seals, approximately .005″ to .010″ (.13 to .25 mm). The larger clearances are needed to compensate for the higher expansion rate of phenolic pistons. However, once they reach operating temperature, both metal and phenolic pistons have similar clearances.

The fixed O-ring seal overcomes the two main disadvantages of stroking seals. First, with the seal located at the outer edge of the caliper bore, and the piston bottomed in the bore when the caliper is new, all of the sealing surfaces that come into play as the brake pads wear are bathed in brake fluid and protected from harm. Although the sealing surface of the piston *outside* the seal may be attacked if the dust boot is damaged, this has little effect on the service life of the caliper. The only potential cause of premature caliper failure is damage to the piston *inside* the caliper as a result of brake fluid contamination.

Some vehicle manufacturers recommend that when the brake pads are replaced in a caliper with fixed seals, the caliper should be rebuilt and the pistons inspected, especially if the dust boot is damaged or torn. Any scoring, rusting, or pitting of the piston will cause the O-ring seals to leak. If pads are installed in a fixed-seal caliper with dirty or damaged pistons, forcing those pistons back through the seals will likely result in a brake fluid leak.

Seal Flex and Piston Retraction

One advantage of fixed seals is that they positively locate the piston in the caliper bore. The O-ring seal fits tightly against the piston and holds it in position at all times. When the brakes are applied, figure 7-23A, the seal flexes outward until the brake pads contact the rotor. When the brakes are released, figure 7-23B, the seal returns to its original shape and retracts the piston from the brake rotor.

The amount of piston retraction provided by the flexing action of a fixed seal is small and constant. Because the piston holds the pad next to the rotor with a small clearance, knockback caused by rotor runout is not as much of a problem, and brake pedal height and travel remain constant. However, excessive rotor runout (more than .005 (.127 mm)) can still cause knockback. Fixed seals also eliminate the need for springs behind the pistons, which improves fuel economy by reducing drag when the brakes are not applied.

Adjustment for Wear

Because the piston seal can only flex so far, it provides automatic adjustment of the clearance between the brake pads and rotor. When the amount

Wheel Cylinders and Brake Caliper Hydraulics

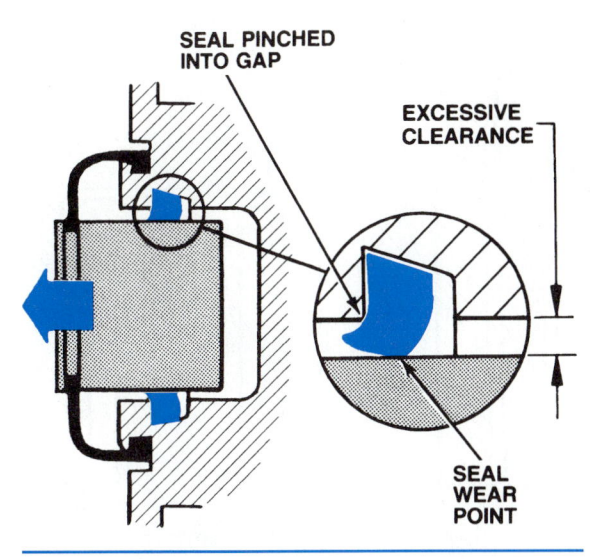

Figure 7-24. Excessive piston-to-bore clearance can cause seal nibble and fluid leaks.

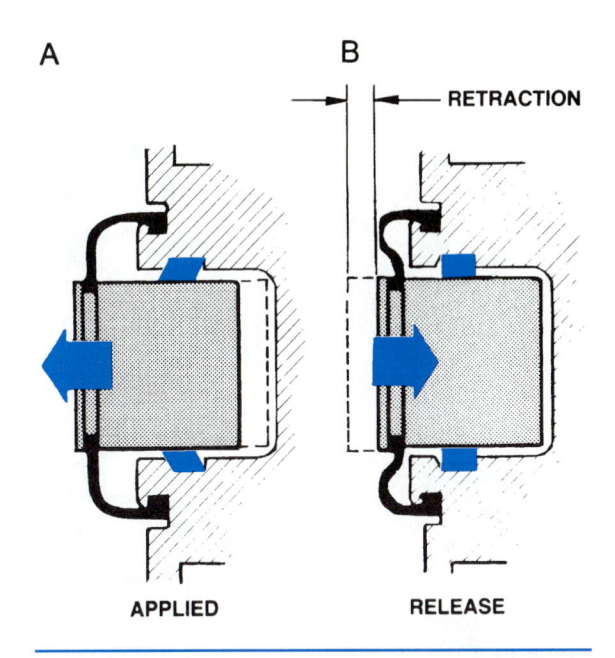

Figure 7-25. Low-drag calipers use special seal groove shapes to provide greater piston retraction.

of piston movement required to apply the brake pads becomes greater than the seal's ability to flex, the piston slides through the inside of the seal until the pads are in full contact with the rotor. When the brakes are released, the seal again retracts the piston but only within the limits of its flexibility; any travel of the piston through the seal remains.

One of the reasons fixed-seal calipers have tighter piston-to-caliper-bore clearances is that excessive clearance can allow a lathe-cut seal to flex so far that its inner edge rolls under and its outer edge is pinched between the piston and bore, figure 7-24. This overflexing causes rapid wear, called seal "nibble," on the face and edges of the seal. Wear in these locations reduces the sealing action of the seal and leads to brake fluid leaks.

Excessive piston-to-bore clearance can also affect the self-adjusting process. If there is too much clearance when the piston must move outward to compensate for pad wear, the seal will simply flex rather than allow the piston to slide through its center. This will cause increased brake pedal travel.

Low-Drag Caliper Seals

Chapter 5 discussed the quick-take-up master cylinders used with low-drag brake calipers. Low-drag calipers have less friction when the brakes are not applied, and this means better fuel economy. To reduce friction, low-drag calipers retract the caliper pistons, and therefore the brake pads, farther away from the rotor.

The extra piston retraction of a low-drag caliper is obtained by modifying the shape of the groove that holds the caliper piston seal. The groove in a conventional caliper is cut to match the shape of the seal, but in a low-drag caliper, the seal is installed in a groove with a beveled outer edge. The bevel allows the seal to deflect farther outward when the brakes are applied, figure 7-25A. And this in turn allows the seal to retract the piston farther inward when the brakes are released, figure 7-25B. Compare this to figure 7-23.

BRAKE CALIPER DUST BOOTS

Although wheel cylinders are shrouded and protected by the brake drums, brake calipers are fully exposed to the elements. Dust boots are used on all brake calipers to prevent moisture, dirt, and other contaminants from entering the caliper bore. If allowed in, these substances will cause physical damage to the caliper body, pistons, and seals; create brake fluid leaks; and contaminate the brake fluid. Brake calipers generally use one dust boot for each piston. In all cases, the opening in the center of the boot fits tightly around the end of the caliper piston. However, the outer edge of the boot that seals against the caliper body can be designed in several different ways.

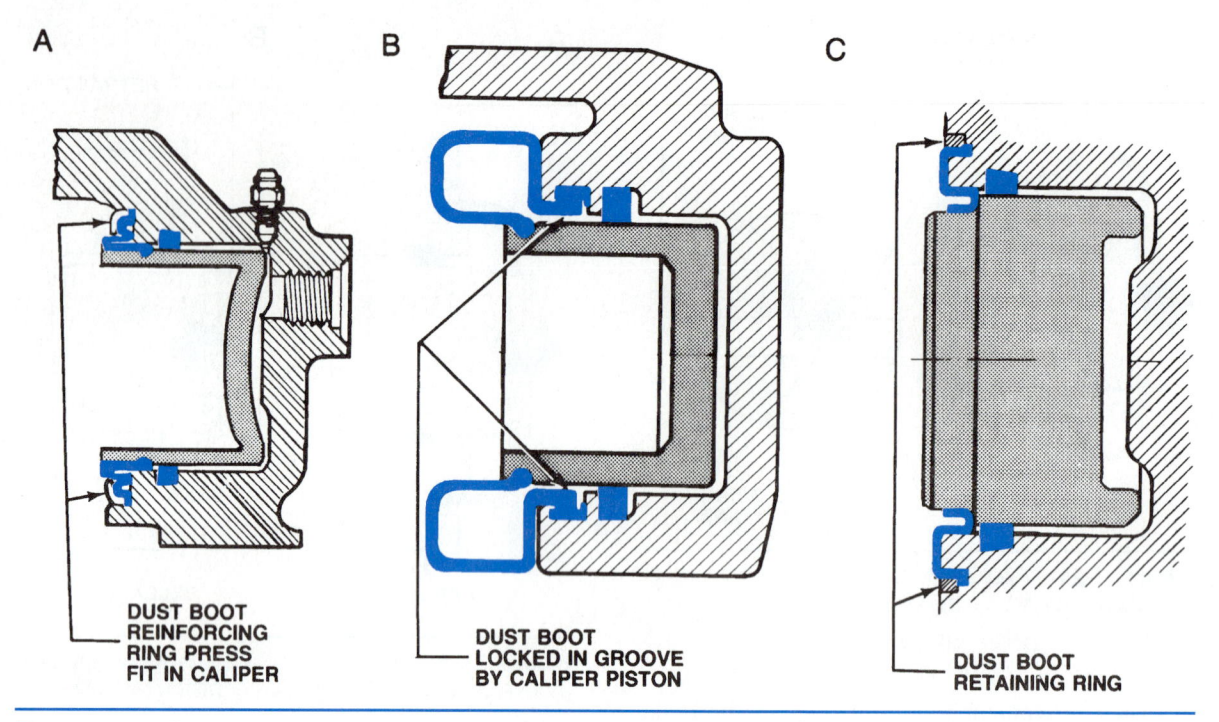

Figure 7-26. Brake caliper dust boots are held in place by a variety of means.

Some dust boots have a metal reinforcing ring around their outer edges, figure 7-26A. The ring is pressed into a groove cut in the caliper body after the piston is installed in the bore. On other calipers, the edge of the dust boot fits into a machined groove in the caliper bore similar to the one that holds the piston seal, figure 7-26B. When the piston is installed in the bore, it locks the boot in place. Other boots fit into a groove in the face of the caliper body and are then held in place by a separate metal ring installed on top of them, figure 7-26C.

BRAKE CALIPER OPERATION

Brake caliper operation is similar to wheel cylinder operation with the exception that the pistons are directing their force inward toward the brake rotor rather than outward toward the brake drum. When the brakes are applied, hydraulic pressure is routed from the master cylinder through the brake lines to the brake calipers. At the calipers, the pressure is routed through internal passages or external lines to the caliper bores where it acts on the backsides of the caliper pistons. The pistons

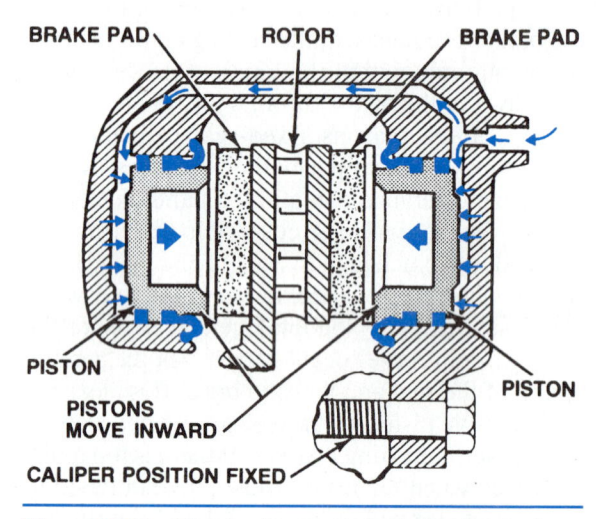

Figure 7-27. Fixed calipers apply the brakes by forcing pistons toward the rotor from both sides.

then move outward in their bores and force the brake pads against both sides of the brake rotor to create the friction needed to stop the car. When the brakes are released, rotor runout and piston seal deflection retract the piston from the pads and rotor.

With a fixed caliper, figure 7-27, the caliper body remains stationary and the pistons act to apply the pads with equal force from both sides of the

Wheel Cylinders and Brake Caliper Hydraulics

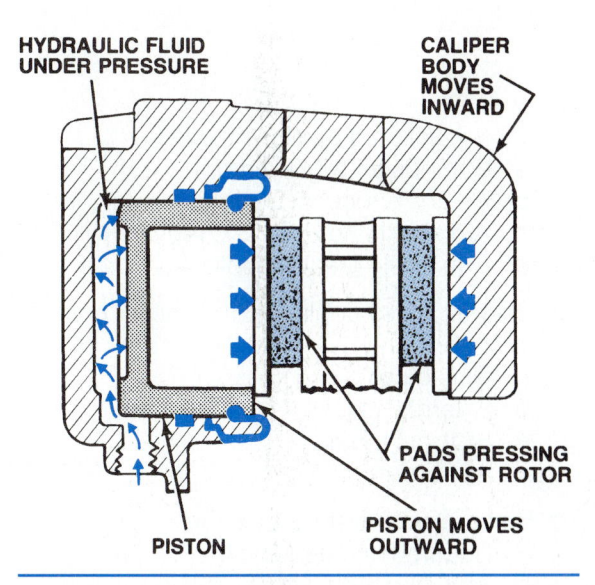

Figure 7-28. Sliding and floating calipers apply the brakes by forcing the piston in one direction and the caliper body in the other.

rotor. With a sliding or floating caliper that has pistons on only one side of the rotor, figure 7-28, the pistons first move the inboard pad into contact with the rotor. Then, because neither the inboard pad nor piston can move further, the caliper body slides or floats on its anchor plate pulling the pad on the outboard side of the caliper into contact with the rotor.

SUMMARY

The wheel cylinders and brake calipers are the last components in the brake hydraulic system. They both convert hydraulic pressure into mechanical force that is used to apply the wheel friction assemblies.

A wheel cylinder actuates a drum brake by forcing the brake shoe linings against the brake drum. There are three types of wheel cylinders now in use: single-piston, dual-piston with a straight bore, and dual-piston with a stepped bore. Dual-piston cylinders with straight bores are the most common. Wheel cylinder pistons are fitted with cup-type seals that prevent fluid leakage. To keep air from bypassing the seals and entering the hydraulic system when the brakes are released, residual pressure check valves or mechanical cup expanders are fitted to the brake system.

A disc brake caliper actuates a disc brake by forcing the brake pads against the brake rotor. Brake caliper bodies are usually made of cast iron in one or two pieces and contain from one to four pistons. Depending on how the caliper is mounted, it may be a fixed, sliding, or floating design. The body of a fixed caliper locates pistons on both sides of the rotor and does not move when the brakes are applied. Sliding and floating calipers usually have a single piston on the inboard side of the caliper, and the caliper body is free to move on an anchor plate to actuate the outboard brake pad.

Brake caliper pistons are made of four materials: aluminum, cast iron, steel, and phenolic plastic. Steel pistons have been the most common type for a long time, but phenolic pistons are used on many newer applications because they insulate fluid in the calipers from the heat of braking to help prevent vapor lock.

Caliper pistons can be sealed to the bore with two different kinds of seals. Many older caliper designs use stroking seals, a form of lip seal that mounts on the piston and moves with it. Stroking seals expose the caliper sealing surfaces to the elements if the dust boot becomes damaged and do not help to locate the piston in the bore. Most newer calipers use fixed, lathe-cut, O-ring seals that fit in a groove near the outer edge of the caliper bore. This design protects the piston sealing surfaces, and the deflection of the seal helps retract the piston from the pads and rotors when the brakes are released.

Review Questions

Choose the letter that represents the best possible answer to the following questions:

1. Both wheel cylinders and brake calipers convert:
   a. Hydraulic force into mechanical pressure
   b. Mechanical pressure into kinetic force
   c. Caliper force into hydraulic pressure
   d. Hydraulic pressure into mechanical force

2. Which of the following is *not* a part of a wheel cylinder?
   a. Dust boot
   b. Piston
   c. Anchor plate
   d. Bleeder screw

3. Wheel cylinder pistons may be fitted with _____ seals.
   a. O-Ring
   b. Cup
   c. Lip
   d. All of the above

4. The spring in a wheel cylinder is used to:
   a. Keep brake pedal travel to a minimum
   b. Maintain a good seal
   c. Prevent pistons from blocking the fluid inlet
   d. All of the above

5. The dust boots on wheel cylinders are used to:
   a. Keep brake fluid in
   b. Keep water out
   c. Both a and b
   d. Neither a nor b

6. Which of the following is *not* a type of wheel cylinder?
   a. Straight-bore dual-piston
   b. Stepped-bore single-piston
   c. Straight-bore single-piston
   d. Stepped-bore dual-piston

7. Split brake calipers are assembled with:
   a. Anchor bolts
   b. Retaining bolts
   c. Security bolts
   d. None of the above

8. Single-piston calipers are used on most late-model cars because they:
   a. Can absorb a great deal of heat
   b. Are the strongest design
   c. Are simpler to build and repair
   d. Can apply greater application force

9. Brake fluid is supplied to the caliper bores through:
   a. Drilled holes
   b. External hoses
   c. External tubes
   d. All of the above

10. Brake caliper pistons must be:
    a. Resistant to corrosion
    b. Able to conduct heat
    c. Both a and b
    d. Neither a nor b

11. The caliper piston material that conducts the most heat is:
    a. Aluminum
    b. Cast iron
    c. Steel
    d. Phenolic plastic

12. Which of the following is *not* true of phenolic caliper pistons.
    a. They provide better retraction
    b. They have superior corrosion resistance
    c. They can be used to replace steel pistons
    d. They shrink as their temperature increases

13. Stroking seals are:
    a. Used only on floating calipers
    b. Typically O-ring seals
    c. Installed on the caliper piston
    d. All of the above

14. A piston-to-bore clearance of .005" to .010" (.13 to .25 mm) would be acceptable for a:
    a. Phenolic piston
    b. Steel piston with a stroking seal
    c. Both a and b
    d. Neither a nor b

15. Piston return springs are installed with stroking seals to prevent:
    a. Knockback
    b. Runout
    c. Nibble
    d. Air leaks

Wheel Cylinders and Brake Caliper Hydraulics

16. A fixed caliper seal uses the _____ as its sealing surface.
    a. Caliper bore
    b. Retaining ring
    c. Piston
    d. Dust boot

17. A seal groove with a tapered outer edge is used in _____ calipers.
    a. Stroking seal
    b. Quick-take-up
    c. European
    d. Lathe-cut

18. The flexing of a fixed seal is used to provide:
    a. Brake adjustment
    b. Piston retraction
    c. Both a and b
    d. Neither a nor b

19. A _____ caliper applies force to both sides of the rotor.
    a. Fixed
    b. Floating
    c. Sliding
    d. All of the above

8
Drum Brake Friction Assemblies

OBJECTIVES

Upon completion and review of this chapter, you will be able to:

- Explain the advantages and disadvantages of drum brakes.
- Describe the four types of brake fade.
- List the parts of a drum brake and describe their function.
- Identify and explain the difference between leading and trailing shoes.
- Describe the operation of non-servo brakes.
- Describe the operation of servo brakes.
- Explain self-energizing action.
- Explain servo action.
- Describe the operation of dual-servo brakes.
- Describe the operation of uni-servo brakes.
- Identify and describe the operation of the four types of manual adjusters.
- Identify and describe the operation of the two types of automatic adjusters.

KEY TERMS

automatic adjusters
brake adjuster
brake fade
double-leading brake
double-trailing brake
dual-servo brake
gas fade
leading shoe
leading-trailing brake
lining fade
manual adjusters
mechanical fade
nondirectional brake
non-servo brake
over-travel spring
pawl
primary shoe
secondary shoe
self-energizing action
servo brake
trailing shoe
uni-servo brake
water fade

DRUM BRAKE ADVANTAGES

The drum brake, figure 8-1, has been more widely used than any other automotive brake design. At one time, its use in the industry was universal; all cars had four-wheel drum brakes. Although the disc brake has proven its superiority in extreme braking conditions and has replaced the drum brake on the front axle of new cars, the drum brake continues to have a number of advantages that contribute to its widespread use on the rear axle of most vehicles.

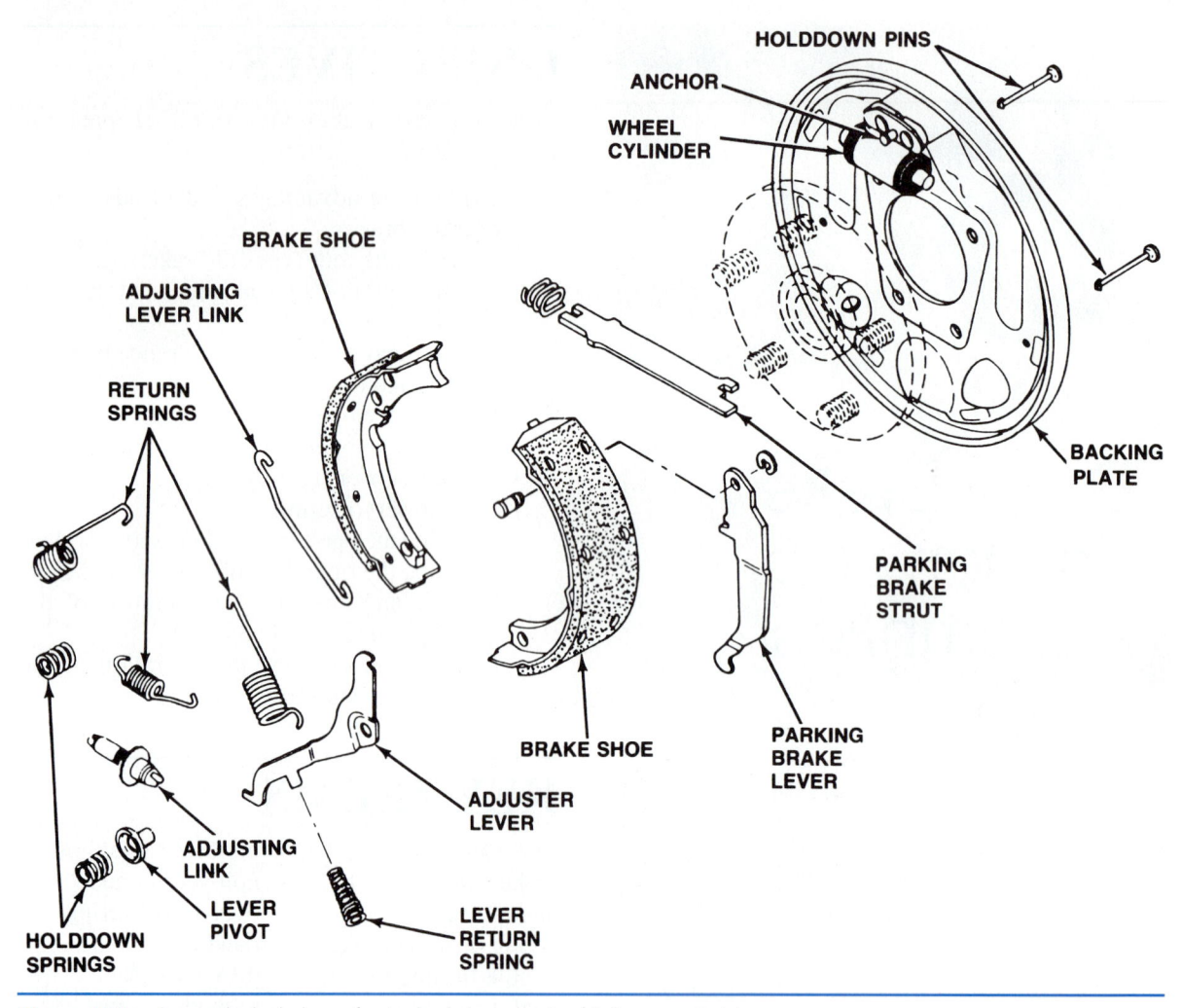

Figure 8-1. An exploded view of a typical drum brake friction assembly.

Self-Energizing and Servo Action

The primary advantage of drum brakes is that they can apply more stopping power for a given amount of force applied to the brake pedal than can disc brakes. This is possible because the drum brake design offers a self-energizing action that helps force the brake linings tightly against the drum. In addition, some drum brake designs use an effect called servo action that enables one brake shoe to help apply the other for increased stopping power. Both self-energizing and servo action are explained in detail later in the chapter.

Parking Brake Service

One significant advantage that results from the superior braking power of drum brakes at low-application forces is that they make excellent parking brakes. A simple linkage fitted to the brake assembly allows relatively low effort on the driver's part to hold a heavy car in place when parked. Disc brakes, which do not benefit from self-energizing or servo action, require a complex set of extra parts to provide enough application force to work well as parking brakes.

DRUM BRAKE DISADVANTAGES

Despite the advantages described previously, the disadvantages of drum brakes (at least for front-axle applications) became significant as cars

Drum Brake Friction Assemblies

grew heavier and highway speeds increased. These disadvantages fall into four areas:

- Brake fade
- Brake adjustment
- Brake pull
- Complexity.

Brake Fade

The greatest drawback of drum brakes is that they are susceptible to fade. **Brake fade** is the loss of stopping power that occurs when excessive heat reduces the friction between the brake shoe linings and the drum. Heat-related fade was introduced in Chapter 3 and takes three forms: mechanical, lining, and gas fade. Drum brakes can also fade if water enters the friction assembly and reduces friction between the linings and drum.

Mechanical Fade

Drum brakes are not very efficient at dissipating heat. The brake drum shrouds the linings, and most of the heat produced when braking must pass through the drum, from the inside out, before it can be carried away by the passing airflow. **Mechanical fade** occurs when the brake drum gets so hot it expands away from the brake linings. The brake shoes then move outward to maintain contact with the drum, causing the brake pedal to drop toward the floor as additional brake fluid moves into the hydraulic system.

The driver can compensate for mechanical fade to a limited degree by pumping the brake pedal to keep the linings in contact with the drum. However, this also results in greater heat and more fade. At the first sign of mechanical fade, a car's speed should be reduced and the brakes allowed to cool, otherwise, a total loss of stopping power may result.

Lining Fade

Lining fade occurs when the friction coefficient of the brake lining material drops off sharply because intense heat makes it "slippery." Unlike mechanical fade, brake pedal travel does not increase when lining fade occurs. Instead, the pedal becomes hard and there is a noticeable loss of braking power. Sometimes the driver can apply more pressure on the brake pedal to regain the lost braking power. However, this often worsens the problem by increasing the heat and causing brake drum distortion that makes it impossible for the linings and drum to stay in complete contact. The portions of the lining that do remain in contact with the drum become extremely hot and mechanical fade of the drum soon results.

Gas Fade

A relatively rare type of brake fade is called **gas fade**. Under extended hard braking from high speeds, a thin layer of hot gases and dust particles can build up between the brake shoe linings and drum. The gas layer acts as a lubricant and reduces friction. As with lining fade, the brake pedal becomes hard, and greater application force is required to maintain a constant level of stopping power.

Gas fade becomes more of a problem as the size of the brake lining increases; gases and particles have a harder time escaping from under a large lining than a small one. Some high-performance drum brake linings are slotted or grooved to provide paths for gas and particles to escape.

Water Fade

Drum brakes are also affected by a problem called **water fade**. A drum brake friction assembly cannot be made waterproof because clearance is necessary between the rotating drum and the fixed backing plate. This clearance allows a small amount of air circulation that helps combat heat fade, but it can also allow water to enter the friction assembly. Water fade occurs when moisture is trapped between the shoes and drum where it acts as a lubricant. This lowers braking efficiency until friction creates enough heat to evaporate the water.

Technically, water fade is not a true form of fade because the brakes do not start out operating at full power then fade into ineffectiveness. Instead, the opposite happens; water fade causes a delay in braking action when the brake pedal is first applied, after which the brakes gradually return to full power. Depending on the amount of water that has entered the brakes, this delay can last up to several seconds.

Adjusting Mechanism

Another disadvantage of the drum brake design is its need for an adjusting mechanism. As the brake shoe lining material wears, the clearance between the linings and drum increases, resulting in longer brake pedal travel. To maintain a high brake pedal, a mechanism must be included in the friction assembly for periodic adjustment of the clearance between the shoe and drum. Early cars had manual brake adjusters that required relatively frequent attention. Most late-model vehicles have automatic adjusters that maintain the proper clearance between the brake linings and drum.

Brake Pull

A further disadvantage of drum brakes is that they sometimes pull the car to one side or the other during braking. Certain designs are more susceptible to this than others, but all drum brakes suffer from it to one degree or another. Brake pull occurs when the friction assemblies on opposite sides of the car have different amounts of stopping power. These differences can be caused by brake fade or misadjustment of the clearance between the brake linings and drum.

Complexity

Finally, when compared to disc brakes, the drum brake assembly is much more complex, both in number of parts and difficulty of assembly. As many technicians know, it is remarkably easy to make mistakes when installing new shoes, along with the various springs, links, clips, and retainers. It is not uncommon for the drum brake assembly to have 30 or more separate parts, whereas the disc brake may have as few as 8 parts.

DRUM BRAKE CONSTRUCTION

Before looking at the different types of drum brake designs, it is important to understand those components that are common to most drum brake friction assemblies. These parts are described in the following sections. Parts that differ from one design to another are detailed in the *Drum Brake Design* section.

Backing Plate

The foundation of every drum brake is the backing plate, figure 8-2, that mounts to the steering knuckle on the front brakes, or to the suspension or axle housing on the rear brakes. The backing plate serves as the mounting surface for all the other friction assembly parts. The backing plate also functions as a dust and water shield to keep

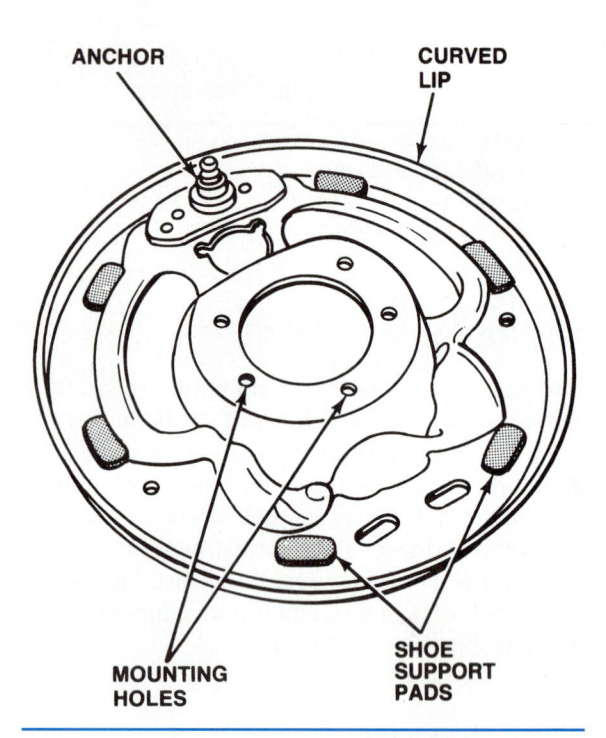

Figure 8-2. The backing plate is the basic building block of every drum brake.

contaminants out of the brake assembly. The edge of the backing plate curves outward to form a lip that strengthens the backing plate and fits inside the brake drum to help prevent water entry. In some cases, the lip fits into a machined groove in the open edge of the brake drum to provide an even better water barrier.

In addition to mounting holes for the various brake parts, the backing plate may also have openings that are used to inspect the wear of the brake linings, or adjust the lining-to-drum clearance. These openings are sometimes sealed with rubber plugs that are removed to make the inspection or adjustment. On brakes using automatic adjusters, the adjustor slots are often sealed with metal plugs that must be knocked out in order to access the star-wheel adjustor. This is necessary when the brakes are badly worn to enable the removal of the brake drum as is explained in the *Shop Manual*. Rubber plugs are available to seal the slots to prevent water entry once the metal plugs have been removed. Other prominent features on backing plates are the shoe anchors, piston stops, and brake shoe support pads.

Shoe Anchors

Shoe anchors prevent the brake shoes from rotating with the drum when the brakes are applied.

Drum Brake Friction Assemblies

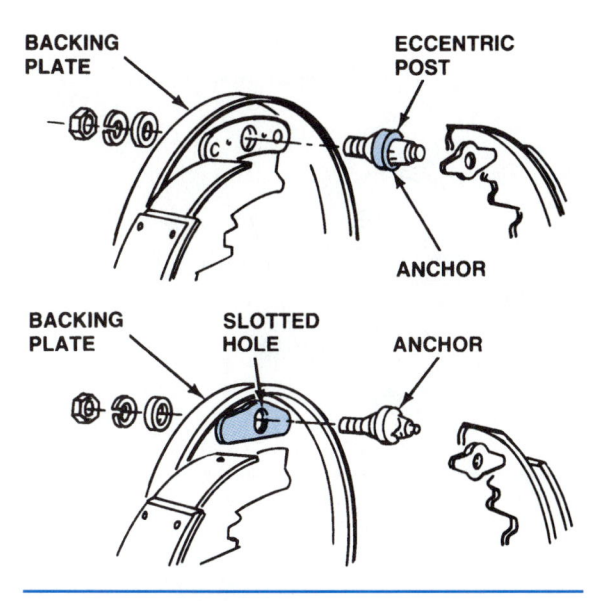

Figure 8-3. Anchor adjusters use either an eccentric post or a slotted backing plate.

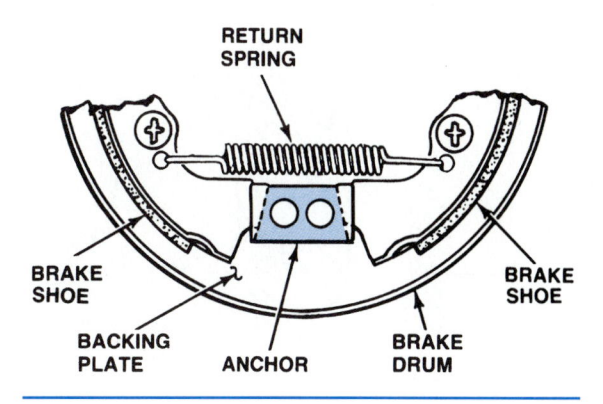

Figure 8-4. A keystone anchor allows the brake shoes to self-center in the drum.

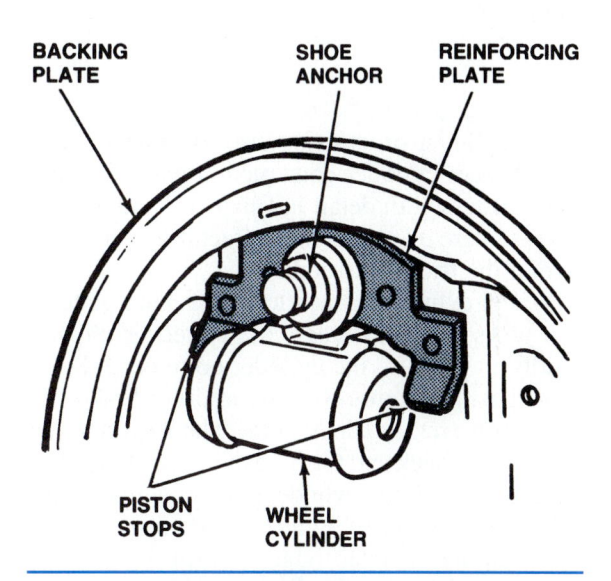

Figure 8-5. Piston stops prevent the wheel cylinder from coming apart unintentionally.

The majority of drum brakes have a single anchor, but some drum brake designs use two or more.

Many anchors are a simple round post that is permanently mounted on the backing plate, figure 8-2. The brake shoes have semicircular cutouts where they contact the anchor, and the anchor positively locates the shoes on the backing plate. Some early brakes used a post-type anchor that could be moved in a slotted mounting hole or had an eccentric shape. Both designs provided adjustment so new brake shoes could be centered within the drum, figure 8-3.

Another type of anchor is the self-centering or keystone anchor, figure 8-4. The ends of the brake shoes that contact this type of anchor are flat or slightly rounded. When the brakes are applied, the shoes slide up or down along the anchor to center themselves for best contact with the brake drum.

Piston Stops

Some backing plates incorporate piston stops that prevent the wheel cylinder pistons from coming out of their bores when the friction assembly is disassembled for servicing. The stops may be part of a reinforcing plate positioned under the anchor, figure 8-5, or they can be stamped directly into the shape of the backing plate itself. When piston stops are used, the wheel cylinder must be removed from the backing plate before it can be taken apart for servicing.

Shoe Support Pads

The shoe support pads, figure 8-2, are stamped into the backing plate and contact the edges of the brake shoes to keep the linings properly aligned with the center of the friction surface inside the brake drum. The support pads are lightly coated with special high-temperature silicone brake grease to minimize wear, prevent rust, and eliminate the squeaking that can occur when the shoes move slightly on the pads during a stop.

Wheel Cylinders

Wheel cylinders convert hydraulic pressure into the mechanical force required to apply the brake linings against the brake drum. The vast majority of current drum brakes use either a single two-piston

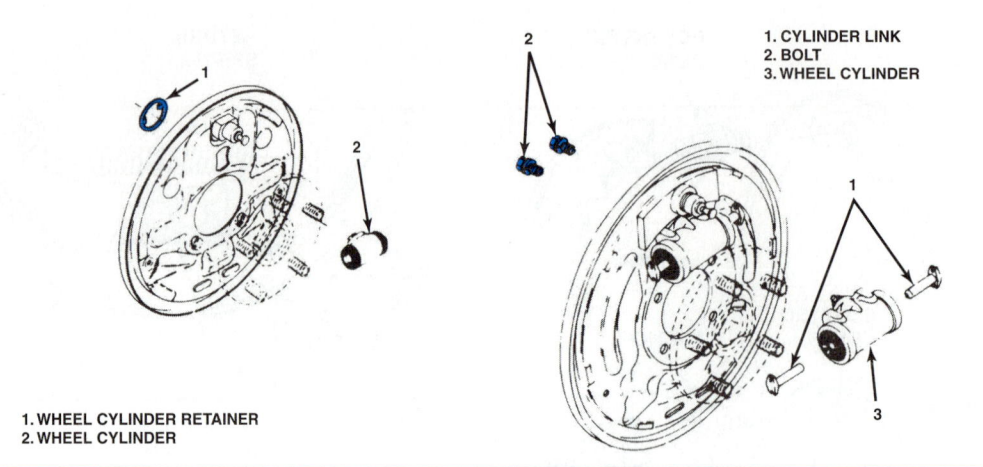

Figure 8-6. The wheel cylinders may be bolted on or held in place with a spring clip. (Courtesy of General Motors Corporation, Service and Parts Operations)

wheel cylinder or two single-piston wheel cylinders. Wheel cylinder construction and operation were discussed in detail in Chapter 7 and are not covered here except as they affect the overall design of the friction assembly.

Wheel cylinders attach to the backing plate in two ways. Most cylinders are bolted or clipped solidly in place so the cylinder body is held in a fixed position figure 8-6. Some drum brake designs, however, require that the wheel cylinder be free to slide back and forth. In this type of installation, figure 8-7, the cylinder is mounted in a slotted hole and retained by one or more spring clips that hold the cylinder tightly against the backing plate while still allowing limited movement.

Brake Shoes

The brake shoes, figure 8-1, are made from welded steel or cast aluminum. The outer portion of the shoe, called the shoe table, is curved to match the contour of the drum; the brake lining friction material is riveted or bonded to this surface. The inner portion of the shoe, called the web, is contacted by the wheel cylinder to apply the brakes. The web contains mounting holes for the shoe return springs, holddown devices, and the linkages for the parking brake and self-adjusting mechanisms. Brake shoes are covered in greater detail in Chapter 10.

Brake Shoe Return Springs

The brake shoe return springs, figure 8-1, retract the shoes to their unapplied positions when the

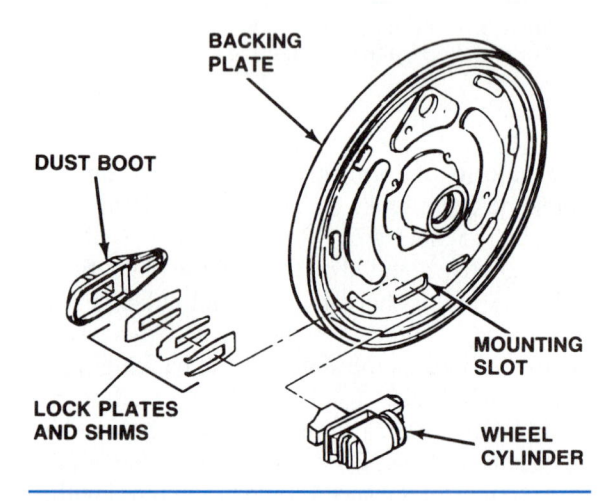

Figure 8-7. This type of mounting allows the wheel cylinder to slide on the backing plate.

brake pedal is released. This helps prevent brake drag and aids the return of brake fluid to the master cylinder reservoir. Most brakes use closed-coil return springs to retract the brake shoes. The coils on these springs are very tightly wound and contact one another when the spring is relaxed. Some imported cars have a single, large, horseshoe-shaped return spring, figure 8-8.

The type, location, and number of return springs varies from one brake design to the next; however, all springs are installed in one of two ways. Some connect directly from shoe to shoe, whereas others connect from one shoe to a fixed support on the backing plate, often the anchor post.

Drum Brake Friction Assemblies

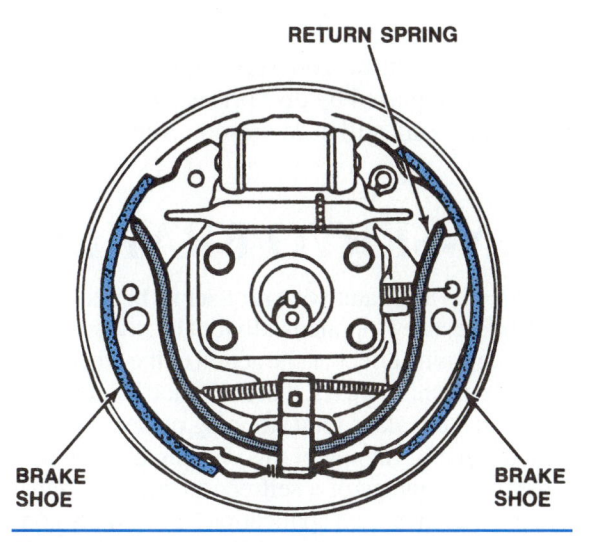

Figure 8-8. A single spring-steel return spring is used on some drum brakes.

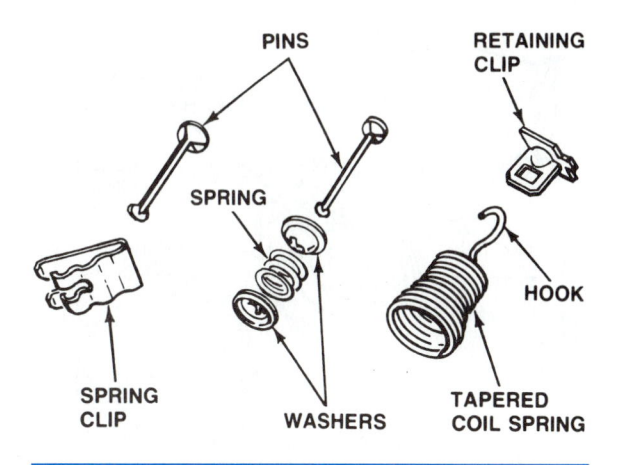

Figure 8-9. Brake shoe holddowns keep the shoes properly positioned against the backing plate.

Brake Shoe Holddowns

Whereas the return springs retract the brake shoes to their unapplied positions, the shoe holddowns keep the shoes securely against the support pads on the backing plate. The holddowns prevent noise, vibration, and wear but still allow the shoes to move out and back as the brakes are applied and released. The holddowns also provide enough freedom of movement to allow adjustment of the shoes outward as the linings wear.

Shoe holddowns take many forms, figure 8-9. The most common design is a steel pin installed through a hole in the backing plate and a corresponding hole in the brake shoe web. A spring fits over the end of the pin against the shoe web, and a special washer compresses the spring and locks onto the flattened end of the pin. Some versions of this holddown use an additional washer between the spring and the shoe web.

A variation of the pin-type holddown uses a simple spring clip between the brake shoe web and the end of the pin. A few brakes carry this design one step further and have a large spring clip mounted directly to the backing plate. When the brake shoe is installed, the web is slipped under the clip, which then holds the shoe against the support pads.

Another type of holddown is a taper-wound coil spring with a hook formed on its end. Because of its shape, this part is sometimes called a "beehive" holddown. The hooked end of the spring is installed through a hole in the brake shoe web and attached into a retaining clip that fits into a corresponding hole in the backing plate.

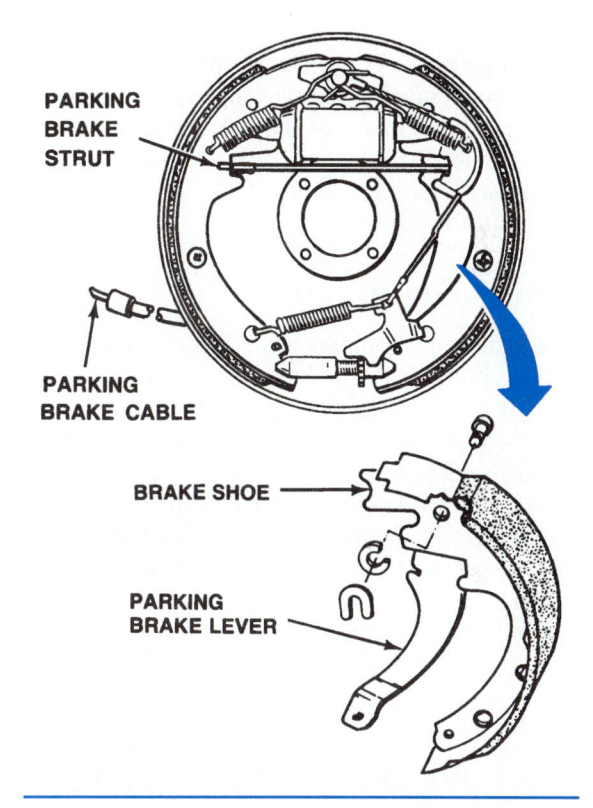

Figure 8-10. A mechanical parking brake linkage is part of most rear drum brakes.

Parking Brake Linkage

Most rear drum brake friction assemblies include a parking brake linkage, figure 8-10. The linkage commonly consists of a cable, lever, and strut system

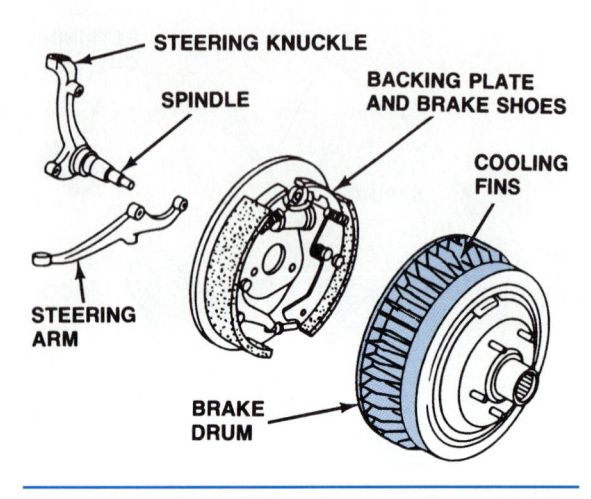

Figure 8-11. The brake drum covers the rest of the friction assembly and rotates on the spindle or axle.

that spreads the brake shoes apart to apply the brake mechanically. The parking brake strut plays a large part in many of the automatic brake adjusters described later in this chapter. Parking brakes are examined in detail in Chapter 12.

Brake Drum

The last major component in a drum brake friction assembly is the brake drum, figure 8-11. Unlike all of the other parts discussed thus far, the brake drum is not connected to the backing plate, but turns with the wheel. The drum mounts on the hub or axle, and covers the rest of the friction assembly. Brake drums are made of cast iron, cast iron and stamped steel, or cast aluminum with a cast-iron liner. Any of these drum types may have ribs or fins on its outer edge to help dissipate heat. Brake drums are examined in detail in Chapter 11.

DRUM BRAKE DESIGN

Not all drum brakes apply the shoes in the same manner. In fact, drum brake designs are classified by the way in which the shoes are applied, and how they react when the linings make contact with the drum. All drum brakes fall into two basic categories:

- Non-servo brakes
- Servo brakes.

Early automotive drum brake friction assemblies were non-servo designs, and non-servo brakes are still used today in certain rear-wheel applications. The more powerful servo drum brakes were developed later and used extensively on both the front and rear axles. Front servo drum brakes were eventually replaced by disc brakes, but rear servo drum brakes are still used on some vehicles.

NON-SERVO BRAKES

The identifying feature of a **non-servo brake** is that each brake shoe is applied individually; the action of one shoe has no effect on the action of the other. This does not mean, however, that each shoe has an equal effect on the brake's total stopping power. As mentioned at the beginning of this chapter, drum brakes have the advantage of a self-energizing action that can provide increased application force. Many non-servo drum brakes use this **self-energizing action** to improve their braking performance.

Self-Energizing Action

The simple drum brake assembly in figure 8-12 shows how the self-energizing process works. As the forward or **leading shoe** contacts the drum, the drum attempts to rotate the shoe along with it. However, the shoe cannot rotate because its far end (relative to drum rotation) is fixed in place by an anchor. As a result, drum rotation *energizes* the shoe by forcing it outward and wedging it tightly against the brake drum.

The drum also attempts to rotate the reverse or **trailing shoe** as it contacts the drum. However, in this case, the far end of the shoe (relative to drum rotation) is not solidly anchored. As a result, drum rotation *de-energizes* the shoe by forcing it inward away from the brake drum.

When this type of brake is applied with the car backing up, the roles of the forward and reverse shoes are switched; the reverse shoe becomes the leading shoe, which is self-energized by drum rotation, whereas the forward shoe becomes the trailing shoe, which is de-energized. A leading shoe is always energized by drum rotation; a trailing shoe is always de-energized by drum rotation.

To identify the leading shoe on a non-servo brake with only one wheel cylinder, point to the wheel cylinder then move your finger in the direction of drum rotation; the first shoe encountered is the leading shoe, the other shoe is the trailing shoe. With multiple wheel cylinder non-servo brakes, shoe identification is slightly more complicated. If the piston of a wheel cylinder moves in the same direction as drum rotation when the brakes are ap-

Drum Brake Friction Assemblies

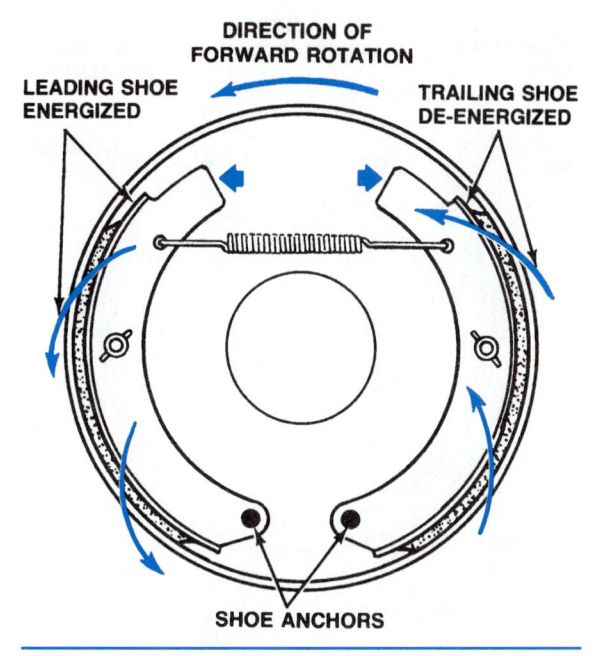

Figure 8-12. Self-energizing action can increase or decrease the stopping power of a brake shoe.

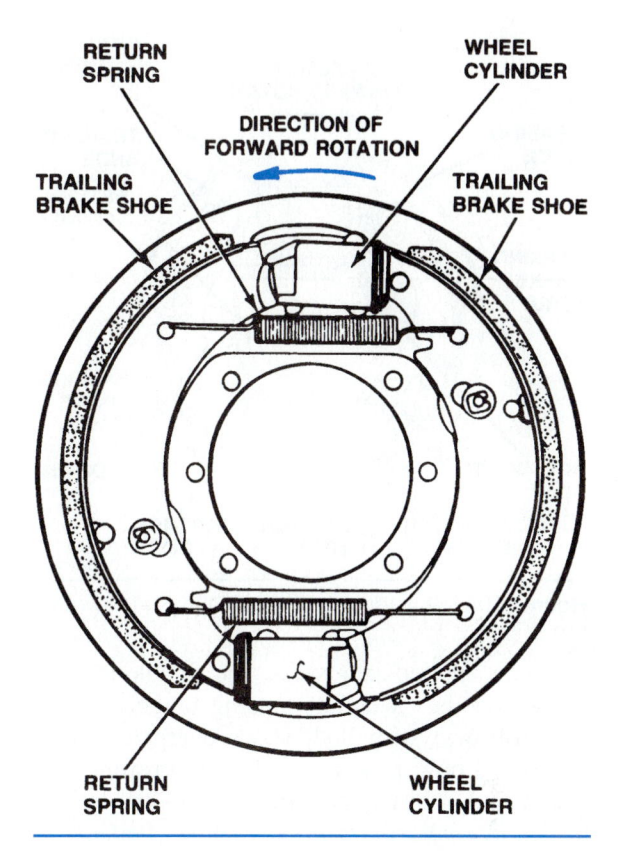

Figure 8-13. A double-trailing non-servo drum brake.

plied, the shoe it actuates is a leading shoe. If the piston moves opposite the direction of drum rotation, the shoe actuated is a trailing shoe.

Leading shoes generally wear at a faster rate than trailing shoes because they are applied with greater force. Where a brake uses one leading and one trailing shoe, like the friction assembly described previously, the leading shoe will sometimes have a thicker lining or one with a larger surface area than that of the trailing shoe. The thicker or larger lining balances the wear between the two shoes so that they will both need replacement at about the same time.

Specific Non-Servo Brakes

Some non-servo brake designs take full advantage of the self-energizing ability of drum brakes, others make only partial use of that ability, and still others use no self-energization at all. There are basically four different non-servo brake designs:

- Double-trailing
- Leading-trailing
- Double-leading
- Nondirectional.

These designs differ in the total amount of braking power they can provide, and also in how effective they are at stopping the vehicle in both the forward and reverse directions. The sections below describe the four types of non-servo brakes, beginning with the least powerful type and progressing to the design with the most stopping power.

Double-Trailing Brake

The least powerful non-servo drum brake is the **double-trailing brake**, figure 8-13. This design has two trailing shoes and does not use any self-energization. Both shoes have identically sized and shaped linings that are applied with equal force by a pair of single-piston wheel cylinders. Each shoe is anchored at the end opposite the wheel cylinder that applies it. In many double-trailing brakes, the backside of one wheel cylinder serves as the anchor for the brake shoe actuated by the other wheel cylinder.

The double-trailing brake is usually found only on the rear axles of vehicles with an extreme forward weight bias. In these applications its relative lack of stopping power aids brake balance and helps prevent rear lockup. Unfortunately, it also makes the double-trailing design a poor parking brake in the forward direction.

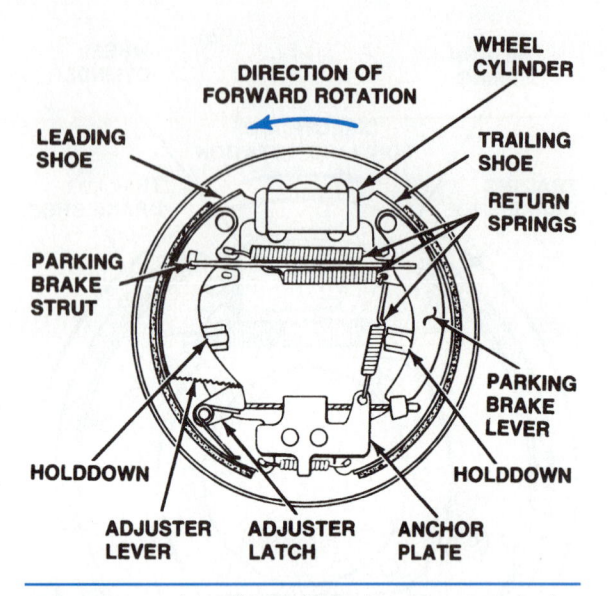

Figure 8-14. A leading-trailing non-servo drum brake.

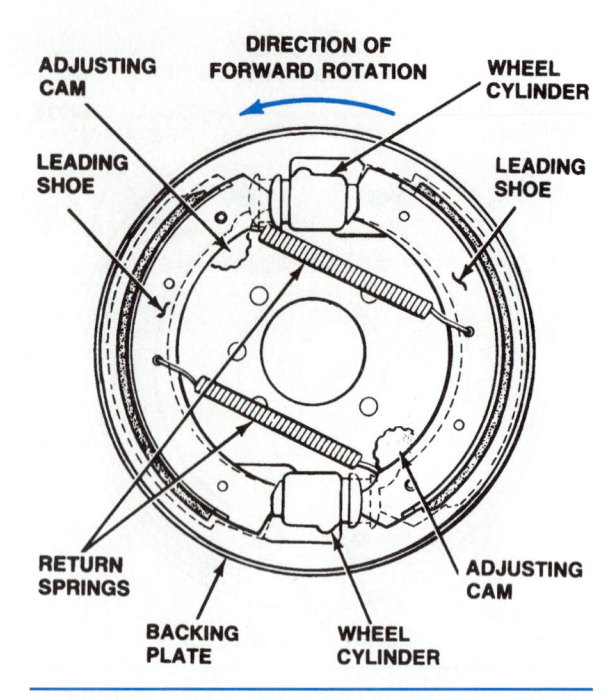

Figure 8-15. A double-leading non-servo drum brake.

When applied in reverse, the double-trailing brake becomes a double-leading brake (discussion follows) with both shoes energized. This makes it a powerful parking brake in reverse but also makes it somewhat prone to lock the lightly loaded rear wheels while slowing in reverse.

Leading-Trailing Brake

The non-servo **leading-trailing brake**, figure 8-14, has one leading shoe and one trailing shoe. Typically, a single, two-piston wheel cylinder is mounted at the top of the backing plate, and the two brake shoes are anchored at the bottom of the backing plate.

The operation of the leading-trailing brake was explained earlier in the description of self-energizing action. In essence, this brake design has one energized and one de-energized shoe regardless of whether it is applied while the vehicle is traveling forward or in reverse. This allows the leading-trailing brake to work equally well in either direction.

Leading-trailing brakes are popular on the rear wheels of many small and FWD cars because, although they are not as powerful as a double-leading or servo brake, they are also less prone to lockup. They have the further benefit of making good parking brakes in both directions.

Double-Leading Brake

The non-servo **double-leading brake**, figure 8-15, takes advantage of self-energizing action on both brake shoes when it is applied in the forward direction. As with the double-trailing design, the brake linings on both shoes are identical and are applied with equal force. The shoes are actuated by a pair of single-piston wheel cylinders, and each shoe is anchored at the end opposite the wheel cylinder that applies it.

The double-leading brake makes a powerful front brake, although not quite as powerful as the servo designs described later. However, it is also not as prone to brake pull or locking. The greatest drawback of the double-leading brake is that it acts like a double-trailing brake in the reverse direction, so it makes a poor parking brake when the vehicle is facing uphill.

Nondirectional Brake

The most powerful non-servo brake design is the **nondirectional brake**, figure 8-16, used on the rear axles of some trucks. A nondirectional brake uses a pair of two-piston wheel cylinders and special brake shoe mounts that allow the shoes to be actuated or anchored at either end. This allows the nondirectional brake to operate like a double-leading brake in both the forward and reverse directions. Although the nondirectional brake is an effective design, its increased cost and complexity compared to a servo brake of similar stopping power make it impractical for automotive applications.

Drum Brake Friction Assemblies

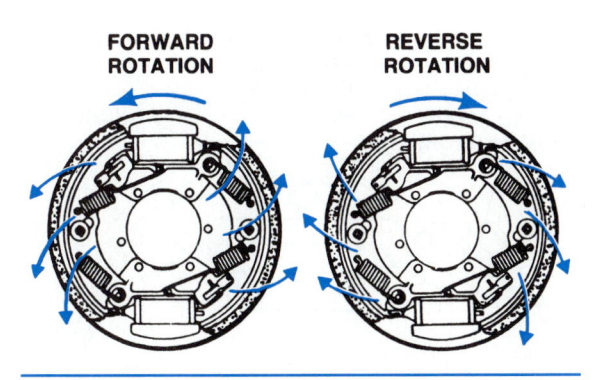

Figure 8-16. A nondirectional non-servo drum brake.

DUAL-SERVO BRAKES

The **servo brake** is the most common drum brake design. It gets its name from the fact that one shoe "serves" the other to increase application force. One version of this brake, the uni-servo design, is used primarily on trucks and supplies additional stopping power in the forward direction only. All servo brakes used on automobiles, however, are of the duo- or dual-servo design that works with equal force in both directions.

The primary advantage of the dual-servo brake is that it is more powerful than any of the non-servo designs. Before the development of practical disc brakes, dual-servo brakes were the obvious choice for front brakes, and dual-servo brakes are still used on the rear axle of some vehicles. Another advantage of the dual-servo brake is that it makes a good parking brake. Dual-servo action not only makes the brake very powerful but it also allows the brake to hold equally well in both directions.

The increased stopping power of the dual-servo brake can be a two-edged sword, however. If there are any problems in a dual-servo brake system, the servo action can magnify any imbalance that results. For example, dual-servo brakes are more susceptible to pull than other brake designs, and their greater application force can lead to faster fade under extreme braking conditions. If the brakes get too far out of adjustment, some dual-servo designs will grab and lock the wheels as the brakes are first applied. And when dual-servo brakes are used on the rear axle of a small or FWD car, they must be very carefully engineered to ensure that their great stopping power does not contribute to wheel lockup.

Dual-Servo Brake Construction

The basic **dual-servo brake**, figure 8-17, uses one anchor and a single two-piston wheel cylinder. The anchor is usually mounted at the top of the backing plate with the wheel cylinder directly beneath it. The tops of the brake shoes are held against the anchor by individual return springs. The bottoms of the shoes are spaced apart by an adjusting link held in position by a third return spring that connects the two shoes.

■ The Creature That Wore Three Shoes

In the pre-disc–brake era some fairly exotic drum brake designs were built in the quest for better braking. An Alfa Romeo design of the 1950s used three single-piston wheel cylinders and three leading shoes. Although this brake did not have any servo action, all three shoes were self-energized. The lack of servo action was not missed because high pedal pressures were not a problem in stopping the small, light Alfa cars of the day.

Compared to servo brakes, the triple-leading-shoe design was less likely to fade or pull. It also had a more linear feel, making it easy for the driver to use the brakes very hard without locking a wheel—just the thing for a quick trip through the Alps or a quick time around a race track. On the debit side, the brake design was heavy, and adjusting it was more complex than most brakes. Still, the triple-leading-shoe brake served Alfa well until effective and reliable disc brakes became available.

Adjusting Link

The adjusting link, figure 8-18, consists of a starwheel that is part of an adjusting screw, a pivot nut that one end of the adjusting screw threads into, and a socket that rotates freely on the opposite end of the adjusting screw. The outer ends of the pivot nut and socket are notched to fit over the brake shoe webs. Some adjusting links have a steel thrust washer or spring washer installed between the socket and the starwheel. These washers allow easier rotation of the starwheel and help reduce brake squeal.

Adjusting links generally have specific left- or right-hand threads and must be installed on the correct side of the car. If the wrong part is used, the starwheel will not align with the adjusting slot

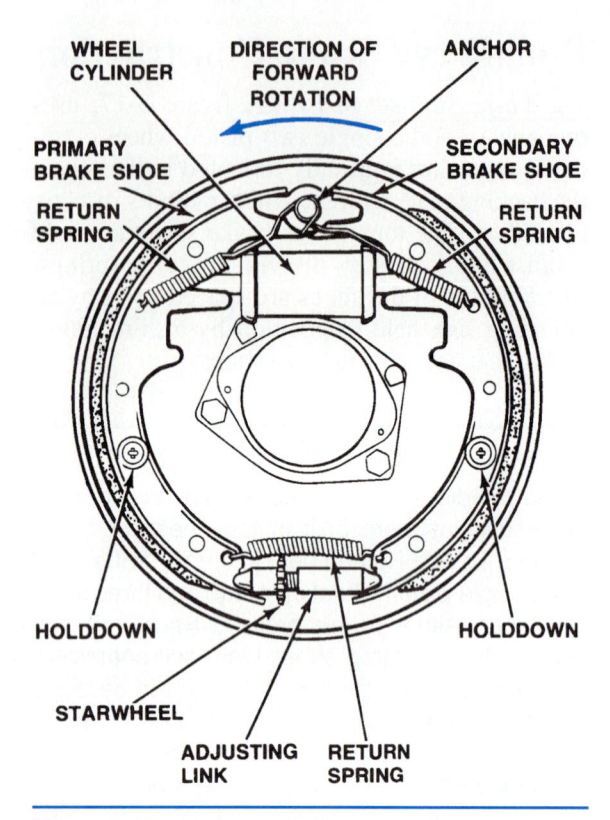

Figure 8-17. A simple dual-servo drum brake.

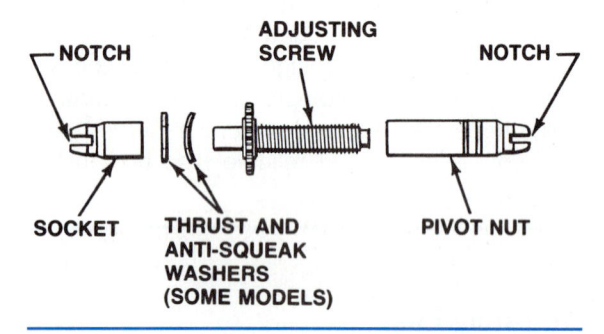

Figure 8-18. A dual-servo brake adjusting link.

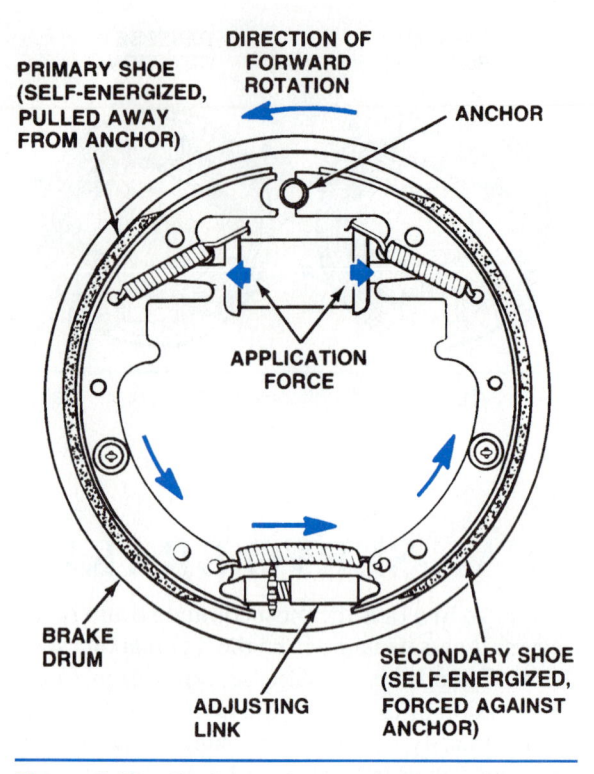

Figure 8-19. Dual-servo brake operation.

in the backing plate, or the automatic adjuster will increase lining-to-drum clearance rather than reduce it.

Primary and Secondary Brake Shoes

Although dual-servo brakes make use of self-energizing action to help provide servo action, the two brake shoes are not called leading and trailing parts as in non-servo brakes. Instead, they are identified as the **primary shoe** and the **secondary shoe**. To identify the primary shoe on a dual-servo brake with a single two-piston wheel cylinder, point to the wheel cylinder then move your finger in the direction of drum rotation. The first shoe reached is the primary shoe; the other shoe is the secondary shoe.

The secondary brake shoe provides approximately 70 percent of the total braking power in a dual-servo brake. For this reason, its lining is usually somewhat larger than that of the primary shoe. In addition, some manufacturers use different types of friction materials on the primary and secondary shoes to help equalize wear.

Dual-Servo Brake Operation

When a dual-servo brake is applied, figure 8-19, the wheel cylinder attempts to force the tops of both brake shoes outward against the drum. As the primary shoe makes contact it rotates with the drum because its far end (relative to the direction of drum rotation) is not directly anchored to the backing plate. As the primary shoe rotates, it forces the adjusting link and the secondary shoe to also rotate until the secondary shoe seats firmly against the anchor.

Although the wheel cylinder attempts to push the top of the secondary shoe outward, the rotational force developed by friction between the brake shoes

Drum Brake Friction Assemblies

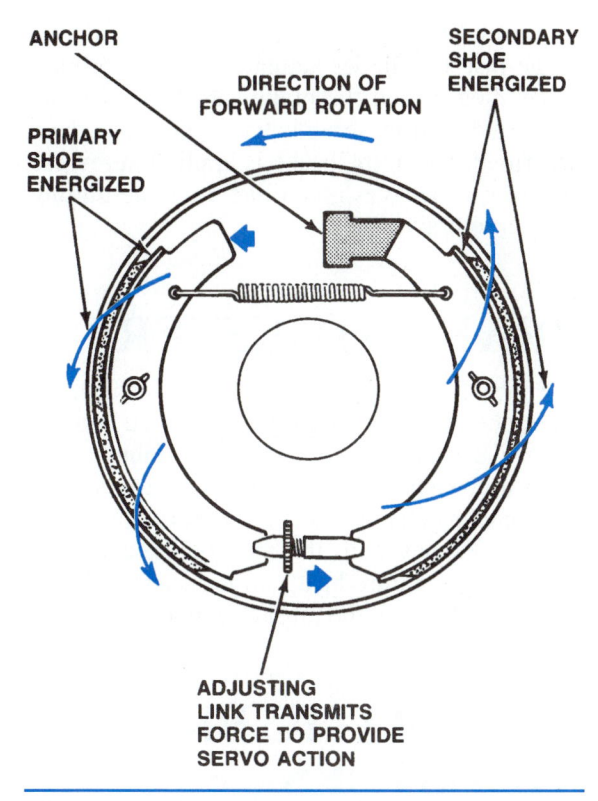

Figure 8-20. Servo action is used in the most powerful drum brake designs.

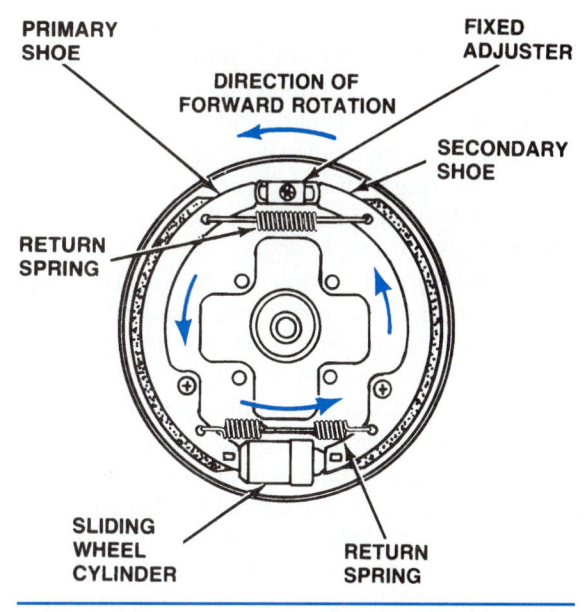

Figure 8-21. In this dual-servo brake, application force is transmitted through the wheel cylinder to the secondary shoe.

and drum is much greater than the application force developed by hydraulic pressure in the wheel cylinder. As a result, the secondary shoe is held solidly against the anchor. In effect, only one-half of the wheel cylinder is used to apply the brakes.

Servo-Action

Once all slack is taken up between the brake shoes, adjusting link, and anchor, both brake shoes become self-energized like the leading shoes in a non-servo brake. The anchor pin prevents the secondary shoe from rotating, and the adjusting link (held in position by the secondary shoe) serves as the anchor for the primary shoe. Servo action then occurs as a portion of the braking force generated by the primary shoe is transferred through the adjusting link to help apply the secondary shoe, figure 8-20. Servo action greatly increases the application force on the secondary shoe and improves overall stopping power.

When a dual-servo brake is applied with the vehicle moving in reverse, the primary and secondary shoes switch roles. The primary shoe is forced against the anchor, whereas the secondary shoe moves outward and rotates with the drum to apply the primary shoe with greater force. Because this occurs only for a short time during relatively low-speed reverse braking, the smaller lining of the primary shoe is able to provide adequate stopping power without overheating or wearing too rapidly.

Although servo action enables a drum brake to provide increased stopping power, it can also cause the brakes to grab and lock if they get too far out of adjustment. As clearance between the shoes and drum increases, the primary brake shoe is allowed a greater range of movement. The farther the shoe moves, the more speed it picks up from the rotating brake drum. At the moment the slack is taken up between the brake shoes, adjusting link, and anchor, the speed of the primary shoe is converted into application force by servo action. If the primary shoe is moving too quickly, it will apply the secondary shoe very hard and fast, causing the brakes to grab and possibly lock the wheels.

Other Dual-Servo Brakes

As explained previously, only half of a two-piston wheel cylinder is used when a dual-servo brake is applied. To reduce the number of parts and simplify the friction assembly, some cars have dual-servo drum brakes that use a sliding, single-piston wheel cylinder, figure 8-21. In this design, a manual brake adjuster solidly mounted near the top of the backing plate serves as the shoe anchor. The wheel cylinder mounts in a slotted opening at the bottom of the backing plate and is free to slide back and forth, much like the adjusting link in the dual-servo design described previously.

To identify the primary and secondary brake shoes in this design, point at the shoe adjuster and move your finger in the direction of drum rotation. The first shoe reached is the primary shoe; the other shoe is the secondary shoe.

When the brakes are applied, the wheel cylinder attempts to move the bottoms of both brake shoes outward against the drum. One shoe is actuated by the wheel cylinder piston, the other by the cylinder body, which slides in its mounting slot in the opposite direction. Pascal's law, discussed in Chapter 3, dictates that equal force is applied in both directions. When all slack is removed from the brake, it functions like any other dual-servo brake. The primary and secondary shoes become self-energized, and the primary shoe provides servo action through the wheel cylinder to increase application force on the secondary shoe.

UNI-SERVO BRAKE

The **uni-servo brake,** figure 8-22, provides servo action only when applied in the forward direction. The construction of the uni-servo brake is similar to that of the dual-servo design except that a single-piston wheel cylinder, solidly mounted to the backing plate, replaces the two-piston part. The single-piston cylinder actuates only the primary brake shoe. When the brakes are applied in the forward direction, servo action occurs as already described. When the brake is applied in reverse, however, it behaves much like a double-trailing, non-servo brake.

BRAKE ADJUSTERS

The job of a **brake adjuster** is to establish, restore, and maintain the proper clearance between the brake shoes and drum as the linings wear. There are two basic types of brake adjusters: automatic and manual. All early adjusters were manual designs. However, in the mid-1950s, automatic adjusters were introduced on domestic cars and have since become the most common type.

AUTOMATIC BRAKE ADJUSTERS

Automatic adjusters use the movement of the brake shoes to continually adjust lining-to-drum clearance as the brakes wear. There are several designs of automatic adjusters, and they operate in a variety of ways. Some adjust the clearance as the brakes are applied; others adjust when the brakes are released. Some work only when the car is moving forward; others work only when the car is traveling in reverse. Still others work when the brakes are applied in either direction. Some automatic adjusters are entirely independent of the service brakes and adjust when the parking brake is applied instead.

There are two basic types of automatic adjusters used on late-model cars:

- Starwheel
- Ratchet.

Starwheel Automatic Adjusters

Starwheel automatic adjusters are found on both servo and non-servo brakes. There are too many variations of the starwheel adjuster design to describe every one in detail, but the operation of the basic designs is explained in the following sections.

Servo Brake Starwheel Automatic Adjusters
Servo brakes use three styles of starwheel adjusters: cable, figure 8-23; lever, figure 8-24; and

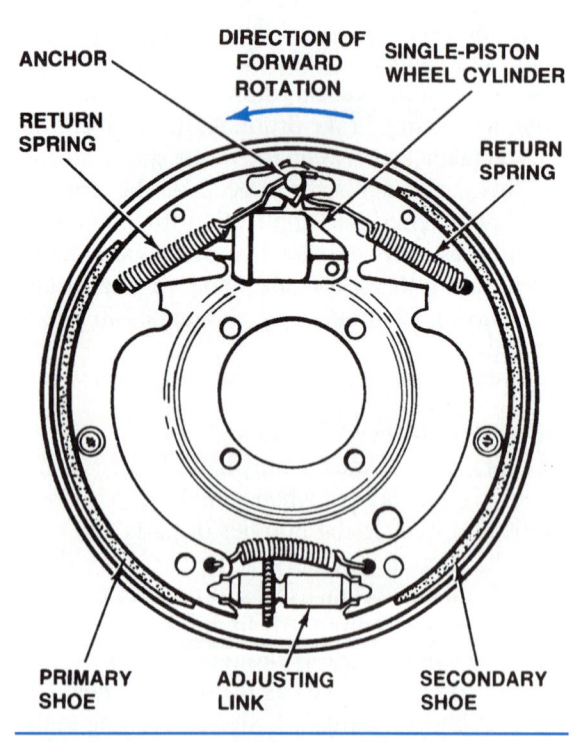

Figure 8-22. A uni-servo drum brake.

Drum Brake Friction Assemblies

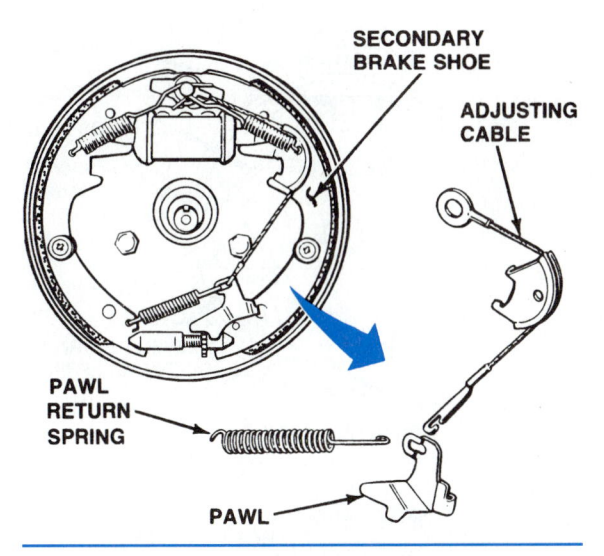

Figure 8-23. The cable-actuated starwheel automatic adjuster is one of the most common designs.

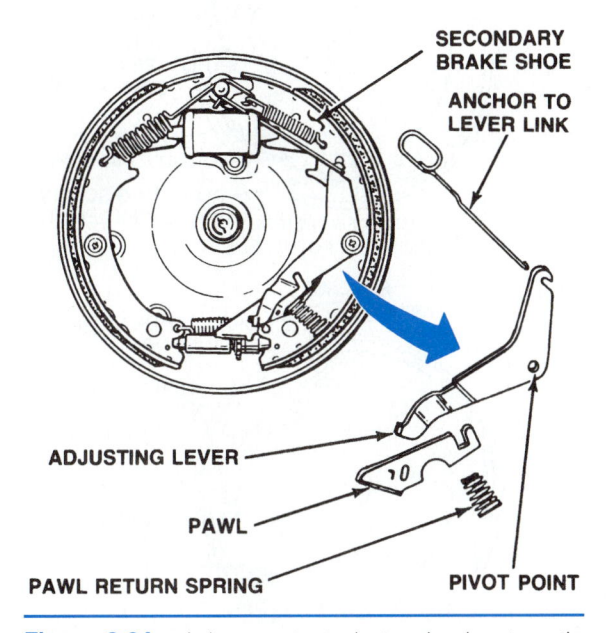

Figure 8-24. A lever-actuated starwheel automatic adjuster.

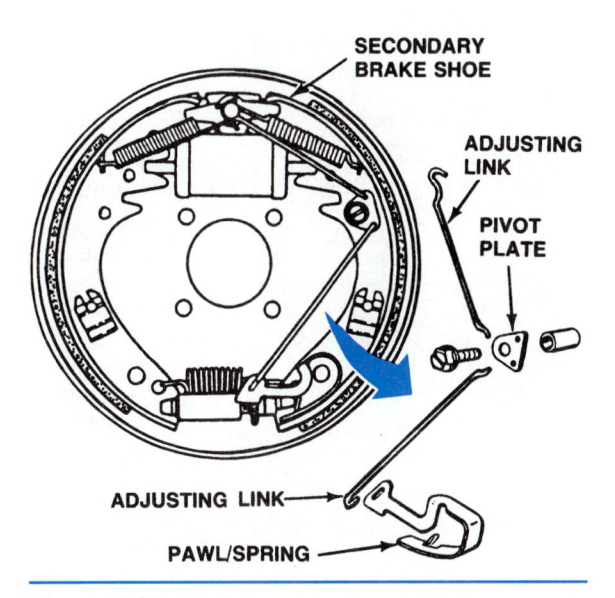

Figure 8-25. A link-actuated starwheel automatic adjuster.

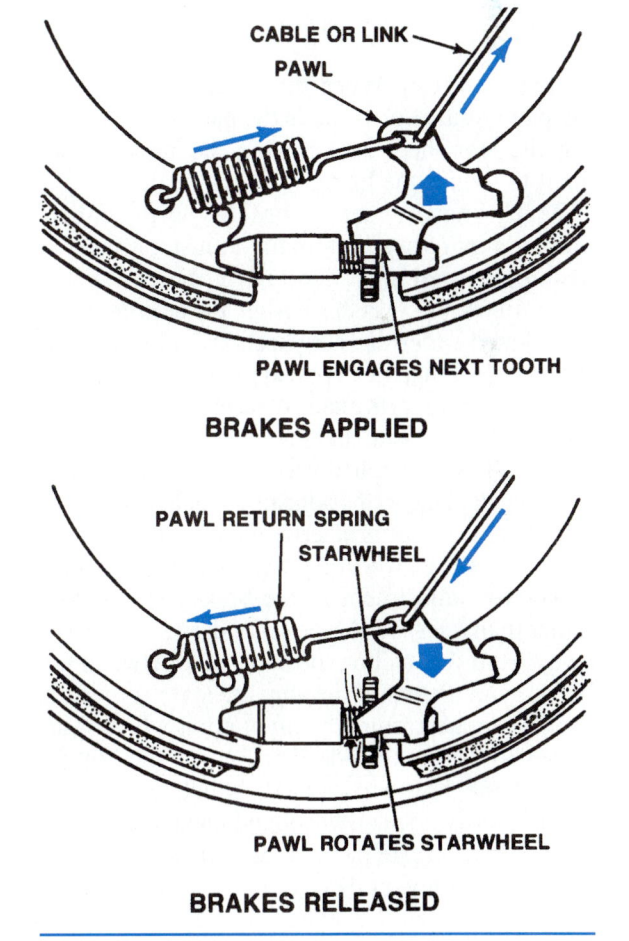

Figure 8-26. The basic operation of a starwheel automatic adjuster.

link, figure 8-25. The cable type was the earliest style, and the lever and link designs were introduced as later refinements. All three variations mount on the secondary brake shoe and adjust only when the brakes are applied while the car is moving in reverse.

As the brakes are applied on a car with a cable or link automatic adjuster, the wheel cylinder and drum rotation combine to move the secondary shoe away from the anchor. Movement of the shoe causes the cable or linkage to pull up on the adjuster **pawl**, figure 8-26. If the brake lining has worn far enough, the pawl engages the next tooth

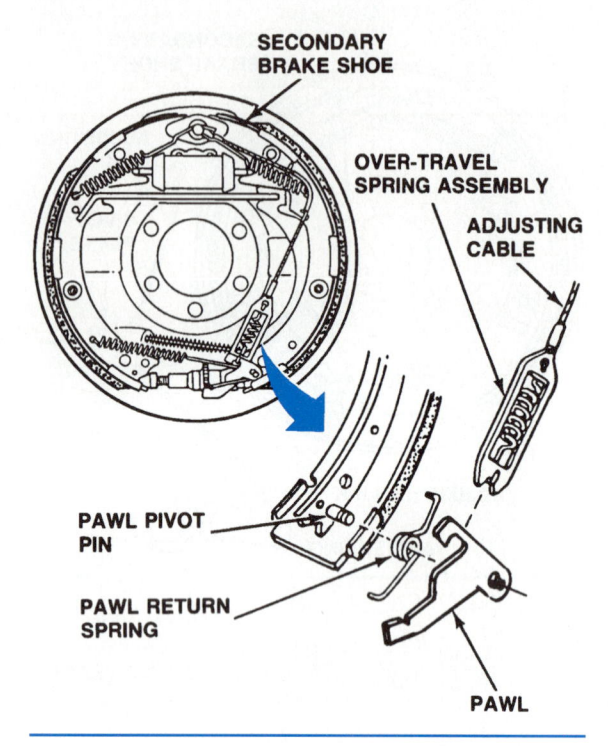

Figure 8-27. A cable-actuated starwheel automatic adjuster equipped with an over-travel spring.

on the starwheel. When the brakes are released, the pawl return spring pulls the pawl down, rotating the starwheel and moving the brake shoes apart to reduce the lining-to-drum clearance.

Many newer servo brakes with cable-actuated starwheel automatic adjusters have an **over-travel spring** assembly on the end of the cable, figure 8-27. In this design, the adjuster pawl is mounted *under* the starwheel, and adjustment is made as the brakes are applied rather than released. The over-travel spring damps the movements of the adjuster mechanism, and prevents overadjustment if the brakes are applied very hard and fast. It also prevents damage to the adjusting mechanism if the starwheel seizes or is otherwise unable to rotate.

The lever starwheel adjuster, figure 8-28, also makes the adjustment as the brakes are applied rather than released. As the secondary shoe moves away from the anchor, the solid link between the anchor and the top of the adjuster lever forces the lever to rotate around the pivot point where it attaches to the brake shoe. This moves the bottom half of the lever downward, which causes the pawl to rotate the starwheel and make the adjustment. The separate pawl piece is free to pivot on the lever to prevent damage if the starwheel will not rotate. When the brakes are released, the return springs lift the lever. If the brakes have worn

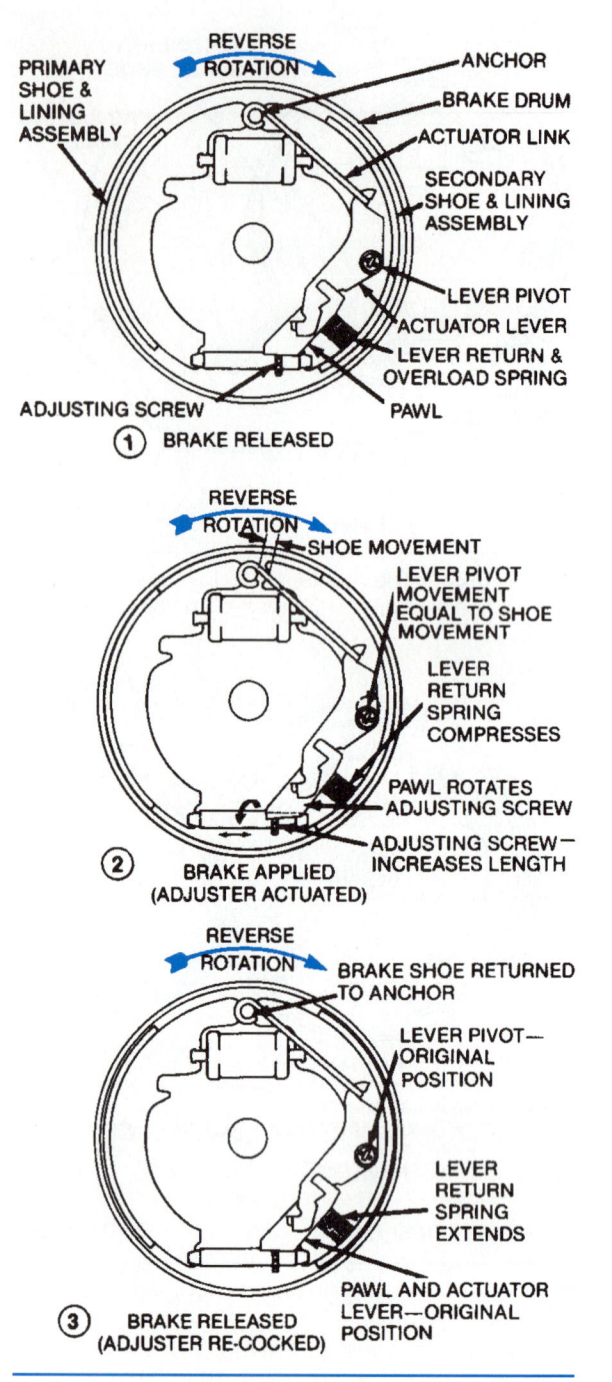

Figure 8-28. Operation of a lever starwheel automatic self-adjuster system. (Courtesy of General Motors Corporation, Service and Parts Operations)

enough, the end of the lever engages the next tooth on the starwheel and additional adjustment will be made the next time the brakes are applied.

Non-Servo Starwheel Automatic Adjusters

The starwheel automatic adjusters used on non-servo brakes may be mounted on either the lead-

Drum Brake Friction Assemblies

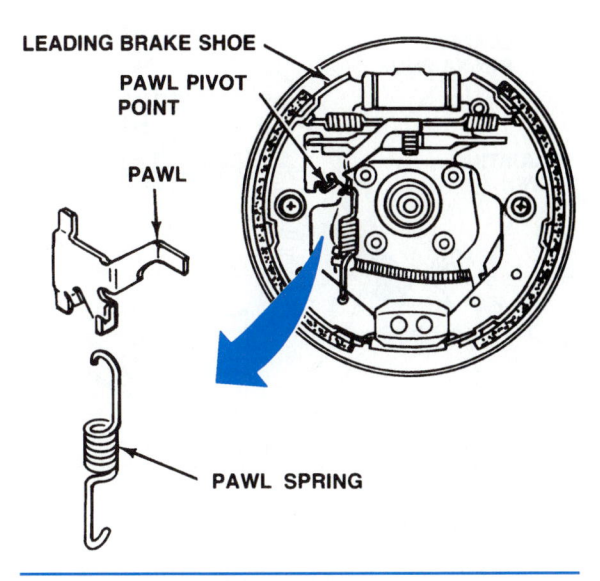

Figure 8-29. A non-servo brake with a lever-actuated starwheel automatic adjuster on the leading shoe.

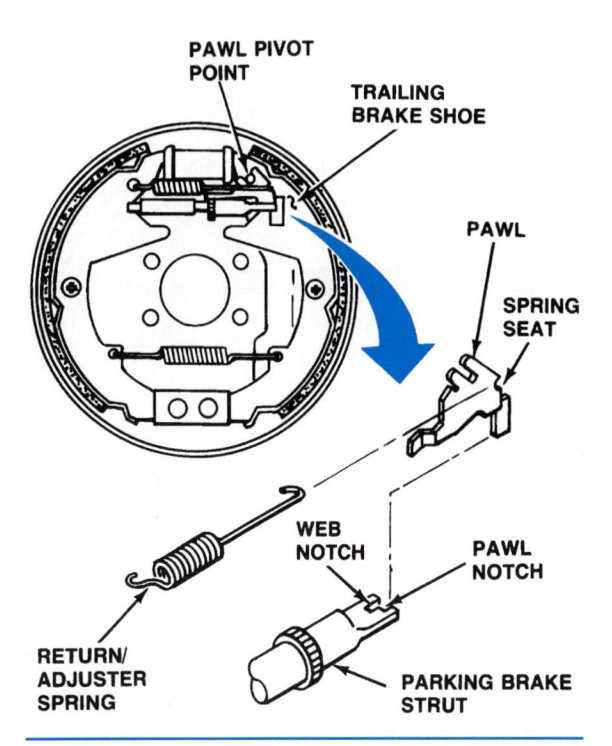

Figure 8-30. A non-servo brake with a lever-actuated starwheel automatic adjuster on the trailing shoe.

ing or trailing shoe. These types of adjusters work whenever the brakes are applied—in either the forward or reverse direction.

A leading-shoe design is shown in figure 8-29. When the brakes are not applied, the adjuster pawl is held in position by the parking brake strut. When the brakes are applied and the primary shoe moves out toward the brake drum (away from the parking brake strut), the pawl spring pivots the pawl downward where it mounts on the brake shoe; this rotates the starwheel to adjust the brake. When the brakes are released, return springs retract the shoes and the pawl is levered back into its resting position by the parking brake strut. If the linings have worn far enough, the lever engages the next tooth on the starwheel and further adjustment will occur the next time the brakes are applied.

The trailing-shoe non-servo starwheel adjuster shown in figure 8-30 works somewhat like the leading-shoe design just described; however, it makes the adjustment as the brakes are released rather than applied. The upper shoe return spring in this design actually serves two purposes: It returns the brake shoes and operates the automatic adjuster. When the brakes are not applied, spring tension holds the trailing shoe and the adjuster pawl tightly against the parking brake strut—each in its own respective notch. When the brakes are applied, the trailing shoe moves out toward the drum and away from the parking brake strut. This allows the adjuster pawl, which is restrained by the return spring, to pivot where it attaches to the brake shoe, causing the adjuster arm to move up-

ward. If the brakes have worn far enough, the arm will engage the next tooth of the starwheel. When the brakes are released, the return spring pulls the brake shoes back together and the parking brake strut levers the adjuster pawl downward to rotate the starwheel and adjust the brakes.

Parking Brake Automatic Adjusters

Parking brake automatic adjusters operate when the parking brake is applied, figure 8-31. The adjusting lever is attached, together with the parking brake lever, to the shoe. The lower end of the adjusting lever is held to the brake shoe with a spring, and the other end of the lever engages the starwheel, pulling it downward.

When the parking brake cable pulls the parking brake lever, it also lifts the adjusting lever. If an adjustment is needed, the adjusting lever drops into the next notch on the starwheel. When the parking brake is released, the adjusting lever turns the starwheel to make the adjustment.

This system works fine on vehicles with manual transmissions, where the owner is most likely to use the parking brake whenever parked. On vehicles with automatic transmissions, the parking brake rarely may be used, causing gradually increasing pedal travel as the shoes wear.

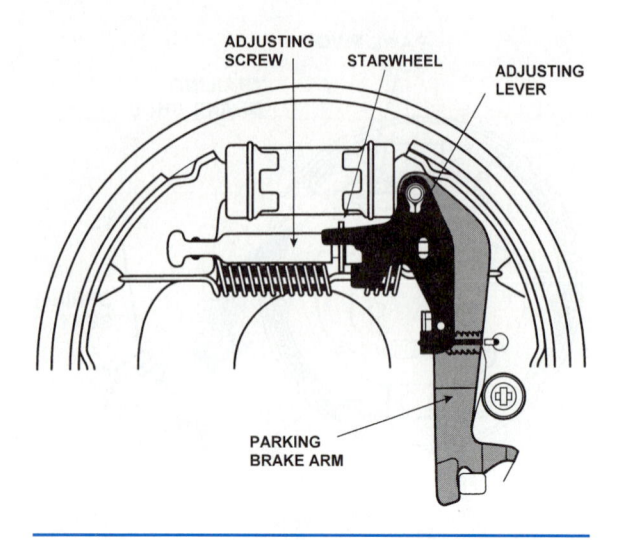

Figure 8-31. This adjuster actuates each time the parking brake is set and released. (Courtesy of Toyota Motor Sales U.S.A., Inc.)

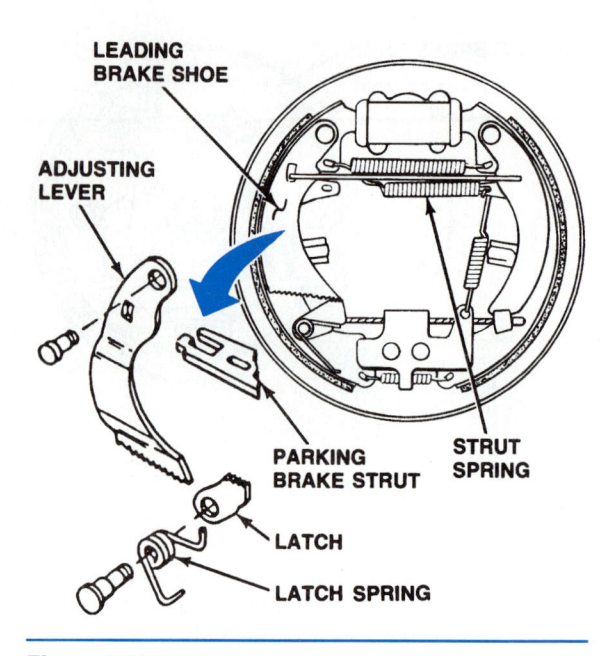

Figure 8-32. A lever-latch ratchet automatic adjuster.

Ratchet Automatic Adjusters

Like the starwheel design, most ratchet automatic adjusters use movement of the brake shoes to adjust the lining-to-drum clearance. However, at least one ratchet adjuster uses the parking brake mechanism to effect the adjustment. Although there are different types of ratchet adjusters, they all share one feature in common: The adjustment is carried out by two parts that have small interlocking teeth. As the adjustment is made, the two toothed elements ratchet across one another. Once adjustment is complete, the teeth lock together to hold the brake shoes in their new positions.

There are three basic variations of the ratchet automatic adjuster:

- Lever-latch
- Strut-rod
- Strut-quadrant.

Lever-Latch Automatic Adjuster

The lever-latch automatic adjuster, figure 8-32, installs on the leading shoe of a non-servo brake and operates whenever the brakes are applied. This design consists of a large lever and a smaller latch with interlocking teeth. A spring on the latch piece keeps it in contact with the lever to maintain the adjustment. One end of the parking brake strut hooks into an opening in the lever; the other end is held against the trailing brake shoe by a strong spring.

As the brakes are applied and the shoes move outward toward the drum, the parking brake strut pulls on the adjuster lever and forces it to pivot inward from where it attaches to the top of the leading shoe. If the brakes are sufficiently worn, the bottom of the lever will ratchet one or more teeth on the latch. When the brakes are released, the parking brake strut, which bottoms against the lever, will hold the shoes farther apart to reduce the lining-to-drum clearance.

Strut-Rod Automatic Adjuster

The strut-rod ratchet automatic adjuster, figure 8-33, is used on some non-servo brakes. This design is incorporated into the parking brake strut and operates whenever the parking brake is applied. The strut assembly mounts between the leading shoe and the parking brake lever. One part of the strut assembly applies the leading shoe; the other connects to the web of the trailing shoe through a hooked rod. The strut sections are held in position by a toothed rack and two spring steel fingers, figure 8-34, that allow the assembly to extend but not retract.

When the parking brake is applied, the parking brake lever moves the trailing shoe outward directly and applies the leading shoe through the strut assembly. This action moves the brake shoes and the two sections of the strut assembly apart. If the brakes are sufficiently worn, the rod attached to the trailing shoe will pull its half of the

Drum Brake Friction Assemblies

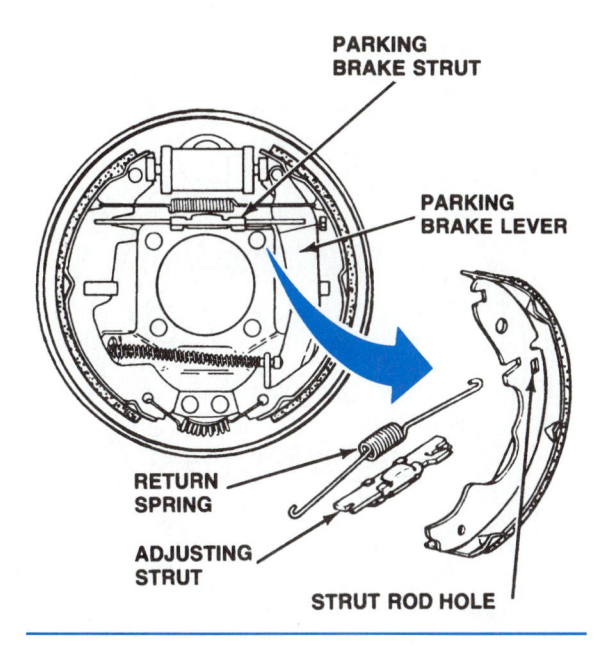

Figure 8-33. A strut-rod ratchet automatic adjuster.

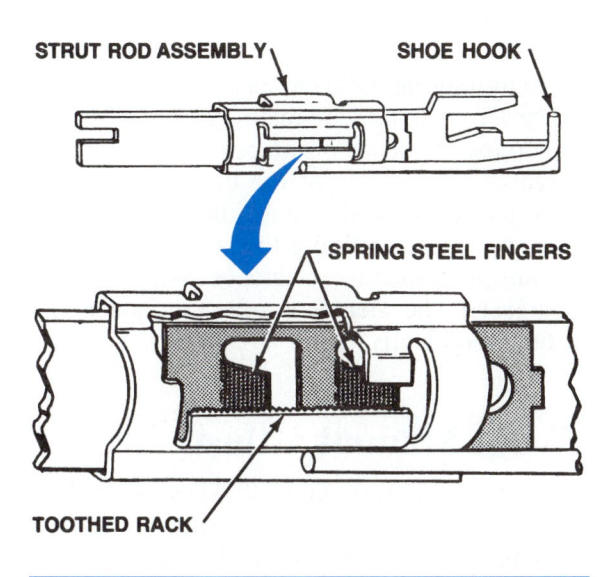

Figure 8-34. The ratchet mechanism of the strut-rod automatic adjuster.

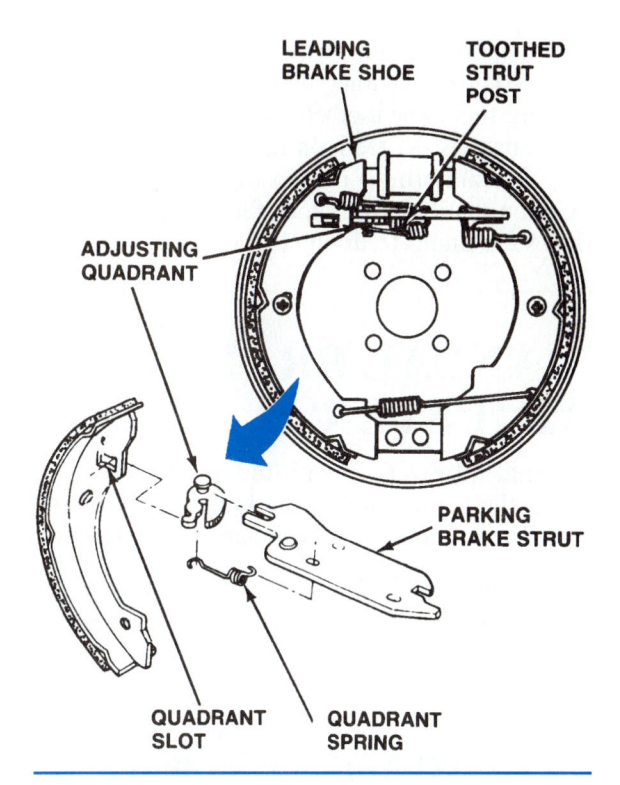

Figure 8-35. A strut-quadrant ratchet automatic adjuster.

strut one or more teeth away from the half of the strut that is applying the leading shoe. When the brakes are released, the strut will remain at its longer length and hold the shoes farther apart, reducing the lining-to-drum clearance.

Strut-Quadrant Automatic Adjuster

The strut-quadrant automatic adjuster, figure 8-35, is a relatively new design used on some non-servo brakes. Like the strut-rod design described previously, the strut-quadrant adjuster is part of the parking brake strut; however, it operates when the service brakes, not the parking brakes, are applied.

The strut-quadrant adjuster consists of three basic parts: the parking brake strut, an adjusting quadrant, and a quadrant spring. The strut has a toothed post solidly mounted on its underside. The adjuster quadrant pivots on a pin that slips into a notch in the end of the strut, and the backside of the quadrant has a toothed, cam-shaped surface that interlocks with the toothed post on the strut. The quadrant also has an arm that extends through an opening in the web of the leading brake shoe. The outer side of this arm serves as the brake shoe stop when the brakes are released; the inner side operates the adjuster as described following. The quadrant spring holds the quadrant in contact with the post to maintain the adjustment.

When the brakes are applied, the leading shoe moves out toward the brake drum. If there is sufficient wear of the brake lining, the edge of the slot in the shoe web contacts the inner side of the adjuster quadrant arm and pulls it outward. When this happens, the toothed section of the quadrant is lifted away from the post on the parking brake strut. The quadrant spring then rotates the quadrant until its pivot pin is bottomed in the slot in the parking brake strut. When the brakes are released, the quadrant returns inward with the leading shoe. The toothed section of the quadrant then engages the teeth on the strut post, causing the quadrant arm to remain in its new extended position that hold the shoes farther apart and reduces the lining-to-drum clearance.

MANUAL BRAKE ADJUSTERS

Manual adjusters require the technician to get under the car and adjust the brakes with a hand tool. Three common types of manual adjusters are:

- Starwheel
- Cam
- Wedge.

Starwheel Manual Adjusters

The starwheel adjuster, shown in figures 8-17 and 8-18, is the most common mechanism for manual brake adjustment. In servo brakes, the starwheel is built into the adjusting link that separates the brake shoes at the bottom of the friction assembly. The spring that holds the adjusting link and brake shoes together also engages the starwheel to prevent it from rotating unless it is positively levered to the next notch. Starwheel adjusters have either right- or left-hand threads, and must be mounted on the proper side of the vehicle.

Turning the starwheel screws one threaded part of the adjuster into or out of a second part. When the adjuster is screwed together, the shoes move inward and lining-to-drum clearance is increased. When the adjuster is screwed apart, the shoes move outward and lining-to-drum clearance is reduced. The starwheel is normally turned with a special adjusting tool or screwdriver inserted through a slot in the backing plate or brake drum.

Cam Manual Adjusters

Cam adjusters, figure 8-36, consist of rotating studs or posts installed through the backing plate. One adjuster is used for each brake shoe, and a cam-shaped portion of the post contacts the brake shoe web or lining table to serve as a shoe "stop."

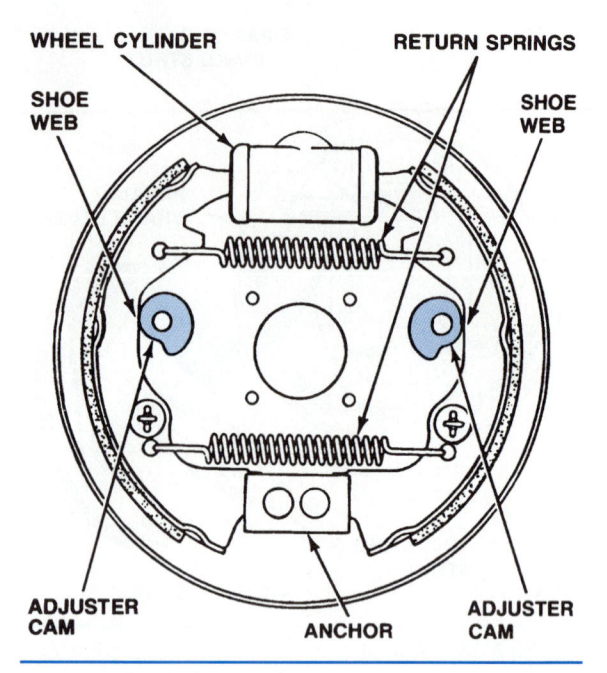

Figure 8-36. A non-servo drum brake with cam adjusters.

The return springs hold the shoes against the cams whenever the brakes are released.

Turning the heads of the posts, which are exposed on the outside of the backing plate, rotates the cams and moves the brake shoes inward to increase lining-to-drum clearance, or outward to reduce the clearance. With this type of adjuster, the clearance of each shoe is adjusted individually.

Wedge Manual Adjusters

The wedge adjuster, figure 8-37, was used on some imported vehicles. It consists of an adjuster housing bolted to the backing plate and two shoe links with angled ends that install in bores on either side of the adjuster body. A wedge bolt screws into the adjuster housing from outside the backing plate and has a square head that can be turned with an open-end wrench or a square socket.

As the wedge bolt is threaded into the adjuster, its tapered inner end contacts the angled faces of the shoe links and forces them apart. The links then move the brake shoes outward to reduce lining-to-drum clearance. To prevent the wedge bolt from turning unintentionally, its tapered tip has four flattened faces. These faces align flush with the angled ends of the shoe links, and pressure from the return springs keep the wedge bolt from turning.

Drum Brake Friction Assemblies

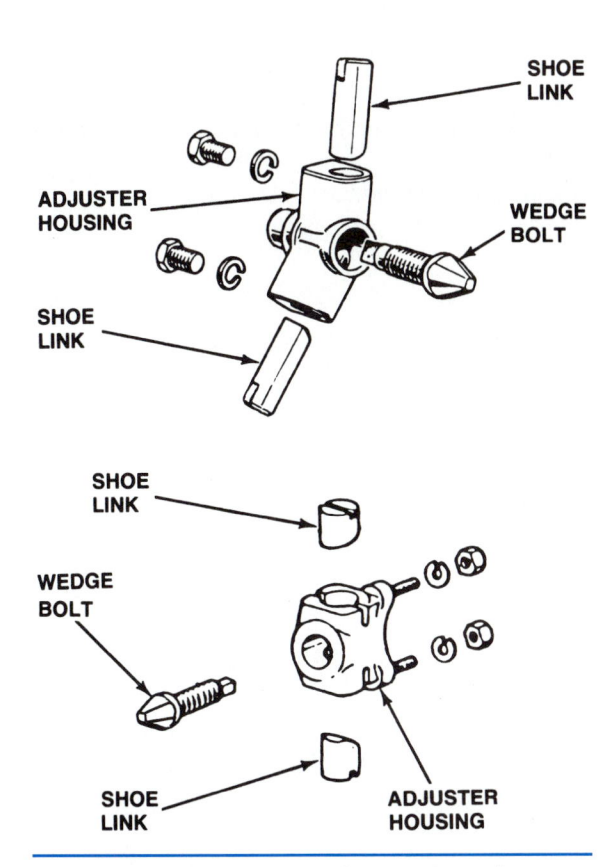

Figure 8-37. Two versions of the wedge adjuster.

SUMMARY

The primary advantage of drum brakes is that they can provide hard stops with relatively low pedal application force. This is possible because drum brake designs can use both self-energizing and servo actions to increase application force. For this reason, drum brakes work very well as parking brakes.

Drum brakes do have a number of disadvantages. They are susceptible to mechanical, lining, gas, and water fade; they require frequent manual adjustment or an automatic adjusting system; and they are susceptible to pull during braking. In addition, drum brakes can be difficult to service due to their complexity. For these reasons, drum brakes are now used only on rear axles of cars and light trucks.

Certain parts are common to many drum brakes: the backing plate, shoe anchors, piston stops, shoe support pads, wheel cylinders, brake shoes, return springs, holddown devices, parking brake linkage, and brake drum. The backing plate supports most of the other brake parts and incorporates the anchor that prevents the shoes from rotating with the drum. The support pads contact the edges of the brake shoes to align the shoes with the drum, and the holddown keep the shoes against the support pads to reduce noise. Piston stops prevent accidental disassembly of the wheel cylinder that provides the application force to move the brake linings into contact with the drum. The shoes consist of a lining table and web and are faced with a special friction material where they contact the brake drum. Return springs move the brake shoes back to their resting positions when the brakes are released. The drum rotates around all the parts on the backing plate, and provides the friction surface for the linings to rub against.

There are two basic types of drum brakes: non-servo and servo. Non-servo brakes apply each brake shoe individually. Each shoe may or may not take advantage of self-energizing action, which uses the rotation of the brake drum to wedge the lining of the shoe tighter against the drum. A shoe that is energized in this manner is called a leading shoe. A shoe that is deenergized is called a trailing shoe.

Ranked from least to most powerful, the four types of non-servo brakes are the double-trailing, leading-trailing, double-leading, and nondirectional. The double-trailing brake does not use self-energization, whereas the leading-trailing brake has one energized and one de-energized shoe. Both the double-leading and nondirectional brakes make use of self-energization on both brake shoes.

Servo drum brakes are more powerful than non-servo brakes of similar size because they take advantage of both self-energizing and servo action. Servo action uses the stopping power of one shoe to increase the application force of the other. The shoe that helps increase application force is called the primary shoe and often has a smaller lining. The shoe that receives the extra application force and does most of the braking is called the secondary shoe. The secondary shoe often has a larger lining than the primary shoe, and the lining may be made of a different friction material that is more resistant to heat.

Dual-servo brakes are the most common type of servo brake; they provide servo action in both the forward and reverse directions. Most dual-servo brakes use a fixed-wheel cylinder—braking force is transferred from the primary shoe to the secondary shoe through an adjusting link. An alternate dual-servo design has a fixed anchor and a sliding single-piston wheel cylinder that both applies the brake and transfers the force from one shoe to the other. A simpler form of the servo brake, called the uni-servo design, provides servo action in the forward direction only.

Brake adjusters are used to set and maintain the proper clearance between the brake lining and drum. The two basic types are automatic and manual adjusters. Automatic adjusters use movement of the brake shoes or parking brake linkage to make the brake adjustment. The two types of automatic adjusters are the starwheel and ratchet designs. The manual type requires adjustment with a hand tool and comes in three varieties: starwheel, cam, and wedge.

Drum Brake Friction Assemblies

Review Questions

Choose the letter that represents the best possible answer to the following questions:

1. Which type of brake fade does *not* result in a hard pedal?
    a. Lining
    b. Mechanical
    c. Gas
    d. Water

2. Drum brakes must be periodically adjusted to compensate for:
    a. Loose wheel bearings
    b. Brake fluid evaporation
    c. Poor design
    d. Lining wear

3. Brake pull may result from:
    a. Misadjustment
    b. Fade
    c. Both a and b
    d. Neither a nor b

4. The backing plate serves as:
    a. A mounting surface for other brake parts
    b. A backup in case of brake failure
    c. A barrier against excess heat
    d. All of the above

5. Holes in the backing plate are used to:
    a. Mount the brake parts
    b. Inspect the linings for wear
    c. Adjust lining-to-drum clearance
    d. All of the above

6. Which of the following keeps the brake linings centered on the drum friction surface?
    a. Piston stops
    b. Shoe support pads
    c. Shoe anchors
    d. All of the above

7. Which of the following converts hydraulic pressure into mechanical force?
    a. Master cylinder
    b. Brake pedal
    c. Brake fluid
    d. Wheel cylinder

8. Brake shoes are made from:
    a. Asbestos or welded steel
    b. Sintered metal or phenolic resins
    c. Welded steel or cast aluminum
    d. Spring steel or cast iron

9. Which of the following is a type of brake shoe holddown?
    a. Pin-type
    b. Spring clip
    c. Coil spring
    d. All of the above

10. Which part of the brake assembly turns with the car's wheels?
    a. Brake shoes
    b. Wheel cylinders
    c. Both a and b
    d. Neither a nor b

11. The term *servo* refers to the action of:
    a. One brake shoe on the other
    b. Drum rotation on a brake shoe
    c. Both a and b
    d. Neither a nor b

12. The term *self-energizing* refers to the action of:
    a. One brake shoe on another
    b. Drum rotation on a brake shoe
    c. Both a and b
    d. Neither a nor b

13. The effect of servo action and self-energizing action is to:
    a. Reduce brake fade
    b. Increase the life of friction materials
    c. Increase stopping ability
    d. Reduce the cost of manufacture

14. Which of the following non-servo brake designs is the best at stopping a vehicle moving in reverse?
    a. Double-trailing
    b. Leading-trailing
    c. Double-leading
    d. None of the above

15. Which of the following is *not* true of dual-servo brakes?
    a. They use a single wheel cylinder
    b. Both linings are the same size
    c. They use an adjusting link
    d. Both shoes have individual return springs

16. Approximately 70 percent of the braking action of a dual-servo brake is supplied at the:
    a. Front axle
    b. Rear axle
    c. Primary shoe
    d. Secondary shoe

17. If a dual-servo brake gets too far out of adjustment, _____ may occur.
    a. Vibration
    b. Wheel lockup
    c. A grinding sound
    d. Wheel bearing failure

18. A single-piston wheel cylinder may be used on:
    a. Uni-servo brakes
    b. Dual-servo brakes
    c. Either a or b
    d. Neither a nor b

19. Which of the following is *not* a manual adjuster mechanism?
    a. Anchor
    b. Crank
    c. Starwheel
    d. Wedge

20. Rotating a starwheel adjuster changes the:
    a. Adjuster length
    b. Lining-to-drum clearance
    c. Both a and b
    d. Neither a nor b

21. Which type of adjuster accepts a special wrench?
    a. Wedge
    b. Anchor
    c. Crank
    d. All of the above

22. Starwheel automatic adjusters:
    a. May operate when the brakes are applied
    b. Are used on some non-servo brakes
    c. May be operated by mechanical links
    d. All of the above

23. On all ratchet-type automatic adjusters, the adjustment is carried out by:
    a. Moving in reverse
    b. Heavy springs
    c. Cam gears
    d. Interlocking teeth

24. Ratchet automatic adjusters use movement of the brake shoes or _____ to adjust the brakes.
    a. Parking brake
    b. Wheel cylinder
    c. Return springs
    d. All of the above

25. Which of the following is *not* a type of ratchet automatic adjuster?
    a. Strut-quadrant
    b. Strut-rod
    c. Lever-screw
    d. Lever-latch

9

Disc Brake Friction Assemblies

OBJECTIVES

Upon completion and review of this chapter, you will be able to:

- Explain the advantages and disadvantages of disc brakes.
- Describe how brake fade affects disc brakes.
- List the parts of a disc brake and describe their function.
- List and describe the function of the three major caliper designs.
- Explain the advantages and disadvantages of the three major caliper designs.
- Explain the advantages and disadvantages of rear disc brake design.

KEY TERMS

anchor plate
brake disc
drum-in-hat
fixed brake caliper
floating caliper
gas fade
inboard disc brakes
lining fade
mechanical fade
natural frequency
sliding caliper
swept area
unsprung weight
water fade
ways

INTRODUCTION

Disc brakes are used on the front wheels of late-model cars and on the rear wheels of an increasing number of automobiles. Disc brakes were adopted primarily because they can supply greater stopping power than drum brakes with less likelihood of fade. This makes disc brakes especially well suited for use as front brakes, which must provide 60 to 80 percent of a car's total stopping power. In fact, the higher levels of braking performance specified in the 1976 revision of Federal Motor Vehicle Safety Standard 105 virtually guaranteed that manufacturers would use only disc brakes on the front axles of new cars.

The first part of this chapter discusses the advantages and disadvantages of disc brakes compared to drum brakes. The second part describes the construction and operation of the three major disc brake designs. Two additional sections look at the special features and problems of four-wheel disc brake systems, and inboard disc brakes.

DISC BRAKE ADVANTAGES

Although increased federal brake performance standards hastened the switch to disc brakes, front drum brakes would eventually have been eliminated anyway because disc brakes are superior in almost every respect. The disc brake friction assembly, figure 9-1, has several significant strong points and only a few relatively minor weak points. The main advantages of the disc brake are:

- Fade resistance
- Self-adjustment
- Freedom from pull.

Fade Resistance

One of the biggest limitations on brake design is size; the friction assembly must fit within the wheel. By the mid-1960s, many drum brakes had grown to fill all of the space available but still could not provide sufficient braking power with adequate resistance to fade. When a disc brake is compared to a drum brake of similar diameter, its biggest advantage is a much greater ability to resist fade. In fact, disc brakes are more resistant to all kinds of fade: mechanical, lining, gas, or water.

The main design feature that helps disc brakes avoid heat-induced fade is their cooling ability; all of the major parts of a disc brake are exposed to the air flowing over the friction assembly. Many brake rotors also have cooling passages cast into them to further reduce operating temperatures.

Swept Area

Another reason disc brakes have greater fade resistance than drum brakes is that they have greater **swept area**. Swept area is the amount of brake drum or rotor friction surface that moves past the brake linings every time the drum or rotor completes a rotation. A larger swept area allows the heat generated in braking to be transferred more rapidly into the drum or rotor for better cooling. A disc brake has swept area on both sides of the rotor; a drum brake has swept area only on the inside of the drum. Although a drum's swept area is around its outer edge and therefore relatively large, a rotor of the same diameter still has substantially more swept area.

Figure 9-2 shows the approximate swept areas for a drum with a 10-in. inside diameter, and a rotor with a 10-in. outside diameter; both with a 2-in.-wide contact area between the lining and the

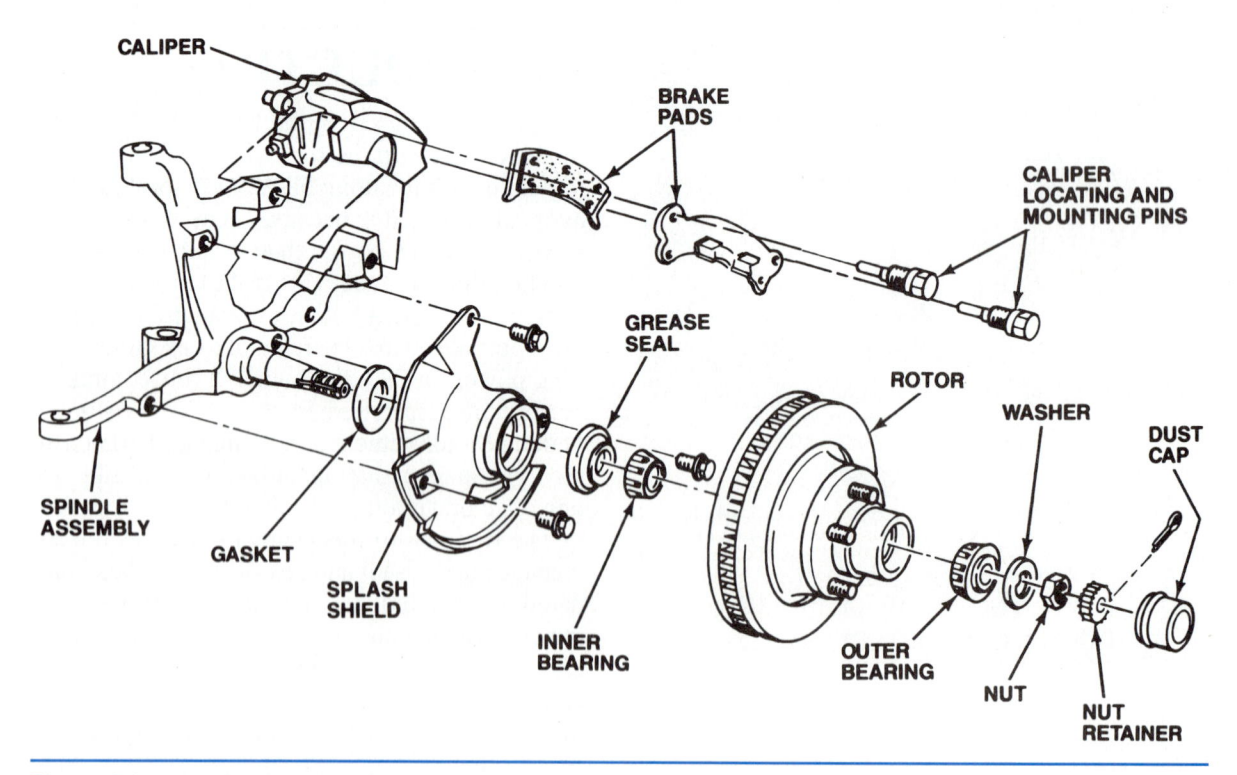

Figure 9-1. A typical disc brake friction assembly.

Disc Brake Friction Assemblies

friction surface. In this example the rotor has more than 50 percent more swept area than the drum. On a car that allows a maximum drum or rotor diameter of 10 in., the difference in swept area is even greater because the thickness of the drum casting reduces the inside diameter of the drum by at least half an inch.

Mechanical Fade

Mechanical fade is not a problem with disc brakes because, unlike a brake drum, the disc brake rotor expands *toward* the brake linings as it heats up rather than *away* from them. This fundamental design difference makes it physically impossible for heat to cause the rotor to expand out of contact with the brake linings. Because of this, there is never the need to move the brake linings out to keep them in contact with the rotor, so brake pedal travel does not increase. If the brake pedal on a car with disc brakes drops toward the floor, it is almost always a sign of vapor lock, a fluid leak, fluid bypassing the seals in the master cylinder, or mechanical fade of the rear *drum* brakes.

Lining Fade

Although mechanical fade is impossible with a disc brake, **lining fade** can and does occur if the brakes become overheated. As discussed in Chapter 3, a little bit of heat brings the brake pads to their operating temperature and actually increases the friction coefficient of the lining material; a warm brake performs better than a cold brake. However, when too much heat is generated by braking, the lining material overheats, its friction coefficient drops, and lining fade occurs.

The primary symptom of lining fade is a hard brake pedal that requires the driver to apply greater force to maintain stopping power. Unlike the similar situation in a drum brake, however, increased application force will not distort the brake rotor because the caliper applies equal force to both sides, figure 9-3. Increased pressure will, however, create even more heat, and if brake lining temperatures continue to increase, gas fade and vapor lock of the hydraulic system can occur.

Because of the disc brake's superior cooling ability, the point at which lining fade becomes a problem occurs much later than in a drum brake. This is one of the reasons a disc brake can supply full stopping power for longer periods than a drum brake. A disc brake's cooling ability also allows it to recover from lining fade much faster. However, if the pads are overheated to the point where the lining material is physically damaged, the brakes will not recover their full stopping power until the pads are replaced.

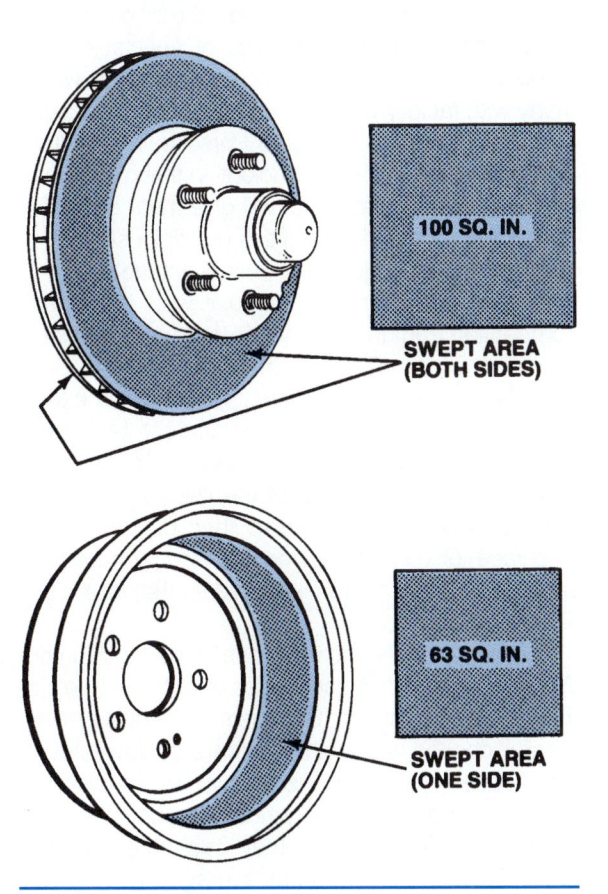

Figure 9-2. One reason a disc brake resists fade better than a drum brake of similar size is because it has more swept area.

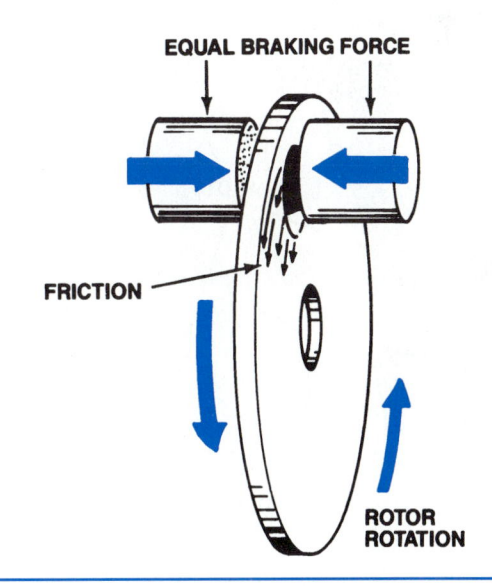

Figure 9-3. Braking force is applied equally to both sides of a brake rotor.

Gas Fade

Gas fade is a problem only under severe braking conditions when hot gasses and dust particles from the linings are trapped between the brake linings and rotor where they act as lubricants. The symptoms of gas fade are the same as those for lining fade: The pedal becomes hard and increased force is required to maintain stopping power.

Even though disc brakes operate at higher temperatures than drum brakes, they have fewer problems with gas fade for a number of reasons. First, disc brakes do not have a drum to contain gasses and particles in the area around the brake linings. Second, the constant flow of air over the brake carries away contaminants that might otherwise build up. And finally, the surface area of the brake lining material in a disc brake is smaller than that of a comparable drum brake; this allows gasses and particles to escape more easily.

To help prevent gas and water fade (see the following discussion), many brake pads have slots cut in the lining material, figure 9-4; these slots allow gasses and dust particles to escape. The holes required in riveted linings also perform this function. For even greater protection against gas fade, high-performance cars and motorcycles sometimes have holes or slots cut in the rotor. These openings allow gasses and water to escape.

Water Fade

Because the disc brake is essentially self-cleaning, it is also excellent at counteracting **water fade.** Two factors help keep water from between the linings and rotor: centrifugal force created by the spinning rotor throws off most moisture, and the brake pads positioned only a few thousandths of an inch away from the rotor continuously wipe it clean. When the brakes are applied, the leading edge of the brake pad lining material wipes the last bit of water from the disk. Once good lining-to-rotor contact is established, water is unable to enter the space between the linings and rotor until the brakes are released.

■ Disc Brakes at Le Mans

Every June, Sarthe, France, hosts a contest of speed and endurance recognized as the most famous race in Europe, and possibly the world—the 24 hours of Le Mans. This race draws hordes of spectators and some of the best equipped race teams in the world, many with factory support from manufacturers such as Ferrari, Jaguar, Ford, and Porsche. These firms employ the very latest technology in trying to fit their cars with special parts that will give them an "unfair advantage" and a victory in this prestigious event. In 1953 Jaguar found just such an advantage.

In the race that year, Jaguar would face stiff competition from Ferrari, Alfa Romeo, the Chrysler-powered Cunningham, and the Cadillac-powered Allard. All of the top cars were capable of more than 150 mph on the long Mulsanne straight, but a very slow hairpin corner at the end of the straight pushed the drum brakes of the day to, and often past, their limits. To conserve their brakes for the long race, drivers would begin to slow hundreds of yards before the hairpin—everyone, that is, but the Jaguar drivers.

Working closely with the brake division of Dunlop, Jaguar had installed disc brakes at all four wheels of their C-type racing car. Because of the disc brake's greater cooling ability, the Jaguars could maintain their speed much farther down the Mulsanne straight. Although the Jaguars were not the most powerful cars in the race, they finished in first and second place. And the winner averaged 105 mph for 24 hours, the first time Le Mans had been won at more than 100 mph! From that date forward, the days of the drum brake in racing were numbered.

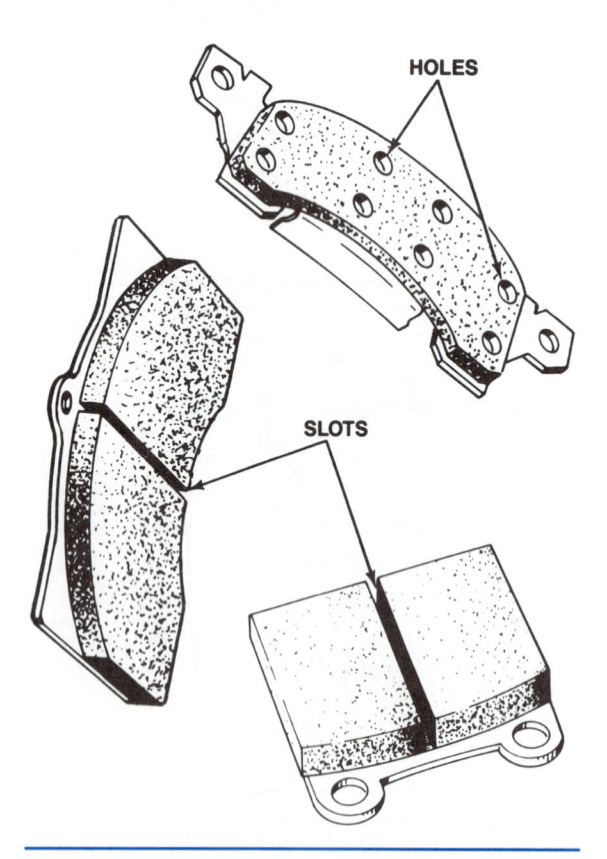

Figure 9-4. Slots and holes in the brake linings help prevent gas and water fade.

Disc Brake Friction Assemblies

Although far more resistant to water fade than drum brakes, disc brakes are not entirely free from its effects. Manufacturers take a number of steps to ensure any delay in brake application is kept to a minimum. Splash shields and the vehicle's wheels help keep water off of the rotor, and the brake lining materials specified for most cars minimize the effects of water fade. As mentioned previously, slotted brake linings and holes or slots in the rotors provide additional paths for water to escape on some vehicles.

Self-Adjusting Ability

In addition to superior fade resistance, disc brakes also have the advantage of being self-adjusting. Unlike drum brakes, disc brakes do not require cables, links, levers, ratchets, or struts to provide either manual or automatic brake adjustment. The brake pads are always right next to the spinning rotor, and any wear of the linings is automatically compensated for by the action of the brake caliper.

As described in Chapter 7, when the brakes are applied the caliper pistons move out as far as needed to force the brake pads into contact with the rotor. When the brakes are released, the piston retracts only the small distance dictated by rotor runout and piston seal flex. Because an automatic adjusting system is not needed, the typical disc brake friction assembly is much simpler than a drum brake.

Freedom from Pull

The last advantage of the disc brake is its freedom from pull. A disc brake will stop straighter under a wider range of conditions than will a drum brake. Two things make this possible. Foremost is the disc brake's self-cleaning ability. If one wheel of a car with drum brakes passes through a puddle, water and other contaminants that enter the drum can seriously reduce the braking power at that wheel and cause a pull toward the opposite side of the car. A disc brake in the same situation will throw off most of the water and perform with little or no problem.

The second reason disc brakes suffer less from pulling problems is that they have no self-energizing or servo action. These actions increase the power of drum brakes but depend on friction between the linings and drum for their effect. This means that even a small loss of lining-to-drum friction causes a large loss of braking power and a significant side-to-side variation in the amount of braking force. Because disc brakes do not use friction between the linings and rotor to increase their braking power, the effects of a loss of friction on one side of the car are far less pronounced than with drum brakes.

DISC BRAKE DISADVANTAGES

The most notable fact about the disadvantages of disc brakes is that there are so few. The weaknesses of disc brakes include:

- No self-energizing or servo action
- Brake noise
- Poor parking brake performance.

No Self-Energizing or Servo Action

The disc brake's lack of self-energizing or servo action is a disadvantage for two reasons: It contributes to poor parking brake performance and requires the driver to push harder on the brake pedal for a given stop. Early manual disc brakes often required very high pedal pressure to make a fast stop. However, the problem of high pedal pressures has been virtually eliminated through the use of brake power boosters. Today, all but the lightest disc-brake-equipped cars have a power booster as standard equipment.

Since the widespread adoption of the power booster, the disc brake's lack of self-energizing or servo action is actually more of a help than a hindrance. As mentioned previously, this lack contributes to straighter stopping. And without the added application force of self-energizing or servo action, a disc brake responds more directly to pressure on the brake pedal. This makes it easier to modulate the brakes for the exact amount of stopping power desired.

Brake Noise

Probably the biggest complaint about disc brakes is that they sometimes make a great deal of noise—usually various squeaks and squeals. These noises can occur both when the brakes are applied and when the brakes are released. As long as the brake linings are not worn down to the backing plate,

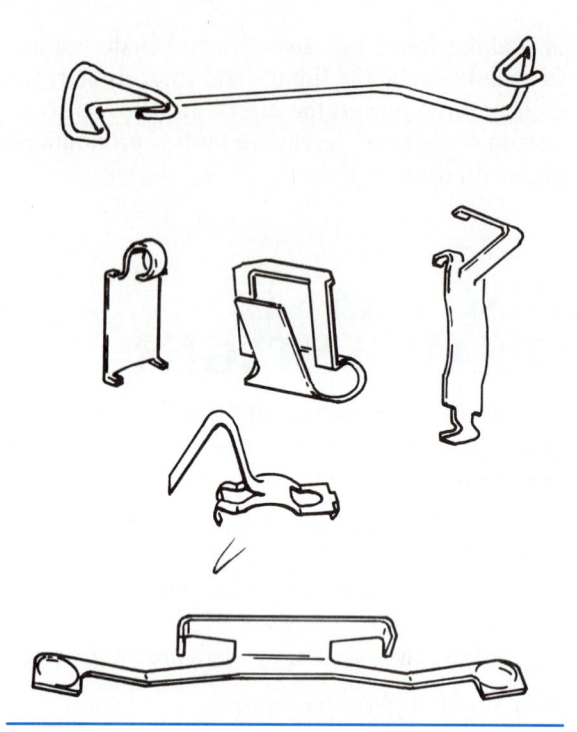

Figure 9-5. Antirattle spring clips reduce brake pad vibration.

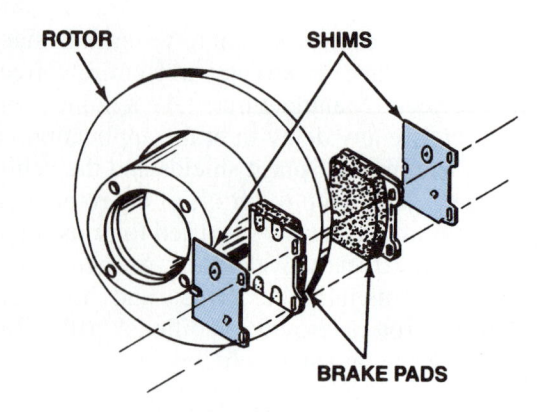

Figure 9-6. Antivibration shims are used behind the pads on some brake calipers.

these noises are usually caused by high-frequency rattling or vibration of the brake pads.

Several methods are used to quiet noisy disc brakes. Manufacturers use specific lining materials that damp vibrations, and most calipers have antirattle clips or springs, figure 9-5, that hold the pads in the caliper under tension to help prevent vibration. Some calipers use special shims, figure 9-6, between the brake pad backing plate and the caliper piston to damp vibrations. These shims may be made of metal, fiber, asbestos, or a plastic-like nylon.

In addition to the mechanical parts used to control vibration, antinoise sprays and brush-on liquids are available from aftermarket suppliers. These products are applied to the back of the brake pad where they set up to form a tacky cushion layer. When the pad is installed, the cushion layer creates a flexible bond between the pad and the caliper piston. The bond lowers the **natural frequency** of the pad, and the cushion layer damps any vibration that may still occur.

Poor Parking Brake Performance

Although noise is the biggest complaint about disc brakes, their only real functional weakness is poor parking brake performance. As already mentioned, the lack of self-energizing and servo action plays a large part in poor disc brake parking performance. In addition, the lining-to-rotor contact area of a disc brake is somewhat smaller than the lining-to-drum contact area of a drum brake. As explained in Chapter 3, this causes the disc brake to have a lower static coefficient of friction and, therefore, less holding power when the car is stopped.

Actually, poor parking brake performance is only a problem on cars with four-wheel disc brakes; most cars with rear drum brakes use the rear brakes as the parking brake. The problems associated with disc parking brakes are covered in greater detail in the *Rear Disc Brakes* section later in this chapter and in Chapter 12, which covers parking brakes exclusively.

DISC BRAKE CONSTRUCTION

A disc brake is relatively simple compared to a drum brake. The major disc brake friction assembly components are the:

- Caliper
- Splash shield
- Brake pads
- Rotor.

Caliper

With the exception of the rotor, the caliper is the largest part of a disc brake friction assembly. The brake caliper uses hydraulic pressure to create the mechanical force required to move the brake pads

Disc Brake Friction Assemblies

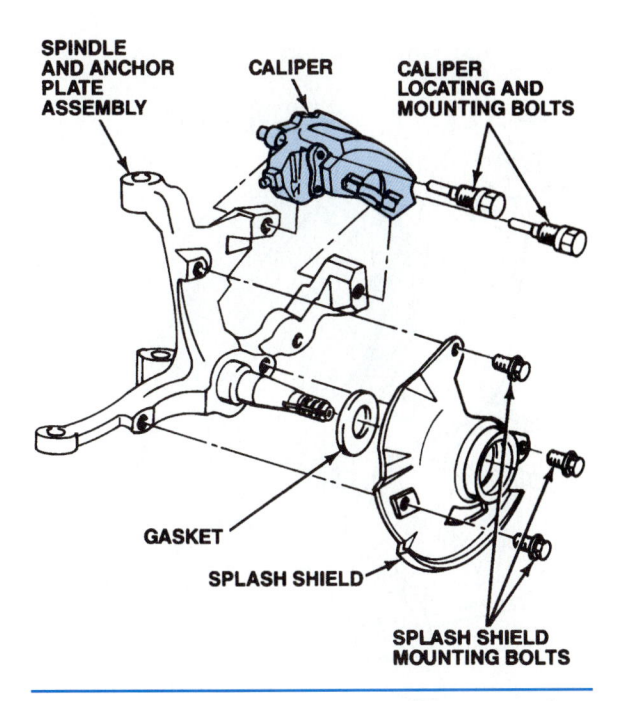

Figure 9-7. This brake caliper attaches to the front spindle.

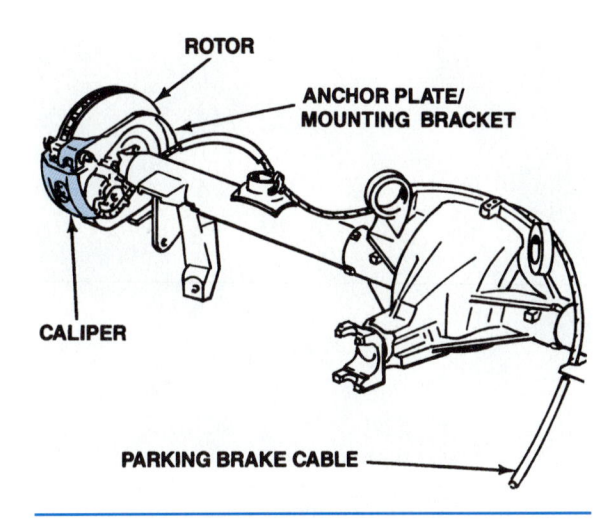

Figure 9-8. This brake caliper attaches to a mounting bracket on the rear axle housing.

into contact with the brake rotor. At the front axle, the caliper mounts to the spindle or steering knuckle, figure 9-7. Rear disc brake calipers mount to a support bracket on the axle flange or suspension, figure 9-8.

The hydraulic operation of brake calipers was covered in Chapter 7 and is not discussed here except as it affects overall disc brake design. The mechanical construction of the caliper is the determining factor of the overall design and is covered in the *Disc Brake Design* section that follows.

Splash Shield

The splash shield, figure 9-7, bolts to the front spindle or steering knuckle, or in rear disc brake applications, to the axle flange or a suspension adapter plate. Its job is to protect the inner side of the brake rotor from water and other contaminants; the outer side of the rotor is protected by the wheel. Most splash shields are made of stamped steel, although some newer cars use splash shields made of plastic to save weight.

Brake Pads

The brake pads in a disc brake do essentially the same job as the brake shoes in a drum brake. The pads contact the spinning rotor to create the friction that converts kinetic energy into heat to stop the car. Two pads are used in each brake caliper, one on each side of the rotor.

Brake pads are manufactured in various shapes and sizes to fit specific applications, figure 9-4, but all pads consist of a metal backing plate to which lining material is bonded or riveted. Some pads have mechanical or electrical wear indicators built into them to signal the driver when the linings have worn to the point where they need replacement. Brake pads are detailed in Chapter 10.

Brake Rotor

The brake rotor, figure 9-1, also called the **brake disc,** provides the friction surfaces for the brake pads to rub against. The rotor is the largest and heaviest part of a disc brake and is usually made of cast iron because that metal has excellent friction and wear properties. Some rotors are made with a stamped steel center and a cast friction surface in order to reduce weight. Where cost is not a factor, rotors made of ceramic-carbon fiber composites afford the greatest weight savings, figure 9-9.

There are two basic types of rotors: solid and vented, figure 9-10. Solid rotors were the first type fitted on automobiles, and they are still used on some vehicles, usually on the rear brakes. Solid rotors have been replaced by vented rotors on almost all vehicles. Vented rotors have radial cooling passages cast between the friction surfaces that allows the rotor to dissipate heat from both the inside and outside of

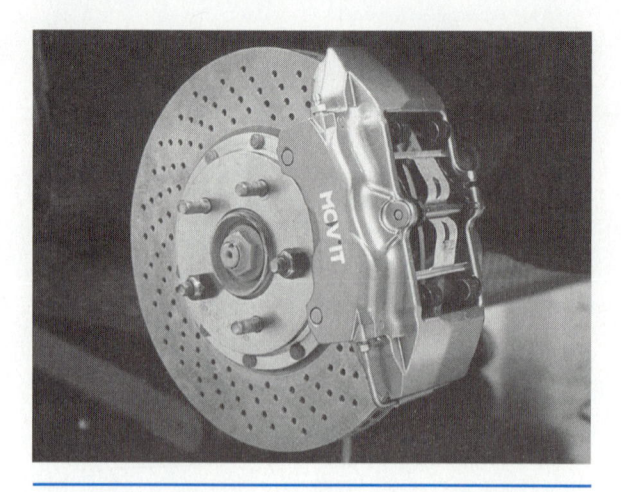

Figure 9-9. Brake rotor made from a carbon fiber-ceramic composite material. (Courtesy of Mov'It GmbH)

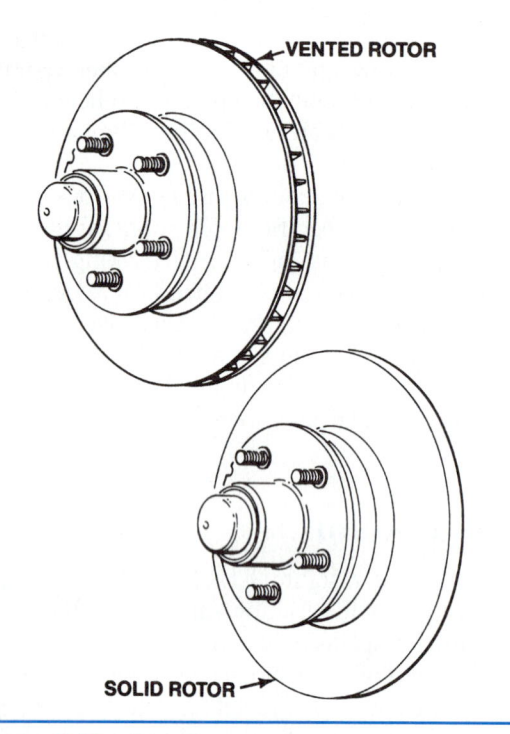

Figure 9-10. Disc brake rotors are either solid or vented.

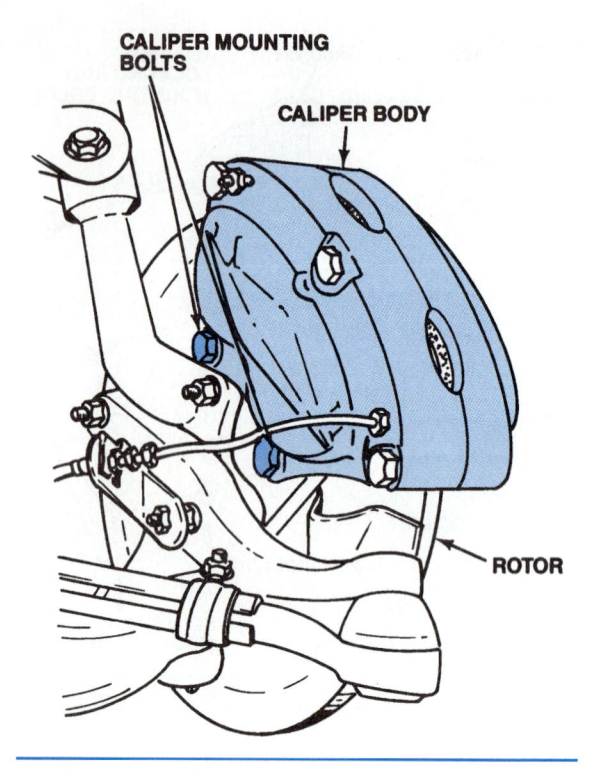

Figure 9-11. A fixed brake caliper bolts solidly to the vehicle suspension.

the rotor, aided by air flow over the cooling fins. Brake rotors are covered in detail in Chapter 11.

DISC BRAKE DESIGN

Although the hydraulic operation of all brake calipers (covered in Chapter 7) is similar, calipers differ in two important areas: how they attach to the vehicle, and how they apply the brake pads to the rotor. The manners in which these tasks are performed determine the design of a disc brake friction assembly. There are basically three types of calipers:

- Fixed
- Floating
- Sliding.

These three caliper types can be further divided into two groups, fixed calipers and all others. Fixed calipers have several unique features, but sliding and floating calipers share many traits. Each caliper design has advantages and disadvantages that make it best suited to certain applications.

Fixed Caliper Design

The **fixed brake caliper,** figure 9-11, is the earliest design and was once used in all disc brake applications. As described in Chapter 7, the fixed caliper has a body manufactured in two halves and uses two, three, or four pistons to apply the brake pads. The fixed caliper gets its name from the fact that the caliper body is rigidly mounted to the suspension. When the brakes are applied,

Disc Brake Friction Assemblies

the pistons extend from the caliper bores and apply the brake pads with equal force from both sides of the rotor. No part of the caliper body moves when the brakes are applied.

Fixed Caliper Advantages

Fixed calipers are relatively large and heavy, which enables them to absorb and dissipate great amounts of heat. This allows the brake rotor and pads to run cooler and reduces the amount of heat transferred to the brake fluid. Compared to other caliper designs, a fixed caliper is able to withstand a greater number of repeated hard stops without heat-induced fade or vapor lock of the hydraulic system.

The size and rigid mounting of a fixed caliper also means it does not flex as much as other designs. A flexing caliper is usually felt by the driver as a spongy brake pedal. Fixed calipers are very strong and provide a firm and linear brake pedal feel.

■ Early Disc Brakes

The disc brakes used today were once called spot disc brakes to separate them from other designs. One early alternative "disc" brake was the unique friction assembly found on the front wheels of the Chrysler Imperial from 1950–55. This design had two full-circle steel pressure plates faced with friction material. The pressure plates were loosely anchored to the backing plate and separated by ball bearings trapped between small ramps cast into the backs of the plates. A modified form of brake "drum" wrapped around the pressure plates to provided the friction surfaces for the linings to rub against.

When the brakes were applied, a hydraulic cylinder rotated one of the pressure plates causing it to ride up the ramps on the ball bearings; this forced the plates apart until they contacted the "drum" to create friction and stop the car. Chrysler's early "disc" brakes worked acceptably well but were costly to build, difficult to service, and almost as likely to overheat as drum brakes. Chrysler abandoned this brake design after six years of limited production and did not offer another disc brake for more than a decade.

The strength and heat-dissipating abilities of fixed calipers make them ideally suited to heavy-duty use. Most race cars use fixed calipers, as do many high-performance and European cars that are designed to be driven at high speeds.

Fixed Caliper Disadvantages

Although the size and weight of fixed calipers are advantages in heavy-duty use, they make fixed calipers undesirable for most cars sold in North America. To obtain better fuel economy, manufacturers want to eliminate as much weight as possible from new cars. In most cases, the added cost and complexity of multipiston fixed calipers cannot be justified at the lower speeds and more moderate braking experienced by U.S. drivers.

Another disadvantage of fixed calipers is their basic construction; multiple pistons and split bodies make service more difficult and allow greater opportunity for leaks. The drilled passages that route fluid through the inside of the caliper body also may contribute to cracking as miles accumulate and the caliper goes through hundreds of thousands of heating and cooling cycles. Where maximum caliper strength and durability are required, fluid is routed to the separate caliper halves through external steel brake lines.

Fixed Caliper Alignment

Because the caliper body is locked in position, a fixed caliper must be centered over the rotor and aligned so the pistons contact the brake pad backing plates parallel to the friction surface of the rotor, figure 9-12. If the caliper is not properly aligned, the pistons will contact the pads at an angle and cause tapered wear of the brake linings. If the misalignment is bad enough, the pistons will cock in their bores, suffer increased wear, and possibly crack.

Misalignment is generally not a problem with modern fixed calipers; manufacturing tolerances are controlled to ensure that any caliper will align properly when bolted to the vehicle suspension. However, some fixed calipers are individually aligned with shims on the mounting bolts between the caliper body and the suspension, figure 9-13. Whenever a fixed caliper is unbolted from its mounts, care should be taken to note the location and quantity of any shims so they can be replaced in the same positions during reassembly. If a different caliper is installed, its alignment must be checked and adjusted as necessary.

Floating and Sliding Caliper Design

The front brakes of most modern cars are fitted with either floating or sliding calipers. These calipers

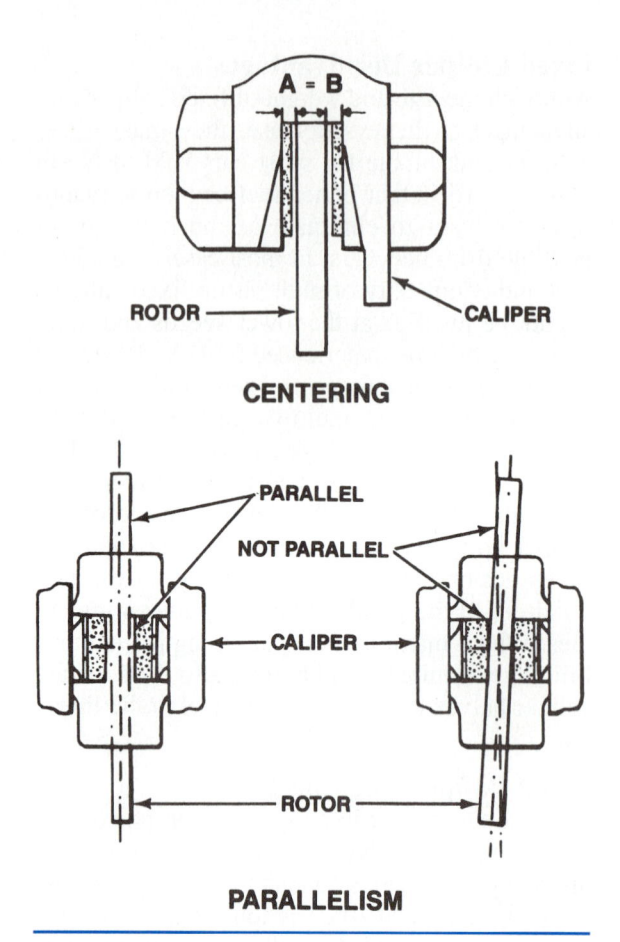

Figure 9-12. Fixed brake calipers must be centered over the rotor with their pistons parallel to the rotor friction surfaces.

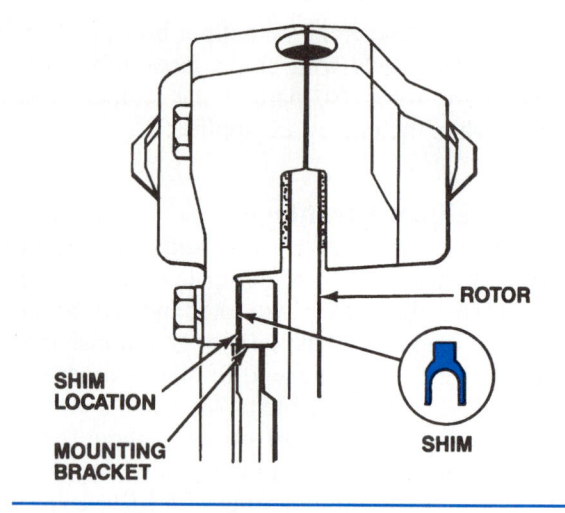

Figure 9-13. Shims are used to align some fixed brake calipers.

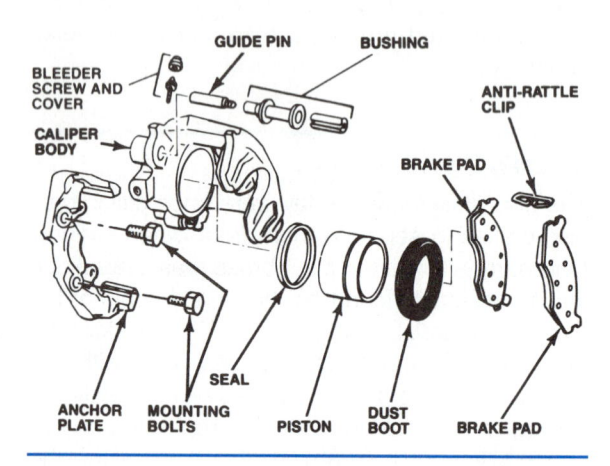

Figure 9-14. This floating caliper mounts on a separate anchor plate that bolts to the vehicle suspension.

share a number of traits, but what sets them apart from fixed calipers is that the caliper body is *not* rigidly mounted. Instead, it is free to move within a limited range on an **anchor plate** that *is* solidly mounted to the vehicle suspension. The anchor plate may be cast into a suspension member (often the front spindle), as in figure 9-7, or it can be a separate piece that bolts to the suspension, figure 9-14.

The movement of a floating or sliding caliper body plays an important role in brake operation. As described in Chapter 7, when the brakes are applied the caliper piston moves out of its bore and applies the inner brake pad. At the same time, the caliper body moves in the opposite direction on the anchor plate and applies the outer brake pad. With a floating or sliding caliper, the caliper body moves every time the brakes are applied.

Floating and Sliding Caliper Advantages

The biggest advantages of floating and sliding calipers are lower cost, simple construction, and compact size. Except for a few two-piston, light-truck calipers, all floating and sliding calipers are single-piston designs. Because they have fewer pieces, floating and sliding calipers are cheaper to build and service, and have fewer places where leaks can develop.

The smaller size of floating and sliding calipers also allows better packaging of the caliper on the vehicle. A single-piston caliper with the piston located on the inboard side of the brake rotor fits easily within the diameter of a small wheel. The inboard position of the caliper piston also contributes to better cooling because the bulk of the caliper body is exposed to the passing airflow.

Like any disc brake, floating and sliding calipers have poor parking brake performance, although they are somewhat better suited for the job than fixed brake calipers. The main reason is that

Disc Brake Friction Assemblies

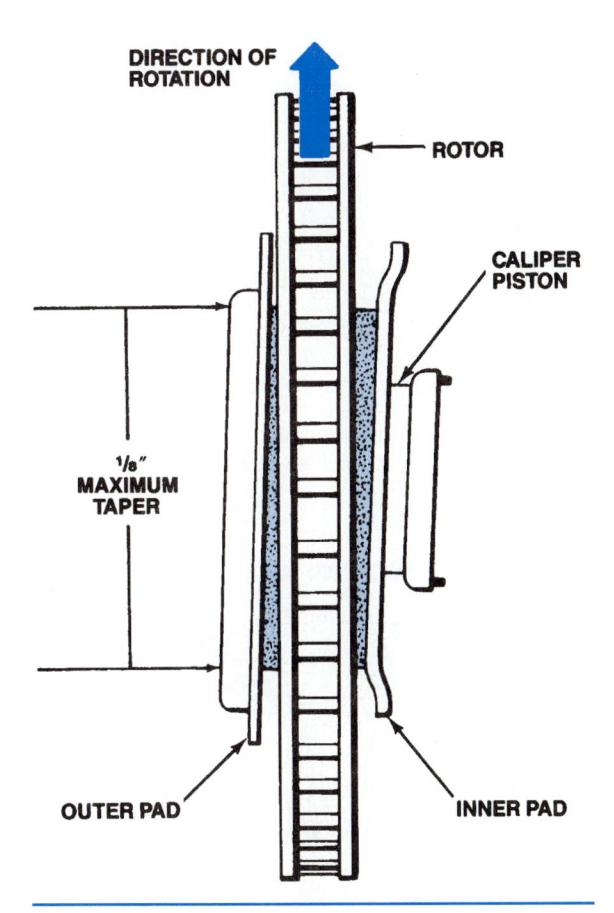

Figure 9-15. Caliper flex can cause tapered wear of the brake linings.

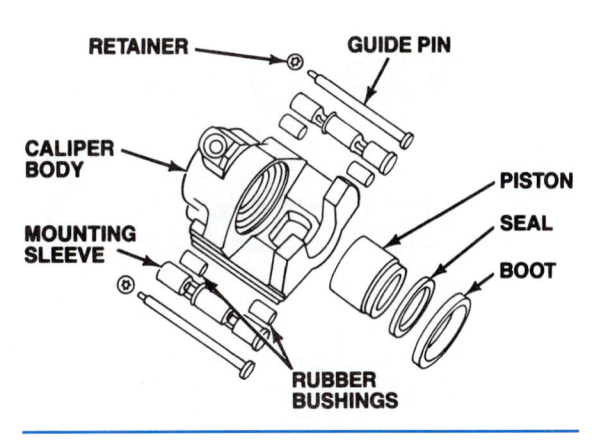

Figure 9-16. A typical floating brake caliper.

the floating or sliding action automatically distributes application force equally between the brake pads on both sides of the rotor. Unlike a fixed caliper, a floating or sliding caliper can be mechanically actuated by applying the single piston with a cable and lever mechanism.

Floating and Sliding Caliper Disadvantages

Floating and sliding calipers are not without their weak points. The movable caliper body allows a certain degree of flex, which can contribute to a spongy brake pedal. Caliper flex also allows the caliper body to twist slightly when the brakes are applied causing tapered wear of the brake pad lining material, figure 9-15.

Although the inboard piston location of floating and sliding calipers provides good cooling, these designs can never absorb as much heat (and therefore have the fade resistance) as a fixed caliper with similar stopping power. Floating and sliding calipers simply do not have the mass of fixed calipers, and their flexible mounting systems slow the transfer of heat from the caliper body to the anchor plate and other suspension components that aid in the cooling process.

Floating Calipers

Although movement of the caliper body is a feature floating and sliding calipers have in common, the manner in which the caliper body mounts to the anchor plate determines the exact kind of caliper. The body of a **floating caliper**, figure 9-16, does not make direct metal-to-metal contact with the anchor plate. Instead, the caliper body is supported by bushings and O-rings that allow it to "float" or slide on metal guide pins or locating sleeves attached to the anchor plate. For this reason, some automakers call the floating caliper a pin-slider caliper.

The bushings that support floating calipers, figure 9-17, are made from a number of materials including rubber, teflon, and nylon; O-rings are generally made from high-temperature synthetic rubber. The guide pins and sleeves, figure 9-18, are made of steel and come in a variety of shapes and sizes for different caliper designs.

Floating calipers depend on proper lubrication of their pins, sleeves, bushings, and O-rings for smooth operation. If these parts become rusted or corroded, the caliper will bind and stick causing a loss of braking power that is usually accompanied by rapid and unusual wear of the brake pads. Special high-temperature brake grease must be used to lubricate these parts any time the caliper is disassembled. Many manufacturers recommend that floating caliper pins, sleeves, bushings, and O-rings be replaced whenever the caliper is serviced. These parts come in a "small-parts kit" available from brake parts suppliers.

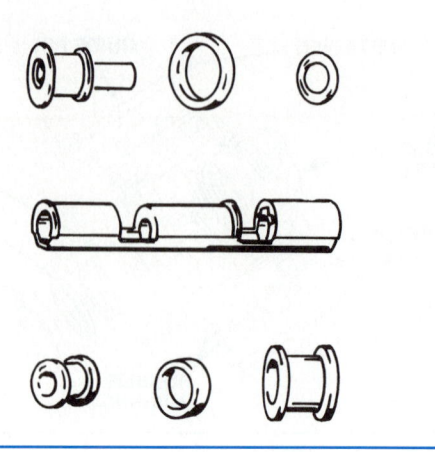

Figure 9-17. Floating calipers are supported by rubber O-rings and plastic bushings.

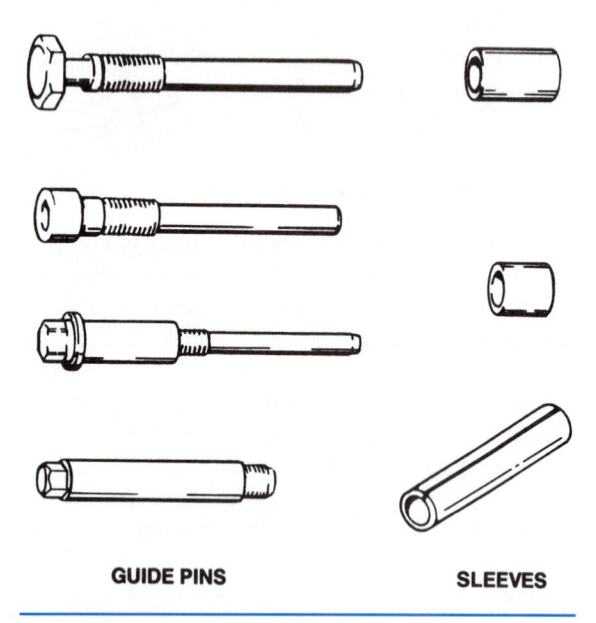

Figure 9-18. Metal guide pins and sleeves are also used to locate floating calipers.

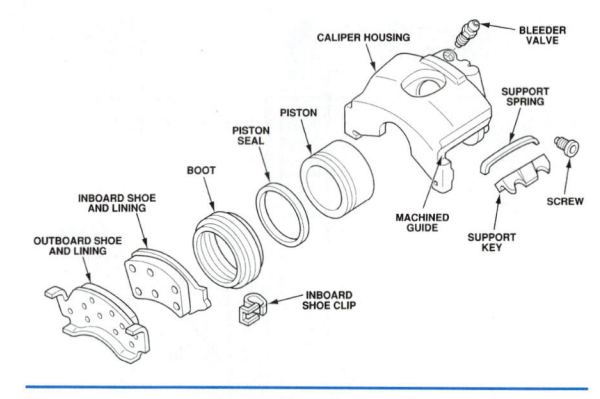

Figure 9-19. A typical sliding caliper. (Courtesy of General Motors Corporation, Service and Parts Operations)

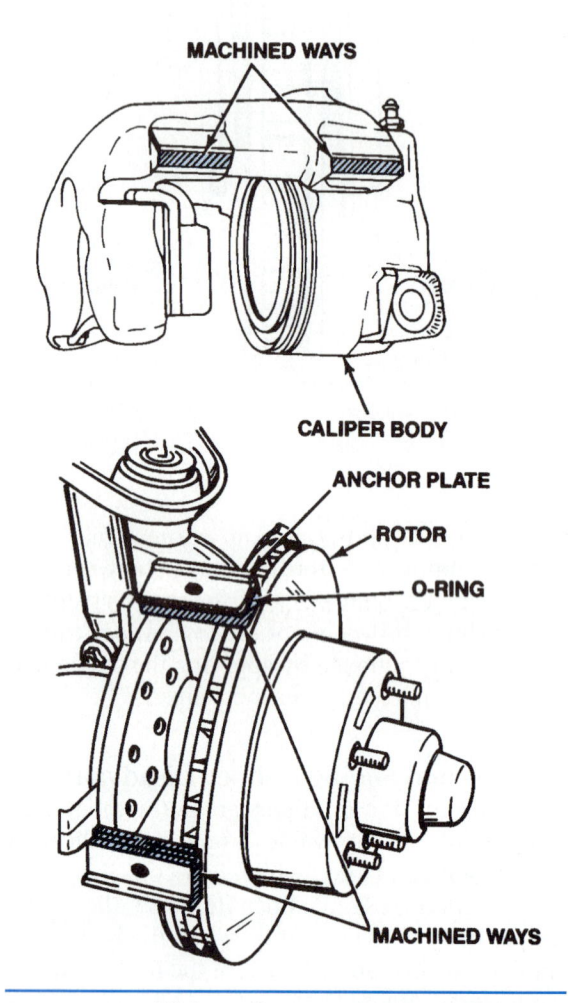

Figure 9-20. Sliding calipers move on machined ways.

Sliding Calipers

Unlike a floating caliper, the body of a **sliding caliper,** figure 9-19, mounts in direct metal-to-metal contact with the anchor plate. Instead of pins and bushings, sliding calipers move on **ways** cast and machined into the caliper body and anchor plate, figure 9-20. Retaining clips and the design of the caliper prevent the body from coming out of the ways once the caliper is assembled. On some calipers, the ways may have to be filed for proper clearance between the caliper body and anchor plate if the caliper is replaced.

Like floating calipers, sliding calipers depend on good lubrication of their ways for proper operation. If not properly coated with high-temperature brake grease, the ways can rust or corrode causing the caliper to drag or seize.

Disc Brake Friction Assemblies

REAR DISC BRAKES

In recent years, four-wheel disc brake systems have become more common. In theory, rear disc brakes offer the same advantages as front discs: improved resistance to fade, automatic adjustment, and freedom from pull. However, in the real world, the benefits of rear disc brakes must be weighed against their increased cost and complexity.

In most rear-wheel applications, drum brakes are adequate to provide the relatively small portion of a car's total braking power required of them. Because rear drum brakes are lightly loaded, fade is a problem only in extreme conditions when the front brakes fade and force the rear brakes to take on a larger part of the braking load. The automatic adjusting ability of disc brakes is also less of an advantage in slow-wearing rear brakes, and because the rear wheels are not steered, a drum brake's susceptibility to pull is not a significant problem on the rear axle.

In practical terms, the main advantage of rear disc brakes is superior stopping power and fade resistance under repeated heavy braking. For this reason, four-wheel disc brakes are usually found on heavyweight luxury cars and trucks, and on performance models capable of high speeds.

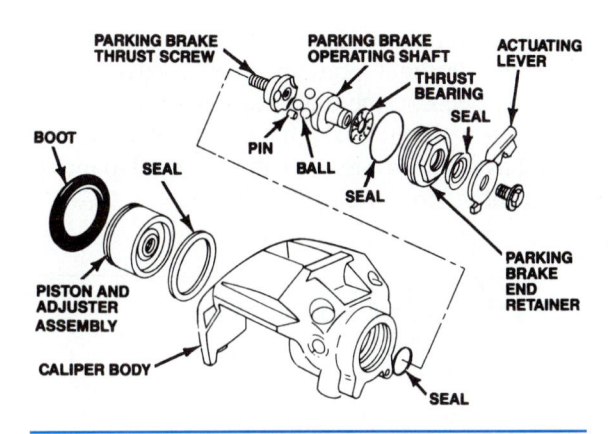

Figure 9-21. This single-piston brake caliper is mechanically actuated to serve as a parking brake.

Rear Disc Parking Brakes

As already mentioned, the most significant drawback of rear disc brakes is their poor parking brake performance. There are two methods of providing parking brakes when rear discs are installed on a car; both increase the complexity of the brake caliper.

Where single-piston floating or sliding rear calipers are fitted, the normal practice is to adapt the disc brake to also function as the parking brake. This is done by installing a series of cables, levers, and internal parts to mechanically actuate the brake caliper, figure 9-21.

When fixed calipers are fitted, mechanical actuation of the caliper becomes impractical because force must be generated on both sides of the rotor to apply the brake. A linkage able to actuate multiple pistons would be very complex, so the parking brake on most cars with fixed calipers is provided by small, mechanically actuated drum brakes inside the rear rotors, figure 9-22. This type of parking brake, also referred to as a **drum-in-hat** parking brake, is also used on some floating caliper-type rear brakes. The operation of disc brake parking brakes is explained in detail in Chapter 12.

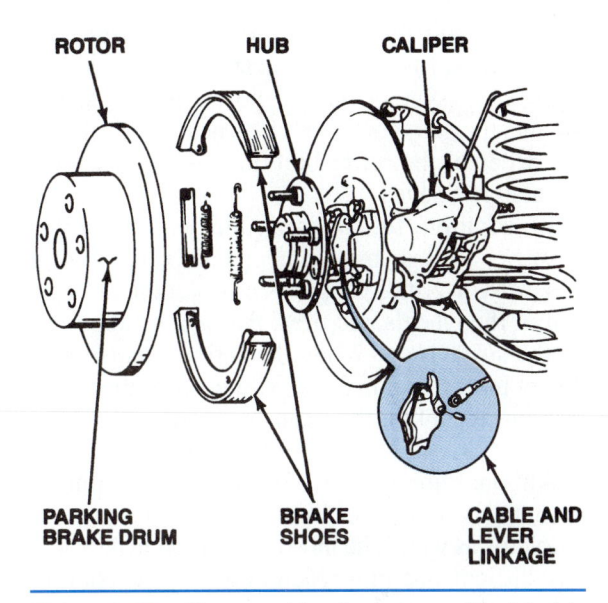

Figure 9-22. Drum parking brakes are fitted inside the rotors on most fixed rear calipers.

Four-Wheel Disc Hydraulic Systems

Although four-wheel disc brakes make the parking brake system more complex, they allow the brake hydraulic system to be simpler than that of a disc/drum brake system. When rear drum brakes are replaced by disc brakes, the metering valve becomes unnecessary because there is no need to delay front brake application until the rear brake shoes have moved into contact with the drums; the disc brakes at all four wheels can be applied simultaneously.

The elimination of rear drum brakes also allows the residual pressure check valve to be deleted. Because there are no longer any wheel cylinders with cup seals that can allow air to enter the hydraulic system, there is no need to maintain positive pressure in the system.

INBOARD DISC BRAKES

It was mentioned earlier that the size of the brake friction assembly is limited because it must fit inside the wheel. Racing cars and a few production vehicles with fully independent suspensions have gotten around this limitation by using **inboard disc brakes**, figure 9-23. There are several variations of this system, but the most common design mounts the calipers on brackets near the differential, whereas the rotors bolt directly to the differential axle flanges.

In addition to allowing the use of larger brakes, inboard mounting reduces **unsprung weight**. Unsprung weight is any weight, such as that of the wheels and tires, that is not supported by the suspension. When the springs and shocks do not have to control the extra mass of the brake friction assembly, they can be tuned to respond more quickly to changes in the road surface.

Although inboard disc brakes provide limited advantages in specific applications, they present problems as well. The inboard location makes service difficult and may result in poor brake cooling unless special air ducts are installed. Another disadvantage of inboard brakes is that they require braking torque to pass through the axles rather than directly into the chassis as with a conventional brake. The added stress on the halfshafts increases universal joint wear and the chances of shaft failure. These drawbacks are acceptable on racing cars but usually too costly and impractical for the street.

Aircraft Disc Brakes

Although most smaller planes use spot disc brakes similar to those on cars, larger aircraft use a type of brake called the multidisc brake. In this design, two sets of alternating discs, much like a hydraulic clutch in an automatic transmission, take the place of the brake rotor and pads. The *driven* discs spin with the wheel and act as rotors. The *stationary* discs, fitted between the driven discs, are keyed to the suspension and act as the brake pads. To apply the brakes, the stack is clamped together and friction between the discs slows the plane.

Multidisc brakes are well suited to aircraft use because they are light and compact. Multidisc brakes do have poor cooling ability, however, and would fade if required to perform the repeated stops common in automotive service. This is not a concern on aircraft, however, because they only stop hard infrequently—when they land—and then have a long cooling period before the brakes are called on again.

SUMMARY

Disc brakes are used on the front wheels of all late-model cars and on the rear wheels of some vehicles. Compared to drum brakes, disc brakes provide greater stopping power with increased resistance to fade. Disc brakes are less susceptible to fade because they have greater swept area, their open construction exposes their components to the passing airflow, and the spinning rotor makes the brake self-cleaning.

Mechanical fade is not a problem with disc brakes because the rotor expands toward the brake linings as it heats up rather than away from them. Lining fade occurs later in a disc brake because of its superior cooling ability. And disc brakes have little trouble with gas fade because they have no drum to contain gasses, the flow of air over the brake carries away particles, and the small surface area of the linings allows both gasses and particles to escape easily. Disc brakes

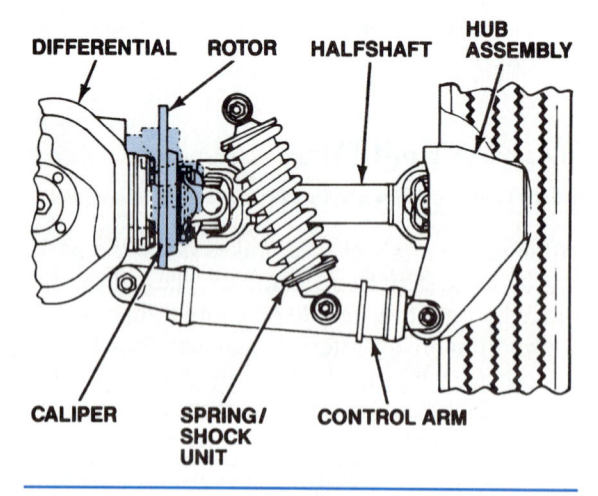

Figure 9-23. Inboard disc brakes were used on a few older imported cars.

Disc Brake Friction Assemblies

counteract water fade well because the centrifugal force of the spinning rotor, and the position of the brake pads next to the friction surface, help keep water from between the pads and rotor.

Disc brakes automatically compensate for lining wear and do not require cables, links, levers, ratchets, or struts to provide brake adjustment. A disc brake can stop straighter under a wider range of conditions than a drum brake because it has no self-energizing or servo actions that affect the side-to-side balance of braking power.

Noise caused by vibration of the brake pads is a common complaint about disc brakes; however, their only functional weakness is poor parking brake performance. Disc brakes make less effective parking brakes because they lack self-energizing and servo action, and their small lining-to-rotor contact area results in a low static coefficient of friction. The lack of self-energizing or servo action also means disc brakes usually require a power booster to keep pedal pressures reasonable.

There are three basic types of calipers: fixed, floating, and sliding. Fixed calipers are relatively large and heavy, and mount solidly to the suspension. They can dissipate large amounts of heat and are quite rigid, which provides a firm and linear brake pedal feel. However, the multiple pistons, split bodies, and drilled fluid passages of fixed calipers make them more difficult to service and present greater opportunity for leaks and cracks to develop.

Most modern cars use floating or sliding calipers. These designs have a caliper body free to move on an anchor plate mounted to the suspension. The advantages of floating and sliding calipers are simple construction, light weight, and compact size. Floating and sliding calipers do flex somewhat. This allows the caliper to twist when the brakes are applied causing a spongy brake pedal and tapered wear of the brake linings. Floating and sliding calipers also have poor parking brake performance but are better suited to the job than fixed calipers because they can be mechanically actuated more easily.

Rear disc brakes offer superior stopping power and fade resistance under repeated heavy braking. However, the benefits may not outweigh the increased cost and complexity in most applications. Four-wheel disc brakes do allow the metering and residual pressure relief valves to be eliminated from the brake hydraulic system.

Two types of parking brakes are used with rear disc brakes. Single-piston floating or sliding calipers are mechanically actuated to function as the parking brake. Cars with fixed caliper rear disc brakes have small, mechanically actuated drum brakes inside the rear rotors.

A few vehicles use disc brakes mounted inboard of the wheel and tire assemblies. This allows the use of larger brakes and reduces unsprung weight. However, servicing inboard brakes is difficult, and they sometimes suffer from inadequate cooling.

Review Questions

Choose the letter that represents the best possible answer to the following questions:

1. Which of the following is *not* an advantage of disc brakes?
   a. Increased resistance to fade
   b. Better brake cooling
   c. Self-energizing action
   d. Less chance of brake pull

2. Brake swept area is a measurement of:
   a. Brake pad area compared to rotor area
   b. Brake shoe area compared to drum area
   c. Both a and b
   d. Neither a nor b

3. A disc brake's resistance to fade depends on its:
   a. Open construction
   b. Centrifugal force
   c. Heat expansion
   d. All of the above

4. A sinking brake pedal on a car with front disc brakes may be a sign of:
   a. Vapor lock
   b. Front brake fade
   c. Improper pedal adjustment
   d. None of the above

5. A hard brake pedal on a car with disc brakes may be a sign of:
   a. Gas fade
   b. Lining fade
   c. Both a and b
   d. Neither a nor b

6. Which of the following is *not* true of disc brakes?
   a. They are more resistant to water fade than drum brakes
   b. Drilled rotors help prevent water fade
   c. Slotted linings provide escape paths for water
   d. The wheels completely shroud the rotor to prevent water fade

7. Disc brakes suffer less from brake pull because they do *not*:
   a. Have any self-energizing effect
   b. Use lining friction to increase application force
   c. Have any servo effect
   d. All of the above

8. Most disc-brake-equipped cars use a power booster because they:
   a. Have poor parking brake performance
   b. Require greater application force
   c. Need it to increase servo effect
   d. Are noisy

9. Which of the following is *not* a method used to control disc brake noise?
   a. Plastic shims
   b. Metal springs
   c. Natural frequency
   d. Antivibration liquids

10. A disc brake's parking brake performance is affected by:
    a. A lack of servo action
    b. Its static friction coefficient
    c. Both a and b
    d. Neither a nor b

11. Large heavy cars can benefit most from using:
    a. Ventilated rotors
    b. Fixed calipers
    c. Both a and b
    d. Neither a nor b

12. Fixed brake calipers:
    a. Can absorb the most pressure
    b. Are popular on race cars
    c. Can cause a spongy brake pedal
    d. Are popular on newer production cars

13. Fixed calipers are used mostly on high-performance vehicles. This is because:
    a. They are expensive to manufacture
    b. They require closer tolerances in manufacturing
    c. They are difficult to service
    d. All of the above

14. Fixed calipers must be aligned so that they are _____ the brake rotor.
    a. Centered over
    b. Parallel with
    c. Both a and b
    d. Neither a nor b

15. The body of a _____ caliper is free to move when the brakes are applied.
    a. Sliding
    b. Fixed-pin
    c. Anchor-slider
    d. All of the above

Disc Brake Friction Assemblies

16. Floating calipers are used on many new cars because they:
    a. Provide good parking brake performance
    b. Have simple construction
    c. Allow a small amount of caliper flex
    d. Offer the best brake cooling

17. Bushings, O-rings, sleeves, and pins are used to support the bodies of _____ calipers.
    a. Fixed
    b. Way-type
    c. Floating
    d. Anchor-slider

18. Sliding caliper ways:
    a. Require special lubrication
    b. Are machined into the anchor plate
    c. Both a and b
    d. Neither a nor b

19. Rear disc brakes are common on:
    a. Heavy luxury cars and trucks
    b. High-performance vehicles
    c. Sport utility vehicles
    d. All of the above

20. Inboard disc brakes:
    a. Allow the use of larger brakes
    b. Generally run cooler than other brakes
    c. Increase unsprung weight
    d. All of the above

10
Brake Shoes and Pads

OBJECTIVES

Upon completion and review of this chapter, you will be able to:

- Describe brake shoe construction and function.
- Describe brake pad construction and function.
- List and describe the composition of friction materials.
- Explain the properties of the four types of friction materials.
- Explain the term "edgebranding."
- List and describe the four lining assembly methods.
- Describe the proper shoe-to-drum fit.

KEY TERMS

anchor eyes
aramid fiber
arcing
binders
bonded linings
brake block
brake pad
brake shoes
core charge
curing agents
edgebrand
fillers
friction material
friction modifiers
glazed lining
hybrid pad sets
lining table
metallic linings
mold bonded linings
nibs
organic linings
oversize brake shoes
pad wear indicators
riveted linings
semimetallic linings
shoe web
sintering
synthetic lining

INTRODUCTION

This chapter deals with drum brake shoes and disc brake pads. Together with the drums and rotors, which are covered in the next chapter, the shoes and pads make up the friction surfaces that convert kinetic energy into heat to stop the car. Shoes and pads operate under the most extreme conditions in the entire brake system and are subject to a great deal of wear. The replacement of worn brake shoes and pads is a common part of brake service.

Although they appear to be simple parts, modern brake shoes and pads are the result of years of engineering development. Often, two shoes or pads that look identical will not perform the same because different friction materials are used for their linings. When a brake system is designed,

179

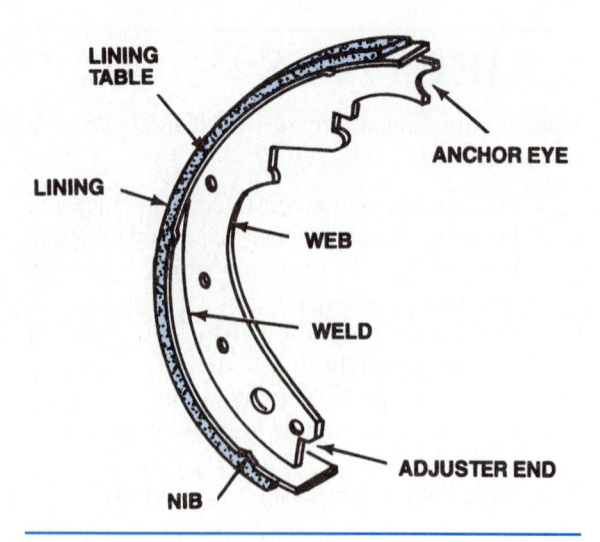

Figure 10-1. Steel brake shoes are made from two stampings welded together.

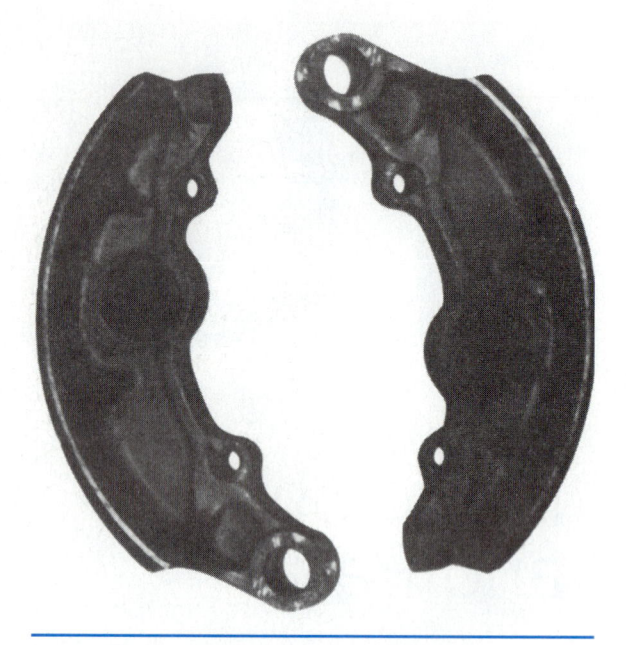

Figure 10-2. A cast-aluminum brake shoe.

engineers test the performance characteristics of many friction materials; they then specify those that will work best in that particular application.

The first part of this chapter covers the physical construction of brake shoes and pads. The second part looks at the various types of brake lining materials, the friction properties of those materials, and how they are attached to brake shoes and pads. The final part of the chapter examines the ways in which brake shoe linings are fitted to the drums.

DRUM BRAKE SHOE CONSTRUCTION

The linings of drum brakes are attached to curved metal assemblies called **brake shoes.** Most shoes are made of two pieces of sheet steel welded together in a T-shaped cross section, figure 10-1. Some shoes are cast from aluminum in the same basic shape, figure 10-2. Although aluminum shoes are lighter than their steel counterparts, they are more expensive to make and not as durable at high temperatures. Regardless of whether a shoe is made of steel or aluminum, its outer edge is lined with a friction material that contacts the brake drum to generate the actual stopping power.

The curved metal piece on the outer portion of the shoe is called the **lining table**; it can also be called the shoe rim or platform. The lining table supports the block of friction material that makes up the brake lining. On some shoes, the edge of the lining table contains small V-or U-shaped notches called **nibs.** The nibs rest against the shoe support pads on the backing plate when the shoe is installed.

The metal piece of the shoe positioned under the lining table and welded to it is called the **shoe web.** All of the application force that actuates the shoe is transferred through the web to the lining table. The web usually contains a number of holes in various shapes and sizes for the shoe return springs, holddown hardware, parking brake linkage, and self-adjusting mechanism.

The inner edge of the web can be shaped in many ways depending on the friction assembly design. One end usually has a notch or protrusion where the wheel cylinder applies the shoe, whereas the other end commonly has a flat or curved surface where the shoe meets an anchor or adjusting link. The upper ends of the webs on dual-servo brake shoes have semicircular **anchor eyes,** figure 10-1, where the webs contact the anchor post on the backing plate.

The brake lining can be any one of several materials and can be fastened to the lining table in many ways. Lining materials and attaching methods are covered in greater detail later in the chapter. The ends of the linings on most brake shoes are tapered to prevent vibration and brake noise, figure 10-3.

Brake shoes are sturdy parts that can be relined and reused many times if the web and lin-

Brake Shoes and Pads

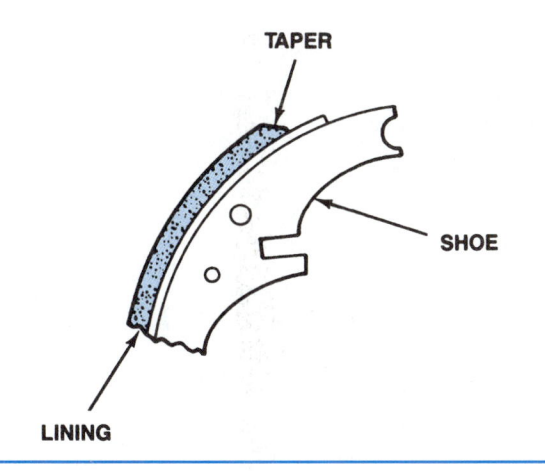

Figure 10-3. Tapered ends on brake shoe linings reduce brake noise.

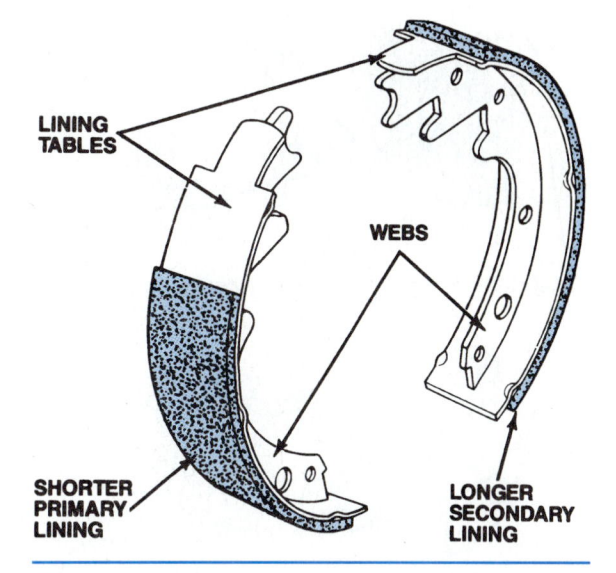

Figure 10-4. The primary and secondary shoes of dual-servo brakes have unique features.

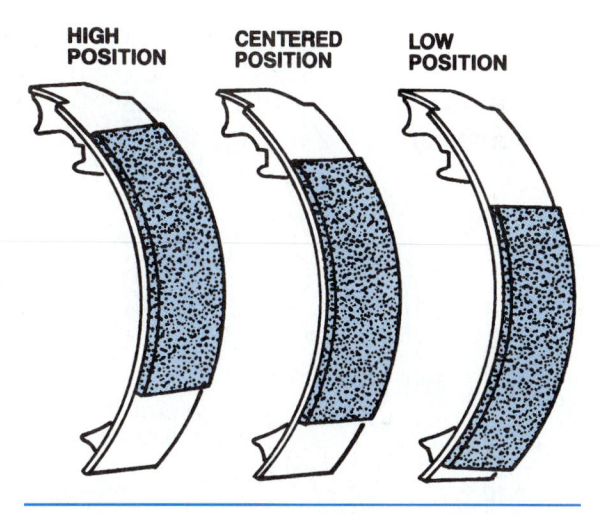

Figure 10-5. Primary shoe lining position may vary in different applications.

ing table are not damaged. In fact, the brake shoes for any given application are usually available in both "new" and "relined" versions from suppliers. Because of their high cost, aluminum shoes are almost always relined. Relined brake shoes are usually sold on an exchange basis. At the time of purchase, a **core charge** is added to the cost of the relined parts. This charge is refunded when the old shoes are returned in rebuildable condition.

Primary and Secondary Brake Shoes

In many drum brake friction assemblies, the two shoes are interchangeable; one is a mirror image of the other. However, as discussed in Chapter 8, the shoes in a dual-servo brake perform very different jobs; the primary shoe is self-energized by drum rotation to create a servo action that forces the secondary shoe more firmly against the drum. Whereas the primary shoe increases application force, the secondary shoe provides most of the braking power. Because of this, the two shoes have definite physical differences and cannot be interchanged.

To help deal with the added friction, heat, and wear it undergoes, the lining of the secondary shoe extends nearly the full length of the shoe lining table, figure 10-4. The secondary shoe lining material also has a high coefficient of friction to provide good stopping power. The primary shoe undergoes far less stress than the secondary shoe, and its lining is often shorter—sometimes less than half the length of the lining table. In addition, the lining material usually has a lower coefficient of friction. This prevents the shoe from engaging the drum too quickly or harshly, which could cause the brakes to grab or lock.

On most dual-servo brake primary shoes, the lining is positioned near the center of the lining table, figure 10-5. However, in some cases, the lining may be positioned above or below the lining table centerline. Higher or lower lining positions provide better braking action, or prevent noise, in certain applications.

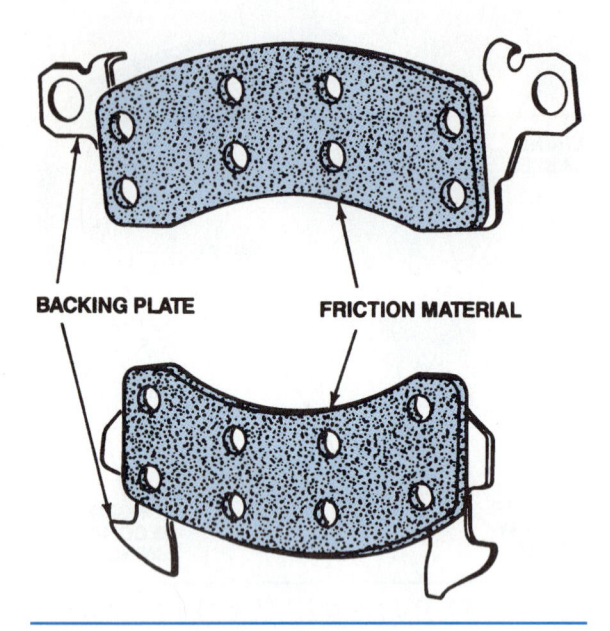

Figure 10-6. Typical disc brake pads.

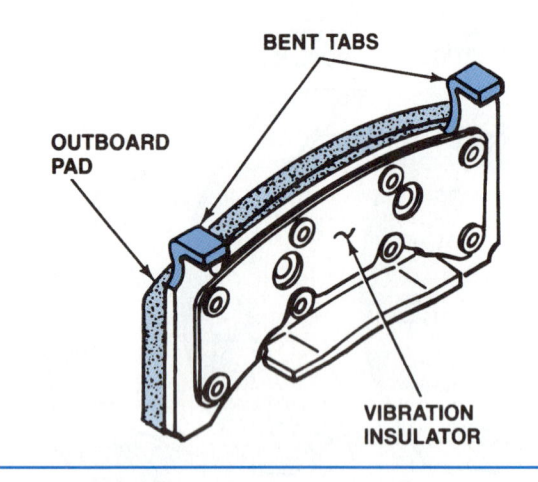

Figure 10-7. To prevent noise, bent tabs on the backing plate clip some brake pads firmly to the caliper.

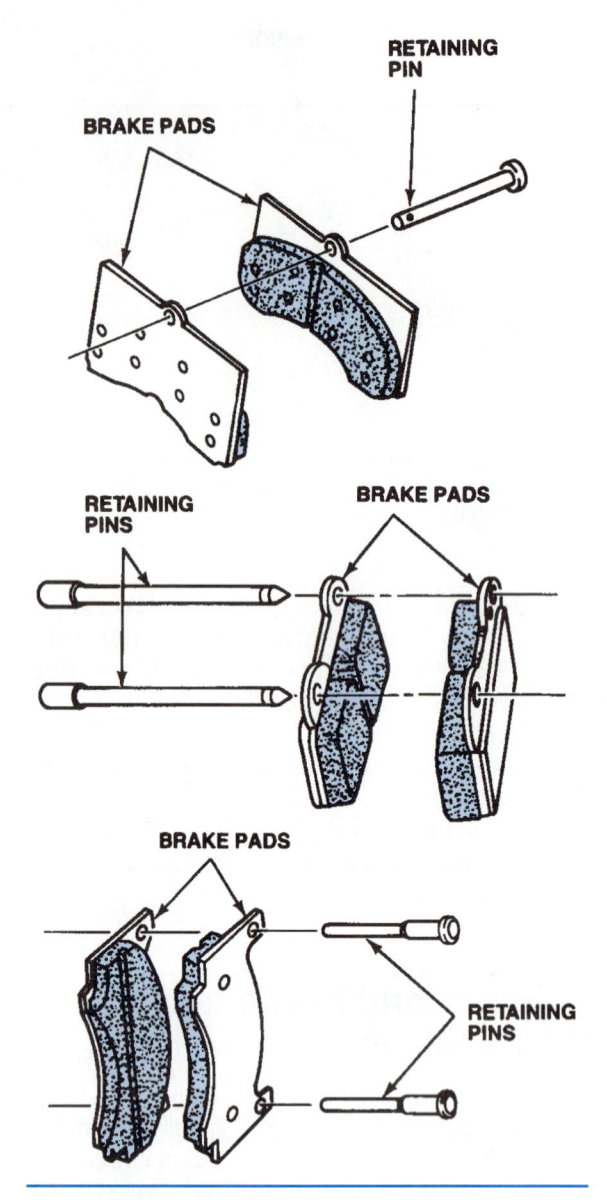

Figure 10-8. Holes in the backing plate are a common method of locating a pad in the caliper.

DISC BRAKE PAD CONSTRUCTION

The lining of a disc brake is part of an assembly called the **brake pad,** figure 10-6. Compared to a brake shoe, a brake pad is a relatively simple part that consists of a block of friction material attached to a stamped steel backing plate. Some pad backing plates have tabs, figure 10-7, that bend over the caliper to hold the pad tightly in place and help prevent brake noise. Other pad backing plates have tabs with holes in them, figure 10-8; a pin slips through the holes and fastens to the caliper body to hold the pads in position. Still other pad backing plates have a retainer spring attached that locates the pad in the caliper by locking it to the caliper piston, figure 10-9.

As with brake shoes, the lining material of a disc pad can be any one of a number of products that can be fastened to the backing plate in several ways. Friction materials and the methods used to attach them are covered later in the chapter.

The edges of the lining material on a brake pad are usually straight cut or slightly chamfered, although a few pads will have tapered edges to help combat vibration and noise. Many pad linings will also have a slot cut into the lining to help dis-

Brake Shoes and Pads

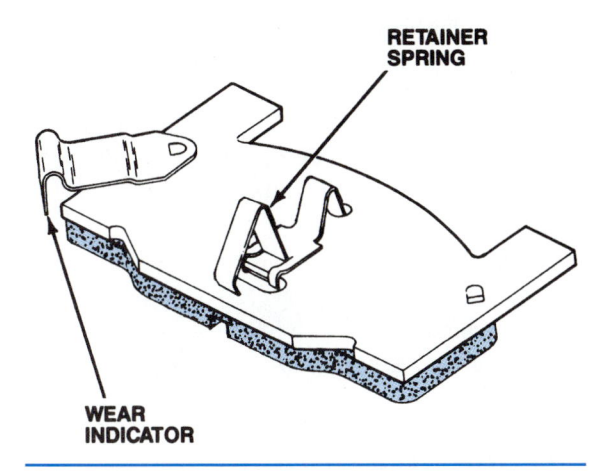

Figure 10-9. Retainer springs lock the pad to the caliper piston to prevent brake noise.

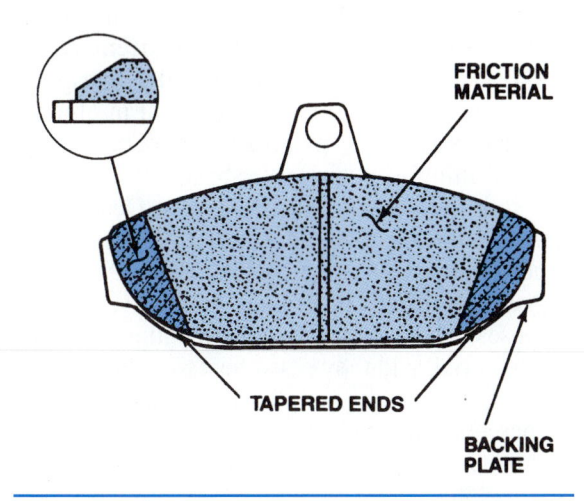

Figure 10-10. The lining edges of a few larger brake pads are tapered to prevent vibration.

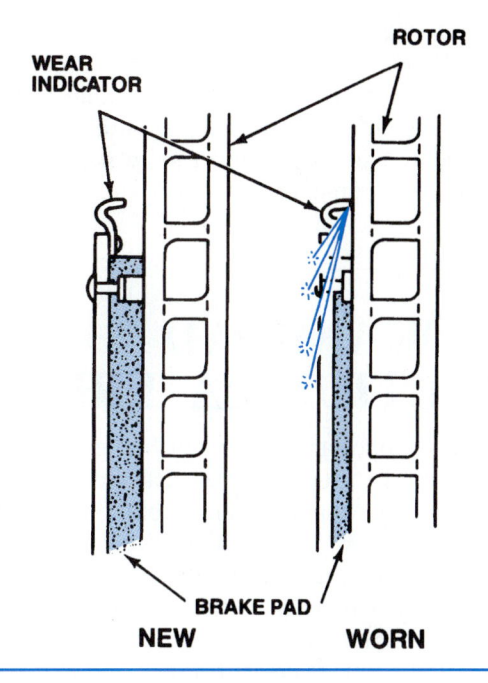

Figure 10-11. Mechanical wear indicators rub against the brake rotor when the pads need replacement.

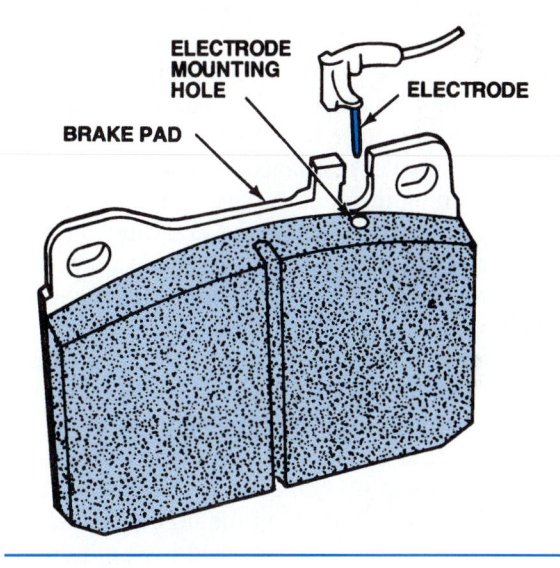

Figure 10-12. Electrical wear indicators ground a warning light circuit when the pads need replacement.

sipate gases generated during high-temperature braking, figure 10-10.

Pad Wear Indicators

Although not required by law, a growing number of vehicle manufacturers are fitting **pad wear indicators** to their brakes for safety reasons. Modern pad wear indicators are either mechanical or electrical and signal the driver when the lining material has worn to the point where pad replacement is necessary.

A mechanical wear indicator, figure 10-9, is a small spring-steel tab riveted to the pad backing plate. When the friction material wears to a predetermined thickness, the tab contacts the rotor and makes a squealing or chirping noise (when the brakes are not applied) that alerts the driver to the need for service, figure 10-11.

Electrical wear indicators, figure 10-12, use a coated electrode imbedded in the lining material to generate the warning signal. The electrode is wired to a warning light in the instrument panel; when the lining wears sufficiently, the electrode grounds against the rotor to complete the circuit and turn on the warning light.

In many applications the slot in the pad lining, figure 10-12, can be considered to be a wear indicator. When the pad is worn to the point that the slot is no longer visible, the pad is ready for replacement. This inspection usually requires that the wheel be removed for a visual inspection of the pads.

SHOE AND PAD FRICTION MATERIALS

The most important element of brake shoes and pads is the **friction material** that makes up the brake linings. Manufacturers of brake linings often refer to raw pieces of friction material as **brake block**. Technically, the brake *lining* is made of *brake block* manufactured from a particular type of *friction material*.

All brake linings create friction, and in theory, the same friction material could be used for both drum brake shoes and disc brake pads. In practice, however, different brake designs require different kinds of friction materials. The most basic example of this is disc brakes, which routinely operate at much higher temperatures than drum brakes. Because of this, disc brake pad friction materials must have greater resistance to heat-induced fade than the lining materials on drum brake shoes.

Another reason there are several kinds of friction materials is that no one type can do everything well under every condition. The perfect friction material would not produce any byproducts that are hazardous to your health; however, it would have a high coefficient of friction that remained constant over a wide range of temperatures. It would neither wear out rapidly nor cause excessive wear of brake drums and rotors. It would be able to withstand the highest temperatures without fade or failure, and it would do all this without making any noise! Unfortunately, all current brake friction materials have both good and bad points, so most modern brake systems are designed around a particular type of lining. This allows the engineer to specify components that take maximum advantage of the friction material's strengths, while keeping the effects of any weaknesses to a minimum.

In the search for the perfect brake lining, engineers have tried many substances and combinations of substances. Modern friction materials are made from mixtures of several elements that are blended together in precise ratios, and then molded under heat and pressure into brake block. The exact formulation of specific friction materials is rarely known because that information is a trade secret. Nonetheless, lining friction materials can be divided into categories based on the substance that makes up the bulk of their content:

- Organic
- Metallic
- Semimetallic
- Synthetic
- Other lining materials.

Organic Brake Linings

Until recently, **organic linings** were the most common type on automobiles. The name *organic* comes from early friction materials that were made of substances that had once been living: wood, leather, or cotton impregnated with asphalt and rubber. Later, the main ingredient in organic linings was asbestos, an inorganic mineral. Today, organic linings contain less asbestos and more of the other ingredients described later. Despite the many changes in content, the term *organic* continues to be used to identify this group of linings.

Asbestos fibers in organic linings serve two purposes: They reinforce and strengthen the friction material, and they resist heat to reduce lining wear. In the past, organic linings contained 50 to 75 percent asbestos fibers. However, since the health hazards of asbestos have been identified (see Chapter 2), the amount of asbestos has fallen below 25 percent in many new organic linings. Because of the dangers, the Environmental Protection Agency (EPA) encourages the elimination of asbestos from brake linings. Many brake pad and shoe manufacturers offer asbestos-free linings as a feature of their higher-priced products.

In addition to asbestos fibers, a typical organic lining material contains one or more of the following: friction modifiers, powdered metal, binders, fillers, and curing agents. **Friction modifiers** are substances that alter the friction coefficient of the brake lining; graphite and oil of cashew-nut shells are common friction modifiers. Powdered metals increase a friction material's resistance to heat fade; lead, zinc, brass, and aluminum are among the many metals added to brake linings. **Binders** are the glues that hold a friction material together; phenolic resin is one of the most common binders used today. **Fillers** are materials added to friction materials in small amounts for specific purposes; rubber chips that

Brake Shoes and Pads

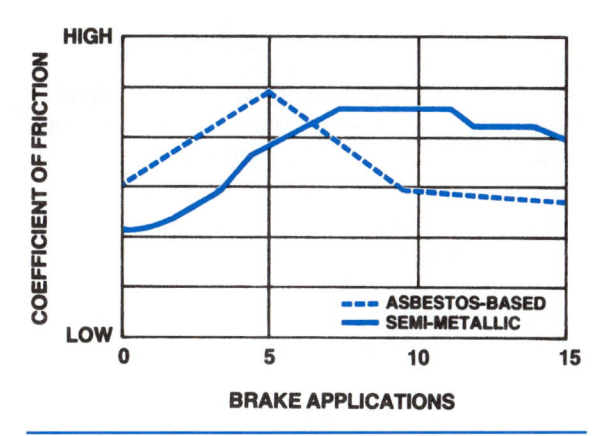

Figure 10-13. The friction coefficient of a brake lining changes as repeated brake applications increase operating temperatures.

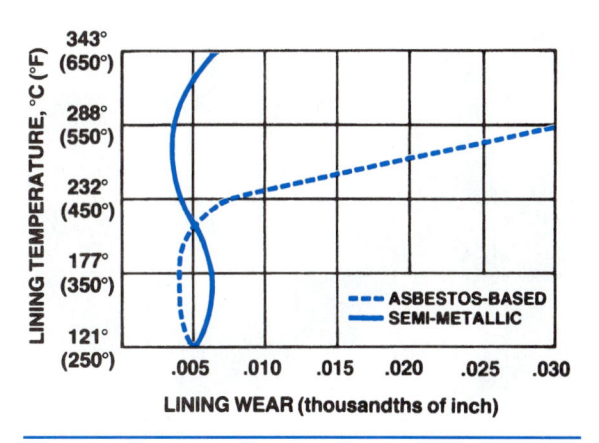

Figure 10-14. Brake lining wear is directly related to operating temperature.

reduce brake noise are a good example of a filler. Finally, **curing agents** ensure that the correct chemical reactions occur between the various substances during manufacture.

The biggest advantages of organic linings are low cost and quiet operation. In addition, organic linings are only mildly abrasive, which extends the life of drums and rotors. Finally, organic linings wear slowly under average braking conditions and have a high coefficient of friction when cold, which keeps brake pedal pressures low during the first few stops.

The main disadvantage of organic linings is that they fade quickly and at relatively low temperatures. Organic linings perform best when kept under 400°F (204°C); above this temperature, they wear rapidly and their coefficient of friction falls off, figure 10-13, as fade becomes a problem. By 550°F (288°C), fade is severe and wear is six times greater than normal, figure 10-14. Because of these traits, organic linings perform best only in large heavy brake systems designed to limit maximum lining temperatures.

High-Temperature Organic Linings

One response to the heat-related problems of organic brake linings was the development of high-temperature organic linings. These linings contain a high percentage of metal particles to improve their wear and friction properties at high temperatures. Many racing-car brakes use high-temperature organic linings, and some of these friction materials work well at temperatures as high as 1250°F (677°C).

The drawbacks of the high-temperature organic linings used on racing cars include rapid lining wear and a poor coefficient of friction at low temperatures. These friction materials are also very brittle, which makes them prone to cracking and difficult to attach to shoes and pads. Finally, high-temperature racing organics are highly abrasive and cause rapid drum and rotor wear.

"Autobahn Formula" Organic Linings

High-performance imported cars often use special high-temperature organic brake linings made from a friction material blend called an "autobahn formula." The name refers to the German highways without speed limits where cars must be able to slow repeatedly from high speed without brake fade. The mixture of substances in these linings includes a high metal content but eliminates many of the drawbacks of racing-type high-temperature organics.

Although they perform better than basic organic linings, "autobahn formula" friction materials will not operate to quite the same temperature extremes as racing organic linings. However, they are not as brittle or abrasive as racing linings, and they have better wear characteristics and a higher coefficient of friction when cold. One drawback of these linings is their relative hardness, which makes them rather noisy. Brake noise is considered normal in Europe, but many North American drivers find it unacceptable.

Nonasbestos Organic Linings

Most brake manufacturers offer a lining material that is completely free of asbestos, often as a deluxe or premium brake pad. These are referred to as nonasbestos organic (NAO) brake pads and are formulated from nonasbestos, organic materials.

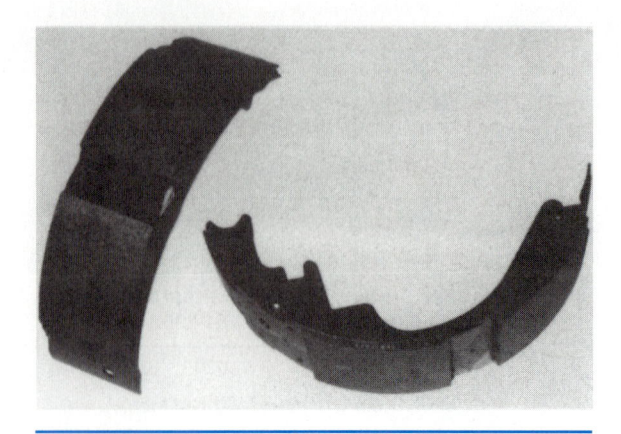

Figure 10-15. Metallic brake linings resist fade even at extreme temperatures.

They feature good heat resistance, smooth braking, low dust, and low squeal, according to one company.

Metallic Brake Linings

Full-metallic brake linings, often called simply **metallic linings,** use friction materials made entirely of powdered metal, figure 10-15. The metals are formed into a brake block by **sintering,** a process in which a powdered mixture is compressed in a mold under high temperature and pressure until the individual elements fuse into a single part. Sintering makes binders unnecessary.

The main advantage of metallic brake linings is excellent fade resistance at high temperatures. This benefit is offset, however, by several significant problems. Metallic linings work very poorly on the first few stops when cold, and even after they reach operating temperature, metallic linings require high pedal pressures to generate acceptable stopping power. Other drawbacks are increased drum and rotor wear, difficulty in attaching the lining material to the brake shoe or pad, and brake noise. In addition, some metallic friction material formulations grab unpredictably.

Metallic linings were used primarily in the heyday of drum brakes. At that time, they were the preferred friction material for heavy-duty use in racing cars, taxicabs, and police vehicles. They have since been replaced by high-temperature organic linings for racing cars and by semimetallic linings (see following discussion) for heavy-duty use in production cars. Metallic linings are generally unavailable for modern automotive applications, although they are still fitted extensively to heavy aircraft. Some motorcycles with stainless steel brake rotors also use metallic pads.

Semimetallic Brake Linings

Semimetallic linings were developed to combat heat better than organic friction materials while avoiding the drawbacks of full-metallic linings. Semimetallic friction materials contain no asbestos and are approximately 50 percent metal; the main ingredients are iron powder and steel fibers similar to chopped steel wool. The remainder of the material consists of organic binding resins, various fillers, and graphite or other friction modifiers.

Semimetallic friction materials have several significant advantages compared to organic linings: They wear longer, are more resistant to heat fade, and suffer less from taper wear when used with floating calipers. The reason for these benefits is that semimetallic linings work best at medium to high temperatures. As shown in figures 10-13 and 10-14, semimetallics suffer little increase in fade or wear up to about 650°F (343°C), far above the maximum operating temperature of an organic lining. The upper temperature limit for semimetallic linings is about 1000°F (538°C); above that, the steel fibers in the friction material can melt and weld to the rotors.

One drawback of semimetallic linings is their cost—two to three times that of organic linings. They are also somewhat brittle and can crack or break if handled roughly. The main functional weakness of semimetallic friction materials is their low coefficient of friction when cold. Cars with original equipment semimetallic linings allow for this in the brake system design. However, when semimetallic linings are installed in a system designed for organic friction materials, increased pedal pressure is required when the brakes are cold.

To provide both good fade resistance and low pedal pressures, some manufacturers fit their disc brakes with **hybrid pad sets.** These include a semimetallic pad for one side of the rotor and an organic pad for the other. The fade-resistant semimetallic pad guarantees good high-temperature stopping power, whereas the superior low-temperature performance of the organic pad keeps pedal pressures reasonable when the brakes are cold.

In rare instances, the iron content of semimetallic linings can allow the brake block to rust to the rotor. This is most likely to occur when the car is parked in a damp or corrosive environment for an extended time. If this happens, part of the brake block may be torn away when the car is first moved,

Brake Shoes and Pads

■ Early Brake Linings

Brake linings at the turn of the twentieth century were made from basic organic materials: wood, metal, and for a higher coefficient of friction, leather. At the 1907 National Automobile Show in New York City, camel hair cloth linings and cotton linings impregnated with asphalt and rubber were displayed as the latest technology. All of these early linings provided erratic braking at best. And if they were used too severely, they had a disturbing tendency to catch fire!

Some say it was the smell of burning leather that prompted inventors to weave asbestos fibers into brake linings. Others claim the invention of internally expanding brakes, and the friction and heat they could generate, forced the development of asbestos linings. Regardless of motivation, asbestos proved to be the material needed to advance the state of the braking art. Sometime around 1908, a lining woven of asbestos and copper wire mesh was introduced to provide char-proof performance with smooth, grab-free brake application. For more than the next half century, asbestos-based organic linings would be used on virtually all cars.

and the brake pedal will pulsate for several stops until any rust and debris on the pad and rotor is worn away. The lining can also rust where rivets attach the brake block to the backing plate. If the rust is severe, the friction material may break loose from the backing plate and the pads will have to be replaced.

Semimetallic Linings and Rotor Wear

Although semimetallic linings are often singled out as a major cause of brake rotor wear those charges are not entirely correct. Although the biggest factor affecting rotor wear *is* the composition of the lining friction material, all semimetallic linings are not alike. As explained earlier, every friction material is made from a unique blend of substances. As a result, some semimetallic pads are more abrasive than others and cause more wear. Early semimetallic linings, like full-metallics, were highly abrasive. Most newer semimetallic friction materials are much improved in this respect and cause no more wear than a typical organic pad.

Another factor affecting rotor wear is cooling ability; like brake pads, rotors wear faster at higher temperatures. The downsized disc brakes on early FWD cars operated at very high temperatures that increased wear regardless of the type of lining installed. In fact, contrary to popular belief, a rotor may actually last *longer* with semimetallic linings than with organic parts because the high metal content of the pads conducts heat away from the rotor much faster.

A third factor affecting rotor wear is the smoothness of the friction surface on the rotor. Wear problems can arise if semimetallic pads are installed against a rotor with too rough a surface finish. This is discussed in greater detail in the next chapter.

Semimetallic Lining Applications

Semimetallic brake linings have been used for many years in heavy-duty applications. In recent years, semimetallic lining applications have increased greatly because manufacturers want to get away from the health hazards of asbestos.

At the present time, semimetallic linings are used exclusively on disc brake pads. Semimetallic linings are not used on drum brake shoes because the lightly loaded rear drum brakes on modern cars provide only a small part of the total braking force and seldom reach the temperatures where semimetallic linings are most effective. Also, rear drum brakes often serve as parking brakes so the superior low-temperature friction coefficient of an organic lining is better suited in this application. Finally, in light of the preceding discussion, the added cost of semimetallic brake shoe linings cannot be justified.

Synthetic Brake Linings

Some of the newest brake shoes and pads use a **synthetic lining** manufactured from man-made, nonasbestos, nonmetallic friction materials. Synthetic linings have been developed because of the health problems associated with asbestos-based organic linings and the resulting safety regulations that have greatly increased organic lining manufacturing costs. The unsatisfactory performance of semimetallic friction materials in drum brakes is also a factor in synthetic lining development.

The research into synthetic brake linings has resulted in two new types of friction materials.

- Fiberglass
- Aramid fiber.

Synthetic friction materials can use either of these substances, and at least one formulation contains both. The performance characteristics of each are explained in the following sections.

Fiberglass Brake Linings

The first type of synthetic lining is made with fiberglass. Like asbestos fibers, glass fibers reinforce a friction material and increase its resistance to heat. The creation of fiberglass linings involves much more than simply replacing asbestos with fiberglass, however; the entire friction material has to be reformulated. Additional ingredients include a large amount of resin binder, barium sulfate, certain types of clay, and crushed pecan shells.

The advantages of fiberglass linings are similar to those of high-quality asbestos: slow wear, quiet operation, and nonabrasiveness to friction surfaces. However, there are two disadvantages to be considered. Fiberglass friction materials have failed in some extreme high-temperature tests, so present use is limited to rear drum brake shoe linings only. Also, fiberglass linings are more expensive than organic friction materials.

Aramid Brake Linings

The most promising new brake lining material is **aramid fiber**, a synthetic product invented by DuPont, which markets it under the trade name Kevlar. Aramid fibers are used as a reinforcing agent in a variety of products from motorcycle helmets to soft body armor. Pound-for-pound, aramid fiber is five times as strong as steel, yet it weighs 43 percent less than an equal volume of fiberglass. As with all friction materials, the exact formulation of aramid-fiber brake block is a closely guarded secret, although the substances used in it are generally similar to those in fiberglass brake block: a large portion of resin binder, barium sulfate, clay, and crushed nut shells.

The most impressive feature of aramid friction materials is their performance in wear tests. Durability studies show aramid brake linings last two to six times longer than organic parts. The longevity of these friction materials is so great that some manufacturers market aramid brake linings as "lifetime" replacements. Aramid linings also cause little wear to drums and rotors, and perform similarly to organic linings in most other respects.

The main drawback of aramid brake linings is their frictional characteristics. Aramid-fiber linings have a coefficient of friction similar to that of a semimetallic lining when cold, and an asbestos lining when hot. The overall performance of aramid linings falls somewhere between that of organics and semimetallics. Because of this, many manufacturers of aramid linings recommended they be used only in applications with original equipment organic linings. At least one manufacturer, however, is marketing aramid linings as a replacement for all other types of friction materials.

Other Lining Materials

We have covered organic, semimetallic, metallic, and synthetic friction materials. Of these, the most commonly used are organic and semimetallic. Today, a variety of friction materials may be used in combinations with different amounts of organic, nonasbestos organic, low-metallic, semimetallic, carbon fiber, and ceramic formulas. Often these materials differ in name only. However, there are very few restrictions on what you can put in friction material, so you may find any number of different raw materials in friction material.

Generally speaking, semimetallic (which is a generic industry term that has virtually no specific meaning) pads are used when a vehicle is likely to generate more heat and the higher metal content will be needed to stop in a reasonable distance while providing acceptable pad life. The ceramic/NAO/low-metallic–type materials are generally used on lighter weight vehicles where heat build-up is not as great.

Ceramic materials have been used in combination with other materials for many years, most often on Asian imports. Ceramic brake pads that contain little or no iron or steel powders are popular for their low-noise and low-dust characteristics. High-performance ceramic pads may contain up to 15 percent copper fibers or dust to enhance heat tolerance. Other ceramic pads may use a certain amount of carbon fiber along with some metallic content, called by one company Carbon Metallic® pads.

LINING MATERIAL COEFFICIENT OF FRICTION

Vehicle manufacturers specify certain lining friction characteristics for each particular car, brake system, and axle. It is common practice, for example, to specify linings with a lower friction coefficient on the rear brakes than on the front to minimize rear brake lockup. The exact friction

Brake Shoes and Pads

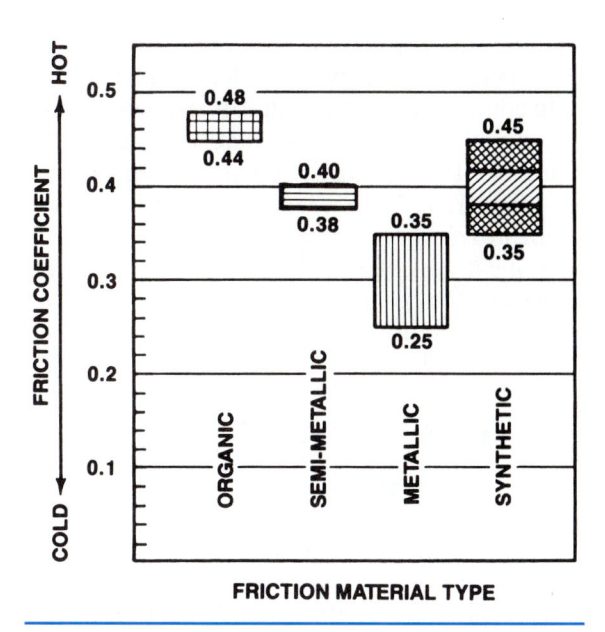

Figure 10-16. Brake lining material friction coefficients.

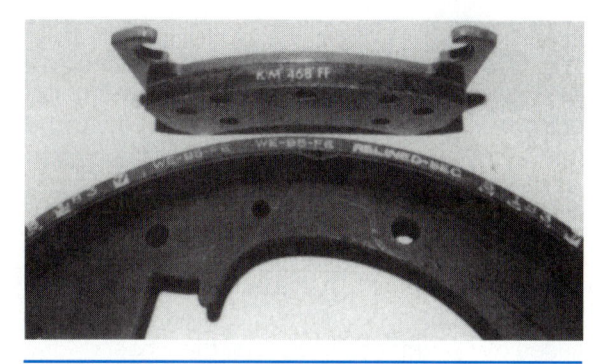

Figure 10-17. Edgebrands provide coded information about the brake lining.

FRICTION CLASS CODE	COEFFICIENT OF FRICTION
C	Not over 0.15
D	Over 0.15 but not over 0.25
E	Over 0.25 but not over 0.35
F	Over 0.35 but not over 0.45
G	Over 0.45 but not over 0.55
H	Over 0.55

Figure 10-18. SAE codes for brake lining friction coefficients.

properties of a brake lining differ with the mixture of substances that go into it, and even linings of the same basic type (organic, semimetallic, etc.) can differ in their coefficients of friction.

Despite the many differences, some generalizations can be made about the friction coefficients of each type of friction material. As shown in figure 10-16, typical organic linings have a cold friction coefficient of around 0.44 that increases to about 0.48 when the linings warm to operating temperature. Semimetallic lining friction coefficients average a bit lower, with cold and hot figures around 0.38 and 0.40, respectively. Sintered metallic linings have the lowest numbers with a typical cold coefficient of friction in the 0.25 range and a hot figure that seldom rises above 0.35. Synthetic linings, both fiberglass and aramid, have coefficients of friction that compare favorably to those of organic and semimetallic linings. When cold they are in the 0.35 to 0.38 range; when hot the figures rise to around 0.42 to 0.45.

Edgebranding

Information on the frictional properties of a brake lining can be found in the **edgebrand** on the side of the brake block, figure 10-17. A typical edgebrand might read R/M 6793–2 EF. The first letters, R/M, identify the manufacturer of the lining. The numbers in the center, 6793-2, identify the specific lining material. The final two letters, EF, describe cold and hot friction coefficients of the lining based on a set of SAE standards, figure 10-18. The first letter gives the cold value; the second gives the hot value. The friction coefficients of brake linings for passenger cars generally fall into the E, F, and G ranges.

Although codes in the edgebrand provide some indication of a lining's friction coefficient, they are *not* a measure of quality. Although the codes can be used to compare the frictional characteristics of various linings, there are many factors of braking performance the codes do not address, such as resistance to fade, service life, and noise. The SAE states that the edgebrand is simply a uniform means of identification. It *does not* represent friction requirements for brake linings, or any significant characteristic of quality. These codes should not influence the selection of a lining for a particular vehicle.

Hard and Soft Brake Linings

When the frictional traits of brake linings are discussed, linings are sometimes referred to as being either "hard" or "soft." These relative terms can

■ **"Mag" Wheels To Go, Hold The Dust!**

Brake dust from front disc brakes was never much of a problem until many cars began to be delivered with alloy "mag" wheels. Unlike hub-cap-covered steel wheels, alloy wheels have an open construction, and sometimes a relatively porous surface, which allows dust to collect on the wheel's outer surface. Although this has no effect on vehicle operation, it is unattractive and in extreme cases can permanently discolor the wheels.

The discoloring elements in brake pads are primarily graphite and carbon black. Graphite is a friction modifier with good high-temperature wear qualities. Carbon black is also a friction modifier, but it is more commonly added to friction materials for cosmetic reasons; many people prefer pads with a darker lining color. To prevent brake dust problems, several friction material manufacturers have released pads formulated to reduce dust. One of these products eliminates graphite and carbon black so the pad produces only white dust. Other manufacturers claim to have solved the problem with their standard brake pads, but the natures of their solutions remain a trade secret.

An alternative to special brake pads is a dust shield that installs between the wheel and the brake friction assembly. These shields are marketed by several firms, but although they reduce dust accumulation on the wheels, they can also reduce the amount of air flowing over the brakes, and thus, increase operating temperatures and the chance of brake fade.

---

be applied to various linings within a single friction material family, or they can be used to compare the different types of friction materials. Generally, a soft lining has a high coefficient of friction, wears more rapidly, fades at a lower temperature, is less abrasive to brake drums and rotors, and operates very quietly. A hard lining has a lower coefficient of friction, wears more slowly, fades at a higher temperature, is more abrasive to drums and rotors, and tends to be noisier.

Traditionally, organic linings have been considered soft, whereas semimetallic linings are harder, and full-metallic linings are the hardest. These overall rankings still hold true, but recent developments in lining technology have made them less meaningful than they once were. For example, aramid synthetic linings are easy on drums and rotors like a soft lining, but they offer the long service life usually associated with a hard lining.

In addition, there are both soft and hard linings within each friction material family; basic organic linings (soft) and high-temperature organic linings (hard) are a good example. The important factor in selecting brake linings is not the hardness or softness of a particular part, but how well it matches the design of the brake system and the car's expected use.

LINING ASSEMBLY METHODS

Before any type of brake lining can do its job, it must be firmly attached to the lining table of a drum brake shoe or the backing plate of a disc brake pad. Several methods are used to mount brake linings; the most common are:

- Riveting
- Bonding
- Mold bonding
- Fusing and brazing.

Riveting

Riveted linings take advantage of the oldest method of lining attachment still in use. In this system, the brake block is attached to the lining table or backing plate with copper or aluminum rivets, figure 10-19. In the past, most organic friction materials were attached using rivets, but since bonding was developed (see following section), rivets have been used primarily to mount semimetallic linings.

The major advantage of riveting is that it allows a small amount of flex between the brake block and lining table or backing plate. This play enables the assembly to absorb vibration, and the result is that riveted linings operate more quietly than bonded linings. Rivets are also very reliable and will not loosen at high temperatures.

Despite the benefit of quiet operation, riveted linings present a number of problems. In order to leave sufficient friction material for the rivets to clamp the brake block securely against the lining table or backing plate, the rivet holes are countersunk only about two-thirds to three-quarters of the way through the lining. This reduces the service life of the assembly because the shoes must

Brake Shoes and Pads

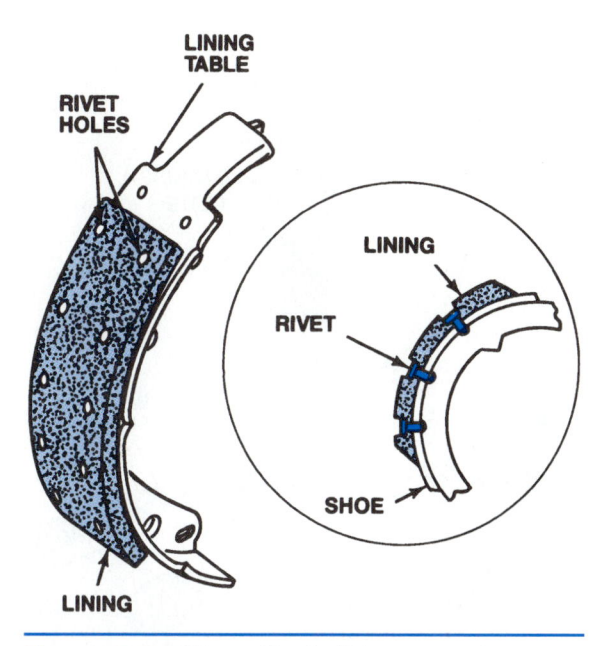

Figure 10-19. Riveted brake linings are quiet and reliable at high temperatures.

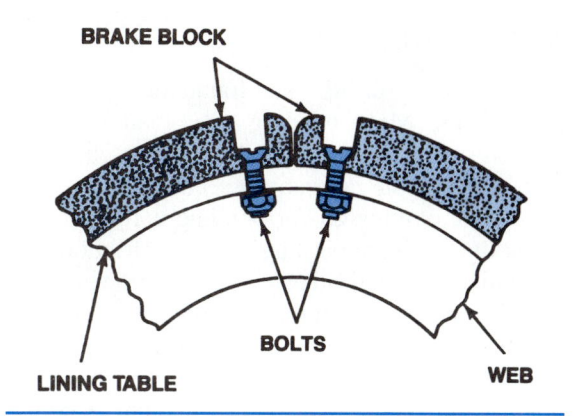

Figure 10-20. Bolted brake linings are found on many large trucks.

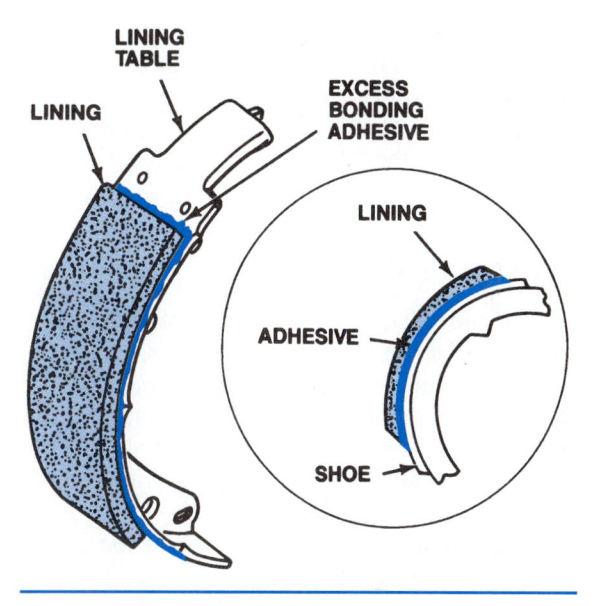

Figure 10-21. Almost all modern organic brake linings are bonded in place.

be replaced before the rivet heads contact and score the drum.

The rivet holes themselves also present some unique problems. First, they trap abrasive brake dust and other grit that can score the drum or rotor. Some rivets are hollow to allow these materials to escape, but not all of them do. It is not uncommon to find a drum that is scored along the path of the rivet holes even though the rivets have not touched the friction surface.

Rivet holes also create stress points in the lining where cracks are likely to develop. Some riveted semimetallic brake pads have a thin layer of asbestos bonded to the back of the brake block to provide a more crack resistant mounting surface.

In a process similar to riveting, heavy-duty truck and trailer brake linings are often bolted in place, figure 10-20. The lining is countersunk as in riveting, but the rivet is replaced by a brass bolt, nut, and washer.

Bonding

Bonded linings result from a fairly modern process that uses high-temperature adhesive to glue the brake block directly to the shoe lining table or pad backing plate, figure 10-21. Heat and pressure are then applied to cure the assembly. Bonding is a common form of shoe and pad assembly and is most often used to mount organic friction materials.

Bonding offers several advantages. Without rivets, bonded linings can wear closer to the lining table or backing plate and provide a longer service life. If the linings wear too far, bonding adhesive is not as destructive to drums or rotors as rivets. Bonded linings also have fewer problems with cracking because they have no rivet holes to weaken the brake block.

The primary disadvantage of bonded linings is a limited ability to withstand high temperatures. If a bonded lining gets too hot, the bonding adhesive will fail and allow the brake block to separate from the lining table or backing plate. Bonded linings are also more prone to be noisy because they do not allow any vibration-absorbing flex between the brake block and lining table or backing plate.

Mold Bonding

Mold bonded linings are found on some disc brake pads. Mold bonding is a manufacturing process that combines the advantages of bonding with some of the mechanical security of riveting. Instead of riveting and/or bonding a cured brake block to a separate backing plate, the friction material in a mold bonded pad is cured on the backing plate during manufacture. Most high-performance disc brake pads are made in this way.

To make a mold-bonded pad, one or more holes are punched in the pad backing plate, and a high-temperature adhesive is applied to it. The backing plate is then installed in a molding machine where uncured friction material is formed onto the plate and forced into the holes, figure 10-22. Once the pad is cured under heat and pressure, the bonding adhesive combines with the portions of the lining that extend into the backing plate holes to solidly lock the brake block in place.

Fusing and Brazing

Fusing and brazing are assembly methods used only on sintered full-metallic linings. Such unusual attaching methods are required because the brittleness of these friction materials and the high temperatures at which they operate make it very difficult to attach them securely to a lining table or backing plate.

Some metallic linings are fused to the lining table or backing plate as part of the sintering process; this method is somewhat similar to the mold bonding process described previously for nonmetallic linings. On occasion, fused linings will also be drilled and riveted to provide an added measure of security.

The alternative method of mounting a full-metallic brake block is to braze it directly to the shoe lining table or pad backing plate. This method ensures secure attachment and is used in the most extreme applications.

BRAKE SHOE-TO-DRUM FIT

A proper fit between the brake shoes and drum is important if the friction assembly is to operate at maximum effectiveness. Full contact between

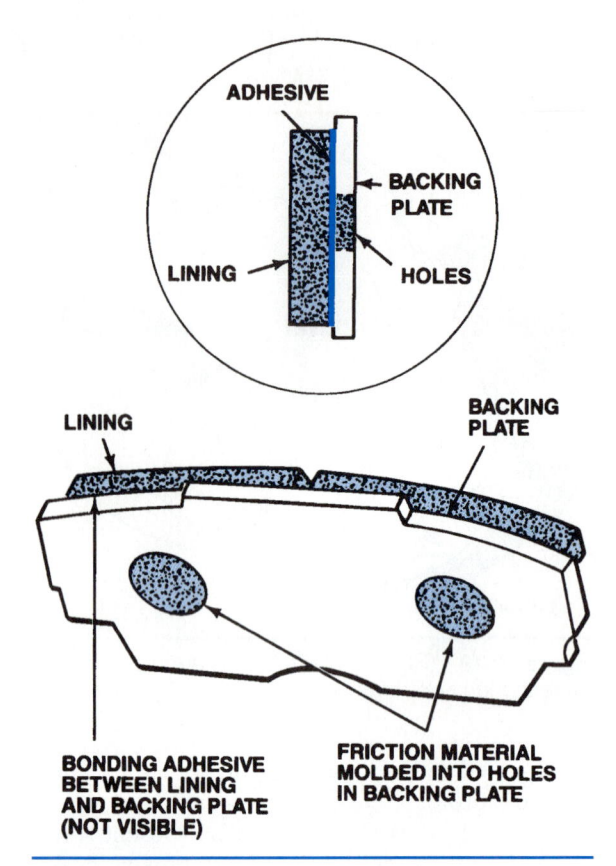

Figure 10-22. Mold bonded linings are common in high-performance disc brake applications.

these surfaces provides quicker break in because the entire lining is burnished at the same time. It also gives better brake pedal feel because it eliminates the sponginess that can be caused by excessive shoe flex. Full contact between the shoe and drum also gives maximum lining life and, at the same time, provides superior fade resistance because more heat is conducted out of the drum.

When only part of a brake lining contacts the drum, or greater force is applied in one area than another, problems can result. Most drum brake noise is caused by binding that occurs when the ends of the lining contact the drum with more force than the center, figure 10-23. When only the center or one end of the lining makes good contact with the drum, figure 10-24, the friction material overheats causing various elements in the friction material to liquify and rise to the surface. As these elements cool and harden, they form a **glazed lining**. Glazed linings have a smooth shiny appearance and a reduced coefficient of friction that decreases stopping power and increases brake noise.

Brake Shoes and Pads

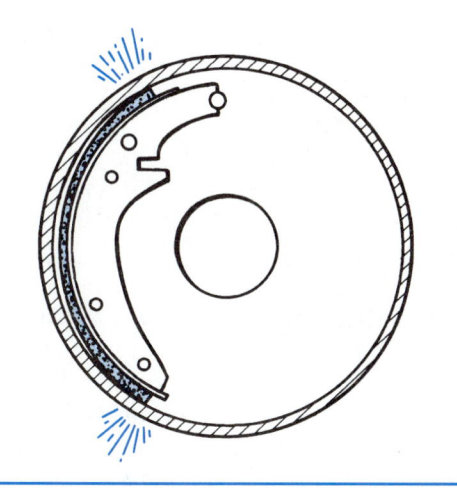

Figure 10-23. The major cause of drum brake noise is binding at the ends of the linings.

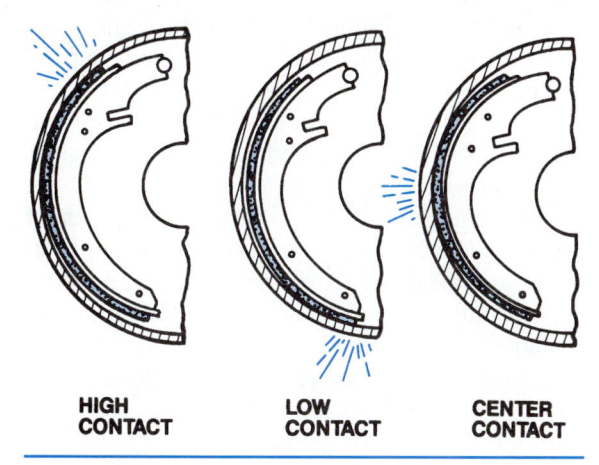

HIGH CONTACT **LOW CONTACT** **CENTER CONTACT**

Figure 10-24. Partial shoe-to-drum contact can cause glazed brake linings.

Proper Shoe-to-Drum Fit

With a disc brake pad and rotor there is little question about the proper fit; the two friction surfaces should be perfectly parallel to one another. In theory, the same holds true for a brake shoe and drum; however, in practice, parallel alignment of the friction surfaces is undesirable. When the brakes are applied, a proper fit allows the center of the lining to contact the drum first while there is still a very slight amount of clearance at each end of the friction material.

The reason brake shoes must fit in this manner involves the way drum brakes behave when they are applied. First, consider a disk brake: The pad is applied directly against the rotor by a caliper piston that moves along the same axis as the pad. Also the piston contacts a broad area of the thick and relatively flex-free pad backing plate. Because of these factors, a disc brake pad has roughly equal application force across its lining.

In comparison, a drum brake shoe is applied indirectly against the drum by an anchor fixed to the backing plate and a wheel cylinder piston that moves along a different axis than the shoe. Self-energizing and servo actions also affect the amount and direction of the brake application force, and the drum distorts to a greater or lesser degree depending on how hard the brakes are applied. As a result of these factors, a brake shoe is applied with greater force at its ends than in the center.

Because the brake shoe web and lining table are long and relatively thin (compared to pad backing plates), the added force at the top and bottom of the lining causes the shoe to flex outward when the brakes are applied. If the lining is fitted parallel to the drum friction surface, shoe flex will cause the ends of the lining to bind against the drum and create a great deal of noise. When the lining is fitted with clearance at the ends, shoe and drum flex combine to allow the lining to conform more precisely to the actual shape of the drum. This creates approximately equal application force along the length of the shoe.

■ Brake Shoe Arcing

Arcing is the process of grinding the friction material on a drum brake shoe to provide the proper amount of clearance at the ends of the lining. There should be just enough clearance to avoid binding; an excessive amount allows the shoes to flex too far, and results in a spongy brake pedal. Non-servo brakes usually require .005″ to .007″ (.15 to .20 mm) of clearance at each end; dual-servo brakes need .010″ to .012″ (.25 to .30 mm) of clearance to compensate for the greater amount of shoe flex caused by their increased application force.

To establish the correct curvature, the shoe is clamped in a grinding machine and moved through an arc of a prescribed radius while an abrasive disc removes excess friction material. Because of the dangers posed by asbestos dust, the shoe arcing machine must be equipped with an approved dust collection system to trap the particles ground from the lining. The machine can be adjusted to fit the lining to different sizes of drums, and in some cases, to specific types of friction assemblies.

Shoe Arcing

When all cars had four-wheel drum brakes, and friction materials were not as forgiving as those available today, extreme brake drum wear was commonplace and drums were routinely machined oversize as a part of brake service. Oversize drums caused problems with shoe-to-drum fit, however, because the radius of the drum increased, whereas that of the brake linings remained the same. This caused only the center of the lining to contact the drum, and brake fade and glazed linings resulted.

To prevent the problems associated with oversize brake drums, most manufacturers offered new and relined **oversize brake shoes** in addition to the standard parts. The standard shoes fit a new or unworn brake drum, whereas oversize shoes had linings that were about .030″ (.75 mm) thicker for use with oversize drums. A few brake parts catalogs still list oversize shoes for selected applications.

In addition to the use of oversize parts, both standard and oversize front brake shoes (and often rear shoes as well) were routinely arced to ensure a proper fit. This had to be done because the front brakes provide most of the braking power. If the shoes were not arced, they could easily overheat and become glazed before they were burnished in.

Today, the situation is different. Modern rear drum brakes provide only a small part of the overall braking effort, and the latest friction materials make it unlikely the drums will suffer wear that requires machining. As a result, oversize linings are unavailable for most modern brake applications, and shoes are rarely arced before installation.

Most modern replacement brake shoes are arced at the factory to fit standard-size drums. Technicians commonly install standard-size linings in rear drums, even if they have been turned oversize. This practice is workable, if not ideal, because rear drum brakes perform such a small part of the total braking job that an exact shoe-to-drum fit is not critical. One negative side effect of using standard shoes in oversize drums is that it takes far longer for the shoes to become completely burnished in.

It remains true in *all* cases that the best braking performance is obtained when standard shoes are used with standard drums, oversize shoes are used with oversized drums, and all linings are arced to fit the drum using the appropriate type of grind for that particular design of friction assembly. If the drum must be turned and oversize shoes or proper arcing equipment is unavailable, the only alternative for maximum braking performance is to fit new drums with standard linings.

SUMMARY

Drum brake shoes are made of welded sheet steel or cast aluminum and consist of a lining table that supports the friction material and a web that supports the lining table and channels application force to it. Non-servo brake shoes are often interchangeable, but the primary and secondary shoes in a dual-servo brake cannot be switched because they often use different friction materials and have their linings placed differently on the lining tables.

Disc brake pads consist of a steel backing plate lined with friction material. The pad backing plate usually has tabs, holes for pins, or a spring clip to hold the pad in the caliper and help prevent vibration. Some pads are fitted with mechanical or electrical wear indicators.

All brake shoes and pads have linings made of friction material formed into brake blocks. Friction materials differ from one another in their exact formulation, although typical ingredients include mineral, metal, or synthetic fibers, friction modifiers, powdered metals, binders, fillers, and curing agents. Friction materials are classified into the following groups: organic, metallic, semimetallic, synthetic, and other.

Organic linings contain asbestos fibers to strengthen the friction material and improve its resistance to heat. Organic linings are inexpensive and operate quietly, but they fade quickly at relatively low temperatures. Fade-resistant high-temperature organic linings are available, but they are noisy, abrasive, and stop poorly when cold.

Full-metallic linings are made entirely of powdered metals sintered into the brake block. Metallic friction materials resist fade well at very high temperatures but perform poorly in almost every other area. Metallic linings are not used on automobiles today.

Semimetallic linings are made primarily of iron powder and steel fibers. These linings last longer and fade less than organics because they operate best at higher temperatures. The drawbacks of semimetallics are increased cost, a slightly lower cold coefficient of friction, and in some cases, increased abrasiveness. Hybrid brake pad sets, with one organic and one semimetallic pad for each ro-

Brake Shoes and Pads

tor, are used on some cars to take advantage of the strengths of both materials.

Synthetic friction materials, reinforced with fiberglass or aramid fiber, offer a long service life combined with very low abrasiveness. But in addition to costing more, fiberglass linings have limited high-temperature strength, whereas aramid blends have a lower friction coefficient than semimetallic linings.

A brake lining with a low coefficient of friction is termed *hard,* whereas one with a high coefficient of friction is called *soft.* A hard lining resists heat fade better, wears longer, but performs worse at low temperatures and requires more pedal pressure. A soft lining has the opposite characteristics. Metallic linings are generally the hardest followed by semimetallic linings and organic linings. The new synthetics are somewhere between semimetallic linings and organic linings. The hot and cold friction coefficients of linings are coded on the side of the brake block in an edgebrand.

Linings are attached to brake shoes and pads by riveting, bonding, mold bonding, fusing, and brazing. Riveting is the most secure method, but it decreases the amount of usable lining and the rivet holes can cause cracks. Bonding glues the friction material to the shoe or pad and leaves more usable friction material, but bonding adhesives occasionally suffer high-temperature failure. Mold bonding cures the brake block and mechanically joins it to the backing plate at the same time. This provides added security without the problems caused by rivets. Fusing and brazing are used only to attach full-metallic linings.

Proper shoe-to-drum fit requires a small amount of clearance at the ends of the linings when they are first applied. This clearance helps prevent noise and glazed linings. Because there is greater application force at the ends of the shoes, they will flex enough to ensure full contact between the linings and the drum. Dual-servo brakes require more clearance than non-servo brakes because the increased force from servo action causes greater shoe flex. The lightly loaded rear drum brakes on modern FWD cars allow standard shoes to be used with oversize drums. Maximum brake system performance is obtained with properly arced linings sized to match the drum.

Brake linings are arced to create the proper clearance at the ends of the linings. Arcing is done at the factory and can be done in the field to provide a custom fit.

Review Questions

Choose the letter that represents the best possible answer to the following questions:

1. A drum brake lining is attached to the _____ of a drum brake shoe.
   a. Web
   b. Platform
   c. Anchor eye
   d. Nib

2. A brake shoe contacts other parts of the friction assembly at the:
   a. Nibs
   b. Anchor eye
   c. Lining
   d. All of the above

3. In a dual-servo brake, the _____ shoe often has a shorter lining.
   a. Secondary
   b. Tertiary
   c. Primary
   d. Reverse

4. A disc brake pad may be held in place by:
   a. Pins through the lining
   b. Bent tabs on the pad backing plate
   c. A spring clip built into the caliper piston
   d. All of the above

5. Wear indicators can cause:
   a. Brake noise
   b. An electrical circuit to open
   c. Both a and b
   d. Neither a nor b

6. The portion of a brake shoe or pad that contacts the drum or rotor is called the:
   a. Lining
   b. Brake block
   c. Friction material
   d. All of the above

7. Which of the following is *not* a type of friction material:
   a. Metallic
   b. Metamorphic
   c. Synthetic
   d. Semimetallic

8. In addition to asbestos, an organic friction material usually contains:
   a. Phenolic resins
   b. Powdered metals
   c. Curing agents
   d. All of the above

9. Standard organic linings work best at temperatures of less than:
   a. 600°F (316°C)
   b. 400°F (204°C)
   c. 1000°F (538°C)
   d. 500°F (260°C)

10. Special high-temperature organic brake linings:
    a. Have a low metal content
    b. Operate quietly
    c. Are relatively hard
    d. All of the above

11. Compared to organic brake linings, semimetallic linings:
    a. Are less expensive
    b. Wear faster
    c. Perform worse at high temperatures
    d. None of the above

12. Increased brake rotor wear is caused by:
    a. High brake temperatures
    b. Specific friction materials
    c. Incorrect surface finish
    d. All of the above

13. Semimetallic friction materials are:
    a. Not used for brake shoe linings
    b. Popular because metallic linings are a health hazard
    c. Used because modern brakes run cooler
    d. All of the above

14. Synthetic brake linings are:
    a. Made with aramid or glass fibers
    b. Used only on drum brakes
    c. More abrasive than semimetallic linings
    d. All of the above

15. The edgebrand on a shoe or pad:
    a. States a friction requirement
    b. Identifies the lining material
    c. Is a shoe part number
    d. Provides a measure of quality

16. A brake lining with a high coefficient of friction generally:
    a. Is called a hard lining
    b. Resists fade better than a lining with a low-friction coefficient
    c. Requires less pedal pressure than a lining with a low-friction coefficient
    d. Wears more slowly than a low-friction coefficient lining

Brake Shoes and Pads

17. Riveted brake linings are used:
    a. In combination with mold bonding
    b. With semimetallic friction materials
    c. Both a and b
    d. Neither a nor b

18. The lining attachment method most likely to fail at high temperatures is:
    a. Fusing
    b. Riveting
    c. Bolting
    d. Bonding

19. Mold bonding:
    a. Makes linings more likely to crack
    b. Does not work for high-performance brake linings
    c. Combines some advantages of both bonding and riveting
    d. Is commonly used on brake shoes

20. Brake shoe arcing:
    a. Eliminates high spots
    b. Shortens the break-in period
    c. Helps prevent brake noise
    d. All of the above

11

Brake Drums and Rotors

OBJECTIVES

Upon completion and review of this chapter, you will be able to:

- Describe brake drum construction and function.
- Describe brake rotor construction and function.
- Identify and explain fixed and floating drums and rotors.
- List and describe the four types of drum and rotor damage.
- List and describe the three types of drum and rotor distortion.
- List and describe the three types of rotor refinishing.
- Explain drum and rotor metal removal limits.

KEY TERMS

axial cooling fins	hard spots
barrel wear	heat checking
bellmouth	lateral runout
bimetallic brake drum	micro-inches
brake lathe	out-of-round
centrifugally cast brake drum	parallelism
	radial cooling fins
composite brake drum	resurfacing
	rotor hat
discard diameter	scoring
discard dimension	solid brake drum
dished rotor	solid brake rotor
drum web	speed nuts
eccentric distortion	taper variation
elastic limit	taper wear
grinding	turning
	vented brake rotor

INTRODUCTION

Brake drums and rotors are the largest and heaviest parts of their respective friction assemblies. Drums and rotors provide friction surfaces for the brake shoes and pads to rub against, and together these parts create the friction that converts kinetic energy into heat and stops the car. Because they provide the largest friction surfaces, the drums and rotors also absorb and dissipate most of the heat generated in braking.

Although brake drums and rotors suffer wear, they are not considered disposable parts like shoes

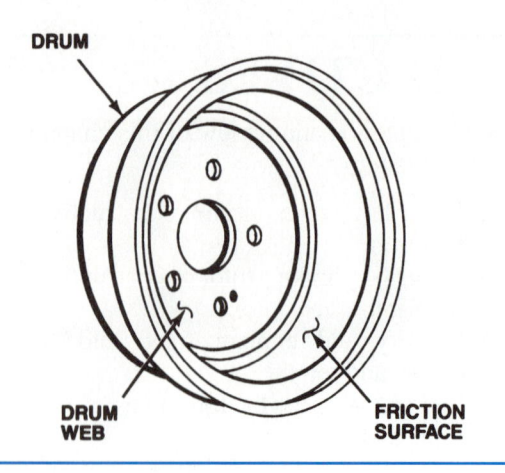

Figure 11-1. A typical brake drum.

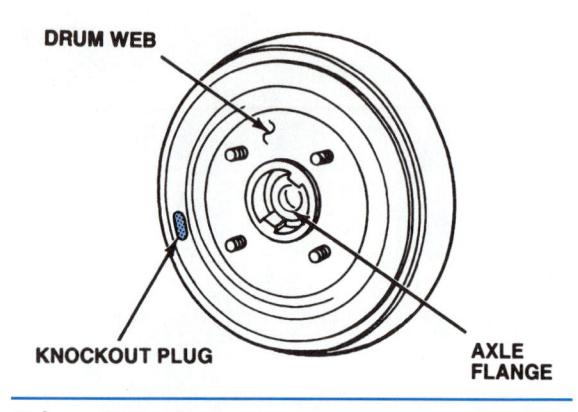

Figure 11-2. Some drum webs contain holes for brake inspection or adjustment.

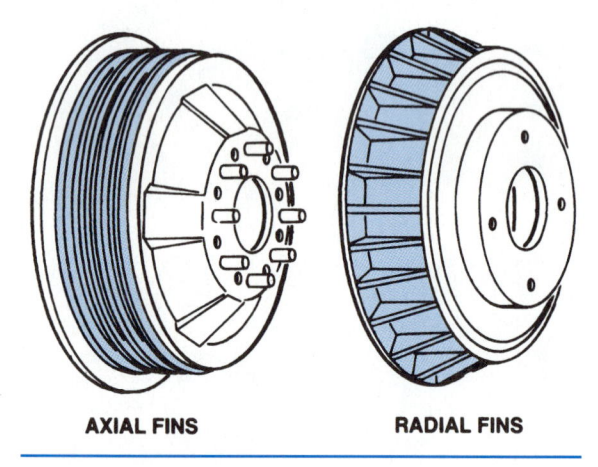

Figure 11-3. Cooling fins increase drum strength and reduce operating temperatures.

or pads. Drums and rotors are made of metals that resist damage from friction and heat, and often have special design features that aid cooling and contribute to a longer service life. The metal used to build a drum or rotor also affects its coefficient of friction, as does the quality of the finish on the friction surface.

This chapter describes the design and construction of brake drums and rotors, and the ways in which they are mounted on the car. It also looks at the most common drum and rotor problems, and provides an introduction to drum and rotor refinishing procedures that are covered in depth in the *Shop Manual*.

BRAKE DRUM CONSTRUCTION

The typical brake drum, figure 11-1, is a round, bowl-shaped casting with a rough-cast exterior and a smooth, machined interior that provides a friction surface for the brake shoe linings. The open edge of the drum installs over the friction assembly so the friction surface inside is centered over the brake shoes on the backing plate. The closed edge of the drum is made up of the **drum web,** which contains the hub, or has holes that allow the drum to be mounted on a separate hub, where it rotates with the wheel and tire. Some drum webs contain an access hole (which may be sealed with a knockout plug), figure 11-2, that is used to inspect lining wear or insert a brake adjusting tool.

A brake drum can have either axial or radial fins cast into its outer edge to improve cooling, figure 11-3. **Axial cooling fins** are continuous raised ridges cast perpendicular to the axle; axial fins are common on large truck brakes. **Radial cooling fins** are many separate raised ridges cast parallel to the axle; radial fins are the most common type on car brake drums. Both axial and radial fins increase the amount of drum surface area exposed to the cooling airflow. In addition, a radially finned brake drum can be designed to work as a centrifugal fan to draw additional cooling air through the space between the wheel and drum.

Some brake drums have a coil spring installed in a groove cast around the drum's outer edge, figure 11-4. The spring damps small vibrations in the brake friction assembly that can be amplified by resonance in the enclosed drum. These vibrations would normally be heard as annoying brake squeal, but the spring changes their frequency to make them inaudible.

Brake Drums and Rotors

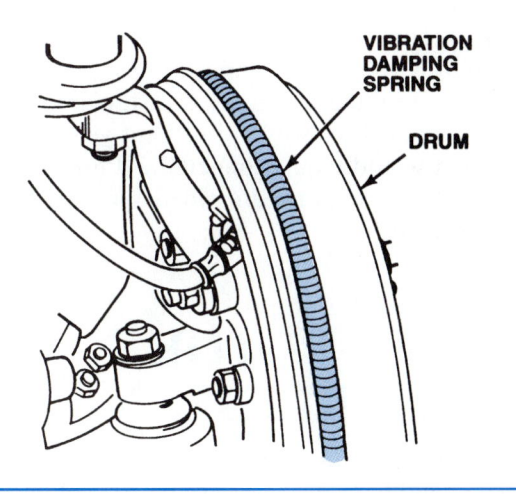

Figure 11-4. A spring around a drum helps reduce brake noise.

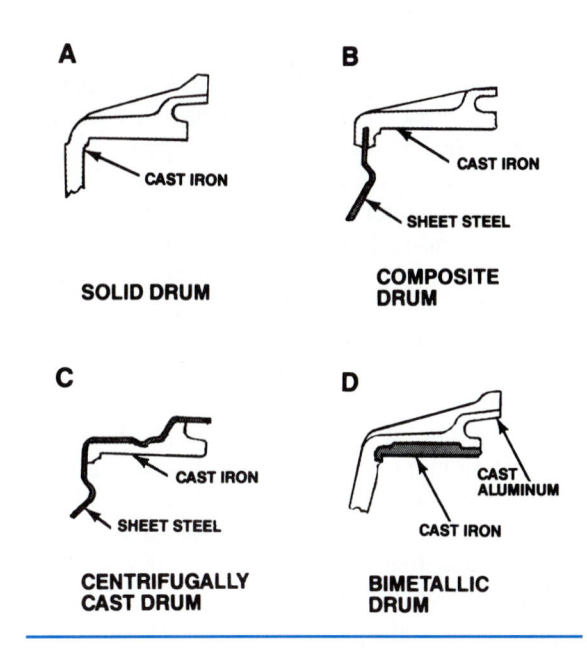

Figure 11-5. There are four types of brake drum construction.

The features described previously apply equally to all brake drums, but not all drums are the same. The differences involve the metals used to make the drum and how those metals are formed. There are two basic types of drums:

- Solid
- Composite.

Solid Drums

A **solid brake drum**, figure 11-5A, is made entirely from gray cast iron. Solid drums are somewhat heavy, and as a result, they can absorb a great deal of heat before they suffer mechanical fade.

One reason cast iron is used for brake drums is its low cost; however, cast iron also has several functional advantages. First, although cast iron is relatively soft, it has excellent wear characteristics. This is possible because the friction, pressure, and heat of braking harden the metal and cause a tough, wear-resistant "skin" to form on the drum friction surface. Despite this surface hardening, the soft core metal in a cast-iron drum makes it easy to machine smooth when it becomes worn.

Another advantage of cast iron is its high coefficient of friction compared to other metals with similar durability. A given amount of brake application force will provide more stopping power with an iron drum than with one made of another metal. Cast iron also contains numerous free graphite particles that act as a lubricant between the brake linings and drum. These particles reduce drum wear and provide a more controlled increase in friction as the brakes are applied harder.

The one significant problem with cast-iron brake drums is that they are brittle and will crack if overstressed. Also, when small cracks first form in cast iron, they are often invisible to the unaided eye. Repeated heavy braking, combined with the continual heating and cooling a drum undergoes, can cause a great deal of drum stress. The thick, heavy walls of cast-iron drums increase strength and rigidity to help prevent cracks.

As mentioned previously, the heavy weight of solid brake drums is a benefit that enables them to absorb large amounts of heat while resisting mechanical fade. However, on late-model cars that have drums only on the lightly loaded rear brakes, fade is rarely a problem. In these applications, the weight of solid drums is a disadvantage because it lowers fuel economy. For this reason, most newer drum brakes are equipped with composite drums.

Composite Drums

A **composite brake drum** is made of two metals joined together. Because the advantages of cast iron are so great, all composite drums use that metal for their friction surface. Where they differ is in the metal used to form the web or outermost portion of the drum. There are two basic types of composite drums:

- Steel-and-iron composites
- Aluminum-and-iron composites.

Steel-and-Iron Composite Drums

Steel-and-iron composite brake drums come in two styles. The most common design has a sheet-steel web combined with a cast-iron drum, figure 11-5B. This is the type of drum most commonly referred to by the general term "composite drum."

The second style of steel-and-iron drum consists of a stamped-steel outer drum with a cast-in iron liner, figure 11-5C. This design is called a **centrifugally cast brake drum** after the process used to form the liner. During manufacture, the steel drum is spun at high speed while molten iron is poured into it. Centrifugal force causes the metal to flow outward and form the liner around the edge of the drum. Centrifugal casting eliminates air gaps between the drum and liner, and the good contact that results improves heat transfer and thus brake cooling.

Steel-and-iron composite drums share a couple of benefits. They are less expensive than solid drums, and their construction saves weight compared to drums made entirely of cast iron. However, the lighter weight of steel-and-iron composite drums also means they are less able to absorb heat and resist fade. And in the case of centrifugally cast drums, the steel shell makes the use of cooling fins impractical.

Because of their weight-versus-fade tradeoffs, steel-and-iron composite drums are most often used on rear brakes. These types of drums are especially common on compact economy cars that are neither heavy nor powerful enough to overtax their cooling abilities.

Aluminum-and-Iron Composite Drums

The second type of composite brake drum has a cast-aluminum outer drum fitted with a cast-iron liner, figure 11-5D. This design is usually called a **bimetallic brake drum.** Because aluminum melts at a much lower temperature than iron, bimetallic drums cannot be formed by centrifugal casting. Instead, the aluminum portion of the drum is cast around a preformed iron liner.

The aluminum drum structure enables bimetallic drums to perform better than any other type. Aluminum is three times lighter than cast iron, so bimetallic drums weigh substantially less than solid drums or steel-and-iron composites. Aluminum also conducts heat four times faster than iron, which means bimetallic drums cool better than other types. The only real problem with bimetallic drums is that they are somewhat costly.

In the past, bimetallic drums were common on high-performance imported cars, and they were used on the rear axles of a few larger domestic models. However, in recent years, many cars that would require bimetallic drums for adequate brake cooling are fitted with rear disc brakes instead.

BRAKE ROTOR CONSTRUCTION

The typical disc brake rotor, figure 11-6, is a circular, metal plate with two machined friction surfaces, one on either side. When the rotor is installed on the car, the brake caliper straddles it and applies the brake pads against the two friction surfaces. With the exception of some high-performance vehicles (see the accompanying sidebar, on exotic brake rotors), all production vehicle brake rotors are cast in one piece and made of gray cast iron for the same reasons this metal is used in drums—low cost, good wear and friction properties, and ease of machining.

On many cars the hub assembly containing the wheel bearings is cast into the center of the rotor. In other applications, particularly FWD vehicles, the center of the rotor is a flat, raised platform with holes in it that allow the rotor to be installed on a separate hub. In these applications, the inner part of the rotor is called the **rotor hat.** On some motorcycle and racing car disc brakes, the hat is a separate part joined to the rotor in a way that reduces heat transfer to the hub.

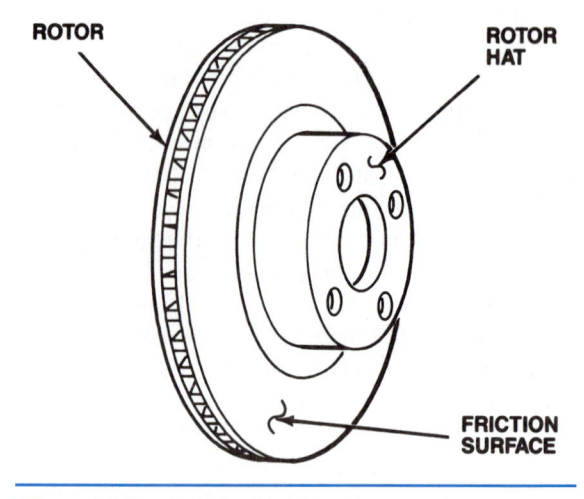

Figure 11-6. A typical brake rotor.

Solid and Vented Rotors

Disc brake rotors come in two styles, solid and vented, figure 11-7. A **solid brake rotor,** as its name states, is one whose friction surfaces are on the opposite sides of a solid piece of metal. Solid rotors do not cool as well as vented ones and are normally used on lighter and less powerful vehicles.

A **vented brake rotor** has radial cooling passages cast between its friction surfaces. These passages allow the rotor to work as a centrifugal fan that draws cooling air into the center of the rotor and forces it out at the edges. Although cooling passages make vented rotors wider and heavier than solid ones, the weight penalty is more than offset by the improved cooling. Lower operating temperatures increase both rotor and brake pad life, and improve fade resistance.

Some vented rotors are large and heavy enough to affect tire and wheel balance if they are not properly balanced themselves. Vented rotors are checked at the factory, and balance weights are sometimes installed in the cooling ducts, figure 11-8. These weights should never be removed when servicing a vented rotor.

Unidirectional Vented Rotors

Some vented rotors are designed as specific right- and left-hand models that have curved cooling passages to improve their air pumping ability, figure 11-9. For proper cooling, the passages must point opposite the direction of forward rotation when the rotor is installed on the car. Although it might seem logical to point the cooling passages forward so they could scoop air *into* the rotor, centrifugal force moving air from the inside of the rotor *outward* is much more powerful than any ram effect that might be gained by reversing the

Figure 11-8. Balance weights are sometimes installed in the cooling passages of vented rotors.

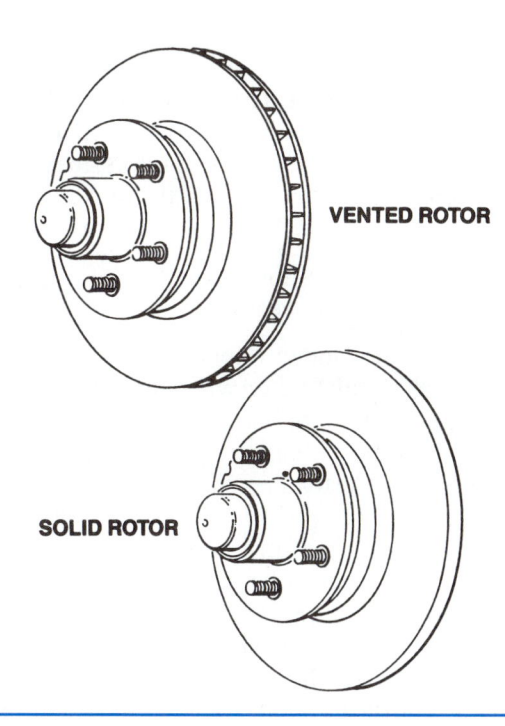

Figure 11-7. Solid and vented brake rotors.

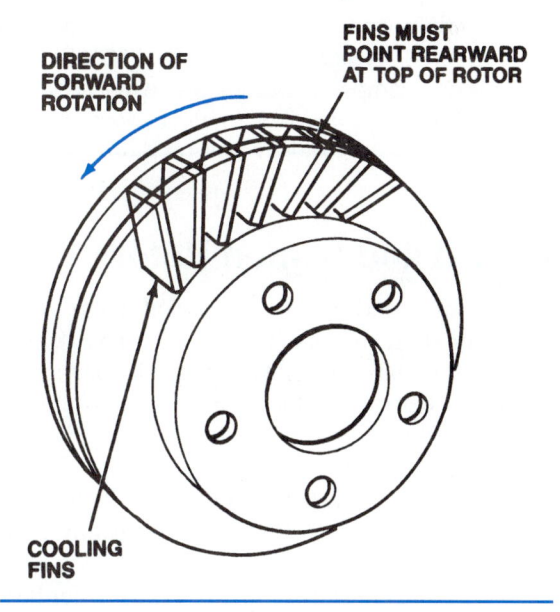

Figure 11-9. A unidirectional vented brake rotor.

direction of the cooling passages. If unidirectional rotors are swapped side-for-side, the increased rotor temperature that result can reduce lining life as much as 75 percent!

■ Exotic Brake Rotors

Although production cars have brake rotors made of cast iron, motorcycles and racing cars often use rotors made of stainless steel, aluminum, or composites. Stainless steel rotors are common on motorcycles because they provide an attractive appearance, resist rust, and do not develop stress cracks from use. However, stainless steel has inferior friction characteristics compared to cast iron and is more likely to gall and warp.

Aluminum rotors weigh only one-third as much as cast-iron parts and conduct heat four times faster. However, because aluminum is relatively soft and brittle, it makes a poor friction surface and must be given a special surface coating to improve its friction and wear properties. This coating can be plasma-sprayed metal, or thin iron or steel plates bonded to the aluminum core. These processes make aluminum rotors costly, and once the surface coating is worn through, the rotor must be discarded.

The rarest material for brake rotors is a synthetic composite made from carbon-graphite and boron. This material was developed for aircraft brakes and is now used on Formula 1 and other high-speed road racing cars. These types of rotors are also used on top-end Porsche and Ferrari automobiles. The advantages of composite rotors are very light weight and the ability to withstand temperatures as high as 1700°F (927°C). On the down side, composite rotors are extremely expensive, and wear rapidly from dirt contamination.

Drilled and Slotted Rotors

A few high-performance vehicles come with drilled or slotted rotors. These rotors have drilled or machined holes or slots, figure 11-10. As discussed in Chapter 8, these features help reduce gas and water fade. Crossdrilling on solid rotors has minimal effect on brake cooling. Crossdrilling on vented rotors can enhance their cooling ability by about 15 to 30 percent, according to some manufacturer's claims.

Rotors that are slotted but not drilled offer better control of hot braking gases and water but are not as expensive to manufacture as the drilled rotor. Depending on the design and angle of the rotor slots, cooling is also increased as compared to a plain rotor.

Figure 11-10. High-performance vehicles may use drilled and slotted rotors to prevent gas and water fade. (Courtesy of Wilwood Engineering)

The technician should be aware that these rotors are designed and engineered from scratch to be slotted and drilled. The drilled holes are carefully laid out, drilled, and chamfered to reduce the possibility of cracking around the holes. Drilling holes in the stock rotor on a drill press is not recommended.

Drilled and Slotted Rotor Concerns

Although there is no doubt about the enhanced performance of drilled or slotted rotors, they have certain characteristics of which the brake technician should be aware. The main concerns are:

- Noise and vibrations
- Cracking under heavy use.

The drilled and slotted rotors may have tendencies to make growling or whirring noises under moderate to heavy braking. This may be heard more or less, depending on the vehicle and brake manufacturer. Customers may perceive the noise as a vibration in the pedal, but this does not normally affect the performance of the brakes. It seems to be normal with this type of rotor, so any attempt to "fix" this noise may not be successful.

Another concern is slight to moderate cracking around the drilled holes. This problem is most likely to appear on "self-drilled" rotors and less likely on properly engineered drilled rotors. Rac-

Brake Drums and Rotors

Figure 11-11. Aftermarket brake kits are popular with vehicle owners looking for enhanced brake performance. (Courtesy of Wilwood Engineering)

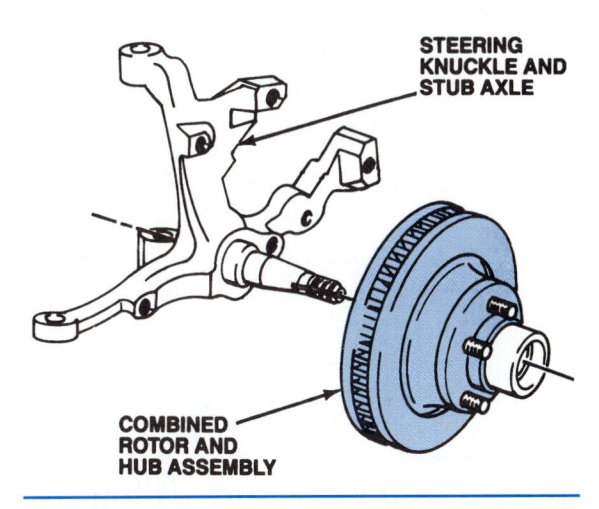

Figure 11-12. Fixed drums and rotors are cast in one piece with the hub assembly.

ing vehicles can show cracking after a few hours of hard use, whereas a street driven vehicle may never have a cracking problem. In any case, vehicles with crossdrilled rotors should be inspected more often and more carefully than those with plain rotors; a cracked rotor that loses a chunk of rotor while driving can be disastrous.

Aftermarket Drilled and Slotted Rotors

A number of companies that specialize in performance brake systems offer aftermarket rotors, calipers, and kits for popular vehicles, figure 11-11. It is important that the brake technician select the proper kits and parts for the vehicle, considering that the brakes are the most important safety system on the vehicle. The technician should receive training or a training/installation manual from the manufacturer before installing or working on these aftermarket systems.

Because these systems are often worked on and driven by nonprofessional mechanics or owners, they require even more rigorous inspection whenever repairs are needed. The manufacturer of the brake rotor/caliper system should supply a required maintenance schedule to follow. Obtain this information and follow it when servicing these types of brake systems.

DRUM AND ROTOR MOUNTING METHODS

Brake drums and rotors are either part of, or attached to, the hubs or axle flanges of the vehicle. All drums and rotors can be grouped into two categories:

- Fixed
- Floating.

With a fixed drum or rotor, the wheel studs, or threaded holes for the wheel lug bolts, are part of the drum or rotor assembly and are removed with it. A floating drum or rotor is separate from the hub that contains the wheel studs or lug bolt holes and can usually be removed from the vehicle independent of it.

Fixed Drums and Rotors

A fixed drum or rotor, figure 11-12, is cast in one piece with the hub assembly containing the wheel bearings. This type of drum or rotor is most often used on nondriven axles; fixed drums can be found on the rear brakes of many FWD cars, whereas fixed discs are common on the front brakes of RWD cars. To remove a fixed drum or rotor, the wheel bearing retaining nut is unscrewed and the drum or rotor assembly is removed as a unit.

Fixed drums and rotors are usually adopted to reduce the number of parts on a car and simplify vehicle assembly at the factory. Fixed drums and rotors do not offer any significant functional advantages. In fact, their one-piece construction can conduct undesirable amounts of heat to the wheel bearings, and they are more expensive to replace than floating drum and rotor designs.

Some trucks and older cars have fixed drums on their driven axle. This type of drum installs over a

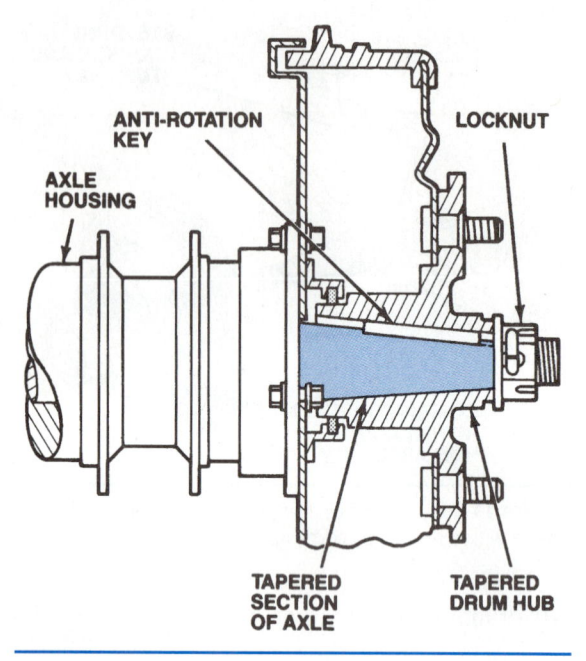

Figure 11-13. Taper-fit fixed drums are used on some driven axles.

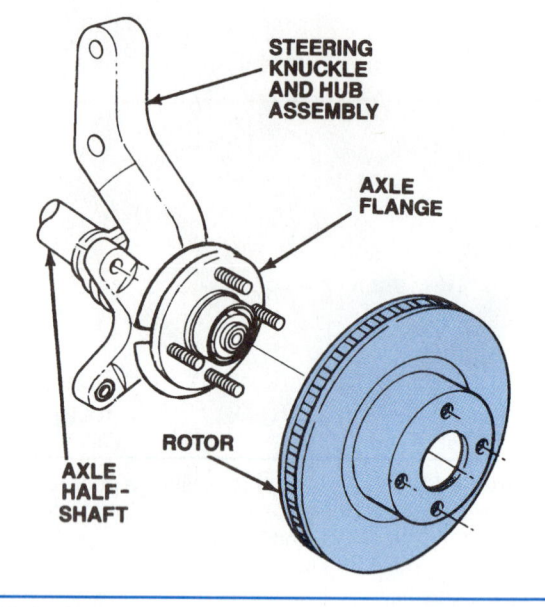

Figure 11-14. Floating drums and rotors are separate from the hub or axle flange.

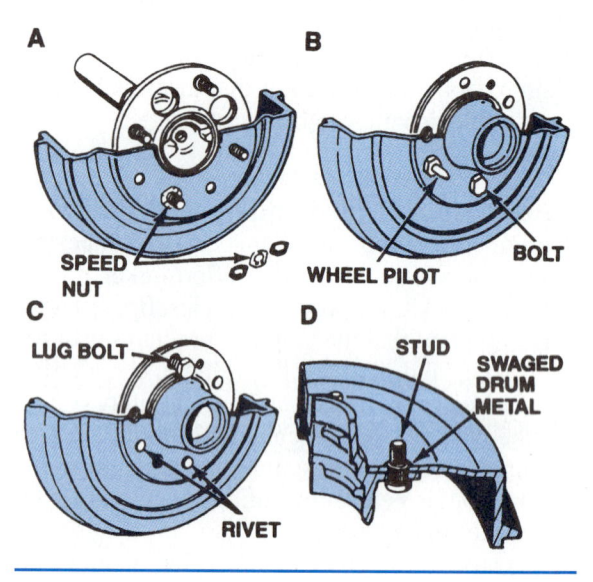

Figure 11-15. Floating drums and rotors are secured by a number of means.

taper on the end of the axle, figure 11-13, and is prevented from slipping by a combination of a woodruff key, or other locating device, and high-installation torque. A locknut and cotter pin on the end of the axle secure the drum in place. These drums become very tightly wedged on the axle, and a special puller is required to remove them.

Floating Drums and Rotors

Floating drums and rotors, figure 11-14, are installed on a separate axle flange or hub assembly, and are common on both driven and nondriven axles. Floating drums and rotors are less expensive, and usually easier to service, than fixed-drum and rotor designs. In addition, the separate castings reduce heat transfer from the drum or rotor to the wheel bearings.

All floating drums and rotors are held in place by the wheel and lug nuts or bolts during normal operation. However, many floating drums and rotors are secured by other means as well. Popular methods include speed nuts, bolts or screws, and riveting or swaging.

Speed nuts, formally called tinnerman nuts, are small, flat, spring-steel clips that thread onto the wheel studs, figure 11-15A. Speed nuts are installed at the factory to keep floating drums and rotors from falling off the car as it moves down the assembly line. On cars with high shoulders on the wheel studs, speed nuts also prevent the drum or rotor from catching on the shoulder and becoming cocked. Speed nuts are not needed for normal operation and are usually discarded the first time the drum or rotor is removed.

Bolts or screws retain the floating drums and rotors on many imported cars, figure 11-15B. These fasteners pass through the drum or rotor and thread into the hub or axle flange. Bolts and screws serve the same purpose as speed nuts, although unlike speed nuts, they should always be

Brake Drums and Rotors

reinstalled after servicing. This is especially important on drum brakes to prevent dirt and other contaminants from entering the friction assembly through the bolt or screw holes.

On some cars with wheel lug bolts, one retaining bolt or screw on each drum or rotor has an extended head. The extension, called a wheel pilot, fits through a hole in the wheel to help hold it in place on the hub while the lug bolts are installed.

A few cars have floating drums or rotors that are riveted or swaged to the separate hub casting, figures 11-15C and 11-15D. These retaining methods create an assembly similar to a fixed drum or rotor, which allows the manufacturer to deal with a single part during vehicle manufacture. Once in the field, however, the drum or rotor can be replaced independently of the hub. To do this, the rivets are sheared off with a chisel, or a special cutter is used to remove the swaged metal so the drum or rotor can be parted from the hub.

BRAKE DRUM AND ROTOR WEAR

The application force, friction, and heat of braking cause drums and rotors to suffer wear, damage, and distortion. Drum and rotor wear is covered in this section, damage and distortion are dealt with in the following two sections.

The most common problem with brake drums and rotors is wear of the friction surfaces. Under ideal conditions, the drum and rotor friction surfaces wear evenly and their finish remains relatively smooth, figures 11-16 and 11-17. Conditions are rarely ideal, however, so friction surface wear is often uneven. There are three common types of wear that affect brake drum and rotor friction surfaces:

- Drum taper wear
- Drum barrel wear
- Rotor taper variation

Unless too much metal must be removed, drum and rotor wear can usually be corrected by machining the friction surface true again.

Drum Taper Wear

The most common type of brake drum wear is **taper wear**, figure 11-18, in which the closed edge of the drum friction surface wears more than the open edge. Taper wear is caused by the excessive application force and heat of repeated

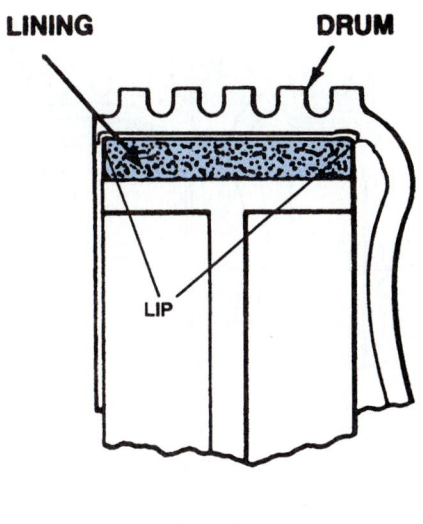

Figure 11-16. Normal brake drum wear will be fairly even across the drum, leaving slightly more wear at each edge and an inner and outer "lip."

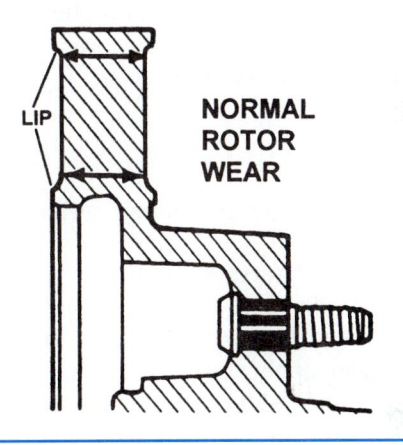

Figure 11-17. Normal rotor wear will leave the rotor evenly worn where the pads make contact, leaving a lip on the inner and outer edges.

heavy braking. These factors cause the open edge of the drum to expand outward, whereas the closed edge of the drum, supported by the drum web, expands less and maintains better contact with the brake linings. The result is greater friction, heat, and wear at the closed edge of the drum.

Brake dust, dirt, and other grit in the drum also contribute to taper wear. These contaminants collect near the closed edge of the drum where they act as abrasives between the linings and the friction surface.

When a drum with taper wear cools, its open edge contracts to a smaller diameter than the closed edge. This will often cause a low brake pedal

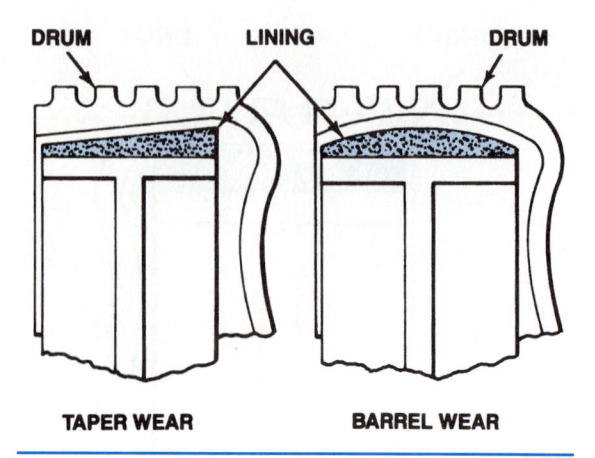

Figure 11-18. Taper and barrel wear are common in brake drums.

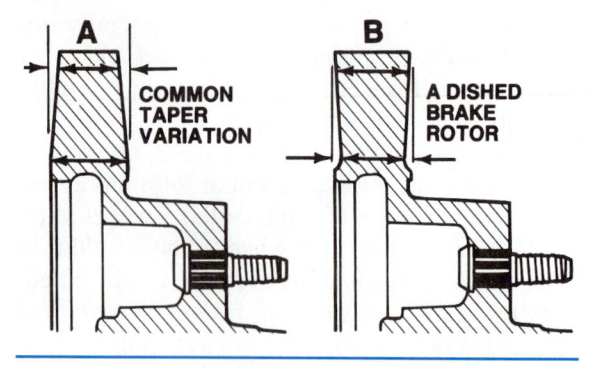

Figure 11-19. Taper variation is a type of wear suffered by brake rotors.

because brake adjustment is based on lining contact with the friction surface near the open edge of the drum where the metal is thickest. When the brakes are applied, extra pedal travel is required to distort the drum and shoes to provide full lining contact. If new shoes are installed in a drum with taper wear, unusual lining wear or permanent shoe distortion can result.

Drum Barrel Wear

When the brake drum friction surface is worn more at the center than at the edges, the drum is suffering from **barrel wear,** figure 11-18. Barrel wear is caused by uneven application force where the linings contact the drum. This occurs because brake shoe application force is transmitted through the shoe web attached along the centerline on the backside of the lining table. During heavy braking, the center of the lining material is applied most firmly against the drum friction surface, whereas the unsupported edges of the lining table deflect slightly downward. The result is greater friction, heat, and wear at the center of the drum.

Generally, drum brakes with rigid cast-aluminum shoes are less susceptible to barrel wear than those with more flexible welded steel shoes. Barrel wear does not present any operating hazards, and it has no obvious symptoms that are relayed to the driver. The only real problem is that it shortens drum service life.

Rotor Taper Variation

A rotor with **taper variation** is usually thinner at the outer edge of its friction surface than near its center, figure 11-19A. This type of wear occurs because the outer edge of the friction surface has a larger diameter than the inner edge. Therefore, the outer edge moves past the brake linings at a higher speed and creates greater friction, heat, and wear.

In some cases, particularly on certain imported-car rotors, taper variation occurs in which the inner edge of the friction surface wears more than the outer edge, figure 11-19B. A rotor with this type of wear is sometimes referred to as a **dished rotor**. Dishing of the rotor occurs when the outer edges of the linings routinely operate at temperatures so high that their coefficients of friction drop off. The cooler-running inner edge of the lining then provides the bulk of the friction, heat, and wear on the rotor.

If new brake pads are installed against a rotor with excessive taper variation, the linings will not contact the rotor evenly across their entire surface. In addition, the pads may cock, placing a side load on the caliper pistons that will cause them to bind in their bores. Most manufacturers allow a maximum taper variation of about .003″ (.08 mm).

BRAKE DRUM AND ROTOR DAMAGE

In addition to wear, drums and rotors often experience damage to their friction surfaces. The most common forms of brake drum and rotor damage are:

- Scoring
- Cracking
- Heat checking
- Hard or chill spots.

Most of these conditions are the result of extremes in brake operation, extremes of wear, stress, temperature, or temperature variation. Because drum and rotor damage is caused by extremes of operation, it is most commonly found

Brake Drums and Rotors

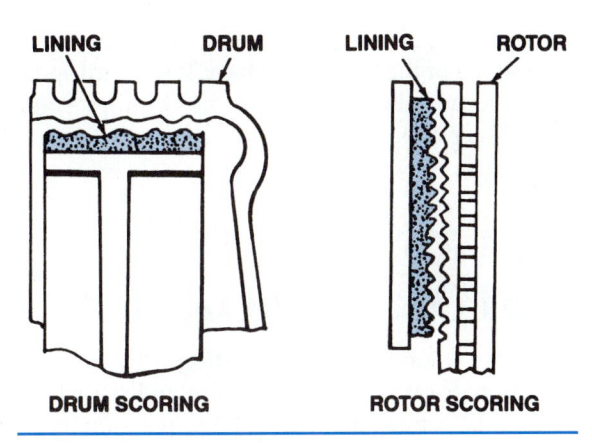

Figure 11-20. Scored drums and rotors often result from metal-to-metal contact.

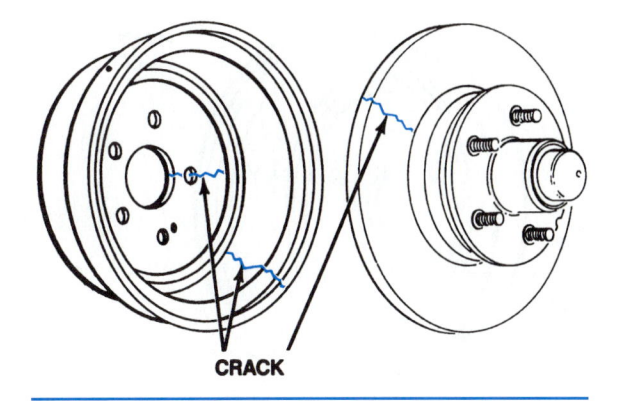

Figure 11-21. Cracked drums and rotors must be replaced.

on front brakes which experience more severe service than rear brakes.

Scoring

Scoring, figure 11-20, is an extreme form of drum and rotor wear consisting of scratches, deep grooves, and a generally rough finish on the friction surface. There are a number of causes for scoring, but the most common is brake linings that have worn to the point where a rivet, lining table, or pad backing plate contacts the drum or rotor. Also, certain friction materials are more likely to score drums and rotors than others, and glazed linings that have hardened from exposure to extreme heat can also cause scoring.

Drum brakes are more likely to become scored than disc brakes because their closed construction holds dirt, sand, and abrasive dust inside the friction assembly. This allows the contaminants to be scrubbed repeatedly between the linings and drum. Severe drum scoring often results when metal parts of the friction assembly fatigue, break loose, and are trapped between the linings and drum.

A scored drum or rotor will cause very rapid lining wear, often accompanied by a growling or grinding noise, particularly if there is metal-to-metal contact between the shoe and drum or pad and rotor. Scoring can be machined out of a drum or rotor as long as the amount of metal removed is within the allowable limits.

Cracking

Cracks in a brake drum or rotor, figure 11-21, are caused by the stress of severe braking or an impact during an accident. Generally, drums and rotors that have been previously machined are more susceptible to cracking than new parts. Cracks can appear anywhere on a drum or rotor, although on drums they are most often found near the bolt circle on the web or at the open edge of the friction surface. Rotors generally crack first at the edges of their friction surfaces.

Small cracks in cast iron are difficult to spot and often will not show up until after a drum or rotor has been machined. Cracks cannot be repaired, and *any* crack, other than the minor heat checking described next, is reason to replace the drum or rotor.

Heat Checking

A lesser form of drum and rotor cracking is called **heat checking,** figure 11-22, which consists of many small interlaced cracks on the friction surface. These cracks typically penetrate only a few thousandths of an inch into the metal; they do not pass through the structure of the drum or rotor. Heat checking is usually caused by a driver who leaves one foot on the brake pedal while applying the accelerator with the other. This practice, called "riding" the brakes, causes a continuous light drag that creates excessive heat. Heat checking can also be caused by repeated heavy braking or numerous panic stops made in rapid succession.

Because the friction surface of a heat-checked drum or rotor is no longer smooth, it causes rapid wear of the brake linings. It may also cause a slight pedal pulsation or brake noise. Light heat checking can often be machined away; in more severe cases the drum or rotor must be replaced.

Hard or Chill Spots

Earlier, it was stated that cast-iron drums and rotors are durable because the friction, heat, and

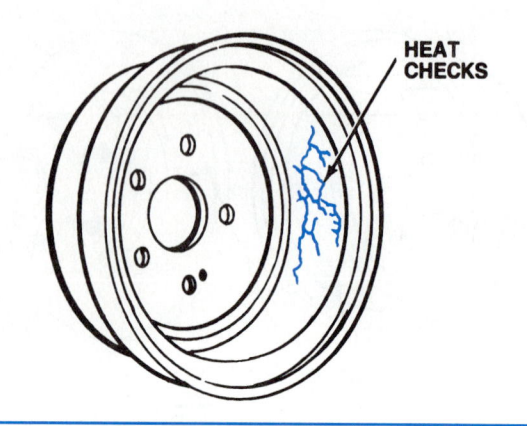

Figure 11-22. Heat checks are a minor form of cracking on the friction surface.

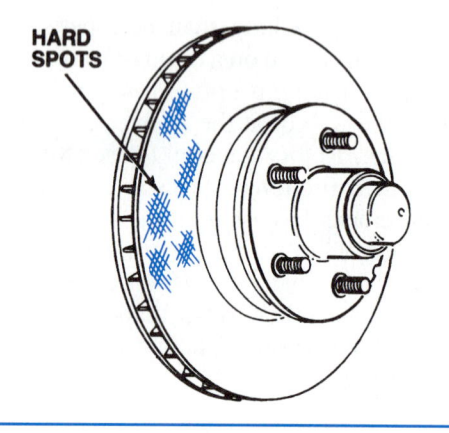

Figure 11-23. Hard spots appear where extreme heat has altered the structure of the iron in the drum.

pressure of braking cause a tough "skin" to form on their friction surfaces. However, if brake temperatures become too great, localized impurities in the metal can be burned away, altering the structure of the metal and causing **hard spots**, also called chill spots, to appear. Hard spots, figure 11-23, are roughly circular, bluish/gold, glassy appearing areas on the friction surface.

Hard spots create a number of problems. First, they are harder than surrounding areas of the friction surface and do not wear at the same rate. Once the spots begin to stand out from the rest of the friction surface, they cause rapid brake lining wear. Second, the friction coefficient of hard spots is less than that of surrounding areas so braking power is reduced or becomes uneven. This can cause the brakes to chatter, or result in a hard or pulsating brake pedal. Finally, a drum or rotor is more likely to crack in the area of hard spots than elsewhere.

If the hard spots in a drum or rotor are not too large and do not penetrate too deeply into the metal, they can sometimes be machined away or ground out of the friction surface. If the hard spots are large and deep, however, the drum or rotor should be replaced.

BRAKE DRUM AND ROTOR DISTORTION

To ensure smooth brake application without pedal pulsation or other problems, drum and rotor friction surfaces must maintain a fixed position in relation to the shoes or pads. In some cases, a position variation of less than a thousandth of an inch will create braking problems. Distortion puts the drum or rotor friction surfaces out of proper alignment with the shoes or pads.

A common cause of distortion in new brake drums and rotors is improper storage. Drums and rotors should always be stored lying flat; they should never be stood on edge. Distortion of replacement drums and rotors is so common that they are routinely checked, and machined true, before installation.

Drums and rotors in service distort for a variety of reasons. And because of their fundamentally different constructions, drums and rotors suffer unique forms of distortion. These conditions are discussed in separate sections below.

Drum Distortion

The friction surface of a brake drum in perfect condition is parallel to the axis of the axle and rotates in a precise circle centered on the axle or hub, figure 11-24. All brake drums suffer some distortion during brake operation, but they usually return to their original shape once the brakes are released. When the friction surface does not return to its proper shape, is no longer parallel to the axle, or does not rotate in a precise circle around the axle, the drum is distorted.

In general, drum brakes are less affected by distortion than disc brakes. In many cases, distortion does not become apparent until the drum is machined in a brake lathe. One reason for this is that most drum brake shoes are free to move on the support pads where they contact the backing plate. Although this can cause wear and noise, it also allows the shoes to shift position to maintain contact with the friction surface and compensate for small amounts of distortion. Drum brake dis-

Brake Drums and Rotors

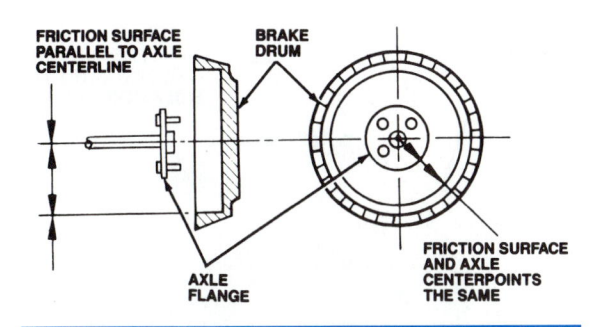

Figure 11-24. Proper alignment of a brake drum friction surface.

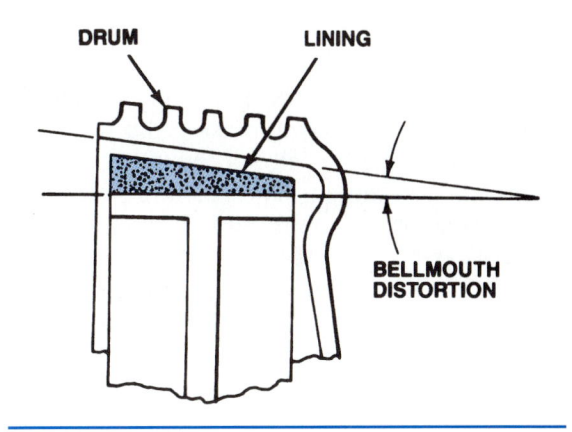

Figure 11-25. Bellmouth brake drum distortion.

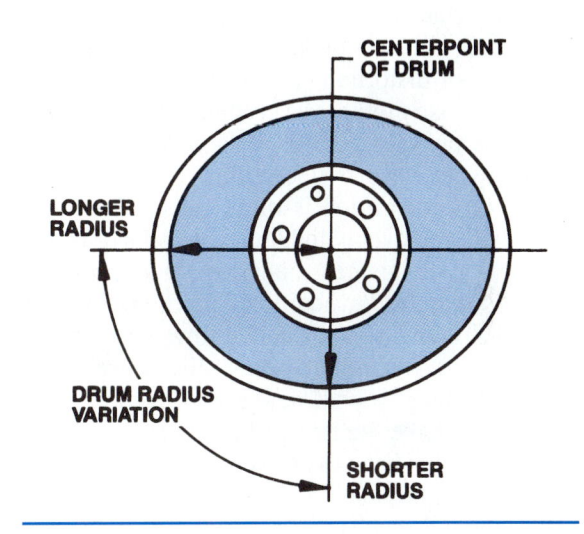

Figure 11-26. Out-of-round brake drum distortion.

tortion is also less likely to create brake pedal pulsation because the relatively small pistons in the wheel cylinders transmit less movement back to the master cylinder. There are three basic types of brake drum distortion:

- Bellmouth drums
- Out-of-round drums
- Eccentric drums.

Bellmouth Drums

When the open edge of a brake drum friction surface has a larger diameter than the closed edge, the drum is suffering from **bellmouth** distortion, figure 11-25. Bellmouth distortion is caused by poor drum rigidity combined with high heat and brake application force. These are the same factors that cause drum taper wear, but in this case they are present at more extreme levels and cause a different effect.

Bellmouth distortion occurs when a drum suffers mechanical fade and its open edge, unsupported by the drum web, expands more than its closed edge. When the brakes are applied harder to compensate for the fade, the shoes distort the open edge of the drum so far outward that the **elastic limit** of its metal is exceeded. Once this happens, the drum will not return to its original shape after it cools. Repeated occurrences of this process eventually cause the drum to take on a bellmouth shape.

Bellmouth distortion is especially common on wide drums, and often takes place when the drum is not strong enough for the application. This commonly occurs when a drum has had too much metal machined from it. If new shoes are installed in a bellmouthed drum, brake fade and unusual lining wear will result. Bellmouth distortion combined with new shoes will also cause a low brake pedal because the shoe lining tables must be distorted to achieve full lining-to-drum contact.

Out-of-Round Drums

Uneven heat distribution can sometimes cause **out-of-round** distortion, figure 11-26, in which the drum radius varies when measured at different points around its circumference. Out-of-round distortion can take place when a car drives through a puddle after a series of hard stops and cold water splashed on the brakes causes rapid and uneven cooling of the drums. It can also result if the parking brake is firmly applied after a series of hard stops before the drums have had a chance to cool. In this case, the shoes extend out against the heat-expanded drum and prevent it from contracting to its original circular shape as it cools. Instead, the extended shoes force the drum into an out-of-round shape. Improper handling of the drum during service can also cause an out-of-round condition. Accidentally dropping the drum on its edge from a height of 1 to 2 feet can result in a .004" to .010" out-of-round condition. Use care when handling brake drums.

The most common symptom of an out-of-round drum is a pulsating brake pedal when the brakes are applied at all speeds. Out-of-round drums can also cause a vibration or brake chatter at speeds above approximately 40 mph. In more extreme cases, an out-of-round drum can result in erratic braking action and possibly cause the brakes to grab with every revolution of the wheel.

Eccentric Drums

Eccentric distortion exists when the geometric center of the circle described by the brake drum friction surface is other than the center of the axle or the bolt circle in the drum web, figure 11-27. This type of distortion causes the drum to rotate with a camlike motion. An eccentric drum will result in a pulsating brake pedal similar to that caused by an out-of-round drum.

Eccentric distortion is often caused by over-tightened or unevenly tightened lug nuts or bolts. Not only will this cause an immediate problem but it also can lead to additional problems later. If a car is driven for an extended time with an eccentric drum, the linings may slowly wear the friction surface "round" again. However, as soon as tension on the lug nuts or bolts is released, the drum will relax back to its original shape creating an out-of-round condition that was not apparent before the wheel was removed.

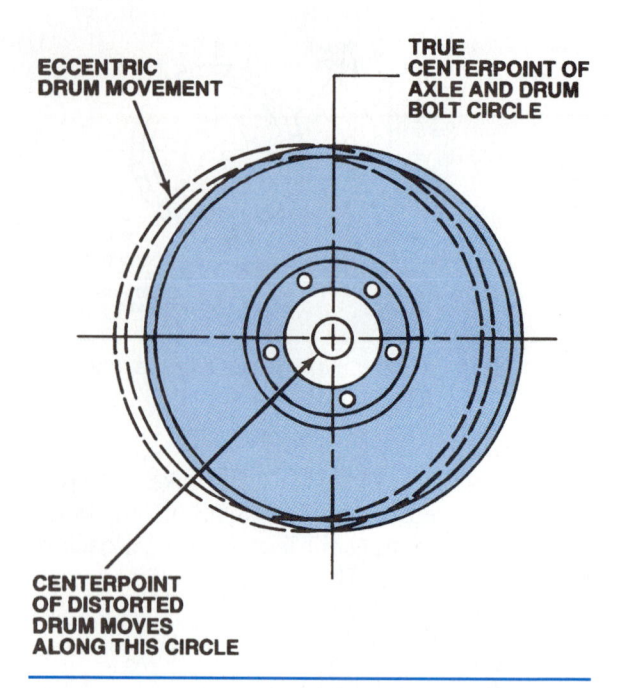

Figure 11-27. Eccentric brake drum distortion.

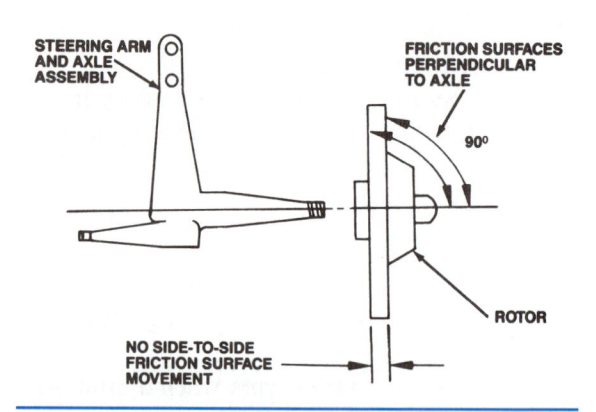

Figure 11-28. Proper alignment of the brake rotor friction surfaces.

Rotor Distortion

The friction surfaces of a rotor in perfect condition are perpendicular to the axle centerline, and have no side-to-side movement, figure 11-28. Unlike brake drums, rotors do not suffer distortion as a routine part of brake operation. However, distortion can occur during braking if there is a problem with the friction assembly, such as a frozen caliper piston that creates unequal application force on the two sides of the rotor.

Friction surface distortion is much more significant in a disc brake rotor than in a brake drum because the design of the friction assembly magnifies the effects of any wear. The hydraulic principles discussed in Chapter 3 dictate that small movements of the large pistons in the brake calipers are converted into large movements of the small pistons in the master cylinder. Even very small amounts of distortion in disc brake rotor can cause large amounts of pedal pulsation. The two most significant forms of rotor wear are:

- Lateral runout
- Lack of parallelism.

Rotor Lateral Runout

Lateral runout is a side-to-side wobble of the rotor as it rotates on the spindle, figure 11-29. As discussed in Chapter 7, a small amount of runout provides caliper piston knockback that reduces drag when the brakes are not applied. However, if the amount of runout is too great, excessive brake pedal travel and front-end vibration will result. In

Brake Drums and Rotors

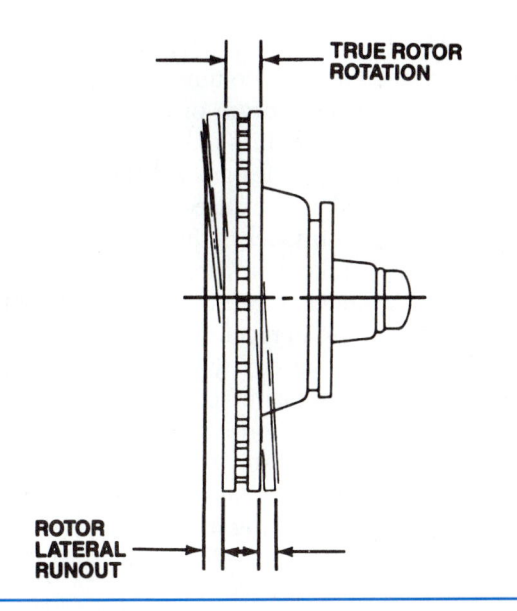

Figure 11-29. Brake rotor lateral-runout distortion.

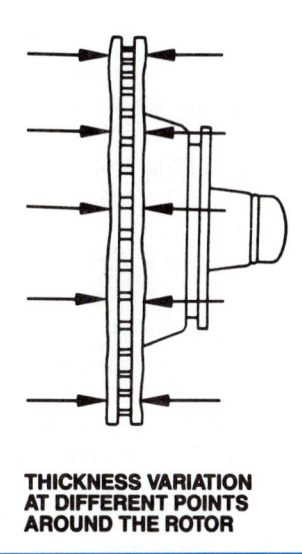

Figure 11-30. Brake rotor lack-of-parallelism distortion.

cases of severe runout, a pulsating brake pedal may also be present.

Lateral runout can be caused by several factors. Overtightened or unevenly tightened lug nuts or bolts are a common source of runout on newer cars with downsized brake rotors. Extreme heat or rapid temperature variations also cause runout. And inaccurate machining at the factory or in the field is a common cause of this distortion.

All rotors have some lateral runout, and disc brakes operate normally in spite of it. The maximum amount of runout allowed varies with the vehicle manufacturer and friction assembly design. Rotors used with fixed calipers are generally allowed less lateral runout than those teamed with floating or sliding calipers. Most maximum values range between .002″ and .008″ (.05 and .20 mm).

Rotor Lack of Parallelism

Lack of **parallelism** is a variation in the thickness of the rotor when it is measured at several places around its circumference, figure 11-30. A rotor with friction surfaces that are not parallel is the most common disc brake cause of a pulsating brake pedal. A lack of rotor parallelism will cause a pedal pulsation when the brakes are applied at all speeds and can also cause front-end vibration during braking.

Next to concerns about braking noise, the most common customer complaint concerning brakes is a pulsating pedal during braking. The pulsation is caused by a variation in the thickness of the rotor. As the pads are applied to the rotor during braking, the caliper piston moves in and out in response to the variations on the rotor surface. This movement is transferred by the hydraulic system to the brake pedal. Lack of parallelism (thickness variation) can be caused by a number of factors. Some are:

- Soft spots in the rotor casting
- Rust buildup on the rotor
- Uneven heating and cooling
- Excessive rotor lateral runout.

Soft spots in the rotor, caused by poor casting or excessive machining, can result in areas on the rotor surface that wear more rapidly than other areas. The uneven wear leaves high and low spots on the rotor.

Rust buildup can occur on a vehicle that is not driven for an extended period of time. The part of the friction surface that is protected by the brake pads does not rust as much as the rest of the rotor. The more rusted areas wear at different rates, which results in a thickness variation when the car is driven again. This can be verified during measurement of the rotor; the area of excessive variation will be shaped like the brake pad.

Uneven cooling of the brake rotor may cause thickness variation on vehicles driven under extreme conditions, such as towing up and down hills or during repeated high-speed stops. If the vehicle is parked without allowing the brakes to cool evenly under light braking, uneven cooling can slightly change the surface composition of the rotor. The pads and caliper assembly will cool much more slowly than the exposed areas of the rotor, allowing this change to occur. Similar to the rusting condition, the areas of excessive variation will take the shape of the brake pad. In addition, the faulty area may be discolored when compared to the rest of the rotor.

Lateral runout can be a significant cause of rotor thickness variation. Excessive lateral runout allows a portion of the rotor to slightly contact the brake pad with each rotor revolution. Over time, the rotor will wear at the point of contact, causing the thickness variation. Eliminating runout is the only way to make sure that the problem does not recur.

Because parallelism variation is much more likely to cause a problem than lateral runout, the maximum amount allowed is much smaller than for runout. Most manufacturers specify that the two friction surfaces of a rotor must be parallel within half a thousandth of an inch, .0005″ (.013 mm).

DRUM AND ROTOR REFINISHING

Some wear of drum and rotor friction surfaces is considered normal. If the finish on the friction surfaces remains smooth, and there is no damage or distortion, new brake linings can usually be installed without the need to machine the drum or rotor. This is desirable because it provides the longest possible component life. A few thousandths of an inch of metal removed from a drum or rotor is equal to thousands of miles of normal wear.

However, if a drum or rotor is worn too severely, or it is damaged or distorted, metal must be cut or ground away to restore a smooth and true friction surface. There are three basic operations used to do this:

- Machining
- Grinding
- Resurfacing.

Drum and Rotor Machining

The most common method of resurfacing drums and rotors is to machine them in a **brake lathe**. The lathe rotates the drum or rotor while a hardened steel cutting bit is slowly drawn across the friction surface to remove a thin layer of metal, figure 11-31. Because the lathe rotates the drum or rotor, technicians commonly refer to this process as **turning** a drum or rotor. Some brake lathes are designed to turn only drums or only rotors, but many newer designs can do both. Lathes that machine brake rotors use two cutters to refinish both friction surfaces at the same time.

Traditional brake lathes are large free-standing machines designed to refinish drums and rotors that are off the car. In recent years concerns about pedal pulsation caused by rotor lateral runout have caused many manufacturers to recommend the use of a portable, on-car brake lathe, figure 11-32. This allows the rotor to be machined while it is still on the vehicle. These machines are especially useful when turning the hard-to-remove rotors found on four-wheel-drive trucks and some

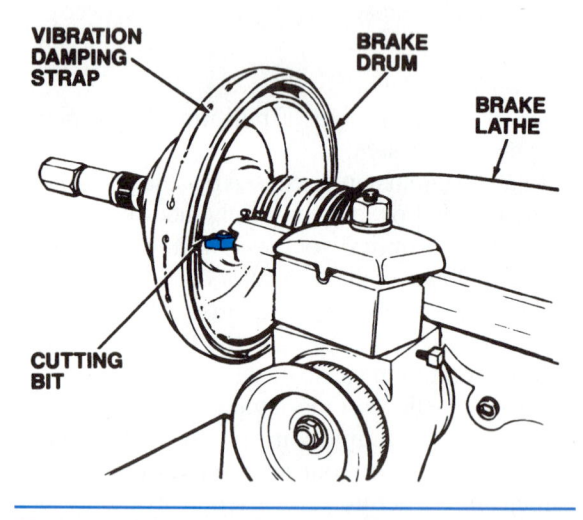

Figure 11-31. A steel cutting bit is used to remove metal when a drum or rotor is turned.

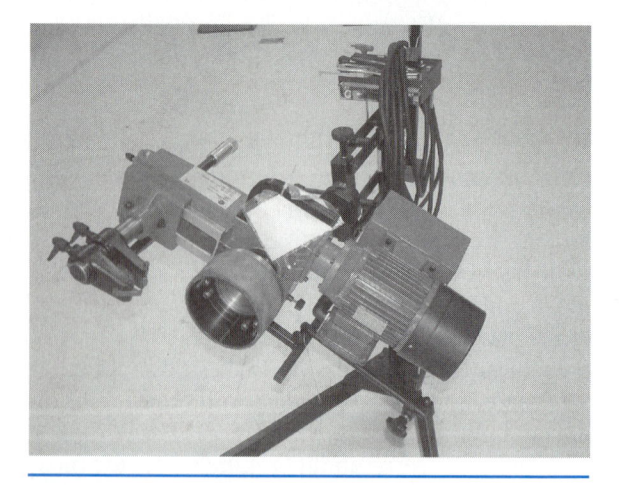

Figure 11-32. An on-car brake lathe.

Brake Drums and Rotors

import vehicles. Turning the rotor while it is still mounted on the vehicle compensates for any runout in the hub or rotor.

There are several variables in the drum and rotor machining process that affect the quality of the final finish. The speed of drum or rotor rotation, the depth of cut, and the rate at which the cutting tool moves across the friction surface are all important and can be adjusted to achieve specific results. The details of machining drums and rotors are explained in the *Shop Manual*.

Drum Grinding

Hard spots in brake drums often cause the cutting bit in a brake lathe to chatter when it passes over them. This makes it difficult, if not impossible, to obtain a good finish on the friction surface. **Grinding** is an alternate resurfacing method used to remove hard spots from brake drums. In this process, the drum is turned in a brake lathe as described previously, then the cutting bit is replaced by a special attachment fitted with a power-driven stone grinding wheel, figure 11-33. The spinning wheel moves across the rotating friction surface and grinds away the hard spots.

Drum grinding was a common process when front drum brakes routinely experienced the extreme temperatures that cause hard spots. Today's lightly loaded rear drum brakes rarely get hot enough for hard spots to form, so grinding is seldom done any more. In addition, once hard spots have been ground from a drum they are more likely to recur because the metal removed increases the drum's operating temperature. In the long run, it is usually better to simply replace a drum if any hard spots cannot be removed by normal machining.

Rotor Resurfacing

Resurfacing is a minor machining operation in which the friction surfaces of a brake rotor are ground smooth with a spinning abrasive disc while the rotor is spun on a lathe, figure 11-34. Resurfacing removes rust, brake lining deposits, and minor damage; it can also true a rotor that has a very small amount of distortion.

Resurfacing is also done to put a nondirectional finish, figure 11-35, on the rotor friction surfaces after they are machined. When a rotor is turned, the movements of the cutting tools leave almost undetectable spiral grooves across the friction surfaces. A nondirectional finish removes these grooves and helps shorten the break-in period for new pads. As discussed later in the chapter, this is especially important when fitting semimetallic brake pads.

Resurfacing removes very little metal and is not intended to be a substitute for rotor machining. Some companies have introduced portable resurfacing machines, similar to the lathes described

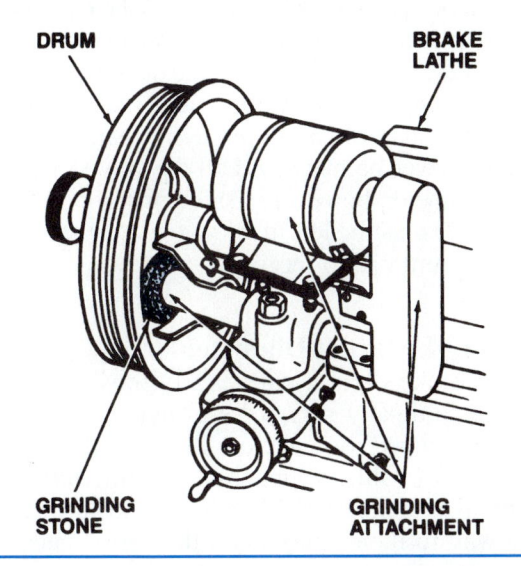

Figure 11-33. Grinding with an abrasive stone can remove hard spots from brake drums.

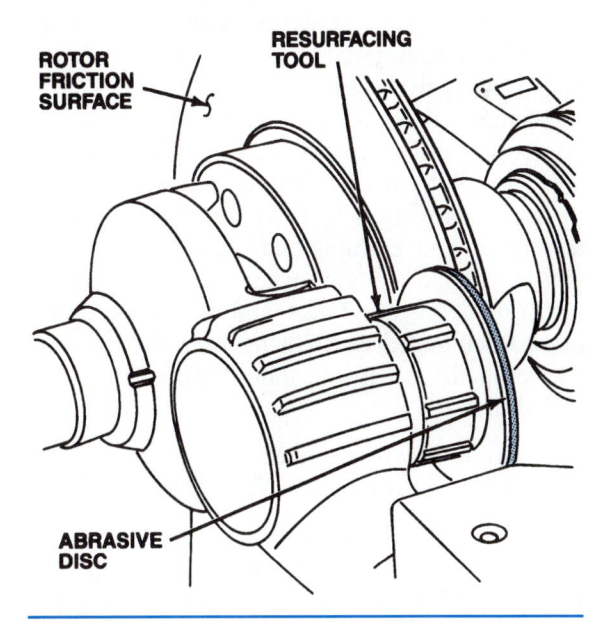

Figure 11-34. Resurfacing with an abrasive disc can refinish a lightly worn friction surface.

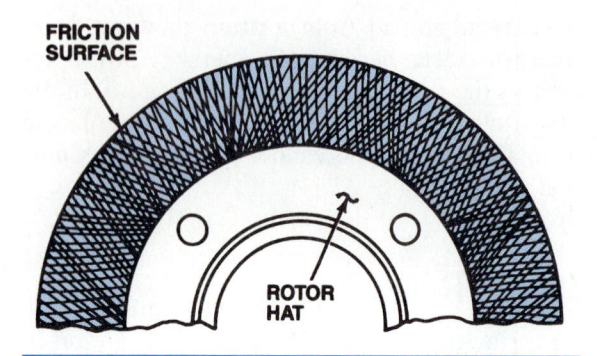

Figure 11-35. A nondirectional finish on a brake rotor friction surface.

earlier, that can resurface rotors on the car. This is an acceptable procedure for rotors with minor wear, but it is not an alternative to machining a badly damaged or distorted rotor.

DRUM AND ROTOR METAL REMOVAL LIMITS

When brake drums and rotors are turned, only the minimum amount of metal necessary to "clean up" the friction surfaces should be removed. The more metal removed from a drum or rotor, the sooner various forms of fade will occur. Metal content also affects strength; when too much metal is removed from a drum or rotor, it will crack and distort more easily. Finally, there are legal and liability issues to consider. As discussed in Chapter 2, the DOT has regulations about maximum allowed wear on drums and discs. Anyone that installs an excessively machined disc or drum is asking for trouble.

In response to these DOT regulations and to help ensure that drums and rotors are not machined beyond the point where they can provide safe operation, all drums and rotors manufactured since the early 1970s have machining limits stamped or cast into them. These limits are called the:

- Drum discard diameter
- Rotor discard dimension.

It is important to recognize that the terms used to express these values, and the manner in which they are interpreted, vary from one manufacturer to another, and sometimes vary among different cars from the same manufacturer! The most common uses are described in

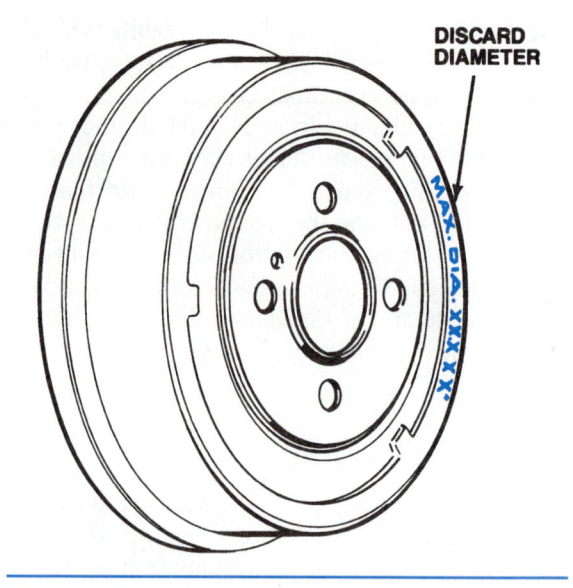

Figure 11-36. Discard diameter and maximum diameter are brake drum machining and wear limits.

the following sections; however, if you have any question about the meaning of the markings on a drum or rotor, consult the factory service manual for details.

Drum Discard Diameter

Most brake drums have a **discard diameter,** sometimes called the maximum diameter, stamped or cast on the outside of the drum, figure 11-36. The discard diameter is the maximum inside diameter allowed. When a drum reaches its discard diameter as a result of machining or wear, it must be replaced.

Traditionally, the discard diameter has been .090″ (2.25 mm) larger than the standard diameter although smaller drums, and modern lightweight units, often allow less wear. A drum should never be machined all the way to its discard diameter, however; a safety margin of .030″ (.75 mm) must be left to allow for wear when the drum is returned to service.

Drum Maximum Diameter

Some brake drums, particularly those on Ford Motor Company cars, have a "maximum diameter" stamped or cast on the outside of the drum. This value is a machining limit, *not* a discard diameter. These drums can be turned all the way to their "maximum diameter" because they contain sufficient metal to allow approximately .030″ (1.5 mm) wear beyond that point. If the friction surface can be restored without

Brake Drums and Rotors

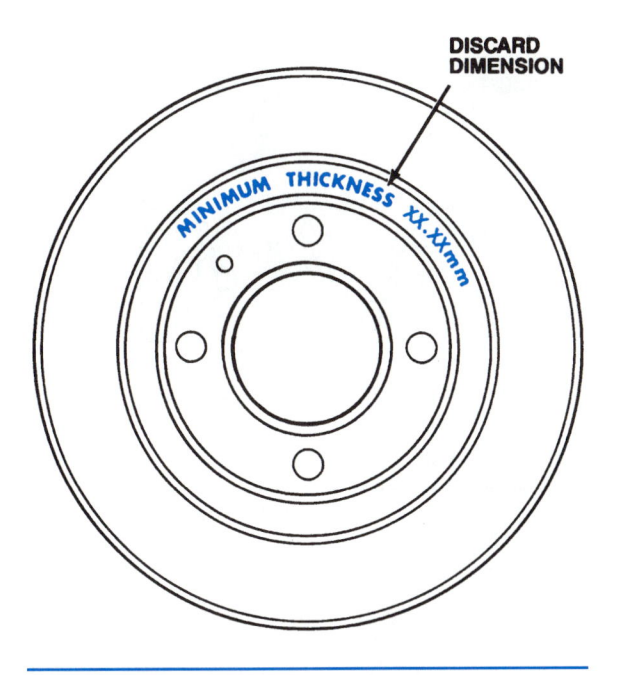

Figure 11-37. Discard dimension and minimum refinish thickness are brake rotor machining and wear limits.

exceeding this limit, the drum can be returned to service.

In keeping with the machining limits already described, the maximum diameter on Ford drums is typically .060″ (1.5 mm) larger than the standard drum diameter.

Rotor Discard Dimension

Most brake rotors have a **discard dimension**, sometimes called a minimum thickness, cast or stamped into the hub of the rotor, figure 11-37. The discard dimension is the minimum rotor thickness allowed. If a rotor reaches its discard dimension as a result of machining or wear, it must be replaced. There is greater variation among rotor discard dimensions than there is among drum discard diameters; however, most rotor discard dimensions are about .060″ (1.5 mm) thinner than the standard thickness.

As with a brake drum, a rotor is never machined all the way to its discard dimension; approximately .030″ (.75 mm) of metal must be left to allow for additional wear in service. To prevent too much metal from being removed, most manufacturers specify a minimum refinish thickness in addition to the rotor discard dimension. The minimum refinish thickness is usually .015″ (.37 mm) to .30″ (.75 mm) thicker than the discard dimension.

Rotor Minimum Thickness

Some brake rotors, particularly those on Ford Motor Company cars, have a "minimum thickness" stamped or cast on the outside of the rotor. Like the "maximum diameter" on some brake drums, the rotor "minimum thickness" value is a machining limit, *not* a discard dimension. The "minimum thickness" is the thinnest dimension to which the rotor can be turned. If the friction surfaces can be cleaned up without going below this figure, the rotor can be returned to service.

SPECIAL DRUM REFINISHING CONSIDERATIONS

Machining metal out of brake drums does two things; it increases the amount of leverage the brake has on the wheel, and it decreases the drum's resistance to mechanical fade. These two factors can have a major effect on side-to-side brake balance. When turning drums, it is very important to keep the diameter of drums on the same axle within .010″ to .020″ (.25 to .50 mm) of one another. If one drum has significantly more metal removed, that friction assembly will fade sooner and cause the brakes to pull.

Because drum brakes act in this manner, if one drum is turned .060″ (1.5 mm), the other drum on the same axle should be turned the same amount regardless of its condition. When a new drum is paired with an older machined drum, the new drum must be turned to match it. If one drum on an axle is worn excessively, it is often less expensive in the long run to replace both drums rather than remove excessive metal from a new drum.

SPECIAL ROTOR REFINISHING CONSIDERATIONS

Although brake rotors do not suffer the fade and pull concerns of brake drums, they do have some unique concerns all their own. These fall into two areas:

- Special metal removal limits
- Rotor surface finish.

Special Rotor Metal Removal Limits

Unlike a brake drum, a rotor has two friction surfaces—one on each side. Whenever a rotor is turned, at least a small amount of metal is removed on *both* sides to minimize runout, maintain parallelism, and ensure a consistent coefficient of friction between the two sides of the rotor. However, in some situations, it is desirable to machine different amounts of metal from each side. This allows the repair of severe wear and damage on one side of the rotor while minimizing the total amount of metal removed.

Although it is possible to remove different amounts of metal from each side of a rotor, this practice is not recommended for all types of friction assemblies. With fixed-brake calipers, the same amount of material must be removed from both sides of the rotor to maintain proper caliper alignment. If more metal is removed from one side than the other, the rotor will no longer be centered in the caliper. And if the maximum allowable amount of metal were machined from only one side of the rotor, the caliper pistons on that side might extend too far out of their bores and cock, jam, or create a leak.

With floating and sliding calipers, however, it is permitted to machine different amounts of metal from the two sides of the rotor. This can be done because the calipers in these friction assemblies automatically center themselves over the rotor. Most floating and sliding calipers can accommodate a moderate difference in the amount of metal removed from the friction surfaces.

Rotor Inner Surface Stock Removal Limit

To prevent the caliper body from shifting too far in one direction, some manufacturers specify an inner surface stock removal limit for their rotors. When the rotor is machined, a depth micrometer, special bracket, and rotor gauge are used to determine how much metal can be removed from the inner friction surface, figure 11-38. If more than the measured amount of metal must be removed to true the inner friction surface, the rotor must be replaced, even if its total thickness is greater than the minimum thickness dimension.

Rotor Surface Finish

A smooth friction surface finish is desirable on any drum or rotor, and the normal machining processes usually provide a good finish for most

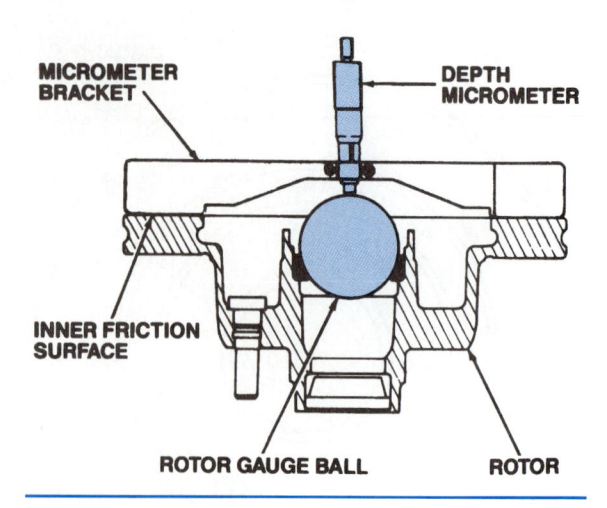

Figure 11-38. Special tools are used to determine how much metal may be removed from the inner friction surface of some rotors.

applications. However, a rotor that runs against semimetallic pads does require a smoother finish than one used with organic pads. Otherwise, slow break in, poor braking performance during break in, and rapid rotor wear may result. These problems result from basic differences in the two friction materials.

Compared to semimetallic friction materials, organic linings are soft and have a high coefficient of friction when cold. These traits allow organic linings to burnish into full contact with a rough rotor very quickly. In addition, organic linings are very unlike a cast-iron rotor and fade at low temperatures; this protects the rotor when it is extremely hot and more susceptible to damage. In other words, organic linings are forgiving; they wear faster, fade, or otherwise absorb damage themselves rather than causing it to the rotor.

Compared to organic friction materials, semimetallic linings are hard and have a low coefficient of friction when cold. This leads to an extended break-in period if the friction surface is too rough to provide good initial contact. In addition, semimetallic linings grip the rotor best at the high temperatures where it is more susceptible to damage. If the finish on the friction surface is too rough, it may provide sufficient "tooth" to allow the lining to actually engage and tear away metal from the rotor. Finally, semimetallic linings are made of iron and steel similar to the cast iron in rotors. If the pads and rotor are not properly mated, extreme heat can develop and weld these parts together momentarily, causing severe damage. In general, semimetallic friction materials

Brake Drums and Rotors

are less forgiving and will create problems if the rotor is not properly finished.

Determining Rotor Surface Finish

The smoothness of the friction surface finish is measured in **micro-inches**, with a lower number indicating a smoother finish. A typical rotor friction surface will be in the 10 to 150 micro-inch range, but for the best braking performance with semimetallic pads, the finish should be less than 50 micro-inches—the lower the better.

Micro-inch measurements require expensive special equipment that is impractical for use in the field. To ensure that a proper finish is obtained, manufacturers specify certain settings for brake lathe turning speed, crossfeed, and depth of cut. If these settings are followed when machining a rotor, the correct finish will result.

Rotors used with semimetallic pads must also be given a nondirectional finish to aid lining break in. This is applied through the resurfacing techniques described earlier in the chapter. To ensure a smooth finish, finer grit resurfacing abrasives are recommended when semimetallic linings are used than when organic friction materials are employed.

SUMMARY

Drums and rotors are the heaviest parts of their respective brake systems and absorb most of the heat generated in braking. The typical drum is a one-piece gray iron casting that may have cooling fins or a vibration damping coil spring around its outer edge. All drums use gray cast iron for their friction surface, but drums not made entirely of cast iron use composite construction: a cast-iron drum with a sheet-steel web, a stamped-steel outer drum with a centrifugally cast-in liner, or an aluminum outer drum with a cast-iron liner.

All production car brake rotors are made of gray cast iron in either solid or vented styles. The vented type is wider and heavier than the solid type, but it cools far better. Some vented rotors, designed for use on one side of the car only, have curved cooling passages to improve airflow. A few high-performance rotors are drilled or slotted to reduce gas and water fade.

Drums and rotors are either fixed or floating. Fixed designs are cast in one piece with the wheel hub and must be removed with it. Floating drums and rotors can usually be removed from the vehicle independently of the hub or axle. Fixed drums and rotors are held in place by the wheel bearing retaining nut, or a taper on the axle combined with a woodruff key and a locknut. Floating drums and rotors are held on by the wheel and lug nuts or bolts during normal operation, although speed nuts, bolts, screws, rivets, or swaging may be used to keep them in place when the wheel is removed.

Brake drums and rotors suffer wear, damage, and distortion. Wear of the friction surfaces is a normal result of brake operation. Brake drums suffer taper and barrel wear, whereas rotors have taper variation. Damage is the result of extremes in brake operation. Drums and rotors both experience four forms of damage: scoring, cracking, heat checking, and hard spots. Distortion is caused by a variety of factors including improper storage. Drums suffer from bellmouth, out-of-round, and eccentric distortion, whereas rotor distortion is confined to lateral runout and a lack of parallelism.

Drums and rotors are refinished using three techniques: machining, grinding, and resurfacing. Machining uses a brake lathe equipped with steel cutting bits to remove metal from friction surfaces. Grinding is a process that uses an abrasive stone to remove hard spots from brake drums. Resurfacing uses abrasive discs to remove minor scoring, rust, and other contaminants from rotor friction surfaces while giving them a nondirectional finish.

Only a limited amount of metal can be removed from a drum or rotor when it is refinished. The amount is indicated by values stamped or cast into drums and rotors. Drums are marked with a discard diameter or maximum diameter. Rotors have either a discard dimension or a minimum refinish thickness.

In addition to maximum metal removal limits, brake drums on the same axle must be kept close to the same size to prevent brake pull. New drums must be turned to match worn parts, or both drums must be replaced.

With a fixed disc brake caliper, equal amounts of metal must be removed from both friction surfaces on the rotor. With floating or sliding calipers, different amounts can be removed from each side as needed. Some brake rotors have an additional limit on the amount of metal that can be removed from the inner friction surface.

Because of the traits of the respective friction materials, rotors used with semimetallic linings require a smoother friction surface finish than those used with organic linings. This finish is obtained by following special rotor machining procedures.

Review Questions

Choose the letter that represents the best possible answer to the following questions:

1. Solid brake drums:
   a. Are made entirely of steel
   b. Are very good at resisting fade
   c. Have a poor coefficient of friction
   d. Contain carbon particles

2. Composite brake drums are made from:
   a. Steel and iron
   b. Iron and aluminum
   c. Both a and b
   d. Neither a nor b

3. A _____ drum is the most common type of composite drum.
   a. Bimetallic
   b. Centrifugal-cast
   c. Radial-cast
   d. None of the above

4. A unidirectional brake rotor is always a _____ rotor.
   a. Solid
   b. Vented
   c. Slotted
   d. None of the above

5. A drum or rotor that is held on the car primarily by the wheel and lug nuts or bolts is a _____ design.
   a. Fixed
   b. Floating
   c. Sliding
   d. All of the above

6. Fixed drums and rotors that become worn:
   a. May be replaced as a unit
   b. Are often machined
   c. Cannot be separated from the hub
   d. All of the above

7. Which of the following is not a type of drum or rotor problem?
   a. Distortion
   b. Wear
   c. Distress
   d. Damage

8. Taper variation is a problem with:
   a. Brake drums
   b. Brake rotors
   c. Both a and b
   d. Neither a nor b

9. Barrel wear of a brake drum is caused by:
   a. Uneven application force
   b. Aluminum brake shoes
   c. The brake lining material
   d. All of the above

10. A _____ brake is more likely to score.
    a. Front disc
    b. Front drum
    c. Rear disc
    d. Rear drum

11. Oftentimes, _____ will not appear until after a drum or rotor is machined.
    a. Heat checks
    b. Cracks
    c. Hard spots
    d. All of the above

12. Brake drum and rotor distortion:
    a. Can be caused by improper storage
    b. Affects suspension alignment
    c. Is the same for both drums and rotors
    d. All of the above

13. High heat, wide drums, and heavy braking can combine to cause _____ distortion.
    a. Out-of-round
    b. Bellmouth
    c. Eccentric
    d. None of the above

14. Overtightening the wheel lug nuts or bolts can cause _____ distortion.
    a. Out-of-round
    b. Bellmouth
    c. Eccentric
    d. None of the above

15. Rotor distortion is more significant than drum distortion because:
    a. Hydraulic feedback to the master cylinder is less in a disc brake
    b. Hydraulic feedback to the master cylinder is greater in a disc brake
    c. Drums routinely distort during operation
    d. Drum brake shoes are fixed on the backing plate

16. Lateral rotor runout:
    a. Is present in all rotors
    b. May be caused by overtightened lug nuts
    c. Both a and b
    d. Neither a nor b

Brake Drums and Rotors

17. A lack of rotor friction surface parallelism:
    a. Is a common cause of brake pedal pulsation
    b. Can cause vibration
    c. Is sometimes caused by rust
    d. All of the above

18. Which of the following means of refinishing brake drums and rotors uses a brake lathe?
    a. Grinding
    b. Resurfacing
    c. Turning
    d. All of the above

19. Resurfacing and turning may be done with the brake drum:
    a. Off the car
    b. Mounted on the car
    c. Either a or b
    d. Neither a nor b

20. Brake rotors are resurfaced:
    a. To remove rust
    b. To eliminate hard spots
    c. To apply a unidirectional finish
    d. All of the above

21. The amount of metal that can be removed from a brake drum is determined by the:
    a. Maximum diameter
    b. Discard diameter
    c. Machining limit
    d. All of the above

22. A brake rotor will be thinnest when it has reached the:
    a. Minimum refinish thickness
    b. Discard dimension
    c. Either of the above, they are the same
    d. None of the above

23. Brake drums on the same axle should be machined to within _____ of one another.
    a. .020" (.50 mm)
    b. .030" (.75 mm)
    c. .040" (1.0 mm)
    d. .060" (1.5 mm)

24. Different amounts of metal may be removed from the friction surfaces of rotors:
    a. On cars with floating calipers
    b. That do not have to be mechanically centered in the caliper
    c. That have inner surface stock removal limits
    d. All of the above

25. Rotors used with semimetallic brake linings will suffer from _____ if they do not have a smooth enough surface finish.
    a. Distortion
    b. Poor break-in performance
    c. Reduced operating temperatures
    d. All of the above

12

Parking Brakes

OBJECTIVES

Upon completion and review of this chapter, you will be able to:

- Describe the construction and operation of the three types of parking brake controls.
- Describe the construction and operation of the automatic parking brake release.
- Describe the construction and operation of parking brake linkages.
- Describe the construction and operation of the two types of drum parking brakes.
- Describe the construction and operation of caliper-actuated disc parking brakes.
- Describe the construction and operation of driveline auxiliary parking brakes.

KEY TERMS

body control module
drum-in-hat
equalizer
intercable adjuster
intermediate lever
jamnuts
ratchet
vacuum servo

INTRODUCTION

The parking brake holds a vehicle stationary while parked and provides a secondary means of stopping the car in an emergency. FMVSS 135 states that a parking brake must be able to hold a car in position on a 20-percent grade with an application force of no more than 150 pounds on a foot pedal or 100 pounds at a hand-operated lever.

The most common type of parking brake applies the service brake friction assemblies at two wheels, usually those on the rear axle. Some older cars, and many trucks, have a single separate friction assembly mounted on the driveline that functions as the parking brake. All parking brakes are operated by mechanical linkages independent of the service brake hydraulic system.

The first part of this chapter covers the pedals, levers, and handles that apply parking brakes. Warning lights and switches are examined next. The second part of the chapter deals with the linkage that connects the pedal, lever, or handle to the parking brake friction assemblies. The final portion of the chapter describes the operation of specific drum and disc parking brakes.

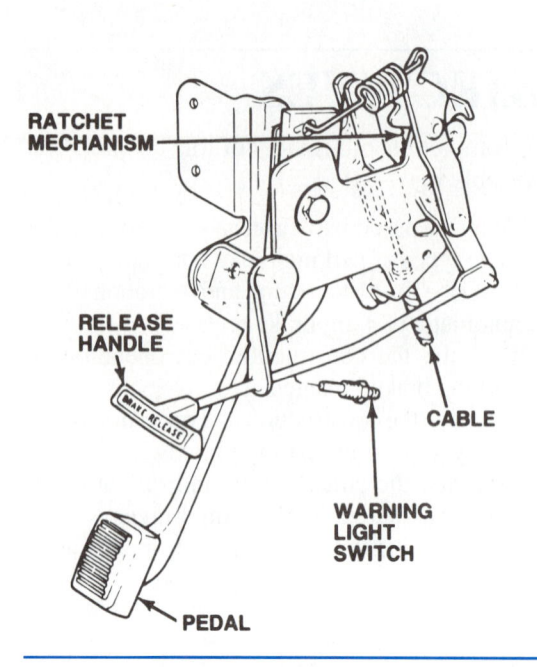

Figure 12-1. A typical parking brake pedal.

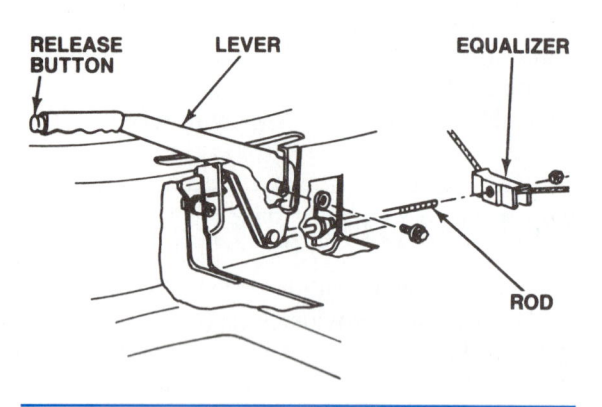

Figure 12-2. A typical parking brake lever.

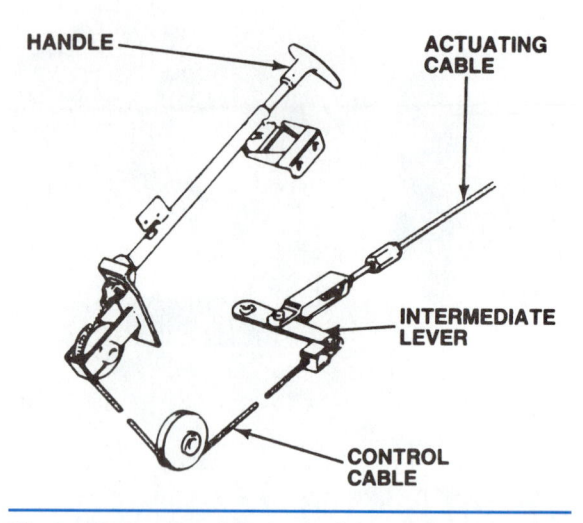

Figure 12-3. A parking brake linkage actuated by an underdash handle.

PEDALS, LEVERS, AND HANDLES

The parking brake is applied by a mechanical linkage actuated by a pedal, lever, or handle (hereafter called the parking brake control) inside the car. Foot pedals and floor-mounted levers, figures 12-1 and 12-2, are the most common means of applying parking brakes, but some trucks and older cars have a handle under the dashboard instead, figure 12-3. All parking brake controls incorporate a **ratchet** mechanism, figure 12-4, to lock the brake in the applied position.

When service brake friction assemblies are used as the parking brakes, the service brake pedal should be depressed while the parking brake control is set. This increases parking brake holding power because the brake hydraulic system provides the actual application force, which is far greater than the force that can be developed mechanically; the parking brake mechanism simply locks the brakes in position. If the parking brake is operated only by the pedal, lever, or handle, holding power will be reduced, and the cables in the linkage will stretch and wear faster.

All parking brakes are applied manually, and most are released in this manner as well. However, the exact application and release procedure varies with the design of the parking brake control.

Parking Brake Pedals

A parking brake pedal, figure 12-1, is applied by depressing it with the foot. The ratchet engages automatically and the pedal remains in the depressed position. Releasing the parking brake is accomplished in one of several ways.

- On older vehicles, the pedal is released by a pull on a small T-handle or lever under the dash. This disengages the ratchet mechanism and allows a return spring to move the pedal to the unapplied position.
- On many vehicles, the release lever is integrated into the underside of the dash, figure 12-5, and connects to the release mechanism through a rod or cable.
- Late model vehicles may not have a release lever. Instead, the parking brake pedal is pushed a second time to release it.

Parking Brakes

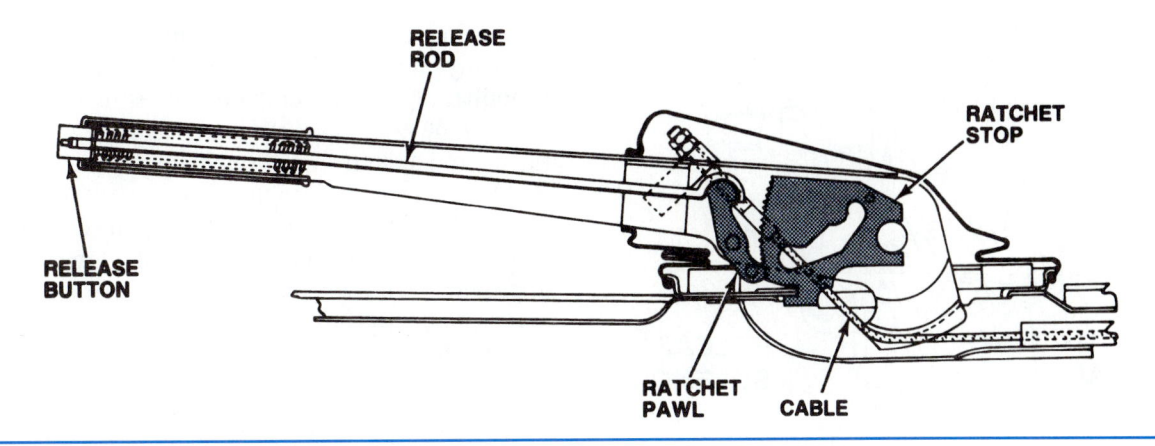

Figure 12-4. A ratchet mechanism is used to lock parking brakes in the applied position.

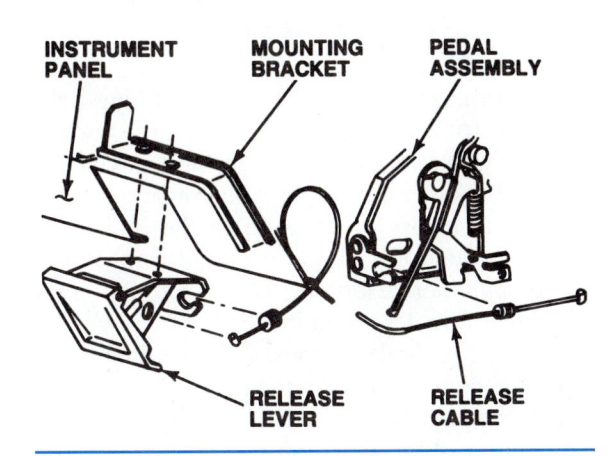

Figure 12-5. A remote-mounted parking brake release lever.

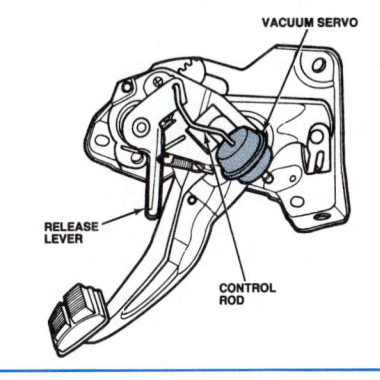

Figure 12-6. Automatic parking brake release mechanisms use a vacuum servo to operate the release lever.

In the mid- to late-1970s, a number of heavyweight luxury models were equipped with a special pedal design. This type of pedal had a high leverage ratio and a special ratchet mechanism that locked the brake in the applied position, but allowed the pedal to return after it had been depressed. Two full pedal strokes, or several partial strokes, were required to fully apply the parking brake. The brake was released by a single pull on a T-handle or release lever as described above.

Automatic Parking Brake Release

Some vehicles with pedal-operated parking brakes have an automatic release mechanism that disengages the parking brake when the shifter is taken out of the park position. Some systems use a **vacuum servo** controlled by an electrical solenoid, figure 12-6, whereas on other systems a solenoid or pair of solenoids operates the release directly.

On vacuum-operated systems a metal rod connects the vacuum servo to the upper end of the parking brake release lever. When the engine is running (to provide vacuum) and the shifter is placed in gear, an electrical contact closes to energize the solenoid and route vacuum to the servo. The servo diaphragm then retracts the rod, which releases the parking brake.

The solenoid-controlled parking brake release system uses a pair of solenoids located on the parking brake apply assembly, figure 12-7. These solenoids are usually controlled by the vehicle electronic **body control module.** When the body control module (BCM) receives data that indicates that the brake pedal is applied and the shifter has moved from the park position, it will ground the solenoids and release the parking brake.

Automatic parking brake release systems supplement the mechanical T-handle or release lever; they do not replace it. A manual release lever is fitted to all parking brake pedal assemblies with an automatic release mechanism, although the

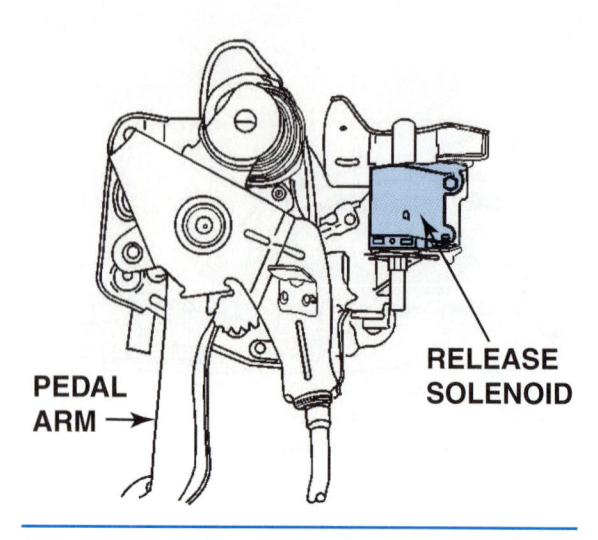

Figure 12-7. The release solenoid is controlled by the body control module. (Courtesy of General Motors Corporation, Service and Parts Operations)

lever may not be visible from the driver's seat because it is seldom used. Automatic parking brake release mechanisms were developed mainly as a luxury convenience; however, they do help prevent the brake damage that can occur when the parking brake is not released before the vehicle is driven away.

Parking Brake Levers

Parking brake levers operate in several different ways. In the most common design, shown in figure 12-4, the ratchet mechanism automatically engages as the lever is pulled upward; once the brake is set, the lever remains in the up position. This design is released by pushing a spring-loaded button on the end of the lever and holding it in while the lever is lowered to the unapplied position.

In a variation of this design described, the parking brake lever drops freely to the floor once the brake has been set. This type of ratchet mechanism is used to allow easier vehicle entry and exit when the lever is mounted outboard between the driver's seat and door. This design is released by pulling the lever upward until resistance is felt, then pushing the release button on the end of the lever and holding it in while the lever is lowered to the unapplied position.

Parking Brake Handles

A parking brake handle, figure 12-3, is applied by pulling it straight back toward the driver. The handle is attached to a ratchet rod that has teeth machined in one side; the ratchet mechanism automatically engages these teeth as the parking brake is applied. A parking brake handle is released by rotating the grip 45 to 90 degrees to disengage the ratchet rod teeth from the ratchet pawl, then feeding the rod forward into the dash.

A few older parking brake handles are attached to a pivot under the dash that allows them to provide additional leverage. These handles are also applied by pulling them back toward the driver. However, they are usually released by squeezing a trigger lever next to the handle grip.

WARNING LIGHTS AND SWITCHES

All modern cars and trucks have a warning light on the instrument panel to alert the driver when the parking brake is applied. In most cases, the parking brake warning light is the same one used to indicate a loss of pressure in the brake hydraulic system or a low brake fluid level in the master cylinder reservoir. These warning light functions were described in Chapter 6.

For parking brake purposes, the warning light is triggered by a switch on the parking brake control mechanism, figure 12-1. The switch is held in the open position when the parking brake is released. When the brake is applied, the pedal, lever, or handle allows the switch to close, completing the warning light circuit. Parking brake warning lights work only when the ignition is on.

PARKING BRAKE LINKAGES

Parking brake linkages transmit force from the pedal, lever, or handle inside the vehicle to the parking brake friction assemblies. Parking brake linkages are made up of four elements:

- Rods
- Cables
- Levers
- Equalizers.

Linkage Rods

Parking brake linkage rods made from solid steel, figure 12-2, are commonly used with floor-mounted actuating levers to span the short dis-

Parking Brakes

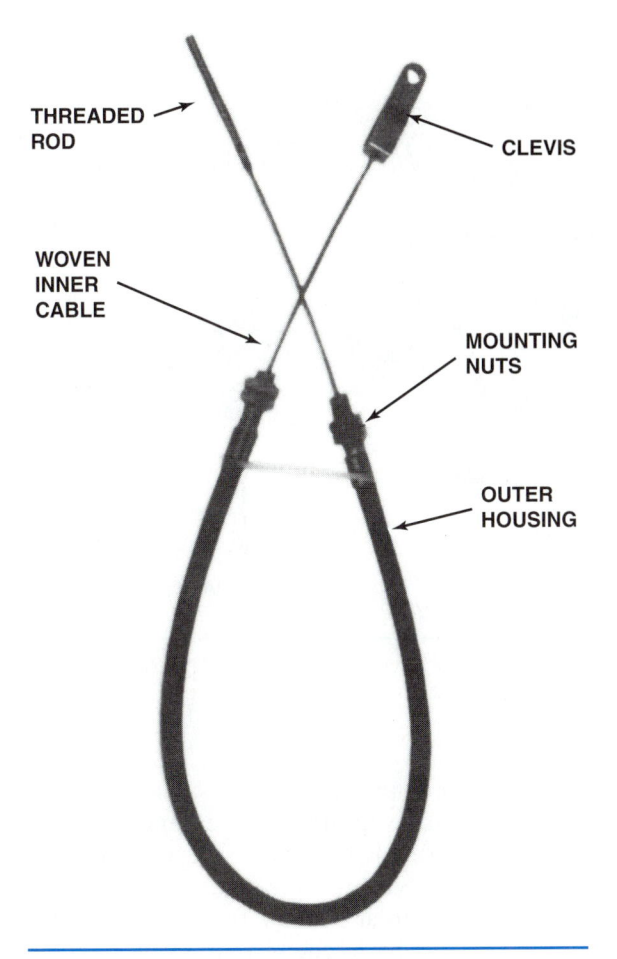

Figure 12-8. A parking brake cable.

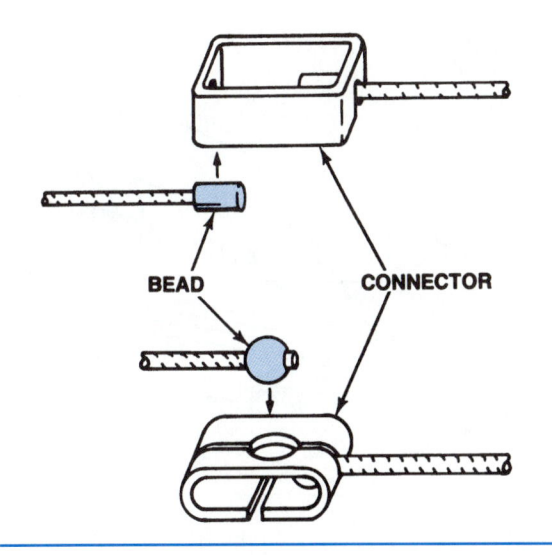

Figure 12-9. Beaded cable ends are common in parking brake linkages.

tance to an intermediate lever or an equalizer (see discussion following). Many large trucks have parking brake linkages that consist entirely of rods and levers. Linkage rods commonly attach to the floor-mounted lever with a pivoting pin; the opposite end is often threaded to accept a pair of nuts that are used to adjust the parking brake.

Linkage Cables

The typical parking brake cable, figure 12-8, is made of woven-steel wire encased in a reinforced rubber or plastic housing. The housing is fixed in position at both ends and is routed under the car through mounting brackets that hold the cable in position yet still allow a small amount of movement. The cable slides back and forth inside the housing to transmit application force, and, depending on the linkage design, the outer housing may play a part in parking brake application as well.

The ends of parking brake cables are fitted with a wide variety of connectors that attach to actuating devices, other linkage parts, or the wheel friction assemblies. The cable in figure 12-8 has a threaded rod on one end and a clevis on the other. Another common type of cable end is a round or cylindrical metal bead designed to fit into a special holder or cable connector, figure 12-9.

Parking brake linkages use control cables, transfer cables, and application cables. Control cables, much like the rods described earlier, attach to the parking brake pedal, lever, or handle inside the car and transmit force to an intermediate lever or equalizer under the car. Transfer cables pass the force from the intermediate lever or equalizer to the application cables. Finally, the application cables use the force passed through the linkage to apply the friction assemblies. Not every parking brake linkage uses all three types of cables; in many systems, the jobs of two or more cables are combined into a single part.

Parking brake cables must transmit hundreds of pounds of application force without jamming or breaking and with minimal stretch. One of the reasons separate control and application cables are used is that they operate at the points in the linkage where stress is greatest and the cable is most likely to break. The use of separate cables in these areas makes replacement simpler and less expensive.

Parking brake cables are also subject to damage from water, dirt, and other debris thrown up under the vehicle by the tires. To help maintain smooth operation, some older cables have grease fittings that allow them to be lubricated, figure 12-10. Modern cables do not require lubrication because they

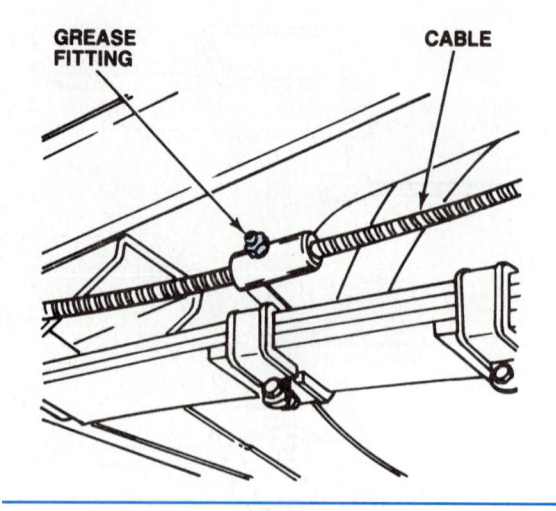

Figure 12-10. Some older parking brake cables are equipped with grease fittings.

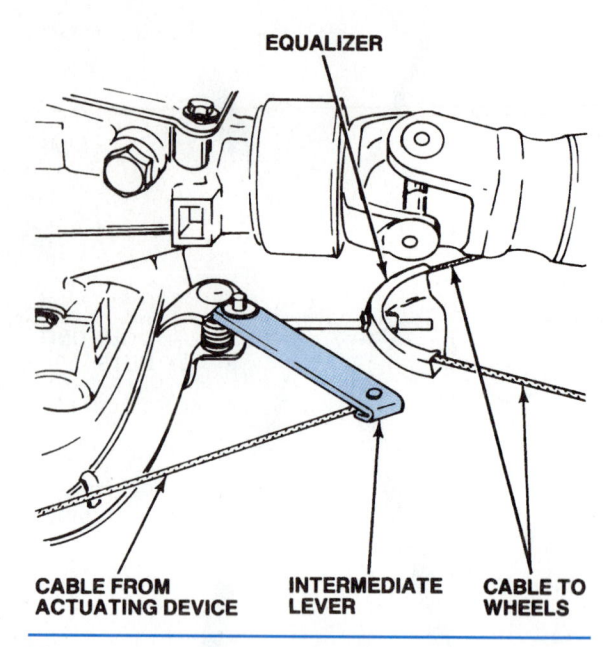

Figure 12-11. Intermediate levers in the parking brake linkage increase application force.

are lined with nylon or teflon and any cable housing ends located under the vehicle are protected by rubber or nylon seals. A few manufacturers do recommend that exposed sections of inner cable be lightly greased.

Linkage Levers

The rods and cables described in the previous section transmit application force in direct proportion; that is, they pass along any force they receive without changing it in any way. If 50 pounds of application force is delivered to one end of a rod or cable, 50 pounds of force will be available at the other end as well. Unfortunately, the amount of physical force a driver can apply to the parking brake control is insufficient for effective parking brake operation. For this reason, all parking brake linkages contain one or more levers that increase application force.

The parking brake pedals, floor-mounted levers, and pivoting underdash handles described earlier are all types of levers used to increase parking brake application force. Straight-pull parking brake handles are not levers themselves, although they are commonly connected to other levers in the linkage. The principles of leverage were covered in Chapter 3, and Chapter 5 dealt with leverage in relation to the service brake pedal; refer to those chapters for detailed discussions of leverage.

A lever in the parking brake linkage under the car is called an **intermediate lever**, figure 12-11. To further increase parking brake application force, intermediate levers provide leverage in addition to that supplied by the parking brake control. Intermediate levers are common in parking brake systems on large, heavy vehicles that require a great deal of force to apply the parking brake hard enough to meet Federal regulations.

Linkage Equalizers

In some parking brake linkages, the rods or cables to the two friction assemblies are adjusted separately. If the adjustments are unequal, one brake will apply before the other, preventing full lining-to-drum contact at the opposite wheel and greatly reducing the holding power of the parking brake. To prevent unequal application, most parking brake linkages use an **equalizer** to balance the force from the parking brake control and transmit an equal amount to each friction assembly.

Equalizers come in many shapes and sizes, but the simplest is a cable guide attached to a threaded rod, figure 12-12. This type of equalizer pivots or allows the inner cable to slide back and forth to even out application force. A variation of this design, figure 12-13, combines the equalizer into an intermediate lever assembly. A transfer cable passes through rollers in the equalizer and connects to short application cables on both sides of the car.

Parking Brakes

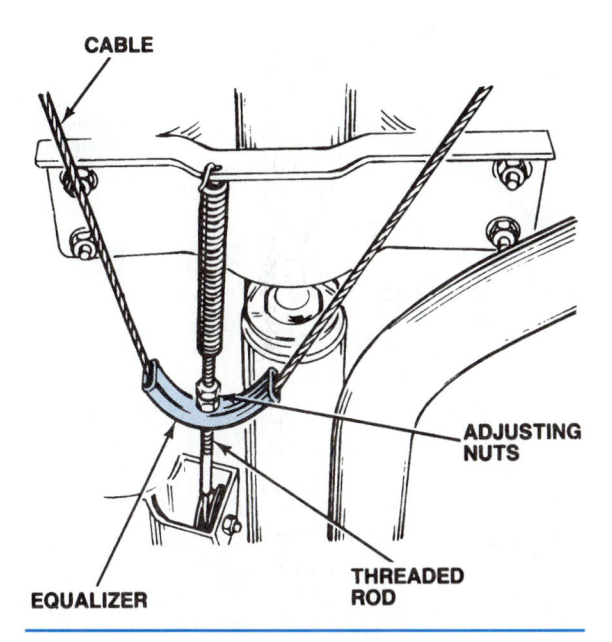

Figure 12-12. A cable guide is a common type of parking brake linkage equalizer.

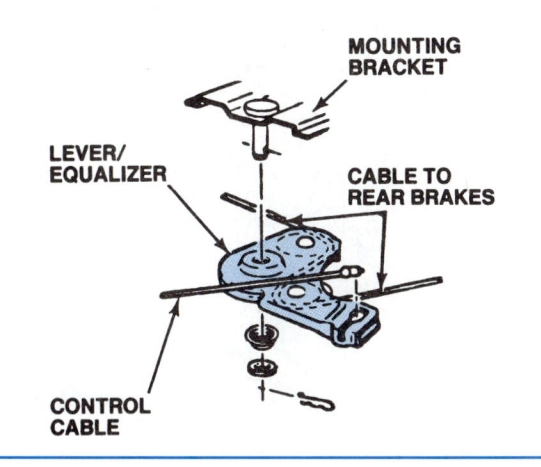

Figure 12-13. This equalizer is designed into an intermediate lever assembly.

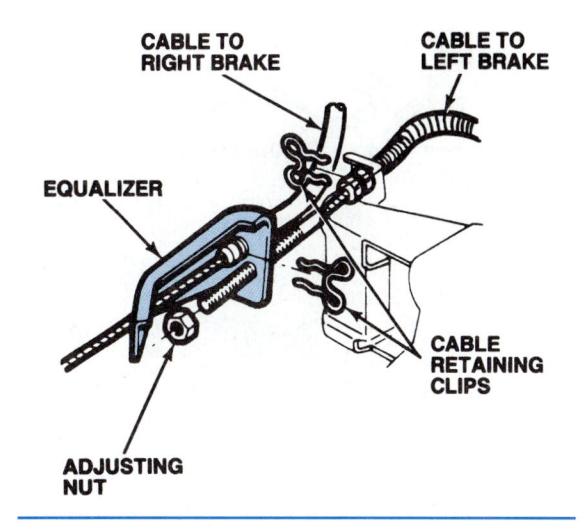

Figure 12-14. Some equalizers install in a parking brake cable.

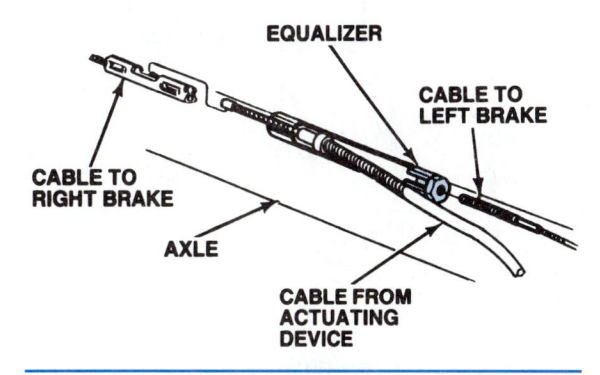

Figure 12-15. Another version of an in-cable equalizer.

Another type of equalizer, figure 12-14, installs in a long application cable that runs from the linkage at the front of the car to one rear brake. A short application cable is routed from the equalizer to the other rear brake. When the parking brake is applied, the long cable actuates its brake directly. Once the shoes make contact with the drum, however, the inner cable can not move any farther so the entire cable assembly, along with the equalizer, is pulled toward the front of the vehicle. Movement of the equalizer then applies the other rear brake through the short cable.

In a variation of this design, figure 12-15, a transfer cable from the linkage at the front of the car attaches to an equalizer near the rear beam axle; short application cables run from the equalizer to the wheel friction assemblies. When the parking brake control is operated, the right brake is applied by the transfer cable through the short application cable on that side of the vehicle. When the shoes contact the drum so the inner cable can move no farther, the transfer cable housing pulls the equalizer toward the right brake applying the left brake through the application cable on that side of the vehicle.

Linkage Design

The number of different parking brake linkage designs is almost as great as the number of car models on the road. Most linkages combine intermediate levers and equalizers in various ways, figure 12-16, and use from one to four cables to actuate the friction assemblies. Some typical linkages are described next.

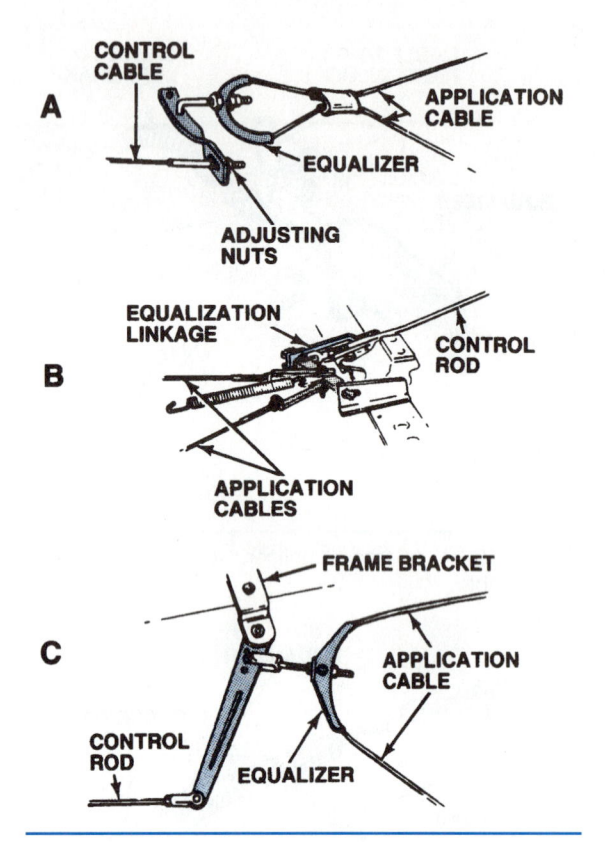

Figure 12-16. Many parking brake linkages use both an intermediate lever and an equalizer.

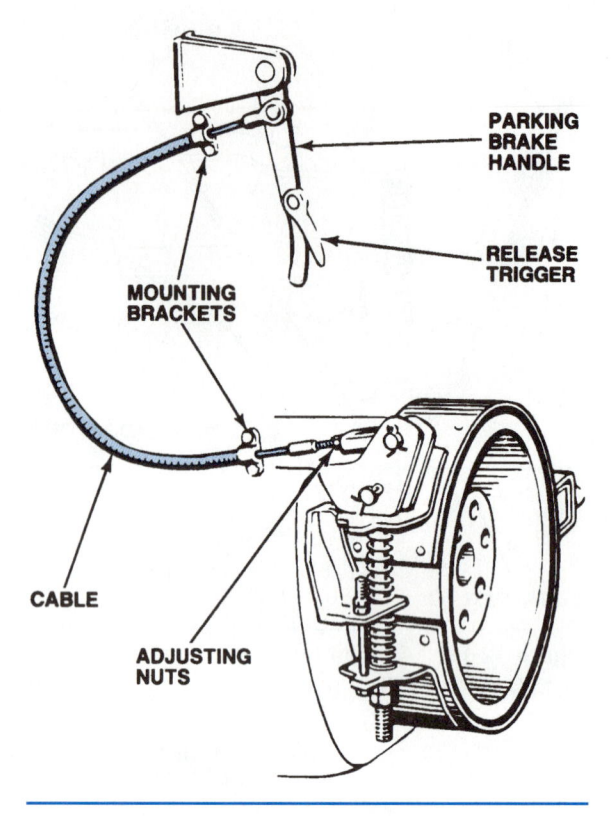

Figure 12-17. A single-cable parking brake linkage.

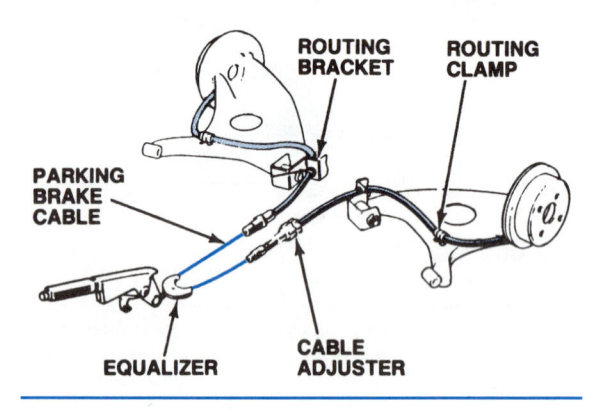

Figure 12-18. A single-cable parking brake linkage that applies both rear wheel friction assemblies.

Single-Cable Linkages

Single-cable parking brake linkages come in a couple of different forms. The most basic version is used to operate driveline-mounted parking brakes on medium and heavy trucks, figure 12-17. In this system, a pedal, lever, or handle transmits force through a single cable to actuate the lone brake. Another type of single-cable parking brake linkage, figure 12-18, is part of a system in which a floor-mounted application lever is connected to an equalizer through a rod; a single cable equipped with dual housings distributes the application force from there to the friction assemblies.

Dual-Cable Linkages

Some parking brake linkages with a floor-mounted actuating lever use two cables to operate the friction assemblies at the wheels, figure 12-19. In this design, the two cables lead directly from the parking brake lever to the wheel friction assemblies. In another type of dual-cable linkage, figure 12-16A, a front cable connects to an equalizer. A single rear cable with dual housings then transmits the application force to the rear wheels. Yet another type of dual-cable linkage, figure 12-16B, uses a rod from the parking brake control to actuate an intermediate lever. Separate cables then run from the intermediate lever to the wheel friction assemblies.

Multiple-Cable linkages

A common linkage design on domestic vehicles, figure 12-20, uses three cables: a control cable from the parking brake pedal, lever, or handle; a long application cable to one rear brake; and a short application cable from an equalizer to the

Parking Brakes

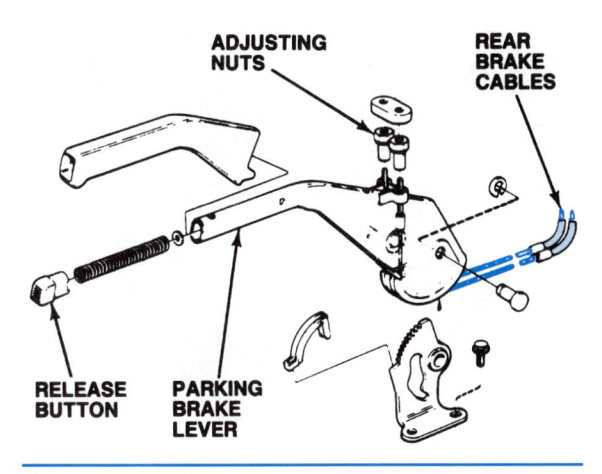

Figure 12-19. Individual rear cables are used in some dual-cable parking brake linkages.

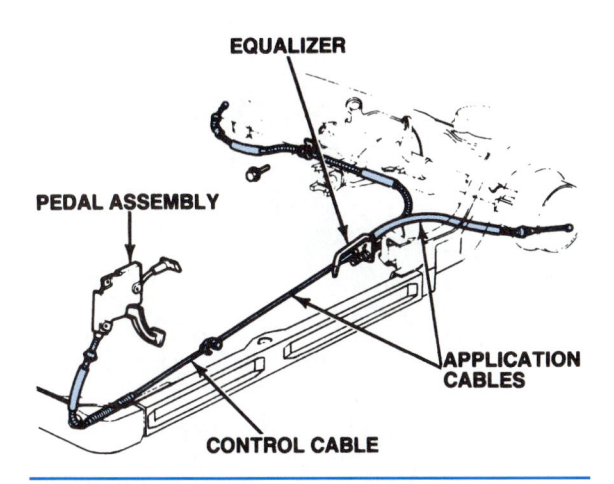

Figure 12-20. A three-cable parking brake linkage.

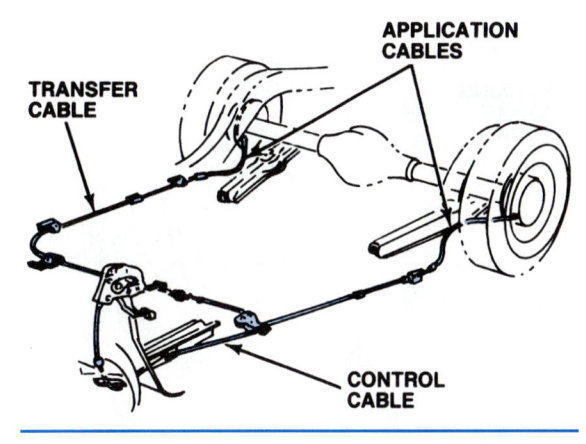

Figure 12-21. A four-cable parking brake linkage.

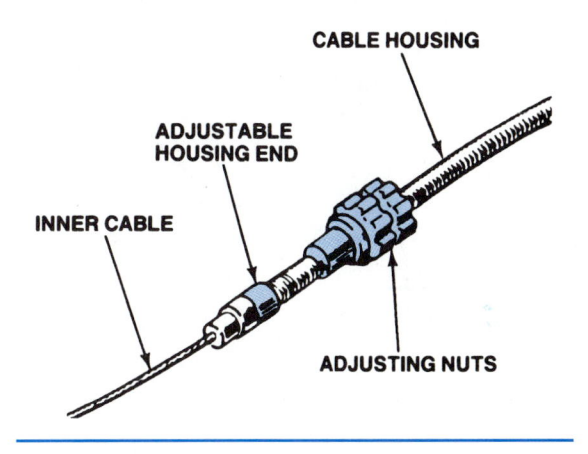

Figure 12-22. An intercable parking brake adjuster.

other rear brake. As described earlier, the equalizer is mounted near the rear axle to keep the second application cable as short as possible.

The parking brake linkage in figure 12-21 uses four separate cables: a control cable from the pedal to the equalizer, a transfer cable from the equalizer to near the rear wheels, and a pair of application cables that operate the rear brakes.

Parking Brake Linkage Adjustment

All parking brake linkages have provision for adjustment to compensate for stretched cables and wear of the linings in the brake friction assemblies. Many linkages adjust with two **jamnuts** where a rod or cable passes through an equalizer, figure 12-12. This type of adjustment is also found where cables pass through intermediate levers, figure 12-16A, and on single-cable systems that actuate a driveline brake, figure 12-17. Dual-cable linkages with floor-mounted levers often have individual adjusting nuts where the ends of the cables pass through the lever assembly, figure 12-19.

The **intercable adjuster** assembly, figure 12-22, is built into the outer housing of some parking brake cables. This type of adjuster allows the length of the cable housing to be altered to take up slack in the linkage.

DRUM PARKING BRAKES

Drum parking brakes are the most common type on cars and light trucks. As discussed in Chapter 8, drum brakes make excellent parking brakes because they have a high static coefficient of friction combined with self-energizing action and, in the case of dual-servo brakes, servo action that increases

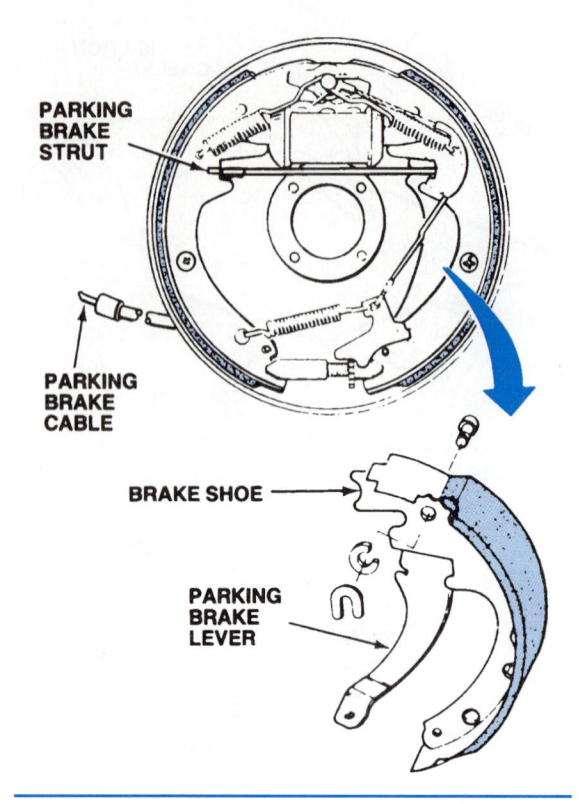

Figure 12-23. An integral drum parking brake.

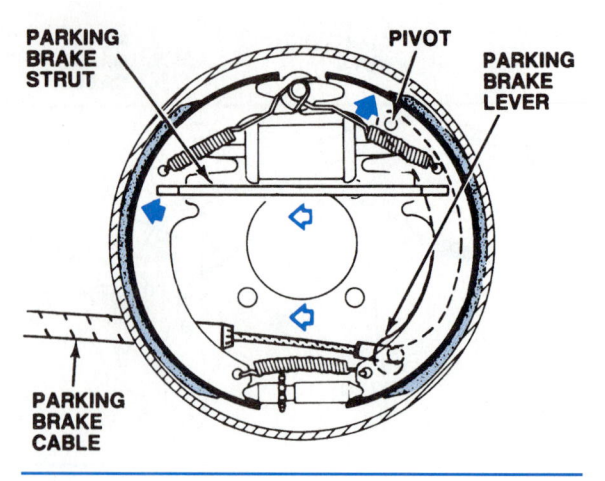

Figure 12-24. Application of an integral drum parking brake.

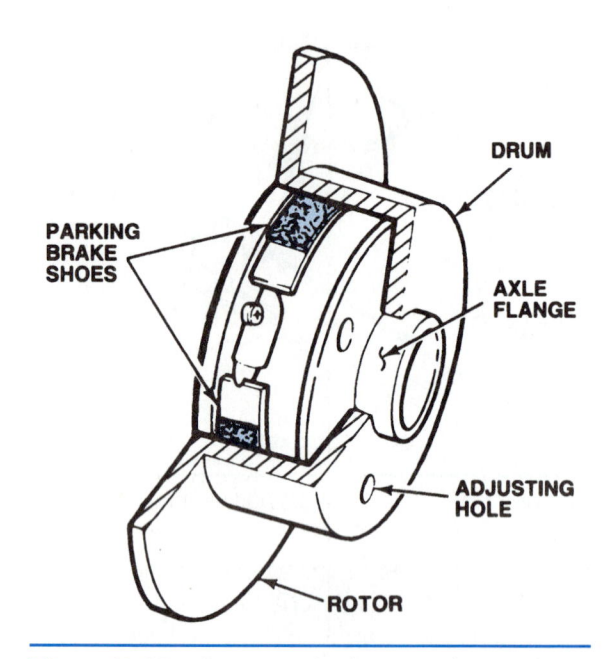

Figure 12-25. Some rear brake rotors incorporate a brake drum for parking brake use.

their application force. There are two basic types of drum parking brakes:

- Integral
- Auxiliary (drum-in-hat)

Integral Drum Parking Brakes

Integral drum parking brakes, figure 12-23, mechanically apply the rear drum service brakes to serve as the parking brakes. Integral drum parking brakes are the most common type not only because of their natural superiority in this application but also because it is simple and inexpensive to design a parking brake linkage into a drum brake.

The typical integral drum parking brake has a pivoting lever mounted on one brake shoe, and a strut placed between the lever and the other shoe. The strut may be fitted with a spring that takes up slack to prevent noise when the parking brake is not applied. The end of the lever opposite the pivot is moved by the parking brake cable, which enters through an opening in the backing plate.

All integral drum parking brakes operate in essentially the same manner, figure 12-24. When the parking brake control is operated, the cable pulls the end of the lever away from the shoe to which it is attached. The lever pivots at the at- taching point and moves the strut to apply the forward shoe. Once the forward brake shoe lining contacts the drum, the strut can travel no farther, the lever then pivots on the strut, and forces the lining of the reverse shoe against the drum.

Rear-Disc Auxiliary Drum Parking Brakes

Rear-disc brakes with floating or fixed calipers may have a parking brake drum formed into the hub of the brake rotor, figure 12-25. Inside the drum is a small dual-servo drum brake friction

Parking Brakes

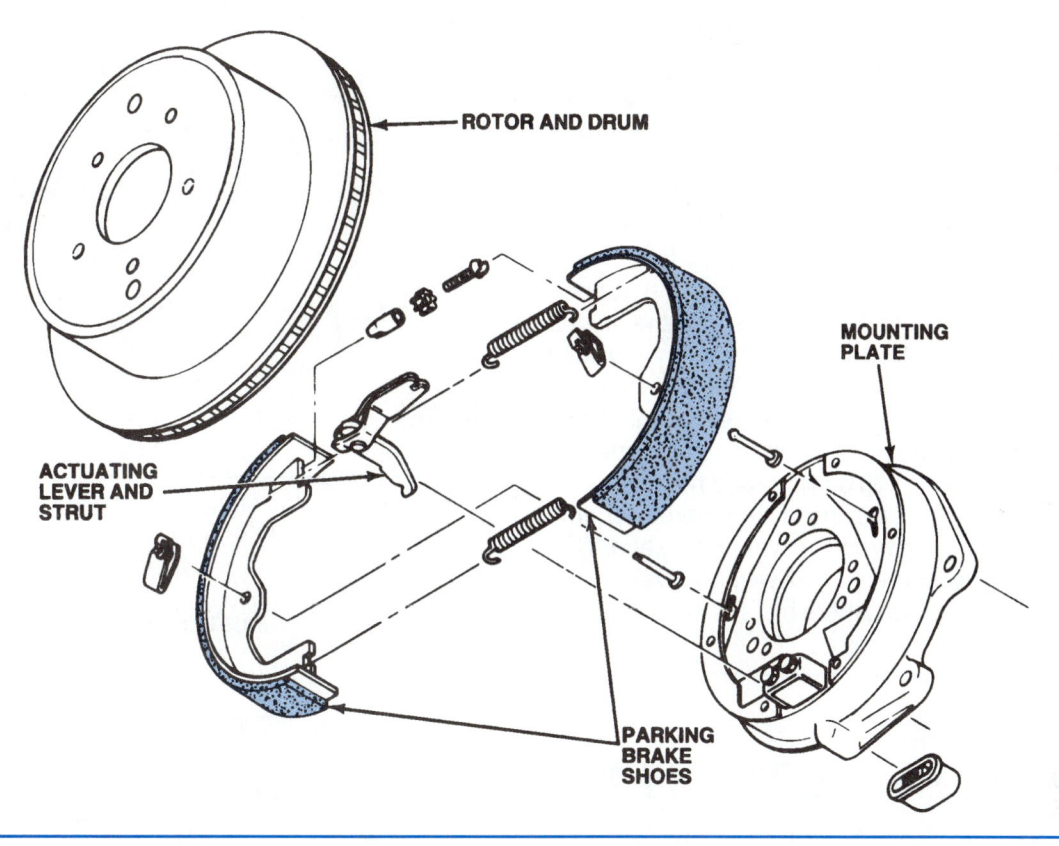

Figure 12-26. A rear-disc auxiliary drum parking brake friction assembly.

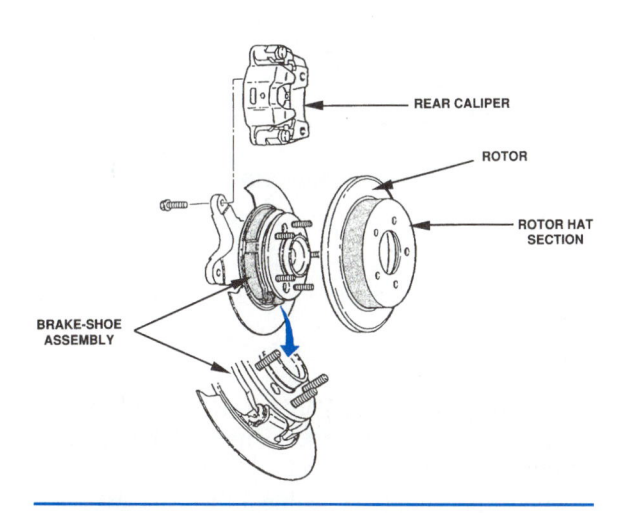

Figure 12-27. Drum-in-hat rear-disc parking brake assembly as used by General Motors. (Courtesy of General Motors Corporation, Parts and Service Operations)

assembly, figure 12-26, that serves as the parking brake. This type of parking brake may also be called a **drum-in-hat** parking brake, figure 12-27.

The rotor splash shield, or a special mounting bracket, provides the backing plate for the friction assembly. Rear-disc auxiliary drum parking brakes use the dual-servo friction assembly design because it provides the most holding power and does so equally in both forward and reverse directions.

Dual-servo parking brake friction assemblies operate in essentially the same manner as their service brake counterparts described in Chapter 8. The main difference is that the wheel cylinder is eliminated and the friction assembly is actuated mechanically. Chrysler rear-disc auxiliary drum parking brakes used in the mid-1970s employed a lever and cam to apply the shoes. Other designs use a lever and strut, see figure 12-26. In both these designs, the lever is actuated by a parking brake linkage cable; the cam or strut then forces the shoes apart against the drum.

All rear-disc auxiliary drum parking brakes are adjusted manually. Most designs have a starwheel adjuster that is reached through an opening in the outside of the drum. Some import cars with wheel lug bolts use a lug bolt hole to access the adjuster. The rear-disc auxiliary drum parking brake on a few imported cars is adjusted by changing the length of the parking brake cable.

■ The Bootleg Turn

The rear axle parking brakes common on American cars made possible the bootleg turn, which gets its name from the "runners" who transported illegal "moonshine" or "bootleg" whiskey during Prohibition. Occasionally, a moonshine runner would use the bootleg turn to get away from a federal agent following in hot pursuit. The runner would slow down a bit, and just as the agent thought he was catching up, the runner would tweak the steering wheel to the left and lock the rear brakes with the parking brake control. If the runner was well practiced in this technique, the car would spin exactly 180 degrees and roar off in the opposite direction, leaving a bewildered federal agent traveling the wrong way at high speed. The bootleg turn was a very effective maneuver on the narrow country roads where most of these dramas were played.

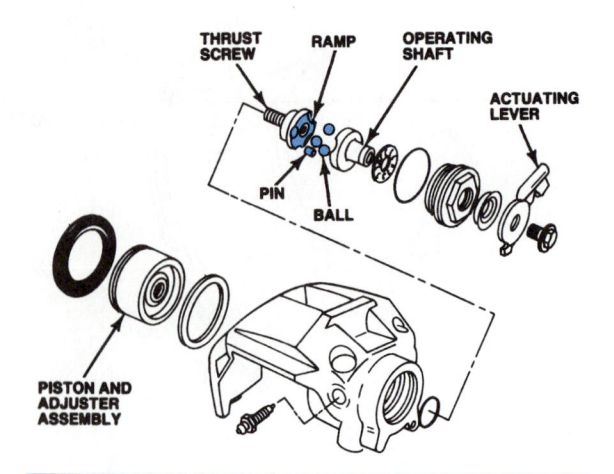

Figure 12-28. A rear brake caliper with ball and ramp parking brake actuation.

CALIPER-ACTUATED DISC PARKING BRAKES

Caliper-actuated disc parking brakes are used on some vehicles whose rear disc brakes are equipped with floating or sliding brake calipers. As discussed in Chapter 9, the single-piston construction of these calipers makes them easier to mechanically actuate than multiple-piston fixed calipers. In this design, a special mechanism in the caliper applies the caliper piston mechanically. The mechanism is operated by a parking brake cable attached to a lever that protrudes from the inboard side of the caliper.

Three different systems are used to mechanically actuate calipers for parking brake service:

- Ball and ramp
- Screw, nut, and cone
- Eccentric shaft and rod.

Ball and Ramp Actuation

The ball and ramp actuating system found in Ford rear brake calipers has three steel balls located in ramp-shaped detents between two plates, figure 12-28. One plate has a thrust screw attached that is threaded into an adjuster mechanism in the caliper piston. The other plate is part

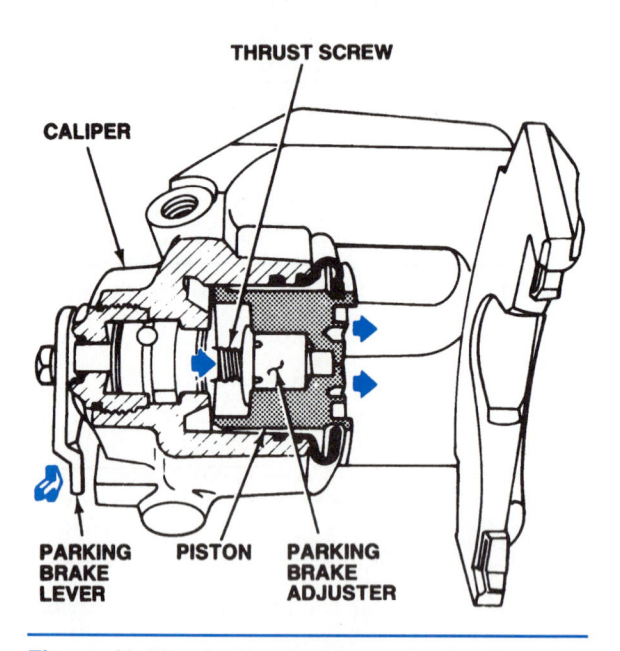

Figure 12-29. Parking brake application in a rear brake caliper.

of the operating shaft that extends out of the caliper; the actuating lever is mounted to the end of this shaft.

As the parking brake cable moves the lever and rotates the operating shaft, the balls ride up the ramps and force the two plates apart. The operating shaft plate cannot move because it butts against the caliper body. Therefore, the thrust screw plate, which is pinned to the caliper body to prevent it from rotating, is driven away from the operating shaft and toward the rotor where the thrust screw moves the caliper piston to apply the brake, figure 12-29.

Parking Brakes

Adjustment of the ball and ramp linkage within the caliper is automatic and takes place during service brake application. When the caliper piston moves away from the thrust screw, an adjuster nut inside the piston, figure 12-30, rotates on the thrust screw to take up any slack created by wear, figure 12-30. A drive ring on the nut prevents it from rotating in the opposite direction when the parking brake is applied.

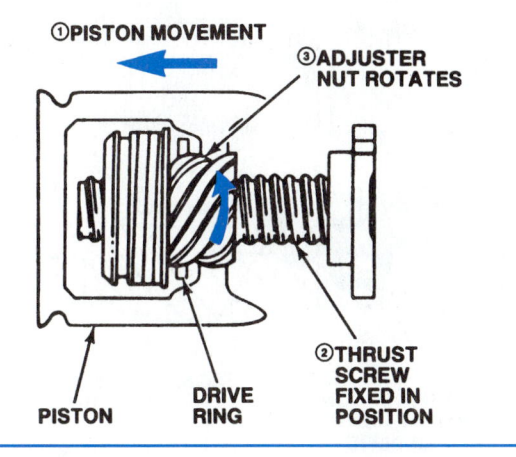

Figure 12-30. Automatic adjustment of a rear-disc brake caliper.

Screw, Nut, and Cone Actuation

General Motors' rear-disc parking brake, figure 12-31, uses a screw, nut, and cone mechanism to apply the caliper piston. In this design, the actuator screw with the parking brake lever attached to it extends through the caliper body. The caliper piston contains a specially shaped nut that threads onto the actuator screw when the piston is installed in the bore. The nut butts against the backside of the cone and is splined to it so that it cannot rotate unless the cone does so as well. The cone is a slip fit in the piston and is free to rotate unless it is held tightly against a clutch surface located near the outer end of the piston bore.

When the parking brake is applied, figure 12-32, the cable moves the lever and rotates the actuator screw. The nut then unthreads along the screw, and jams the cone against the clutch surface of the caliper piston; this prevents the cone from rotating because the caliper piston is keyed to the brake pad, which is fixed in the caliper. Because the cone cannot rotate, movement of the nut along the actuator thread forces the cone and piston outward against the inboard pad to apply the brake.

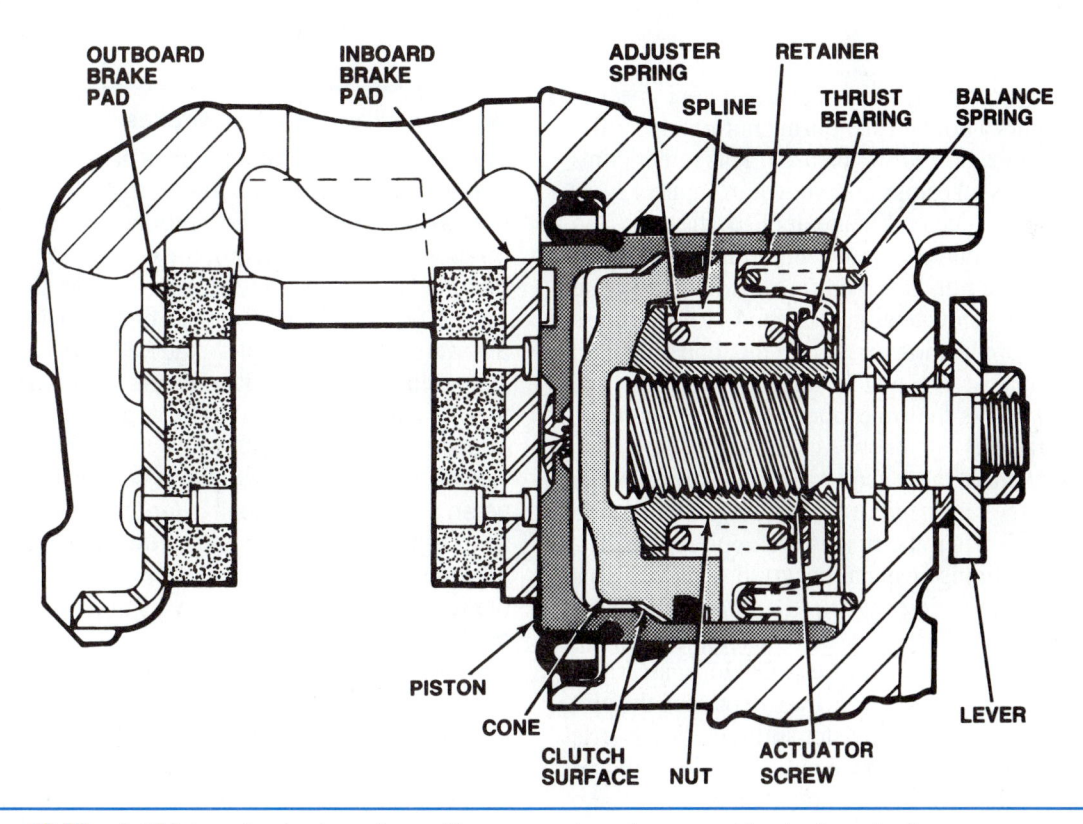

Figure 12-31. A GM rear-disc brake caliper with screw, nut, and cone parking brake actuation.

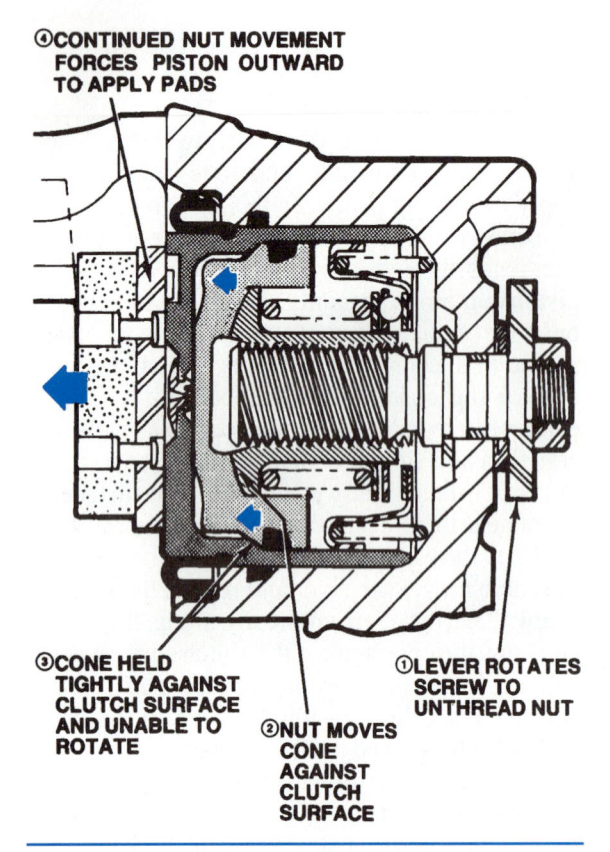

Figure 12-32. Parking brake application of a GM rear brake caliper.

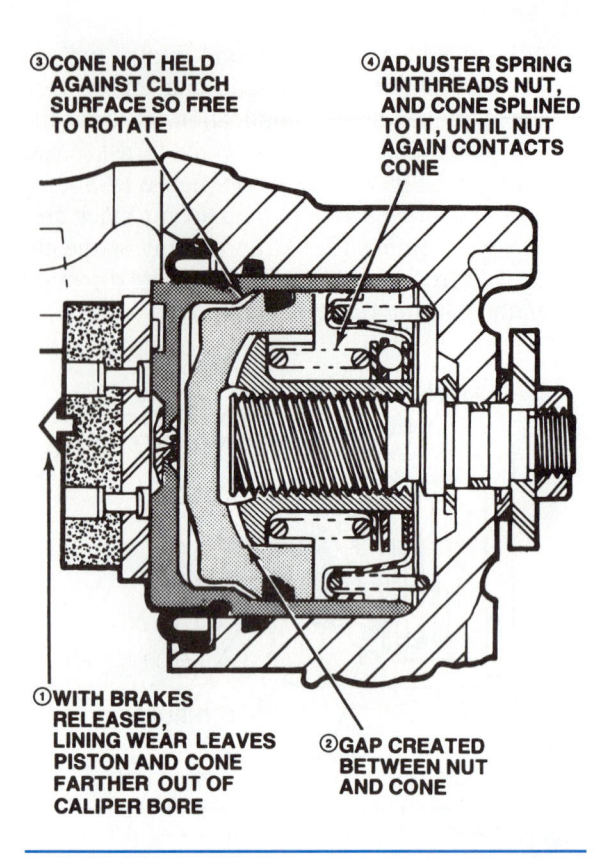

Figure 12-33. Automatic adjustment of a GM rear-disc brake caliper.

Adjustment of the screw, nut, and cone mechanism occurs automatically during normal operation as the service brakes are released. This rather complex process takes place as shown in figure 12-33. When the service brakes are applied, the cone and piston move outward in the bore under hydraulic pressure. The nut, however, remains fixed because the actuator screw does not rotate. As long as there is brake application pressure, the cone is held tightly against the clutch surface of the piston, which prevents the cone, and the nut splined to it, from rotating.

The result of these actions is that a gap develops between the outer end of the nut and the backside of the cone when the brakes are applied. If sufficient brake lining wear has occurred, a gap remains after seal deflection retracts the piston and cone when the brakes are released. Once the brakes are released, the cone is no longer held against the clutch surface and becomes free to rotate in the piston. At this point, the adjuster spring, which exerts strong axial pressure on the nut, causes the nut and cone to unthread along the actuator screw and take up any clearance between the cone and piston.

■ Front-Wheel Parking Brakes

The vast majority of automobiles have the parking brakes incorporated into the drum brakes on their rear axles. However, two automakers, Sweden's Saab and Japan's Subaru, have used front-wheel parking brakes on many of their models. On these cars, a brake lever on the floor between the seats actuates a mechanical linkage that applies the pistons of the front brake calipers. The reason for this unusual layout is simple. A parking brake is equally effective on either the front or rear axle. However, if the parking brake is called on to serve as an emergency brake, it is far more effective on the front. This is especially true on FWD cars, such as the Saab and Subaru, where the front brakes are called on to supply as much as 80 percent of the total stopping power.

The balance spring between the piston and the caliper bore is there for two reasons. First, it prevents excessive piston retraction when the brakes

Parking Brakes

are released. Second, and more importantly, it counterbalances the pressure of the adjuster spring. Keep in mind that the outer end of the nut is in constant contact with the cone whenever the service brakes are not applied. If the automatic adjusting system fails, the tension of the adjuster spring against the thrust bearing at the back of the piston will retract the cone and piston from the rotor until the cone does contact the nut, resulting in a low brake pedal. The balance spring is intended to prevent excessive retraction from taking place.

It is extremely important with this caliper design that the parking brake be used regularly to keep the actuator screw threads clean, and the thrust bearing rotating freely. If either of these parts should stick or freeze up, the automatic adjusting mechanism will not work properly, and the balance spring may be unable to compensate. This will cause the parking brake to become inoperative, and the car will suffer from a low brake pedal.

Eccentric Shaft and Rod Actuation

Some disc brake calipers use an eccentric shaft acting on an actuator rod to apply the caliper piston for parking brake service, figure 12-34. When this type of parking brake is applied, the cable moves a lever and rotates a shaft that is offset in the end of the caliper bore. A connecting link, installed between a notch in the eccentric shaft and a similar notch in the end of the actuator rod, then forces the rod outward against spring tension to apply the piston. In the caliper shown in figure 12-34, a stack of cone springs returns the actuator rod when the parking brake is released; other calipers use a conventional coil spring for this purpose.

Like other designs, this type of disc parking brake mechanism adjusts automatically during service brake operation. As the brake linings wear and the piston moves farther out in the caliper bore, a nut inside the piston unscrews along the threaded actuator rod to take up the slack. This holds the piston out farther in its bore and keeps the clearance between the pads and rotor to a minimum.

DRIVELINE AUXILIARY PARKING BRAKES

Driveline auxiliary parking brakes are completely separate from the service brake system and are sometimes called transmission brakes because

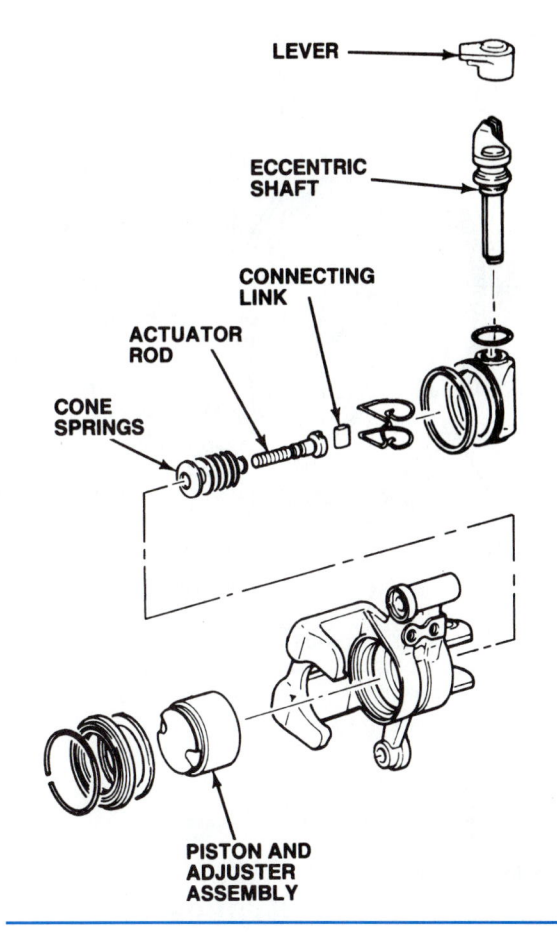

Figure 12-34. A brake caliper with eccentric shaft and rod parking brake actuation.

they usually mount on the rear of the gearbox. Driveline auxiliary parking brakes have not been used on cars for many years. However, they are still common on medium and heavy trucks, and many types of industrial equipment.

A single transmission or driveline parking brake is able to provide sufficient holding power because the differential gearing multiplies the braking torque applied to the wheels. Or, to look at it from the opposite point of view, the differential reduces the amount of torque the wheels can exert on the parking brake friction assembly. The drawbacks to driveline auxiliary parking brakes include the constant strain they exert on the driveline whenever the vehicle is parked on an incline and the additional cost of an independent friction assembly. Driveline auxiliary parking brakes are manufactured in three basic designs:

- External contracting-band
- Internal expanding-shoe
- Disc.

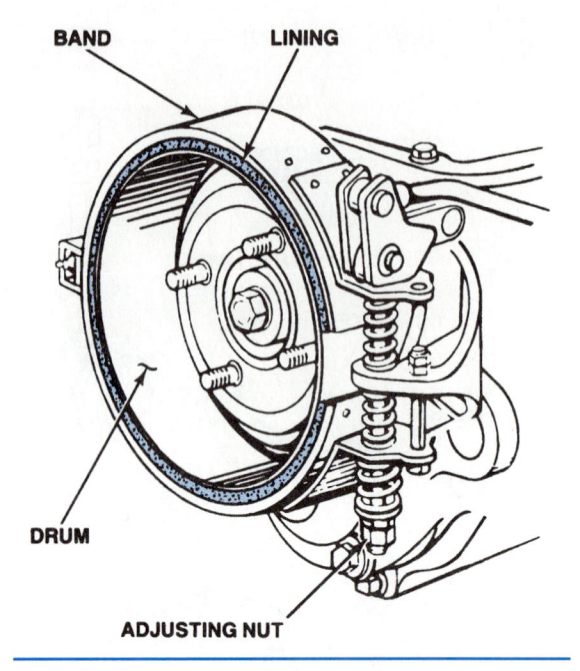

Figure 12-35. A transmission-mounted external contracting-band parking brake.

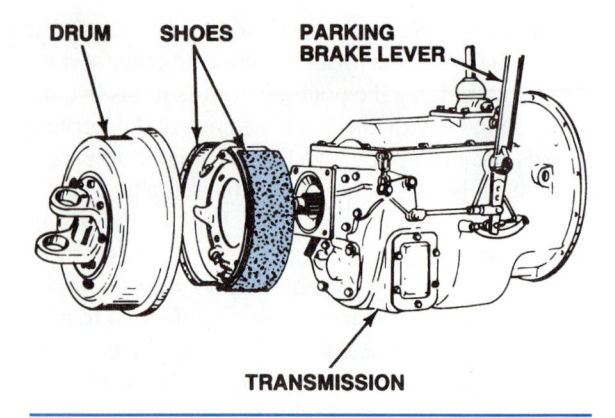

Figure 12-36. A transmission-mounted internal expanding-shoe parking brake.

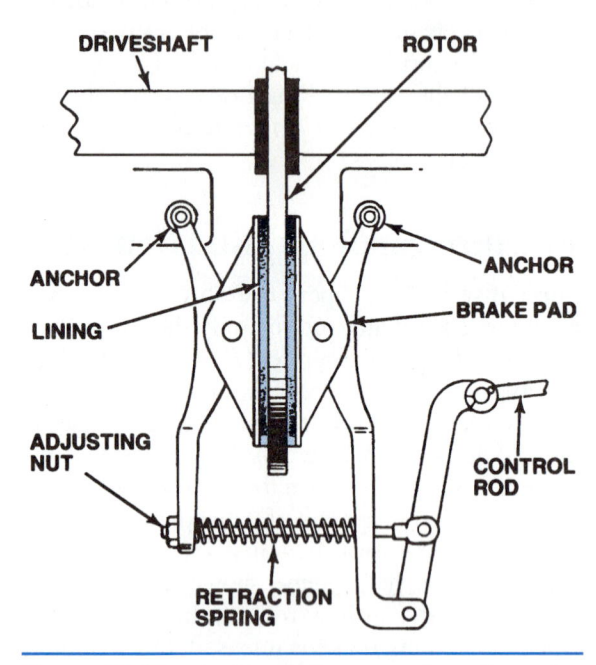

Figure 12-37. A driveshaft-mounted auxiliary disc parking brake.

External Contracting-Band Driveline Auxiliary Drum Parking Brakes

The external contracting-band parking brake, figure 12-35, consists of a brake drum that spins with the driveshaft and a contracting band that is solidly mounted to the rear of the transmission case. A cable or rod controlled by a pedal, lever, or handle inside the vehicle applies the band through a spring-loaded cam or lever linkage. The linkage squeezes the brake band tightly around the drum to hold the driveshaft stationary and lock the rear wheels.

Internal Expanding-Shoe Driveline Auxiliary Drum Parking Brakes

The internal expanding-shoe parking brake, figure 12-36, is similar to a conventional drum brake friction assembly. The brake drum rotates with the driveshaft, and a backing plate securely mounted to the rear of the transmission supports two shoes lined with friction material. The brake may be either a dual-servo or leading-trailing design. A hand lever and linkage operates a cam or lever to apply the shoes of either type of brake against the drum.

Truck Auxiliary Disc Parking Brakes

The auxiliary disc parking brakes used on heavy trucks have a single, mechanically actuated friction assembly built into the driveline between the transmission and differential. With auxiliary disc parking brakes, figure 12-37, the rotor mounts on the driveshaft where it spins between a rod- or cable-actuated clamping assembly fitted with special parking brake pads.

When the parking brake is applied, the linkage rod or cable actuates the levers of the clamping assembly to force the pads against the rotor. When

Parking Brakes

the parking brake is released, a retraction spring helps return the levers to the unapplied position. Auxiliary disc parking brakes used on trucks are adjusted manually. A threaded rod and adjusting nut on the clamping mechanism are used to set the clearance between the rotor and the friction material.

SUMMARY

Parking brakes hold the car stationary when parked and provide a secondary means of stopping the vehicle if the service brake system fails. FMVSS 135 sets standards for parking brake performance.

Parking brakes may be applied by a pedal, lever, or handle. Each of these controls has a ratchet system to lock the brake in the applied position. Pedals are released with a T-handle or lever, and may also be fitted with an automatic release mechanism powered by a vacuum servo. Most parking brake levers release by pushing a button on the end of the lever. Handles are released by twisting the grip or pulling a trigger lever next to the grip. A warning light in the instrument panel alerts the driver when the parking brake is applied.

Parking brake linkages are made up of rods, cables, intermediate levers, and equalizers. Rods and cables transmit force directly; intermediate levers increase application force; and an equalizer balances the pulling force from the actuating device so the same amount is transmitted to each parking brake friction assembly.

Parking brake linkages come in a wide variety of configurations that use one, two, or several cables. Linkages are generally adjusted in one of three ways: jam nuts at the equalizer, individual adjusting nuts at the ends of the cables, or an intercable adjuster built into a cable.

There are two types of drum parking brakes: integral and rear-disc auxiliary (drum-in-hat). Integral drum parking brakes mechanically operate the rear drum service brakes. Rear-disc auxiliary (drum-in-hat) designs are used with rear disc brakes and consist of separate drum brake friction assemblies that fit inside small drums formed into the rear brake rotors.

Disc-type parking brakes are divided into two groups: caliper actuated and auxiliary. Caliper-actuated disc parking brakes are used with single-piston sliding or floating calipers, and are divided into three groups based on the method they use to actuate the caliper piston. Ford designs use a ball and ramp form of actuation. General Motors' calipers have a unique screw, nut, and cone actuating mechanism. Some vehicles use an eccentric shaft and rod to apply the brake caliper piston for parking brake purposes.

Driveline auxiliary parking brakes may use a single external contracting-band, an internal expanding-shoe friction assembly, or a disc-type brake assembly that is usually mounted to the back of the transmission. Automotive applications use the service brake for the parking brake, but heavy equipment installations usually have a separate drum or rotor mounted on the vehicle driveshaft.

Review Questions

Choose the letter that represents the best possible answer to the following questions:

1. Parking brakes may be applied by:
   a. Handles inside the car
   b. Hydraulic pressure
   c. The service brake pedal
   d. None of the above

2. A parking brake pedal may be released by a:
   a. T-handle
   b. Lever
   c. Vacuum servo
   d. All of the above

3. If a parking brake pedal must be depressed several times to set the brake:
   a. The ratchet mechanism is defective
   b. The pedal has a high leverage ratio
   c. Both a and b
   d. Neither a nor b

4. Parking brake handles may be released by:
   a. Pulling a trigger
   b. Twisting the handle
   c. Both a and b
   d. Neither a nor b

5. The warning light for the parking brake:
   a. May also indicate hydraulic pressure loss
   b. Operates off a trigger on the vacuum servo
   c. Is normally in the closed position
   d. None of the above

6. Which of the following is *not* used in parking brake linkages?
   a. Eccentric rod
   b. Equalizer
   c. Woven-steel cable
   d. Intermediate lever

7. Parking brake linkage equalizers:
   a. Are sometimes cable guides
   b. May move both the inner cable and the outer housing
   c. Often attach near the rear axle
   d. All of the above

8. Transmission-mounted parking brakes usually use a _____ parking brake linkage.
   a. Single-cable
   b. Dual-cable
   c. Multiple-cable
   d. Any of the above

9. The most common dual-cable parking brake linkage on domestic cars:
   a. Uses one long and one short cable
   b. Has an equalizer mounted in the center of the short cable
   c. Both a and b
   d. Neither a nor b

10. Which of the following is *not* a form of parking brake adjustment?
    a. Jamnuts
    b. Intercable adjuster
    c. Intermediate lever location
    d. Individual cable-end nuts

11. Integral drum parking brakes:
    a. Are applied with a lever and cam
    b. Are rarely found on cars
    c. Use a strut as a pivot point
    d. All of the above

12. Rear-disc auxiliary drum parking brakes:
    a. Use a leading-trailing friction assembly
    b. Have equal holding power in both directions
    c. Are adjusted through a hole in the backing plate
    d. Are usually applied by a cam

13. Driveline parking brakes are able to hold a large vehicle in place because:
    a. The differential provides torque multiplication
    b. The transmission provides torque reduction
    c. Both a and b
    d. Neither a nor b

14. Caliper-actuated disc parking brakes:
    a. Are used with fixed calipers
    b. Are usually single-piston calipers
    c. Apply the caliper piston hydraulically
    d. None of the above

15. The piston in a caliper with ball and ramp parking brake actuation is applied by the:
    a. Cone nut
    b. Thrust screw
    c. Compensating spring
    d. None of the above

Parking Brakes

16. General Motors' rear-disc parking brakes use a(n) _____ to help prevent a low brake pedal.
    a. Adjuster spring
    b. Balance spring
    c. Both a and b
    d. Neither a nor b

17. Adjustment of General Motors' rear-disc parking brakes takes place when:
    a. The service brakes are applied
    b. A gap exists between the nut and cone
    c. The cone contacts the piston clutch surface
    d. The actuator screw rotates

18. Rear-disc parking brakes actuated by an eccentric shaft:
    a. Are applied through a cable and lever
    b. Use an actuator rod to move the piston
    c. May contain cone springs
    d. All of the above

13

Power Brakes

OBJECTIVES

Upon completion and review of this chapter, you will be able to:

- Explain the need for power brakes.
- List the methods used to increase application force.
- Define the terms "vacuum" and "pressure differential."
- Describe the construction and operation of integral vacuum boosters.
- Describe the construction of multiplier boosters.
- Describe the construction and operation of the Hydro-Boost.
- Describe the construction and operation of the Powermaster.
- Describe the construction of dual-power brake systems.

KEY TERMS

accumulator
atmospheric pressure
atmospheric-suspended power chamber
booster holding position
booster vacuum runout point
dual-power brake system
electro-hydraulic (E-H) boosters
integral vacuum booster
mechanical-hydraulic boosters
multiplier vacuum booster
power chamber
pressure differential
spool valve
tandem booster
vacuum
vacuum-suspended power chamber

INTRODUCTION

Power brakes is the popular expression used to identify a brake system fitted with a power booster. The booster is used to increase brake application force, thus reducing the amount of foot pressure required at the brake pedal. Power boosters are standard on most cars and light trucks, and optional on a few. Because of their large size and heavy weight, all medium-duty trucks, and commercial vehicles have power-assisted brakes.

This chapter examines the reasons power brakes are needed. It then looks at several methods that can be used to increase braking force and

243

discusses why power brake boosters are the best alternative. Finally, the types of power boosters used on modern vehicles are described, along with the principles that make them work.

THE NEED FOR POWER BRAKES

Power brakes were developed for three reasons. The first was the dramatic increase in vehicle weights and speeds during the 1950s and 1960s. Faster and heavier cars and trucks required more brake application force to convert their increased kinetic energy into heat. As time passed, the amount of pedal pressure required to create this force became greater than the average driver could comfortably supply.

Second, the introduction and widespread adoption of disc brakes made power brakes a necessity on heavier cars. Most drum brakes have self-energizing or servo action that increases their application force. Disc brakes have neither and, by nature, require greater pedal pressures to operate.

Finally, marketing reasons made power brakes important. Manufacturers looking to attract customers noted that power brakes made a car easier to drive. Power brakes first appeared on large, expensive luxury cars and were soon made available across all model lines in response to public demand.

WAYS TO INCREASE BRAKING POWER

There are essentially four methods that can be used to reduce pedal pressures or increase application force in a brake system. These methods are:

- Pedal force
- Mechanical advantage
- Hydraulic advantage
- Power boosters.

Pedal Force

The most fundamental method of increasing braking power is for the driver to apply the brake pedal harder. Beyond a certain point, this method is unacceptable because weaker drivers may not be able to push the pedal hard enough to take advantage of all the stopping power the brakes have to offer. In effect, the physical strength of the driver becomes the limiting factor in how well the car stops.

Mechanical Advantage

The next method of providing additional brake application force is to increase the mechanical advantage of the pedal assembly—use a higher pedal ratio to achieve more leverage. This makes it easier for the driver to create a given application force, but it also increases the amount of pedal travel required to do so. As a result, there is less pedal reserve available to compensate for brake fade and wear.

Another factor working against increased mechanical advantage is that the brake pedal and linkage would have to be made somewhat larger to create a significant increase in leverage. This is impractical in many cases because there is limited room in the driver's foot well to place these items.

Hydraulic Advantage

The third method of increasing application force, hydraulics, is already used to some degree in every brake system. If the wheel cylinder or caliper piston area is increased in relation to master cylinder bore size, it is easier for the driver to provide a given amount of application force. However, just as with increased mechanical leverage, increased hydraulic advantage requires more brake pedal travel and reduces pedal reserve. In addition, if the amount of hydraulic advantage built into the system is too great, the brakes become hard to modulate in a panic stop.

Power Boosters

The fourth and final alternative to increase brake application force is to install a power booster in the system, figure 13-1; this is by far the best method. Generally, when a power booster is fitted, the brake pedal ratio is decreased and the master cylinder bore size is increased. The combined effect of these changes is to reduce pedal effort, while greatly increasing pedal reserve.

Power boosters do not alter the hydraulic system in any fundamental way; they still allow braking even if the booster fails or its power supply is cut off. All boosters have a power reserve

Power Brakes

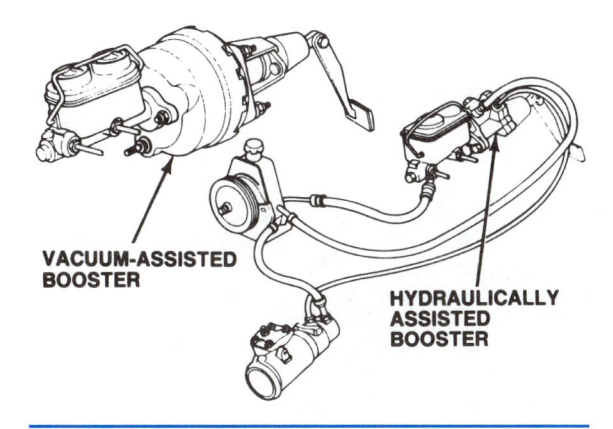

Figure 13-1. Typical brake power boosters.

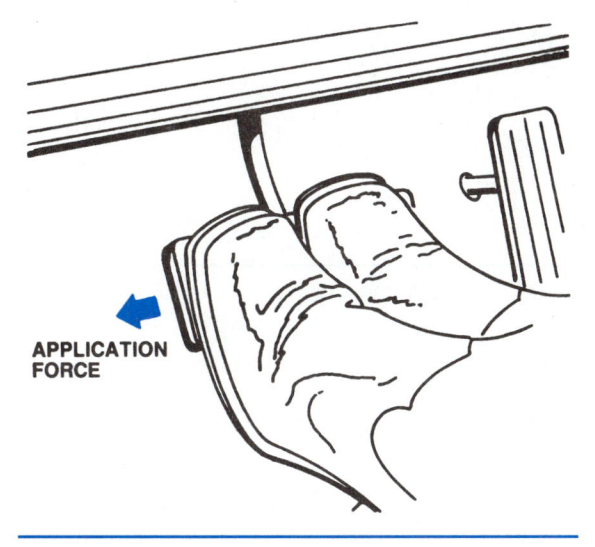

Figure 13-2. Wide power brake pedals allow two-foot braking if power assist is lost.

that provides assist for at least one hard stop, and sometimes several light brake applications, even after power is lost. However, because power brake systems are designed with the added force of the booster taken into account, the amount of brake pedal pressure required to slow or stop the car is much higher than in a nonboosted system once the reserve is used up. For this reason, some cars with power brakes have a brake pedal that is wide enough to allow two-foot braking should the booster fail, figure 13-2.

Two types of power boosters are found on today's automobiles: those that use vacuum to increase application force and those that use hydraulic pressure. The principles governing hydraulics were discussed in Chapter 3; the principles of vacuum, or low air pressure, are explained next.

AIR PRESSURE— HIGH AND LOW

Most vacuum-powered brake boosters get their vacuum supply from the engine intake manifold. An engine is essentially a big air pump; the pistons move up and down in the cylinders to pump in air and fuel, and pump out exhaust. They do this by creating differences in air pressure. Air, both inside and outside an engine, has weight and exerts pressure.

As a piston moves downward on an intake stroke with the intake valve open, figure 13-3, it creates a larger area inside the cylinder for air to fill. This lowers the air pressure within the cylinder, and, as a result, the higher-pressure air outside the engine flows in through the intake manifold in an attempt to fill the low-pressure area. Although it may seem as though the low pressure is pulling air into the engine, it is really the higher pressure outside that forces air in. The difference in pressure between two areas is called a **pressure differential.**

Because throttle valves and manifold shape restrict intake airflow, high-pressure air from outside the engine is almost never able to move into the cylinders fast enough to fill the space created. As a result, gasoline-powered internal-combustion engines normally operate with a low-pressure area, or partial vacuum, in the intake manifold. **Vacuum** is a technical term that means a total lack of pressure (0 psi); however, the word *vacuum* is commonly used to refer to any pressure lower than **atmospheric pressure.** Atmospheric pressure varies with altitude, figure 13-4, but is approximately 14.7 psi at sea level.

Measuring Vacuum

Vacuum is measured in inches or millimeters of mercury (Hg), a figure that indicates how far a column of mercury in a tube will rise when a vacuum is applied at one end and atmospheric pressure at the other. Or, put another way, inches or millimeters of mercury is a measurement of the pressure differential between the lower pressure inside the tube and the higher pressure outside it.

A perfect vacuum will pull roughly 30 inches of mercury (762 mm Hg); however, perfect vacuum occurs only in space and is never achieved in an engine's intake manifold. Manifold vacuum varies with throttle position. The lowest manifold vacuum (highest pressure) is commonly less than

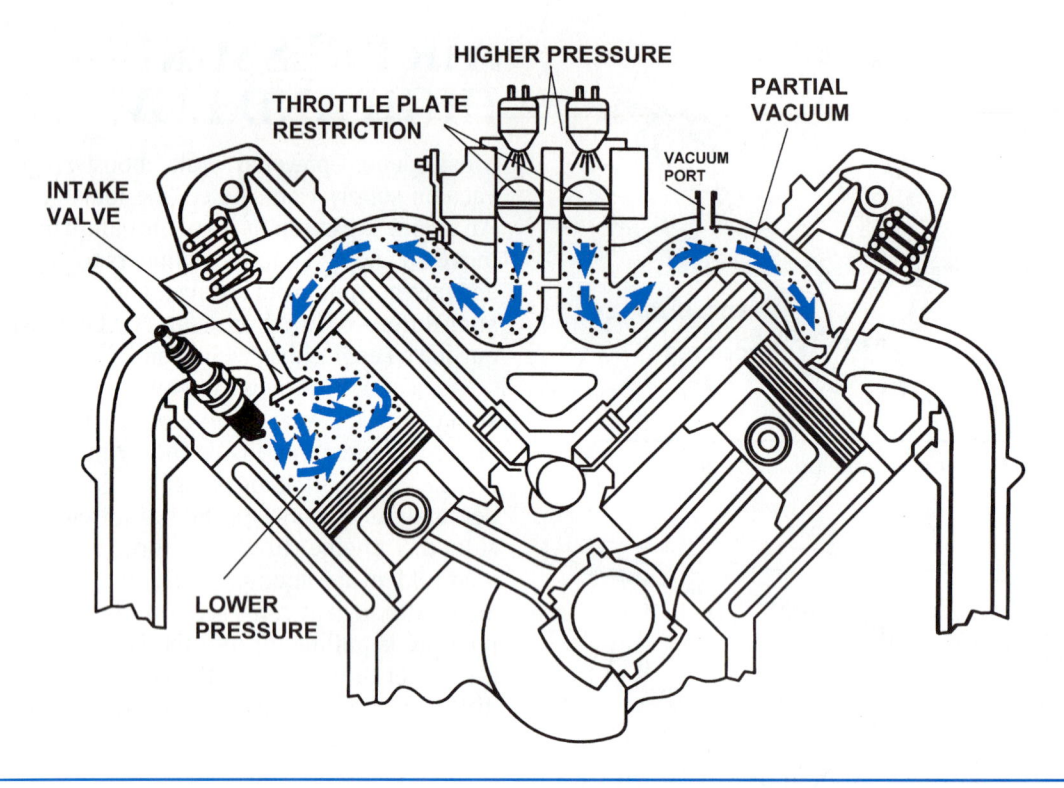

Figure 13-3. Airflow through a gasoline engine creates a partial vacuum in the intake manifold. (Courtesy of General Motors Corporation, Service and Parts Operations)

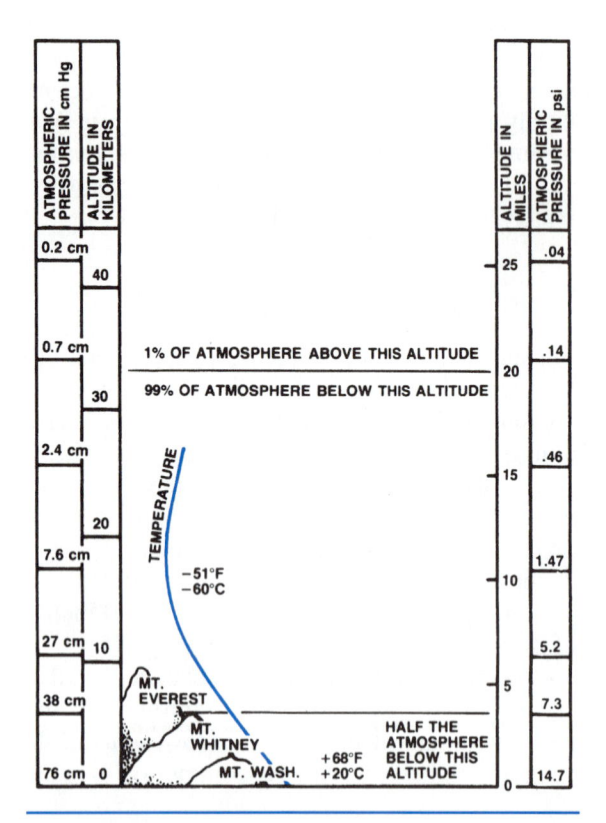

Figure 13-4. Atmospheric pressure varies with altitude.

5 inches of mercury (127 mm Hg) and occurs when the throttle is wide open with the engine under load. The highest manifold vacuum (lowest pressure) may be as much as 24 inches of mercury (610 mm Hg) when the car is rolling rapidly downhill in gear with the throttle closed. Manifold vacuum at idle typically falls between 15 and 20 inches of mercury (381 and 508 mm Hg), and most vacuum brake boosters are designed to operate with vacuum levels in this range.

Booster Vacuum Supply

As discussed previously, most vacuum boosters get their vacuum supply from the engine intake manifold, figure 13-5. Diesel engines, however, run unthrottled (engine speed is controlled strictly by the amount of fuel injected) and have little or no intake manifold vacuum. If a vehicle with a diesel engine is equipped with a vacuum-powered brake booster, it must also be fitted with an auxiliary vacuum pump.

Auxiliary vacuum pumps come in several forms and are powered in a number of ways. Belt-driven add-on pumps, figure 13-6, are one solution; however, many diesel engines are designed with vac-

Power Brakes

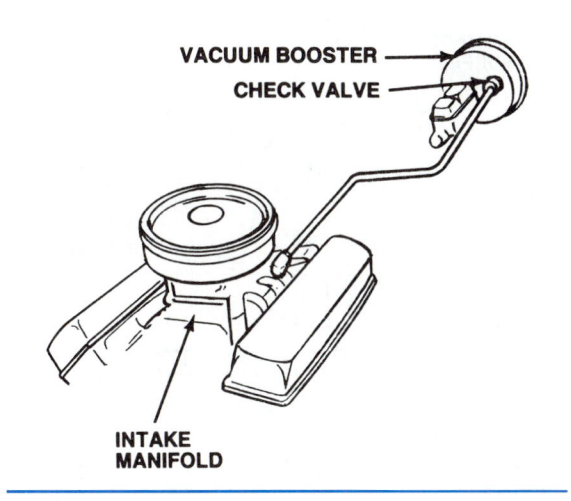

Figure 13-5. The intake manifold is the most common source of vacuum for power boosters.

uum pumps that mount directly to the engine and are driven internally, figure 13-7. An electrically powered vacuum pump, figure 13-8, is used on some cars and trucks. These pumps are turned on and off by a pressure switch on the booster, which means they operate only when needed and thus reduce power drain on the engine.

Supplemental Brake Assist Unit

Some manufacturers use a supplemental brake assist (SBA) unit to supply vacuum in case of a low-vacuum condition. The unit contains a circuit board and pressure sensor, a check valve, and an electrically operated vacuum pump. It mounts in series with the booster vacuum line on the booster, figure 13-9. During normal operation, the engine supplies vacuum through the check valve to the booster. At this time the SBA unit continuously monitors the vacuum in the booster.

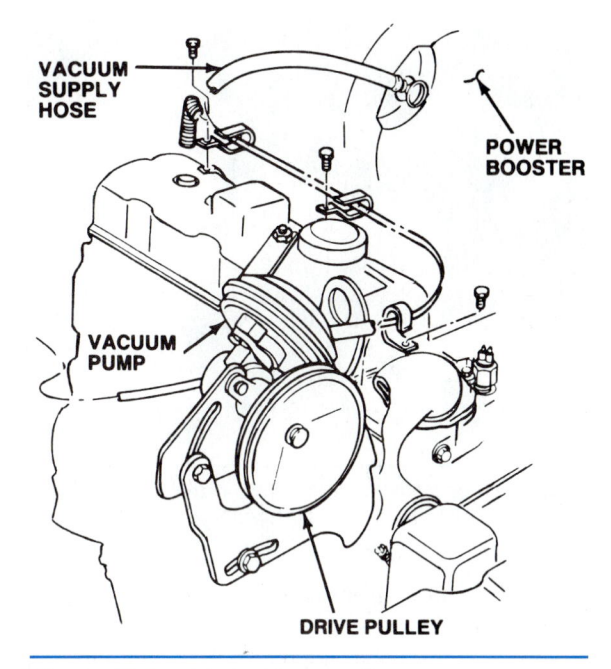

Figure 13-6. A belt-driven auxiliary vacuum pump.

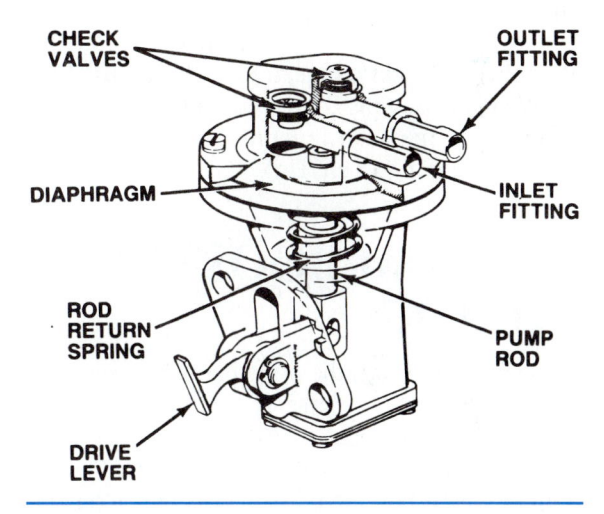

Figure 13-7. A vacuum pump driven directly off the engine.

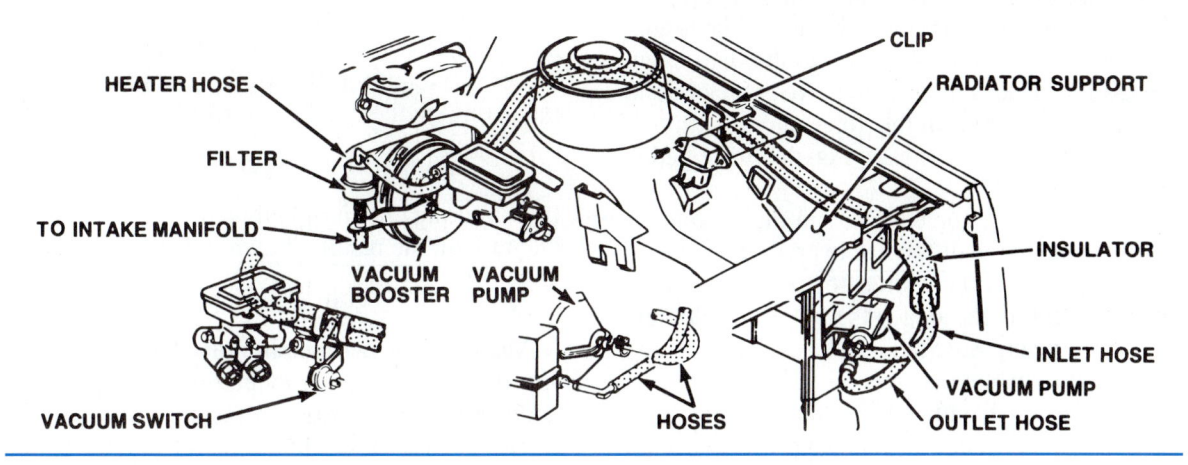

Figure 13-8. An electrically powered vacuum pump.

Figure 13-9. The supplemental brake assist (SBA) unit is mounted on the brake booster.

If the vacuum drops below 7 in./Hg (8 cm/Hg), the SBA starts a 5-second timer and turns on the pump if vacuum does not increase. The pump will turn off when vacuum is more than 9 in./Hg (23 cm/Hg). If vacuum drops to below 6 in./Hg (15 cm/Hg), the pump will come on immediately. Whenever the pump runs, a REDUCED BRAKE POWER message is displayed on the instrument panel. This message remains until vacuum exceeds 10 in./Hg (25 cm/Hg).

VACUUM BOOSTER THEORY

Vacuum booster use the principle of pressure differential to increase brake application force. The typical vacuum booster has a **power chamber** separated into two smaller chambers by a flexible diaphragm. When air pressure is greater on one side of the diaphragm than the other, figure 13-10, a pressure differential is created. In an attempt to equalize pressure in the two chambers, the higher pressure exerts a force that moves the diaphragm toward the lower pressure area. Rods attached to the diaphragm transmit this force, along with that from the driver's foot pressure on the brake pedal, to the master cylinder.

The amount of force created in this manner is porportional to the difference in pressure between the two sides. In other words, the greater the pressure differential, the greater the force. To calculate the force, the pressure differential is multiplied by the diaphragm surface area. For example, if a power booster diaphragm has atmospheric pressure (14.7 psi) on one side, and a perfect vacuum (0 psi) on the other, the pressure differential is:

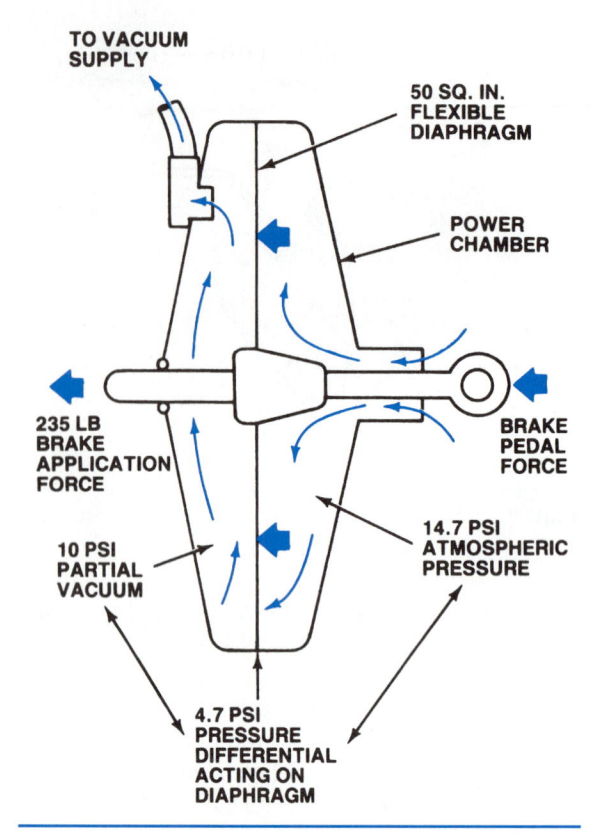

Figure 13-10. Vacuum boosters operate on the principle of pressure differential.

$$14.7 \text{ psi} - 0 \text{ psi} = 14.7 \text{ psi}$$

If this pressure differential is applied to a diaphragm with 50 square inches of surface area, the resulting force would be:

$$14.7 \text{ psi} \times 50 \text{ sq. in.} = 735 \text{ pounds of force}$$

This equation assumes that one side of the diaphragm is acted on by a perfect vacuum; however, as already discussed, a perfect vacuum never occurs in an engine intake manifold. With a more typical intake manifold vacuum of 20 inches of mercury (10 psi), the pressure differential acting on the diaphragm would be:

$$14.7 \text{ psi} - 10 \text{ psi} = 4.7 \text{ psi}$$

If we once again multiply this times the area of the diaphragm, the result (shown in figure 13-10) is:

$$4.7 \text{ psi} \times 50 \text{ sq. in.} = 235 \text{ pounds of force}$$

Vacuum booster diaphragms are sized to fit specific applications and provide the necessary application force. Modern vacuum boosters are capable of providing hundreds of pounds of application force.

Vacuum Booster Suspension

All power chambers operate in one of two ways. The difference lies in the way they create the pressure differential between the two halves of the chamber. The two types of power chambers used on vacuum brake boosters are called:

- Atmospheric suspended
- Vacuum suspended.

Atmospheric-Suspended Power Chambers

In an **atmospheric-suspended power chamber**, there is atmospheric pressure on both sides of its diaphragm when the brakes are not applied. When the brake pedal is depressed, vacuum is admitted to one side of the chamber to create the pressure differential needed to cause power boost.

The primary problem with atmospheric-suspended power chambers is that, when the engine is not running, there is no reserve vacuum in the booster for braking. To get around this problem, most atmospheric-suspended boosters use a separate small tank to store enough vacuum for at least one hard stop, and sometimes several light brake applications, should the engine stall. The atmospheric-suspended power chamber has not been used on cars since the mid-1960s.

Vacuum-Suspended Power Chambers

Virtually all modern vacuum boosters use a **vacuum-suspended power chamber** that has vacuum on both sides of its diaphragm when the brakes are not applied. When the brake pedal is depressed, atmospheric pressure is admitted to one side of the chamber to create the pressure differential needed to cause power boost.

The primary advantage of the vacuum-suspended power chamber is that it always has a vacuum reserve stored in its power chamber to provide assist should the vacuum source to the booster be interrupted. All of the vacuum boosters described in this chapter have vacuum-suspended power chambers.

Vacuum Check Valves and Filter

To maintain a supply of vacuum in the storage tank or power chamber, vacuum-assisted power brake systems are equipped with one-way check valves. The check valve seals the vacuum passage to the booster whenever pressure is *higher* in the supply line than it is in the booster. The valve reopens as soon as the pressure in the supply line is *lower* than in the booster. In this way, the lowest possible pressure (highest vacuum) is maintained in the booster at all times.

Most systems have a single check valve built into the inlet fitting on the booster power chamber, figure 13-5, although some cars and trucks use a separate valve installed in the vacuum supply line, figure 13-11. Vehicles that use an auxiliary vacuum reservoir to supplement the vacuum

■ Assist Power Brakes

The assist power brake used in the 1950s had a conventional (for the time) atmospheric-suspended, bellows-type power chamber. What set this design apart from all others was a unique linkage that hinged on a "power lever" that pivoted on the floor. A pushrod from the brake pedal attached near the bottom of the backside of the lever, whereas the power booster attached near the top of the lever's forward side. A pushrod near the center of the forward side actuated the master cylinder. When the brakes were applied, movement of the brake pedal pushrod applied both the master cylinder and the booster through the lever. As the booster came into play, it pulled the top of the lever forward to help apply the brakes.

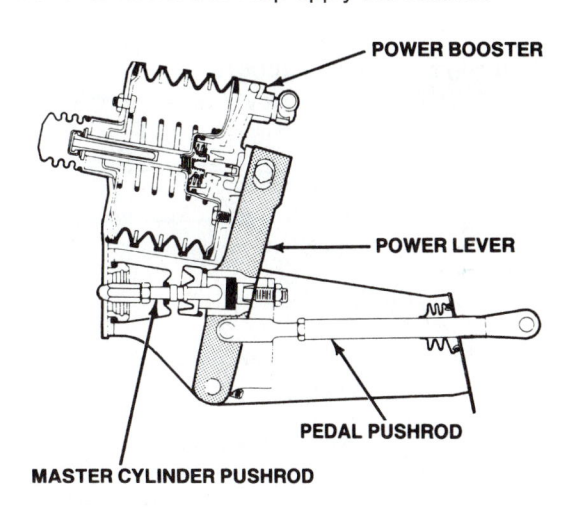

Several factors led to the demise of assist power brakes. The biggest was a desire to move the master cylinder higher in the car. This kept it out of the way of road splash, facilitated system bleeding, and made the cylinder easier to reach for service. In addition, once the master cylinder was in a higher location, a suspended brake pedal and mounting the master cylinder and booster in tandem greatly simplified the pedal linkage.

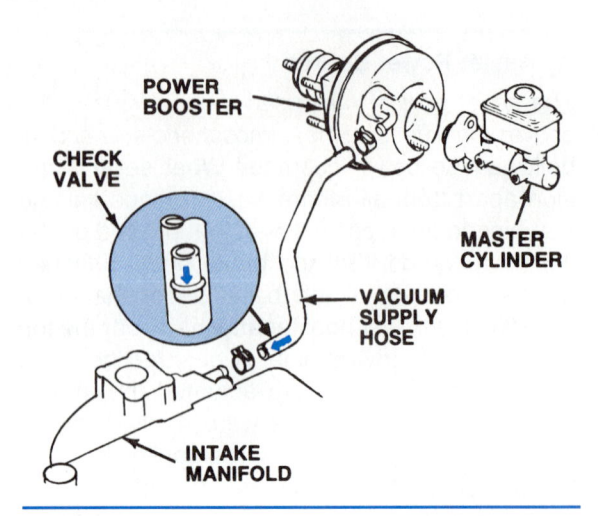

Figure 13-11. An inline vacuum check valve.

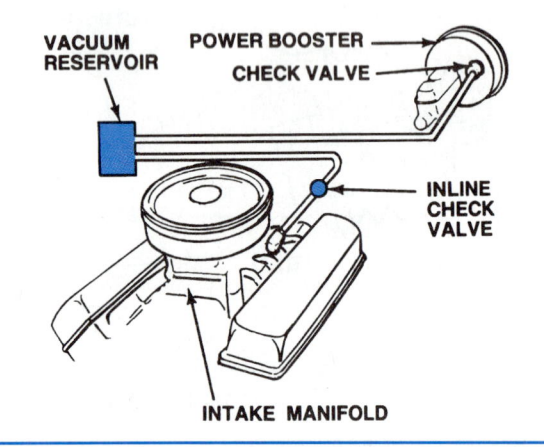

Figure 13-12. A brake system with an auxiliary vacuum reservoir and dual check valves.

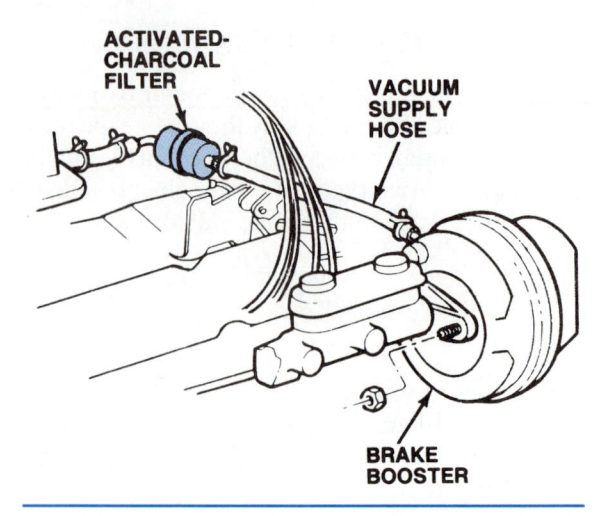

Figure 13-13. A vacuum supply line filter is used to trap fuel vapors.

stored in the booster have two check valves, figure 13-12, one in the conventional location on the booster, and another between the intake manifold and vacuum reservoir.

In power brake systems that use manifold vacuum from a gasoline engine, the check valve serves an additional purpose: It keeps fuel vapors out of the booster. These fumes can attack the rubber diaphragms and seals in boosters, and lead to premature failure. The check valve allows air to flow only from the booster to the engine. If a check valve were not used, low pressure in the booster would allow some of the air-fuel mixture to enter the booster when the throttle was opened and manifold pressure increased.

In addition to a check valve, some cars have an activated charcoal filter, figure 13-13, in the vacuum supply line. This filter traps and absorbs any fuel vapors that may leak toward the booster as a result of a weak or defective check valve. Whenever manifold pressure is less than booster pressure, the fuel vapors are drawn out of the filter and into the engine to be burned.

INTEGRAL VACUUM BOOSTERS

Two types of vacuum boosters are used today: integral and multiplier. The integral boosters described in this section are far and away the most common type. Multiplier boosters are less common and are covered later in the chapter.

An **integral vacuum booster**, figure 13-14, mounts on the bulk head between the brake pedal and the master cylinder. The outside of the booster is a round, two-piece, stamped-steel housing called the power chamber. Some boosters have a reinforced plastic power chamber to reduce weight. The power chamber mounts to the bulk head at the rear and the master cylinder at the front. An input pushrod at the rear of the boosters attaches to the brake pedal. An output pushrod at the front of the booster actuates the master cylinder primary piston.

The master cylinder side of the power chamber is equipped with an inlet fitting where vacuum is supplied to the booster. On vehicles with auxiliary electric vacuum pumps, a pressure switch is installed on the vacuum supply side of the power chamber to control vacuum pump operation. A low-vacuum switch may also be used to illuminate a warning light on the instrument panel if vacuum in the booster falls (pressure rises) to an unsafe level.

Power Brakes

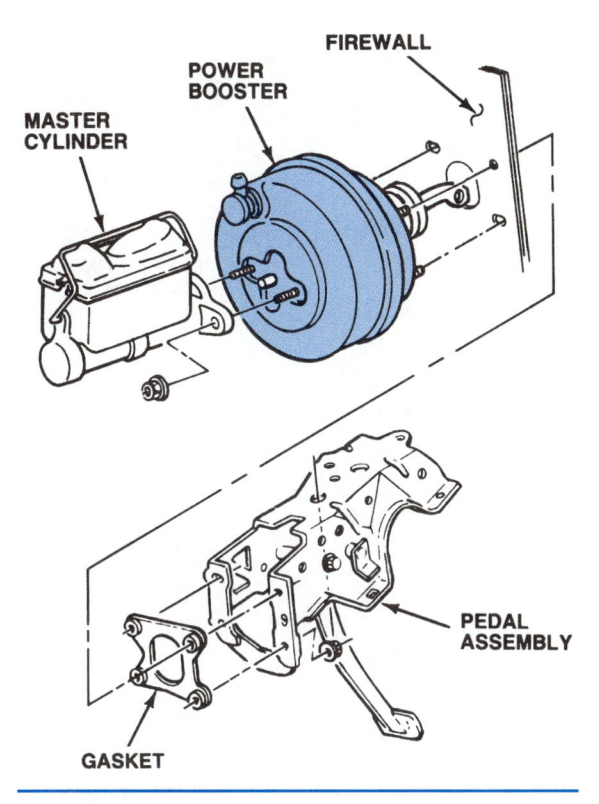

Figure 13-14. An integral vacuum booster installation.

Inside the power chamber is the diaphragm that is acted on by atmospheric pressure to increase brake application force. Most vacuum boosters have a single diaphragm, however, a dual-diaphragm **tandem booster,** figure 13-15, is used in some applications. The use of two diaphragms allows the power chamber to be smaller in diameter (though slightly longer) so it will fit on cars with limited underhood space. Despite its reduced diameter, a tandem booster usually has more diaphragm surface area, and, therefore, supplies greater power assist, than a single-diaphragm booster with a larger diameter. In all other respects, single- and dual-diaphragm boosters are the same.

Also inside the power chamber between the input and output pushrods is the power piston assembly that controls booster operation, figure 13-16. The power piston, along with a support plate, is attached to the center of the diaphragm and moves back and forth with it. A return spring holds the diaphragm, support plate, and power piston rearward in the power chamber when the brakes are not applied.

The typical vacuum-suspended integral power booster has five phases of operation:

- Brakes not applied
- Brake application

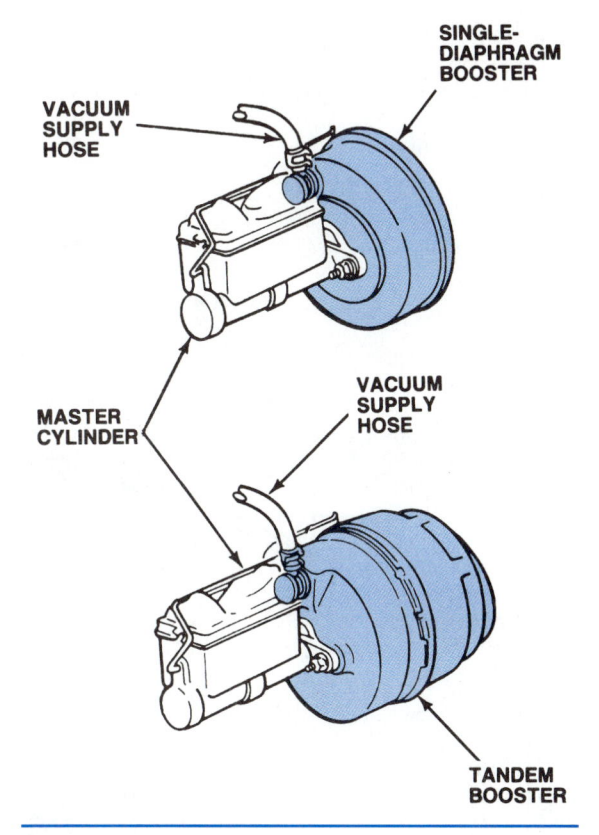

Figure 13-15. Single and tandem vacuum power boosters.

- Brakes holding
- Brakes fully applied
- Brake release.

Most of the following operational descriptions are based on illustrations of a power booster that uses a rubber reaction disc to provide brake pedal feel. Some boosters use a reaction plate and levers for this purpose. Although the two designs differ in their physical construction, the same basic control valves are used in both. Where there is a significant variation in the way the two designs operate, both systems are explained.

Integral Booster—Brakes Not Applied

When the brakes are not applied, figure 13-17, the input pushrod return spring holds the pushrod and valve plunger rearward in the power piston. This does two things. First, the backside of the plunger seats against the floating control valve and closes the atmosphere control port; this prevents atmospheric pressure from entering the rear half of the

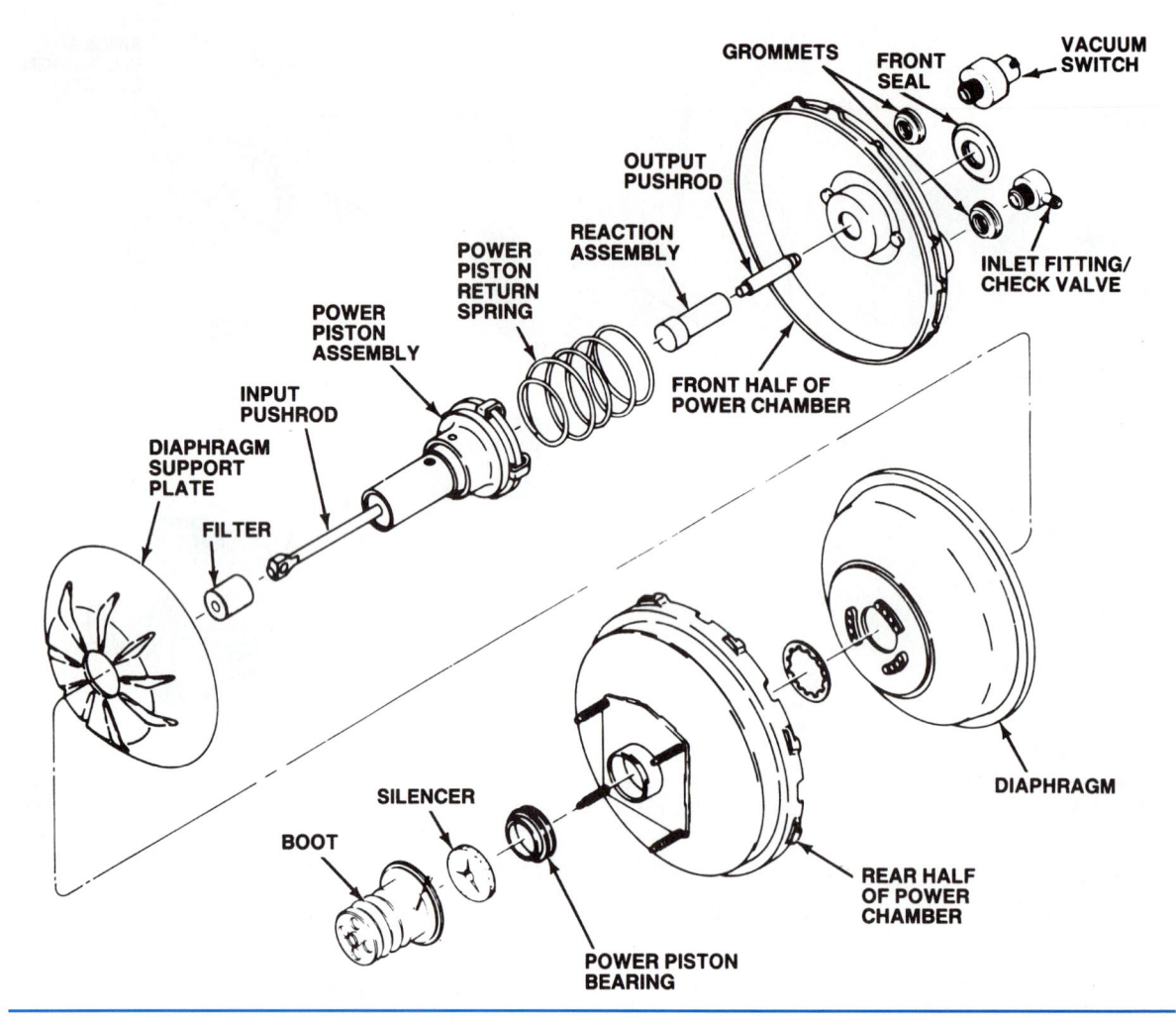

Figure 13-16. An integral vacuum power booster.

power chamber. Second, the plunger compresses the floating control valve against spring tension and opens the vacuum control port; this clears a passage between the front and rear halves of the power chamber so there is equal vacuum in both. Because the diaphragm and power piston assembly is held rearward by the power piston return spring, no pressure is placed on the master cylinder at this time.

Integral Booster— Brake Application

When the driver applies the brakes, figure 13-18, pedal pressure overcomes input pushrod return spring pressure and moves the pushrod and valve plunger forward in the power piston. This also does two things. First, it allows the floating control valve to extend under spring pressure and seal the vacuum control port; this prevents vacuum from entering the rear half of the power chamber. Second, the valve plunger moves away from the end of the floating control valve and opens the atmosphere control port; this lets atmospheric pressure enter the rear half of the power chamber through a filter installed where the input pushrod enters the chamber. A slight sound of rushing air may be heard at this time, although many boosters have a silencer that eliminates most of the noise.

As a result of these actions, there is low pressure in the front half of the power chamber and higher pressure in the rear half. This pressure differential moves the diaphragm and power piston assembly forward against power piston return spring pressure. The application force created in this manner is applied to the master cylinder through the body of the power piston, the reaction disc, and the output pushrod.

Power Brakes

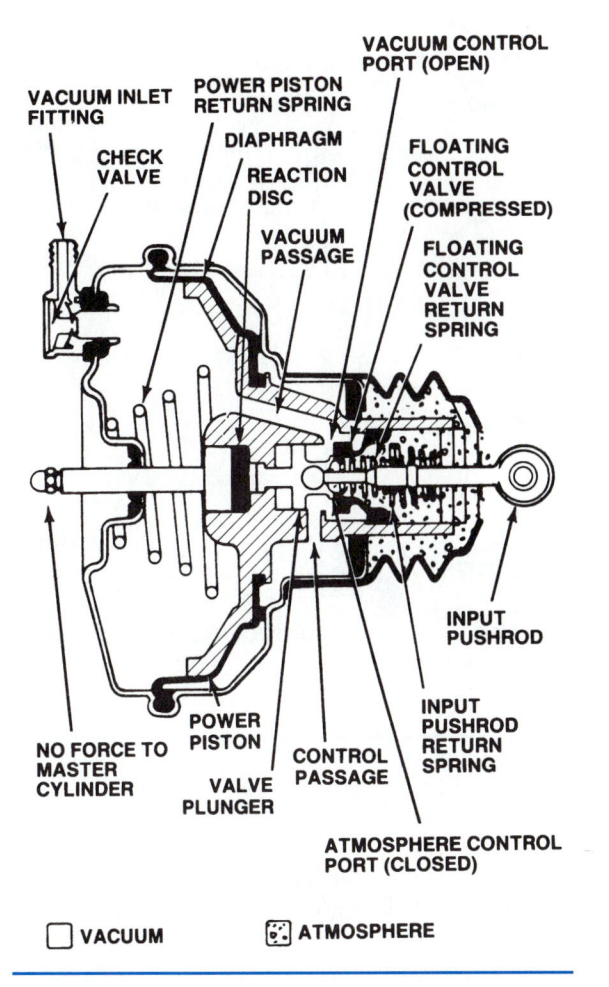

Figure 13-17. An integral vacuum booster in the unapplied position.

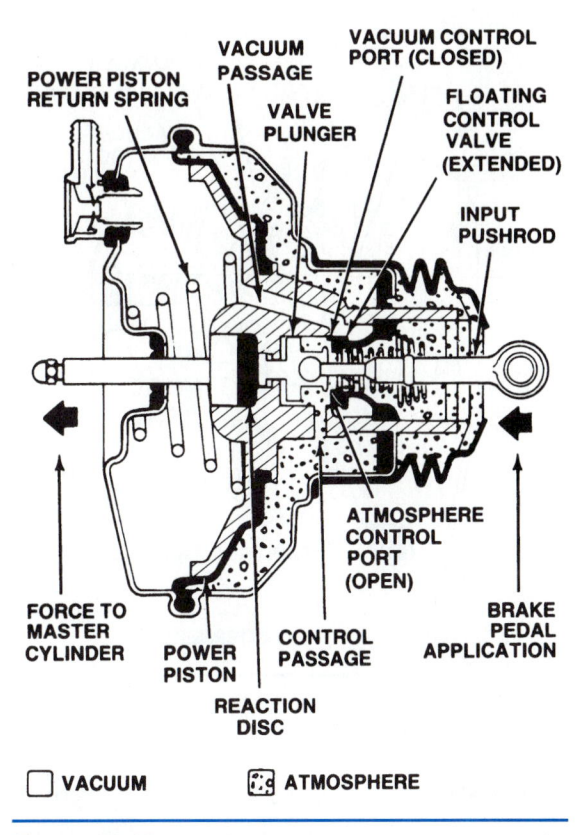

Figure 13-18. An integral vacuum booster as the brakes are applied.

Integral Booster— Brakes Holding

Once the desired rate of deceleration is achieved, no further pressure is applied to the brake pedal, and the input pushrod and valve plunger are held in a fixed position. As long as the atmosphere control port remains open, pressure increases in the rear half of the power chamber, and the diaphragm and power piston move forward. When the power piston moves far enough, the floating control valve (attached to the power piston) seats against the backside of the valve plunger and closes the atmosphere control port, figure 13-19.

At this point, *both* the atmosphere control port and the vacuum control port are closed. As a result, there is a fixed pressure differential between the front and rear halves of the power chamber, so the level of power assist and braking force remains constant. This is called the **booster holding position**. The booster always seeks the holding position whenever brake pedal pressure is constant.

Integral Booster— Brakes Fully Applied

Under very heavy braking, the brake pedal may be pushed hard enough to force the valve plunger so far forward that it bottoms in the power piston assembly, figure 13-20. At this point, the vacuum control port is closed, the atmosphere control port is wide open, and full atmospheric pressure exists in the rear half of the power chamber. This means the maximum possible pressure differential is present in the power chamber, and the booster is supplying as much power assist as it can. This is called the **booster vacuum runout point**.

The vacuum runout point can be felt by the driver as a distinct hardening in the feel of the brake pedal. This occurs for two reasons. The main one is that once the booster reaches vacuum runout, any additional braking force must be supplied entirely by the driver through foot pressure on the brake pedal.

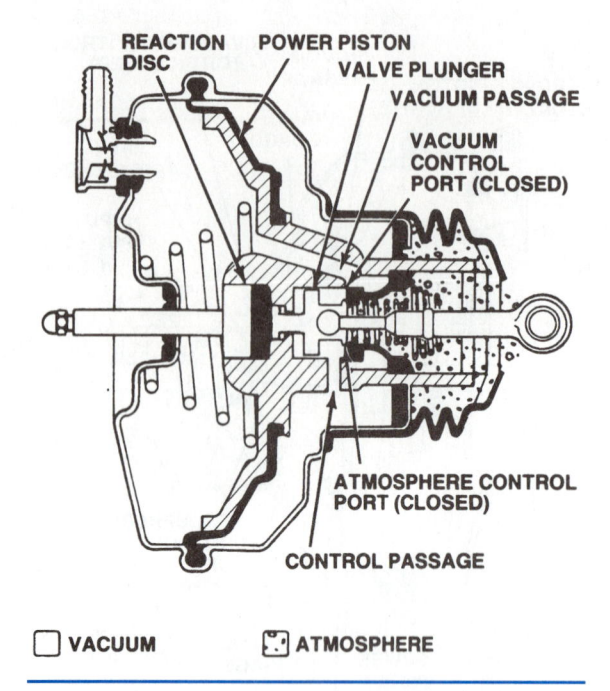

Figure 13-19. An integral vacuum booster in the holding position.

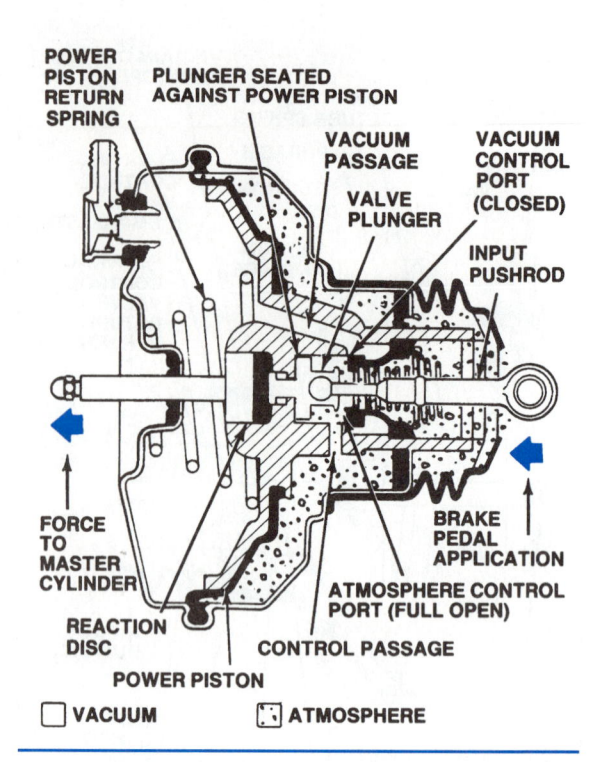

Figure 13-20. An integral vacuum booster in the fully applied position.

The second reason the brake pedal becomes harder during vacuum runout is that the device used to provide pedal feel is bypassed at this time. When the valve plunger seats against the power piston in a booster with a rubber reaction disc, the disc is compressed by the extension on the front of the valve plunger to the point where it is virtually solid. In a booster that uses a reaction plate and levers to provide feedback, the intake pushrod applies the valve plunger with such force that the levers collapse and the forward end of the valve plunger makes metal-to-metal contact with the output pushrod. In both types of boosters, the result is a solid connection between the brake pedal and output pushord.

Integral Booster— Brake Release

When the driver releases the brake pedal, the input pushrod return spring moves the pushrod and valve plunger rearward in the power piston. This does two things. First, the backside of the valve plunger closes the atmosphere conntrol valve; this prevents any further increase in pressure in the rear half of the power chamber. Second, the valve plunger compresses the floating control valve against spring pressure and opens the vacuum control port; this clears the passage to the vacuum-filled front half of the booster. At this time, any residual air pressure in the rear half of the booster is drawn out through the connecting passage and vacuum inlet fitting, and into the intake manifold. When pressure is again equal on both sides of the diaphragm, the booster will be in the brakes-not-applied position shown in figure 13-17.

Brake Pedal Feel

If a power booster applied the brakes strictly as described, the amount of power assist would increase or decrease in sudden steps as the various ports in the booster opened and closed. And once the booster reached the holding position, only the foot pressure applied by the driver would be felt at the brake pedal. To ensure smooth, controllable stops, a brake system must provide the driver with feedback on how hard the brakes are being applied. This feedback, commonly called the pedal feel, is the increasing resistance to further pedal pressure that occurs as the brakes are applied harder.

The hundreds of pounds of application force created by a power booster cannot all be fed back through the brake pedal or it would force the pedal, against driver foot pressure, into the unapplied position. Instead, power boosters have mechanisms that transmit a fixed proportion of the braking force back

Power Brakes

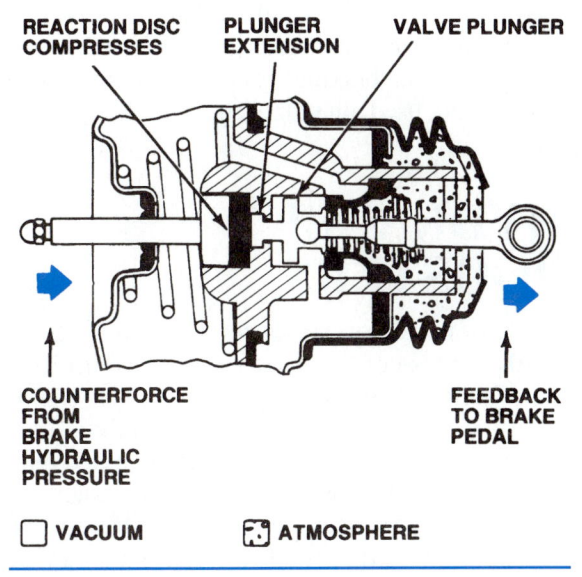

Figure 13-21. Pedal feel provided by a rubber reaction disc.

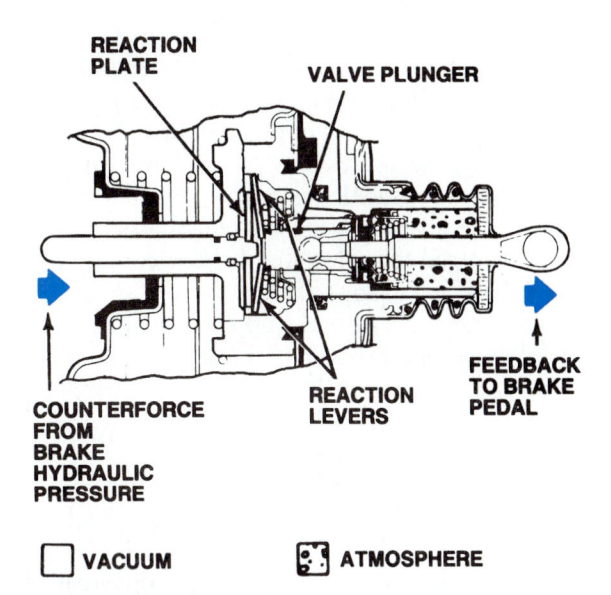

Figure 13-22. Pedal feel provided by a reaction plate and levers.

to the brake pedal to provide pedal feel. Generally, 20 to 40 percent of the force applied to the master cylinder is relayed back to the driver. As mentioned earlier, this is done through either a rubber reaction disc or a reaction plate and series of levers.

The process used to provide pedal feel relies on a basic law of physics, Sir Isaac Newton's third law of motion, which states that for every action there is an equal and opposite reaction. This means that once braking begins and force is applied to the master cylinder, a counterforce is also created. As governed by Newton's third law, the counterforce is equal to the hydraulic pressure in the master cylinder and is transmitted in the opposite direction back into the booster through the output pushrod. Because the force and counterforce must always be equal, the harder the brakes are applied, the stronger the counterforce will be.

Pedal Feel With Reaction Disc

In the power booster shown in figure 13-21, the counterforce acts on a rubber reaction disc in the power piston. The reaction disc compresses under the force, flexes outward into the space occupied by the extension on the end of the valve plunger, and moves the plunger rearward. This seats the backside of the plunger against the floating control valve to close the atmosphere control port and prevents further pressure increase in the rear half of the power chamber. At the same time, the reaction disc acting on the valve plunger transmits feedback to the brake pedal through the input pushrod.

If more braking power is needed, further pressure on the brake pedal moves the valve plunger forward, opens the atmosphere control port, and increases the pressure differential between the two halves of the power chamber. This increases brake application force, but it also increases the counterforce, causing the reaction disc to deform and again push back against the extension on the front of the valve plunger. The plunger again moves rearward to seal the atmosphere control port, restore the booster to the holding position, and transmit feedback to the brake pedal. This cycle occurs repeatedly as the brake pedal is pushed harder and harder.

Whenever the brakes are applied, the reaction disc is compressed. The harder the brakes are applied, the more the disc is compressed and the greater the amount of feedback it transmits to the brake pedal. Engineers specify both the thickness of the reaction disc and the hardness of the rubber it is made of to provide a pedal feel that is proportional to brake application.

Pedal Feel With Reaction Plate and Levers

In the boosters shown in figure 13-22, the counterforce acts on a reaction plate mounted to the rear end of the output pushrod. The reaction plate then pushes against a series of levers that swing around their pivots and extend rearward to contact the end of the valve plunger. The levers push the valve plunger rearward and seat the backside of the plunger against the floating control valve; this closes the atmosphere control port and prevents further pressure increase in the rear half of the power chamber. It also transmits feedback to the brake pedal through the plunger and input pushrod.

Just as with the rubber reaction disc described previously, the reaction plate and lever mechanism cycles repeatedly as braking pedal pressure is increased. And just as the reaction disc is always compressed when the brakes are applied, the levers are always deflected to a greater or lesser degree during brake application to provide pedal feel proportional to brake application force.

MULTIPLIER VACUUM BOOSTERS

The **multiplier vacuum booster,** sometimes called an auxiliary booster, is found on some older imported cars, as well as motor homes and medium-duty trucks. The multiplier booster, figure 13-23, installs in the hydraulic lines between the brake master cylinder and the wheel friction assemblies. Hydraulic pressure from the master cylinder controls the vacuum and atmosphere ports in the booster, and the booster then actuates the wheel brakes through a secondary master cylinder attached to it.

Because a multiplier booster is a self-contained assembly without external levers or linkages, it can be installed almost anywhere on the vehicle. Imported cars sometimes use a location in the engine compartment however, most motor homes and medium-duty trucks install the multiplier booster along the frame rails under the vehicle. On vehicles with dual braking systems, two multiplier boosters are used, one for each hydraulic circuit.

HYDRAULIC BOOSTERS

The hydraulic booster was introduced in the early 1970s because of the crude emission controls on many larger engines. In the days before engine control computers and three-way catalytic converters, vehicle manufacturers used retarded ignition timing and camshafts with large amounts of valve overlap to meet emission standards. These measures reduced intake manifold vacuum to the point where vacuum booster operation became marginal.

Hydraulic boosters have several factors in their favor. They are generally smaller than vacuum boosters and, therefore, easier to fit into the engine compartments of downsized cars. In addition, much greater assist is available from a hydraulic booster than from a vacuum booster. This makes hydraulic boosters especially well suited for applications that require high master cylinder pressure, such as cars with four-wheel disc brakes, and medium-duty trucks. Finally, hydraulic boosters are ideally suited for use with diesel engines that have no intake manifold vacuum.

Hydraulic brake boosters use an existing power supply—hydraulic pressure from the power steering pump. However, some hydraulic boosters are powered by a dedicated electric motor and pump assembly attached to the booster itself. Modern hydraulic power boosters are divided into two catagories:

- Mechanical-hydraulic
- Electro-hydraulic.

MECHANICAL-HYDRAULIC BOOSTERS

The Bendix Hydro-Boost I and II units are the most common **mechanical-hydraulic boosters.** These boosters mount on the bulkhead between the brake pedal and the master cylinder, figure 13-24, in the same location as a vacuum booster. Mechanical-

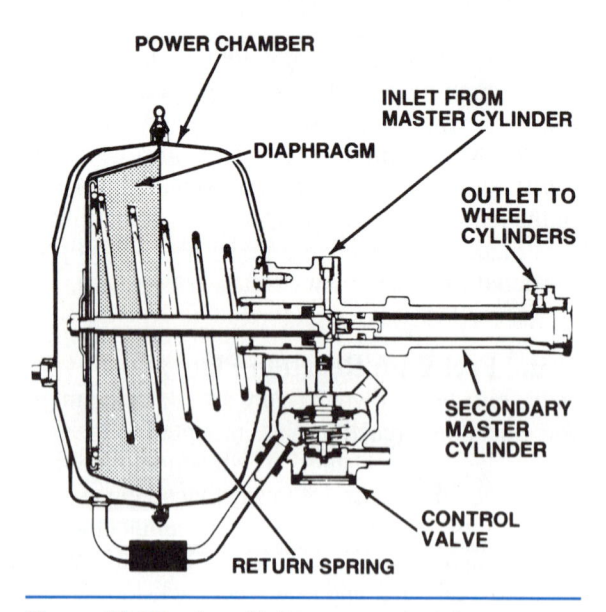

Figure 13-23. A multiplier vacuum booster.

Power Brakes

hydraulic boosters do not change the brake or steering systems in any substantial way from those systems used with a vacuum booster. Although the capacity of the power steering pump is increased, the steering box itself is not changed.

To provide power, three fluid lines are routed to the power booster, figure 13-24. One supplies hydraulic pressure from the power steering pump to the booster. The second routes hydraulic pressure from the booster to the power steering gear. The third is a low-pressure fluid return line from the booster to the power steering pump reservoir. The overall system is designed so the power brake and steering systems do not interfere with one another, whether in use or at rest.

A hydraulic power booster, figure 13-25, is a relatively simple mechanism that consists of input and output pushrods, and four basic operating

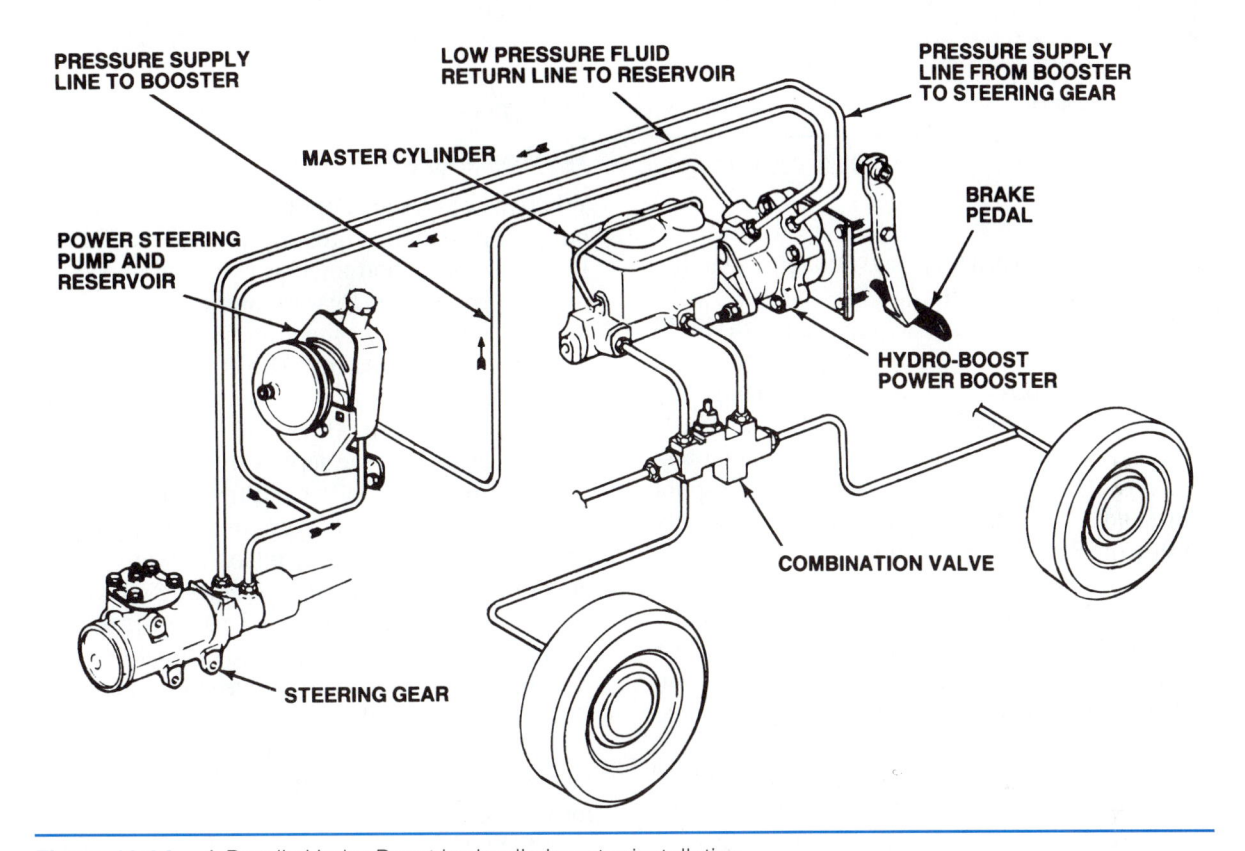

Figure 13-24. A Bendix Hydro-Boost hydraulic booster installation.

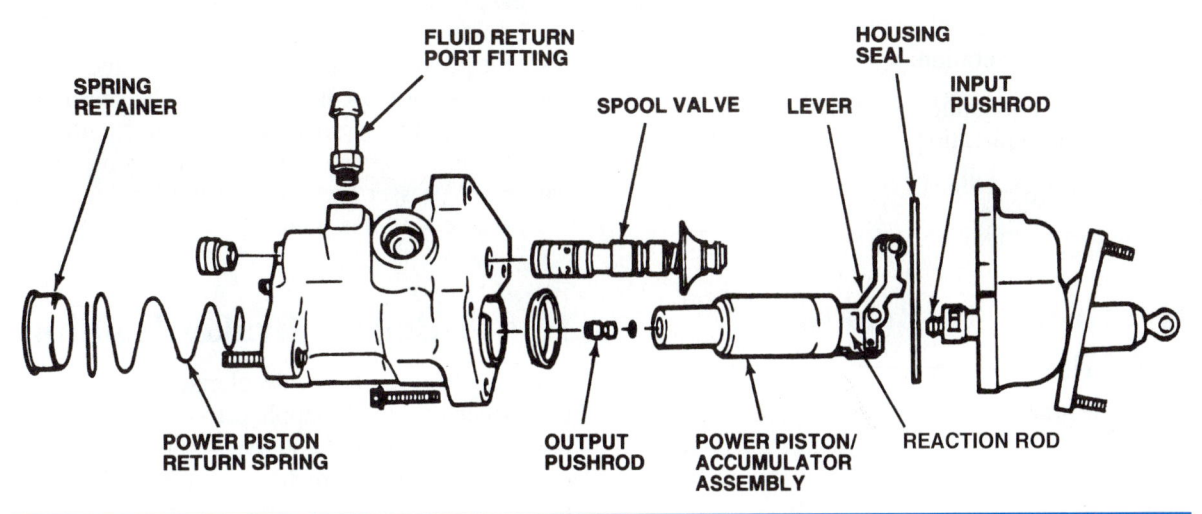

Figure 13-25. A Bendix Hydro-Boost II hydraulic power booster.

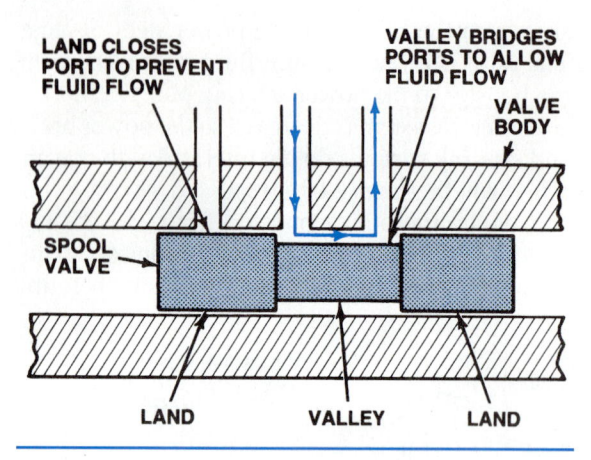

Figure 13-26. The basic operation of a spool valve.

parts: a spool valve, a lever and linkage, a reaction rod, and a power piston.

The **spool valve** is a machined rod with raised "lands" and cut-away "valleys." As the spool valve changes position in its bore, the lands and valleys align with various ports in the valve bore to prevent or facilitate the flow of hydraulic pressure through the booster, figure 13-26. The spool valve and its bore are very finely finished and rely on this finish to provide sealing.

The movement of the spool valve is governed by the lever. The bottom of the lever pivots on the power piston, while a lever pin near the center of the lever is attached to the reaction rod. The reaction rod slides in a bore machined into the rear of the power piston; movement of the reaction rod causes the lever to pivot and thus changes the position of the spool valve in its bore. The power piston slides in a bore in the booster housing and is normally held in the rearward position by a return spring.

Mechanical-hydraulic power boosters have five phases of operation:

- Brakes not applied
- Brakes application
- Brakes holding
- Brake release
- Reserve braking.

Hydro-Boost—Brakes not Applied

When the brakes are not applied, figure 13-27, the spool valve is held in a rearward position by its return spring. Approximately 100 to 150 psi of hydraulic pressure enters the booster from the power steering pump at this time; however, a valley on the spool valve allows the pressure to pass straight through to the steering box, and the lands on the spool valve prevent any pressure from entering the power chamber. To ensure that any pressure leakage past the spool valve does not create application force, the power chamber is open to the low-pressure return line through a vent port that leads to the hollow center of the spool valve.

Hydro-Boost— Brake Application

When the brakes are applied, figure 13-28, the input pushrod applies pressure to the reaction rod. Because it is not bottomed in the power piston, the reaction rod moves forward and causes the lever to pivot on the power piston, moving the spool valve forward in its bore. Movement of the spool valve does two things: It closes the power chamber vent port, and it allows fluid pressure to pass through a valley on the spool valve into the hollow center of the spool valve and from there into the power chamber. The pressure in the power chamber then moves the power piston forward to help apply the output pushrod to the master cylinder.

At the same time the spool valve moves forward to allow fluid pressure into the power chamber, the valley shape on the valve begins to restrict fluid flow through the booster to the power steering gear. The further the valve moves, the greater the restriction. Because the power steering pump moves the same volume of fluid with each revolution, the restriction in flow demands that pressure increase so all of the fluid can be moved through the booster. As system pressure increases, so too does the pressure in the power chamber and, therefore, the amount of power assist. The restriction provided by the spool valve can cause hydraulic pressure to rise as high as 1450 psi in some applications. A pressure relief valve prevents pressure increase beyond this point.

Hydro-Boost—Brakes Holding

Once the desired rate of deceleration is achieved, no further pressure is applied to the brake pedal, and the input pushrod and reaction rod are held in a fixed position, figure 13-29. As pressure in the power chamber increases, the power piston moves

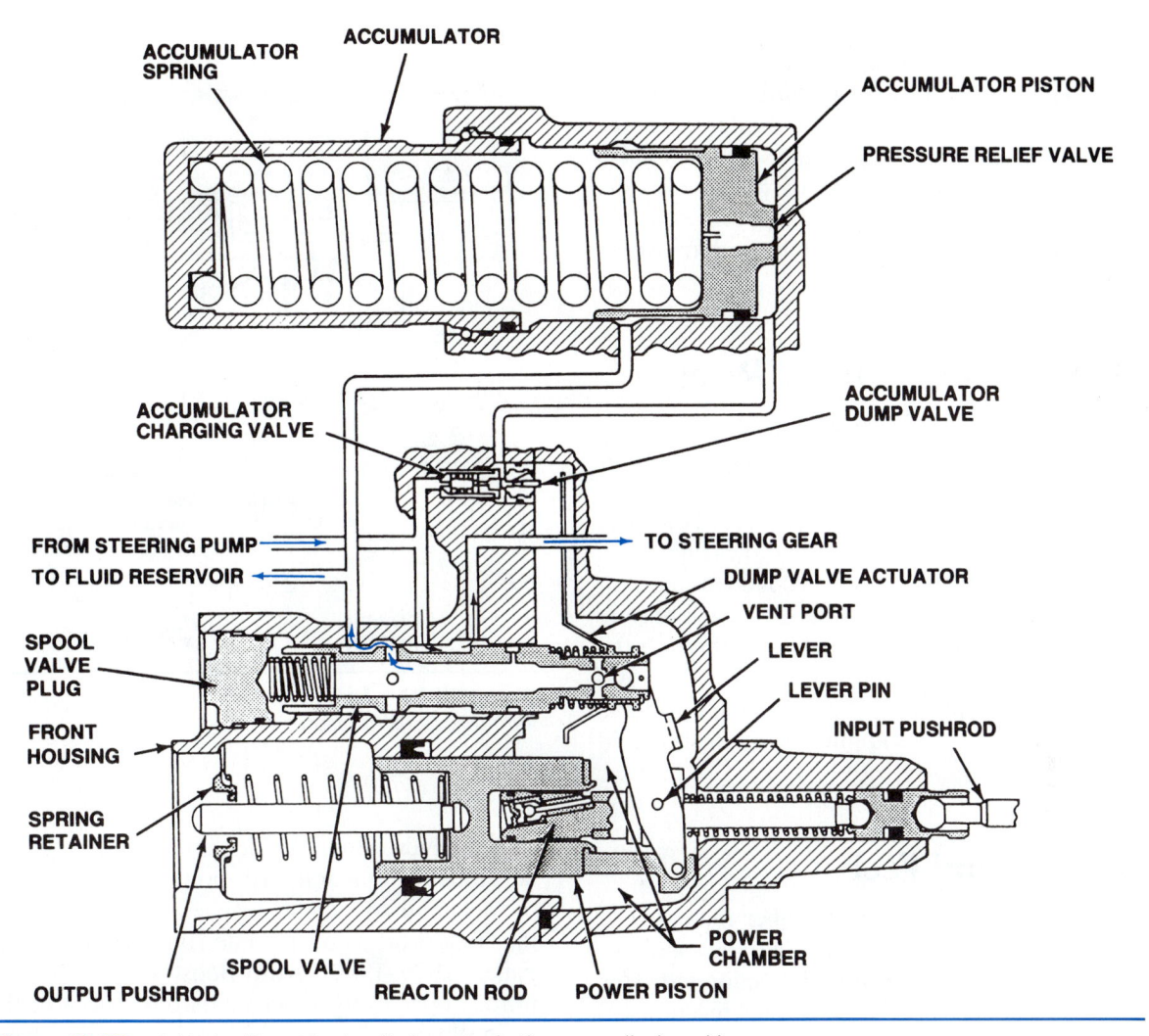

Figure 13-27. A Hydro-Boost hydraulic booster in the unapplied position.

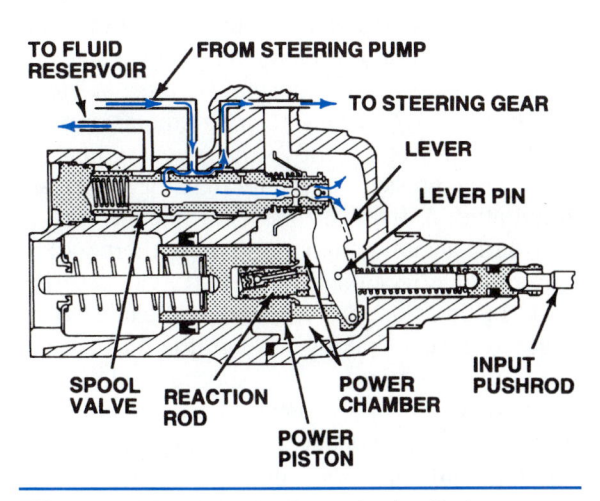

Figure 13-28. A Hydro-Boost hydraulic booster as the brakes are applied.

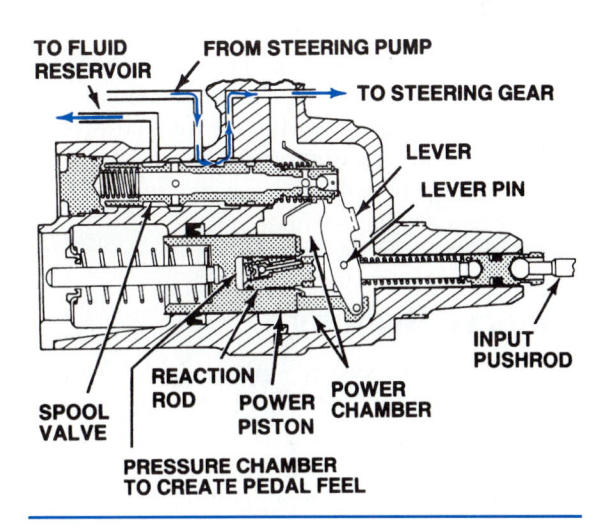

Figure 13-29. A Hydro-Boost hydraulic booster in the holding position.

forward, causing the lever to pivot around the lever pin and move the spool valve rearward in its bore. This closes the port that allows additional pressure into the power chamber, and because the vent port near the back end of the spool valve also remains closed, the amount of pressure in the power chamber (and, therefore, power assist) is held constant. This is the holding position of the power booster.

Hydro-Boost—Brake Release

When the brakes are released, the input pushrod and reaction rod travel rearward causing the lever to pivot on the power piston and move the spool valve all the way rearward in its bore. This does two things. It closes the port that allows fluid pressure into the power chamber, and it opens the vent port near the back end of the spool valve. Pressure in the power chamber then escapes through the hollow center of the spool valve into the low-pressure return line to the reservoir. The booster is then in the brakes-not-applied position shown in figure 13-27.

Brake Pedal Feel

As pressure in the power chamber increases, it passes through a small passage into the space between the power piston and the reaction rod. This creates a counterforce that moves the reaction rod and lever rearward, causing the spool valve to also move rearward in its bore and thus moderate application force. The counterforce also acts directly through the reaction rod and input pushrod to provide the driver with pedal feel. Because the hydraulic pressure in front of the reaction rod is acting on only a small part of the total power piston area, the amount of force fed back to the pedal is less than, but still proportional to, the total braking force being applied.

Reserve Braking

All Hydro-Boost systems are equipped with an **accumulator** that stores hydraulic pressure to provide reserve stopping ability in the event power is lost to the booster. This can be caused by an engine stall, a broken pump drive belt, a burst hose, or a pump mechanical failure. Hydro-Boost I systems have two or three stops in reserve, Hydro-Boost II systems have one or two.

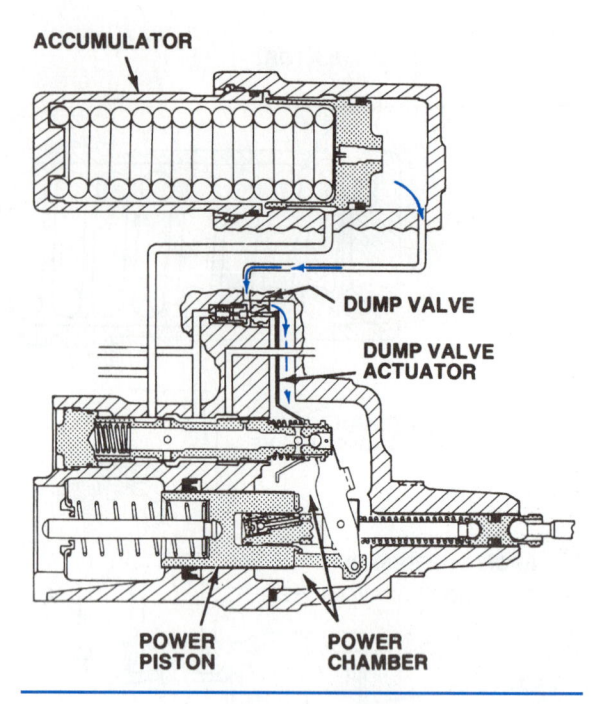

Figure 13-30. An accumulator gives the Hydro-Boost hydraulic booster reserve braking power.

The Hydro-Boost I units have a spring-loaded accumulator similar to that shown in figure 13-30; the accumulator can be attached to the outside of the booster or remote mounted on the inner fender panel. 1979–1980 Hydro-Boost I units have a nitrogen-gas-charged accumulator in place of the spring-loaded design. With the introduction of Hydro-Boost II in 1981, the gas-charged accumulator was integrated into the power piston inside the booster, figure 13-25. This makes Hydro-Boost II units both smaller and lighter than earlier models.

The accumulator is charged by power steering pump pressure during normal braking or steering, figure 13-27. Pressure enters the accumulator through a check valve that opens when the supply line pressure is greater than that in the accumulator. A fixed orifice controls the rate at which the accumulator is charged so there will be a minimal effect on fluid flow to the steering gear. In addition, the accumulator can also be charged during braking through the backside of the dump valve, which is exposed in the boost pressure chamber. If pressure in the chamber during braking rises above that in the accumulator, the dump valve will unseat and open a direct passage to the accumulator. A relief valve vents the accumulator to the low-pressure fluid return line if pressure rises above approximately 1400 psi.

Power Brakes

■ Gimme a Brake, Jake!

The Jacobs brake, commonly called the Jake brake, is often used on diesel engines in large trucks. Although it is not actually a brake system power booster, the Jake brake does create additional braking force at the driving wheels. It does so by changing the engine from a power *producing* air pump into a power *absorbing* air pump.

To accomplish, this, the Jake brake opens the exhaust valve as the piston nears the top of the compression stroke. When this happens, the potential energy in the hot compressed air is released through the exhaust system, and any fuel injected into the cylinder cannot ignite. As a result, the engine produces no power, and the force required to turn the engine and pump air through it is absorbed from the motion of the vehicle through the driveline. In this manner, the Jake brake is able to use up to 100 percent of the engines's rated horsepower to slow the driving wheels.

The Jake brake is especially effective when descending mountain roads because it provides constant braking and will not fade. Because this takes part of the load off the service brake system, the Jake brake can extend the life of brake linings as much as two to three times. The smoother braking provided by a Jake brake also improves tire life by about 5 percent and gives better control on snow and ice.

If braking is attempted and power steering pump pressure is unavailable, extra pedal travel causes an actuator on the spool valve sleeve to mechanically open the accumulator dump valve, figure 13-30. This releases accumulator pressure directly into the power chamber to actuate the power piston and apply the brakes. Once accumulator pressure is depleted, the reaction rod will bottom in the power piston; this creates metal-to-metal contact through the booster to allow full manual braking.

ELECTRO-HYDRAULIC BOOSTERS

Although mechanical hydraulic boosters are effective, they are bulky, heavy, noisy, and provide additional opportunities for fluid leaks to develop. **Electro-hydraulic (E-H) boosters,** figure 13-31,

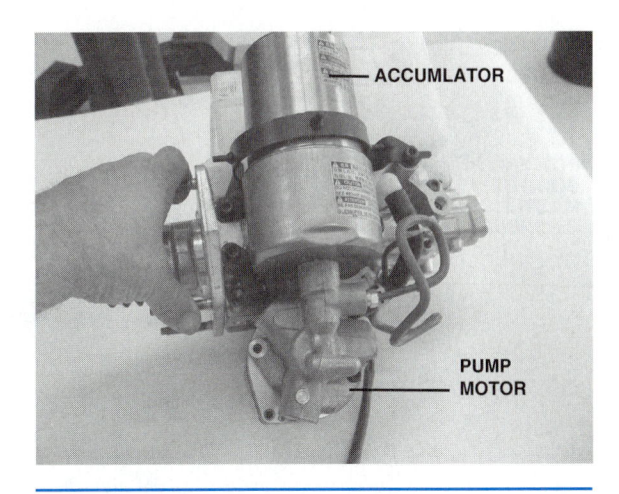

Figure 13-31. An electro-hydraulic power booster showing the pump motor and accumulator.

are neat, compact, self-contained systems that do not require external hydraulic lines because they use an electric pump to provide hydraulic pressure for the booster. E-H boosters also use brake fluid as their operating fluid, which eliminates any chance of power steering fluid and brake fluid becoming accidentally mixed. Finally, E-H boosters provide an excellent reserve of braking power in the event power to the booster is lost.

There are actually two types of E-H boosters: basic models used with conventional brake systems and those that are integrated into an antilock braking system. Basic E-H power boosters are covered in the next section; the antilock system designs are dealt with in Chapter 14.

THE POWERMASTER BRAKE BOOSTER

One popular E-H booster is the General Motors Powermaster, figure 13-32. The Powermaster mounts to the firewall in the conventional location, but its resemblance to other boosters ends there. The Powermaster is not a separate power booster assembly, it is a completely integrated brake actuating system that incorporates the hydraulic booster into the master cylinder. The front part of the Powermaster unit contains a conventional dual master cylinder; the rear part contains the hydraulic booster assembly. Attached to the body of the Powermaster are the supplemental components that complete the system: the fluid reservoir, electric pump, dual pressure switch, and accumulator.

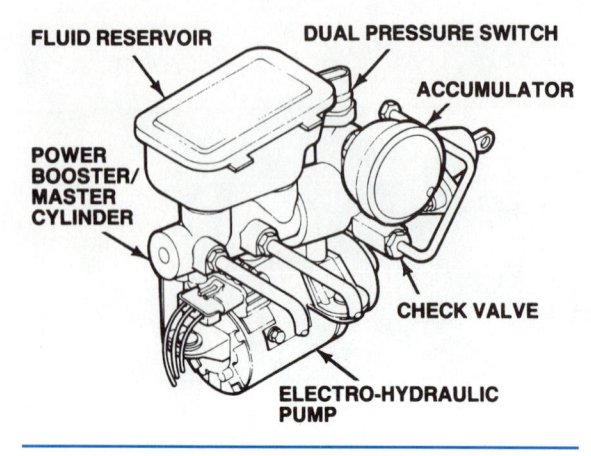

Figure 13-32. A Powermaster power brake system.

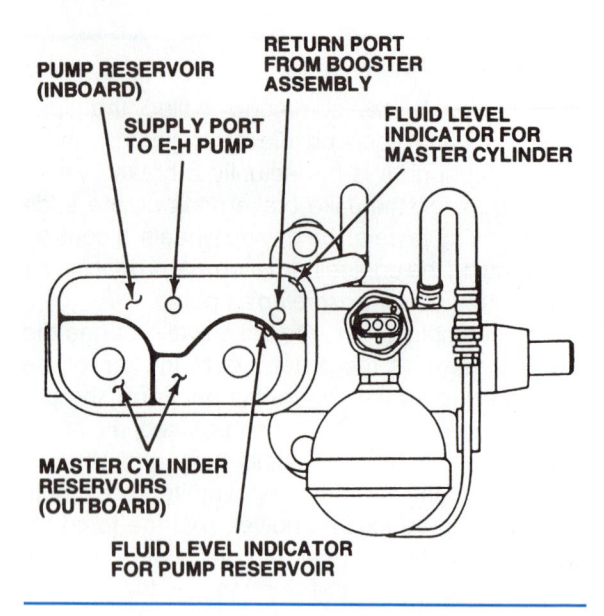

Figure 13-33. The Powermaster brake fluid reservoir.

Powermaster Fluid Reservoir

Because the Powermaster uses brake fluid as its operating fluid, the booster and master cylinder share a common reservoir, figure 13-33. The outboard side of the reservoir is divided into two chambers that serve the primary and secondary circuits of the brake hydraulic system. The inboard side of the reservoir is one large chamber that holds the fluid required to charge the accumulator and provide power assist. There are two ports in the bottom of the booster reservoir. The forward port supplies fluid through a short hose to the electric pump. The rear port allows fluid return from the booster assembly. Because the accumulator is normally charged, the fluid level in the reservoir will appear low under most conditions. Fluid returns to the reservoir as the power booster operates during brake application.

Although a single reservoir housing is used for the entire Powermaster unit, fluid is not allowed to flow from one chamber to another and is never mixed internally between the booster and brake hydraulic systems. The reason for this is that the reservoir cover has a rubber diaphragm to isolate only the master cylinder brake fluid from the air. Because fluid flows in and out of the booster reservoir frequently and very rapidly, that chamber is vented directly to the atmosphere. Moisture absorbed into the booster fluid is not a problem because that fluid is never exposed to extreme temperatures that would cause it to boil. However, for this reason, fluid should never be moved from the booster side of the reservoir into the master cylinder fluid chambers.

Powermaster Pump and Accumulator

The E-H booster pump bolts to the underside of the Powermaster and consists of a vane-type pump driven by a powerful series-wound electric motor. The motor requires a great deal of current and is protected by a 30-amp fuse. The E-H pump runs at irregular intervals to keep the accumulator charged. Pump operation is controlled by the dual-pressure switch that maintains accumulator pressure between 510 and 675 psi. The switch turns the pump on when accumulator pressure drops below 510 psi and shuts the pump off when pressure reaches 675 psi. A check valve retains pressure in the accumulator after the pump shuts off.

The accumulator, figure 13-34, threads into the Powermaster body and contains a diaphragm that is pressurized on one side with nitrogen gas. Unlike Hydro-Boost systems, Powermaster boosters use accumulator pressure to supply *all* power assist pressure, not only reserve pressure.

POWERMASTER OPERATION

The booster section of the Powermaster, figure 13-35, consists of the power piston and the reaction body. The power piston is at the back of the Powermaster bore and receives the force applied by

Power Brakes

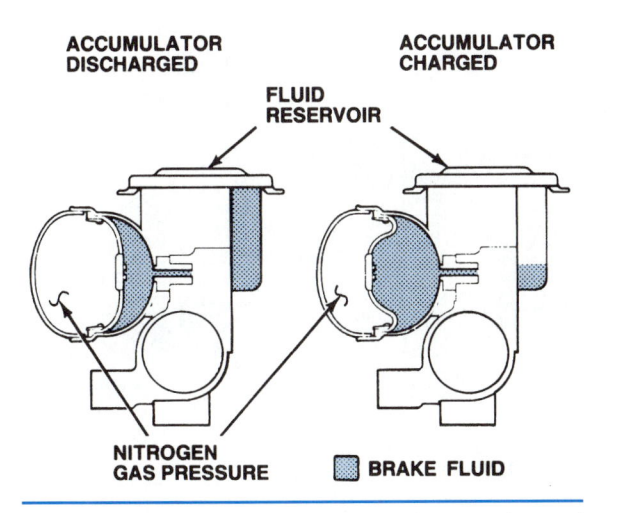

Figure 13-34. The Powermaster nitrogen-charged accumulator.

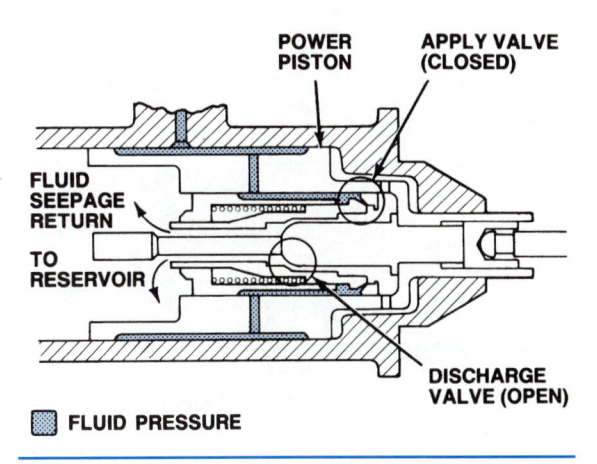

Figure 13-36. A Powermaster hydraulic booster in the unapplied position.

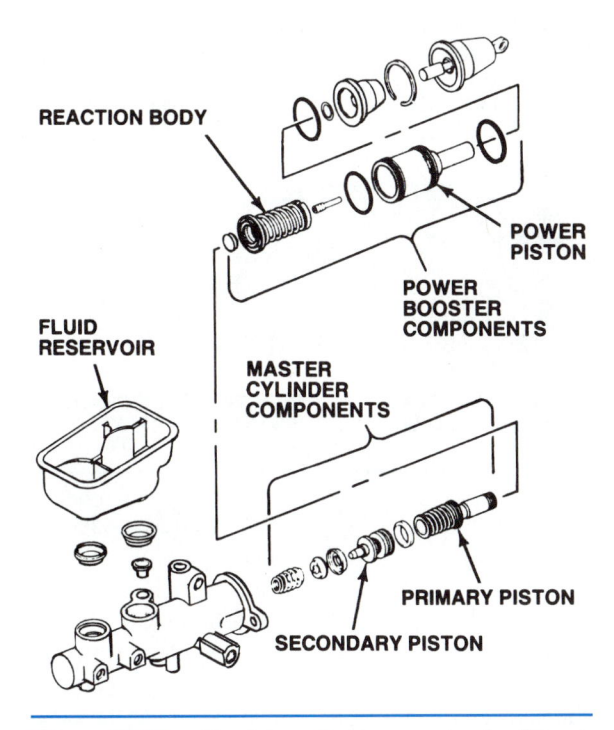

Figure 13-35. The internal components of a Powermaster system.

the brake pedal pushrod. The reaction body mounts between the power piston and master cylinder primary piston, and provides pedal feedback. The reaction body also prevents the brakes from being applied too hard.

The Powermaster booster has four phases of operation:

- Brakes not applied
- Brake application
- Brakes holding
- Brake release.

Powermaster—Brakes Not Applied

When the brakes are not applied, figure 13-36, high-pressure fluid from the accumulator enters the power piston through a port in its surface. However, because the apply valve is held closed by spring tension, the pressure cannot pass through to the backside of the piston and create brake application force. Any fluid that does seep past the apply valve passes out of the booster and back to the fluid reservoir through the discharge, valve which is held open by spring tension.

Powermaster—Brake Application

When the brakes are applied, figure 13-37, the pedal pushrod first moves the discharge valve forward against spring tension and seats it inside the apply valve body; this prevents pressure from escaping the booster assembly. Once the discharge valve seats, the apply valve body moves forward and opens the apply valve; this allows hydraulic pressure to pass through additional ports to the backside of the power piston. The pressure then forces the power piston forward to help apply the brakes.

Powermaster—Brakes Holding

Once the desired rate of deceleration is achieved, no further pressure is applied to the brake pedal, and the input pushrod, discharge valve, and apply valve body are held in a fixed position, figure 13-38. As long as pressure continues to enter the backside of the power piston, the piston moves forward. When it has moved far enough, the apply

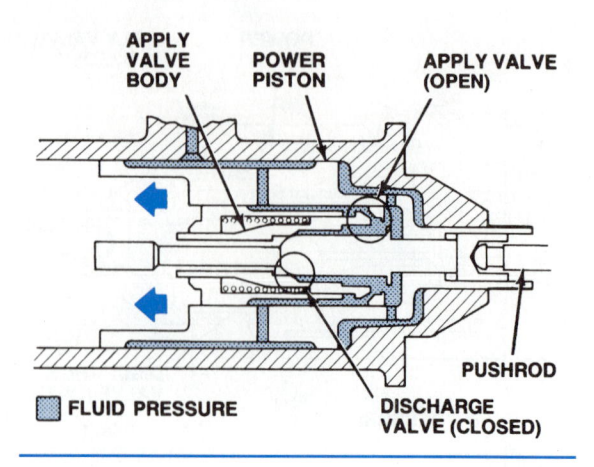

Figure 13-37. A Powermaster hydraulic booster as the brakes are applied.

valve seats and prevents further pressure from entering. This is the Powermaster holding position in which both the discharge and apply valves are closed.

Powermaster—Brake Release

When the brakes are released, spring tension closes the apply valve; this prevents further pressure from entering the booster. Spring tension also opens the discharge valve; this allows the pressure in the booster to bleed back into the fluid reservoir. Once all pressure is released, the booster is in the brakes-not-applied position as shown in figure 13-36

Brake Pedal Feel

Application force from the power piston is transmitted to the master cylinder primary piston through the reaction body, figure 13-39, which seats against a ridge inside the power piston. The reaction body consists of two pieces that are locked together and held at a fixed length by a strong spring. The center of the reaction body contains a rubber reaction disc that contacts the end of the master cylinder primary piston and a reaction piston that transmits force from the reaction disc to the forward end of the discharge valve.

As the brakes are applied, the discharge valve moves forward with the power piston, and hydraulic pressure helps to keep it seated. At the same time, the rubber reaction disc that rides against the master cylinder primary piston transmits a counterforce through the reaction piston against the forward end of the discharge valve. As the brakes are

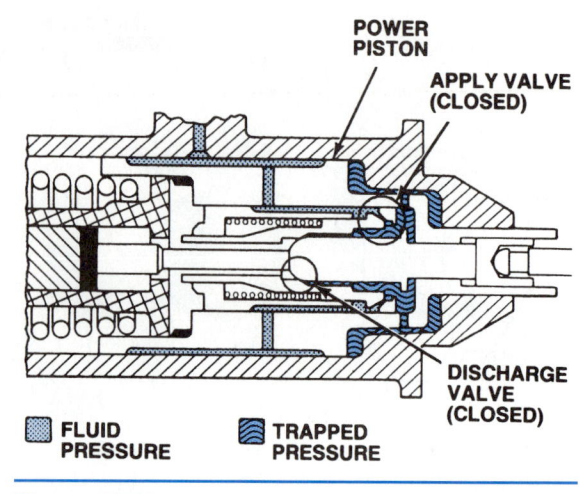

Figure 13-38. A Powermaster hydraulic booster in the holding position.

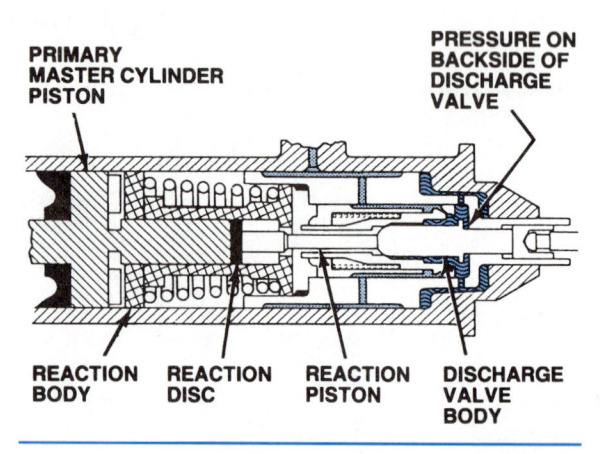

Figure 13-39. Pedal feel provided by the reaction disc and piston in a Powermaster hydraulic booster.

applied harder, the reaction piston unseats the discharge valve and allows pressure to bleed from behind the power piston until the apply valve seats and the booster regains the holding position. At the same time, pedal feedback is transmitted through the discharge valve and input pushrod to the brake pedal.

When the brake pedal is pushed harder, the apply valve reopens and greater pressure is allowed behind the power piston to provide brake application force and hold the discharge valve closed more tightly. However, once the counterforce becomes strong enough, it again opens the discharge valve to stabilize braking and transmit pedal feedback to the driver. Because hydraulic pressure on the discharge valve is acting on only a small area compared to that of the total power piston, the amount of force fed back to the pedal is less than, but proportional to, the total braking force being applied.

Power Brakes

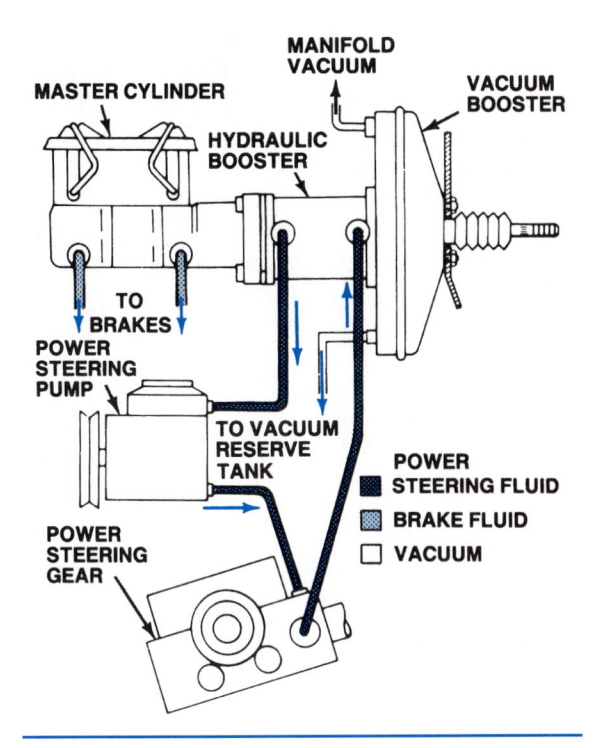

Figure 13-40. A dual-power brake system.

Whenever the brakes are applied, the rubber reaction disc is compressed to some degree to apply pedal feedback. However, if the brakes are applied so hard that there is a possibility of mechanical damage to the system, the two pieces of the reaction body will compress together against spring tension. This shortens the length of the reaction body and allows the reaction piston to unseat the discharge valve to vent excess pressure to the fluid reservoir.

DUAL-POWER BRAKE SYSTEMS

A **dual-power brake system**, figure 13-40, combines a vacuum power booster and a hydraulic booster in tandem. The vacuum booster bolts to the engine bulkhead and is actuated by the brake pedal; the hydraulic booster bolts to, and is actuated by, the vacuum booster. The hydraulic booster then actuates the master cylinder to apply the wheel friction assemblies. The vacuum and hydraulic boosters in a dual-power system are identical to those used alone in single systems.

Dual-power systems are standard on many school buses, and optional on may medium-duty trucks. A dual-power system provides about twice the power assist of either system alone, and it considerably lightens the pedal pressure needed to stop large, heavy vehicles equipped with hydraulic brakes. If more power is needed than even a dual-power system can supply, as on heavy-duty tractor-trailer rigs, air brakes are used in place of hydraulic brakes.

SUMMARY

Power brakes are brake systems fitted with a power booster that reduces the amount of pedal pressure needed to apply the brakes and stop the car. Power brakes were developed because the weight and speed of cars increased, disc brakes were introduced, and people wanted cars that were easier to drive.

Power boosters are not the only method used to reduce pedal pressures or increase braking force. In addition to simply pushing harder on the brake pedal, greater mechanical or hydraulic advantage can also achieve these ends. The problem with these methods is that they reduce brake pedal reserve. A power booster actually increases pedal reserve as it decreases the required pedal force.

There are two kinds of power boosters: vacuum boosters and hydraulic boosters. Vacuum boosters create brake application force with a power chamber that is divided in half by a flexible diaphragm. Low pressure on one side of the diaphragm and atmospheric pressure on the other creates a pressure differential that moves the diaphragm to the low-pressure side and creates brake application force.

Vacuum booster power chambers come in two types. The atmospheric-suspended design has atmospheric pressure on both sides of the diaphragm when the brakes are not applied. The vacuum-suspended design has vacuum on both sides under the same conditions. All modern booster power chambers are of the vacuum-suspended variety. Vacuum can be supplied to the power chamber from the engine intake manifold, or a separate mechanical or electric pump. A check valve contains the vacuum in the booster, and a filter may be used in the vacuum supply line to trap fuel vapors.

There are two types of vacuum boosters: integral and multiplier. The integral type mounts on the bulkhead behind the master cylinder and is the most common design. Integral vacuum boosters are actuated by the brake pedal. The multiplier vacuum booster installs in the hydraulic lines between the master cylinder and the wheel brakes. Multiplier boosters are actuated by hydraulic pressure. A tandem vacuum booster of either type uses two diaphragms for a more compact package and increased power assist.

Hydraulic boosters replace vacuum boosters in applications that have limited space or an inadequate vacuum supply. Hydraulic boosters can produce much higher assist levels than vacuum boosters. Hydraulic boosters operate off power steering pump pressure or a separate electrically powered hydraulic pump. All hydraulic boosters store pressure in an accumulator. Those boosters served by the power steering pump use the accumulator for reserve braking only. Boosters charged by an electric pressure pump rely on the accumulator for all boost pressure.

Dual-power brake systems use a vacuum booster and a hydraulic booster in tandem. This arrangement provides extra power assist for heavy vehicles.

Review Questions

Choose the letter that represents the best possible answer to the following questions:

1. Power brakes were developed because:
   a. Larger cars needed more kinetic energy
   b. Friction assemblies became heavier
   c. Disc brakes had no servo action
   d. Luxury cars were unpopular

2. Which of the following is *not* a way to reduce brake pedal pressure or increase brake application force?
   a. Increase the pedal leverage
   b. Use wheel cylinders with a larger bore
   c. Install larger brake shoes
   d. Use a master cylinder with a smaller bore

3. A power booster in a brake system will generally lead to:
   a. Increased brake pedal reserve
   b. An immediate increase in pedal pressure if power to the booster is interrupted
   c. Both a and b
   d. Neither a nor b

4. The intake manifold of a gasoline engine:
   a. Has no manifold vacuum
   b. Usually contains low pressure
   c. Contains higher pressure when the throttle is closed than when it is open
   d. None of the above

5. A value of 14.7 psi is most closely related to:
   a. A pressure differential
   b. Vacuum
   c. Atmospheric pressure
   d. Manifold pressure

6. A diesel engine is likely to have:
   a. An auxiliary vacuum pump
   b. Less than 5 in. of vacuum at idle
   c. Both a and b
   d. Neither a nor b

7. A vacuum booster uses a _____ to create brake application force.
   a. Pressure differential
   b. Power chamber
   c. Diaphragm
   d. All of the above

8. When the brakes are not applied, a vacuum-suspended power booster has:
   a. Atmospheric pressure in both halves
   b. A partial vacuum in both halves
   c. Atmospheric pressure in the front half and vacuum in the rear half
   d. Vacuum in the front half and atmospheric pressure in the rear half

9. A(n) _____ can always be found somewhere in the vacuum supply line between the intake manifold and the power booster.
   a. Vacuum reservoir
   b. Activated-charcoal filter
   c. One-way check valve
   d. All of the above

10. An Integral vacuum brake booster:
    a. Installs between the brake pedal and master cylinder
    b. Has a larger diameter if ft is a tandem booster
    c. Both a and b
    d. Neither a nor b

11. Which of the following does *not* take place during brake application with an integral vacuum booster?
    a. The floating control valve closes the vacuum control port
    b. Air enters the rear half of the power chamber
    c. The input pushrod opens the vacuum control port
    d. The valve plunger opens the atmosphere control port

12. When an integral vacuum booster is in the holding position:
    a. The reaction disc is bypassed
    b. A pressure differential exists between the two halves of the power chamber
    c. The atmosphere control port is open and the vacuum control port is closed
    d. All of the above

13. Brake pedal feel in an integrated vacuum booster may be provided by:
    a. Compression of a rubber disc
    b. Levers acting on the floating control valve
    c. Both a and b
    d. Neither a nor b

14. When a vacuum booster is applying the maximum amount of power assist possible:
    a. It is at the run-in point
    b. Pressure is equal in the two halves of the power chamber
    c. Brake pedal feel becomes harder
    d. None of the above

15. Multiplier vacuum power boosters:
    a. Apply the wheel brakes hydraulically
    b. Are actuated hydraulically
    c. Do not require external mechanical linkages
    d. All of the above

16. Hydraulic brake power boosters:
    a. Can provide greater brake application force than a vacuum booster
    b. Are larger than vacuum boosters
    c. Are not suited to use with diesel engines
    d. Are never used with four-wheel disc brakes

17. Which of the following is *not* one of the three fluid lines attached to a mechanical-hydraulic booster?
    a. High-pressure line from steering pump
    b. High-pressure line to steering gear
    c. Low-pressure line to steering gear
    d. Low-pressure line to fluid reservoir

18. Hydraulic pressure in a Hydro-Boost booster is controlled by the:
    a. Reaction rod
    b. Pivoting lever
    c. Power piston
    d. All of the above

19. Pedal feel in the Hydro-Boost booster is provided by:
    a. Hydraulic pressure
    b. The power piston and reaction rod
    c. Both a and b
    d. Neither a nor b

20. During reserve braking, a Hydro-Boost booster:
    a. Obtains accumulator pressure through the hollow center of the spool valve
    b. Accumulator is charged by power steering pump pressure
    c. Both a and b
    d. Neither a nor b

21. Electro-hydraulic power boosters:
    a. Provide excellent pedal reserve
    b. Use the accumulator for all power assist
    c. Are lighter than mechanical-hydraulic boosters
    d. All of the above

22. Which of the following is *not* true of the Powermaster fluid reservoir?
    a. The largest chamber is vented directly to the atmosphere
    b. It is divided into two chambers
    c. The fluid level is usually lower in one of the chambers
    d. Fluid cannot pass between the chambers

23. When the brakes are not applied with a Powermaster booster:
    a. Fluid pressure enters the power piston
    b. The discharge valve returns fluid to the accumulator
    c. The apply valve is closed by hydraulic pressure
    d. None of the above

24. When a Powermaster booster is in the holding position:
    a. The discharge valve is open
    b. There is pressure on the backside of the power piston
    c. Both a and b
    d. Neither a nor b

25. Dual-power brake systems:
    a. Provide four times the assist of a single power system
    b. Commonly have the booster mounted on the frame rail under the vehicle
    c. Use a vacuum booster to actuate the master cylinder
    d. None of the above

14
Antilock Brake Basics

OBJECTIVES

Upon completion and review of this chapter, you will be able to:

- Explain the relationship between braking and tire slip.
- Explain antilock brake limitations.
- Describe the major antilock brake system configurations.
- Explain the difference between integral and nonintegral antilock systems.
- Explain how wheel sensors operate.
- Describe the role the ABS control module plays in the antilock system.
- Explain the ABS control pressure strategy.
- Describe pump motor and accumulator functioning.
- Explain how traction control works.
- Describe other functions that the ABS may control.

KEY TERMS

accumulator
active digital wheel speed sensors
air gap
control module
flash codes
integral
isolation solenoid
magneto-resistive wheel speed sensors
nonintegral
release solenoid
scan tool
select-low prinicple
solenoid
throttle relaxer
tire slip
traction
wheel speed sensors
yaw movement

INTRODUCTION

This chapter discusses the basics of antilock brake systems (ABS). We cover the advantages of ABS, its history, a general description of its theory of operation, and typical ABS components.

ABS CHARACTERISTICS

Antilock brakes increase safety because they eliminate lockup and minimize the danger of skidding, allowing the vehicle to stop in a straight line. ABS also allow the driver to maintain steering

269

control during heavy braking so the vehicle can be steered to avoid an obstacle or another vehicle.

ABS can optimize braking when road conditions are less than ideal, as when making a sudden panic stop or when braking on a wet or slick road. ABS do this by monitoring the relative speed of the wheels to one another. This information is used to modulate brake pressure as needed to control wheel slippage and maintain traction when the brakes are applied.

ABS and Tire Traction

Preventing brake lockup is important because of the adverse effect a locked wheel has on tire **traction**. The brakes slow the rotation of the wheels, but it is friction between the tire and road that stops the car and allows it to be steered. If tire traction is reduced, stopping distances increase, and the directional stability of the car suffers.

Traction is defined in terms of **tire slip**, figure 14-1, which is the difference between the actual vehicle speed and the rate at which the tire tread moves across the road. A free-rolling wheel has nearly zero tire slip, whereas a locked wheel has 100 percent tire slip. When the brakes are applied, the rotational speed of the wheel drops, and tire slip increases because the tread moves across the road slower than the actual vehicle speed. This slip creates friction that converts speed. This slip creates friction that converts kinetic energy into braking and cornering force.

Tire Slip and Braking Distance

On dry or wet pavement, maximum braking traction occurs when tire slip is held between approximately 15 and 30 percent, figure 14-1. On snow- or ice-covered pavement, the optimum slip range is 20 to 50 percent. In each case, if tire slip increases beyond these levels, the amount of traction decreases. A skidding tire with 100 percent slip provides 20 to 30 percent less braking traction on dry pavement, and this is generally true on slippery roads as well. In nearly all cases, the shortest stopping distances are obtained when the brakes are applied with just enough force to keep tire slip in the range where traction is greatest.

Tire Slip and Vehicle Stability

A tire's contact patch with the road can provide only a certain amount of traction. When a car is stopped in a straight line, nearly all of the available traction can be used to provide braking force; only a small amount of traction is required to generate the lateral force that keeps the car traveling in a straight line. However, if a car has to stop and turn at the same time, the available traction must be *divided* to provide both cornering (lateral) and braking force. As shown in figure 14-1, cornering traction decreases as the amount of traction used for braking increases.

No tire can provide full cornering power and full braking power at the same time. When a brake is locked and the tire has 100 percent slip, all of the available traction is used for braking, and none is left for steering. As a result, a skidding tire follows the path of least resistance. This means that if the rear brakes lock, the back end of the car will want to swing around toward the front causing a spin. If the front brakes lock, steering control will be lost and the car will slide forward in a straight line until the brakes are released to again make traction available for steering.

ABS Relation to Base Brakes

An important point to keep in mind about antilock brakes is that on most vehicles, ABS is an "add-on" to the existing base brake system. ABS only comes into play when traction conditions are marginal or during sudden panic stops when the tires lose traction and begin to slip excessively. The rest of the time ABS has no effect on normal driving, handling, or braking.

ABS also makes little difference in the maintenance, inspection, service, or repair of conventional brake system components. A vehicle with ABS brakes uses the same brake linings, calipers, wheel cylinders, and other system components as a vehicle without ABS brakes, the only exception

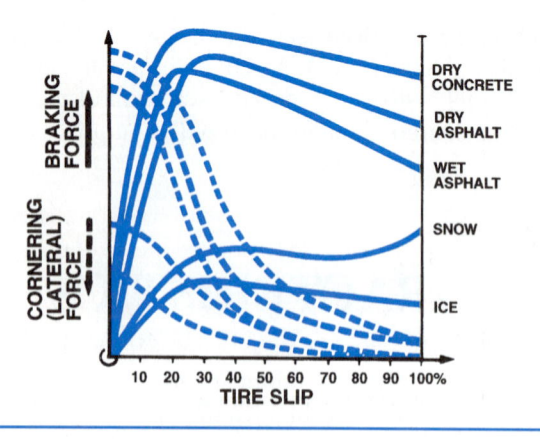

Figure 14-1. Traction is determined by pavement condition and tire slip.

Antilock Brake Basics

being the master cylinder on certain applications. Service and repair procedures for systems are covered in Chapters 12 and 13 of the *Shop Manual*.

All ABS are also designed to be as "fail-safe" as possible. Should a failure occur that affects the operation of the ABS, the system will deactivate itself and the vehicle will revert to normal braking. Thus, an ABS failure will not prevent the vehicle form stopping. Diagnosing ABS problems is also covered in Chapters 12 and 13 of the *Shop Manual*.

ABS Limitations

Antilock brake systems modulate brake application force several times per second to maintain tire slip at a controlled level. On dry pavement, ABS can generally hold tire slip between 5 and 20 percent, figure 14-2. However, not all antilock systems apply and release the brakes at the same rate or with the same degree of control. Although the latest antilock brake systems are very good, they are not perfect and cannot do the impossible. The limitations of antilock brakes fall into two categories:

- System limitations
- Physical limitations.

System Limitations

There are two situations in which an antilock brake system will not provide the shortest stopping distances. The first involves straight stops made on smooth, dry pavement by an *expert* driver. Under these conditions, a skilled driver can hold the tires consistently closer to the ideal slip rate than the antilock system can, figure 14-2. This is possible because current antilock systems may allow the amount of tire slip to drop as low as 5 percent, which is somewhat below the point where maximum tire traction is achieved. However, for the average drive, or under less-than-ideal conditions, antilock brakes will almost always stop the car in a shorter distance.

The other situation in which antilock brakes may not provide the shortest stops is when braking on loose gravel or dirt, or in deep, fluffy snow. Under these conditions, a locked wheel may stop the car faster because loose debris builds up and forms a wedge in front of the tire that helps stop the car, figure 14-3. An antilock system can prevent this wedge from forming.

Physical Limitations

In addition to system limitations, it is important to realize that no matter how good an antilock brake system is, it cannot overcome the laws of physics. The weight and speed of a moving vehicle give it a great deal of kinetic energy, and only so much of that energy can be converted into braking or cornering force at any given time. The limiting factor in this conversion is the traction between the tires and road.

Although a car with four-wheel antilock brakes will stop in very nearly the shortest possible distance, this will still not prevent an accident if the brakes are applied too late to bring the car to a complete stop before impact.

However, because steering control is retained with four-wheel antilock brakes, it may be possible to drive the car around a potential accident while in the process of braking.

Another situation where antilock brakes cannot defy the laws of physics occurs when a car enters a corner traveling faster than it is physically possible to negotiate the turn. In this situation, antilock

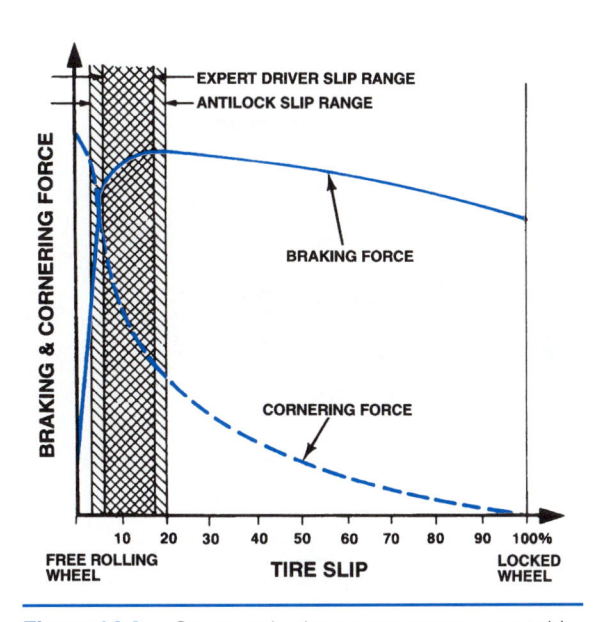

Figure 14-2. On smooth, dry pavement an expert driver can control tire slip more accurately than an antilock system.

Figure 14-3. A wedge of gravel or compressed snow in front of a locked wheel can help stop a car faster.

brakes will not prevent the vehicle from leaving the road. However, they will allow the car to be slowed and steered in the process, thus reducing the severity of the eventual impact, and perhaps allowing the driver to choose what the car is going to hit!

An added consideration in both of these examples is the condition of the road surface. A car without antilock brakes has longer stopping distances and cannot corner as hard on wet and icy roads; *the same is true of a car with antilock brakes.* Once again, although the antilock system provides added braking and steering control, it cannot overcome the physical reality that there is less traction available on a slippery road.

ABS OPERATION

One thing that all ABS controls have in common is their ability to control tire slip by monitoring the relative deceleration rates of the wheels during braking. Wheel speed is monitored by one or more **wheel speed sensors.** If one wheel starts to slow at a faster rate than the others, or at a faster rate than that which is programmed into the an-

■ The Origins of ABS

Within a few years after World War II, the development of heavier aircraft and faster landing speeds led to the development of antilock braking, which would eventually lead to automotive applications. The first such system appeared in 1947 on a B-47 bomber. In 1954, Dunlop Tire in England introduced the first automotive antilock brake system, which was offered on a limited number of Lincolns equipped with power brakes.

In 1968, Ford offered "Sure Track" Electronic Skid Control as an option on the Lincoln Mark III. This was a two-wheel vacuum-actuated antilock system designed to improve the car's straight-line stopping ability. The system worked, but it cycled so slowly that it increased stopping distances on dry pavement. Shortly thereafter, Cadillac introduced its "Brake-Track Master" system and Chrysler offered its "Sure Brake" system on the Imperial. However, all these early systems suffered from slow cycle times and the limited technology that was available at the time.

These early ABSs all used analog computers, which are relatively slow and unreliable compared to today's digital computers. They also had vacuum-actuated pressure modulators, which cannot cycle the brakes as quickly as today's electric solenoid-actuated modulators. Consequently, ABS performance was less than ideal. Even so, these early systems did demonstrate the potential for improved braking under certain driving conditions.

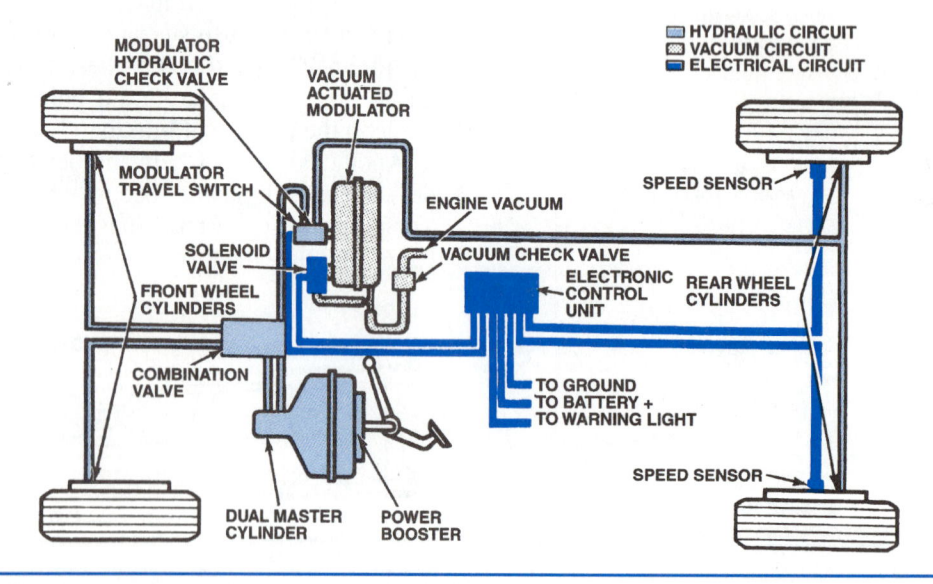

An early Cadillac Brake-Track Master two-wheel antilock system. It used a vacuum-actuated modulator and an analog electronic control module.

Antilock Brake Basics

tilock **control module**, it indicates a wheel is starting to slip and is in danger of losing traction and locking. The antilock system responds by momentarily reducing hydraulic pressure to the brake on the affected wheel or wheels. This allows the wheel to speed up momentarily so it can regain traction. As traction is regained, brake pressure is reapplied to again slow the wheel. The cycle is repeated over and over until the vehicle stops or until the driver eases pressure on the brake pedal.

Electrically operated **solenoid** valves (or motor-driven valves in the case of Delco ABS-VI applications) are used to hold, release, and reapply hydraulic pressure to the brakes. This produces a pulsating effect, which can be felt in the brake pedal during ABS braking. The rapid modulation of brake pressure in a given brake circuit reduces the braking load on the affected wheel and allows it to regain traction to prevent lockup. The effect is much the same as pumping the brakes, except that the ABS does it automatically for each brake circuit and at speeds that would be humanly impossible—up to dozens of times per second depending on the system (some cycle faster than others).

Once the rate of deceleration for the affected wheel catches up with the others, normal braking function and pressure resume, and antilock reverts to a passive mode.

SYSTEM CONFIGURATIONS

Antilock brake systems are configured in different ways, depending on the desired control strategies and drivetrain layout. The system will have a number of channels that allow the ABS hydraulic unit to control the wheels individually or in groups. It is important to note that "channel" always refers to the number of separate or individually controlled ABS hydraulic circuits in an antilock brake system not the number of wheel speed sensor electrical circuits. The number and location of the speed sensors depends on the type of system:

- Four-channel systems have four speed sensors
- Three-channel systems may use three or four speed sensors
- One-channel systems use only one speed sensor.

Four-Channel ABS

On vehicles with fully independent suspensions, each wheel can be controlled individually by the ABS controller. Each wheel is equipped with its own speed sensor. This type of arrangement is called a "four-wheel" or "four-channel" system because each wheel speed sensor provides input for a separate hydraulic control circuit, figure 14-4.

The four-channel ABS is the most complex type of system, requiring dedicated sensors and brake lines for each wheel. However, the precise control of each wheel makes the four-channel system the most effective system, especially when combined with traction and stability control systems.

Three-Channel ABS

A three-channel antilock system controls the front wheels individually and the rear wheels as a pair,

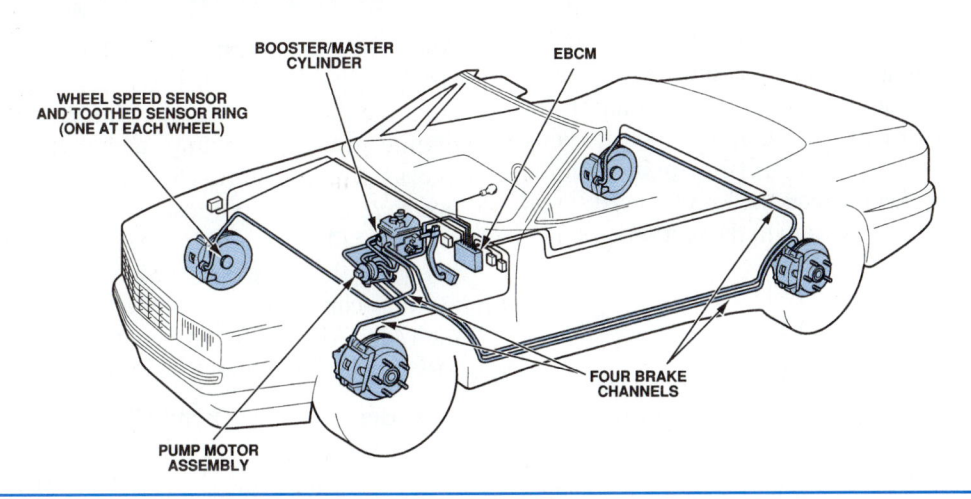

Figure 14-4. Four-channel ABS (Bosch III). (Courtesy of General Motors Corporation, Service and Parts Operations)

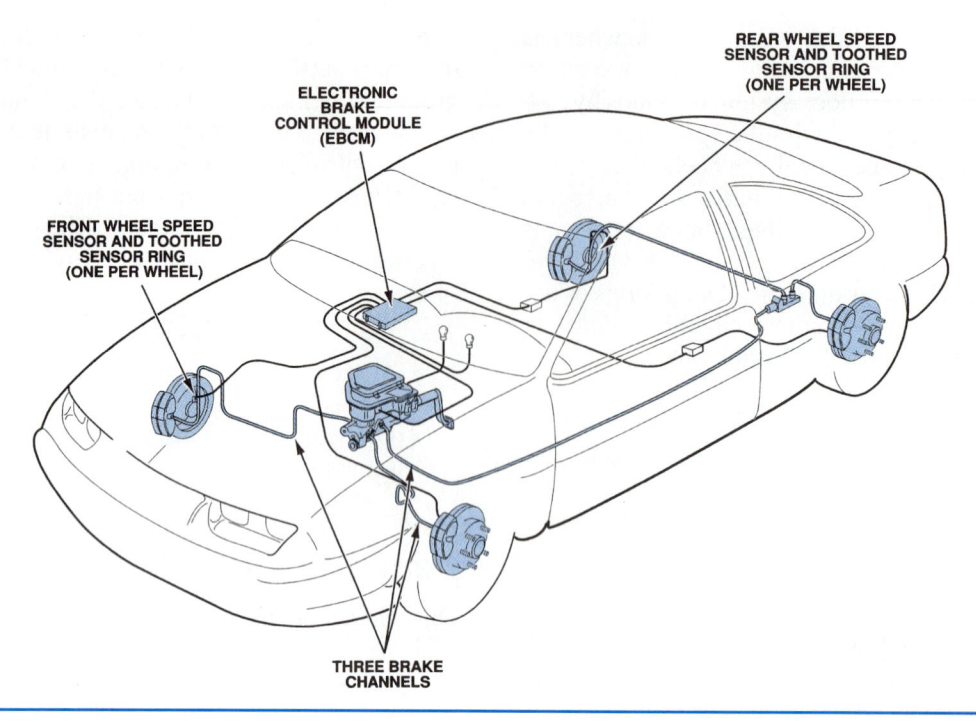

Figure 14-5. Three-channel ABS (Delco Moraine III). (Courtesy of General Motors Corporation, Service and Parts Operations)

figure 14-5. Found on both front- and rear-wheel drive vehicles, the system may have three or four speed sensors.

On rear-wheel drive vehicles, the ABS will have a separate wheel speed sensor for each front wheel but use a common speed sensor for both rear wheels. The rear-wheel speed sensor is mounted in either the differential or the transmission and reads the combined or average speed of both rear wheels. This type of setup saves the cost of an additional sensor and reduces the complexity of the system by allowing both rear wheels to be controlled simultaneously.

On front-wheel drive and some rear-wheel drive vehicles, there are four wheel speed sensor circuits. Although the rear wheels are controlled hydraulically as a pair, the rear wheel speeds are individually monitored by the ABS control module. Using a strategy known as the **select-low principle,** the control module then selects the lowest speed signal to modulate brake pressure to the rear wheels.

One-Channel ABS

Another variation is one-channel, rear-wheel-only ABS that is used on many rear-wheel drive pickups and vans. Ford's version is called "Rear Antilock Brakes" (RABS), whereas GM and Chrysler call theirs "Rear Wheel Anti-Lock" (RWAL), figure 14-6. The front wheels have no speed sensors, and only a single speed sensor mounted in the differential or transmission is used for both rear wheels. Rear-wheel antilock systems are typically used on applications where vehicle loading can affect rear-wheel traction, which is why it is used on pickup trucks and vans. Because the rear-wheel antilock systems have only a single channel, they are much less complex and costly than their multichannel, four-wheel counterparts.

Integral and Nonintegral

Another distinction between ABS is whether they are **integral** or **nonintegral.** Integral systems combine the brake master cylinder and ABS hydraulic modulator, pump, and accumulator into one assembly, figure 14-7. Integral systems do not have a vacuum booster for power assist and rely instead on pressure generated by the electric pump for this purpose. The accumulators in these systems can contain more than 2700 psi. Most of the older ABS applications are integral systems. Integral ABS include:

- Bendix 10 and Bendix 9 (Jeep)
- Bosch 3
- Delco Moraine Powermaster III
- Teves Mark II.

Antilock Brake Basics

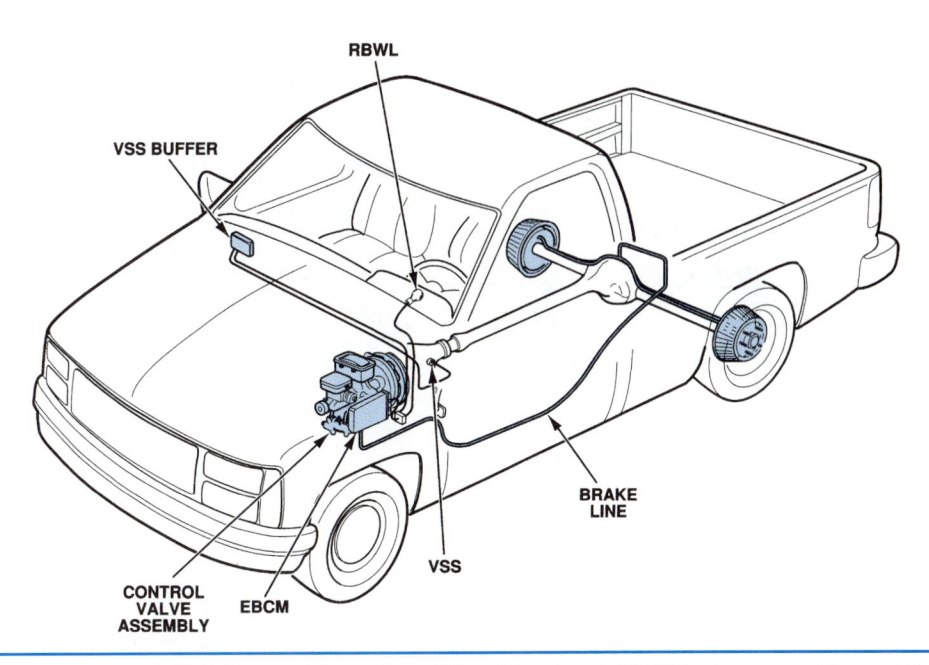

Figure 14-6. One-channel ABS (Kelsey-Hayes Rear Wheel Anti-Lock, RWAL). (Courtesy of General Motors Corporation, Service and Parts Operations)

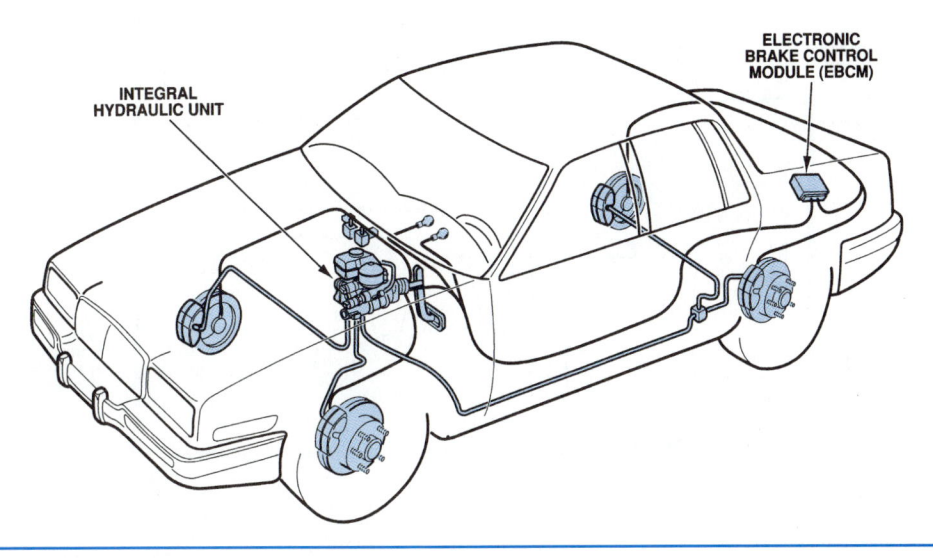

Figure 14-7. Integral ABS (Teves Mark II). (Courtesy of General Motors Corporation, Service and Parts Operations)

Nonintegral ABS, which are sometimes referred to as "add-on" systems, have become the predominant type of ABS because of their lower cost and simplicity, figure 14-8. Nonintegral ABS have a conventional brake master cylinder and vacuum power booster with a separate hydraulic modulator unit. Some also have an electric pump for ABS braking (to reapply pressure during the ABS hold-release-reapply cycle) but do not use the pumps for normal power assist. Nonintegral (add-on) systems include:

- Bendix 6, ABX-4, LC-4, Mecatronic, Mecatronic II
- Bosch 2, 2S Micro, 2U, 2E, 5, 5.3, ABS/ASR, Rear Wheel ABS
- Delco (Delphi) ABS-VI
- Delphi DBC-7
- Honda ABS
- Kelsey-Hayes, RABS/RWAL, 4WAL, EBC-5, EBC-10, EBC 310, EBC325, EBC410
- Nippondenso ABS
- Sumitomo 1,2

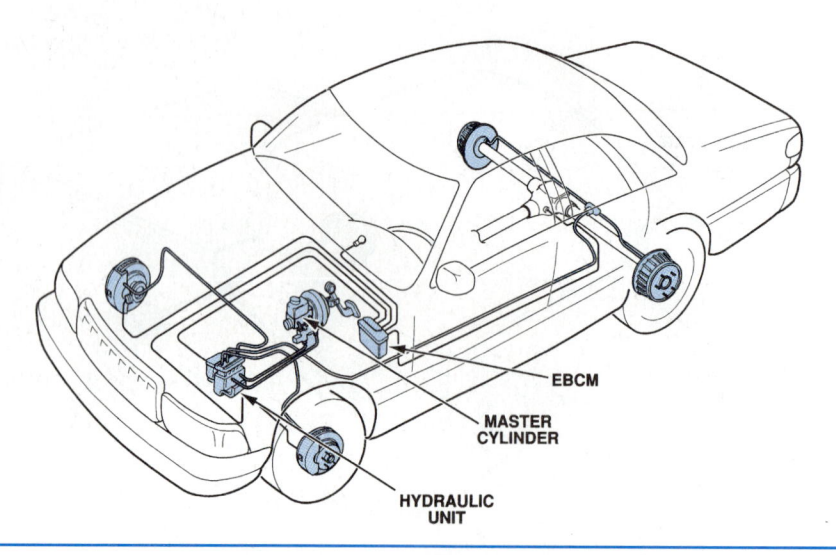

Figure 14-8. Nonintegral ABS (Bosch 2U). (Courtesy of General Motors Corporation, Service and Parts Operations)

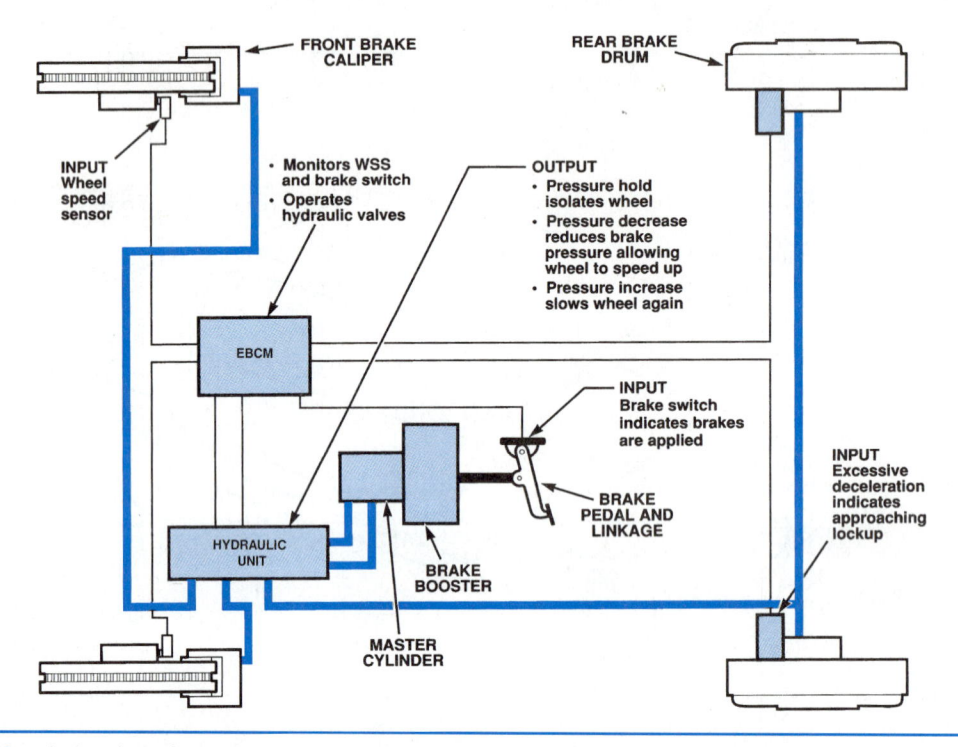

Figure 14-9. A drawing of a typical antilock brake system. (Courtesy of General Motors Corporation, Service and Parts Operations)

- Teves Mark IV, Mark 20
- Toyota Rear-Wheel ABS, Four-Wheel ABS.

ABS COMPONENTS

Basic components that are common to all antilock brake systems, figure 14-9, include:

- Wheel speed sensors
- ABS control module
- ABS self-diagnostics and warning lamp
- Hydraulic modulator assembly with electrically operated solenoid valves (or motor-driven valves in the case of Delco ABS-VI)
- Some systems also have an electric pump and accumulator to generate hydraulic pressure for power assist as well as ABS braking.

Wheel Speed Sensors

Most wheel speed sensors consist of a magnetic pickup and a toothed sensor ring (sometimes called a "tone" ring or "rotor"). The sensor may be mounted

Antilock Brake Basics

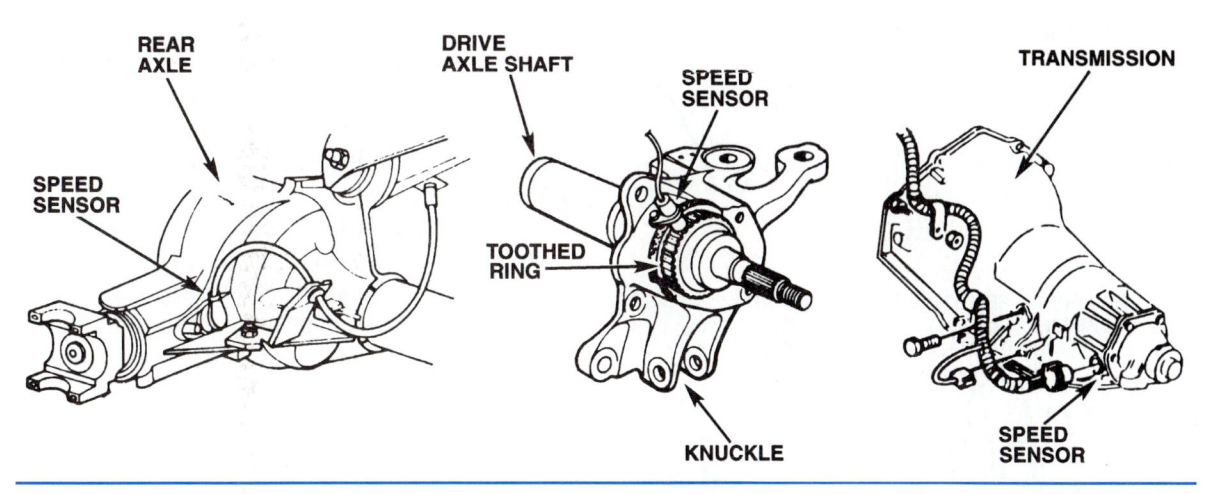

Figure 14-10. Wheel speed sensors for the rear wheels may be located on the rear axle, on the transmission, or on the individual knuckle.

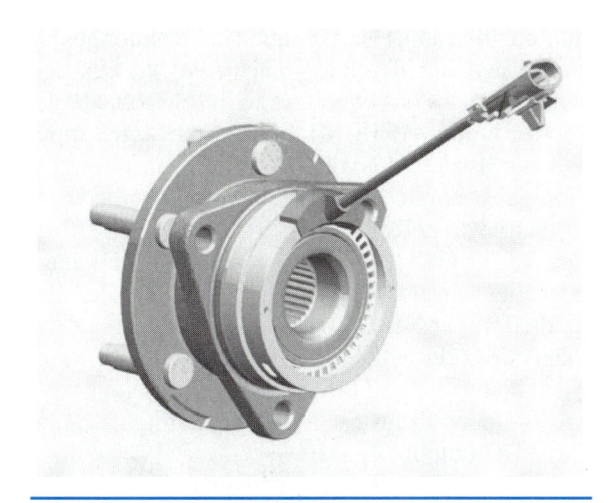

Figure 14-11. The active digital wheel speed sensor used by GM is part of the wheel bearing assembly. (Courtesy of General Motors Corporation, Service and Parts Operations)

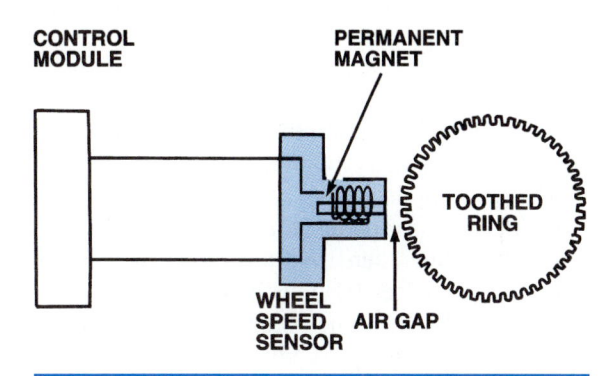

Figure 14-12. Schematic drawing of a wheel speed sensor.

in the steering knuckle, wheel hub, brake backing plate, transmission tailshaft, or differential housing, figure 14-10. On some applications, the sensor is an integral part of the wheel bearing and hub assembly. The sensor rings may be mounted on the axle hub behind the brake rotors, on the brake rotors or drums, on the outside of the outboard constant velocity joints on a front-wheel drive car, on the transmission tailshaft, or inside the differential on the pinion gear shaft.

Another type of wheel speed sensor used by some manufacturers is considered an "active" speed sensor. Called **active digital wheel speed sensors** (GM), figure 14-11, or **magneto-resistive wheel speed sensors** (DaimlerChrysler), these sensors produce a digital signal. In systems using a magnetic sensor (passive sensor), the electronic brake control module (EBCM) receives an analog signal from a "passive" speed sensor at each wheel, but the EBCM has no role in producing the signal. An "active" wheel speed sensor needs a separate 12-volt reference circuit supplied by the EBCM and chassis ground before it can produce an output signal.

Sensor Operation

The sensor pickup, figure 14-12, has a magnetic core surrounded by coil windings. As the wheel turns, teeth on the sensor ring move through the pickup's magnetic field. This reverses the polarity of the magnetic field and induces an alternating current (AC) voltage in the pickup's windings. The number of voltage pulses per second induced in the pickup changes in direct proportion to wheel speed. The result is a voltage signal that changes frequency, figure 14-13. The frequency of the signal is, therefore, proportional to wheel speed. The higher the frequency, the faster the wheel is turning.

WHEEL SPEED SENSOR OUTPUT - LOW SPEED

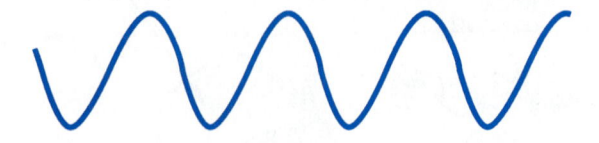

**WHEEL SPEED SENSOR OUTPUT - HIGHER SPEED
(FREQUENCY OF AC SIGNAL INCREASES
IN PROPORTION TO SPEED)**

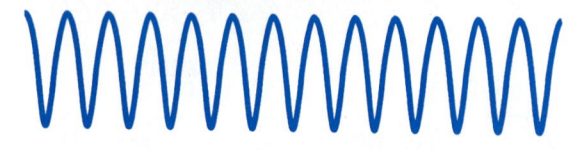

Figure 14-13. Wheel speed sensors produce an alternating current (AC) signal with a frequency that varies in proportion to wheel speed.

The signals are sent to the ABS control module (or an intermediate module in some GM rear-wheel ABS applications), where the AC signal is converted into a digital signal for processing. The control module then monitors wheel speed by counting the pulses from each of the wheel speed sensors. If the frequency signal from one wheel starts to change abruptly with respect to the others, it tells the module that wheel is starting to lose traction. The module then applies antilock braking if needed to maintain traction.

Sensor Air Gap
The distance or **air gap** between the end of the sensor and its ring is critical. A close gap is necessary to produce a strong, reliable signal. But metal-to-metal contact between the sensor and its ring must be avoided because this would damage both. The air gap must not be too wide or a weak or erratic signal (or no signal) may result. The air gap on some wheel speed sensors is adjustable and is specified by the vehicle manufacturer. The gap will vary from one application to another, so always refer to the exact specifications for the vehicle when adjusting the sensor. Wheel speed sensor adjustment is covered in Chapter 13 in the *Shop Manual*.

Sensor Applications and Precautions
Wheel speed sensor readings are affected by the size of the wheels and tires on the vehicle. A tire with a larger overall diameter will give a slower speed reading than one with a smaller diameter. Because the ABS is calibrated to a specific tire size, vehicle manufacturers warn against changing tire sizes. A different tire size or aspect ratio could have an adverse effect on the operation of the ABS.

Wheel speed sensors are also magnetic, which means they can attract metallic particles. These particles can accumulate on the end of the sensor and reduce its ability to produce an accurate signal. Removing the sensor and cleaning the tip may be necessary if the sensor is producing a poor signal.

Active Digital Wheel Speed Sensor Operation
The active wheel speed sensor sends a direct current (DC) square wave signal to the brake control module instead of an AC signal and cannot generate its own signal voltage. The active wheel speed sensor harness has two wires to the eletronic brake control module (EBCM), one for the signal input and one for the 12-volt reference. The reference voltage is supplied to an internal semiconductor in the active wheel speed sensor called a Hall Effect sensor. The Hall Effect sensor creates a small magnetic field around the sensor. A rotating metallic toothed ring inside the wheel bearing interrupts the magnetic field as the wheel spins. When a tooth passes near the hall effect sensor, the signal voltage switches from low to high, creating a square wave DC output.

The frequency of the DC square wave signal output increases with wheel speed but does not increase in voltage the way a magnetic sensor does. The EBCM uses the frequency to interpret wheel speed of ABS operation.

The advantage of a digital active wheel speed sensor is the accuracy of the signal input. Because of the increased signal accuracy, the ABS can react faster to wheel slip. Also, the speed at which the ABS can be activated can occur at slower vehicle speeds.

Magneto-Resistive Wheel Speed Sensors
These wheel speed sensors use an electronic principle known as magneto-resistance to help increase performance, durability, and low-speed accuracy. The sensors convert wheel speed into a small digital signal. A toothed-gear tone wheel serves as the trigger mechanism for each sensor.

The ABS control module sends 12 volts to power an integrated circuit (IC), in the sensor. The IC supplies a constant 7 mA signal to the

Antilock Brake Basics

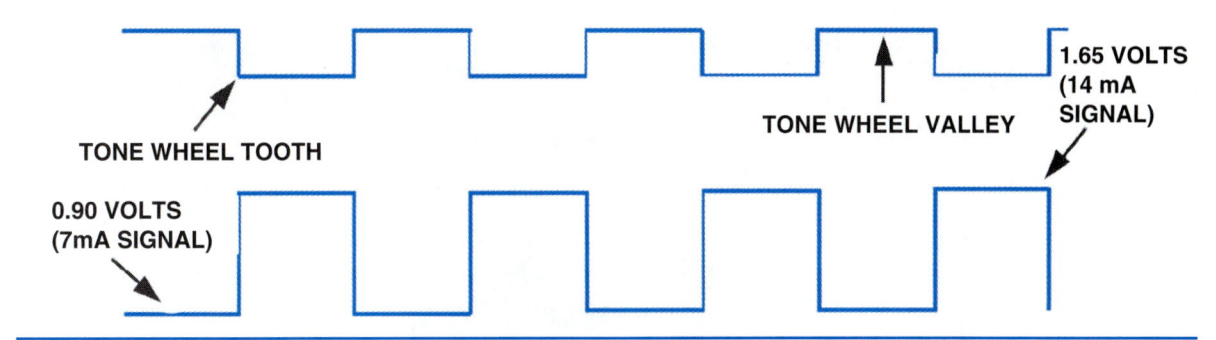

Figure 14-14. The square wave signal produced by a magneto-resistive wheel speed sensor. (Courtesy of Daimler Chrysler Corporation. Used with permission)

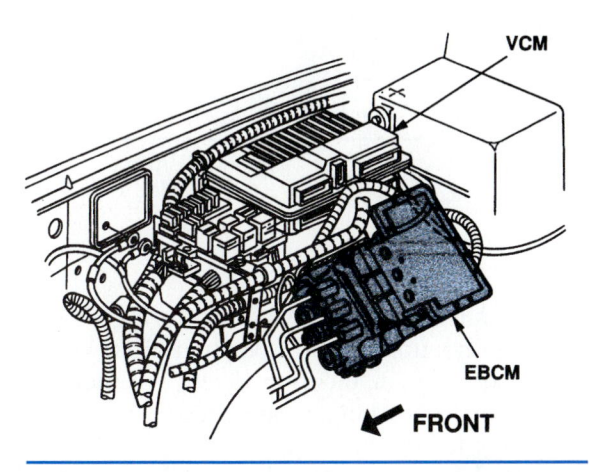

Figure 14-15. This ABS control module (EBCM) is located on the hydraulic modulator assembly mounted under the hood of this SUV. (Courtesy of General Motors Corporation, Service and Parts Operations)

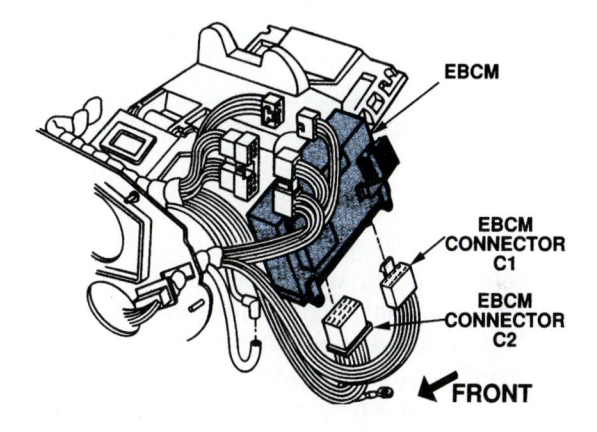

Figure 14-16. This ABS control module (EBCM) is located under the instrument panel inside the vehicle. (Courtesy of General Motors Corporation, Service and Parts Operations)

module. The relationship of the tooth on the tone wheel to the permanent magnet in the sensor signals the IC of the sensor to toggle a second 7 mA power supply on or off. The output of the sensor sent to the ABS module is a DC voltage signal with changing current levels. A square wave is produced as the tone wheel rotates from one tooth to the next, figure 14-14. The number of these square waves that are produced over a specific time is a measurement of frequency known as Hertz (Hz). The module monitors the changing amperage (digital signal) from each wheel speed sensor. The resulting signal is interpreted by the module as the wheel speed.

ABS Control Module

The ABS electronic control module, which may be refered to as an EBCM electronic brake module (EBM), or controller antilock brakes (CAB) module, is a digital microprocessor that uses inputs from its various sensors to regulate hydraulic pressure during braking to prevent wheel lockup. The module may be located on the hydraulic modulator assembly, figure 14-15, or it may be located elsewhere in the vehicle, such as in the trunk, in the passenger compartment, or under the hood, figure 14-16.

Module Inputs

The key inputs for the ABS control module are ignition on, brake pedal switch, and wheel speed sensors, figure 14-17. All antilock brake systems use at least these basic inputs.

- **Ignition on:** The ignition on input from the ignition switch, through a fuse, signals the ABS control module to begin operating.
- **Brake switch:** The brake pedal switch signals the control module when the brakes are

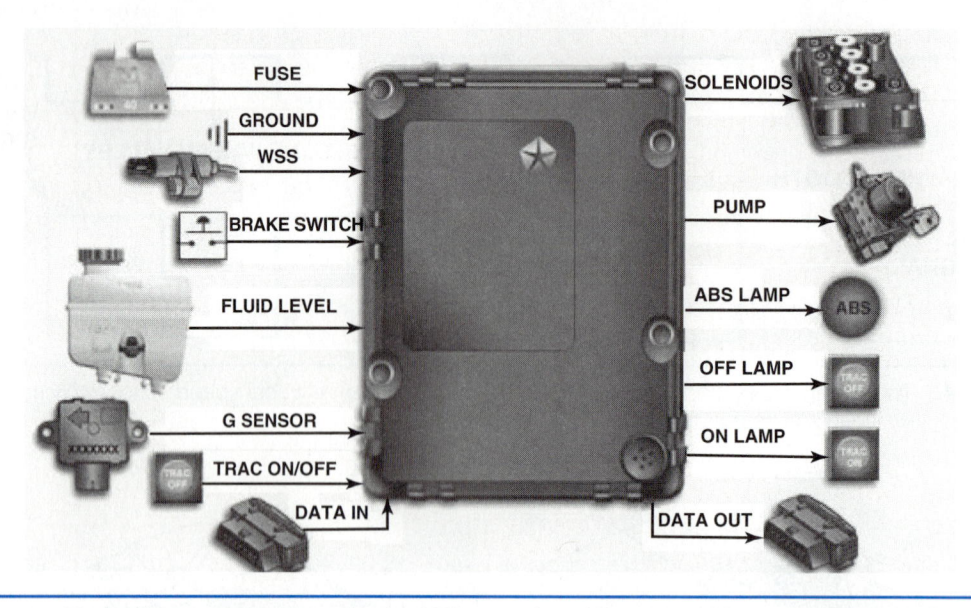

Figure 14-17. Typical inputs and outputs for the ABS control module. (Courtesy of DaimlerChrysler Corporation. Used with permission)

being applied, which causes it to switch from a standby mode to an active mode.
- **Wheel speed sensors (WSS):** The wheel speed sensors provide information about what is happening to the wheels while the brakes are being applied.

Depending on the vehicle, the ABS control module may have additional inputs. Some of these are briefly described here.

- **Vehicle speed sensor:** On vehicles with rear-wheel ABS only, the vehicle speed sensor input used by the speedometer is shared with the ABS module in place of wheel speed sensors.
- **Brake fluid level:** Low fluid level can cause the ABS control module to disable ABS and turn on the amber ABS warning lamp.
- **Lateral acceleration sensor (G sensor):** The G sensor signals the ABS control module to modify ABS operation during hard cornering.
- **Crank sensing:** Crank sensing disables the ABS controller while the engine is cranking to prevent damage from fluctuating system voltage.
- **Battery voltage sensing:** Because the module and solenoids cannot operate correctly below a specified voltage (usually 9.5 volts), the ABS module will disable ABS when voltage falls below this level.

- **Pump and pump relay monitoring:** There are monitors for the operation of the pump and relay in order to detect whether the pump is working, running too long, or running too often.
- **Red brake warning lamp:** The red brake warning lamp may be used to notify the ABS control module of a base brake malfunction.
- **Vehicle data communications:** Both an input and an output, the vehicle data communications allows the ABS control module to "communicate" with other modules in the vehicle and with a diagnostic scan tool.

Other sensor inputs related to traction and stability control systems are discussed later in the chapter.

Module Outputs

Although ABS control module outputs vary with each specific system, as discussed in Chapter 15, there are some outputs that are common to all systems. These include solenoid or motor controlled valves, diagnostic data, and the amber ABS warning lamp. Other outputs, depending on the vehicle, include the ABS power relay, pump control, and the red brake warning lamp.

The main outputs of the control module are the valve solenoids or motors. The solenoids or motors (Delphi ABS VI) in the hydraulic control module are controlled by the ABS control module

Antilock Brake Basics

to produce a three-mode cycle that is common to all systems:

- **Pressure hold:** Isolates brake pressure to the wheel circuits
- **Pressure decrease:** Reduces brake pressure to the wheel circuits
- **Pressure increase:** Resumes brake pressure to the wheel circuits.

Other ABS control module outputs, some related to traction and stability controls, are discussed later in the chapter and as needed in Chapter 15.

Module Operation

If the control module detects a difference in the deceleration rate between one or more wheels while braking, or if the overall rate of deceleration is too fast and exceeds the limits programed into the control module, it triggers the ABS control module to momentarily take over. The control module cycles the solenoid valves in the modulator assembly to modulate hydraulic pressure in the affected brake circuit (or circuits) until its sensor(s) tell it deceleration rates have returned to normal and braking is under control. At that point, normal braking action resumes. When the brake pedal is released or when the vehicle comes to a stop, the control module returns to standby mode until it is again needed.

ABS Self-Diagnostics and Warning Lamp

The ABS control module has a self-diagnostic program that checks the ABS for faults. The self-diagnostic system starts when the ignition switch input is received by the ABS control module (key on). The module then performs a complete self-test of all internal circuits and electrical components in the system. Some manufacturers may call this test a static test.

When vehicle speed reaches a predetermined speed, approximately 10 to 15 mph (16 to 24 km/h), the ABS control module performs a dynamic test that momentarily cycles the solenoid valves and pump motor to check each respective electrical system. During this test, the driver may hear or feel the operation of the hydraulic actuator. This is a normal condition and should not be perceived as a problem. The electrical components are continuously tested for the entire key cycle. On some vehicles the ABS control module may continue to cycle the solenoids periodically.

If any ABS component does not test "good" during these tests or fails during a key on cycle, the ABS control module illuminates the ABS warning indicator lamp and traction control off indicator lamp (if equipped), and stores the appropriate diagnostic trouble code (DTC).

Every ABS has an amber indicator lamp on the instrument panel that warns the driver when a problem occurs with the ABS. The lamp comes on when the ignition is first turned on for a bulb check. The lamp then turns off by the ABS control module at the completion of the static test sequence (assuming no faults are found), which may take from 1 to 5 seconds. If the warning light remains on or comes on while driving, there is a fault in the ABS that requires further diagnosis, which is covered in Chapter 14 of the *Shop Manual*. On most applications, the ABS disables if the ABS warning light comes on and remains on. This should have no effect on normal braking—unless the red brake warning light is also on. The ABS warning light may also be used for diagnostic purposes when retrieving **flash codes** (trouble codes) from the ABS module.

DTCs are provided on most ABSs to help technicians isolate faults to a particular circuit or component, figure 14-18. Some ABSs put the ABS

	None	Blink Codes	Bi-directional Scan Tool	Driver Information Center
Bosch 2U/2S	*	X	X	
Bosch III		X		X
Delco III			X	
Delphi Chassis VI			X	
Kelsey-Hayes RWAL		*	X	
Kelsey-Hayes 4WAL		*	X	
Teves Mark II	*	*		*
Teves Mark IV			X	
Delco/Bosch 5			X	*
Bosch 5			X	
Bosch 5.3			X	*

* Depending on year and application

Figure 14-18. The ABS control module may output diagnostic trouble codes (DTCs) in a variety of ways. (Courtesy of General Motors Corporation, Service and Parts Operations)

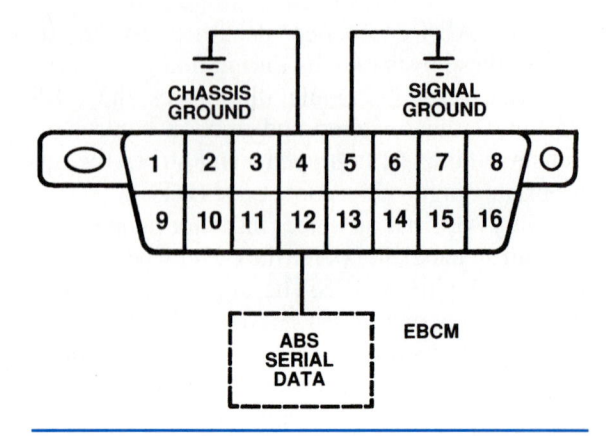

Figure 14-19. The scan tool plugs in to this diagnostic connector located under the dash. (Courtesy of General Motors Corporation, Service and Parts Operations)

module into a special diagnostic mode that causes the ABS warning lamp to flash a numerical sequence that corresponds to particular codes. Most late-model vehicles require the use of a **scan tool** to retrieve ABS DTCs, figure 14-19.

Hydraulic Modulator Assembly

The modulator valve body is part of the master cylinder assembly in integral antilock systems but separate in nonintegral systems. It contains solenoid valves for each brake circuit, figure 14-20, (in Delco ABS-VI applications, however, motor-driven valves are used instead of solenoids). The exact number of valves per circuit depends on the ABS and the application. Some use a pair of on–off solenoid valves for each brake circuit, figure 14-21, whereas others use a single valve that can operate in more than one position.

ABS Solenoids

A solenoid consists of a wire coil with a movable core and a return spring. When current from the ABS control module energizes the coil, it pulls on the movable core. Depending on how the solenoid is constructed, this may open or close a valve that's attached to the movable core. When the control current is shut off, the solenoid snaps back to its normal or rest position (which may be normally open or closed, depending on what is designed to do).

Some solenoids are designed to do more than just switch on or off to open or close a valve. Some pull a valve to an intermediate position when a certain level of current is applied to the coil, then pull the valve to a third position when additional current is provided, figure 14-22. This design allows a single solenoid to perform the same functions as two or even three single-position solenoids.

The solenoids in the hydralic modulator assembly are used to open and close passageways between the master cylinder and the individual brake circuits. By opening or closing the modulator valves to which they're attached, brake pressure within any given circuit can be held, released, and reapplied to prevent lockup during hard braking.

ABS Control Pressure Strategy

The standare ABS control strategy that is used is a three-step cycle:

- The first step is to hold or isolate the pressure in a given brake circuit by closing an **isolation solenoid** in the modulator assembly, figure 14-23. This blocks off the line and prevents any further pressure from the master cylinder reaching the brake.

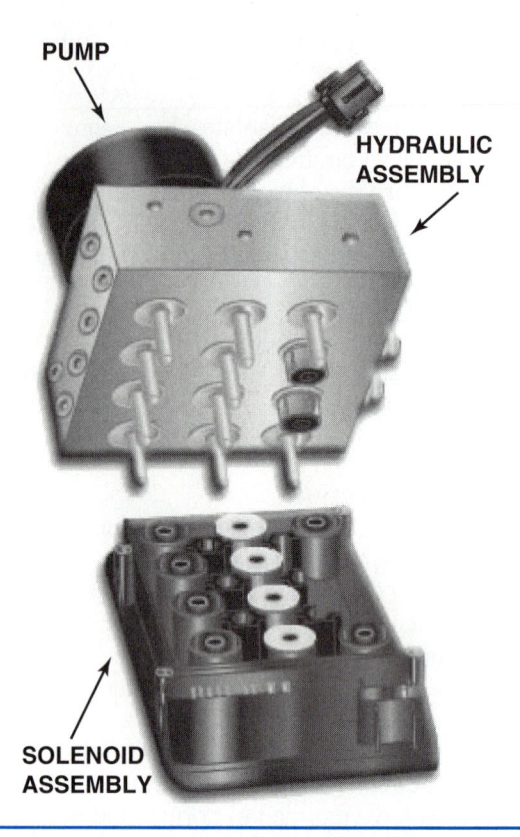

Figure 14-20. Major parts of a nonintegral ABS hydraulic modulator assembly. (Courtesy of DaimlerChrysler Corporation. Used with permission)

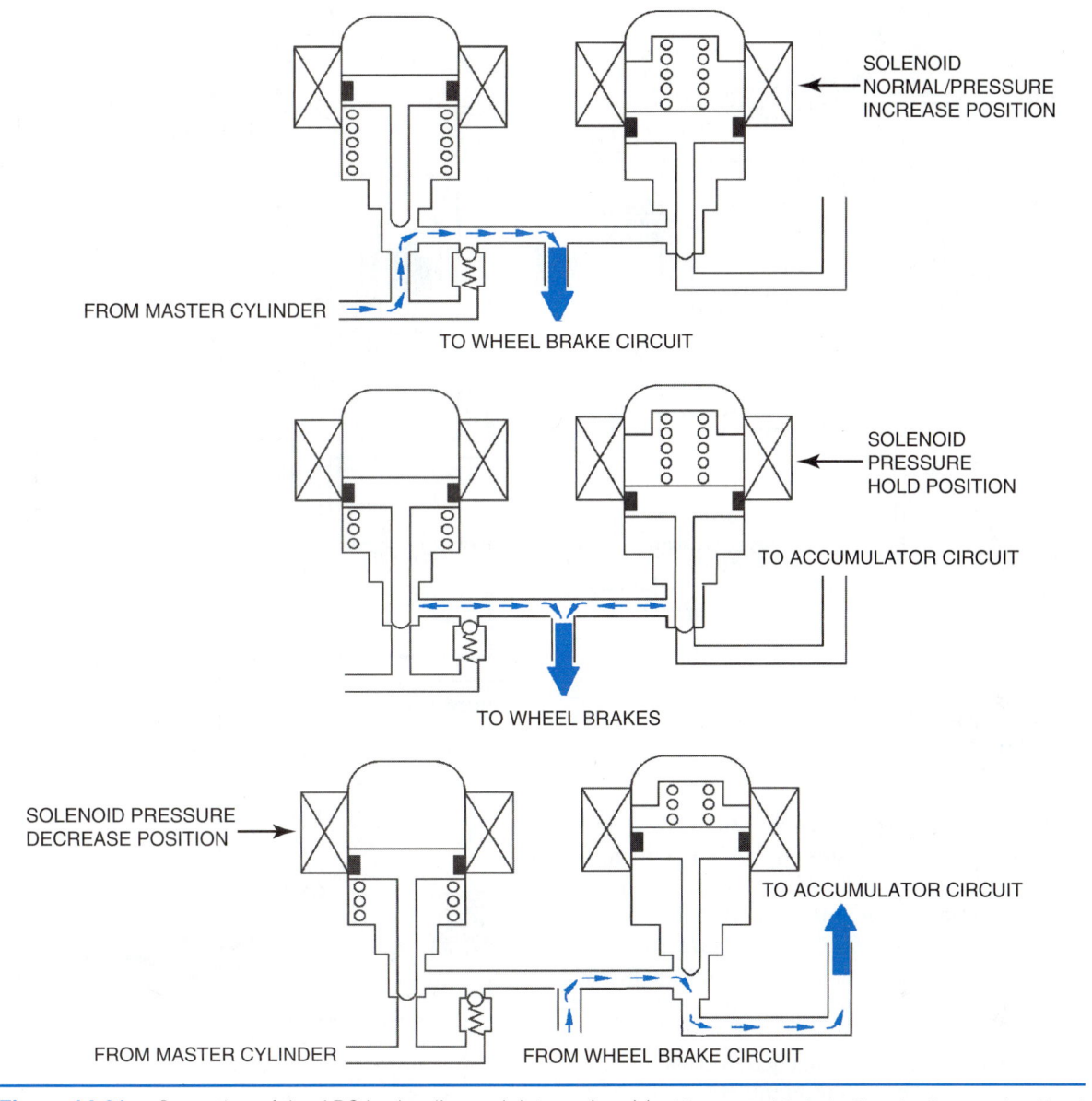

Figure 14-21. Operation of the ABS hydraulic modulator solenoids. (Courtesy of DaimlerChrysler Corporation. Used with permission)

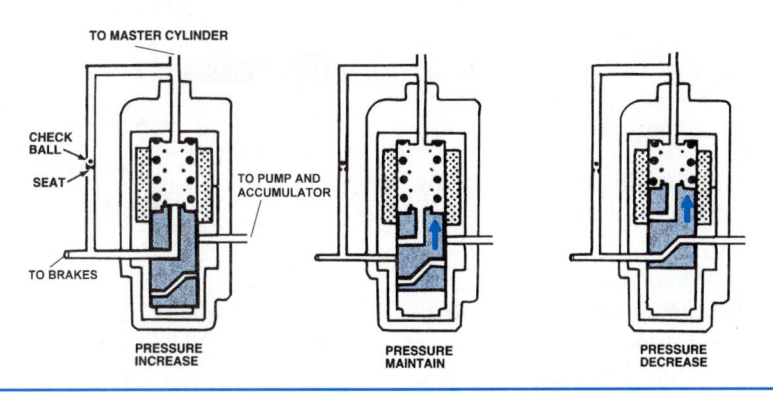

Figure 14-22. An ABS three-way solenoid can increase, maintain, or decrease brake pressure to a given brake circuit.

283

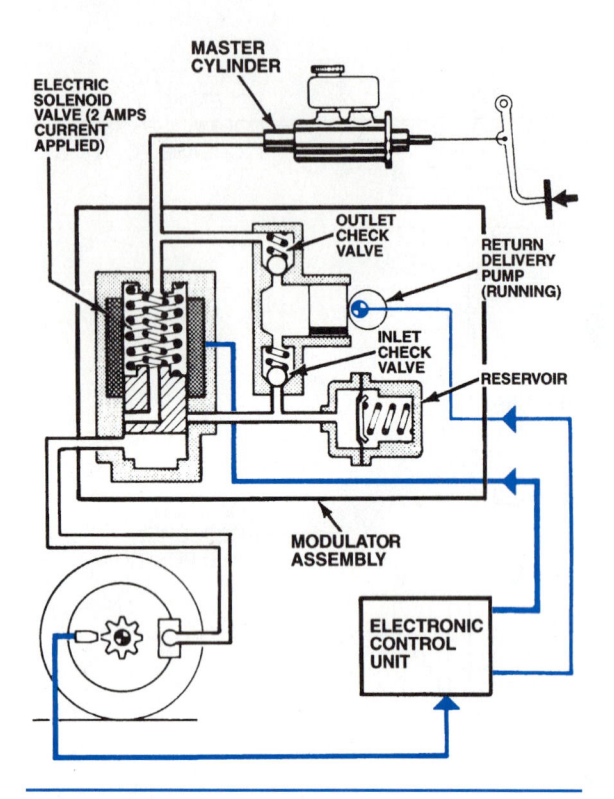

Figure 14-23. The isolation or hold phase of an ABS system. A Bosch 2 system is shown.

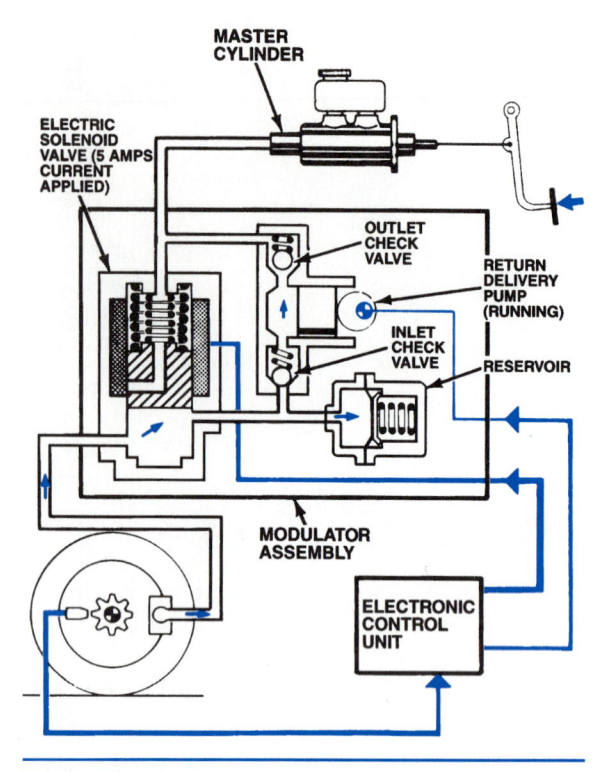

Figure 14-24. During the release phase, pressure is vented from the brake circuit so the tire can regain traction.

- If the wheel speed sensor continues to indicate the wheel is slowing too quickly and is starting to lock, the same solenoid or a second **release solenoid** is energized to open a vent port that releases pressure from the brake circuit, figure 14-24. The fluid is usually routed into a spring-loaded or pressurized storage reservoir (called an "accumulator") so it can be reused as needed. Releasing pressure in the brake circuit allows the brake to loosen its grip so the wheel can speed up and regain traction.
- The release or isolation solenoid(s) are then closed or the additional solenoid energized so pressure can be reapplied to the brake from the master cylinder or accumulator to reapply the brake, figure 14-25.

The hold-release-reapply cycle repeats as many times as needed until the vehicle either comes to a halt or the driver releases the brake pedal. The speed at which this occurs depends on the particular ABS that is on the vehicle but can range from a few times per second up to dozens of times per second.

The hydraulic modulator is not a serviceable component, so you cannot take it apart and replace internal components if it is defective. Replacement

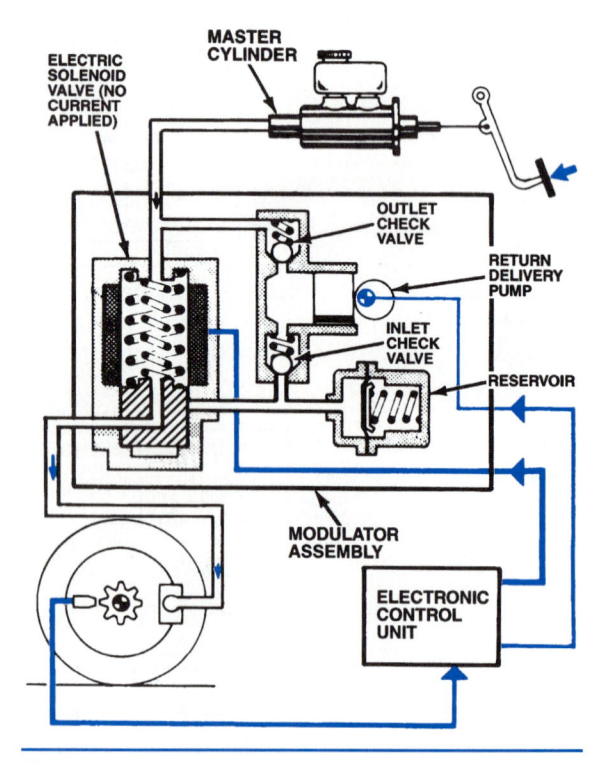

Figure 14-25. The control module reapplies pressure to the affected brake circuit once the tire regains traction.

Antilock Brake Basics

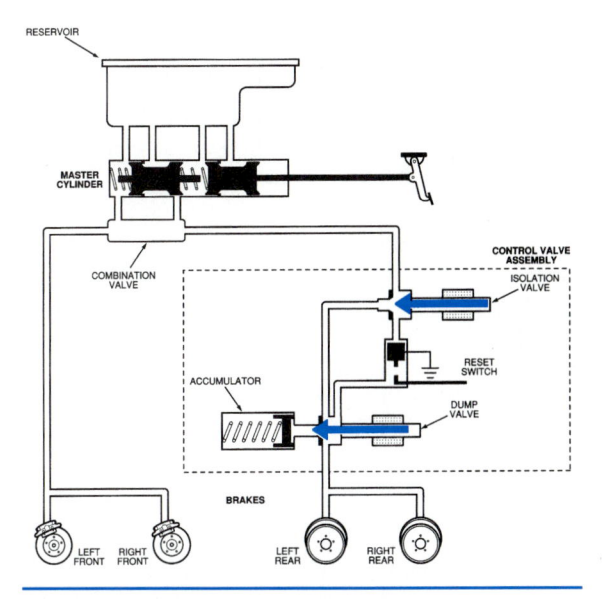

Figure 14-26. The rear-wheel antilock brake system (RWAL) does not use a pump. (Courtesy of General Motors Corporation, Service and Parts Operations)

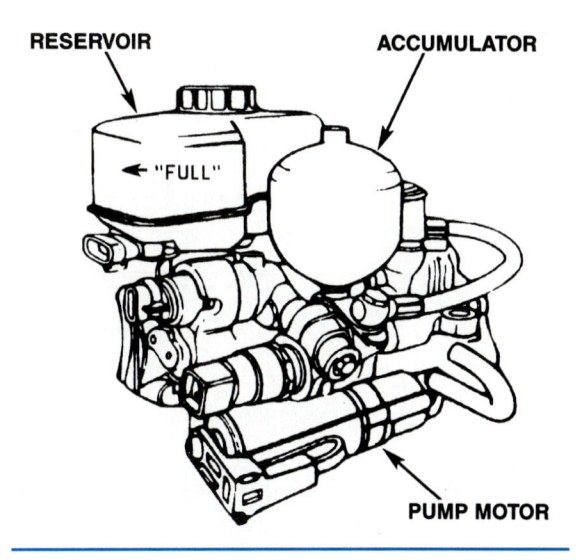

Figure 14-27. An integral ABS unit with a pump motor to provide power assist during all phases of braking and brake pressure during ABS braking.

as an assembly is the only recommended repair. On systems that have a pump and accumulator, these parts can usually be replaced separately.

Pump and Accumulator

Many antilock brake systems use a pump and **accumulator** as part of the hydraulic modulator unit. The pump and motor operation and function varies, depending on the system. A high-pressure electric pump is used in some ABSs to generate power assist for normal braking as well as the reapplication of brake pressure during ABS braking. In other systems, it is used only for the reapplication of pressure during ABS braking.

In some systems the pump serves to only return fluid back to the reservoir and has no role in providing pressure to the ABS. There are also ABS designs that do not use a pump but use accumulator pressure alone to return fluid to the master cylinder reservoir, figure 14-26. Some ABS designs have more than one accumulator.

The accumulator on ABSs where the hydraulic modulator is part of the master cylinder assembly consists of a presure storage chamber filled with nitrogen gas, figure 14-27. A thick rubber diaphragm forms a barrier between the nitrogen gas and brake fluid. As fluid is pumped into the accumulator, it compresses the gas and stores pressure. When the brake pedal is depressed, pressure from the accumulator flows to the master cylinder to provide power assist. A pair of pressure switches mounted in the accumulator circuit signals the ABS control module to energize the pump when pressure falls below a preset minimum, then to shut the pump off once pressure is built back up.

The pump motor is energized via a relay that is switched on and off by the ABS control module. The fluid pressure that is generated by the pump is stored in the accumulator. Should the pump fail (a warning lamp comes on when reserve pressure drops too low), there is usually enough reserve pressure in the accumulator for 10 to 20 power-assisted stops. After that, there is no power assist. The brakes still work but with greatly increased effort.

On ABS designs that have a conventional master cylinder and vacuum booster for power assist, a small accumulator or pair of accumulators may be used as temporary storage reservoirs for brake fluid during the hold-release-reapply cycle, figure 14-28. This type of accumulator typically uses a spring-loaded diaphragm rather than a nitrogen-charged chamber to store pressure.

Accumulator Precautions

A fully charged accumulator in an integral ABS system can store up to 2700 psi of pressure for power assist braking and for reapplying the brakes during the hold-release-reapply cycle of antilock braking. This stored pressue represents a potential hazard for a brake technician who is servicing the brakes, so the accumulator should be depressurized

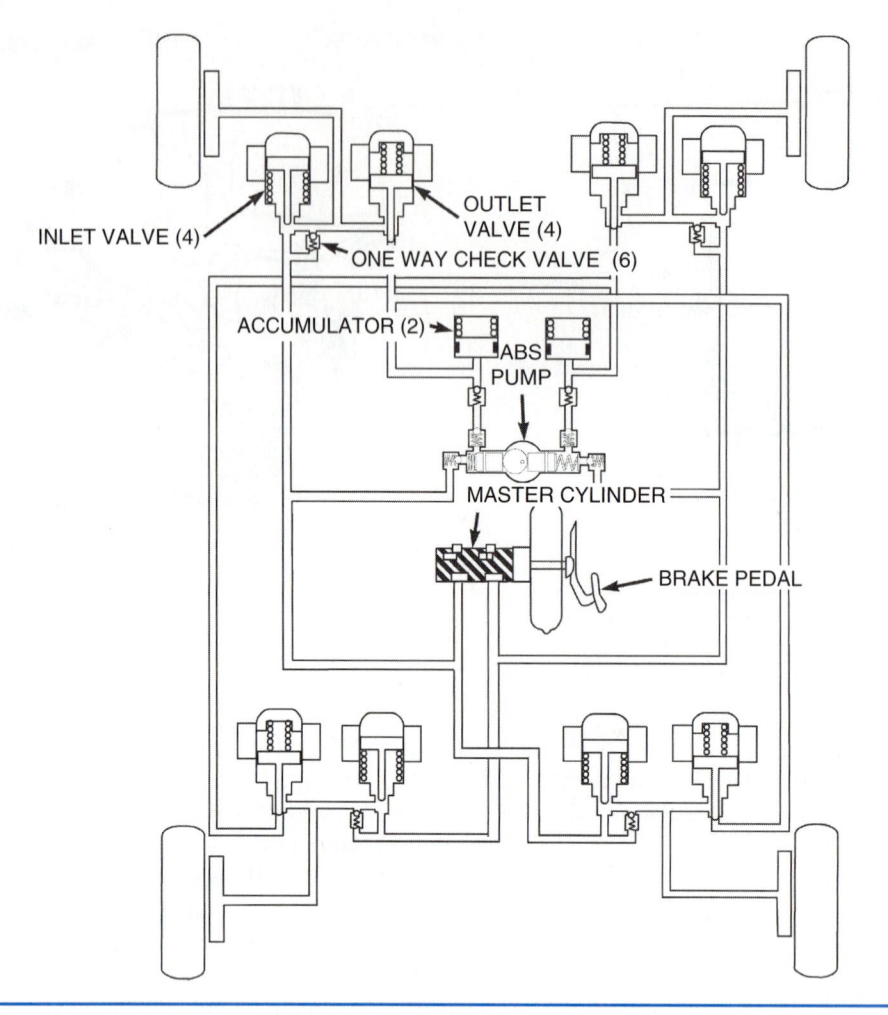

Figure 14-28. This diagonal split ABS hydraulic system features a pump and two accumulators. (Courtesy of DaimlerChrysler Corporation. Used with permission)

prior to doing any type of brake service work. This can be done by pumping the brake pedal 25 to 40 times while the ignition key is off.

In nonintegral ABS systems where an accumulator is used to temporarily hold fluid during the release phase of the hold-release-reapply ABS cycle, the accumulator consists of a spring-loaded diaphragm. This type of accumulator does not have to be depressurized prior to performing brake service.

ADVANCED ABS FUNCTIONS

Much of the same equipment that allows an ABS to control wheel lockup during braking can be adapted to control other vehicle dynamics. Some of these advanced ABS functions are:

- Traction control
- Electronic front/rear proportioning
- Stability control.

Traction Control

These systems may be called traction control (TC), traction assist, traction control system (TCS), or acceleration slip regulation (ASR), depending on the application and vehicle manufacturer, figure 14-29.

The combination of traction control and ABS greatly enhances all-weather traction. High-powered cars and those with low-profile-performance tires (which provide great dry traction but may do poorly on wet or slick surfaces) are the ones that often benefit most from the addition of traction control.

Antilock Brake Basics

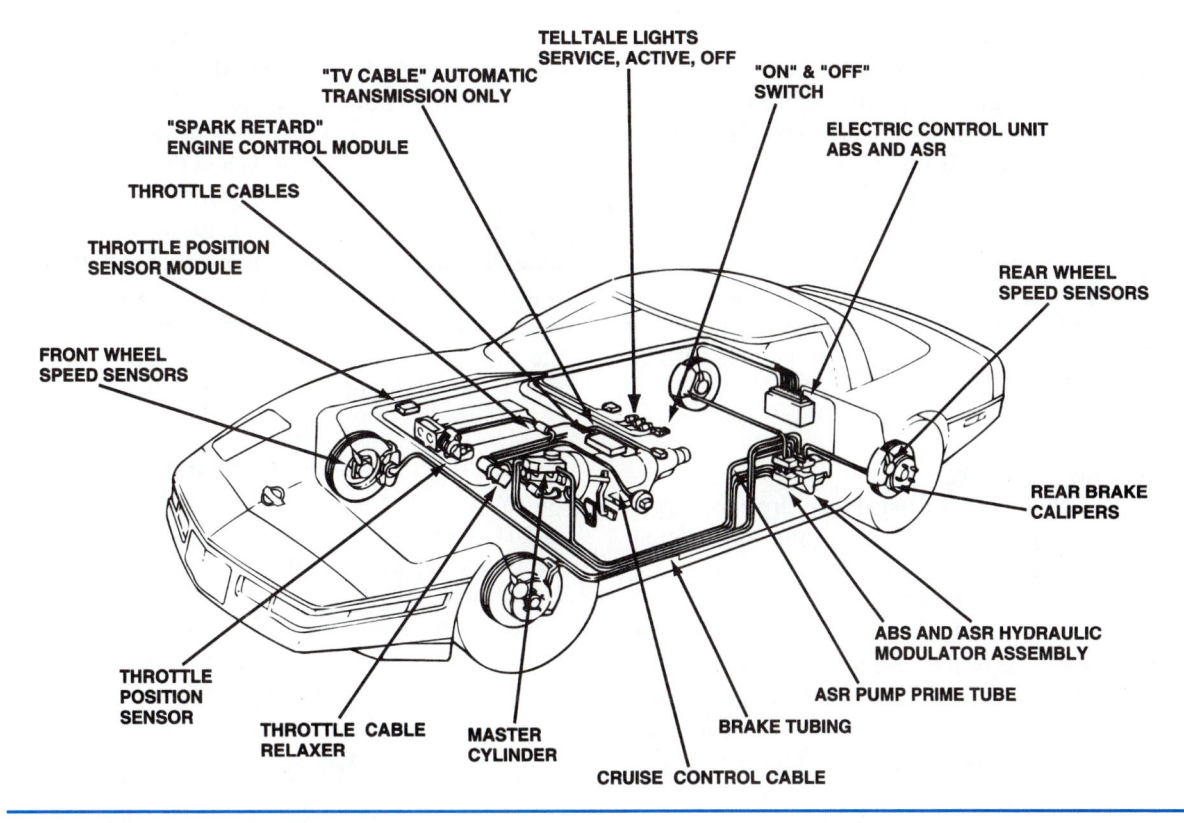

Figure 14-29. The components of the Corvette's ABS/ASR system.

But traction control is not an electronic substitute for a limited-slip differential or all-wheel drive. It is designed primarily to improve traction and vehicle stability on wet or slick surfaces, not for traction in deep snow or mud, or for racing.

Traction Control Operation

Traction control uses the same wheel speed sensors as ABS but requires additional programming in the control module. Traction control also requires additional solenoids in the hydraulic modulator so the brake circuits to the drive wheels can be isolated from the nondrive wheels when braking is needed to control wheel spin, figure 14-30. An ABS system with traction control capability must also have a pump and accumulator to generate and store pressure for traction control braking.

The traction control/ABS system monitors wheel speed during acceleration as well as deceleration (braking). If a wheel speed sensor detects wheel spin in one of the drive wheels during acceleration, the control module energizes a solenoid that allows stored fluid pressure from the

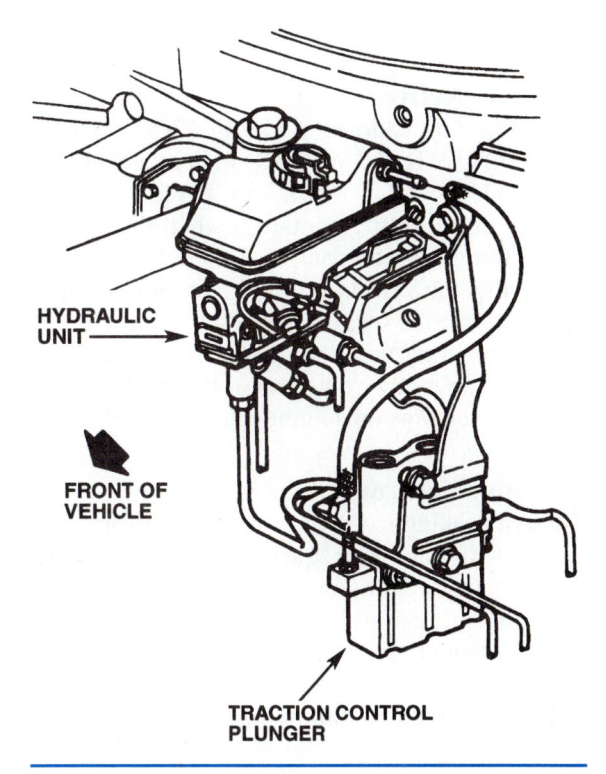

Figure 14-30. The Cadillac Allante's ABS/traction control system uses an additional plunger assembly. It applies pressure to the front brakes when needed to reduce wheel spin.

accumulator to apply the brakes on the wheel that is spinning. This slows the wheel that is spinning and redirects engine torque through the differential to the opposite drive wheel to restore traction. It works just as well on front-wheel drive as it does on rear-wheel drive.

Most traction control systems are only functional at speeds up to about 30 to 35 mph. At higher speeds, they are deactivated because of the adverse effects braking could have on handling and steering stability. Others, however, remain active at all speeds.

Thermal Limiting

Some traction control systems are also programmed to turn themselves off after a predetermined period of constant use to prevent excessive heat buildup in the brakes. This formula is based on a calculation that takes into account the number of times the brakes are applied compared to the total length of time applied. Traction control is disabled temporarily if the ABS control module calculates potentially damaging high-brake pad temperature. Such might be the case if both drive wheels were buried in mud or snow and the driver kept gunning the engine in a vain attempt to get free.

Traction Active Lamp

On most applications, a TRAC CNTL, TRAC ON, LOW TRAC, or ASR indicator light or TRACTION CONTROL ACTIVE message flashes on the instrumentation when the system is engaging traction control. This helps alert the driver that the wheels are losing traction. In most applications, the message does not mean there is anything wrong with the system unless the ABS warning lamp also comes on or the traction control light remains on continuously.

Traction Control Strategies

Though braking is the primary means of limiting wheel spin on many traction control systems, some manufacturers use additional means. Some traction control systems use a throttle close-down strategy that closes the throttle when commanded by the ABS/TCS controller. This may be done by a **throttle relaxer** or throttle adjuster assembly. On vehicles with electronic throttle controls the powertrain control module (PCM) controls the throttle opening, using information from the ABS/TCS controller. By reducing the throttle opening, engine power is reduced to regain traction.

Most applications also link the ABS/traction control computer to the PCM using the vehicle data communication system. This allows the PCM to retard ignition timing or deactivate fuel injectors to reduce power. When the traction control system determines that cutting power is necessary to reduce wheel spin, the engine control module retards timing or shuts off up to half of the injectors momentarily. Another strategy is to communicate with the transmission control module and direct it to upshift the automatic transmission to a higher gear to reduce engine torque.

Traction Control Deactivation Switch

Many vehicles with traction control have a switch on the dash or in the glove box that allows the driver to deactivate the system when desired (for example, when driving in deep snow). An indicator light shows when the system is on or off, and may also signal the driver when the traction control system is actively engaged during acceleration.

Electronic Proportioning

Electronic front-to-rear proportioning control makes it possible to eliminate the mechanical proportioning valve on some vehicles. This is called dynamic proportioning or electronic variable brake proportioning (EVBP), depending on the manufacturer. Using ABS/TCS control module software and the hydraulic modulator assembly, optimum front-to-rear brake balance can be maintained at all times.

As the vehicle decelerates under braking, the ABS/TCS module monitors the wheel speed sensors. If rear-wheel speeds show a slightly greater deceleration rate than the front wheels, the ABS/TCS controller will pulse the rear brakes until the front/rear-wheel speeds equalize. Failures in the electronic proportioning system will cause the red brake warning light to come on.

Stability Control

Some vehicles may be equipped with stability control systems (may be called ABS Plus or Stabilitrak®), sophisticated software extensions of the ABS control module that senses (using the WSS input) when a vehicle is braking in a turn. The software balances brake forces at wheels (side-to-side) to counteract a **yaw movement** and improve the vehicle stability while braking in cornering maneuvers, figure 14-31. ABS plus is active during all braking—not only during ABS events.

Antilock Brake Basics

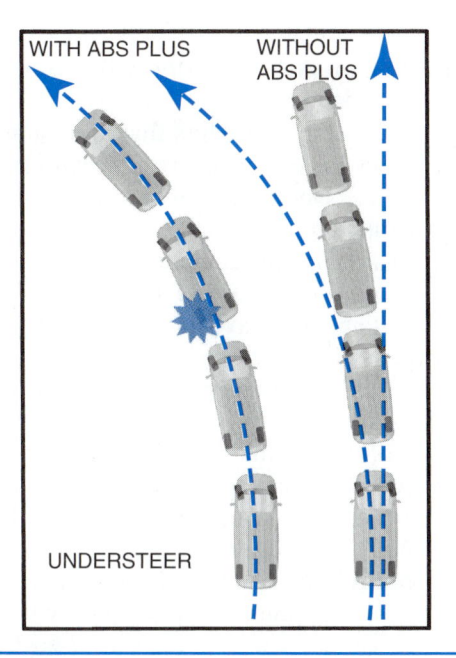

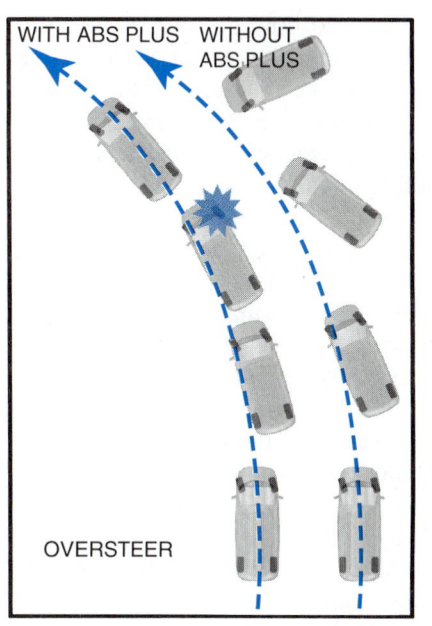

Figure 14-31. How an ABS with stability control activates a single brake (star) to help maintain vehicle control when braking. (Courtesy of DaimlerChyrsler Corporation. Used with permission.)

Figure 14-32. This push button can be used to disable both the traction control and the stability control system. (Courtesy of General Motors Corporation, Service and Parts Operations)

The ABS controller calculates the outside wheel rate of speed compared to the inside wheel rate of speed based on wheel speed sensor input to determine when ABS plus activation may be beneficial. It then reduces the front inside wheel brake pressure relative to the front outside wheel braking pressure. Reducing the inside wheel braking pressure provides less oversteer, reducing the tendency for yaw during the cornering maneuver.

Some vehicles may be equipped with a vehicle stability enhancement system called Stabilitrak® (GM) or by a similar name by other manufacturers. This is an advanced ABS/TCS system that assists with directional control of the vehicle in difficult driving conditions. These systems are active even when there is no braking taking place. Stabilitrak® activates when the computer senses a discrepancy between the intended path and the direction the vehicle is actually traveling. Stabilitrak® selectively applies braking pressure at any one of the vehicle's brakes to help steer the vehicle in the desired direction. The system can be disabled by the driver with the traction control switch, figure 14-32.

SUMMARY

Antilock brake systems prevent lockup by modulating application pressure in the hydraulic system. Preventing lockup is important because it affects tire traction. Traction is defined in terms of tire slip. A free-rolling tire has almost zero slip; a locked wheel has 100 percent slip. Maximum traction on dry pavement occurs when slip is held between 15 and 30 percent.

The performance of antilock systems is limited by their components. The faster and more accurately an antilock system can control the pressure release-apply cycle, the closer to the optimum slip range it will be able to hold the tire. An expert driver can still outbrake the latest systems on smooth dry pavement, and stopping distances on gravel and in deep snow are shorter with locked wheels because of the wedges of loose debris that form in front of the tires.

The laws of physics also limit what an antilock system can do. A car that is braked too late or enters a corner too fast may still have an accident, even if it is equipped with an antilock brake system. However, the systems may allow the driver to avoid the accident or reduce its severity.

All antilock brake systems use speed sensors to detect wheel rotation. If the signal from a sensor slows too quickly during braking—indicating impending lockup—an electronic control unit instructs a modulating device to release and apply the brake several times per second. This keeps the wheel from locking and holds the tire at a controlled amount of slip. Antilock brake systems do not operate unless a wheel is about to lock. In addition, an antilock system failure does not affect normal operation of the brake system.

Antilock brake systems can be divided into two groups: integral or nonintegral. Integral systems combine the master brake cylinder and ABS hydraulic modulator, pump, and accumulator into one assembly. Nonintegral ABS designs, which are sometimes referred to as "add-on" systems, have a conventional brake master cylinder and vacuum power booster with a separate hydraulic modulator unit.

Basic components that are common to all antilock brake systems include wheel speed sensors, an electronic control module, an ABS warning lamp, a hydraulic modulator assembly with electrically operated solenoid valves (or motor-driven valves in the case of Delco ABS-VI), and some systems also have an electric pump and accumulator to generate hydraulic pressure for power assist as well as ABS braking.

Much of the same equipment that allows an ABS to control wheel lockup has been adapted to control wheel spin during acceleration. Traction control uses the same wheel speed sensors as ABS, but requires additional programming in the control module so the system monitors wheel speed continuously and not only when braking.

Though braking may be used to limit wheel spin on traction control systems, additional strategies include a throttle relaxer that decreases the throttle opening, reducing power to regain traction. Some systems also link the ABS/traction control computer to the engine computer electronically to retard ignition timing or deactivate fuel injectors to reduce power. Another strategy is to command the automatic transmission to shift to a higher gear, reducing engine torque.

Review Questions

Choose the letter that represents the best possible answer to the following questions:

1. Preventing brake lockup is important because it affects:
   a. Tire traction
   b. Tire slip
   c. Both a and b
   d. Neither a nor b

2. The optimum tire slip range on icy pavement is:
   a. 15 to 30 percent
   b. 20 to 50 percent
   c. 100 percent
   d. 20 to 30 percent less than on dry pavement

3. A locked and skidding tire:
   a. Has zero slip
   b. Is devoting all of its traction to braking
   c. Both a and b
   d. Neither a nor b

4. A vehicle spin during braking is most likely if:
   a. The front wheels lock
   b. The rear wheels lock
   c. All four wheels lock
   d. The driver attempts to turn

5. Which of the following Is not a part of basic antilock system operation?
   a. Monitoring the vehicle speed
   b. Reapplying the brakes
   c. Modulating brake hydraulic pressure
   d. Computing the rate of wheel rotation

6. An antilock system:
   a. Does not provide antilock action until after a brake locks
   b. Does not affect normal brake application
   c. Both a and b
   d. Neither a nor b

7. During ABS operation, solenoid valves:
   a. Hold hydraulic pressure to the brakes
   b. Release hydraulic pressure to the brakes
   c. Reapply hydraulic pressure to the brakes
   d. Any of the above, when necessary

8. Four-wheel antilock systems have:
   a. Three-channel configurations
   b. Four-channel configurations
   c. Either a or b
   d. Neither a nor b

9. Integral antilock systems:
   a. Combine the master cylinder and hydraulic modulator booster
   b. Use a vacuum booster
   c. Have low-pressure, spring-loaded accumulators
   d. All of the above

10. A four-wheel antilock system will not provide the shortest stopping distance:
    a. When stopping on gravel surfaces
    b. When the car is stopping in deep snow
    c. Both a and b
    d. Neither a nor b

11. Nonintegral antilock systems:
    a. Use a conventional master cylinder
    b. Have an electric pump rather than a vacuum booster
    c. Have nitrogen-charged accumulators
    d. All of the above

12. Wheel speed sensors:
    a. Send a DC signal to the control module
    b. Send an AC signal to the control module
    c. Receive but do not send signals
    d. Work on the principle of high resistance

13. The key inputs for the ABS control module come from:
    a. The wheel speed sensors
    b. A brake pedal switch
    c. Both a and b
    d. Neither a nor b

14. Traction control requires:
    a. Several additional accumulators
    b. Dedicated wheel speed sensors
    c. Special brake calipers
    d. Additional solenoids in the hydraulic modulator

15. In addition to applying the brakes, traction control systems may also:
    a. Reduce the throttle opening
    b. Retard timing
    c. Shut off fuel Injectors
    d. All of the above

15
Antilock Brake Systems

OBJECTIVES

Upon completion and review of this chapter, you will be able to:

- Explain the operation of the major components, such as the accumulator, pressure switches, pump motor, control modules, and hydraulic assembly, for specific antilock brake systems.
- Be able to identify the type of ABS system after visually inspecting the vehicle.
- Know which systems are integral and which are nonintegral.
- Know which systems are four-channel, three-channel, and single-channel.
- Explain the operation of a system's solenoid valves through the isolation, release, and reapply modes.
- Explain the operation of the traction control components for systems that include traction control.
- Explain the operation of optional ABS functions, including electronic brake proportioning, tire inflation monitor, and stability control systems.

KEY TERM

microprocessors

INTRODUCTION

Chapter 14 covered ABS basics, so this chapter details the most popular domestic and import antilock brake systems. The systems covered include:

- Bendix
- Bosch
- Delphi Chassis (Delco Moraine)
- Kelsey-Hayes
- Nippondenso
- Sumitomo
- Teves
- Toyota.

BENDIX ABS

Bendix currently has two integral ABS designs Bendix 9 ABS and Bendix 10; and four nonintegral systems: Bendix 6, Bendix LC4, Bendix ABX-4, and Bendix Mecatronic II.

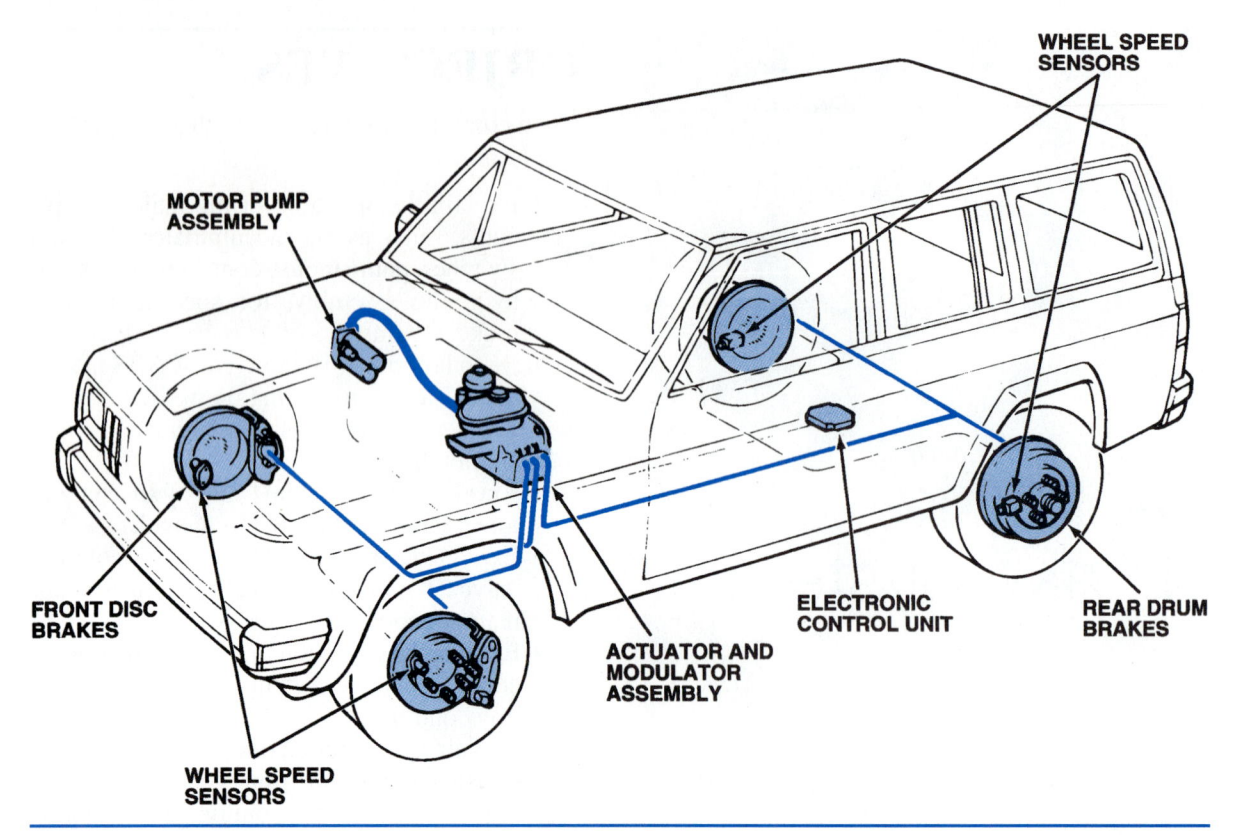

Figure 15-1. Bendix 9 ABS is used on 1989–1991 Jeep Cherokees and Wagoneers.

Bendix 9 ABS

The Bendix 9 ABS (Jeep) system on the 1989–91 Jeep Cherokees and Wagoneers was the first of the "new generation" of Bendix ABSs, figure 15-1. It is an integral ABS that combines the master brake cylinder with the hydraulic control modulator and two accumulators. The pump and motor are a separate assembly but work with the master cylinder and hydraulic modulator assembly to provide power assist as well as antilock braking.

System Description

The Bendix 9 system is a four-wheel, three-hydraulic-channel ABS system. During normal braking, the vehicle's brake hydraulics are split front and rear. During antilock braking, both front brakes are controlled separately whereas the rear brakes are cotrolled as a pair. The modulator assembly contains nine solenoid valves: as isolation valve, a decay valve, and a build valve for each of the three ABS circuits.

Four-wheel speed sensors (one for each wheel) are used. Front sensors are on the steering knuckles with the sensor tone rings mounted on the axle shafts. Rear sensors are on the rear brake backing plates with the sensor rings mounted on the axles.

The Bendix 9 system control module is under the back seat attached to the floorpan. On 1989 and 1990 models, the mounting angle of the module is important because the module contains s mercury switch that monitors vehicle deceleration. This allows the module to adjust antilock braking to better suit slippery driving conditions when the vehicle is being operated in four-wheel drive. If the module is removed or replaced, it must be reinstalled at exactly the same angle as before so the mercury switch will function correctly. On 1991 models, the module is mounted on a bracket that holds it at the proper angle.

Pressure Switches and Valves

The integral master cylinder, hydraulic control modulator, and accumulator assembly contains a boost pressure differential switch mounted on the pressure modulator. It has a single terminal, is self-grounding, and is normally open. This switch functions the same as a differential pressure switch in a conventional brake system; it illuminates the BRAKE and ABS warning lights in the event of fluid loss from the brake system. The relays for both warning lights are located on the

Antilock Brake Systems

driver's side of the engine compartment near the master cylinder/modulator assembly.

The modulator assembly also has a brake proportioning valve and switch, and an accumulator low-pressure switch for signaling the controller if accumulator pressure is lost. The accumulator switch is a single terminal. Self-grounding switch that remains closed during normal operation. If pressure in the accumulator drops below 1050 psi (7240 kPa), the switch opens and the controller illuminates the yellow ABS light. If accumulator pressure does not increase enough to close the switch within 20 seconds, the red BRAKE warning light will also come on.

Dual Accumulators

Bendix 9 uses two accumulators. A high-pressure accumulator, located on the master cylinder booster assembly, is nitrogen precharged to 1000 psi (6895 kPa) and capable of storing fluid at 1700 to 2000 psi (11,722 to 13,790 kPa). A low-pressure accumulator, precharged to 350 psi (2413 kPa), but also capable of storing fluid at 1700 to 2000 psi, serves primarily as a fluid reservoir. Both accumulators work together to provide additional fluid pressure during antilock braking as well as power-assisted brake applications should the booster pump and motor fail.

CAUTION: Before doing any work on the master cylinder and hydraulic modulator assembly, accumulator pressure must be discharged. This is accomplished by pumping the brake pedal 25 to 40 times with the ignition off.

The remote pump and motor assembly are on the passenger side of the engine compartment. The pump supplies system pressure for normal power-assisted braking as well as antilock braking. Power is supplied to the motor through a relay and pressure switch located next to the pump and motor assembly.

The pressure switch is designed to keep system operating pressure between 1700 and 2000 psi (11,722 to 13,790 kPa). When pressure drops below 1700 psi, the switch closes and causes the pump to run until it builds sufficient pressure to recharge the accumulators. At 2000 psi, the switch opens and the pump shuts off. Normal pump time for a completely discharged system (zero psi) is about 60 to 80 seconds. The pump runs only when the ignition is on.

To protect the system from serious damage caused by too much pressure, a pressure relief valve and thermal fuse are included as part of the motor and pump assembly. The pressure relief valve is a simple blow-off valve that opens if pump pressure exceeds 3000 psi (20,685 kPa). The relief valve does not reseat and must be replaced if pressure forces it open. The thermal fuse will shut the pump off if the pump's operating temperature exceeds approximately 385°F (177°C).

System Operation

The Bendix 9 Jeep ABS is unusual compared to other ABSs because it does not become functional until a vehicle speed reaches 12 to 15 mph. It is also designed to disengage at speeds below 3 to 5 mph, which may allow the wheels to chirp or skid slightly just before the vehicle completes an ABS stop. This is a normal condition for Bendix 9 systems and does not indicate a problem. It was designed to function this way to make it compatible with Jeep's Selec-Trac 4WD.

When a driver brakes, the brake pedal switch signals the ABS module to compare the relative deceleration rates of all four wheels. If ABS braking is needed, the module energizes the appropriate ABS isolation solenoid to seal off the affected wheel brake. If either rear brake needs ABS braking, a single solenoid isolates both lines simultaneously.

The system then releases pressure in the isolated brake circuit(s) by energizing the normally closed ABS decay solenoid. Pressure is reapplied by energizing the ABS build solenoid. This opens a passageway to allow pressure from the pump and accumulators to enter the line. Cycling of the decay and build solenoids continue until the vehicle slows to 3 to 5 mph, when the control module deenergizes the isolation solenoid and discontinues antilock braking. The wheels may then lock up slightly or chirp as the vehicle comes to a complete stop.

Bendix 10 ABS

Bendix 10 ABS is used by Chrysler on the 1990–93 New Yorker, Imperial, and Dodge Dynasty cars and the 1991–93 Dodge Caravan and Plymouth Voyager minivans. Like the Bendix 9 system, Bendix 10 is an integral ABS with a combined master cylinder and hydraulic modulator assembly, and a separate pump and motor for

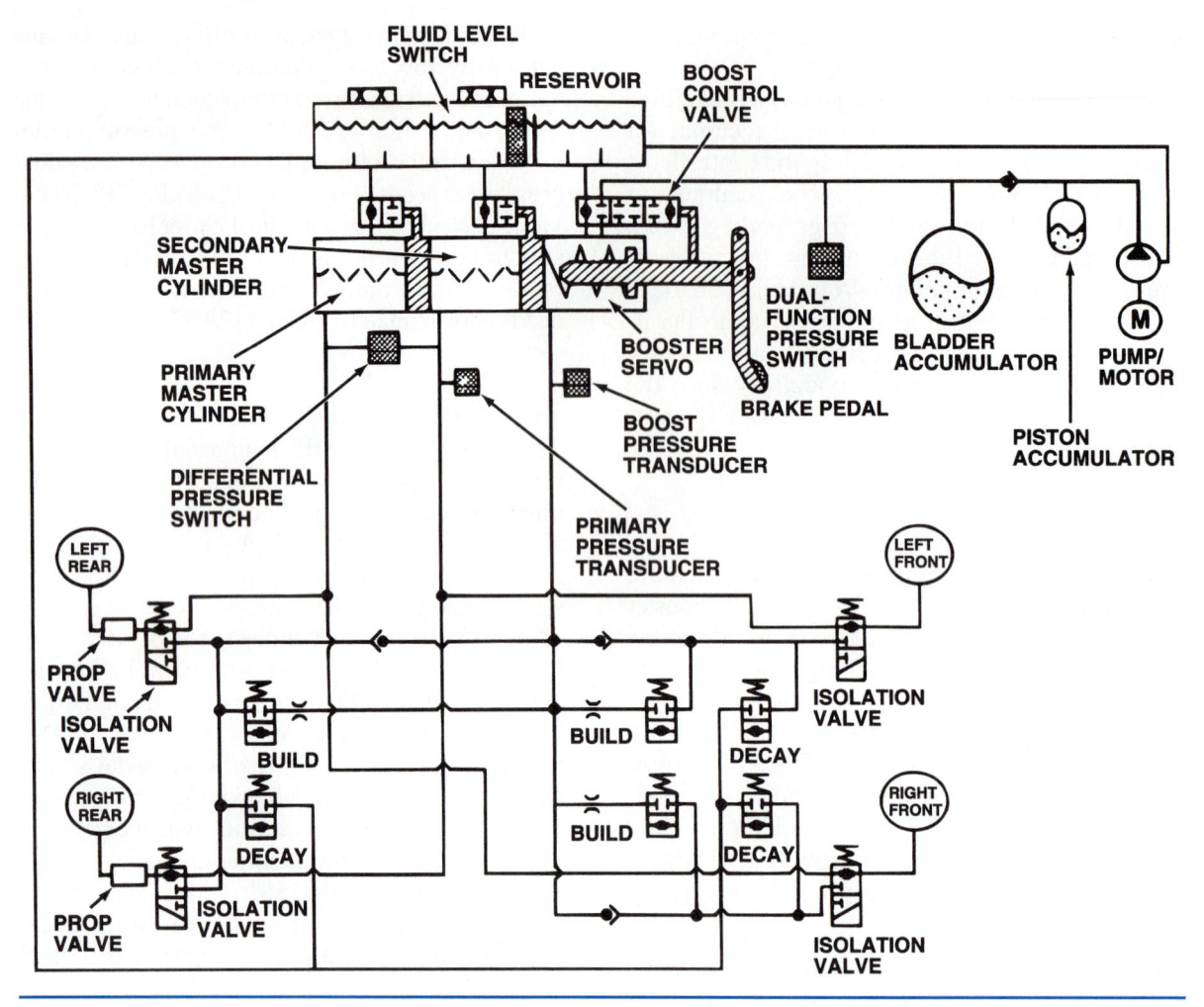

Figure 15-2. The Bendix 10 system uses four isolation valves, three decay valves and three build valves.

power assist and antilock braking. It is called Bendix "10" because the hydraulic modulator contains 10 ABS solenoid valves, figure 15-2. There are four isolation solenoid valves (one for each wheel), a decay valve and build valve for each front brake and a decay valve and build valve controlling the rear brake channel. The extra isolation solenoid in this system helps reduce pedal pulsation and feedback during antilock braking.

System Description
Like the Bendix 9 Jeep applications, Bendix 10 ABS controls all four wheels but divides the brake system into three channels for antilock braking. The system controls the front brakes independently and the rear brakes as a pair.

The system has four nonadjustable wheel speed sensors. The sensor rings for the front wheel speed sensors are located on the outboard constant velocity joint housings.

The control module is usually located under the battery tray. Power to the module is provided through a relay in the power distribution center located on the left side of the engine compartment.

Dual Accumulators
The Bendix 10 has two accumulators. The high-pressure accumulator, figure 15-3, located on the master cylinder booster assembly, is precharged with nitrogen to 1000 psi (6895 kPa), with a normal operating pressure range of 1600 to 2000 psi

Antilock Brake Systems

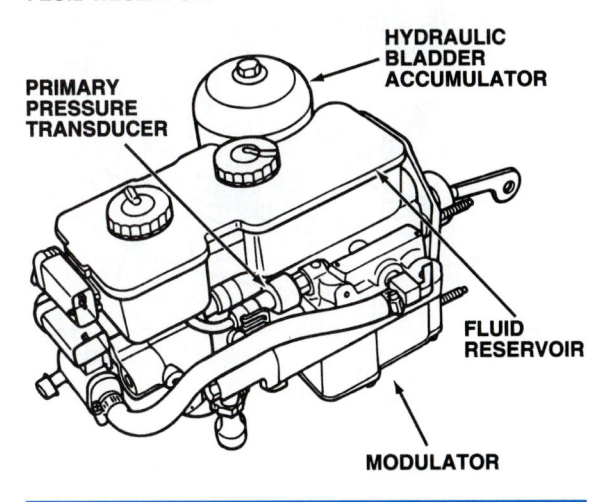

Figure 15-3. The Bendix 10 hydraulic control unit.

(11,032 to 13,790 kPa). The low-pressure accumulator, located on the pump/motor assembly, is precharged to 460 psi (3172 kPa) with a sliding piston design rather than a diaphragm. The accumulators provide additional brake pressure during ABS braking as well as boost pressure for power-assisted normal stops.

CAUTION: Before doing any work on the master cylinder and hydraulic modulator assembly, accumulator pressure must be discharged. This is accomplished by pumping the brake pedal 25 to 40 times with the ignition off.

Pressure Switches and Transducers

The modulator has a dual function pressure switch located on the bottom of the modulator. It turns the remote pump motor on and off by grounding the pump motor relay through the starter solenoid. The switch maintains accumulator pressure by turning the pump on when pressure drops below 1600 psi (11,032 kPa) and off when pressure reaches 2000 psi (13,790 kPa). Like the Jeep system, this switch also warns the control module if accumulator pressure drops below 1000 psi (6,895 kPa). When the switch opens, the red brake and yellow ABS warning lights are illuminated, and the ABS is disabled. A differential pressure switch detects pressure differences of more than 300 psi (2069 kPa) between the primary and secondary sides of the brake hydraulics. A problem here illuminates both warning lights.

Two pressure transducers detect pressure problems in the Bendix 10's modulator assembly. Each generates a voltage between 0.25 and 5.0 volts that is proportionate to the amount of pressure. One transducer monitors boost pressure and is located on the bottom of the modulator assembly. The other transducer monitors primary pressure in the master cylinder and is located on the left side of the hydraulic assembly.

A fluid level switch on the reservoir monitors the fluid level. The switch consists of a float and magnetic reed switch that closes when the fluid level is too low, illuminating both warning lights and disabling the ABS system.

Bendix 6 ABS

The Bendix 6 system is on 1991–93 Chrysler Lebaron, Dodge Daytona, Dodge Spirit, Plymouth Acclaim, Plymouth Laser, Eagle Talon, and Eagle Premier models. Unlike the Bendix 9 (Jeep) system or Bendix 10, the Bendix 6 system is a nonintegal ABS with a conventional master brake cylinder and power booster. The hydraulic modulator, pump/motor and accumulator assembly is located on the frame rail, figure 15-4.

System Description

Bendix 6 is a four-wheel, three-channel nonintegral ABS with four nonadjustable wheel speed sensors. The sensor rings for the front wheel speed sensors are located on the outboard constant velocity (CV) joints. Rear sensors are located on the brake backing plates.

The system is called Bendix "6" because the modulator assembly contains six ABS solenoids, figure 15-5. There are two isolation solenoid valves: one for the primary side of the hydraulics and one for the secondary side. The control module activates the isolation valves when either front wheel needs antilock braking. The isolation valves are necessary for these applications because the vehicles are front wheel drive with diagonally split brake hydraulics.

There are four combination build/decay solenoid valves (one for each wheel). These are two-position solenoids that release fluid pressure from an affected wheel circuit in the decay mode or

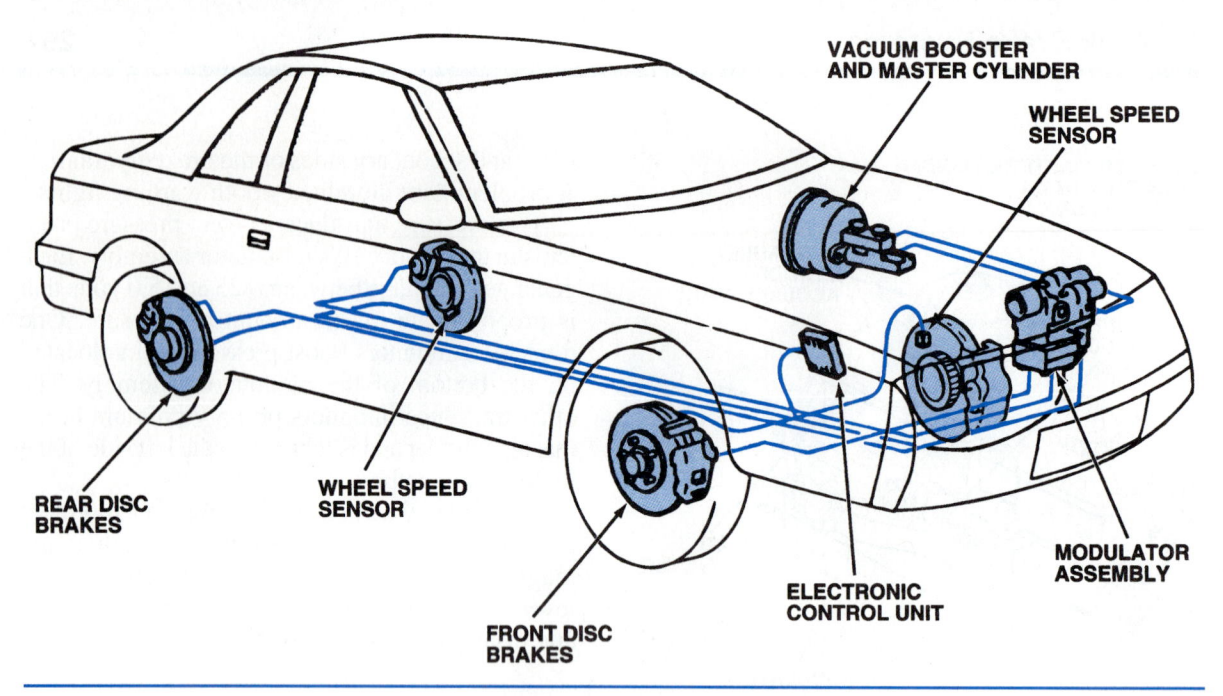

Figure 15-4. The Bendix 6 ABS system is a nonintegral system.

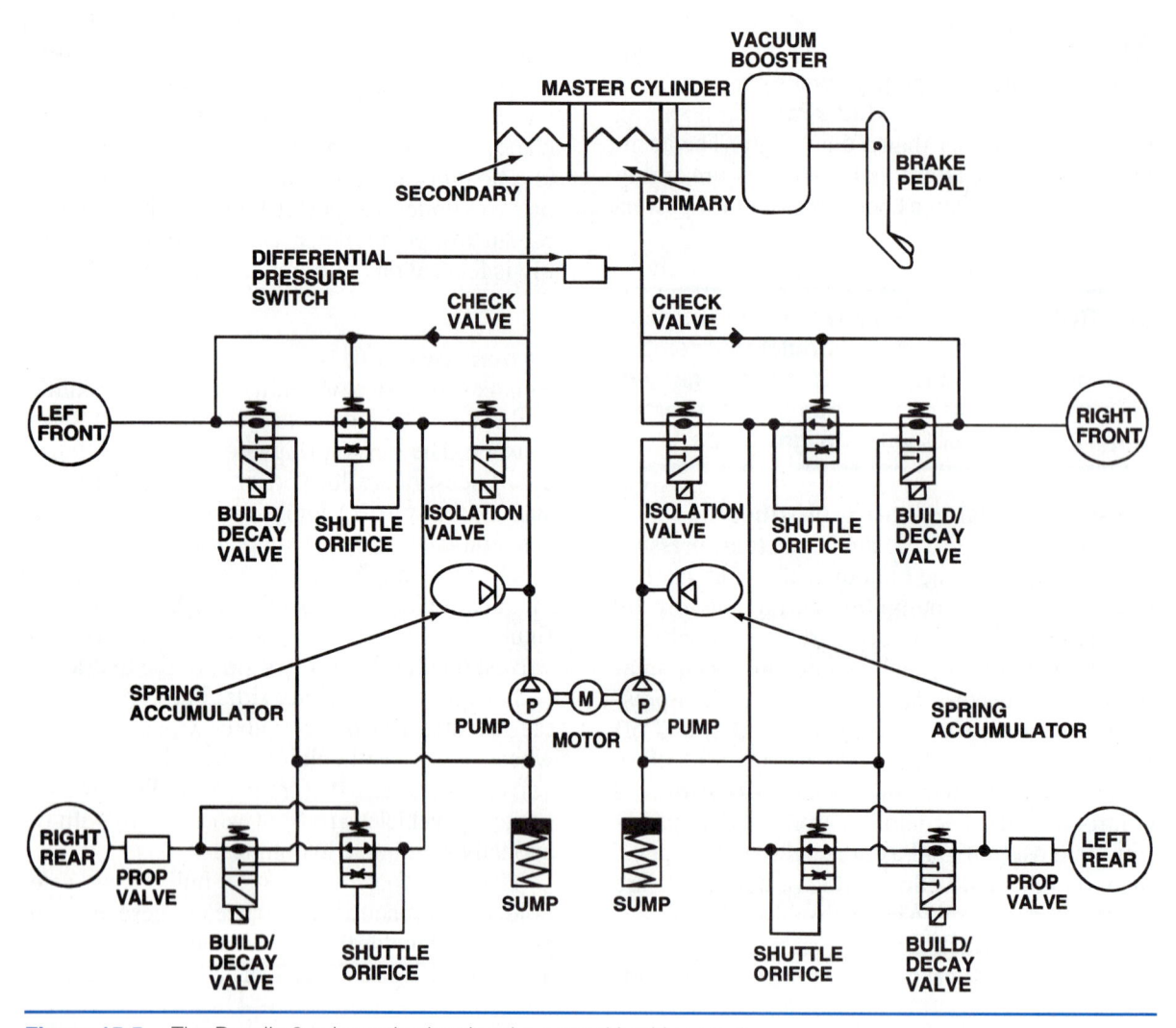

Figure 15-5. The Bendix 6 schematic showing the normal braking mode.

298

Antilock Brake Systems

increase it in the build mode. The front brakes are controlled individually, whereas the build/decay solenoids from both rear brake circuits are controlled simultaneously as a pair, giving three-channel ABS braking.

The modulator also contains four shuttle orifice valves, one for each wheel. These are hydraulically actuated valves that restrict fluid flow in a circuit between the isolation valve and build/decay valve when the build/decay valve is in use. This provides a more gradual build rate during an antilock stop for smoother ABS operation. The shuttle orifice valve in the affected wheel circuit remains in the restricted position until the ABS stop has been completed. When the build/decay valve returns to its rest position, pressure equalizes in the circuit and spring pressure opens the shuttle orifice valves.

Dual Accumulators

The modulator assembly has two accumulators, one each for the primary and secondary sides, figure 15-6. Both are piston- and spring-type accumulators and do not have a gas precharge. Each accumulator also has a small fluid sump that holds fluid temporarily during an ABS stop as fluid returns through the build/decay valve from the brake circuit. The pressure within the sumps is typically about 50 psi (345 kPa). When the ABS function ceases, the fluid in the accumulators is pumped back to the master cylinder fluid reservoir.

Pressure Switch

A differential pressure switch, located inside the modulator assembly, illuminates the red BRAKE warning light if there is a difference of 70 to 225 psi (483 to 1550 kPa) between the primary and secondary sides of the hydraulic system when the brake pedal is depressed, or if there is pressure differential of more than 225 psi (1550 kPa) at any other time.

Pump Motor and Control Module

The pump motor is part of the modulator assembly, and operates only during antilock stops. An electric motor drives two pistons, one for the primary side of the system and one for the secondary side. Power is provided by a relay located in the power distribution center (or on the left shock tower on vehicles without a power distribution center). Each pump takes

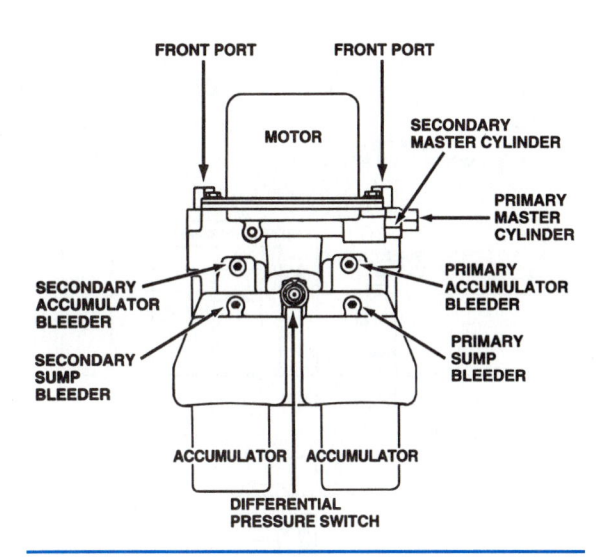

Figure 15-6. The Bendix 6 modulator with dual accumulators.

low-pressure brake fluid from its sump and delivers it at high pressure to one of the spring-loaded accumulators.

The ABS control module is located in the right side of the engine compartment, and has a 60-pin connector. It can cycle the solenoids 5 to 7 times a second during an ABS stop, not quite as fast as the Bendix 10 system, which cycles up to 20 times a second, but fast enough to provide antilock control. The pedal on the Bendix 6 ABS drops slightly and feels harder than ordinary during an ABS stop, which is normal for this system.

Bendix LC4 ABS

The Bendix LC4, which Bendix calls Bendix III but Chrysler refers to as LC4 (for "low cost" system), is a lighter, more compact version of the Bendix 6 nonintegral ABS it supersedes. LC4 is used on 1994–96 Chrysler, Dodge, and Plymouth minivans, Chrysler LeBaron, Dodge Spirit and Shadow, and Plymouth Acclaim and Sundance. It is a nonintegral ABS with a conventional master cylinder and vacuum booster, and is functionally similar to the Bendix 6 system.

System Description

LC4 is a four-wheel, four-channel ABS with four wheel sensors and four combination build/deacy solenoid valves in the modulator assembly, figure 15-7. Unlike Bendix 6, however, it uses no isolation solenoids because each brake

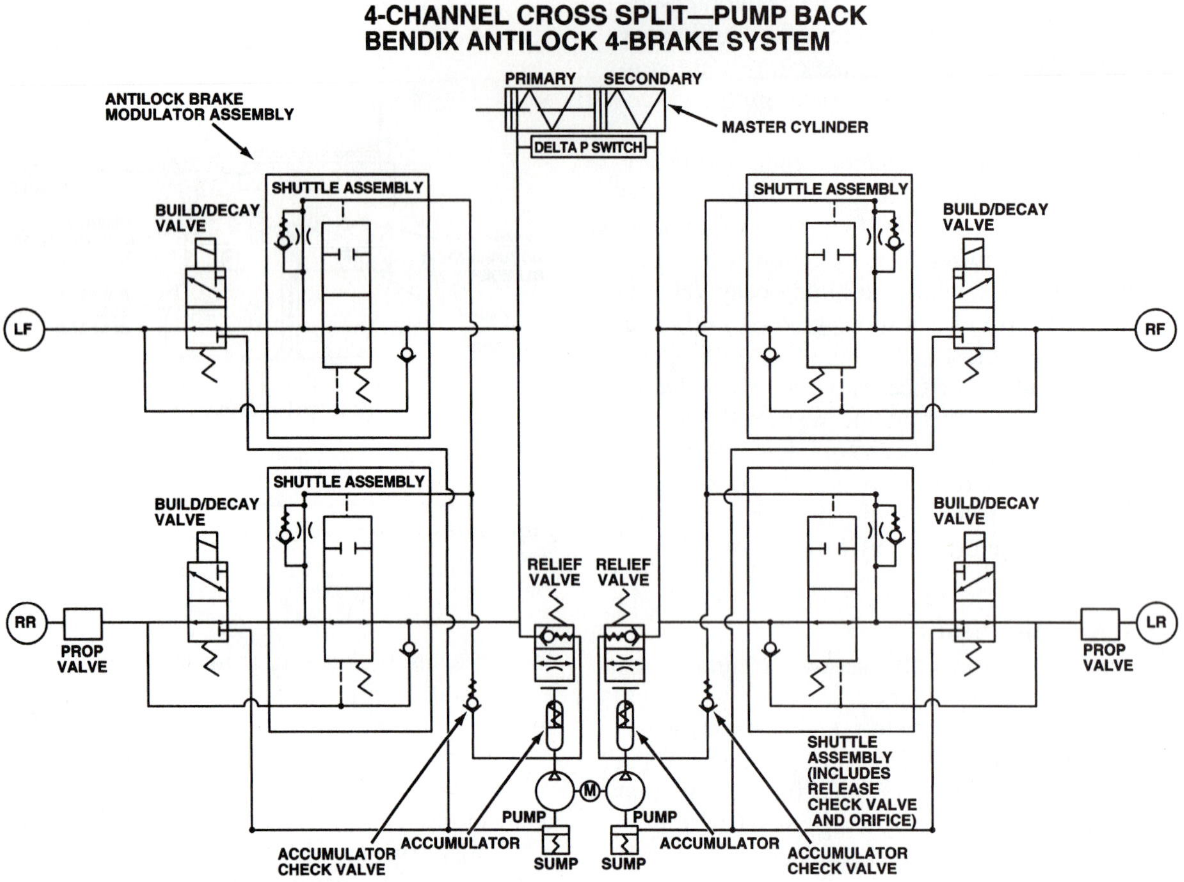

Figure 15-7. The Bendix LC4 system schematic.

is controlled separately. The brake system is split diagonally so the left front and right rear wheel are on the primary circuit, whereas the right front and left rear wheels are on the secondary circuit.

System Operation
In the released position, the ABS build/decay solenoids provide a direct fluid path from the master cylinder to the brakes. In the actuated (decay) position, they allow fluid to flow from the brakes back to the fluid sumps in the hydraulic modulator assembly. There are two sumps, one for the primary and one for the secondary brake circuits. The build/decay valves are spring loaded in the normally released (build) position.

Like Bendix 6, the LC4 modulator contains shuttle orifice valves for each wheel circuit. These valves restrict fluid flow between the isolation valve and build/decay valve when the build/deacy valve is in use. This provides a more gradual build rate during an antilock stop for smoother ABS operation. The shuttle orifice valve in the affected wheel circuit will remain in the restricted position until the ABS stop ends. When the build/decay valve returns to its rest position, pressure equalizes in the circuit and spring pressure returns the shuttle orifice valves to the normally open or unrestricted position.

Dual Accumulators
The LC4 modulator has two accumulators, one for the primary side and one for the secondary side, figure 15-8. Both are spring loaded with no gas precharge. The accumulators store pressurized fluid during ABS braking only and provide supplemental pressure when needed. During normal braking, no pressure is stored in the accumulators.

Each accumulator also has a small fluid sump that holds fluid temporarily during an ABS stop as fluid returns through the build/decay valve from the affected brake circuits. Pressure within the sumps is typically about 50 psi (345 kPa). When ABS ceases, the fluid is pumped back to the master cylinder fluid reservoir.

Antilock Brake Systems

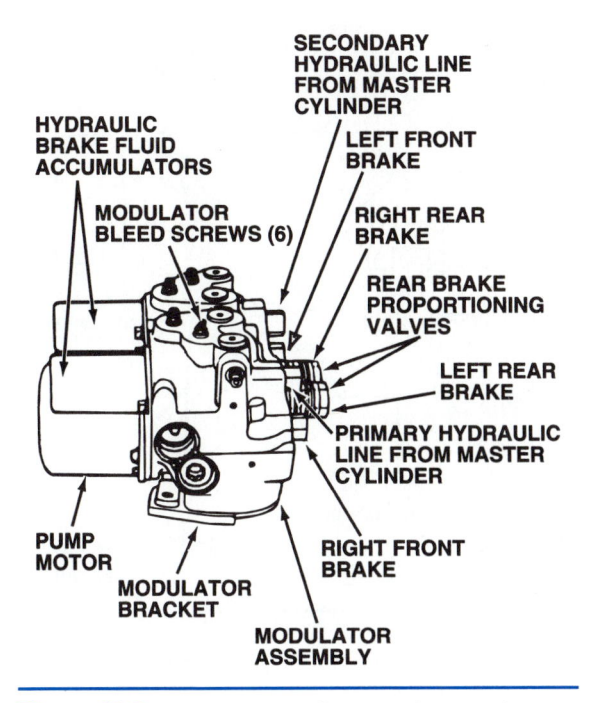

Figure 15-8. The Bendix LC4 modulator and pump assembly.

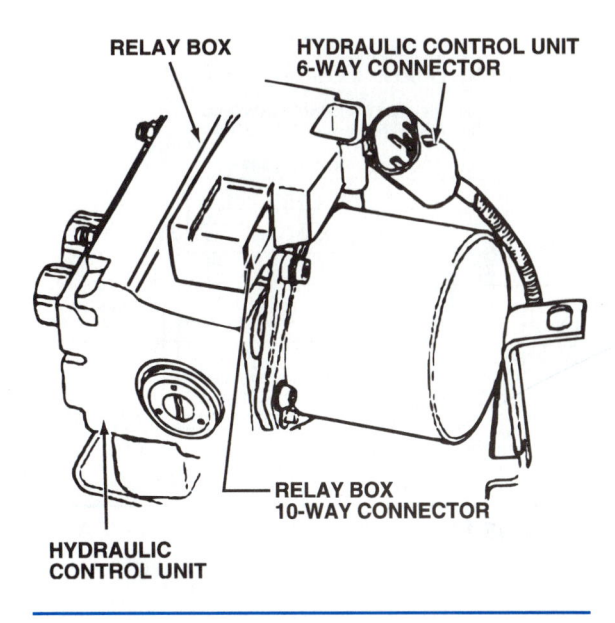

Figure 15-9. The Bendix ABX-4 hydraulic control unit.

Pump Motor and Control Module

The modulator also has a pump motor with a dual-piston pump. Each piston pumps fluid from its respective sump to its accumulator as needed during ABS braking. Power for the motor is routed through a relay mounted on either the left front inner fender shield or the front of the left strut tower. The ABS control module energizes the pump motor relay. The control module is mounted on the right front frame rail and has a 60-pin connector.

Bendix ABX-4 ABS

The Bendix ABX-4 antilock brake system is used on 1995–97 Dodge and Plymouth Neon models as well as the 1995–97 Chrysler Cirrus and Dodge Stratus. It is a four-wheel, four-channel ABS with a more, compact design than earlier Bendix systems.

Though ABX-4 is a nonintegral ABS with a conventional master cylinder and vacuum power booster, the master cylinder and booster on ABS-equipped Neon models are different than those without ABS. On models without ABS, the master cylinder has four outlets, two for the primary brake circuit lines and two for the secondary brake circuit lines. On models with ABS, the master cylinder has only two outlets (one for the primary brake circuit and one for the secondary brake circuit). Both lines are routed to the hydraulic modulator assembly, which Chrysler refers to as the hydraulic control unit (HCU).

Inside the master cylinder are more differences. Those without ABS use a standard compensating port design with two screw-in proportioning valves. For those with ABS, the master cylinder has a center valve design and the proportioning valves are located on the hydraulic control unit rather than the master cylinder.

The power booster has a coventional vacuum diaphragm. On models with ABS, though, a special seal between the master cylinder and booster just inside the booster opening seals vacuum. You must replace this seal if either the booster or master cylinder are removed or replaced. Failure to do so may cause a vacuum leak and a loss of power assist.

System Description

The ABX-4 system is functionally similar to the Bendix LC4 system and consists of a compact hydraulic modulator and pump assembly, figures 15-9 and 15-10. Inside the modulator are four ABS build/decay solenoids, one for each brake circuit. However, the system works like a three-channel system because both rear ABS solenoids are cycled simultaneously to improve vehicle stability.

As with Bendix 6 and LC4, shuttle orifice valves for each brake circuit restrict fluid flow be-

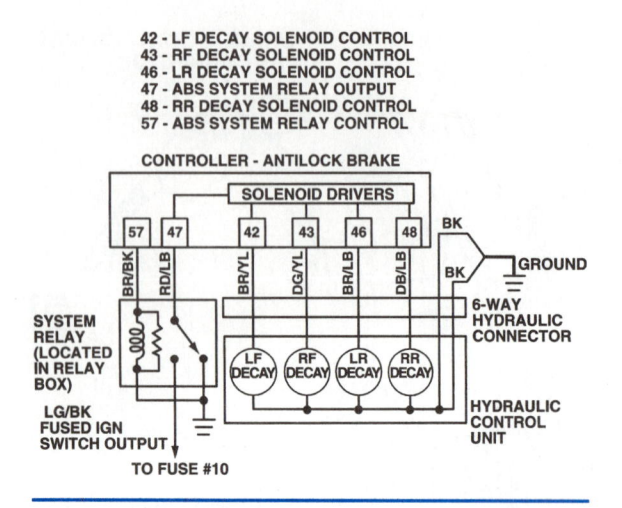

Figure 15-10. Wiring schematic for the Bendix ABX-4 hydraulic control unit.

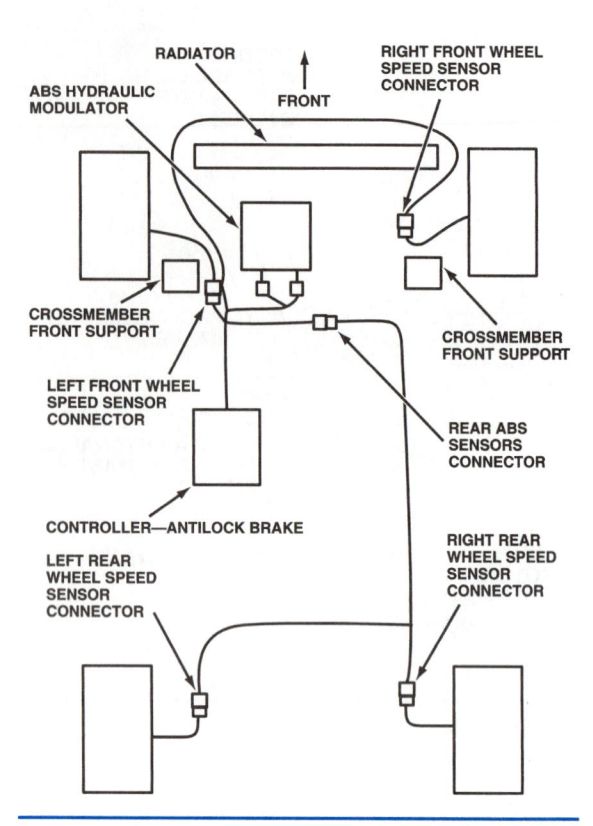

Figure 15-11. The Bendix ABX-4 wiring harness and component locations.

tween the pump and decay solenoid, providing a gradual build rate during an antilock stop for smoother ABS operation.

The 60-pin control module, which Chrysler refers to as the controller antilock brakes (CAB) on these applications, receives inputs from four nonadjustable wheel speed sensors, figure 15-11, and the brake pedal switch. The sensor rings for the front wheel speed sensors are located on the outboard CV joints. The rear wheel speed sensors bolt to the brake support plates and read sensor rings on the rear hubs. The module is located in the kick panel on the driver's side.

No Accumulators

Unlike the Bendix systems previously described, ABX-4 has no accumulators. There are two fluid sumps in the hydraulic modulator for the primary and secondary hydraulic circuits. The fluid sumps temporarily store brake fluid vented from the brake circuits during ABS braking. The fluid is then delivered to the pump to provide additional build pressure. The typical pressure in the sumps is 50 psi (345 kPa) during ABS braking only.

Like the Bendix LC4 system, a single electric pump motor drives a dual-piston pump (one piston for each hydraulic circuit). The motor runs only during ABS braking to provide build pressure. When the system needs additional build pressure, the control module grounds the pump motor relay, which is also mounted on the modulator, to start the pump.

Bendix Mecatronic II ABS

The 1995–98 Ford Contour and Mercury Mystique use the Bendix Mecatronic II ABS. It is a four-wheel four-channel, nonintegral ABS with four wheel speed sensors and a conventional master cylinder and power booster.

The Mecatronic II ABS also provides traction control. Below 31 mph it uses braking alone to control wheel spin. Above 31 mph, a throttle relaxer backs off the throttle opening to reduce engine power. If the driver steps on the brake pedal while traction control is operating, the system uses the brake switch to deactivate traction control, causing the system to go into ABS brake mode.

System Description

The hydraulic modulator assembly has four ABS solenoids, plus two additional solenoids on models equipped with traction control, figure 15-12. The control module, relay, and pump are also mounted on the compact "unitized" modulator assembly.

During ABS braking, Mecatronic II uses "closed-loop" hydraulic control. Fluid pressure relieved from the brake channels is held temporarily within an internal low-pressure reservoir in the

Antilock Brake Systems

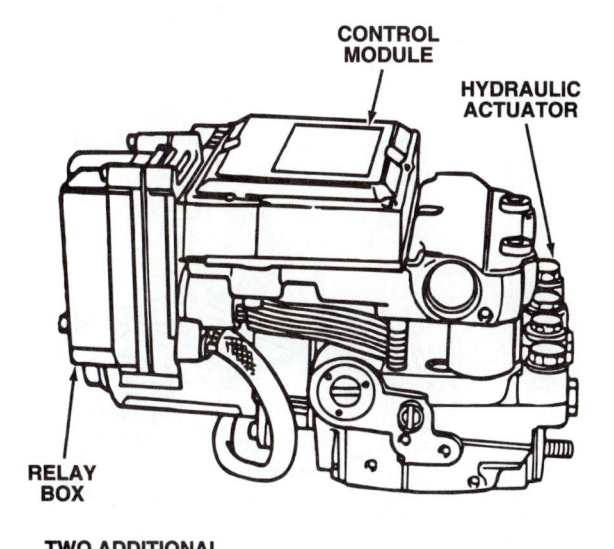

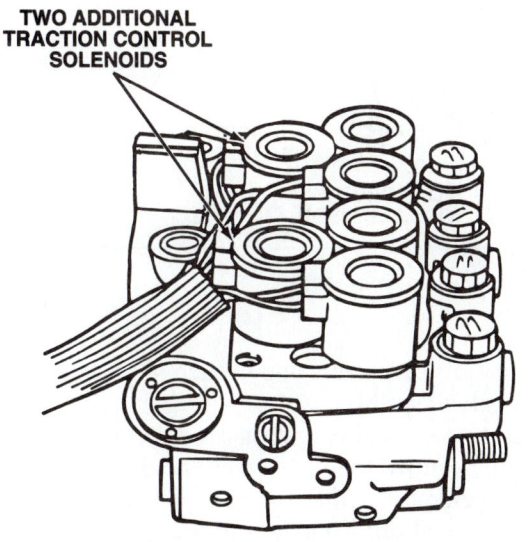

Figure 15-12. The Bendix Mecatronic modulator controls ABS and drive wheel traction.

modulator. Because there are no accumulators, the pump provides additional build pressure as needed. Like the Bendix LC4 system, this system uses a single pump motor with a dual-piston pump that is capable of supplying hydraulic pressure to either brake circuit.

The front and rear wheel speed sensors are similar to those on the 1994 Ford Taurus and Mercury Sable. The front sensors attach to the knuckle assembly and read a sensor ring on the outboard CV joint. Rear sensors mount on the brake backing plates and read sensor rings on the rear hubs. Sensors are nonadjustable.

Dual Microprocessors

The Mecatronic control module contains two **microprocessors.** The first controls the function of the ABS solenoids and pump motor. The second monitors and compares inputs from the wheel speed sensors, overseeing the overall operation of the system.

The relay box, which is located on the modulator next to the control module, includes the ABS main relay and a second relay for the pump motor. Also included are diodes to protect the electronics from polarity reversal and power surges when the system energizes the main ABS relay. The main ABS relay provides power to the ABS solenoids and the pump motor relay. The pump motor relay, however, is not activated by the control module until ABS braking is needed. If the main ABS relay fails, the ABS warning light comes on and deactivates the system.

Traction Control

Applications with traction control, figure 15-13, have two additional solenoids in the modulator. These solenoids create a hydraulic path that allows the system to reverse the ABS control process. When the control module detects wheel spin during acceleration, it energizes the pump to create hydraulic pressure. At the same time, it energizes the appropriate ABS solenoid to route brake pressure to the spinning wheel. The appopriate traction control system (TCS) solenoid is also energized to prevent brake pressure from entering the brake reservoir. This isolates the spinning drive wheel brake so it is the only brake receiving hydraulic pressure during traction control.

A warning indicator on the instrument panel signals the driver when the traction control system is operating. When traction control is no longer active, the light goes out.

Throttle Actuator

The Mecatronic traction control system uses a throttle actuator to back off the throttle, reducing engine power to help control wheel spin, figure 15-14. The throttle actuator works only at speeds more than 31 mph (50 km/h), whereas traction control *braking* is deactivated above 31 mph to preserve handling and steering stability.

The throttle actuator, located in the right front corner of the engine compartment, is attached to the throttle cable linkage. A potentiometer on top of the throttle actuator informs the ABS/TCS control module about the exact location of the throttle. The actuator consists of an electric motor and two drive disks, each of which are connected to

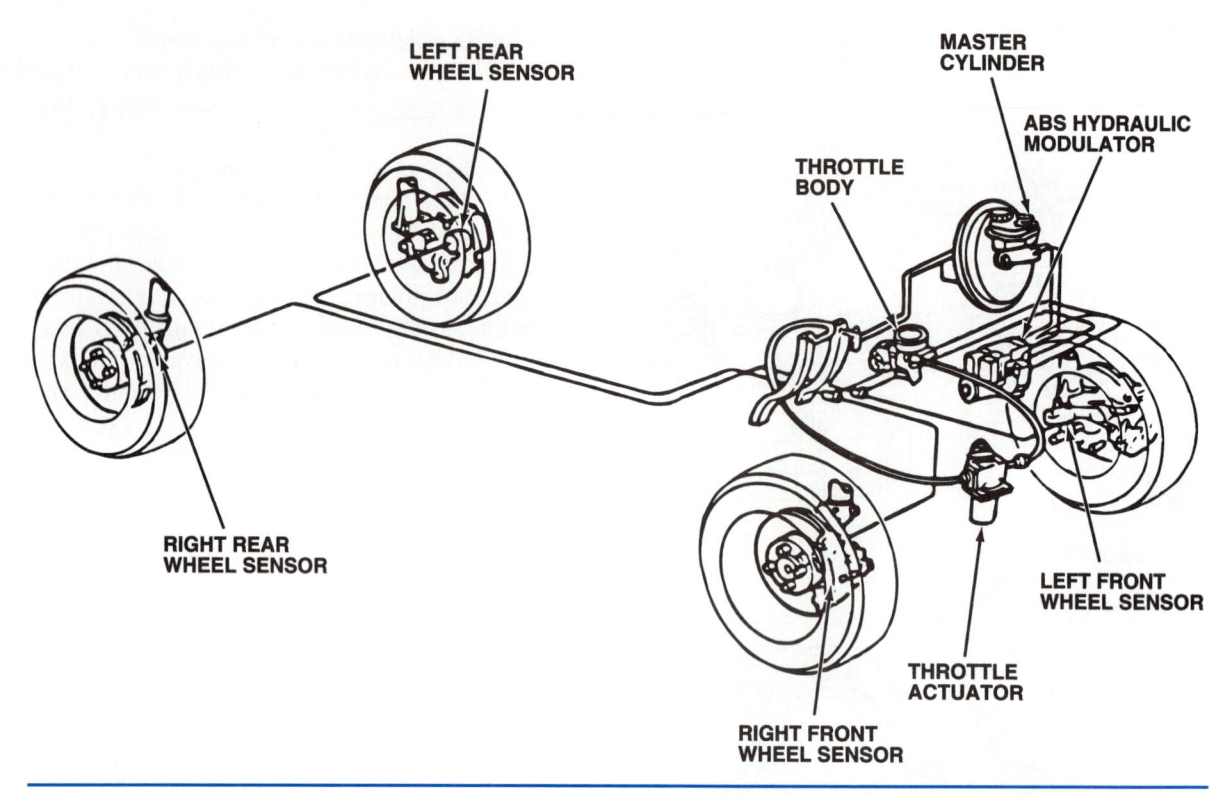

Figure 15-13. A Bendix Mecatronic system with traction control.

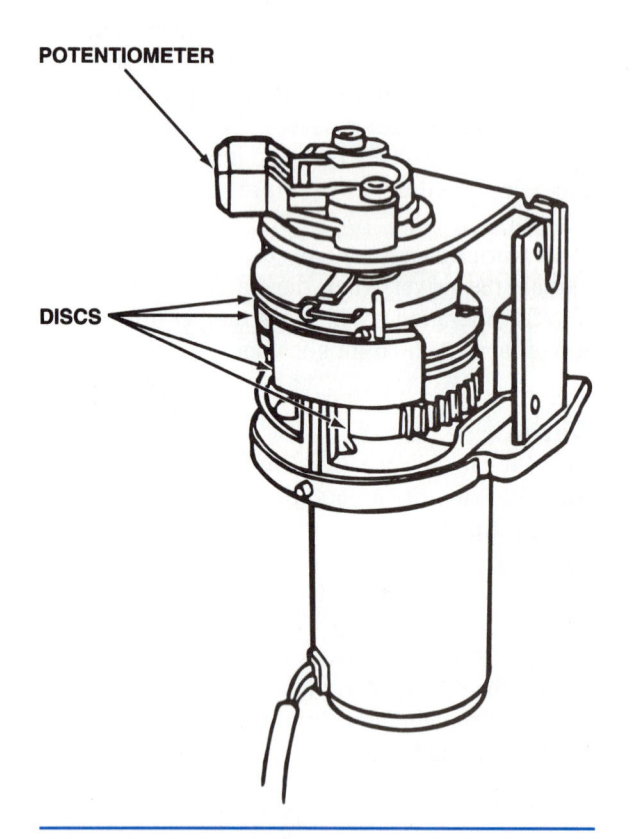

Figure 15-14. The Bendix Mecatronic throttle actuator pushes back the throttle linkage to reduce wheel spin.

the split throttle cable and to each other by means of a helical spring.

During normal operation (no traction control), the disks move together when the driver presses the accelerator pedal and have no effect on engine operation. When throttle reduction is needed to control traction, the electric motor rotates the drive disks, causing the throttle cable to lengthen or shorten. When the system reduces the throttle opening, the driver feels the accelerator pedal pushing back against his or her foot. If the driver continues to push down on the pedal, the system will continue to offer resistance but will allow a more gradual increase in power. When traction control is no longer needed, the actuator allows a smooth transition to occur from the TCS mode to prevent any sudden surge in vehicle speed.

BOSCH ABS

Bosch has one integral ABS, the Bosch 3; and a variety of nonintegral ABS designs including the Bosch 2, 2S, 2E, 2U, 2U Micro, Bosch 2U ABS/ASR, Bosch 5, 5.3, 5.7, and Bosch 8.0.

Antilock Brake Systems

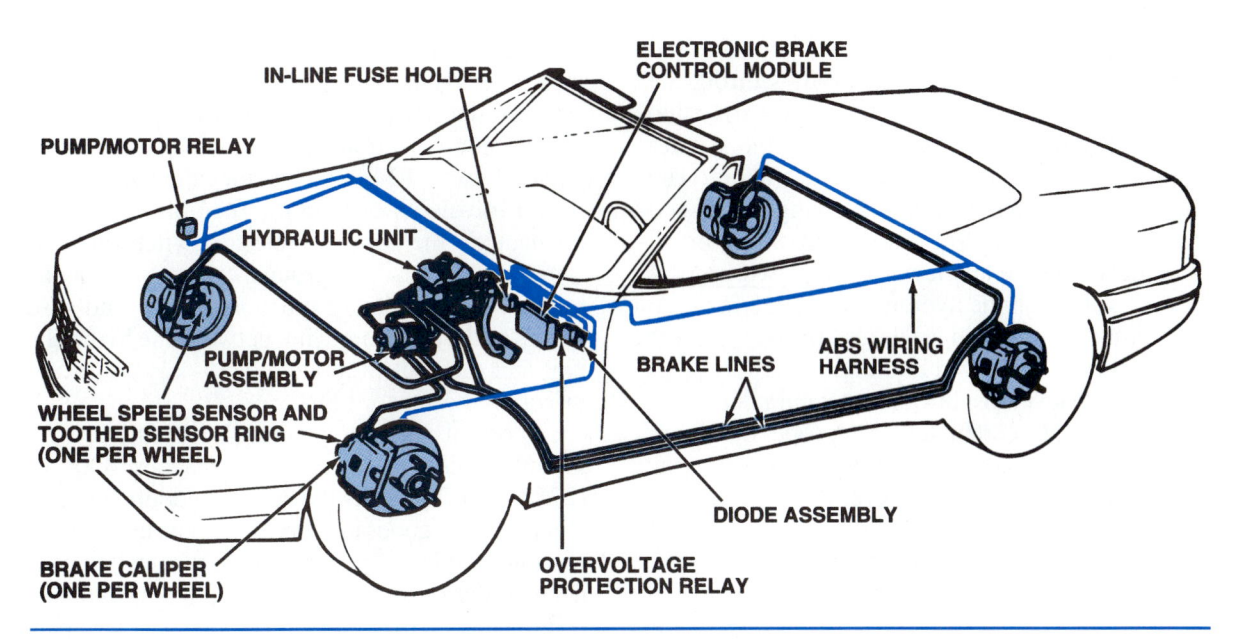

Figure 15-15. A Bosch 3 system overview on the Cadillac Allante.

Bosch 3 ABS

The Bosch 3 ABS is found on 1987–92 Cadillac Allante, figure 15-15, and 1988 to early 1990 Chrysler Fifth Avenue, Imperial, New Yorker, and Dodge Dynasty models. It is an integral ABS with a combined master cylinder and hydraulic modulator assembly. A high-pressure pump and accumulator provide power assist, rather than a conventional vacuum booster.

System Description

Bosch 3 is a four-channel ABS, with a separate three-way ABS solenoid valve for each brake circuit. This permits independent control of each front and rear brake. The system uses four wheel speed sensors. The front wheel sensors have their sensor rings on the outer CV joint housings. The rear sensor rings are an integral part of the rear hub bearing assembly. On Chrysler cars, front and rear sensors are nonadjustable. On Allante, the rear sensors are nonadjustable but the front sensors are mounted in brackets that require a sensor air gap adjustment of 0.020″ (0.51 mm).

The Bosch 3 control module is referred to as an antilock brake control module (ABCM) by Chrysler, whereas Cadillac calls it an electronic brake control module (EBCM). Except for the 1990–92 Allante with traction control, the Chrysler and Cadillac modules both have 35-pin wiring connectors and use essentially the same

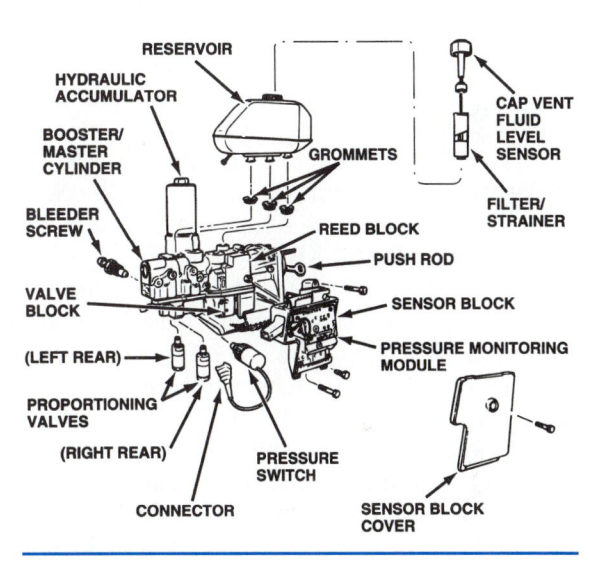

Figure 15-16. A Bosch 3 integral hydraulic control unit used on a Chrysler.

electronics. The Allante application with traction control, however, has a different module with a 55-pin connector. Chrysler locates the module on the rear seat bulkhead in the trunk, and Cadillac mounts it under the dash to the left of the steering column on the Allante.

Hydraulic Assembly

The combination modulator/master cylinder/booster assembly, figure 15-16, provides power-assisted braking as well as antilock braking. The master cylinder is in the upper portion of the unit.

When the driver presses the brake pedal, stored pressure from the accumulator routes through a booster control valve into the master cylinder. This pressure pushes the piston in the master cylinder forward for power-assisted braking. The master cylinder has standard metering valves and check valves to control the movement of brake fluid during brake application. Also, two proportioning valves on the hydraulic unit, one for each rear brake circuit, reduce brake pressure to the rear brakes for balanced braking.

The fluid level inside the master cylinder reservoir must be at a certain level for proper operation of the ABS, so the reservoir filler cap has a built-in fluid level sensor. If the level drops too low, a switch in the sensor closes to signal the control module. The control module then illuminates the ABS warning light and deactivates the ABS.

Reed and Valve Blocks

Inside the hydraulic assembly is a reed block, containing two switches to keep the control module informed about the position of the booster piston. These switches are named, S2A and S2B.

Attached to the bottom of the master cylinder is the valve block that contains four three-way ABS solenoid valves, one for each brake circuit. As with other Bosch ABSs, the three-way solenoid valves are normally in the open position to allow hydraulic pressure from the master cylinder to flow to each of the brake circuits. The system energizes the solenoids with a solenoid valve relay located in the sensor block on the modulator assembly. The control module monitors the status of the relay through a feedback circuit. If the relay's actual position and the commanded position do not agree, the control module sets a fault code and disables the ABS.

Replenishing Valve and Sensor Block

The valve block also contains a replenishing valve, which helps maintain pedal height during hard braking or antilock braking. The replenishing valve's solenoid, called the pilot solenoid, is energized by the control module via the solenoid valve relay. Opening the valve redirects brake fluid from the booster servo circuit back into the master cylinder. This may produce a slight "clunk" that can be heard and felt. A replenishing valve status switch allows the control module to monitor the valve's operation. If the switch fails to open when the replenishing valve is energized, the control module will log a fault code and illuminate the ABS warning light.

Connected to the left side of the valve block is the sensor block, which can be replaced as a separate component. The sensor block provides all electrical connections for the solenoids in the valve block (a 12-pin connector), the two booster piston travel switches in the reed block (a 2-pin connector), the control pressure switch (another 2-pin connector), the ground connection for the hydraulic unit (single-pin connector), and also serves as the electronic link between the hydraulic unit and the control module (a 15-way connector on top of the sensor block).

A control pressure switch signals the control module when boost pressure reaches 400 psi (2758 kPa), This tells the control module whether the pump is generating pressure during power-assisted braking. The control pressure switch connects to the sensor block by a 2-pin connector.

In addition, the sensor block contains the relay that powers the four three-way ABS solenoid valves and repelnishing valve. It also contains the pressure monitoring module that control the operation of the pump. The pressure monitoring module has two redundant switches that turn the pump motor relay on and off. The switches ground the pump relay and turn the pump on when pressure inside the accumulator drops below 2100 psi (14,480 kPa). The switches open and turn the relay and pump off when accumulator pressure reaches 2600 psi (17,927 kPa). If both switches do not open and close at the same time, the control module logs a fault code, illuminates the ABS warning light, and deactivates the ABS.

Accumulator

The accumulator mounts upright on the right side of the unit. It stores pressure generated by the pump for power-assist braking as well as antilock braking. On 1990–92 Allantes, it also provides pressure when braking is required for traction control. A sliding piston inside the accumulator has a nitrogen precharge of 800 psi (5516 kPa). As fluid is pumped into the accumulator, it pushes the piston up, compressing the nitrogen gas to achieve a working pressure of 2100 to 2600 psi (14,480 to 17,927 kPa), which is sufficient to provide 10 to 20 power-assisted stops without recharging from the pump.

CAUTION: As with all integral ABSs, the pressure inside the accumulator must be discharged prior to doing any brake work. Pump the brake pedal 25 to 40 times with the ignition off.

The accumulator, master cylinder, and other hydraulic components that make up the hydraulic assembly are not serviceable. If may one is defective, the entire hydraulic assembly must be replaced. The only parts that can be replaced separately are the plastic fluid reservoir, fluid level sensor (part of the fluid reservoir cap), sensor block, control pressure switch, and two rear brake proportioning valves.

Pump/Motor Assembly

The Bosch 3 pump/motor assembly is separate from the hydraulic assembly. On the Chryslers and the Allante, it is located on a transaxle bracket on the left side of the engine compartment below the hydraulic unit. Chrysler also covers it with a heat shield.

A relay, controlled by the two pressure switches in the sensor block pressure monitoring module, turns the pump/motor on and off. The pump relay on Chrysler applications is located on the left inner fender shield just ahead of the battery. On the Allante, the relay is mounted on the right fender rail behind the coolant reservoir. When the ignition is turned on, the relay coil receives voltage from the ignition circuit. But it will not close its contacts to complete the pump motor circuit unless pressure in the accumulator is less than 2100 psi (14,480 kPa). If accumulator pressure is low, the two pressure switches in the sensor block close and energize the relay, which then feeds current to the pump. The pump runs continuously until pressure in the accumulator reaches 2600 psi (17,927 kPa), up to 30 seconds or longer, depending on how long the vehicle was idle. When the two pressure switches open again, the relay is de-energized, and the pump shuts off.

To protect the ABS control module and sensor block from dangerous spikes, voltage spikes, an overvoltage protection relay is included in the system's wiring harness. This relay also functions as the main system power relay that routes voltage to the ABS when the ignition is on. Chrysler locates the overvoltage protection relay just aft of the pump/motor relay on the left fender side shield. On Allantes, it is next to the ABS diode under the dash.

The ABS control module monitors input from the brake light switch to confirm whether the brakes are applied when the hydraulic unit is called on to provide power assist. The input serves as a diagnostic check only and plays no role in the operation of the pump/motor.

System Operation

The Bosch 3 integral ABS provides power-assisted braking as well as antilock braking. The system programing allows no more than 12 to 30 percent wheel slippage before the control module starts ABS braking. To initiate ABS braking, the control module energizes the ABS solenoid valve relay and activates the appropriate ABS solenoid valve to prevent further pressure increase to the brake that is locking. It moves the three-way ABS solenoid valve to the hold position, which isolates the brake circuit and prevents any increase in pressure to the brake. The replenishing valve also opens to reroute brake fluid from the booster servo circuit back into the master cylinder. This helps to maintain pedal height. The ABS solenoid valve is then cycled to the release position so pressure can be vented from the circuit, allowing the wheel to regain traction.

The Bosch 3 system does not have spring-loaded accumulators like other Bosch ABS, so the fluid flows through the solenoid back to the reservoir where it is recycled by the pump to the accumulator and master cylinder. The ABS solenoid is then de-energized and allowed to return to the rest or open position so hydraulic pressure can be reapplied to the brake. The replenishing valve remains open as long as ABS braking occurs. The hold-release-apply cycle repeats up to 10 times a second as long as needed or until the car's speed drops below 3 to 5 mph. At that point, the ABS system reverts back to standby mode and normal braking resumes. This may result in a momentary locking of the brakes.

Traction Control (1990–92 Allante Only)

To handle the additional task of braking the drive wheels for traction control, the Allante traction control system (TCS) has a separate TCS plunger assembly mounted next to the hydraulic unit. The assembly contains two TCS plunger solenoid valves and two plungers that are connected to the front brake circuits. The TCS solenoids are powered by the same relay in the sensor block that powers the ABS solenoids in the hydraulic unit. You can replace the TCS plunger assembly as a separate unit if it is defective, but it contains no serviceable parts inside.

The traction control system also communicates with the engine control module to further control wheel spin. The engine module will temporily deactivate up to four fuel injectors, one at a time in 3-second intervals. At the same time, the ABS module commands the hydraulic unit to brake on or both drive wheels to further reduce wheel slippage. The braking and power reduction will continue until it is no longer needed, at which point the injectors are reactivated one at a time in 1-second intervals to give a smooth transition back to normal.

When traction control is needed, the control module energizes one or both TCS solenoid valves to brake one or both drive wheels. Power for the TCS solenoids comes from the same solenoid valve relay on the sensor block that supplies the ABS solenoids. Energizing the TCS solenoid valve opens a hydraulic circuit that allows pressure from the accumulator to enter a spring-loaded plunger. The plunger, in turn, acts like a single circuit minimaster cylinder to apply pressure to its brake. One plunger is connected to the circuit for the right front caliper, whereas the other plunger is connected to the left front caliper.

The control module provides two levels of ground to the TCS solenoids. A low-resistance ground allows 5 amps to travel through the TCS solenoid circuit. This pulls the TCS solenoid to a position that blocks the line from the master cylinder and opens the circuit through which pressure from the accumulator and pump can flow to the spring-loaded plunger that applies the brake. The higher-resistance ground provided by the control module cuts the solenoid current to 2 amps, causing it to move to the hold position. This maintains steady brake pressure at the wheel until braking is no longer needed. At that point, the control module cuts power to the TCS solenoid and it returns to its normal position. Residual pressure in the brake circuit pushes the spring-loaded plunger back to its normal position, blocking further pressure from the accumulator and reopening the normal brake line to the master cylinder. Cycling of the TCS solenoids can occur several times a second as needed to control wheel spin during acceleration.

Bosch 2 and 2S ABS

Mercedes used the first Bosch 2 ABSs in Europe back in 1978. But it was not until 1985 that Mercedes and BMW offered Bosch 2 ABS as standard equipment on export models for the North American market. Since then, Bosch 2 ABS has been used on a wide variety of domestic and import vehicles, including Audi, BMW, Chevrolet Corvette, Infiniti, Lexus, Mercedes, Porsche, Mazda, Mitsubishi, Nissan, Rolls-Royce, Subaru, and Volvo.

Bosch 2 is a nonintegral ABS with a conventional master brake cylinder and vacuum power booster. The hydraulic modulator assembly is a separate unit.

System Description

Bosch 2 ABS provides four-wheel ABS braking with three control channels. The front wheels are controlled independently; the rear brakes share a common ABS circuit.

On most applications, figure 15-17, only three wheel speed sensors are used: one for each front wheel and a common sensor for both rear wheels. On some applications, such as the Corvette, separate sensors are used for each rear wheel.

In applications where each rear wheel has its own speed sensor, the select-low principle of control is used. The control module monitors the

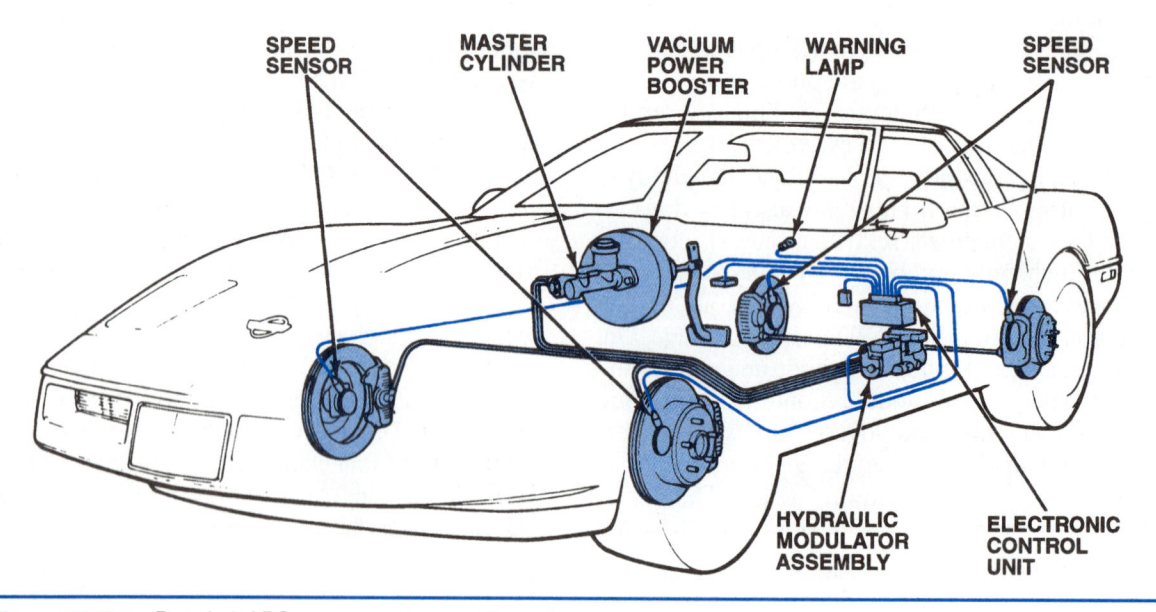

Figure 15-17. Bosch 2 ABS components on a Corvette.

Antilock Brake Systems

■ The Growth of ABS Applications

The first production vehicle equipped with antilock brakes as standard equipment was the 1972 Jensen Intercepter. That same year, the first electronic skid control system for truck and bus air brakes was introduced. The obvious safety aspects of antilock brakes quickly became apparent, and in 1975 the Federal Motor Vehicle Safety Standard 121 called for the installation of antilock brakes on most heavy trucks and buses. The standard was withdrawn in 1978, however, because of the limitations of ABS technology available at the time. Interest in antilock brakes in the United States waned for a number of years—but not in Europe.

In 1978, Mercedes-Benz and BMW made antilock brakes available for the first time on a limited production basis. The "new" ABS they offered was supplied by Bosch and featured a much faster and more reliable digital computer as well as electric ABS solenoid valves that allowed brake pressure to be cycled more than 10 times per second. The improvement in braking performance over the earlier ABSs was significant, as was the improvement in braking performance compared to vehicles with standard brakes. The new ABS technology was immediately recognized as a major safety innovation and was soon adopted by other European vehicle manufacturers on their vehicles.

In 1985, the first Audis, BMWs, and Mercedes with antilock brakes were exported from Europe to the North American market. The following year Ford became the first domestic vehicle manufacturer to offer a state-of-the-art antilock system on the Lincoln Mark VII and Continental. General Motors soon followed, offering ABS on a number of full-size models.

Within a few short years, ABS was offered either as an option or standard equipment on a growing number of luxury model passenger cars as well as sports cars, pickup trucks, and vans. By 1990, a quarter of all new cars and light trucks had ABS either as standard equipment or as an option. By the 1995 model year, advancements in ABS design, reduced manufacturing costs, and increased public demand swelled ABS use to more than 90 percent of all new models.

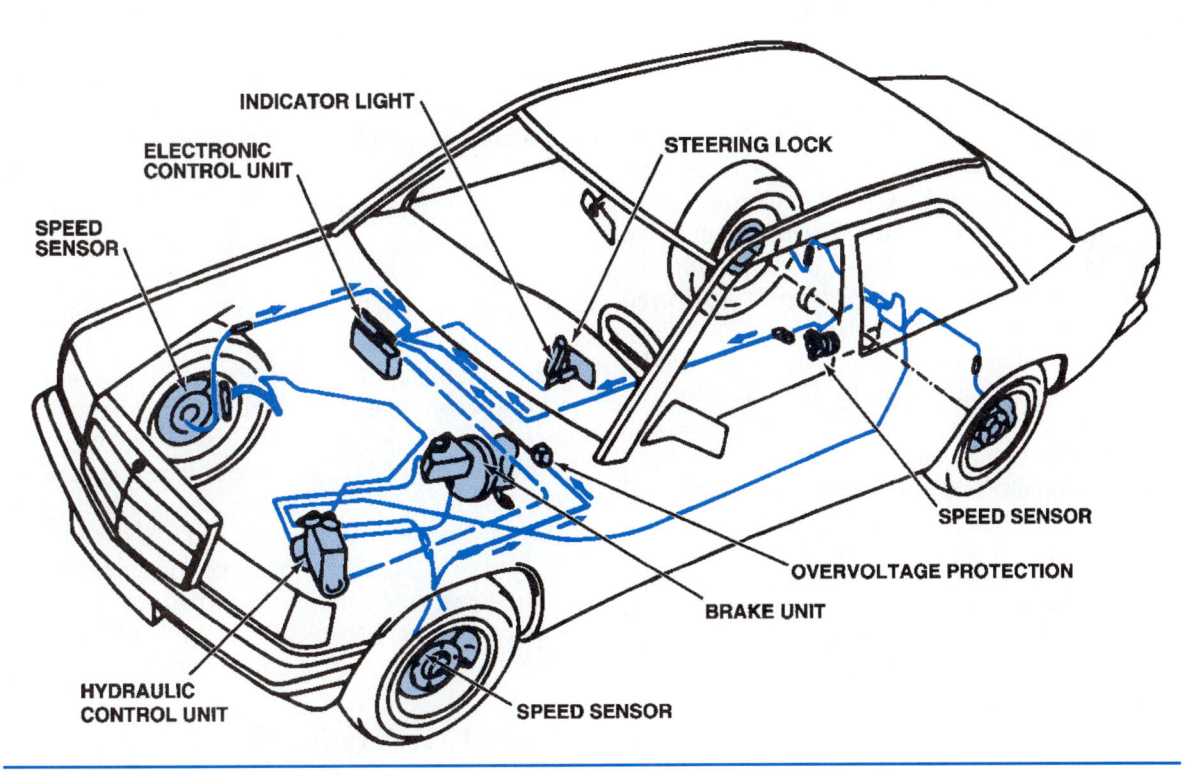

A Mercedes-Benz ABS.

speed of both rear wheels and reacts to the one that is slowing (decelerating) the fastest. Consequently, if either rear wheel starts to lock up during a hard stop, the module activates the rear brake ABS circuit and cycles the pressure to both rear brakes to prevent wheel lockup.

The sensor rings for the wheel speed sensors may be mounted inside the front wheel hub assembly, on the brake rotor or drum, on the axle shaft or constant velocity joint housing. On applications where the rear wheels share a common sensor, the sensor ring is mounted on the pinion input shaft in the differential. Air gaps are nonadjustable on some applications: Corvette, Infiniti, Lexus, Mercedes, Porsche, and Toyota; but adjustable on others: Audi, BMW, Mazda, Mitsubishi, Nissan, and Volvo.

Hydraulic Modulator

The hydraulic modulator assembly, figure 15-18, has three three-position ABS solenoids (one for each front brake and one for the combined rear brake circuit), a high-pressure return pump, two fluid accumulators, and pump and solenoid relays. The modulator assembly is located in the engine compartment on most vehicles, but on Corvettes it is located behind the driver's seat under an access panel.

Each of the ABS solenoid valves in the modulator is normally in the open position. Voltage is supplied through a relay on the modulator assembly. When a small current (1.9 to 2.3 amps) is applied to one of the ABS solenoids, it pulls the valve up to its second or intermediate position, which isolates the brake circuit and prevents any additional pressure from reaching the brake. When current to the solenoid is increased (4.5 to 5.7 amps), it pulls the valve to the third position which opens a port and releases pressure from the brake circuit.

The fluid that flows back through the release port routes into a spring-loaded accumulator. There are two accumulators in the modulator, one for the front brakes and one for the rear brakes. The accumulators hold the fluid that is released from the brake circuits until the return pump can pump in back to the master cylinder (which produces the pulsations that are felt in the brake pedal during ABS braking). Each of the accumulators can hold about 2 cc of brake fluid, and operates at pressures of less than 148 psi (1020 kPa).

The return pump receives its power through a pump relay mounted on the modulator. The ABS control module switches the pump relay on when the return pump is needed during ABS braking. The pump draws about 45 amps and is capable of

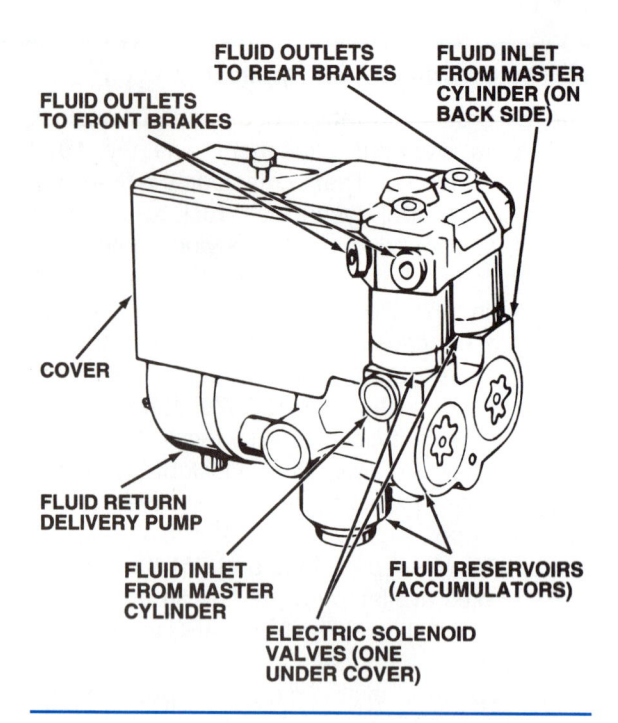

Figure 15-18. A Bosch 2 hydraulic modulator.

moving 3.5 cc of brake fluid per second at 2900 psi (19,996 kPa). High pump pressure is needed to overcome brake pressure from the master cylinder.

Of all the components in the modulator assembly, only the relays and pump motor are serviceable. If one of the solenoid valves is defective or if the modulator assembly is leaking, the entire modulator assembly must be replaced.

Control Module

The control module may be located in the engine compartment, passenger compartment, or trunk depending on application. It receives inputs from each of the wheel speed sensors, the brake pedal switch, and a lateral acceleration switch (Corvette only) or deceleration sensor (Toyota All-Trac only). Outputs go to the pump motor and solenoid relays, the three three-way ABS solenoid valves, and the ABS warning light. The system is capable of cycling the ABS solenoids continuously from 4 to 10 times per second until ABS braking is no longer needed or until the vehicle comes to a stop. At that point, the controller de-energizes the solenoid and pump relays and reverts back to standby mode.

Lateral Acceleration Switch (1986–89 Corvette Only)

The lateral acceleration switch signals the control module when the car is cornering hard at

Antilock Brake Systems

high G-loads (side forces). Hard cornering produces body roll and/or yaw in the chassis which changes the loading on the tires. To compensate, the control module changes its ABS braking characteristics slightly when it receives a signal from the lateral acceleration switch.

The switch itself actually contains two mercury switches connected in series. One switch opens when making hard left turns (more than 0.6 G) and the other when making hard right turns. Both switches are normally closed. When the car goes into a hard turn, the appropriate switch open and causes a voltage loss to module terminal 13. The control module then switches to modified ABS programming in case antilock braking is needed. This improves overall braking performance and handling stability in such situations.

Deceleration Sensor (Toyota All-Trac Only)

Toyota's special deceleration sensor compensates for changes in suspension loading during deceleration and acceleration. The sensor, mounted near the ABS control module in the trunk, is a position-sensitive mercury switch. It functions similarly to the lateral acceleration sensor used on the Corvette except it reacts to G-loads during deceleration and acceleration rather than cornering. This signals the ABS module to modify its control strategy.

Bosch 2E ABS

Bosch 2E is a variant of Bosch 2 and is also a non-integral, four-wheel, three-channel ABS with a conventional master cylinder and brake booster. It is used on 1991 and later Plymouth Laser, Eagle Talon, Dodge Stealth, and Mitsubishi 3000 GT, Eclipse, and Galant models.

Each wheel has its own adjustable wheel speed sensor with a specified air gap of 0.012″ to 0.035″ (0.3 to 0.9 mm). But on all-wheel drive applications, rear sensors are not adjustable.

The hydraulic modulator contains three, three-way ABS solenoids: one for each front brake and a common solenoid for both rear brakes. The control module governs the rear brakes with the select-low principle. Relays for the pump motor and ABS solenoids are located on the modulator.

Stealth and Mitsubishi 3000 GT all-wheel drive applications have a G-force sensor located under the back seat to provide lateral acceleration information to the controller. The sensor is essentially an on–off switch that signals the controller when G-forces exceed about 0.6 G.

Bosch 2U ABS

The Bosch 2U ABS system, figure 15-19, functionally similar to the Bosch 2 and 2E systems, is on 1991 and later full-size GM rear-wheel drive passenger cars, such as the Buick Riviera, Reatta, and Roadmaster wagon; Cadillac Brougham RWD and Eldorado Touring Coupe (through 1992 only); Chevrolet Caprice; and Oldsmobile Toronado. It is also on the 1993–98 Mercury Villager and Nissan Quest minivans, and the 1994–98 Ford Mustang. The differences that set this system apart from other Bosch 2 systems are:

1. The Bosch 2U modulator assembly has a different design and appearance but functions in exactly the same way as the one used in Bosch 2 applications. It contains three, three-way ABS solenoid valves, one for each of its three control channels (each front brake and a combined control channel for both rear brakes) and uses the select-low principle for the rear brakes. Because of physical differences, the Bosch 2U modulator is not interchangeable with the Bosch 2 modulator.

2. As a rule, Bosch 2U in rear-wheel drive (RWD) applications uses only three wheel speed sensors. Front-wheel drive (FWD) applications use four wheel speed sensors. The single wheel speed sensor for the rear brakes on the RWD GM applications is located in the differential, where it monitors the average speed of both rear wheels. Front-wheel drive GM applications have a separate speed sensor at each rear wheel. The front sensors are mounted in the steering knuckles, and the sensor rings are pressed into the backs of the hub and rotor assembly. Rear sensors are mounted on the rear brake backing plates.

3. The Bosch 2U control module for General Motors applications has the ability to provide diagnostic fault codes. This eliminates the need for a special dedicated ABS tester and allows a technician to access the codes using either a manual flash code procedure or plugging a scan tool into the diagnostic connector.

Bosch 2S Micro ABS

The 2S Micro ABS, figure 15-20, is yet another Bosch variant. It is functionally similar to all of the Bosch systems described thus far but is only

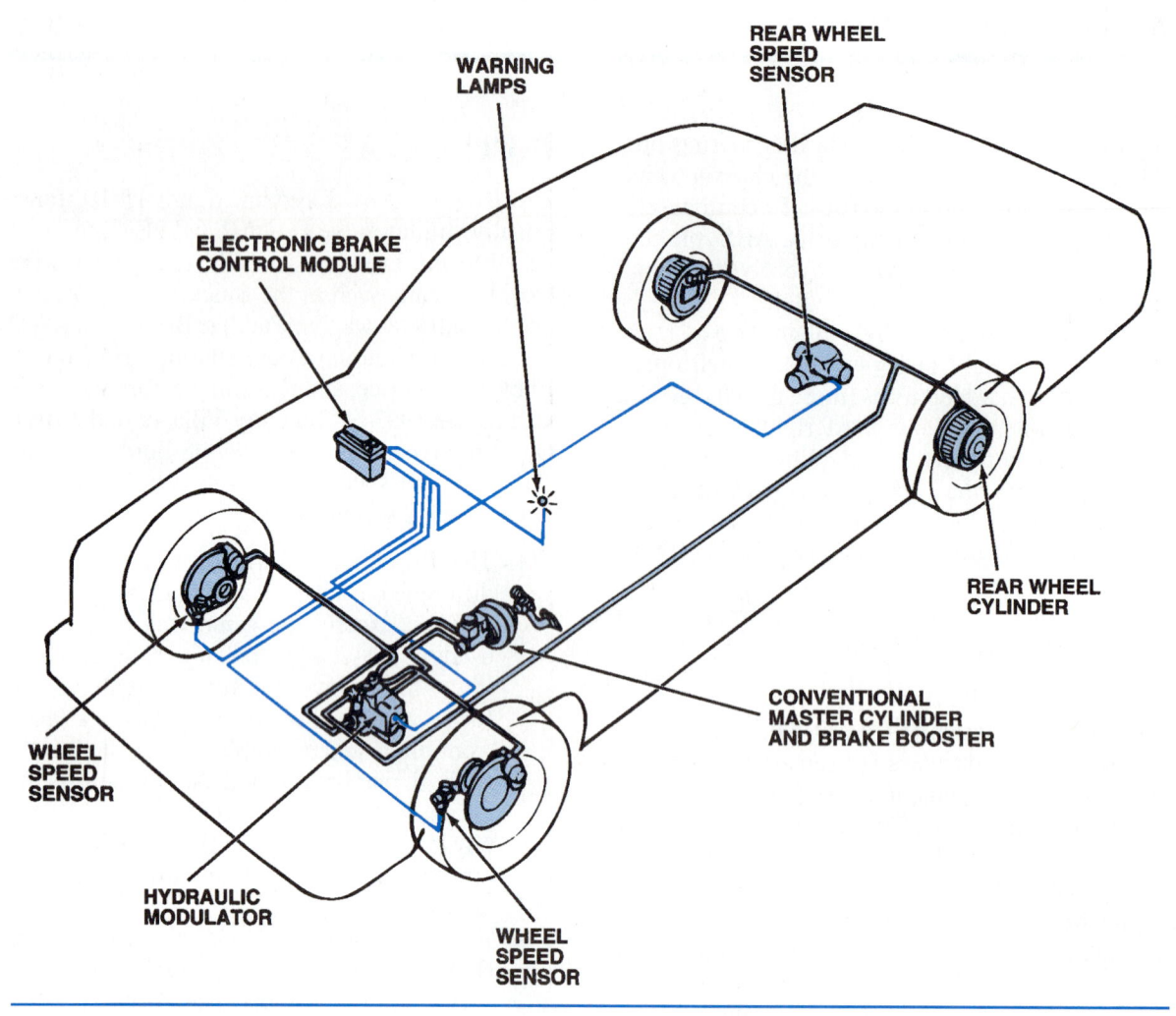

Figure 15-19. Bosch 2U ABS components.

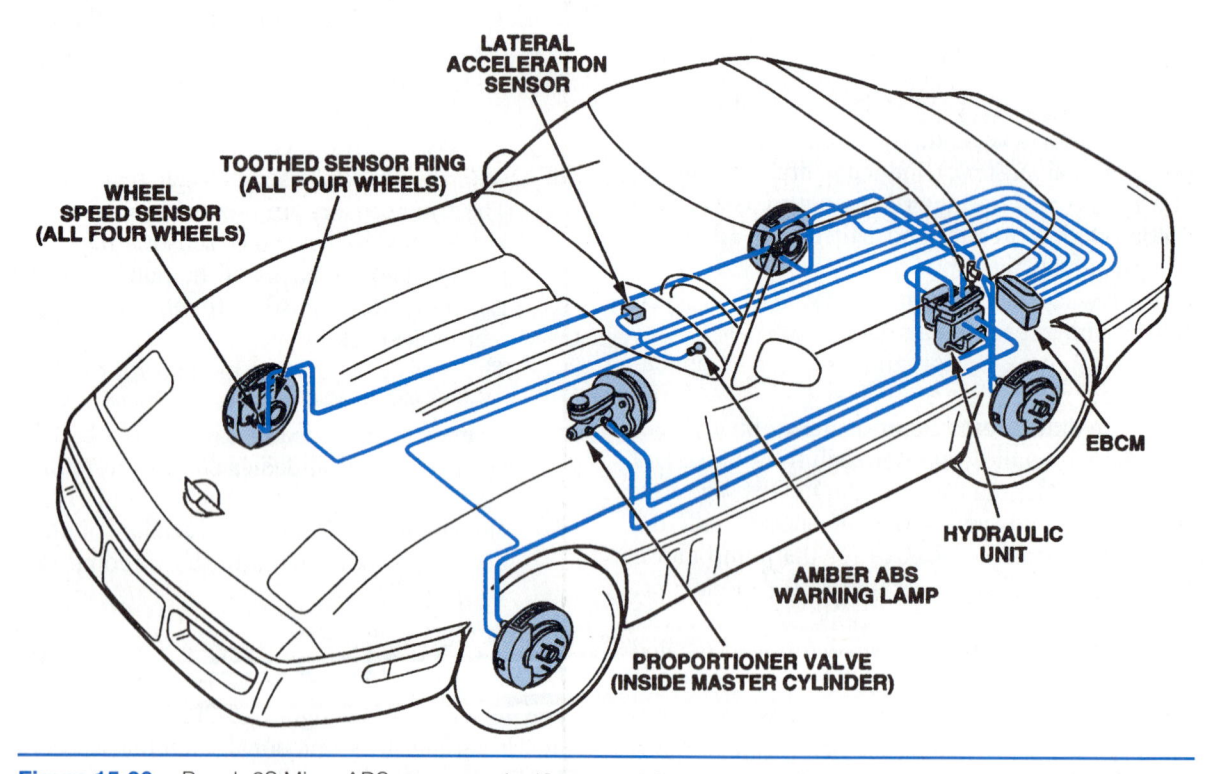

Figure 15-20. Bosch 2S Micro ABS components. (Courtesy of General Motors Corporation, Service and Parts Operations)

Antilock Brake Systems

on the 1990–91 Corvette. The main differences between 2S Micro and the earlier Bosch 2 system used on 1986–89 Corvette are:

1. The electronic brake control module (EBCM), located in the small compartment behind the driver's seat, provides diagnostic fault codes. As with the other GM Bosch 2U applications, a technician can access the codes using either a manual flash code procedure (grounding pin "H" on the ALDL connector) or plugging a scan tool into the ALDL connector.
2. Four wheel speed sensors are used with three control channels, one for each front brake and a shared channel for both rear brakes. The rear wheel speed sensors on the Corvette are mounted on the rear suspension knuckles. Left and right sensors are different, so replacement sensors must be for the appropriate side. The sensor rings are pressed onto the ends of the driveshaft spindles and are not serviceable. If damaged, you must install a new driveshaft spindle. The front wheel speed sensors on the Corvette are part of the wheel bearing and hub assembly and are not adjustable or replaceable.
3. A lateral acceleration sensor replaces the switch used in the earlier Corvettes. It is located on the instrument panel carrier, figure 15-20, beneath the radio and modifies ABS braking strategy during hard cornering when body roll loads the outside wheels more heavily than the ones on the inside. The sensor is a Hall-Effect switch that provides an analog reading of progressively higher side loads, starting at 0.4 gravities (G).

Bosch 2U ABS/ASR

The Bosch ABS/ASR system, figures 15-21, 15-22, and 15-23, is used on the 1992–94 Chevrolet Corvette and on 1993 and newer Cadillac Allantes, Broughams, Sevilles, and Eldorados. This system adds traction control capability, which Bosch refers to as acceleration slip regulation (ASR). The ASR system, figure 15-24, uses both braking and engine torque management to limit wheel spin during acceleration.

Like the Corvette's earlier Bosch 2S Micro ABS system, the ABS/ASR system is a nonintegral, four-wheel, three-channel ABS system, but

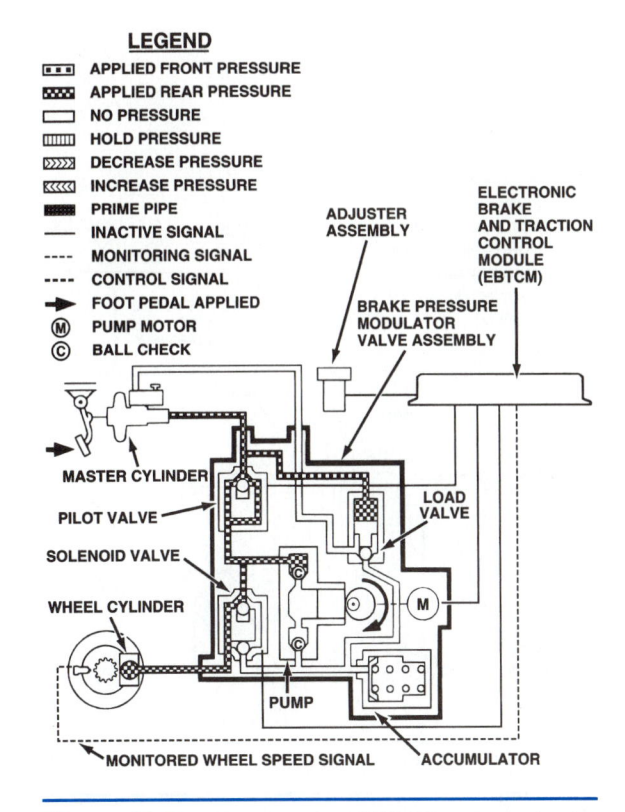

Figure 15-21. The Bosch ABS/ASR system in the normal braking mode.

only in the antilock brake mode. The hydraulic modulator assembly in the 1992 and later ABS/ASR applications contains an additional three-way solenoid valve so the ASR can operate the rear brakes independently. The return pump is also used to provide hydraulic pressure for rear braking during ASR. The controller is different, too, because modifications were necessary to give it traction control ability. In addition to these differences, the ABS/ASR modulator contains two other valves:

- **Load valve:** The load valve is a hydraulically operated valve used to isolate the master cylinder prime pipe from the pump during brake apply and ABS control. The load valve is spring-loaded in the open position.
- **Pilot valve:** The pilot valve is a two-way valve that is electronically operated to isolate the master cylinder from the pump when ASR is needed. When the normally open valve is closed, the pump routes fluid to the rear brake circuit. The control module also uses this valve to regulate fluid volume in the brake circuits when switching from ASR to ABS.

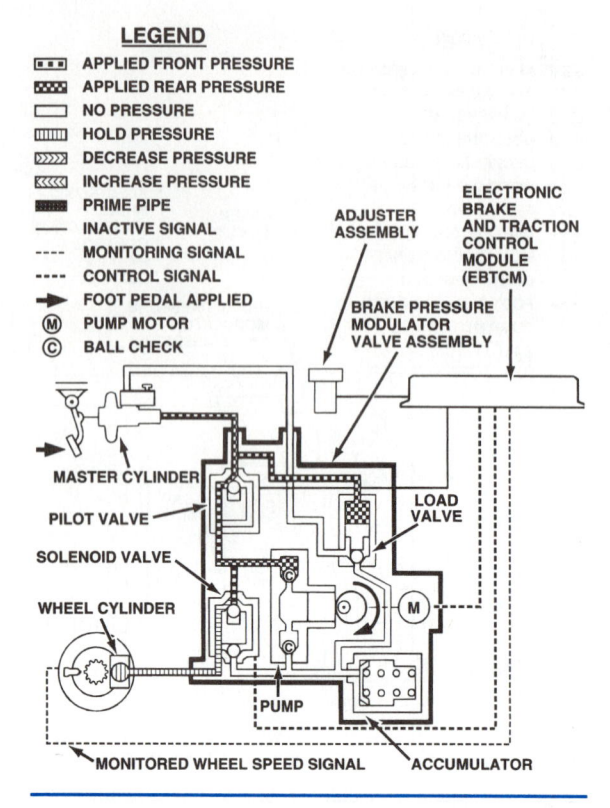

Figure 15-22. The Bosch ABS/ASR in the pressure hold mode.

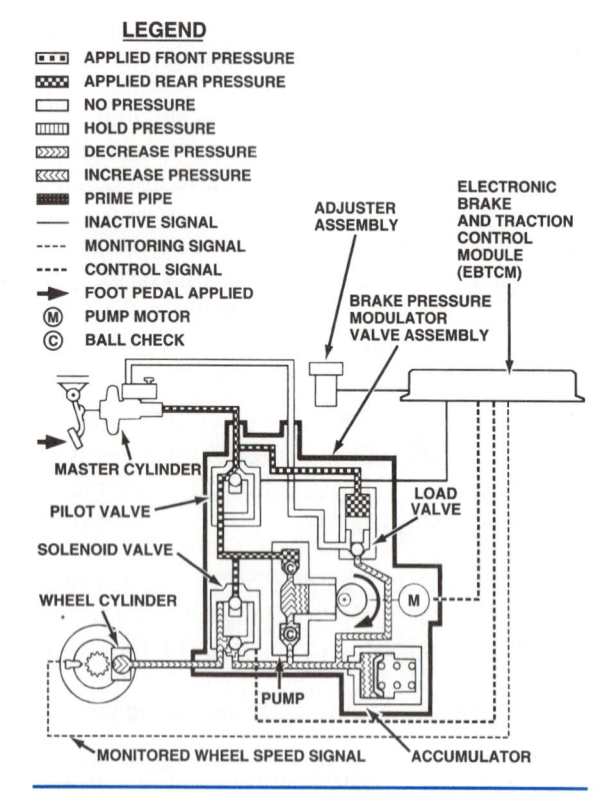

Figure 15-23. The Bosch ABS/ASR in the pressure decrease mode.

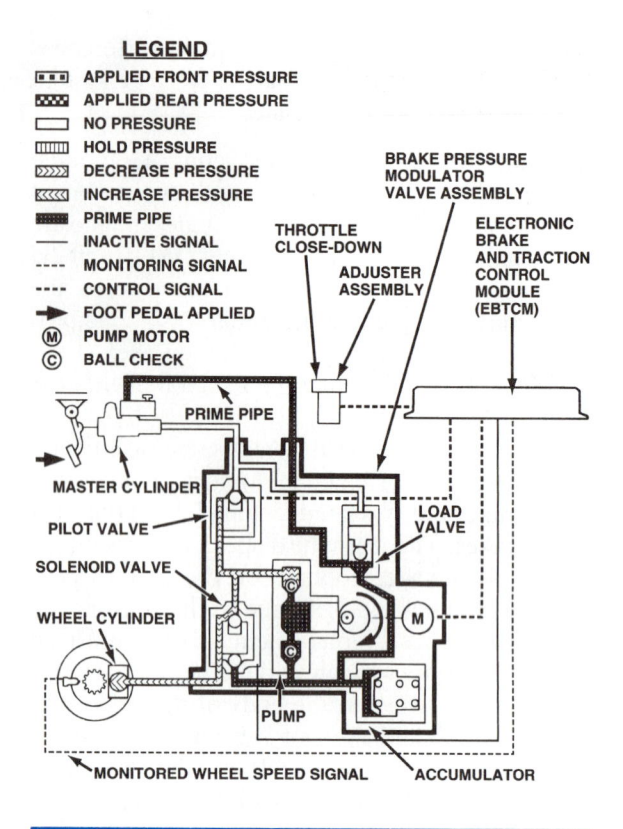

Figure 15-24. The Bosch ABS/ASR in the traction control mode.

The ABS/ASR module is linked electronically to the powertrain control module (PCM) so it can reduce engine power if needed to control wheel spin during acceleraton. This allows the ASR system to apply either rear brake as needed or to reduce engine power by reducing the throttle opening (via a throttle relaxer on the throttle linkage), retarding ignition timing, or deactivating some of the fuel injectors depending on application.

The ABS/ASR system on the Corvette uses braking and throttle reduction to control wheel spin; the Cadillac uses braking, throttle reduction, timing retard, and injector deactivation. Rear braking and/or torque management are both used at speeds below 50 mph. Above 50 mph, only spark retard, injector disabling, or throttle reduction are employed to control wheel spin. Braking is not used above 50 mph because the sudden application of only one rear brake could upset vehicle stability. Thus, the system gives priority to directional control over traction at high speed.

An ASR warning light (SERVICE ASR) is part of the system as is an indicator light to signal the driver when the ASR system is preventing wheel

Antilock Brake Systems

spin (ASR ACTIVE). The driver can deactivate the ASR system with a dash switch.

Bosch 5 Series

The Bosch 5 series ABS is found on many vehicles built in the past 15 years. The type of system depends on the desired features or the year built. Major features of each system are:

- Bosch 5.0
  - Three- or four-channel ABS functions
  - Option for traction control function
  - Sixteen-kilobyte memory (Bosch 2 systems have 8 kilobyte)
  - Diagnostic system by way of flash codes or special ABS checker tool
  - Weight 6 to 7 lbs (2.8 to 3.3 kg).
- Bosch 5.3
  - Three- or four-channel ABS functions
  - Traction control
  - Electronic stability control
  - Electronic brake proportioning
  - Twenty-four-kilobyte memory
  - Diagnostic system by way of flash codes, ABS checker, or communication over vehicle diagnostic system with a scan tool
  - Weight 6 lbs (2.6 kg).
- Bosch 5.7
  - Three- or four-channel ABS functions
  - Traction and stability control
  - Electronic brake proportioning
  - Electronic four-wheel traction system for off-road use
  - Forty-eight-kilobyte memory
  - Diagnostic system by way of communication over vehicle communication network using a scan tool
  - High-speed controller area network (CAN) bus communication system (see explanation under *Delphi DBC 7.2 and 7.4 ABS,* page 332.)
  - Weight 5.5 lbs (2.5 kg).

Bosch 5.0 is on the 1995–96 Corvette and 1995–2001 Porsche 911. In 1996 the system was added to the Ford Taurus and Mercury Sable. A Delco(Delphi)/Bosch hybrid system, using a Delphi control module, is found on 1996–99 the Pontiac Bonneville, Buick Park Avenue, Buick Le Sabre, Oldsmobile 98 and 88, and 1996–2001 Corvette.

The Bosch 5.3 ABS is used on Subaru and Toyota models beginning in 1997. The 5.3 system is also used on 1998 and later Camaro, Firebird, Pontiac Grand Prix, Oldsmobile Intrigue, and Cadillac Catera. Ford, Mazda, Mercury, Mitsubishi, and Nissan vehicles use the 5.3 system starting in 1999, depending on the model. The Bosch 5.7 system is used, beginning in 1999, on some Mercedes and other European vehicles.

System Description

Like the Bosch ABS/ASR system they supersede, the Bosch 5 series systems are nonintegral, four-wheel ABS with traction control. However, they are four-channel, rather than three-channel, systems with a compact "unitized" design, figure 15-25. The hydraulic modulator or brake pressure modulator (BPM) valve body and control module are both part of the electronic brake traction control module (EBTCM) assembly. Nonadjustable wheel speed sensors at each wheel provide inputs to the control module. On some models, a lateral accelerometer keeps the control module informed about G-forces.

System Operation

The brake pressure modulator valve assembly contains a pair of ABS valves for each brake circuit (one normally open inlet solenoid and one normally closed outlet solenoid) for a total of eight. When the control module senses wheel lockup, the inlet solenoid is energized to close and isolate the brake circuit. This prevents any further pressure increase to the brake. If wheel speed does not increase and match the reference speed of the other wheels, the control module energizes the outlet solenoid to open it so fluid pressure can escape from the circuit. This relieves brake pressure and allows the wheel to regain traction. When the reference speed is obtained, the control module opens the inlet valve, allowing pressure to be reapplied to the brake. The cycle is repeated continuously as long as needed on until the vehicle stops.

The EBTCM also contains the pump motor (used for fluid return during ABS braking and to generated rear brake pressure for traction control), pump motor relay brake pressure modulator (BPM) valve relay (provides power to the ABS and traction control solenoids), and a pair of additional solenoid valves used for traction control.

Bosch 5.3 ABS

The Bosch 5.3 system is a more compact, lighter unit than the Bosch 5.0, but its operation is identical to the 5.0 system, figure 15-26. It costs less to manufacture and also offers additional features not found in the 5.0 system.

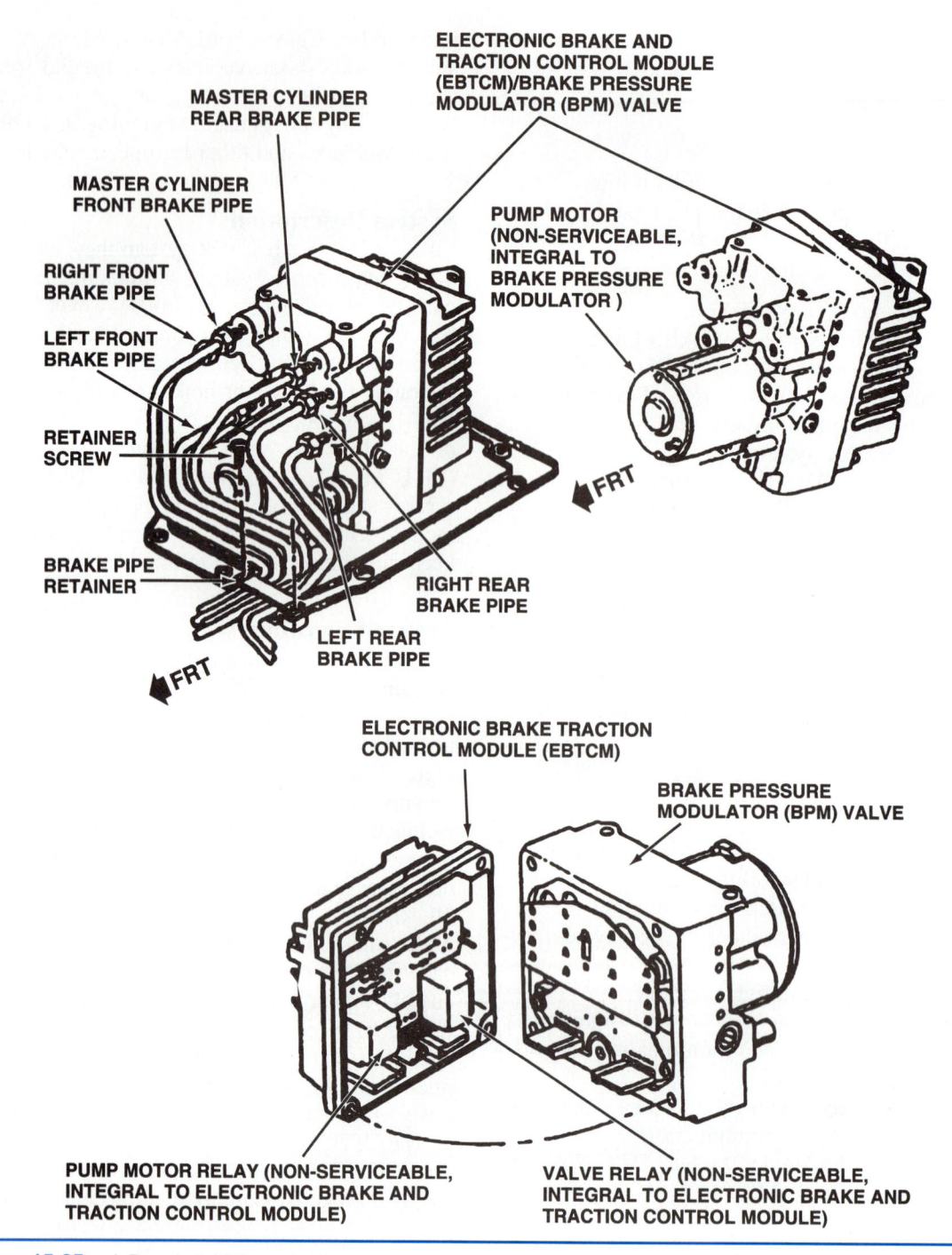

Figure 15-25. A Bosch 5 ABS modulator.

One enhancement, used on Firebird, Camaro, and others, is an electronic proportioning function. Called dynamic rear proportioning (DRP) by GM or electronic brake proportioning (EBP) by Bosch, these vehicles use electronic proportioning and do not have proportioning valves in the hydraulic system. Instead, they use computer control of the rear inlet valves to the brake pressure modulator valve. When the EBTCM detects that the rear wheels are decelerating just slightly faster than the front wheels, it pulses the rear brakes until the rear wheels' rate of deceleration matches the front wheels. A problem in this system will illuminate the red brake lamp but will not set a code.

Antilock Brake Systems

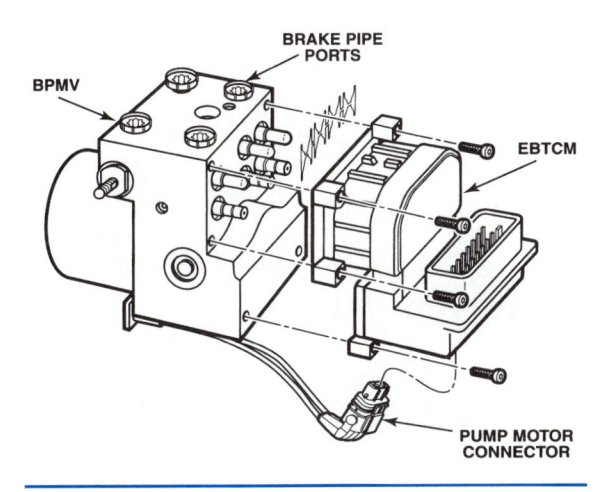

Figure 15-26. A Bosch 5.3 ABS modulator. (Courtesy of General Motors Corporation, Service and Parts Operations)

Bosch 5.7 ABS

The Bosch 5.7 ABS, introduced on some Mercedes models in 1998, is a development of the Bosch 5 series. It functions the same as other 5 series systems but with a smaller and more compact modulator/controller assembly. A faster computer processor and more memory make it easier to include advanced ABS functions such as:

- Traction control
- Vehicle stability control
- Faster communication with other vehicle modules.

Traction Control

Like the earlier Bosch ABS/ASR system, Bosch 5 also uses braking in conjunction with throttle reduction and spark retard to reduce wheel spin when traction control is needed. Braking is only used at speeds below 50 mph and primarily at slow speeds when wheel spin is excessive.

When traction control braking is needed, the control module in the EBTCM energizes the pump and two solenoids: the prime line solenoid valve, which opens a circuit that allows fluid to flow to the return pump so it can generate brake pressure, and the traction control solenoid, which closes to prevent pump pressure from returning to the master cylinder. A relief valve in the traction control solenoid valve also limits the maximum pressure that the pump can apply to the rear brakes when controlling wheel spin.

In addition to braking, a combination of spark retard and throttle reduction via a throttle adjuster assembly attached to the throttle linkage allows the system to quickly and easily control engine torque as well as traction. The throttle adjuster assembly consists of a small electric motor and a control cable cam assembly. The unit is attached to three cables: the accelerator pedal cable, the cable to the throttle body, and the cruise control cable. The adjuster only moves in one direction, which is to pull the throttle closed. When the unit is engaged to reduce power, the driver feels the accelerator pedal pushing back against his or her foot.

Like the Bosch ABS/ASR system, an indicator light signals the driver when the traction control system is active. The Corvette has a switch to deactivate the traction control system.

Vehicle Stability Control

The later Bosch 5.3 and 5.7 systems can also include a feature called the electronic stability program (ESP). The addition of a steering angle sensor, and a yaw-rate and lateral acceleration sensor to the other ABS components allows the ABS control module and hydraulic modulator to help maintain vehicle stability during situations where the driver may otherwise lose control of the vehicle.

The ESP system prevents possible skidding by using the brakes, engine torque control, or transmission shift control to prevent understeer or oversteer. The steering angle sensor measures how far, how fast, and in which direction the steering wheel is being turned. The yaw-rate and lateral acceleration sensors signal to the ESP computer that under- or oversteer is about to occur. These sensors and the ABS sensors are monitored 25 to 150 times per second (5.3 or 5.7 ABS) under all driving conditions. By selectively applying braking pressure at the wheels, the ESP "steers" the vehicle back to the desired direction.

As an example, suppose a vehicle is exiting the expressway at a high a rate of speed. As the vehicle reaches the tighter curve of the exit ramp, the driver turns the wheel more to the right but the vehicle continues straight, a condition called understeer. In this case the ESP knows, from the steering angle sensor, that the driver wishes to turn more sharply; the ESP also knows, from the yaw and lateral acceleration sensors, that the vehicle is not turning as the driver desires. ESP reacts instantly by quickly and sharply applying the brakes on the rear wheel on the inside of the curve and, if necessary, reducing engine power. This action helps prevent the understeer condition and brings the car safely under control, figure 15-27.

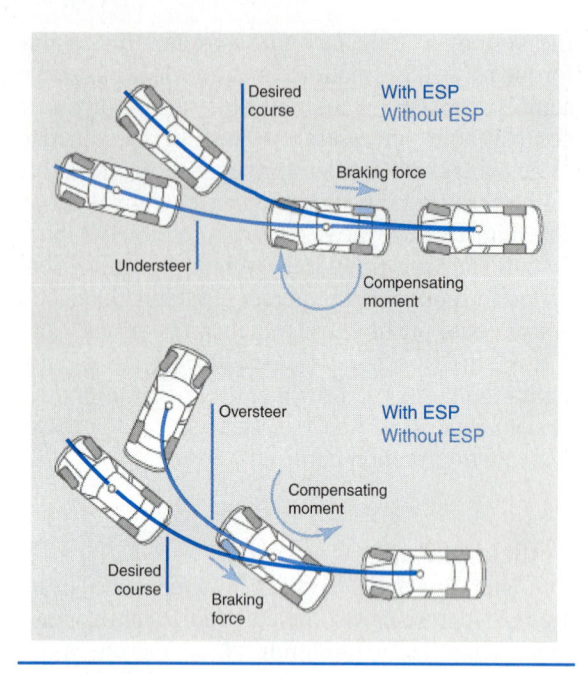

Figure 15-27. Operation of the electronic stability program (ESP). (Courtesy of Continental Teves)

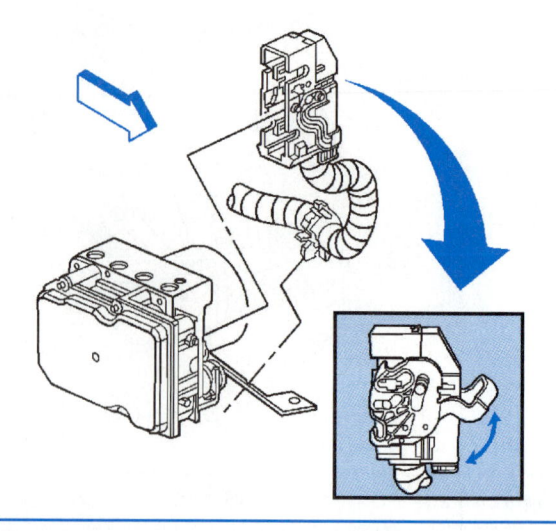

Figure 15-28. The Bosch 8.0 ABS modulator and electrical connector. (Courtesy of General Motors Corporation, Service and Parts Operations)

ESP is active under all driving conditions and functions, in addition to ABS, during an ABS braking event. When ESP is controlling the vehicle, an indicator lamp on the dash lights up to inform the driver. The system can be disabled by the driver, if desired.

Bosch 8.0 ABS

The BOSCH 8.0 antilock braking system is the newest generation of ABS from Bosch. Introduced in 2001 and used on the 2003 Saturn Ion and other vehicles, the memory is increased to 128 kbyte, and the weight of the controller/modulator is reduced to 3.5 lbs (1.6 kg). The systems increased processing power makes future expansion easier. Bosch 8.0 ABS offers the following braking functions:

- Antilock brake system
- Electronic brake proportioning or dynamic rear proportioning
- Traction control system
- Brake assist system
- Enhanced vehicle communications using CAN (see page 332).

Bosch 8.0 systems are nonintegral, four-wheel ABS systems with traction control and additional features, depending on the vehicle. The electronic brake control module (EBCM) and the brake pressure modulator valve (BPMV) are combined into one assembly but can be replaced separately, if needed. The EBCM controls the system functions, detects failures, and also contains the solenoid power relay and pump motor relay. The nonadjustable wheel speed sensors are replaceable only as part of the wheel hub and bearing assemblies.

The BPMV contains the hydraulic valves and pump motor that are controlled electrically by the EBCM, figure 15-28. The Bosch 8.0 system BPMV uses a four-circuit configuration and directs fluid from the reservoir of the master cylinder to the front and rear wheels as needed. The BPMV contains the following components:

- Pump motor
- Inlet valves (one per wheel)
- Outlet valves (one per wheel).

Other system inputs include the traction control switch and the stop lamp switch. The TCS is manually disabled or enabled using the traction control switch. The EBCM uses the stop lamp switch as an indication that the brake pedal is applied.

System Operation

The Bosch 8.0 operation is similar to the 5.3 and 5.7 systems. The EBCM performs one self-test each ignition cycle. The self-test occurs when the vehicle speed is greater than 10 mph (16 km/h). The test sequence cycles each solenoid valve and the pump motor, as well as the necessary relays for approximately 1.5 seconds to check component operation. The EBCM sets a DTC if any error is detected. The initialization sequence may be

Antilock Brake Systems

heard and felt by the driver while it is taking place and is considered part of normal system operation.

When wheel slip is detected during a brake application, the ABS enters antilock mode. During antilock braking, hydraulic pressure in the individual wheel circuits is controlled to prevent any wheel from slipping. A separate hydraulic line and specific solenoid valves are provided for each wheel. The ABS can decrease, hold, or increase hydraulic pressure to each wheel brake. The ABS cannot, however, increase hydraulic pressure more than the amount that is transmitted by the master cylinder during braking.

Dynamic Rear Proportioning

The dynamic rear proportioning (DRP) is a control system that replaces the hydraulic proportioning function of the mechanical proportioning valve in the base brake system. The DRP control system is part of the operation software in the EBCM. The DRP uses active control with the existing ABS in order to regulate the vehicle's rear brake pressure.

Traction Control System

The traction control system (TCS) compares front wheel speeds to rear wheel speeds to determine whether drive wheels lose traction. The TCS activates when drive wheel speeds exceed the speed of non drive wheels by a calibrated value. This allows the driver to maintain acceleration and directional stability while accelerating on low-tractions surfaces.

The TCS limits wheel slip during acceleration when one or more of the drive wheels lose traction. The brake switch must be off for TCS to operate. During a traction event, the EBCM sends a requested torque value to the powertrain control module (PCM) over the serial data link. The PCM initiates an engine torque reduction routine to slow down the drive wheels. This routine consists of ignition spark timing reduction, fuel injector cutoff, and transmission shift control. The PCM also sends a torque-delivered value to the EBCM over the serial data link. The driver can disable TCS, if desired, by pressing the TCS switch. Whenever the TCS switch light-emitting diode (LED) is illuminated, TCS is available. This will default to on every time the vehicle is started.

Brake Assist System

The brake assist system (BAS) is available as a function of Bosch 8.0 (and some 5.7) ABS. Testing has shown that up to 90 percent of drivers fail to brake fast enough and hard enough in emergency situations. BAS applies full power-assisted braking during emergency stops by monitoring the accelerator pedal and brake pedal acceleration to detect an emergency braking situation. Under control of the electronic controller, the vacuum booster is opened to atmospheric pressure to apply full power to the brakes. This can result in up to a 45 percent shorter stopping distance in an emergency.

System Indicators and Warning Lights

The instrument panel cluster illuminates the ABS indicator when the EBCM detects a malfunction with the antilock brake system. The ABS indicator also illuminates for approximately 3 seconds each time the key is turned on. The red brake indicator lamp lights when any of the following occurs:

- At the start of each ignition cycle for approximately 3 seconds
- The body control module (BCM) detects that the park brake is engaged
- The BCM detects that the master cylinder brake fluid reservoir level is low
- In the event of a DRP failure.

A LOW TRACTION message on the driver information center or TRAC lamp will come on whenever the TCS is in operation. The message or lamp will remain on for 3 to 4 seconds after the TCS event is complete. The message does not operate during an ABS event.

DELPHI CHASSIS (DELCO MORAINE) ABS

Delco makes one integral ABS (Powermaster III) and one nonintegral system (ABS-VI).

Delco Moraine Powermaster III

The Delco III Powermaster system is used only on 1989–91 Buick Regal, Oldsmobile Cutlass, and Pontiac Grand Prix and GTU models. It is an integral ABS system that uses a high-pressure pump and accumulator for power assist as well as antilock braking.

System Description

Powermaster III is a four-wheel, three-channel ABS system, figure 15-29. The front brakes are

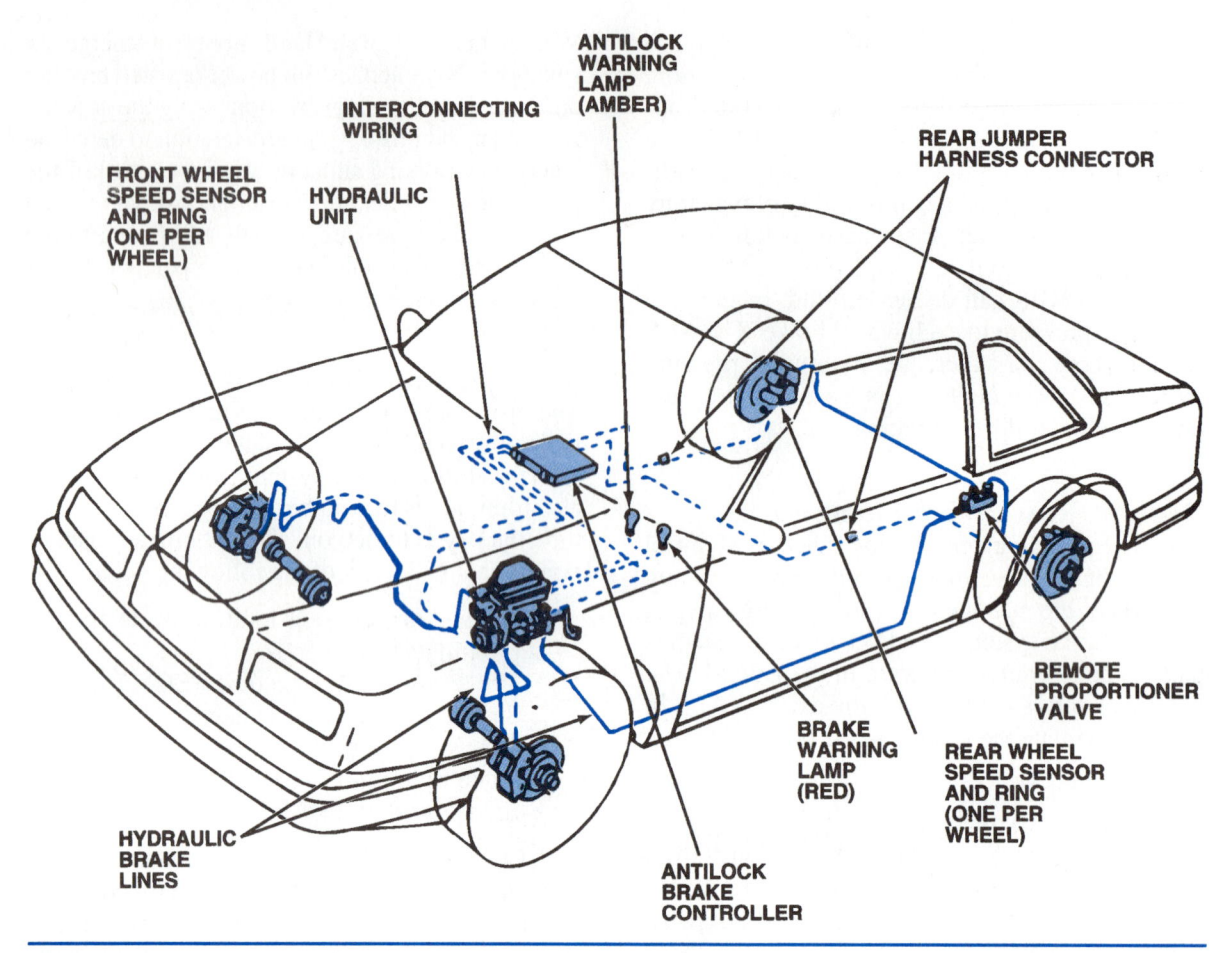

Figure 15-29. The Delco Powermaster III integral ABS system.

controlled individually, but the rear brakes are controlled as a pair by the select-low principle. Each wheel has its own nonadjustable speed sensor. Front sensors are mounted on the spindle assembly with the sensor ring located on the outboard constant-velocity joint. Rear sensors and rings are an integral part of the rear wheel bearing and hub and cannot be serviced separately.

The hydraulic modulator assembly, figure 15-30, contains the master cylinder, pump and motor assembly, pressure switch, relief valve and accumulator, and three two-way, ABS solenoids (one for each front brake circuit and one for both rear brakes). The fluid reservoir contains a level sensor. The brake system is split front and rear, with the primary side of the master cylinder serving the front wheels and the secondary side serving the rear wheels.

When the ignition is turned on, the pump motor runs to build to pressure in the accumulator for power assist as well as antilock braking. A pressure switch causes the pump to kick in when accumulator pressure drops below 2200 psi (15,169 kPa) and to shut off when it reaches 2700 psi (18,617 kPa). A relief valve prevents accumulator pressure from exceeding 3400 psi (23,443 kPa).

Power for the pump motor is provided by the ABS power center, located in a black plastic rectangular container on the left front inner fender near the strut tower. Inside are two relays for the solenoids; two fusible links for the pump motor and front enable relay; and two fuses for the controller and the rear enable relay and solenoid. There is also an "electrical center" by the left strut tower. It has a rectangular plastic box with a red cover. Inside are fusible links for ABS power center circuits. The Powermaster III wiring harness uses a single ground connector located on a stud near the battery tray. The ABS control module is located inside the vehicle under the passenger seat.

ABS Operation

Antilock operation is similar to other integral ABS systems, but there are some differences. Each ABS solenoid in the Powermaster III system

Antilock Brake Systems

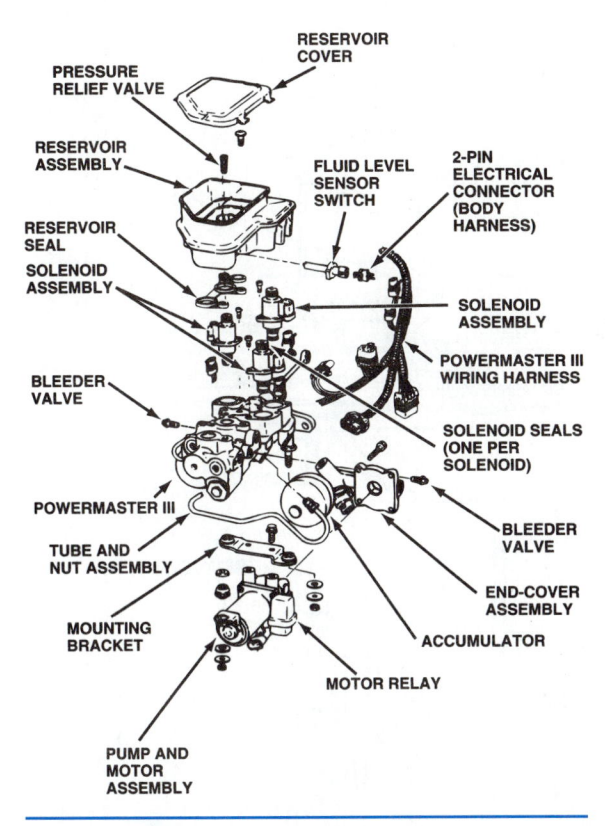

Figure 15-30. A Delco Powermaster III hydraulic modulator assembly.

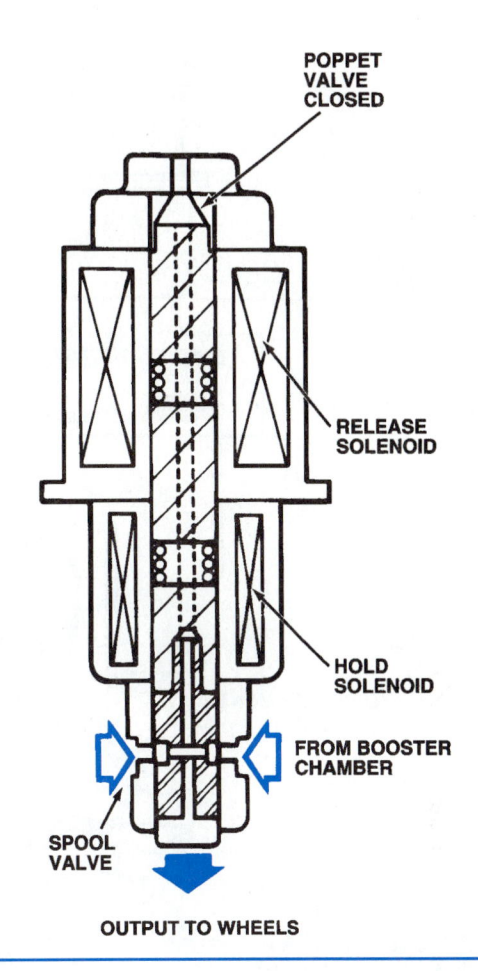

Figure 15-31. A Delco Powermaster III ABS solenoid for rear wheels.

has two valves: a spool valve at the bottom of the solenoid that is normally open and an internal poppet valve at the top that is normally closed, figure 15-31. Each solenoid also contains two sets of windings to control the spool valve's position.

When hydraulic pressure in the rear brakes needs to be held during antilock braking, the control module energizes the hold windings in the rear ABS solenoid to close the normally open spool valve. This closes off the rear brake circuit, isolating the rear brakes from the rest of the system and preventing any further increase in brake pressure.

To reduce pressure and prevent wheel lockup, the control module then energizes the second set of windings in the rear ABS solenoid, which opens the upper poppet valve. This allows line pressure to be released so it can escape back to the master cylinder reservoir.

To reapply pressure to the rear brakes, the control module de-energizes both sets of windings in the rear ABS solenoid. Spring pressure closes the upper poppet valve and reopens the lower spool valve, allowing pressure from the master cylinder to again reach the rear brake circuit and reapply the brakes.

Up front, the same process occurs during an ABS stop except each front brake circuit also contains a displacement cylinder and isolation valve, figures 15-32, 15-33, and 15-34. These are used to prevent the master cylinder chambers from running out of fluid during the release mode. These components are not needed for rear ABS braking because the rear wheel circuits receive brake fluid directly from the boost chamber.

Delphi Chassis (Delco Moraine) ABS-VI

Introduced in 1991 as an option on Saturn, Buick Skylark, Oldsmobile Cutlass Calais, and Pontiac Grand Am models, the Delphi Chassis (Delco Moraine) ABS-VI antilock brake system was used until 2000 on numerous General Motors front-wheel drive models, including the Chevrolet Lumina and APV minivan; Chevrolet Beretta, Corsica, and Cavalier; Pontiac Grand Prix and Sunbird; and Oldsmobile Cutlass Supreme and

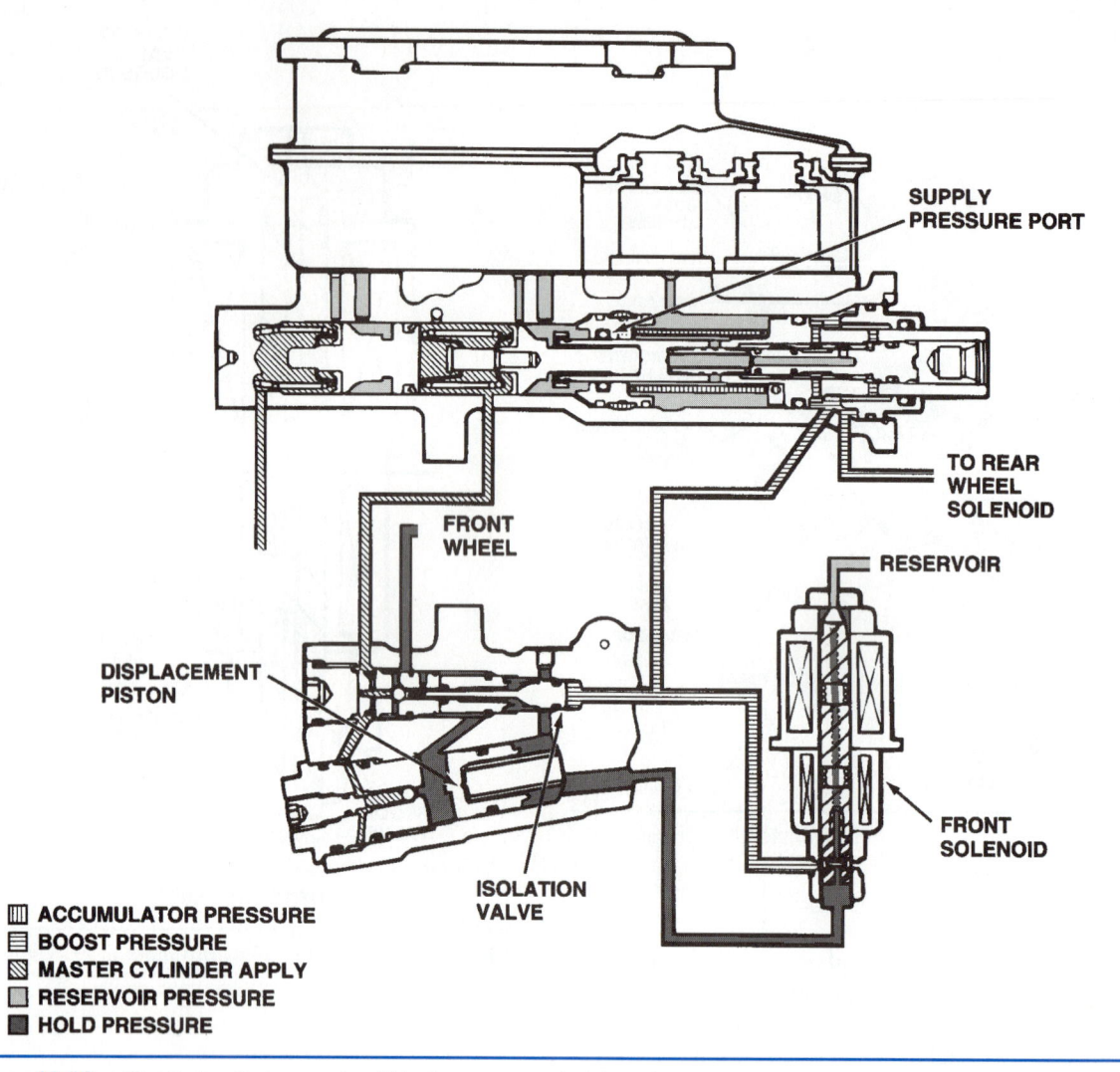

Figure 15-32. The Delco Powermaster III in the pressure hold mode.

Buick Regal. It is a nonintegral ABS with a conventional master brake cylinder and power booster. The brake systems on these applications are split diagonally so the primary (rear) master cylinder piston operates the right front and left rear brakes and the secondary (front) piston operates the left front and right rear brakes.

System Description

Delphi Chassis ABS-VI is a four-wheel, three-channel ABS, figure 15-35. It controls the front brakes separately, but the rear brakes share a common ABS circuit. Each wheel, however, has its own nonadjustable wheel speed sensor. Front sensors are mounted on the steering knuckle assembly with the sensor rings located on the outside of the constant-velocity joints. Front wheel speed sensors are plastic rather than stainless steel, which improves corrosion resistance and lowers manufacturing costs. Rear wheel speed sensors and sensor rings are loated inside the rear wheel bearing and hub assemblies and must be replaced as a complete assembly.

The ABS-VI control module is located on the firewall in the engine compartment in most applications. The control module receives inputs from the four wheel speed sensors as well as a brake pedal switch.

The ABS enable relay supplies power to the ABS when the ignition is turned on. The relay is under the hood in the relay cluster. The enable relay switches power to the DC motor pack and ABS isolation solenoids.

Hydraulic Modulator and Motor Pack

The hydraulic modulator and motor pack assembly is next to the master cylinder, figure 15-36. The modulator is held by two banjo bolts at the

Antilock Brake Systems

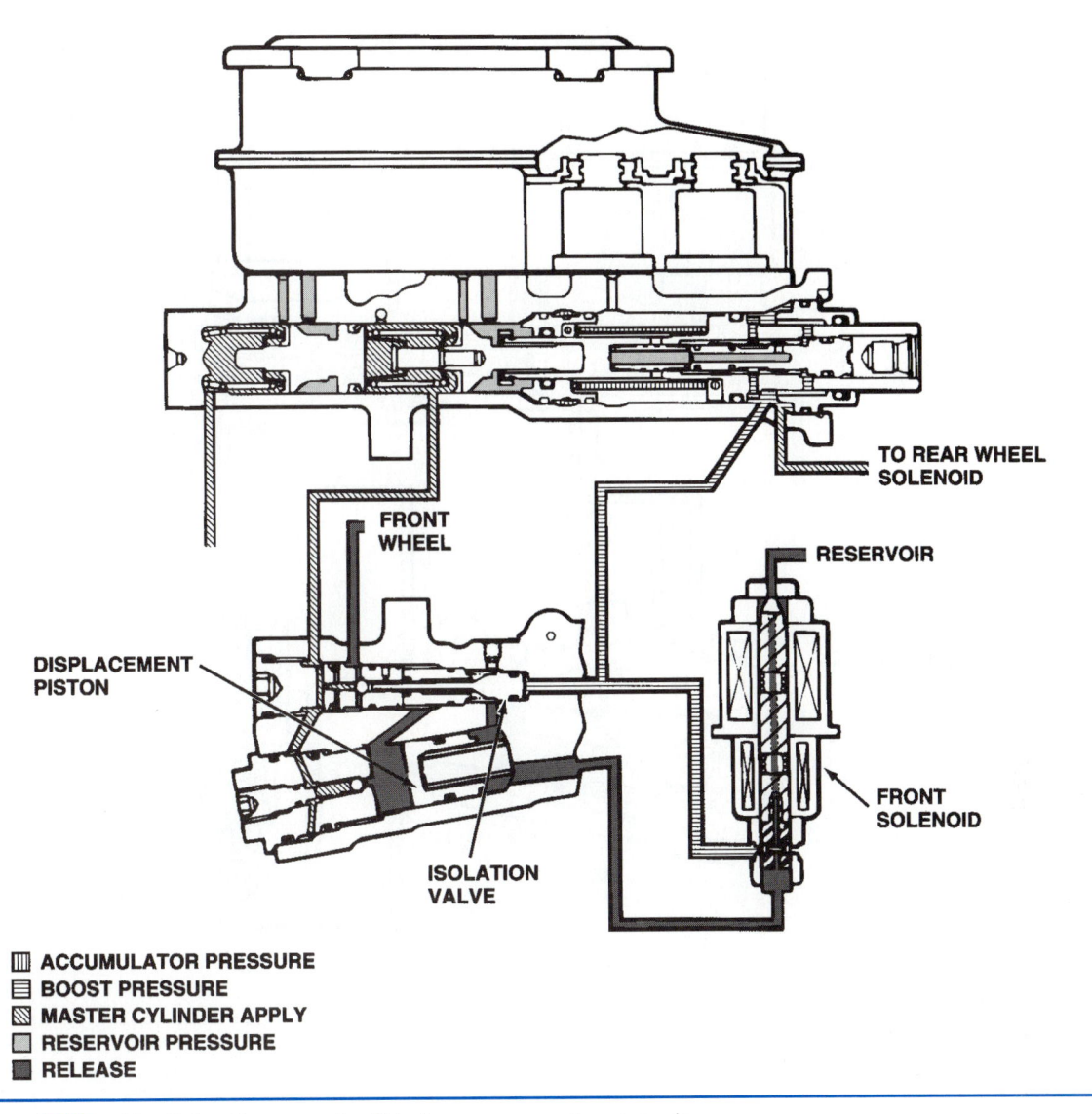

Figure 15-33. The Delco Powermaster III in the pressure release mode.

master cylinder's two upper outlet ports and two transfer tubes at the two lower outlet ports. You must replace the two lower transfer tubes and O-rings if the master cylinder and modulator are disconnected from one another to prevent leaks that might result in brake failure.

The modulator, figure 15-37, has fluid chambers for all four brakes, two ABS isolation solenoid valves, four check balls, and a motor pack, figure 15-38. The motor pack contains three bidirectional direct current motors with electromagnetic brakes (EMBs) or expansion spring brakes (ESBs), figure 15-39, three ball screw assemblies, four pistons, and a gear drive set and gear cover. The modulator motor pack, isolation solenoid valves, gear cover, and individual gears are all serviceable parts and can be replaced separately.

The hydraulic modulator and motor pack assembly on the ABS-VI system is unique. The modulator contains two isolation solenoid valves to block off the front brake circuits when rear antilock braking is required. But there is no isolation solenoid valve for the rear brakes. Isolation is provided by a check ball for each rear brake circuit.

The most important difference between the ABS-VI system and other ABSs is the method by which brake pressure is modulated during antilock braking. On most other four-wheel ABSs, pressure is held in a given brake circuit by closing an isolation solenoid valve, then opening a decay valve to release pressure. In the ABS-VI system, however, brake pressure is modulated during antilock braking by positioning a small piston up and down inside a fluid chamber.

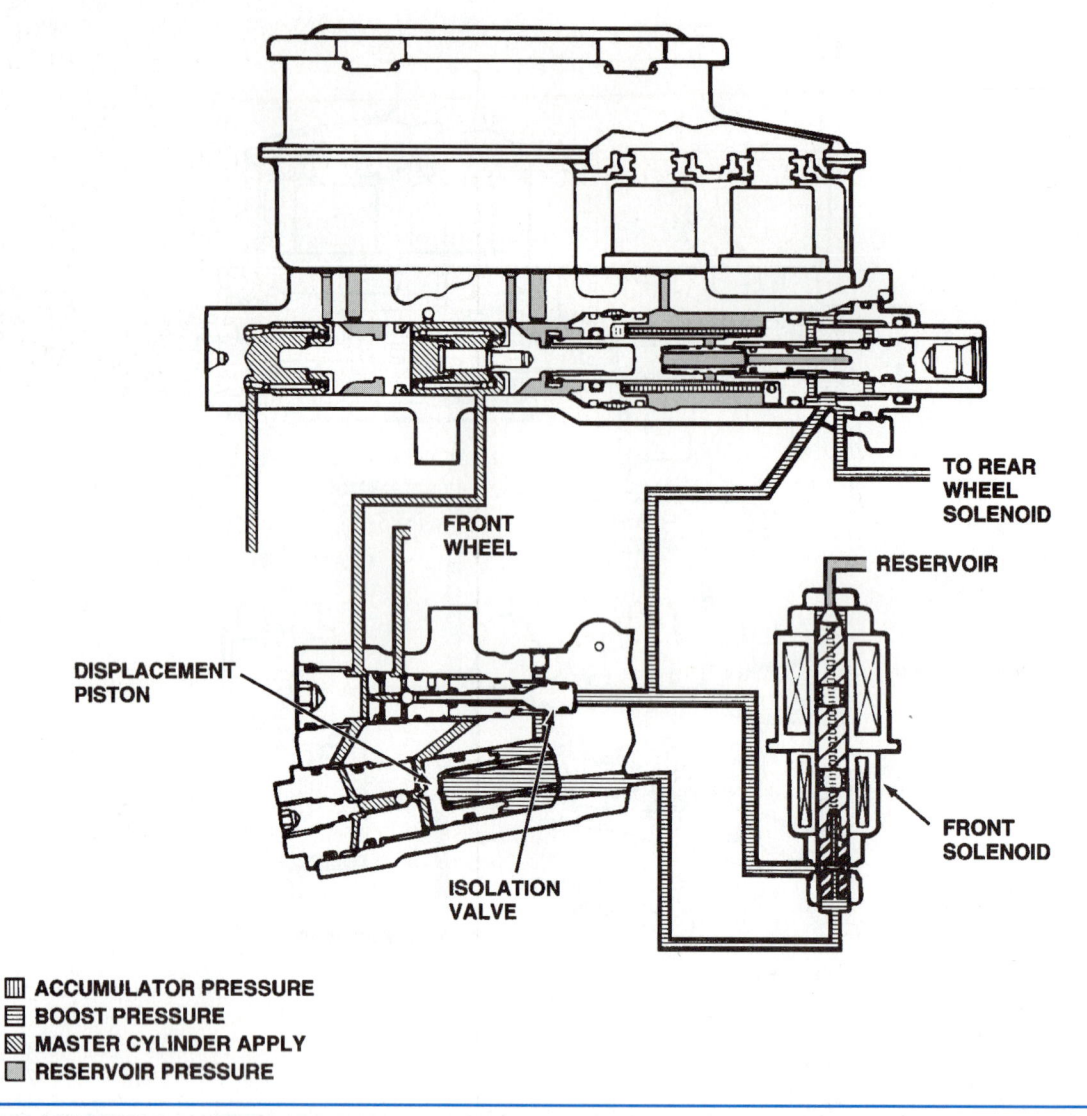

Figure 15-34. The Delco Powermaster III in the pressure apply mode.

Legend:
- ACCUMULATOR PRESSURE
- BOOST PRESSURE
- MASTER CYLINDER APPLY
- RESERVOIR PRESSURE

Labels: TO REAR WHEEL SOLENOID, FRONT WHEEL, RESERVOIR, DISPLACEMENT PISTON, ISOLATION VALVE, FRONT SOLENOID

The modulator is divided into three circuits. The right and left front brakes are controlled individually, whereas the rear brakes are controlled as a pair. The system uses the select-low principle of operation for the rear brakes. For each of the front brake circuits, there's an isolation solenoid valve, a check ball, a ball screw and piston, DC motor, gear drive, and electromagnetic motor brake (EMB). The EMBs are disc style brakes located on top of the front motor assemblies. When no voltage is applied to the EMB, a plate and spring pushes down against the pads, keeping the motor from turning. When voltage is sent to the EMB, the electromagnet pulls up on the plate, disengaging the brake. The motor is now free to turn.

During normal braking, fluid pressure from the master cylinder passes through the modulator and on to all four brakes. The ball screw and piston in each circuit is at its highest or home position. This holds the check ball open so fluid can go through the upper passageways to the brakes. The ball screws and pistons for the front brake circuits are locked in this position by the de-energized electromagnetic brakes that prevent the DC motors from turning. The two solenoid valves for the front brakes are also open when de-energized, which allows fluid to pass through their passageways in the modulator to the brakes. The front brakes provide about 80 percent of the total braking effort. The modulator is designed with two passageways to the front brakes in the event of a

Antilock Brake Systems

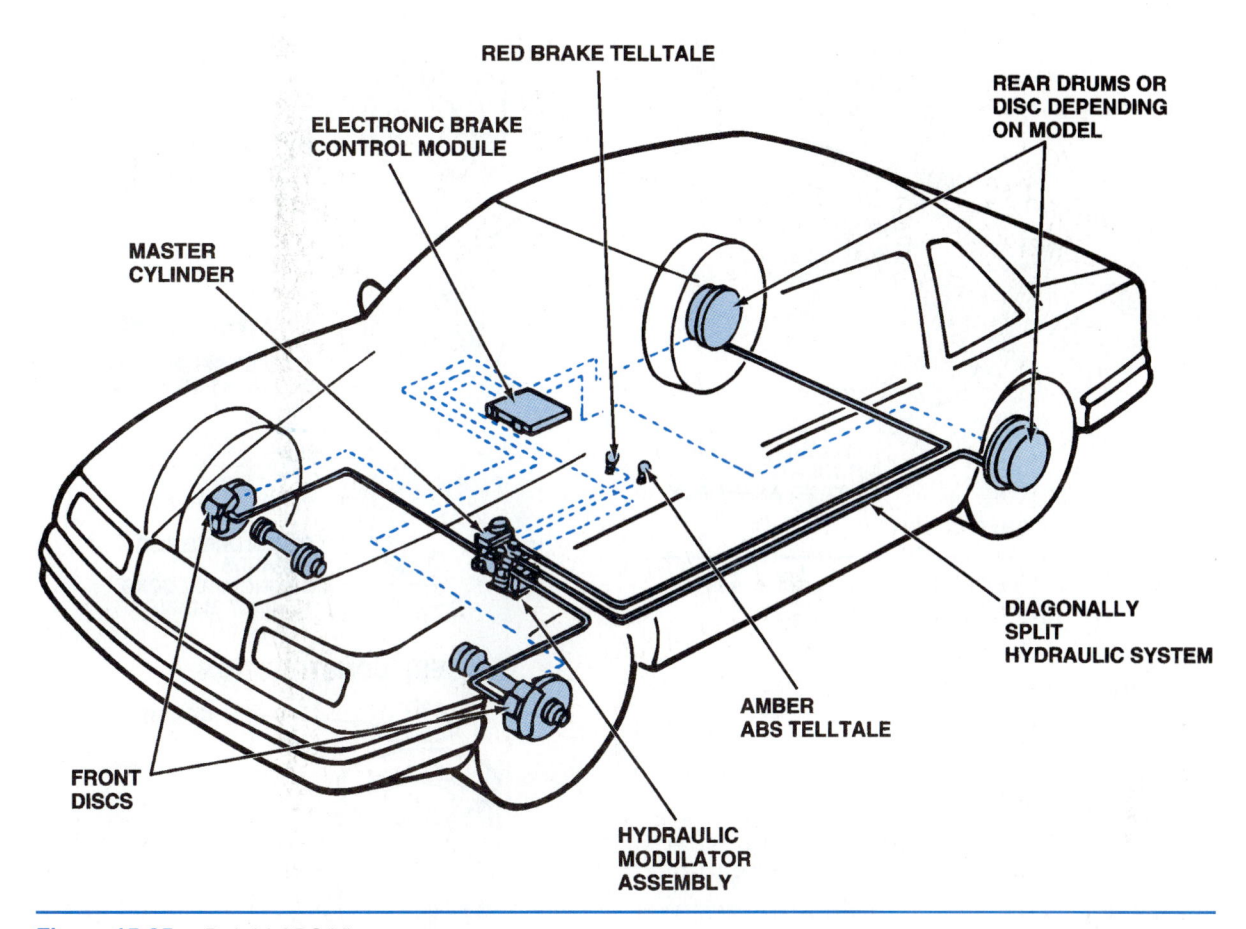

Figure 15-35. Delphi ABS-VI components. (Courtesy of General Motors Corporation, Service and Parts Operations)

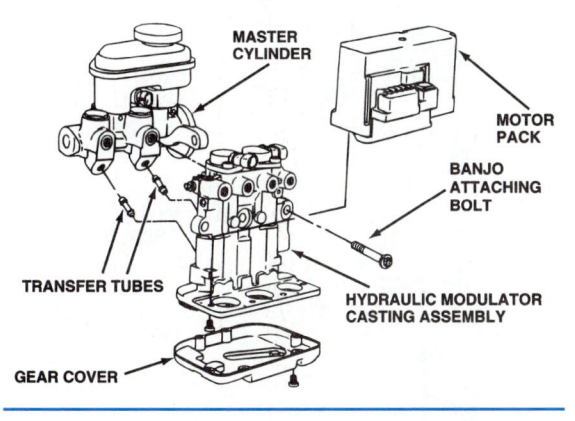

Figure 15-36. A Delphi ABS-VI modulator, motor pack, and master cylinder.

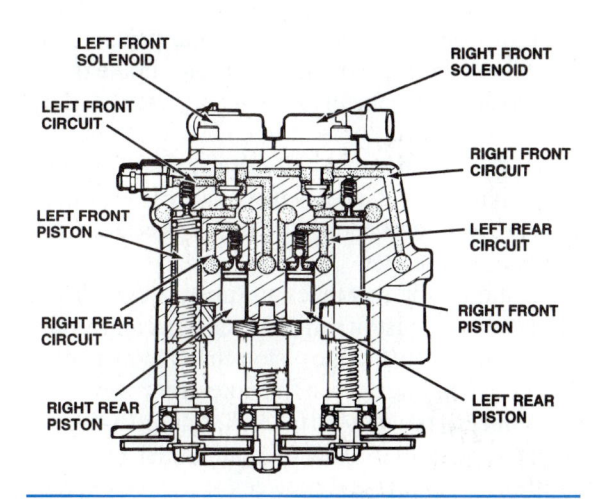

Figure 15-37. A Delphi ABS-VI hydraulic modulator.

failure of either the isolation solenoid or ball screw assembly.

Each rear brake circuit also has its own piston chamber, but the two pistons share a common ball screw and motor. During normal braking, the ball screw for the rear circuit is also at its highest or home position. This holds each check ball for the rear brakes open so fluid can pass through to each rear brake. An expansion spring brake prevents the rear motor from turning during normal braking. The ESB will apply braking action to the rear motor assembly when current to the motor is removed. The ESB has no direct electrical connection to the controller. It is a mechanical brake that

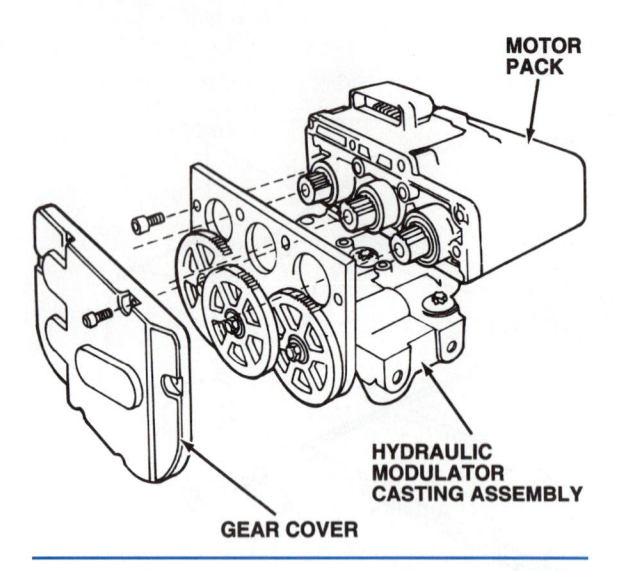

Figure 15-38. A Delphi ABS-VI motor pack assembly.

works similar to a window crank mechanism or overrunning clutch.

System Operation

When the ABS controller detects that a front or rear wheel is about to lock, it initiates ABS braking. If a front brake is involved, the first step is to energize the right or left isolation solenoid valve. This blocks the flow of fluid pressure through the isolation solenoid passage to the calipers. At the same time, the electromagnetic motor brake is energized to free up the motor and ball screw so they can turn. The motor will then draw the piston down by turning the ball screw, allowing the check ball to seat and isolate the brake circuit. This prevents any additional brake pressure from reaching the caliper.

As the motor continues to turn backward, the ball screw and piston move lower. This allows the volume area on top of the piston chamber to receive the corresponding drop in fluid pressure away from the caliper to prevent wheel lock, figure 15-40.

Pressure is then held or maintained when an equilibrium is reached between the force exerted by the motor against the ball screw and piston equals that in the brake circuit itself. When this point is reached, the piston stops moving downward and pressure is held steady. The controller provides current to the motor to accomplish this function. Pressure can then be reapplied as needed by running the piston back up, figure 15-41. To do this, the controller increases current to the motor.

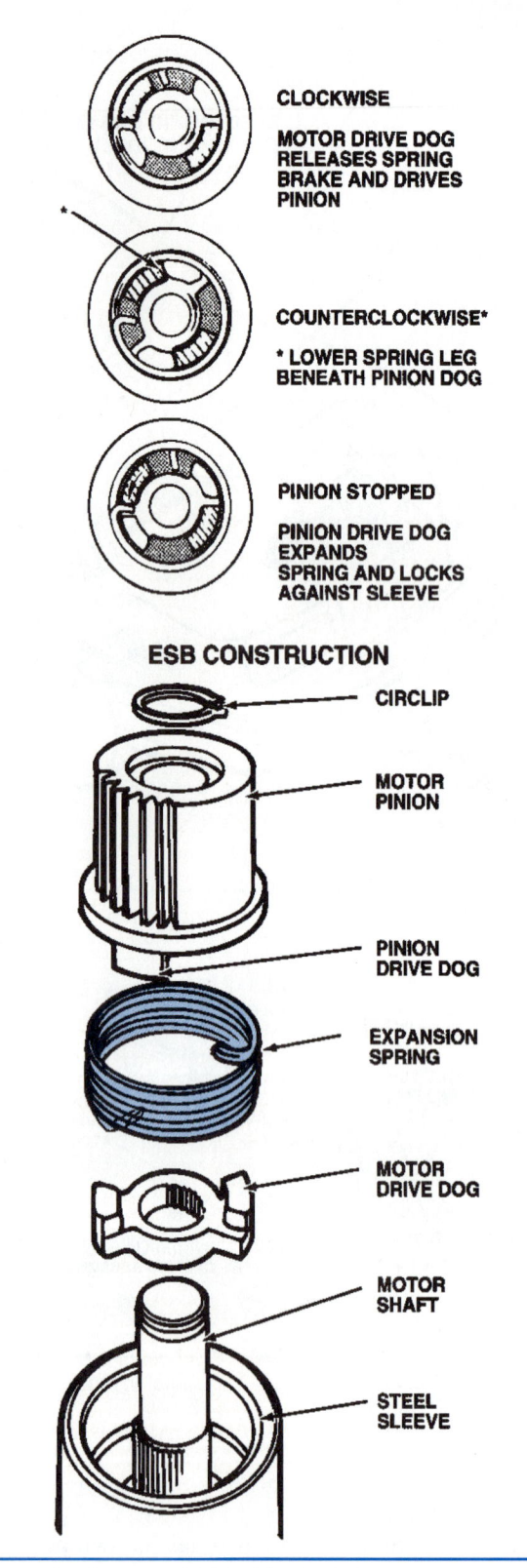

Figure 15-39. Delphi ABS-VI expansion spring brake (ESB) construction and operation.

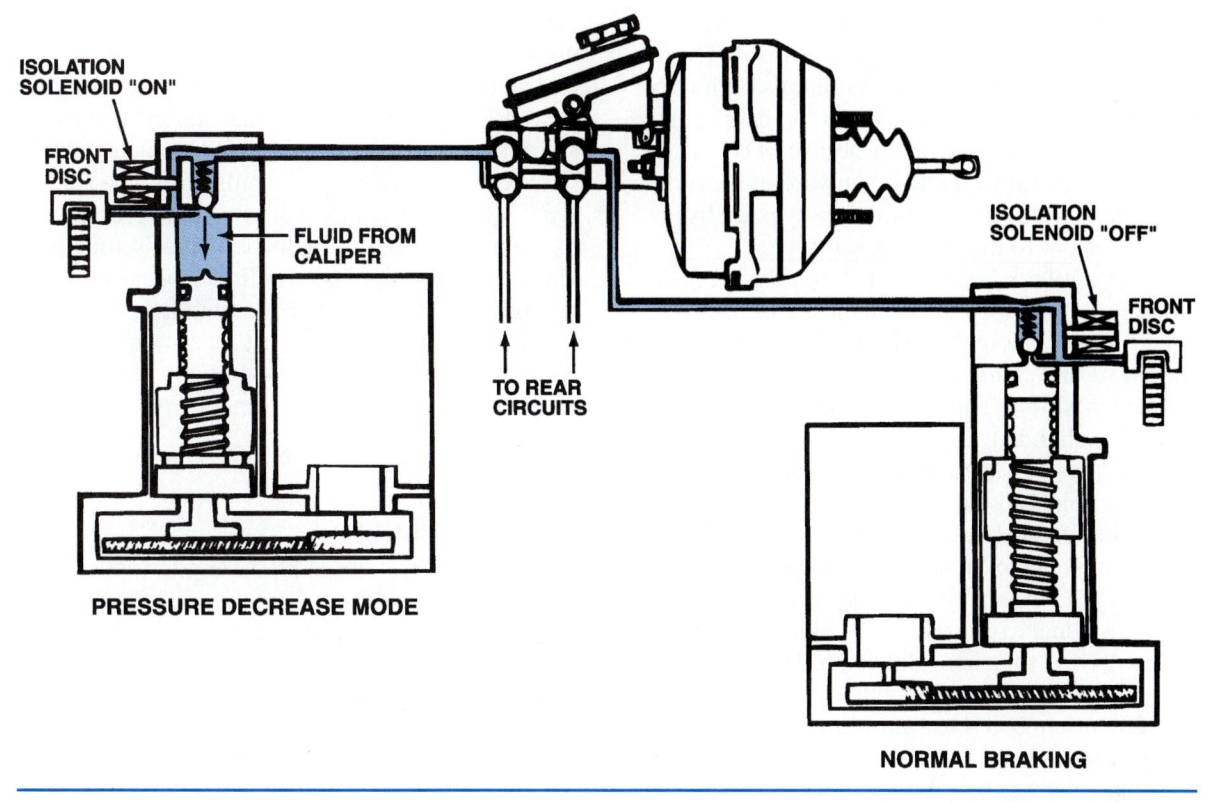

Figure 15-40. A Delphi ABS-VI system in pressure decrease mode for one front brake circuit.

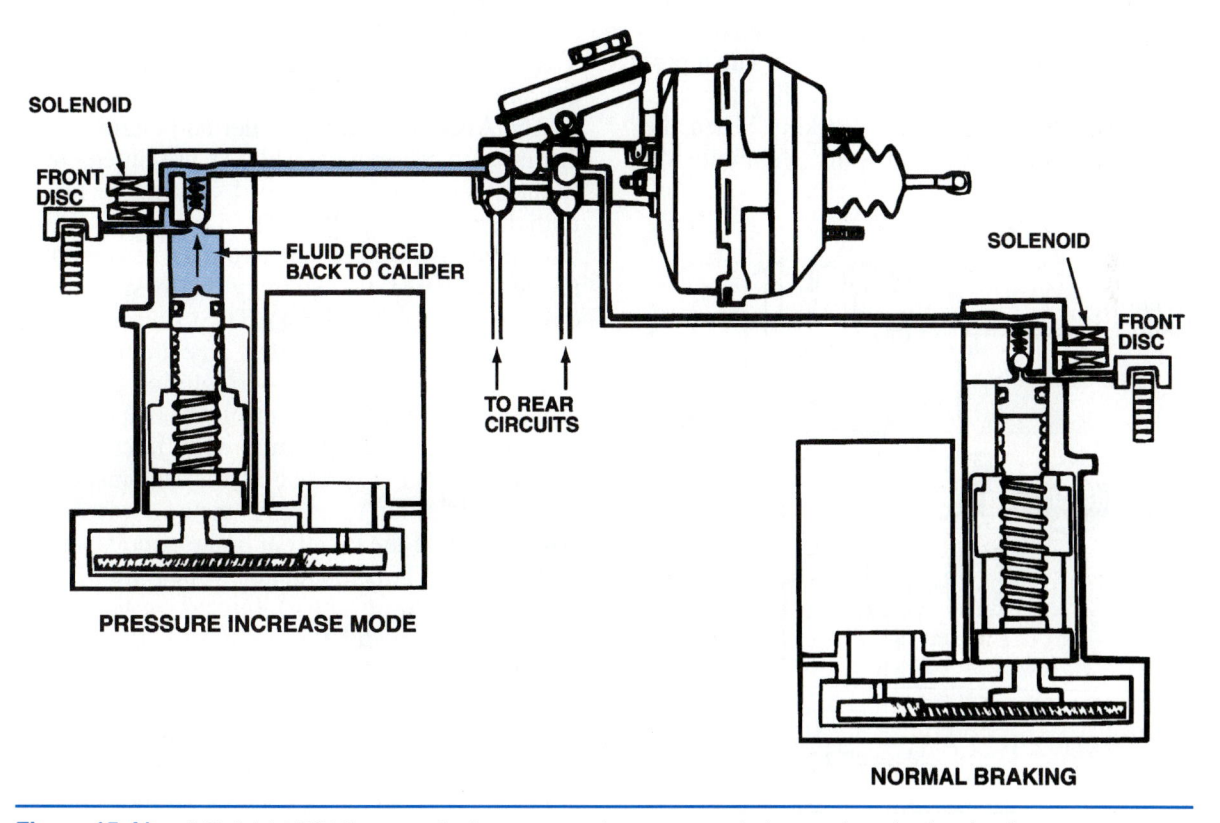

Figure 15-41. A Delphi ABS-VI system in the pressure increase mode for one front brake circuit.

327

This turns the ball screw in the opposite direction and reverses the direction of piston travel. This is the pressure increase phase. As the piston moves back up, it pushes fluid back into the caliper line and increases pressure at the brake. The piston can be moved upward and downward in the modulator bore and held in position at any point by applying current to the motor assembly.

When antilock braking is no longer needed, the motor is commanded to return the ball screw to its uppermost or "home" position. The EMB is de-energized and holds the motor, keeping it in the "home" position. This piston then returns to the top of the chamber and unseats the check ball, reopening that passageway for the brake fluid. The isolation solenoid valve is also de-energized and opens the second passageway to the calipers.

In the rear brake circuit, things happen somewhat differently. During antilock braking, the brake system is no longer split diagonally. The rear wheels are controlled together. The controller commands the rear motor to draw down the ball screw, figure 15-42. The rear ball screw now turns allowing both rear pistons to back down the bore of the modulator. This seats the check balls for the rear brake circuits and isolates the lines from the master cylinder.

At this point, the apply and release phases are much the same as for the front brakes. The motor allows the pistons to back off and reduce pressure at the rear brakes to prevent lockup. When the system needs to reapply pressure, the motor reverses the ball screw and forces the pistons back up, sending more pressure to the rear brakes.

The up and down cycling of the pistons in the ABS-VI system is a somewhat "softer" approach to antilock braking than a solenoid-actuated ABS, so the driver feels less pedal pulsation. The Delphi ABS-VI system cannot increase brake pressure on its own above that which the driver's foot provides through the master cylinder because it has no high-pressure pump or accumulator. The maximum cycling frequency of this system is about half the rate of some ABSs—but is still sufficient to prevent wheel lockup under most circumstances.

Traction Control

On certain applications, ABS-VI also has added traction control capability, referred to as ABS-IV/TCS. On 1994 and later applications, ABS-VI/TCS uses a combination of braking and engine torque management to limit wheel spin at all speeds. Torque management involves retarding spark timing or shutting off up to half of the engine's injectors temporarily until traction is resumed.

On 1993 and earlier applications, ABS-VI/TCS uses braking only to control wheel speed and disengages at speeds above 30 mph. A switch is provided to disable the traction control for unusual conditions, such as rocking the car in deep snow.

The ABS-VI/TCS controller and engine powertrain control module (PCM) are interactive. Signals from the wheel speed sensors are fed to both the PCM and ABS-IV/TCS controller. When wheel spin is detected, the two modules communicate and agree on a strategy that best suits the circumstances. The ABS-IV/TCS control module may apply braking to control wheel spin, or the PCM may momentarily retard ignition timing to reduce engine torque to control wheel spin.

To generate brake pressure when traction control is needed, a TCS modulator is used instead of a pump and accumulator as is the case with other ABSs. Each of the two front brake circuits has its own separate TCS modulator, which are located in separate add-on units mounted on the frame rail. The TCS modulator/motor pack assembly is connected hydraulically to the ABS modulator with lines and has its own electrical connector.

Each traction control channel consists of a motor, piston, ball screw, spring, and poppet valve. Under normal operating conditions, the piston is located in the downward or home position, and the poppet valve is unseated. This is accom-

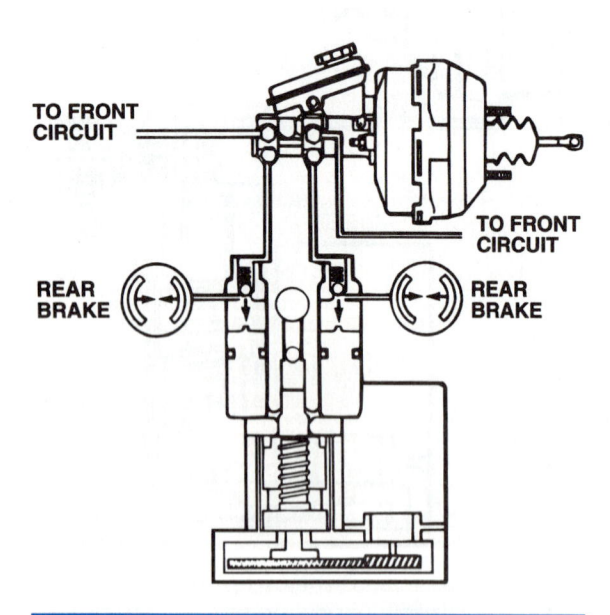

Figure 15-42. A Dephi ABS-VI system in the pressure decrease mode for a rear brake.

Antilock Brake Systems

plished by turning the ball screw via the motor to drive the nut downward. Brake fluid enters the inlet port of the TCS modulator, flows past the unseated poppet valve, and goes out the outlet port to the brake. There are no solenoid valves or ESBs within the TCS modulator.

When traction control braking is needed, the ABS-IV/TCS control module energizes the appropriate TCS modulator. The motor drives the ball screw and pistons up from their home position. This unseats the poppet valve and isolates the circuit from the master cylinder. Now, running the bidirectional motor either way to change the volume within the TCS modulator piston chamber controls the amount of brake pressure applied to the drive wheel brake.

Traction control cannot occur if the driver is applying the brakes. If traction control is operating and the driver touches the brake pedal, the system will immediately disengage traction control, return the TCS ball screws and pistons to the home position, and revert to normal or ABS braking. Normal brake pressure from the master cylinder will also overcome the poppet valve spring in the TCS modulator to apply pressure immediately to the brakes while the TCS modulator is returning to the home position.

The prolonged application of traction control can overheat the brakes, so the ABS-VI/TCS system has a built-in thermal shut-down mode that discontinues traction control after a predetermined amount of application time. The TCS warning lamp will come on and TCS will remain disabled until the brakes have cooled down. The system will then automatically reenable traction control as needed.

Delphi Brake Control DBC 7 Series

In 1999 General Motors began phasing in a new ABS option, the Delphi Chassis DBC 7, as a replacement for the Delphi ABS-VI system. DBC 7 is used on 1999 and and newer Chevrolet Trackers and Buick Regals and Centurys. In 2000 it was introduced on some models of Chevrolet, Pontiac, and Oldsmobile, including the Impala, Malibu, Cutlass, and Cavalier/Sunfire models. The 2001 models include Pontiac Grand Am and Oldsmobile Alero. An upgraded system, Delphi DBC 7.2, is used on the Cadillac CTS, SRX, SLR, and on some Chevrolet models beginning in 2003.

DBC 7 is a nonintegral system that has the electronic controller mounted to the brake pressure modulator valve (BPMV). On vehicles without a TCS, the unit is called an electronic brake control module (EBCM), and on vehicles with traction control, it is referred to as an electronic brake traction control module (EBTCM). The ABS control relay that controls power to the pump motor and solenoids is mounted inside the module.

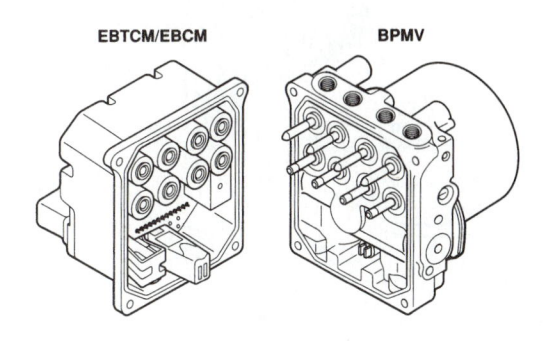

There are three versions of DBC 7 ABS, figure 15-43. On the Chevrolet Tracker a three-channel version is used. Most DBC 7 applications are

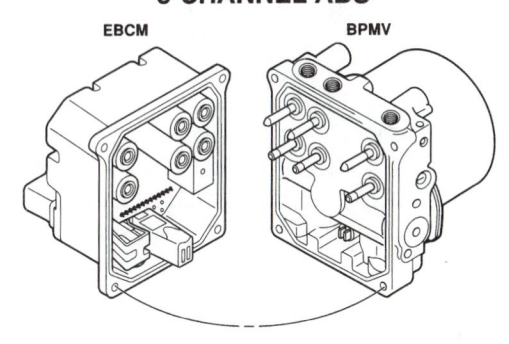

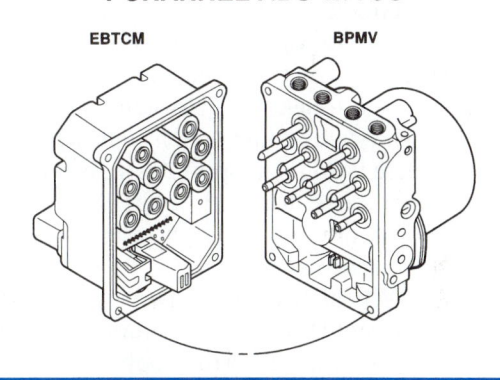

Figure 15-43. Delphi Chassis DBC 7 ABS is available in three versions. (Courtesy of General Motors Corporation, Service and Parts Operations)

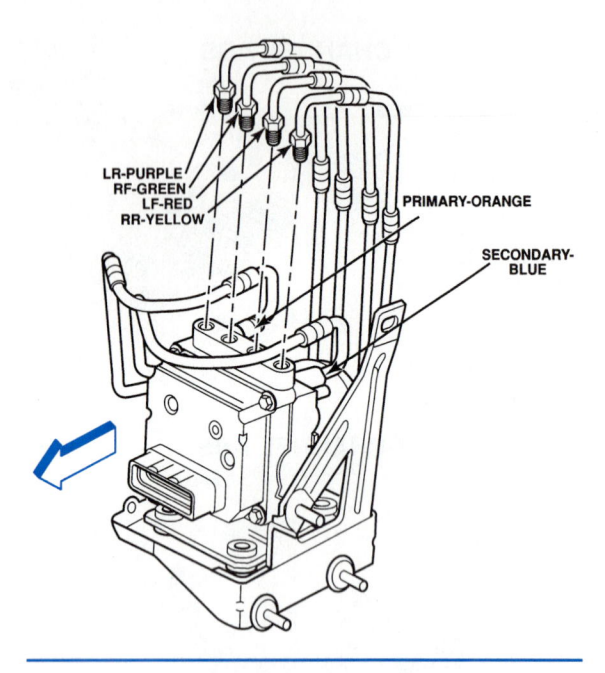

Figure 15-44. The Delphi DBC 7 ABS modulator and brake lines are color coded. (Courtesy of General Motors Corporation, Service and Parts Operations)

four-channel systems with TCS, although there are a few models that have a four-channel system without traction control.

The number of valves in the BPMV depends on the version. On three-channel systems there are six valves; four-channel systems have eight valves, and systems with traction control have ten valves. There is an inlet (apply) valve and an outlet (release) valve for each hydraulic channel. When equipped with a TCS, there is also a TCS valve for each drive wheel.

The BPMV also contains two accumulators and a pump motor. The pump returns fluid from the accumulators back to the apply circuits and also provides pressure to the drive wheels for traction control. The inlet and outlet lines are color coded, figure 15-44.

Wheel sensors at each wheel provide wheel speed input to the EBTCM. Other sensors include the brake switch, longitudinal accelerometer (Tracker only), and four-wheel drive switch (Tracker only).

System Operation
With normal (non-ABS) braking, the inlet and outlet valves are electrically off. The inlet valves are hydraulically open, allowing fluid to pass to the wheels, and the outlet valves are hydraulically closed, figure 15-45.

When wheel lockup is detected, the EBTCM initiates a pressure hold cycle by energizing the inlet valve, which closes the hydraulic circuit on the locking wheel. This stops any further pressure increase from the master cylinder. A pressure release cycle follows, with the inlet valve remaining closed and the outlet valve energized, thus releasing pressure into the accumulators; the pump also runs to decrease fluid in the channel, reducing the pressure. During the ABS pressure increase cycle, the EBTCM turns off both valves. This closes the outlet valve and opens the inlet valve, allowing pressure from the master cylinder to increase in the channel. This "hold-release-increase" process repeats rapidly until wheel slip is within an acceptable range.

Dynamic Rear Proportioning
The Delphi DBC 7 has a function, dynamic rear proportioning (DRP), that eliminates the need for a mechanical proportioning valve in the hydraulic circuits to the rear wheels (see Chapter 6). While slowing during a non-ABS stop, the EBTCM will limit brake pressure to the rear wheel if the deceleration rate of the rear wheels is slightly greater than that of the front wheels. Proper brake system balance requires a pressure bias for the front wheels, and DRP will hold or release pressure to maintain the proper front-to-rear balance. If sufficient releases occur or the system enters ABS braking, the pump will run to empty the accumulators.

DRP is functional at all times unless there are faults in the system. DRP will be disabled if:

- There are multiple EBTCM failures
- Two wheel speed sensors on the same axle fail.

DRP functions will be partially disabled if:

- Either inlet solenoid fails
- The pump fails
- Any wheel speed sensor fails when braking in a turn.

Any fault that disables or degrades the DRP will turn on the red brake warning light.

Tire Inflation Monitoring System
Some applications of the DBC 7 system use software and the vehicle wheel speed sensors to detect low tire pressure. The feature may be called a tire inflation monitoring system (TIMS) or a tire

Antilock Brake Systems

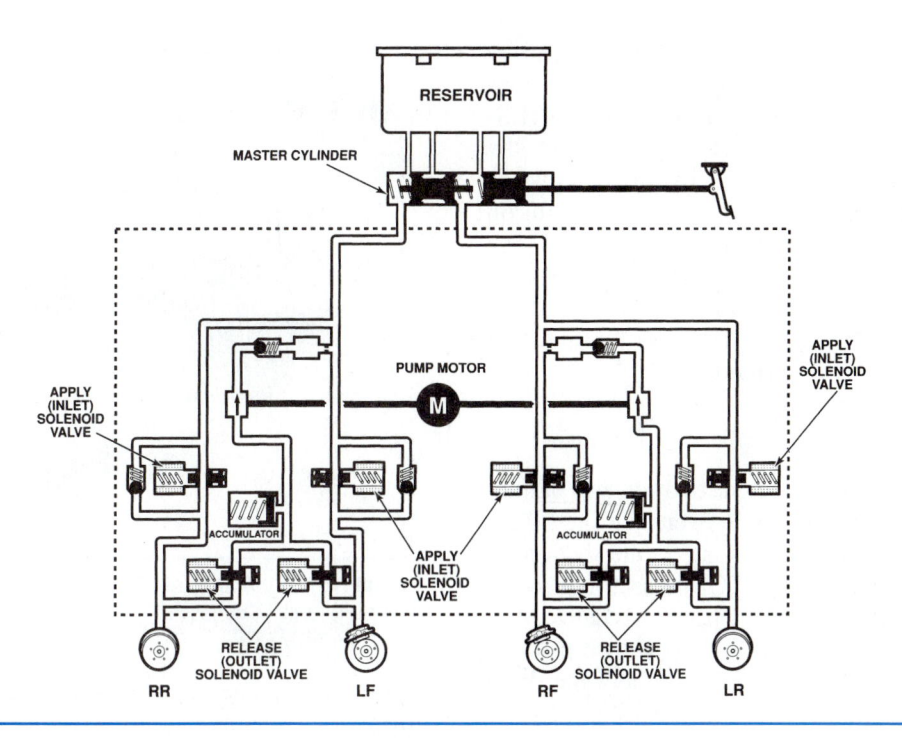

Figure 15-45. The Delphi DBC 7 ABS hydraulic schematic system. (Courtesy of General Motors Corporation, Service and Parts Operations)

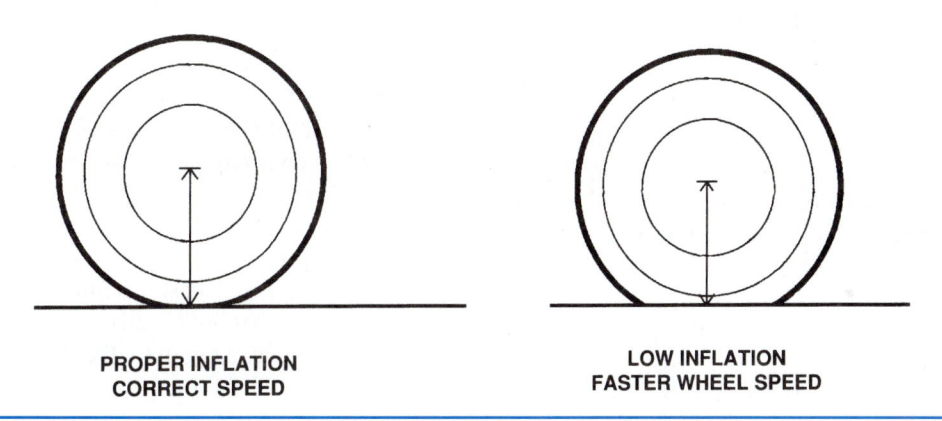

Figure 15-46. The ABS wheel sensors are able to detect a tire with low pressure. (Courtesy of General Motors Corporation, Service and Parts Operations)

pressure monitoring (TPM) system, depending on the vehicle. Any tire that becomes underinflated will also have a smaller rolling radius when compared to the other wheels, figure 15-46. The system can detect a pressure difference of 10 psi (69 kPa).

The EBCM software requires approximately 30 minutes of straight-line driving in each of the three different speed ranges to complete a calibration process in order to have full ability to detect a tire pressure condition. The speed ranges are as follows:

- 15 to 40 mph (24 to 64 km/h)
- 40 to 70 mph (64 to 113 km/h)
- 70 to 90 mph (113 to 145 km/h).

The EBCM learns the tire pressure calibration for each speed range independently. Once the TPM system is calibrated the EBCM monitors the ABS wheel speed sensor inputs, which will all be the same as the learned calibration if the tire pressures stay the same. If the pressure increases or decreases in a tire, so too will that tire's circumference and radius, which causes that wheel speed sensor input to change. The EBCM recognizes this change in wheel speed as a tire pressure condition and will turn on the LOW TIRE PRESSURE warning lamp or set a message on the vehicle message center in the instrument cluster.

The low tire pressure warning will remain on until a specific reset procedure is performed. See the *Shop Manual* for details. After the reset procedure, the TPM system is not fully functional until it has again completed the self-calibration process previously described.

Traction Control

Delphi DBC 7 systems include two versions of traction control, depending on the vehicle. The (TCS) operates as described previously in the section on Bosch 5.3 and 5.7 systems. TCS uses various methods to reduce engine torque when slip is detected. If torque reduction is not enough to reduce positive wheel slip, the EBTCM can also apply the drive wheel brakes to further control wheel slip.

A second version, enhanced traction control (ETC) is used on some applications. ETC achieves traction control using torque reduction alone to control positive drive wheel slip. This is done by retarding spark timing, selective cylinder cutout, leaning out the air/fuel mixture, or forcing a transmission upshift. No brake application is involved.

Either system can be disabled by the driver using a switch. When the system is disabled, the ETC OFF of TCS warning lamp is on. The system will revert to an enabled condition with the next key cycle. When the traction control function is active, the LOW TRAC lamp will illuminate and stay on for a few seconds after the system returns to an inactive mode.

Delphi DBC 7.2 and 7.4

The DBC 7.2 ABS is functionally identical to the DBC 7 system but has expanded feature content. For example, the Cadillac XLR DBC 7.2 ABS features:

- An antilock brake system
- Engine drag control
- Dynamic rear proportioning
- Traction control system
- Vehicle stability enhancement system
- Enhanced vehicle communication using the controller area network communication protocol.

The DBC 7.4 ABS module is a refinement of the DBC 7.2 and is lighter, quieter, and more compact,

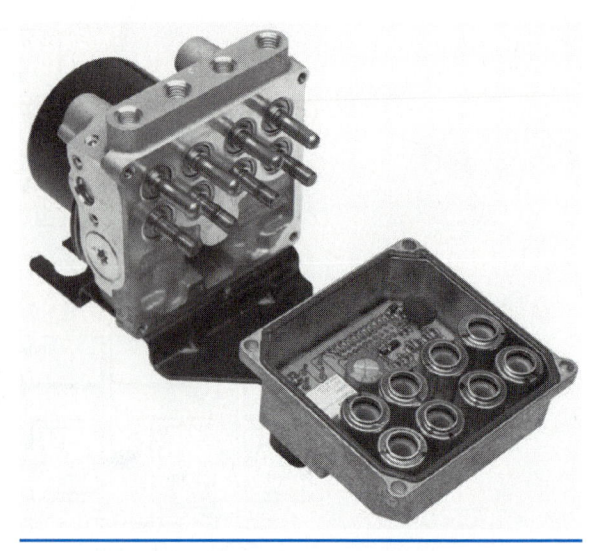

Figure 15-47. A Delphi DBC 7.4 control module and hydraulic unit. (Courtesy of Delphi Corporation)

figure 15-47. The larger memory (256 kbyte) allows for even more features in future applications.

The operation of ABS, dynamic rear proportioning, and traction control function are the same as discussed previously. Some additional features of the DBC 7.2/7.4 system are briefly discussed here.

Engine Drag Control

On vehicles with manual transmissions, when the driver releases the throttle and the drag from the engine overcomes the frictional force between the tire and the road, skidding or loss of control may occur. Under these conditions engine drag control (EDC) becomes active. The EBCM sends a torque request signal to the PCM, which increases the torque at the wheels. This stabilizes the wheels by reducing the slip at the driven wheels.

Vehicle Stability Enhancement System

The vehicle stability enhancement system (VSES) adds an additional level of control to the EBTCM. By sensing the steering wheel position, the vehicle speed, and the lateral acceleration of the vehicle, VSES can apply differential braking to the appropriate wheel to help maintain vehicle control during aggressive driving and in turns. The system is similar to the Bosch 5.3 and Bosch 5.7 vehicle stability control systems. Refer to those sections for operating details.

Controller Area Network

Although is not exactly a braking function, the way that the EBTCM communicates with other

Antilock Brake Systems

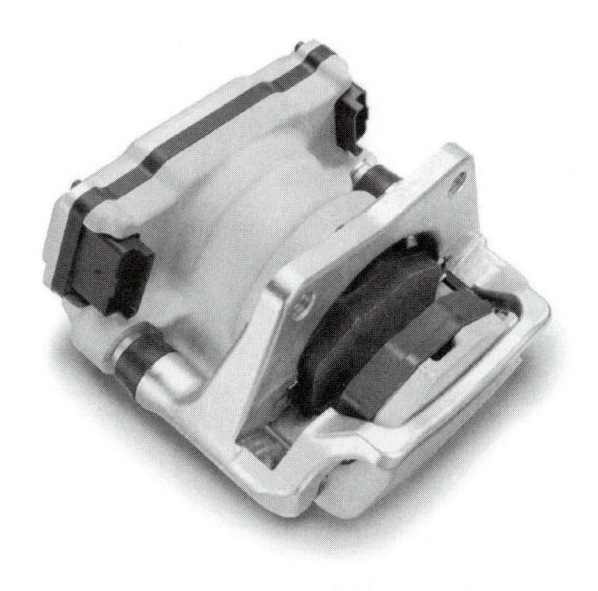

Figure 15-48. An electromechanical brake caliper for use on a CAN-bus-controlled electric brake system. (Courtesy of Delphi Corporation)

vehicle modules has become increasingly important. Manufacturers have been using on-board computers to control various vehicle functions since the 1980s. In the beginning various data communication systems were used to control engine functions. Beginning in the early 1990s, these data systems evolved to permit numerous modules on the vehicle to communicate with each other over a common data line.

With each system change, it has been possible to increase the speed of data transmission to accommodate an ever-growing number of functions. While the domestic automakers were developing their own data protocols, the European industry was developing a system that operates on the controller area network (CAN) protocol. Also used in the Bosch 5.7 ABS, the Delphi DBC 7.2 is able use the CAN data protocol, called local area network by GM GMLAN.

The major advantage of the CAN protocol is the speed of communication. Using the CAN protocol, GMLAN operates at two different data rates, low speed and high speed.

- **Low-speed bus:** The low-speed data system is typically used for functions controlled by the vehicle operator (for instance, door lock, window motor, etc.) where the system response time requirements are on the order of 100 to 200 msec.
- **High-speed bus:** The high-speed bus is typically used for sharing real-time data. Systems with these requirements are primarily powertrain and chassis devices (engine, transmission, brakes). This data system operates at 500 kbps.

Use of a high-speed data system allows real-time electronic control of many vehicle functions that traditionally had been mechanically or hydraulically controlled, such as throttle control, braking, steering, and suspension. For example, Bosch, Continental/Teves, Delphi, and others are developing electric braking systems (no brake fluid) that depend on CAN for operation, figure 15-48.

KELSEY-HAYES ABS

Kelsey-Hayes manufactures rear-wheel-only ABSs (EBC2 RABS, RABS II, and RWAL) as well as a family of nonintegral, four-wheel, three- and four-channel ABSs, including EBC4 (4WAL), EBC5H, EBC5U, EBC310, EBC325, EBC410, and EBC430.

Kelsey-Hayes was bought by Lucas Varity in 1997 and was subsequently acquired by TRW in 1999. The various systems may be called Kelsey-Hayes systems, Kelsey-Hayes (Lucas Varity) systems, Lucas ABS, or TRW ABS, depending on the source. Because the "EBC" prefix for each system is used by all three companies, this text uses the EBC prefix as a system identifier in the following descriptions.

EBC2 RABS, RABS II, and RWAL

Kelsey-Hayes EBC2 rear-wheel antilock brake systems have been in use since 1987 on Ford F series trucks, as well as on later model Ranger, Bronco, Bronco II, and Explorer trucks, and Aerostar and Econoline vans. Ford calls their version of EBC2 rear-wheel antilock brake system (RABS). Beginning in 1993–94, the system is called RABS II, the main difference being in the brake control module self-diagnostic and trouble code software.

On General Motors applications, the EBC2 system is called rear wheel anti lock (RWAL). It is on 1988 and later Chevrolet C and K series pickups, on 1989 M series (Astro) minivans and S and T series pickups, some S series Blazers, and on 1990–92 R and V series light trucks and G series vans.

Dodge has also used the RWAL system since 1989 on its D and W 150-350, Dakota, and Ram pickups. Geo, Isuzu, Mazda, Nissan, and Subaru have used the system since 1991.

A zero pressure rear wheel antilock (ZPRWAL) variant of the basic EBC2 system is found on

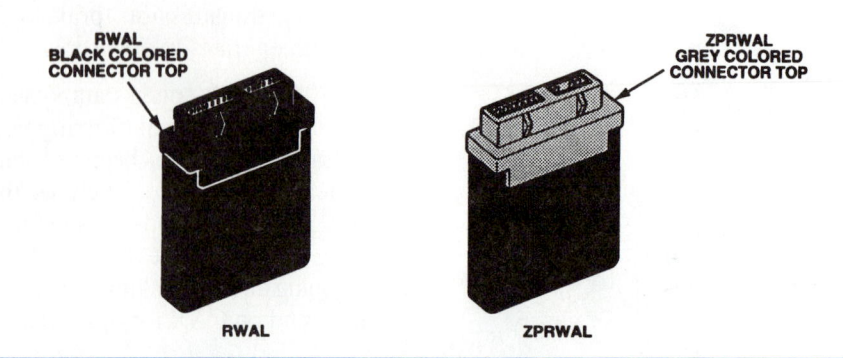

Figure 15-49. The controllers for the RWAL and ZPRWAL look the same but have different colored electrical connectors. (Courtesy of General Motors Corporation, Service and Parts Operations)

1992–94 3500 HD model C and K series GM trucks. All other 1992–96 C and K series truck applications still use either the RWAL or the 4WAL system. The ZPRWAL system works in essentiallly the same manner as the RWAL system, but the components are not interchangeable. The ZPRWAL system is designed to handle the larger volume of fluid that is found in the rear disc brake calipers on the 3500 HD models. The electronic controllers for the RWAL and ZPRWAL systems appear similar, the only external difference being a gray electrical connector on the ZPRWAL controller, compared to a black connector on the RWAL controller, figure 15-49. The two are not interchangeable.

Rear-wheel-only antilock braking cannot provide the same benefits as four-wheel ABS braking, but it is a major improvement over standard brakes on trucks and vans where vehicle loading has a major influence on brake balace and traction. In trucks and vans, the tendency of the rear wheels to lock up and skid is greatly influenced by the payload in the vehicle. The lighter the load over the rear wheels, the greater the tendency for the rear wheels to lock on wet surfaces or when braking hard. Rear-wheel antilock braking became standard on most pickups from 1988 to 1991.

System Description

The EBC2 RABS, RWAL, and ZPRWAL systems are all nonintegral, rear-wheel-only antilock brake systems, figure 15-50. On four-wheel-drive applications, rear-wheel antilock braking is only used when in the two-wheel-drive mode.

The conventional master brake cylinder and power booster supplies brake pressure to a dual solenoid control valve for the rear brakes. This unit contains only two solenoid valves: a normally open isolation valve to block pressure from the master cylinder to the rear brakes during antilock braking and a normally closed dump valve for relieving pressure in the rear brake circuits, figure 15-51. The dual solenoid control valve also contains a pressure accumulator for storing fluid pressure during the dump or release phase of operation, and a reset switch which allows the system to maintain proper brake pressure.

When the ABS control module detects a difference in the average speed of the rear wheels compared to the vehicle's overall speed, it initiates antilock braking. The control module energizes the ABS isolation solenoid, figure 15-52, to prevent any further increase in brake pressure at the rear wheels. Then, the ABS dump solenoid valve is opened to release pressure from ther rear brake cirucits so the wheels can regain speed and traction, figure 15-53. Pressure is reapplied when both solenoids are de-energized and return to their normal positions. The cylce is repeated continuoulsy for as long as ABS braking is needed or until the vehicle stops.

The control module that regulates operation of the ABS solenoid valves is separate from the control valve and is located next to the master cylinder on some applications. On Dodge trucks, it is on the passenger side cowl panel under the dash. On Ford Bronco IIs, it is located under an access panel by the driver's door pillar.

The control module receives a speed signal from a single vehicle speed sensor. On Ford and Dodge applications, the speed sensor is in the differential and the sensor ring is on the ring gear. On GM applications, the speed sensor is located in the transmission tailshaft, and the sensor ring is on the transmission output shaft.

Digital Ratio Adapter Controller

On GM 1988–95 applications, the speed sensor signal first passes through an intermediate module on its way to the antilock controller. This module is called the digital ratio adapter con-

Antilock Brake Systems

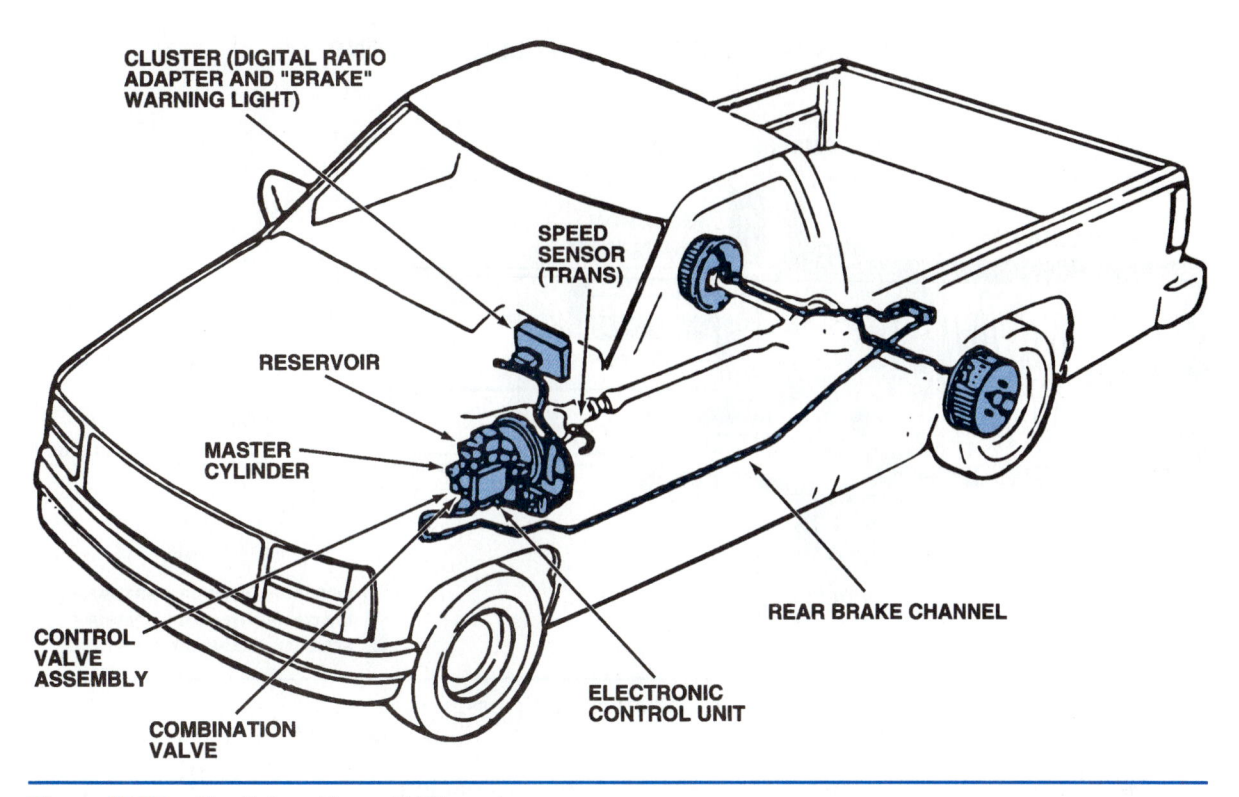

Figure 15-50. The Kelsey-Hayes RWAL system.

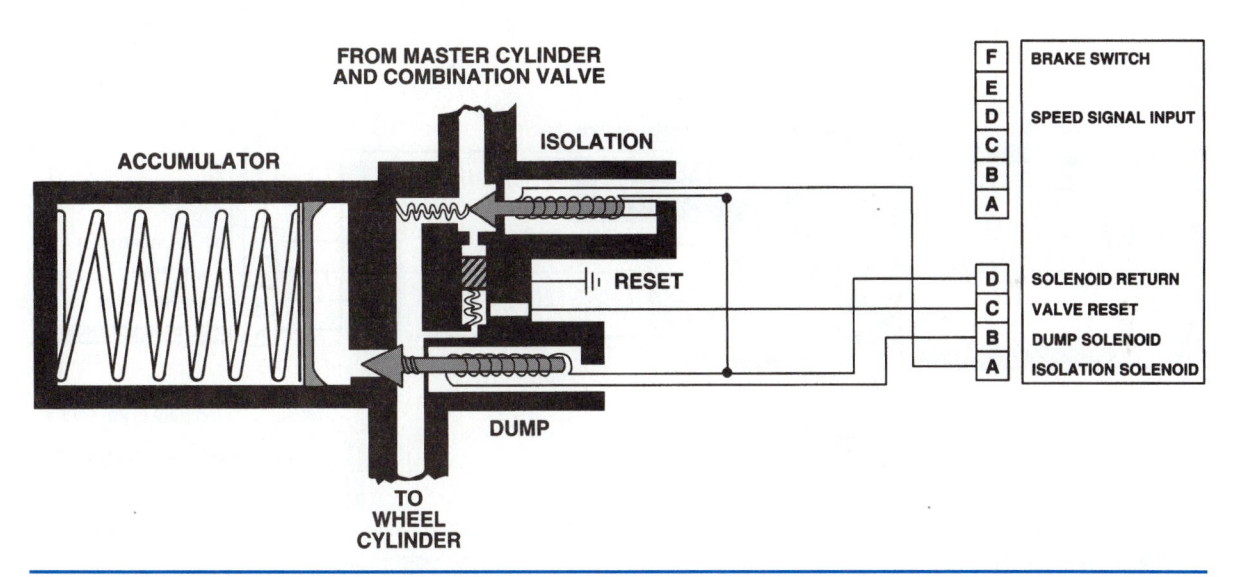

Figure 15-51. The Kelsey-Hayes RWAL system modulator in the normal brake mode.

troller (DRAC). The DRAC module translates the analog speed sensor signl into a digital signal that can be processed by the ABS control module. The DRAC module also divides the basic speed sensor signal into three separate frequencies of signals that are used by other vehicle systems:

- A 4000-pulse-per-mile frequency that is used by the cruise control system and electronic instrument cluster on some vehicles
- A 2000-pulse-per-mile frequency that is used by the engine control module for various emission and torque converter lockup functions
- A 128,000-pulse-per-mile frequency that is used by the ABS controller.

The DRAC module is calibrated to the final drive ratio and original equipment tire size of the vehicle. Replacing the original tires with ones of a different

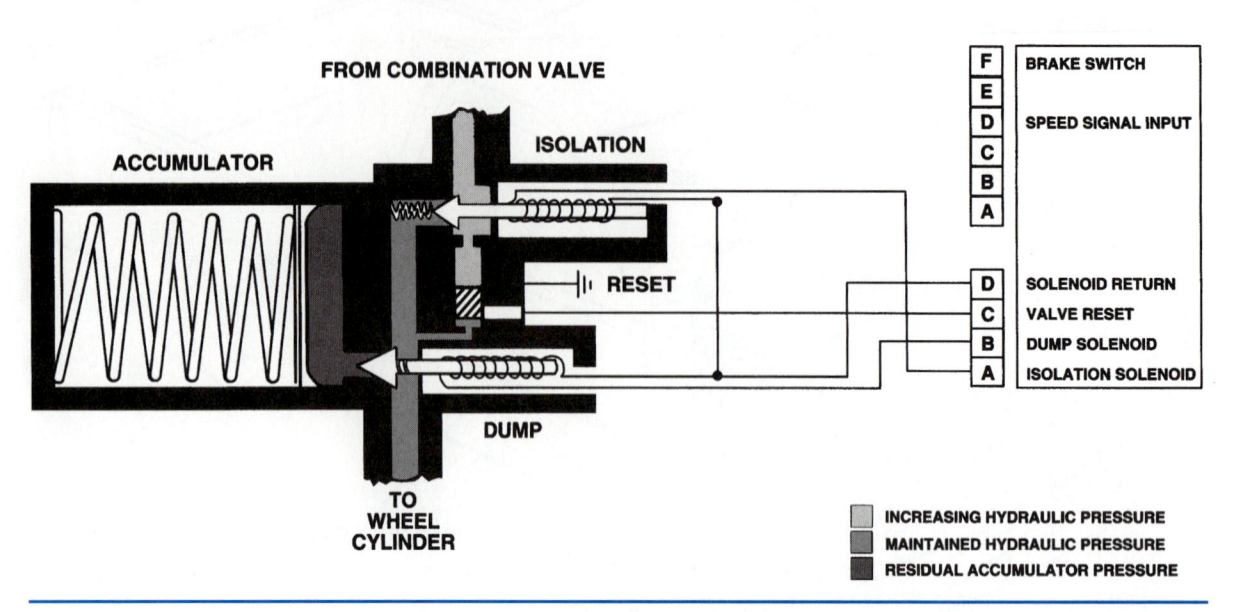

Figure 15-52. The Kelsey-Hayes RWAL system modulator in the pressure hold mode.

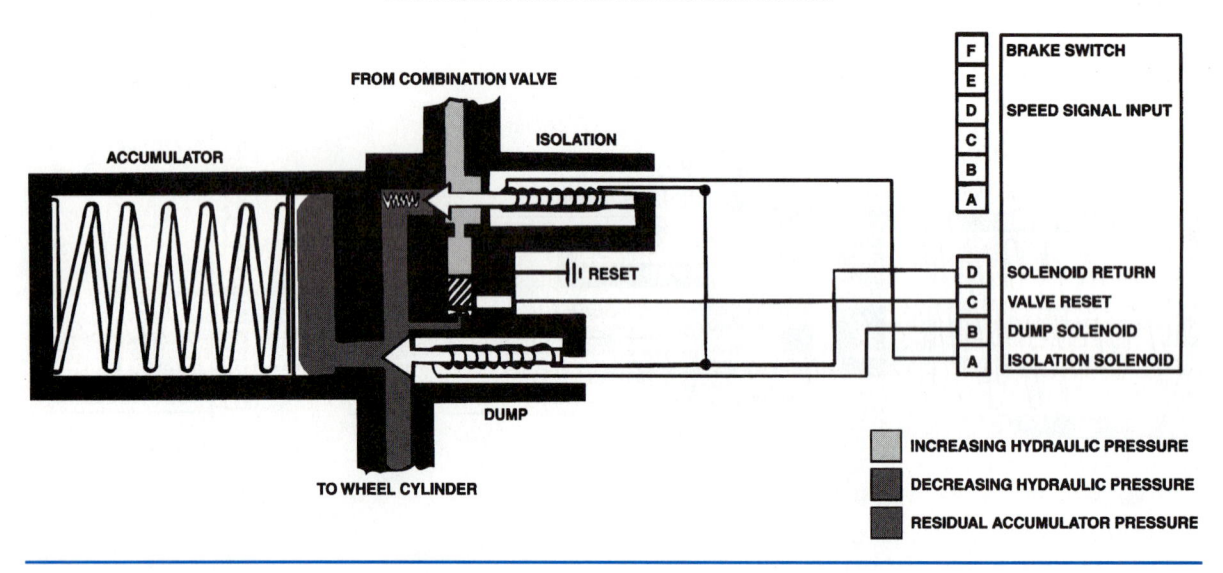

Figure 15-53. The Kelsey-Hayes RWAL system modulator in the pressure decrease mode.

size or aspect ratio will change the speed sensor signal, which in trun can adversely affect the operation of the ABS system as well as torque converter lockup and the accuracy of the speedometer and odometer.

Installing nonstock tires on a GM vehicle with a DRAC module requires the DRAC module to either be replaced or recalibrated. On 1991 and earlier C and K pickups, the DRAC modules can be recalibrated for different tire sizes and axle ratios by changing the configuration of an eight-pin connector that plugs into the instrument panel circuit board connector. A DRAC recalibration kit from GM is necessary for this procedure. By referring to a speedometer calibration chart in the factory service manual, you can determine the correct pin positions for any tire size and gear ratio combination. The speci-

Antilock Brake Systems

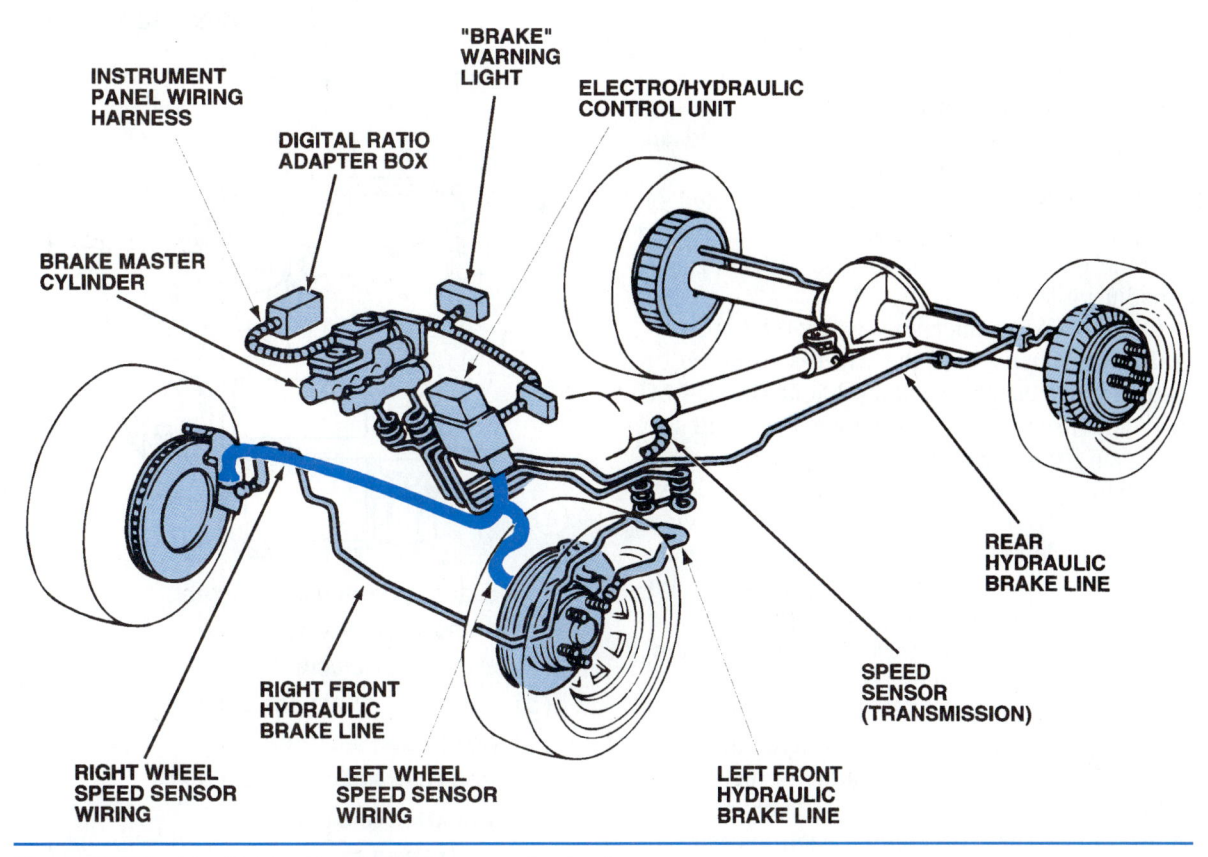

Figure 15-54. The Kelsey-Hayes 4WAL system for the Astro van.

fied pins are then broken off the connector. This alters the DRAC circuits when it is replaced in the instrument cluster to recalibrate the vehicle speed signal.

The S and T Blazers and vans as well as 1992–95 C and K pickup truck applications have a DRAC that is a sealed plug-in module referred to as a vehicle speed sensor buffer. It is matched to the original gear ratio and tire size of the vehicle, and you cannot recalibrate it. Any changes in tire size or axle gearing require replacing the DRAC or buffer with one that is correctly calibrated for the tire size application.

EBC4 (4WAL)

In 1990, Kelsey-Hayes introduced a new EBC4 four-wheel antilock braking system on General Motors M and L series minivans (Chevrolet Astro and GMC Safari). General Motors refers to the system as 4WAL. The following year 4WAL was offered on the S and T series trucks (Chevy Blazer, GMC Jimmy four-door) and the GMC Syclone. In 1992, the four-wheel antilock braking system was added to the two-door S and T trucks, Suburban, full-size Blazer/GMC Yukon, Geo Tracker, and the limited-production GMC Typhoon.

The 4WAL system, figure 15-54, is a nonintegral, four-wheel, three-channel ABS with a conventional master cylinder and brake booster. Each front brake circuit is controlled independently, but both rear brakes are controlled as a pair. The system remains active in four-wheel drive, unlike some earlier RWAL applications where ABS was deactivated in four-wheel drive.

System Description

Most 4WAL applications use four wheel speed sensors, one for each wheel. On the 1992 Suburban, however, only the front wheels have individual speed sensors. A transmission-mounted vehicle speed sensor provides a common signal for both rear wheels.

The front wheel speed sensors on two-wheel drive trucks are located on the splash shield behind the brake rotor, with the sensor ring behind the rotor. Four-wheel drive trucks have the front sensors mounted behind the splash shield and can be replaced without having to pull the brake rotor. The tone ring, however, is part of the wheel bearing assembly. Air gaps on both types of front wheel speed sensors are nonadjustable and are factory set at 0.050″ (1.3 mm).

The rear wheel speed sensors are located on the rear brake backing plates with rings pressed onto the ends of the axle shafts. The air gap is nonadjustable.

On all 4WAL applications except the 1992 Suburban, the signals from the wheel sensors go directly to the electronic control unit (no DRAC module is used). On the Suburban, the speed sensor signal for the rear wheels first goes through an electronic "buffer" that's similar to the DRAC module in RWAL applications. The buffer divides up the vehicle speed signal into various frequencies that are used by the cruise control system, engine control module, and electronic speedometer and odometer, as well as the antilock brake system.

The ABS control module is mounted on the hydraulic modulator assembly, figure 15-55, which General Motors calls the electro-hydraulic control unit (EHCU). One variation is the EBC4-VCM system on 1994 and later Chevy S and T series trucks. The ABS control electronics are integrated with the engine control module so a single module controls both systems.

The ABS control module receives input from the wheel speed sensors and a brake pedal switch, figure 15-56. On trucks with automatic transmissions, the brake switch circuit contacts also control power to the transmission torque converter clutch. The switch is normally closed and provides a 12-volt electrical signal to the ABS controller. With the brake pedal applied, the switch opens and supplies less than 1 volt to the controller.

The 4WAL system also has a 4WD switch and indicator light. The 4WD switch changes the ABS control logic when the vehicle is shifted into four-wheel drive. The switch is mounted on the front axle in T series trucks and is open in two-wheel drive, supplying less than 1 volt to the controller. The switch is closed in four-wheel drive and supplies 12 volts to the controller. The switch is part of the indicator light circuit and is mechanically operated by the shift linkage in the transfer case.

For diagnostic purposes, there is both a red BRAKE and an amber ANTILOCK warning light, instead of the single red BRAKE warning light on Kelsey-Hayes RWAL systems. An assembly line diagnostic link (ALDL) diagnostic connector is provided for manual fault code retrieval and scan tool diagnostics.

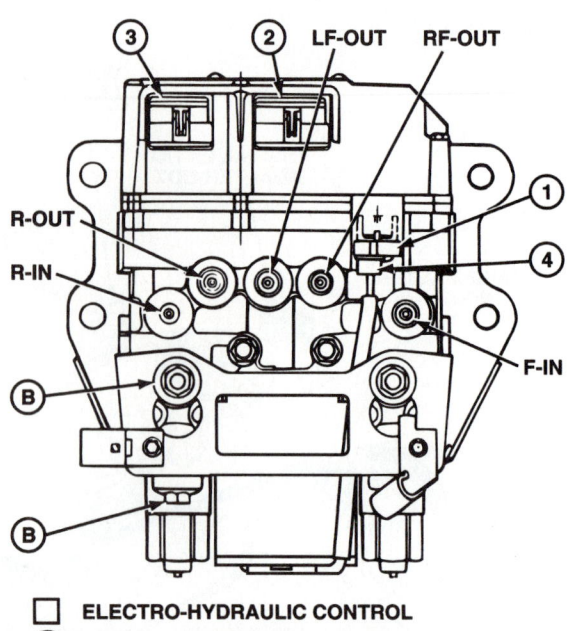

☐	ELECTRO-HYDRAULIC CONTROL
①	BATTERY POWER CONNECTOR
②	SWITCH POWER/CIRCUIT CONNECTOR
③	WHEEL SPEED SENSOR CIRCUIT CONNECTOR
④	MOTOR CONNECTOR
☐	MOUNTING BRACKET FEATURES
B	ISOLATION GROMMETS (SIX)
R-IN	REAR INLET FITTING
R-OUT	REAR CHANNEL OUTLET FITTING
F-IN	FRONT INLET FITTING
LF-OUT	LEFT CHANNEL FITTING
RF-OUT	RIGHT CHANNEL OUTLET FITTING

Figure 15-55. The Kelsey-Hayes 4WAL electro-hydraulic control unit (EHCU).

Electro-Hydraulic Control Unit

The electro-hydraulic control unit (EHCU) contains six ABS solenoid valves: an isolation valve and a modulation valve for each of the three individual brake circuits. The EHCU also has a single pump motor that drives two separate pumps, one for the front brakes and one for the rear brakes. The relay for the pump motor is inside the EHCU. Also inside the EHCU are four pressure accumulators, one front and one rear low-pressure accumulator, and one front and one rear high-pressure accumulator. There are also three reset switches, one for each

Antilock Brake Systems

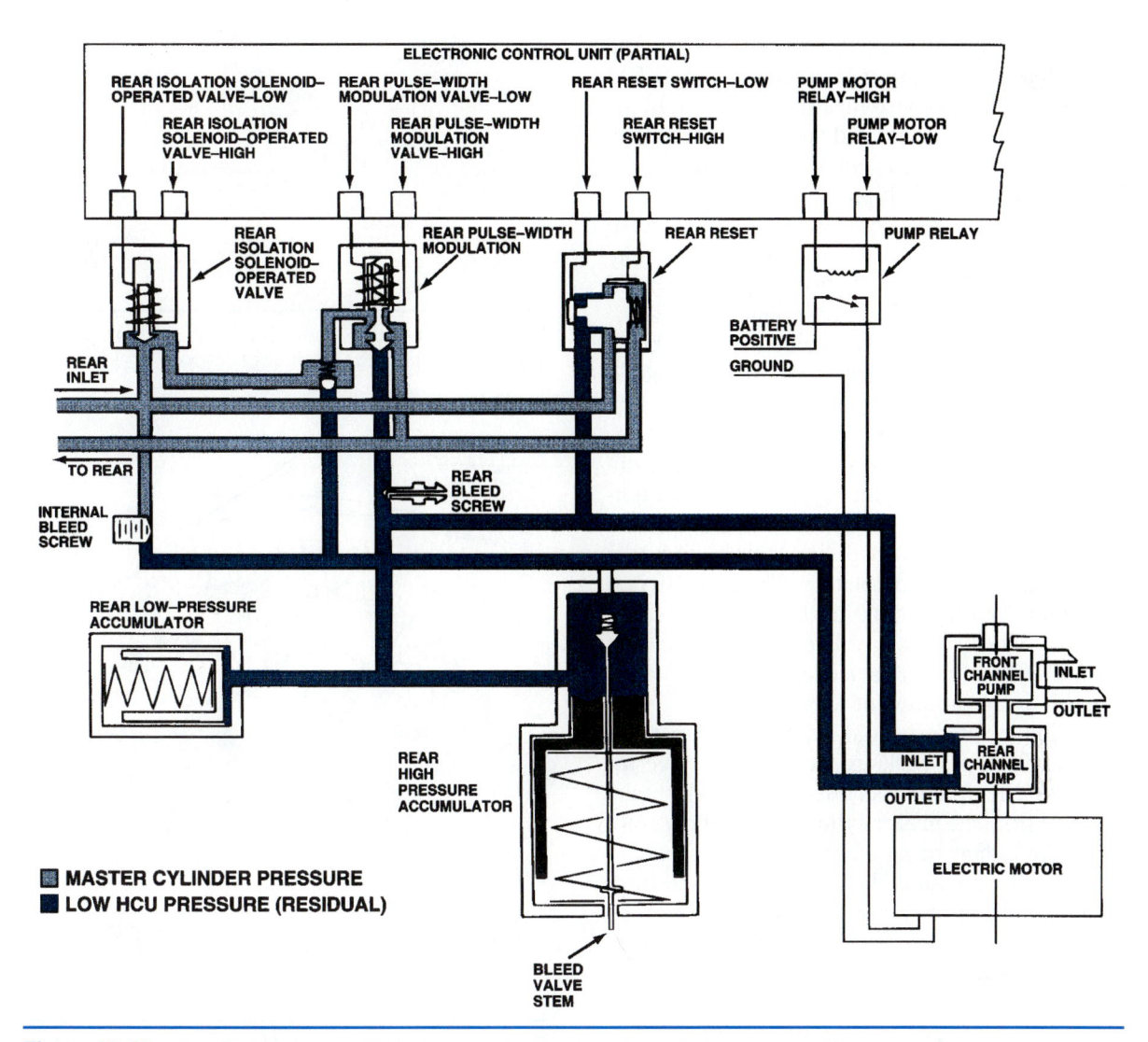

Figure 15-56. The Kelsey-Hayes RWAL system hydraulic schematic in the normal brake mode.

front brake and one for the combined rear brake circuit. The EHCU is located under the master cylinder in the engine compartment on the Astro and Safari vans, and on the left wheel well on S and T Blazers.

System Operation
During normal braking, fluid pressure from the master cylinder passes through the normally open isolation and modulation solenoid valves in the EHCU to the individual brakes. When the controller senses that wheel lockup is about to occur, it activates antilock braking.

The EHCU begins ABS operation in the usual way, by energizing the ABS isolation solenoid for the affected wheel to block any further pressure increase to the brake. At the same time, the pump motor relay is energized to start the pump. This primes the system and builds pressure in the high-pressure accumulator for the pressure increase phase of antilock braking. Next, the system energizes the decay ABS solenoid, which is referred to as the pulse-width modulation (PWM) valve, to bleed pressure from the brake. The 4WAL control module can vary the amount of pressure decrease by varying the valve's open or "on" time. It does this by varying the "on time" or duration of voltage to the solenoid.

When the PWM valve is energized and closes, it opens a passage that allows pressure to be released from the brake circuit. The pressure dumps into one of the two low-pressure accumulators. The spring-loaded low-pressure accumulator (LPA) serves as a temporary storage reservoir to hold the fluid until it can be rerouted back into the system. It also maintains sufficient fluid pressure in the EHCU to keep the high-pressure pump primed.

As the system reduces pressure in the affected brake circuit, the reset switch for that circuit may trip, signaling the ABS control module that fluid pressure has dropped to the point where additional pressure is needed to maintain proper braking action. At this point, the control module de-energizes the PWM valve and initiates the pressure increase mode.

When the PWM reopens, fluid pressure from the high-pressure accumulator (HPA) and high-pressure pump enters the brake circuit and builds brake pressure. This reapplies the brake as the wheel continues to decelerate. The reset switch also returns to its normal position as pressure is restored. The ABS controller then continues to cycle the PWM valve as needed to prevent wheel lockup for the duration of the stop.

During the pressure hold, decrease, and increase phases of antilock braking, the isolation valve for the brake circuit remains closed. This blocks brake pressure from the master cylinder so it cannot enter the brake circuit being controlled. The reapplication of brake pressure, therefore, must come from the high-pressure pump and accumulator.

Once the need for ABSs braking has passed, the system de-energizes the isolation and PWM valves and normal braking resumes. The residual pressure that remains in the high-pressure accumulator bleeds into the master cylinder reservoir and the system returns to the passive mode until it is needed.

EBC5H

Later Kelsey-Hayes four-wheel ABSs include the EBC5 and EBC10. The EBC5H system is on 1993 and later Dodge Dakota trucks, and 1994 and later Dodge Ram and Ram Van. It is a nonintegral, four-wheel hybrid three-channel system that uses a separate hydraulic modulator for the front wheels and a RWAL modulator for the rear wheels, figure 15-57. The RWAL valve body for the rear wheels is mounted just under the master cylinder, and the ABS valve body for the front wheels mounts on the driver side fender panel, figure 15-58.

This system uses three nonadjustable wheel speed sensors, one on the inboard side of each front disc brake rotor hub and a common sensor for both rear wheels mounted on top of the differential housing. The signals route directly to the control module or controller antilock brakes (CAB), which is next to the ABS valve body on the driver side fender panel.

The RWAL valve body for the rear wheels is the same as that used on rear-wheel ABS applications and contains two ABS solenoids: an isolation valve and a dump valve. The ABS modulator for the front brakes contains four ABS solenoid valves, including an isolation solenoid and dump solenoid for each front brake circuit, and a pump to generate pressure as additional brake force is needed during ABS braking. The pump only runs if antilock braking occurs at either front wheel and does not operate if the rear wheels only are involved.

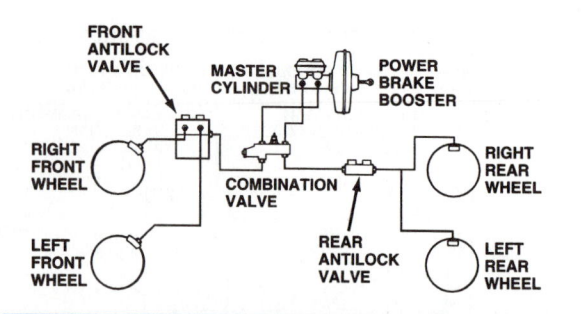

Figure 15-57. The Kelsey-Hayes EBC5H "hybrid" four-wheel ABS for Dodge truck applications.

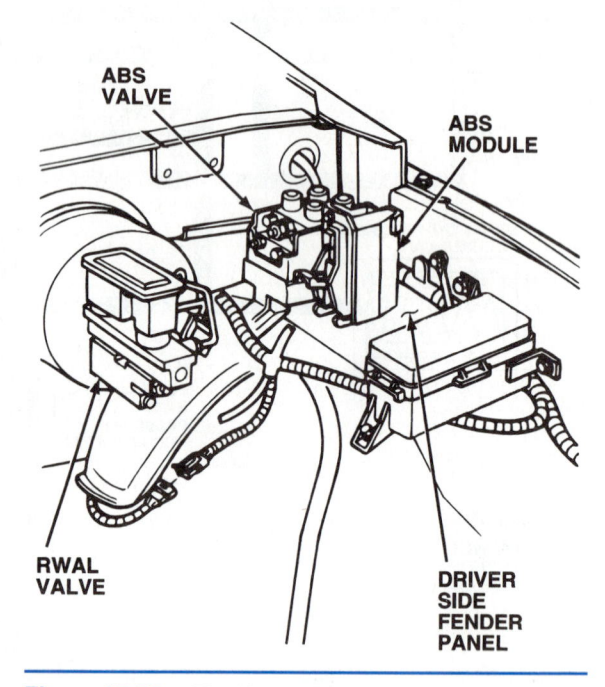

Figure 15-58. The Kesley-Hayes EBC5H component locations on Dodge trucks.

Antilock Brake Systems

The front modulator unit also contains a LPA to temporarily store fluid under pressure during pressure release, and a HPA to store hydraulic fluid under pressure for reapplying the front brakes. The LPA is normally empty and only fills when brake fluid vents from either front brake as its dump valve opens. The HPA fills as soon as the pump starts to run.

A pump outlet check valve separates the master cylinder and pump, preventing unnecessary fluid flow to the HPA from the master cylinder during a high-pressure brake application. The valve is normally closed and opens only when pump pressure exceeds a certain valve as pressure is reapplied to either front brake. This allows fluid to flow from the master cylinder reservoir to the pump, causing a slight drop in the brake pedal.

A hydraulic check valve minimizes pedal feedback during ABS braking and prevents pump pressure from exceeding pedal pressure. It is a normally open valve that closes when pump pressure is applied during an ABS stop.

Other components include a brake return check valve to speed up fluid return after an ABS stop, and a reset swtich that acts like a differential pressure switch to detect pressure imbalances between the brake circuits.

EBC5U

The 1994 and later Ford Econoline vans use a different version of the EBC5 system named EBC5U, which stands for "unitized." This system has a single hydraulic control unit (HCU) for all four wheels instead of the separate front and rear ABS modulators used by Dodge. It is a nonintegral, four-wheel, three-channel system with the ABS control module mounted on the modulator.

The HCU, which is mounted inboard on the left frame rail just ahead of the fuel tank, has six ABS solenoids: three normally open isolation valves (one for each front brake circuit and one for the rear brakes) and three normally closed dump valves. An accumulator on the HCU temporarily stores fluid that is released from a brake circuit through its dump valve. A return pump routes fluid back to the main fluid reservoir. The valve body, pump, and pump motor are not serviceable separately.

The ABS control module is mounted on a bracket just behind the HCU on the left frame rail. The module has a 40-pin connector and receives inputs from a brake pedal switch and three nonadjustable wheel speed sensors, one at each front wheel and a common sensor on ther rear axle housing for the rear wheels. The main ABS relay is in the power network box.

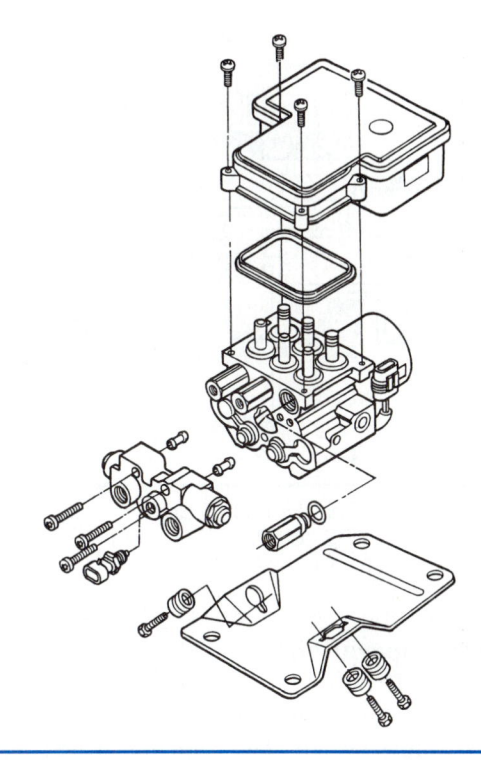

Figure 15-59. The EBC310 ABS electro-hydraulic control unit (EHCU). (Courtesy of General Motors Corporation, Service and Parts Operations)

EBC310 and 325

EBC310 and EBC325 are improved members of the EBC family of ABS units. EBC310 is used on 1995–2000 GM C and K series trucks, Blazer, and S/T four-door trucks. EBC310 or EBC325 is used beginning in 1997 on some DaimlerChrysler, Kia, and Isuzu trucks, and on 2000 and later GM trucks.

EBC310 and EBC325 are nonintegral four-wheel, three-channel unitized systems with the controller mounted on the modulator. The combined assembly is called a controller, antilock brake (CAB) by DaimlerChrysler or the electro-hydraulic control unit (EHCU) by GM. The modulator contains six ABS valves: three isolation valves and three dump valves, plus two pumps and two accumulators, figure 15-59. The system has the following features:

- Three-channel antilock braking
- Dynamic rear proportioning (1998 and later, some models)
- Traction control (EBC325).

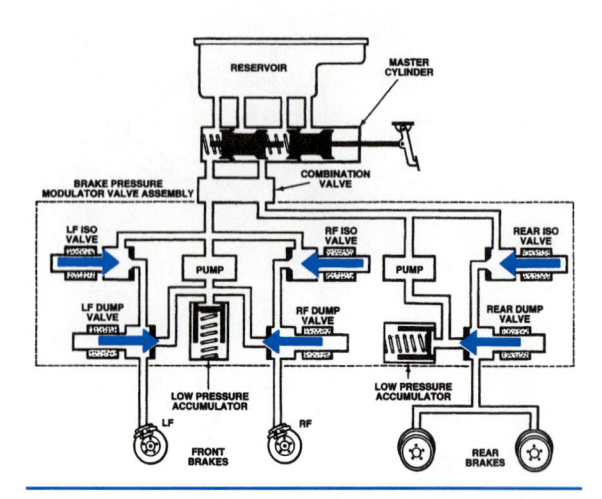

Figure 15-60. The hydraulic circuits of a Kelsey-Hayes EBC310 ABS. (Courtesy of General Motors Corporation, Service and Parts Operations)

System Operation
The EHCU controls each front wheel individually and the rear wheels as a pair, making it a three-channel system. The main inputs are the left and right front wheel speed sensors, the vehicle speed sensor (on transmission or transfer case), and the brake switch. Outputs are the isolation solenoids, dump solenoids, and the pump motor, figure 15-60.

Nonbraking
Each time the key is turned on the EBCM checks the input and output circuit electrically. If a fault is detected, the EBCM will turn on the amber ABS light, store a DTC, and disable ABS braking. When the vehicle reaches 4 mph for the first time each key cycle, the EBCM will cycle the solenoids and pump motor for a functional check.

Nonantilock Braking
When the brake pedal is depressed and the EBCM does not detect wheel lock, the EBCM continues to monitor for faults. The brake pedal switch signals the controller to prepare for ABS brake control, if necessary.

Antilock Braking
When wheel slip is detected during braking the EBCM begins to control the slipping wheel by closing the isolation valve to begin the pressure hold mode. This prevents any additional pressure increase in the slipping wheel.

If the pressure hold mode fails to prevent the wheel from locking, the EBC310 system will enter the pressure decrease mode by leaving the isolation valve closed and opening the dump valve, thus releasing pressure in the affected channel and allowing the wheel speed to increase. At the same time the pump is turned on to empty the accumulators and return fluid to the reservoir. The pump stays on until the end of the ABS braking event.

If the wheel speed begins to increase too rapidly, the system enters pressure increase mode by turning both valves off. Pressure from the pump, master cylinder, and accumulator then causes an increase in channel pressure to bring the wheel speed back down.

These cycles repeat rapidly until the slipping wheel(s) regain their grip or the driver releases the brake pedal. The pump will continue to run briefly at the end of the cycle to empty the accumulators.

The DRP and TCS operate as described on page 328.

EBC410
Another Kelsey-Hayes ABS system, called EBC410, is used on 1995 and later Ford Windstar minivans. It is similar to the EBC310 system except the EHCU has eight ABS solenoids, and an isolation and dump valve for each brake circuit. The two extra solenoids and allow the system to control each rear brake circuit separately, making it a four-channel system, figure 15-61. Independent control of the rear brakes is necessary on the Windstar because it has front-wheel drive and a diagonally split hydraulic system.

The EBC410 has a compact unitized design with the ABS control module mounted on the modulator. The EHCU is under the battery tray in the left front corner of the engine compartment. You can service the ABS control module separately from the modulator, but the EHCU valve body, pump, pump motor, and accumulator are not serviceable separately. The control module receives inpits from four nonadjustable wheel speed sensors and the brake pedal switch.

EBC430
Beginning with the 2003 Chevrolet Avalanche, some vehicles are equipped with the EBC430 ABS, figure 15-62. EBC430 ABS includes the following vehicle functions:

- Antilock brake system
- Dynamic rear proportioning

Antilock Brake Systems

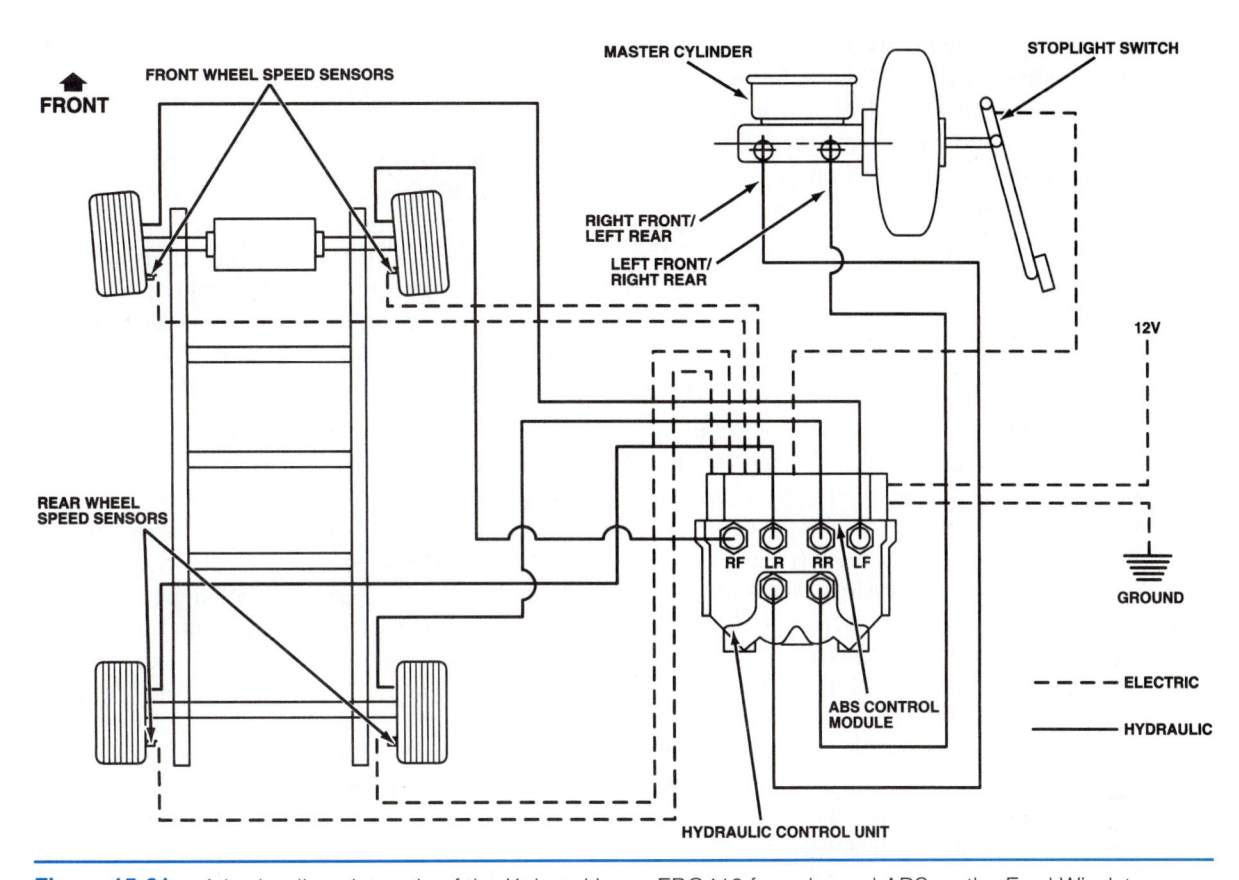

Figure 15-61. A hydraulic schematic of the Kelsey-Hayes EBC410 four-channel ABS on the Ford Windstar.

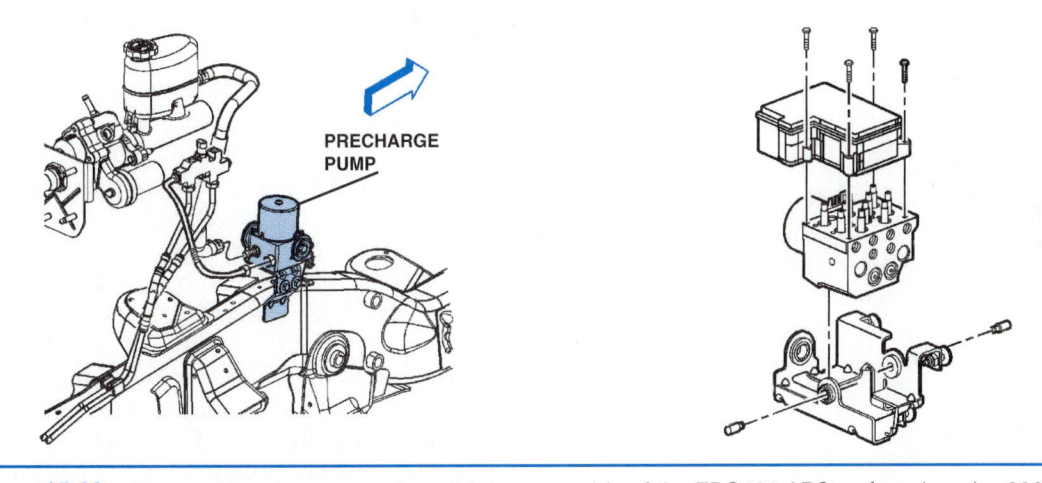

Figure 15-62. The precharge pump and modulator assembly of the EBC430 ABS as found on the 2003 Chevrolet Avalanche. (Courtesy of General Motors Corporation, Service and Parts Operations)

- Traction control system
- Vehicle stability enhancement system.

EBC430 components are:

- **EBCM:** The EBCM controls the system functions and detects failures; it also contains the system relay, which supplies battery positive voltage to the valve solenoids and to the ABS pump motor; the valve solenoids are also in the EBCM
- **Longitudinal accelerometer:** The EBCM uses the longitudinal accelerometer to determine the actual straight-line acceleration of the vehicle

- **BPMV:** The BPMV uses a four-channel configuration to control hydraulic pressure to each wheel independently. The BPMV contains the following components.
  - ABS pump motor and pump
  - Four isolation valves
  - Four dump valves
  - Two traction control isolation valves
  - Two traction control supply valves
  - A front and rear low-pressure accumulator
  - Wheel speed sensors (WSS)
  - Traction control
- **Lateral accelerometer:** The EBCM uses the lateral accelerometer to determine the sideways acceleration of the vehicle. The lateral accelerometer and yaw rate sensor are in the same housing
- **Master cylinder pressure sensor:** The master cylinder pressure sensor is located within the BPMV and senses hydraulic fluid pressure in the front brake circuit at the master cylinder
- **Yaw rate sensor:** The EBCM uses the yaw rate sensor to determine the rate of rotation along the vehicle's vertical axis
- **Steering wheel position sensor.**
- **Precharge pump motor and pump:** The precharge pump assembly is used to assist the ABS pump in rapidly building hydraulic pressure needed during certain TCS or VSES events.

System Operation
EBC430 operation is identical to the four-channel EBC410 system. Refer to the EBC310 and 325 section and the EBC410 section for descriptions of system operations. The dynamic rear proportioning, traction control system, and vehicle stability enhancement system have also been previously described, however, EBC430EV has some unique features.

Traction Control System
EBC430 traction control uses engine torque reduction as the initial mode when controlling drive wheel spin. The system also uses brake pressure application to control traction by transferring torque through the driveling to wheels that are not slipping. The precharge pump motor, ABS pump motor, and appropriate valve solenoids are commanded on and off to apply brake pressure to the slipping wheels. Brake pressure application is used in an attempt to maintain equal WSS signals at the driven wheels.

Vehicle Stability Enhancement System
As discussed previously, VSES provides added stability during aggressive maneuvers. When braking during VSES activation, the driver may notice pedal pulsations. The brake pedal pulsates at a higher frequency during VSES activation than during ABS activation. Both TCS ans VSES activate a fluid precharge and prefill mode when needed.

Precharge Pump
EBC430 uses a precharge pump that runs if the EBCM anticipates that the TCS or VSES may be activated. The TCS and VSES depend on the precharge pump to assist the ABS pump in building hydraulic pressure to perform brake pressure application. The precharge pump inlet is fed by a flexible hose that draws fluid directly from a port on the master cylinder reservoir. When the precharge pump is activated, fluid is pumped into the combination valve, slightly pressurizing the front and rear brake circuits between the master cylinder and the BPMV.

System Prefill
As discussed in Chapter 7, the brake caliper seals retract the pads away from the rotor surface when the brakes are released. If the EBCM determines that a brake application is likely to be needed, the ABS pump motor runs momentarily to take up any clearances between the brake pads and the rotor. By monitoring the master cylinder pressure sensor feedback signal, the EBCM can determine when the brake pads are contacting the rotor. The EBCM then holds this small amount of pressure in the system. A brake application may or may not occur after prefill is complete. If the EBCM determines that a brake application is no longer pending, the prefill pressure is released and the system returns to normal.

NIPPONDENSO ABS

Many Asian imports use one of several different types of Nippondenso ABS. These are usually derivatives of Bosch systems and can be three- or four-channel systems. Some have a G-sensor or deceleration sensor along with four wheel speed sensors.

Antilock Brake Systems

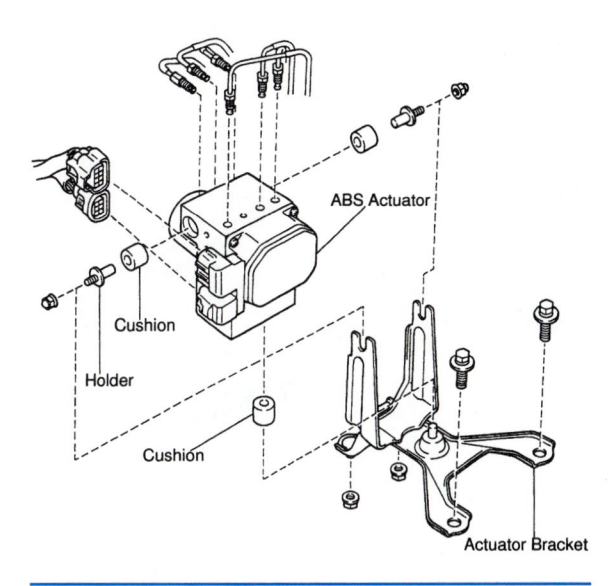

Figure 15-63. The Nippondenso 5.3 ABS modulator assembly as used on the Toyota Camry. (Courtesy of Toyota Motor Sales, U.S.A., Inc.)

Nippondenso 2L

The 2L ABS is used on Subaru 1990–93 Legacy, 1992–97 SVX, and some other Asian imports. Nippondenso 2L is based on the Bosch 2 system. It is a nonintegral system with the hydraulic modulator in the engine compartment and the electronic control unit in the passenger compartment. Refer to the Bosch 2 section for system operation.

Nippondenso 5.3

The Nippondenso 5.3 system, used on some 1996 and later Asian imports, is based on the Bosch 5.0 and 5.3 ABS design. It is a nonintegral four-wheel, three- or four-channel system with the electronic control unit and the hydraulic modulator combined in one unit, figure 15-63. Refer to the sections covering Bosch 5 series ABS for system operation.

SUMITOMO ABS

The first appearance of Sumitomo ABS, Sumitomo 1 (also called Mazda Old Generation), was in 1988 on the Mazda RX7, 626, MX-6, and 1989–91 Ford Probe. In 1990, Honda offered it on the Prelude Si, and then the Accord EX and Civic EX in 1991, as well as the Acura NSX, Legend, Integra GS, and Vigor. The Honda versions may also be referred to as Honda ABS. Sumitomo 1 ABS is a nonintegral, four-wheel, three-channel ABS.

Sumitomo 2 (Mazda New Generation), a four-channel, nonintegral system, is used from mid-1991–97 on Mazda MX-3, MX-6, 626, Miata, and on Ford Probe. It is also used on some models of 1994–96 Ford Escort/Mercury Tracer.

Sumitomo 1 System Description

The Sumitomo 1 ABS has four wheel speed sensors, figure 15-64. Sensors are adjustable on all applications: 0.016″ to 0.039″ (0.41 to 0.99 mm) for Honda; 0.012″ to 0.043″ (0.30 to 0.76 mm) on Mazda and Ford.

The modulator contains three ABS solenoid valves as well as four modulator control pistons, one for each brake circuit, figure 15-65. Honda versions use three pistons, one for each front brake and one for both rear brakes.

A pump and nitrogen-charged accumulator generate ABS brake pressure. On Hondas, the accumulator is also used for power-assisted normal braking. The pump and accumulator are part of the modulator assembly in all applications, except Honda where they are separate components mounted ahead of the modulator in the engine compartment. You can replace the pump separately on all applications.

When ABS braking is needed, the control module, located under the dash or center console, energizes the normally open solenoid outlet valve in the modulator. This closes the reservoir outlet circuit valve and opens the pump supply to the modulator. Pump pressure enters the modulator and works against the control piston, which closes a cutoff valve in the brake circuit. Pressure is then released from the brake circuit so it can flow back to the reservoir, which also produces pedal feedback for the driver. Pressure is reapplied to the brake when the solenoid is de-energized. This allows the control piston to return to its normal position, opening the brake circuit cutoff valve.

Honda

Honda antilock brake (ALB) versions use a conventional master cylinder with the Sumitomo ABS. However, the brake booster is hydraulic rather than vacuum assited. The ABS pump generates pressure for normal power-assisted braking as well as ABS braking and stores it in the accu-

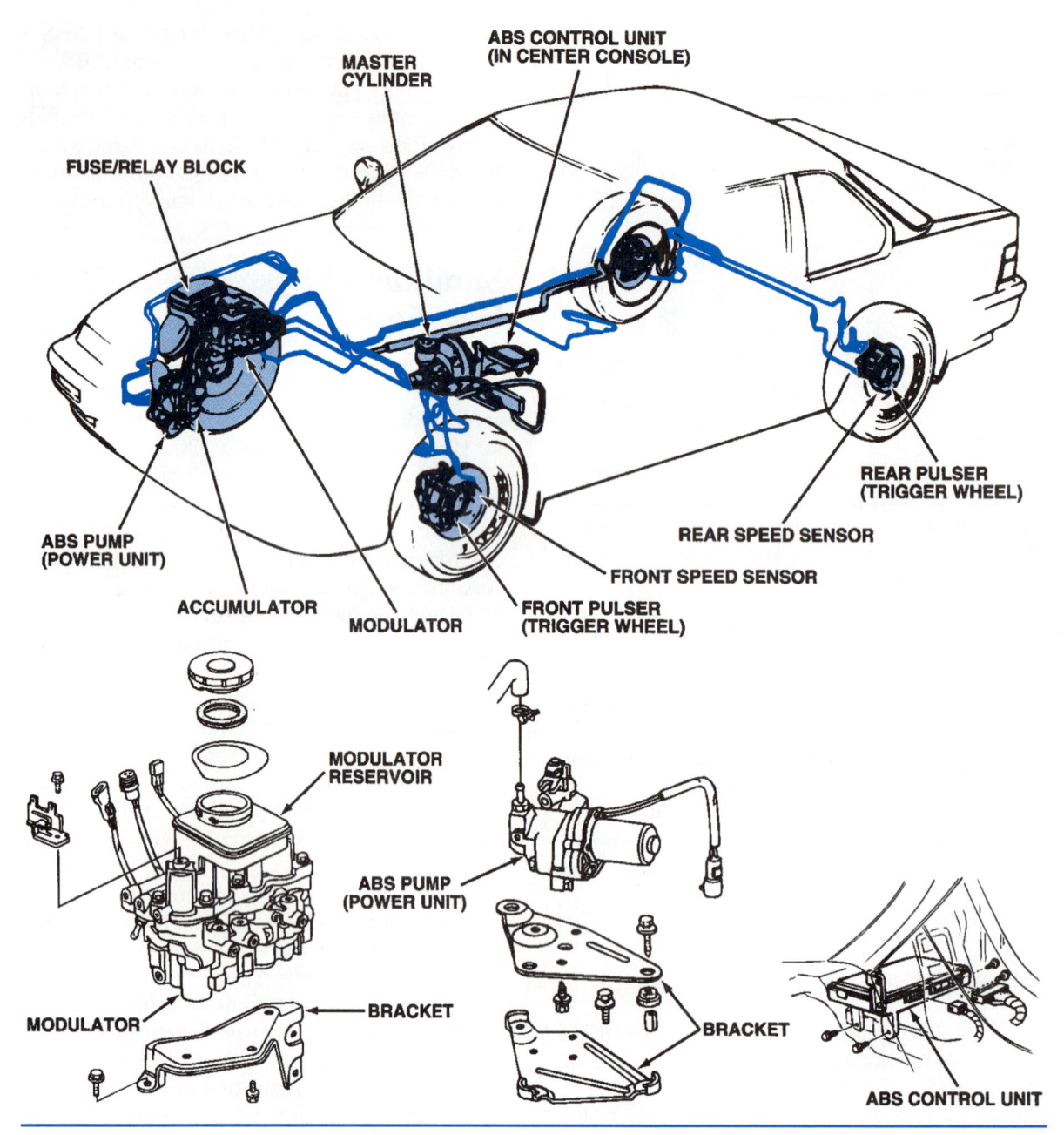

Figure 15-64. The Sumitomo ABS used on the Honda Prelude.

mulator. Both pump and accumulator are located in the left front area of the engine compartment.

When the ignition is on, the ABS control module energizes the pump relay to start the pump if the accumulator pressure switch indicates low pressure, less than 1721 psi (11,866 kPa). The pump draws fluid from the master cylinder reservoir and feeds it into the accumulator. When accumulator pressure reaches its maximum limit of 3271 psi (22,554 kPa), the pump shuts off. The pressure stored in the accumulator is sufficient to provide several power-assisted stops or ABS applications should the pump fail. A pump failure will cause the ABS warning lamp to come on.

CAUTION: Before servicing this system, the accumulator must be discharged by pumping the brake pedal 25 to 40 times with the ignition off.

Antilock Brake Systems

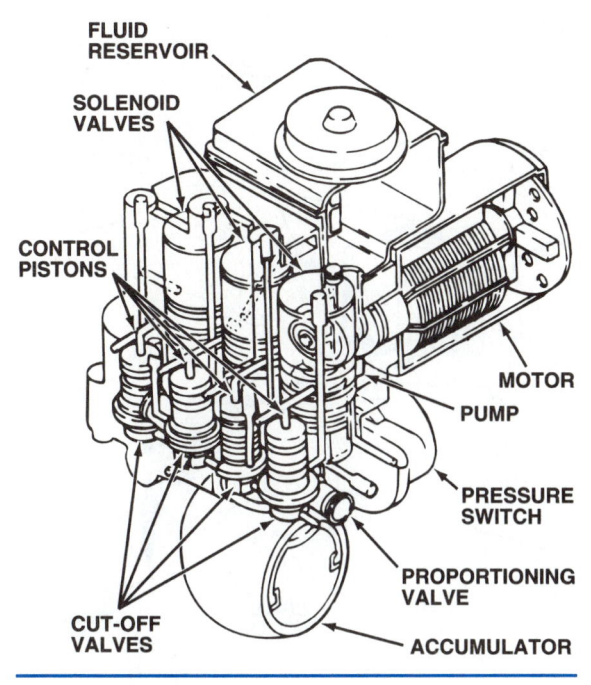

Figure 15-65. A Sumitomo modulator.

Sumitomo 2 System Description

Sumitomo 2 ABS is used in 1997–98 models of Ford and Mazda. The system may be called Sumitomo 2, Compact Sumitomo, or Sumitomo (Mazda New Generation), depending on the manufacturer. The HCU has four solenoids, four control valves, a pump and pump motor, an accumulator (buffer chamber), and a damper chamber. The ECU is mounted separately from the HCU.

The main difference between the Sumitomo 1 and 2 ABS is that Sumitomo 1 uses a high-pressure nitrogen-charged accumulator, whereas Sumitomo 2 uses a more conventional low-pressure, spring-type accumulator.

System Operation

Instead of individual isolation and dump valves, Sumitomo 2 ABS uses a control valve in each hydraulic channel. The control valve combines the functions of the isolation and dump valve into a single, three-position valve.

- **Open, solenoid off:** Allows normal braking
- **Hold, solenoid low-current flow:** Does not allow pressure increase to the caliper
- **Reduction, solenoid high-current flow:** Releases pressure from the caliper into the buffer chamber (accumulator).

During ABS braking, the ECU commands the solenoids to operate the control valves in a "pressure hold–pressure reduction–open (pressure increase)" cycle to control the slipping wheel. The pump will run as needed to return fluid from the buffer chamber back to the master cylinder. The damper chamber helps to reduce pedal feedback and noise in the ABS.

TEVES ABS

Teves, now called Continental/Teves, has one integral ABS (Mark II) and a number of nonintegral ABS systems, including the Mark IV, Mark 20, Mark IV G, Mark 25, and Mark 60.

Teves Mark II

The Teves Mark II ABS was used on the 1985–92 Lincoln Mark VII, 1985–89 Continental, and 1987–92 Thunderbird and Cougar. GM used the system 1986–90 on certain models of Buick, Cadillac, Oldsmobile, and Pontiac. Teves Mark II can also be found on some European vehicles, such as the 1990–94 Jaguar and Saab 9000/900 models.

Teves Mark II is an integral four-wheel, three-channel ABS with a combination master cylinder, hydraulic modulator, and pump motor and accumulator assembly, figure 15-66. The pump and accumulator provide power-assisted braking as well as ABS braking. A booster control valve, located in a parallel bore above the master cylinder, is operated by a lever mechanism connected to the brake pedal rod.

System Description

The integral HCU, figure 15-67, contains six ABS solenoids: an isolation and dump valve for each front brake circuit and one for both rear brakes, plus a main valve solenoid on the end of the master cylinder portion of the unit.

The main valve solenoid connects the boost pressure chamber with the master cylinder internal reservoir, figure 15-68. When energized by the control module, it opens and allows accumulator pressure to fill the area around the master cylinder pistons and in front of the reaction sleeve. During ABS operation, the main valve solenoid opens a path to the master cylinder internal reservoirs, providing a continuous flow of pressurized brake fluid to replace fluid returning to

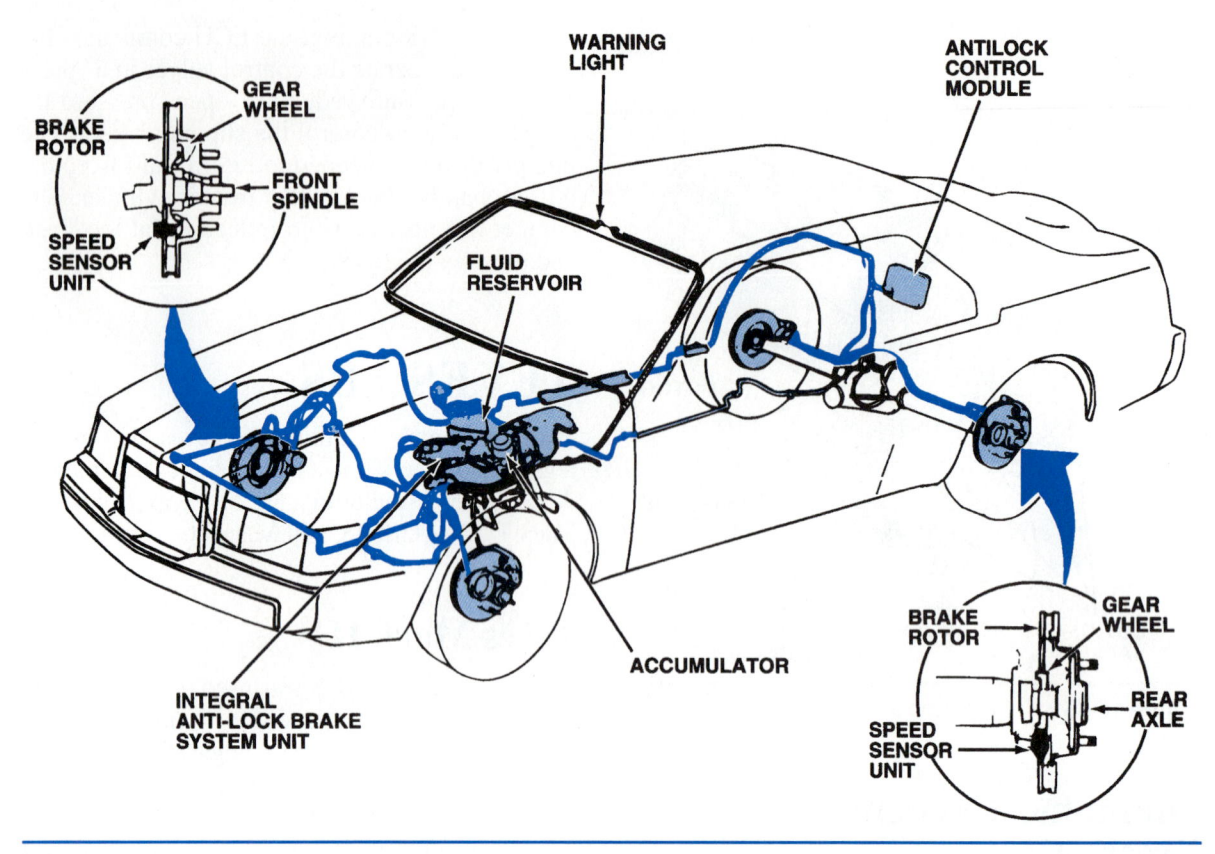

Figure 15-66. The Lincoln Mark VII's Teves Mark II integral antilock brake system.

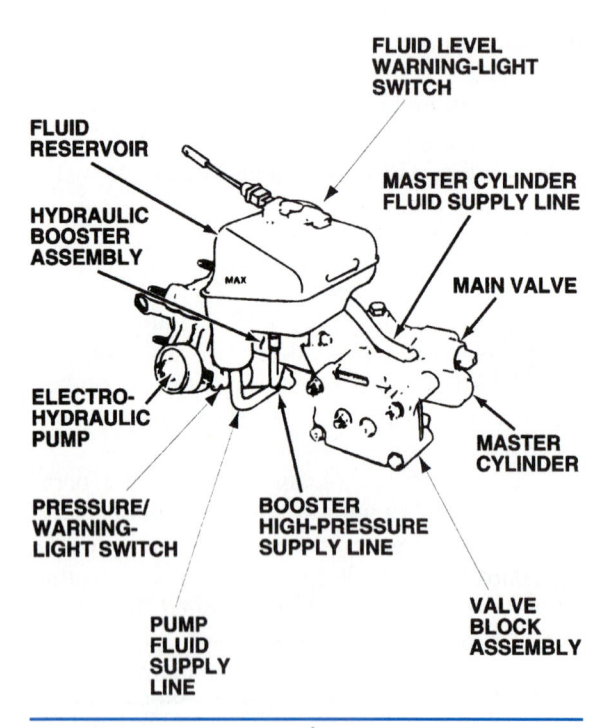

Figure 15-67. The Teves Mark II integral ABS modulator assembly.

the reservoir. When ABS braking is no longer needed, the main valve closes to block the flow of accumulator pressure into the front brake circuits.

Each wheel has its own speed sensor, but the controller uses the select-low principle to initiate ABS at the rear wheels depending on which rear wheel is slowing the fastest. The wheel speed sensors are nonadjustable on most applications. The location of the sensors and sensor rings vary depnding on application. On Fords with independent rear suspensions, the rear wheel speed sensor tone rings are on the inboard CV joints next to the differential.

Wheel speed sensors on 1987 and later GM cars differ from earlier versions. The number of teeth on the speed sensor rings changed from 98 to 47, making the sensors less vulnerable to signal variations caused by rough roads. Therefore, early and late wheel speed sensors are not interchangeable. The EBCM was also changed in 1987 to accommodate the different signal inputs from the revised sensors. An extra pin was added to the EBCM wiring connector to prevent accidental interchange with the earlier EBCM. The EBCM is under the dash on most

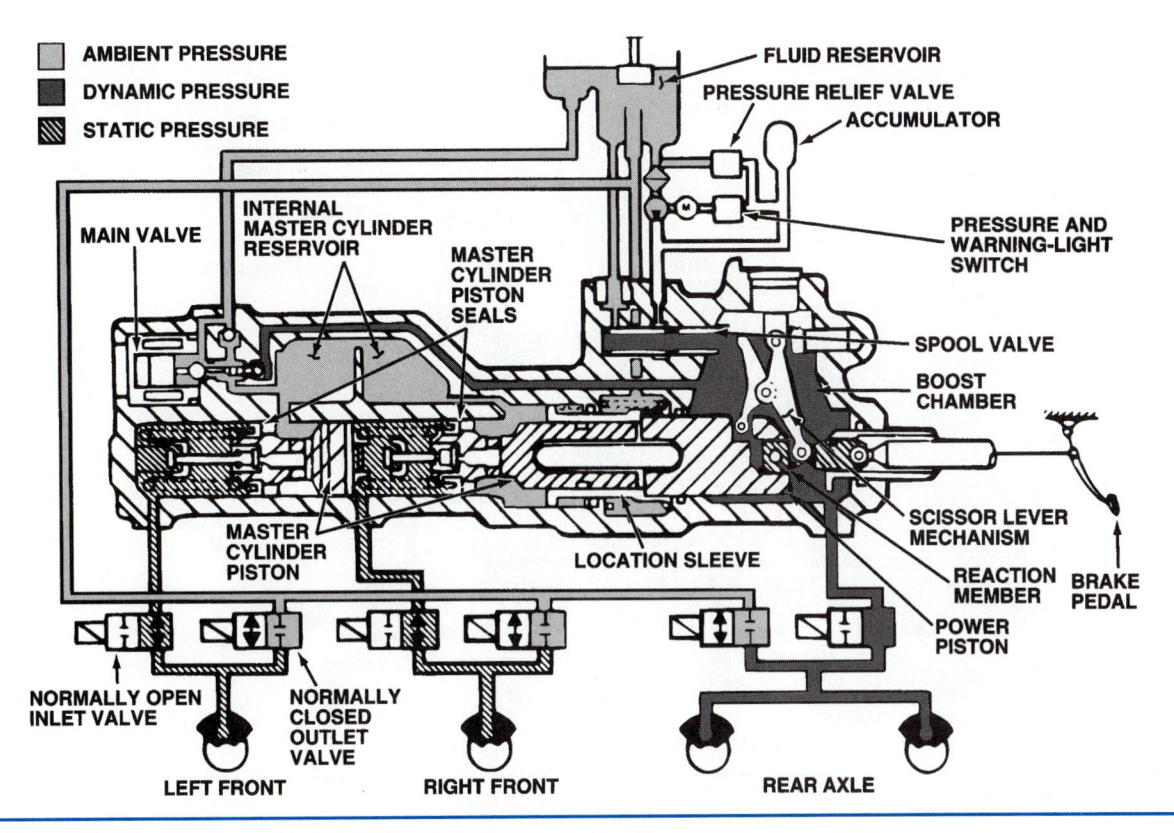

Figure 15-68. The Teves Mark II ABS unit during normal braking.

applications.

System Operation

System operation is similar to other three-channel ABSs. When ABS braking is required, the normally open isolation solenoid valve closes to prevent any further pressure increase in the brake circuit, figure 15-69. The normally closed dump solenoid valve opens to release pressure in the circuit. The two solenoids are cycled open and closed as needed to hold, release, and reapply the brake until ABS braking ceases.

Operation of the Teves Mark II master cylinder is rather unusual in that the master cylinder's dual pistons only apply the front brakes. The primary piston works the right front brake, and the secondary piston works the left front brake. Pressure for both rear brakes comes directly from the accumulator. When the driver presses the brake pedal, the booster control valve opens a passage that routes presure from the accumulator into both rear brake circuits. At the same time, the booster control valve allows pressure from the accumulator to enter the front brake circuits for normal power-assisted braking.

Accumulator Pressure

A pressure control switch monitors pressure in the accumulator and commands the pump motor to run anytime the ignition is on and pressure drops below 1500 psi (10,343 kPa). When accumulator pressure drops below that point, the switch closes, turns the pump relay on, and the pump builds pressure until it reaches 2610 psi (17,996 kPa). Then, the pressure control switch opens, turns off the relay, and the pump stops. Should the pump fail to maintain pressure because of a motor, relay, or switch failure, there is enough reserve pressure designed into the accumulator for a dozen or so power-assisted stops. A warning light comes on if reserve pressure drops below 1500 psi (10,342 kPa). The pump also has a built-in relief valve that opens and remains open if the pump remains on too long because of a faulty pressure switch and accumulator pressure exceeds 3340 psi (23,029 kPa). If the relief valve has opened, the pump must be replaced.

CAUTION: Remember to relieve accumulator pressure by pumping the brake pedal 25 to 40 times with the ignition off before working on the brakes.

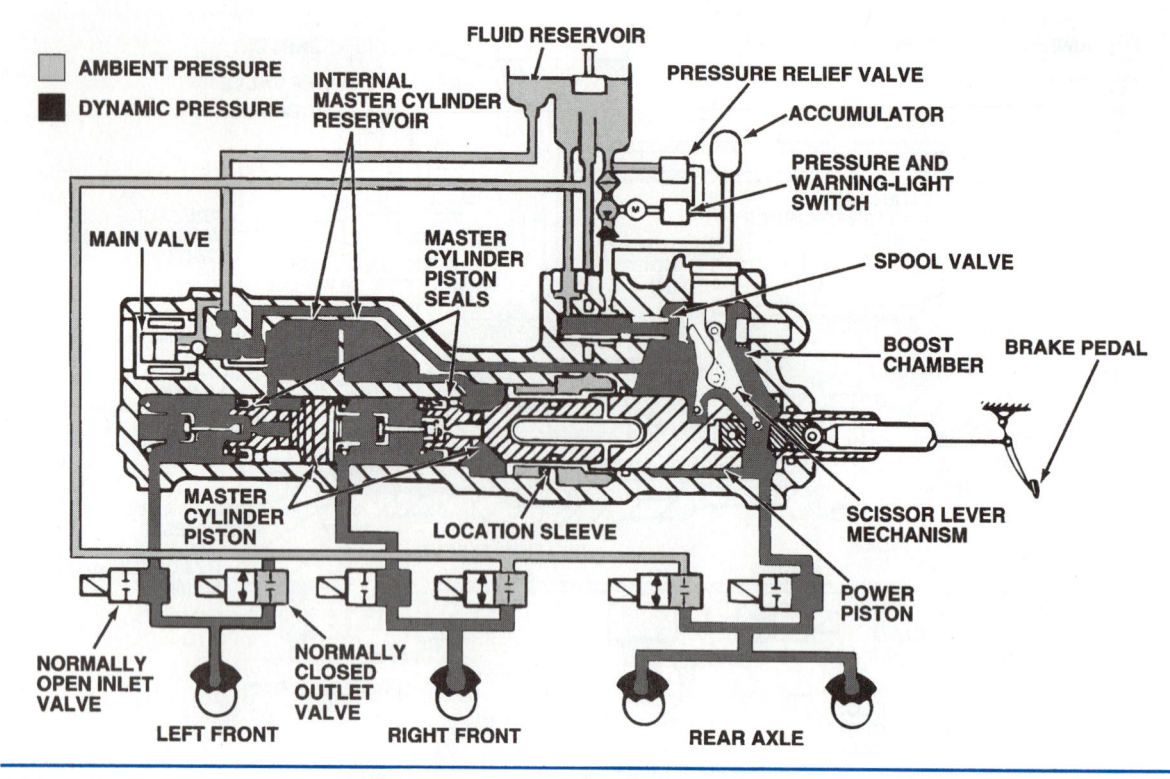

Figure 15-69. The Teves Mark II ABS unit during ABS operation (pressure decrease phase).

Teves Mark IV

The Mark IV antilock brake system first appeared in 1990 on the Lincoln Continental and Ford Taurus SHO, and was offered as an option on the Town Car. Since then, it has been added to the Ford Taurus and Mercury Sable, and the 1992 and later Crown Victoria and Grand Marquis. General Motors adopted the Mark IV system in 1991 for Buick Park Avenue and LeSabre, Cadillac DeVille, Pontiac Bonneville, and Oldsmobile 98, 88, and Royale. It is also used on 1993 and later Chrysler LH cars: the Concorde, New Yorker, Dodge Intrepid, and Eagle Vision.

The Teves Mark IV is a nonintegral, four-wheel ABS system that uses a standard master brake cylinder and booster with a separate hydraulic modulator and pump assembly.

System Description

The Mark IV is a four-channel ABS with individual wheel speed sensors and separate ABS solenoids for each brake circuit, figure 15-70. It controls the front wheels independently, but both rear wheels are controlled as a single output with separate feed lines during ABS braking, similar to a three-channel ABS. On applications with traction control, the rear brake circuit solenoids can provide independent braking for either rear wheel.

The cap on the fluid reservoir on Fords with Teves Mark IV contains a fluid level switch, which will illuminate the red BRAKE warning lamp if the level becomes too low. The reservoir also contains a low-pressure hose that supplies brake fluid to the hydraulic modulator minireservoir. The valve body, pump and motor, and reservoir on the modulator assembly can all be serviced separately.

Some versions of the Teves Mark IV ABS use a brake pedal position sensor. On Jeep it is mounted on the brake vacuum booster, figure 15-71, and on Ford the sensor is mounted on the brake pedal. The pedal position sensor input is used by the control module to determine when to run the pump during ABS braking.

The hydraulic modulator, figure 15-72, is located in the front of the engine compartment. Ford service literature refers to the modulator as the hydraulic control unit (HCU), whereas General Motors calls it the pressure modulator valve (PMV) assembly.

The modulator consists of a valve block manifold with four inlet and four outlet ABS solenoid

Antilock Brake Systems

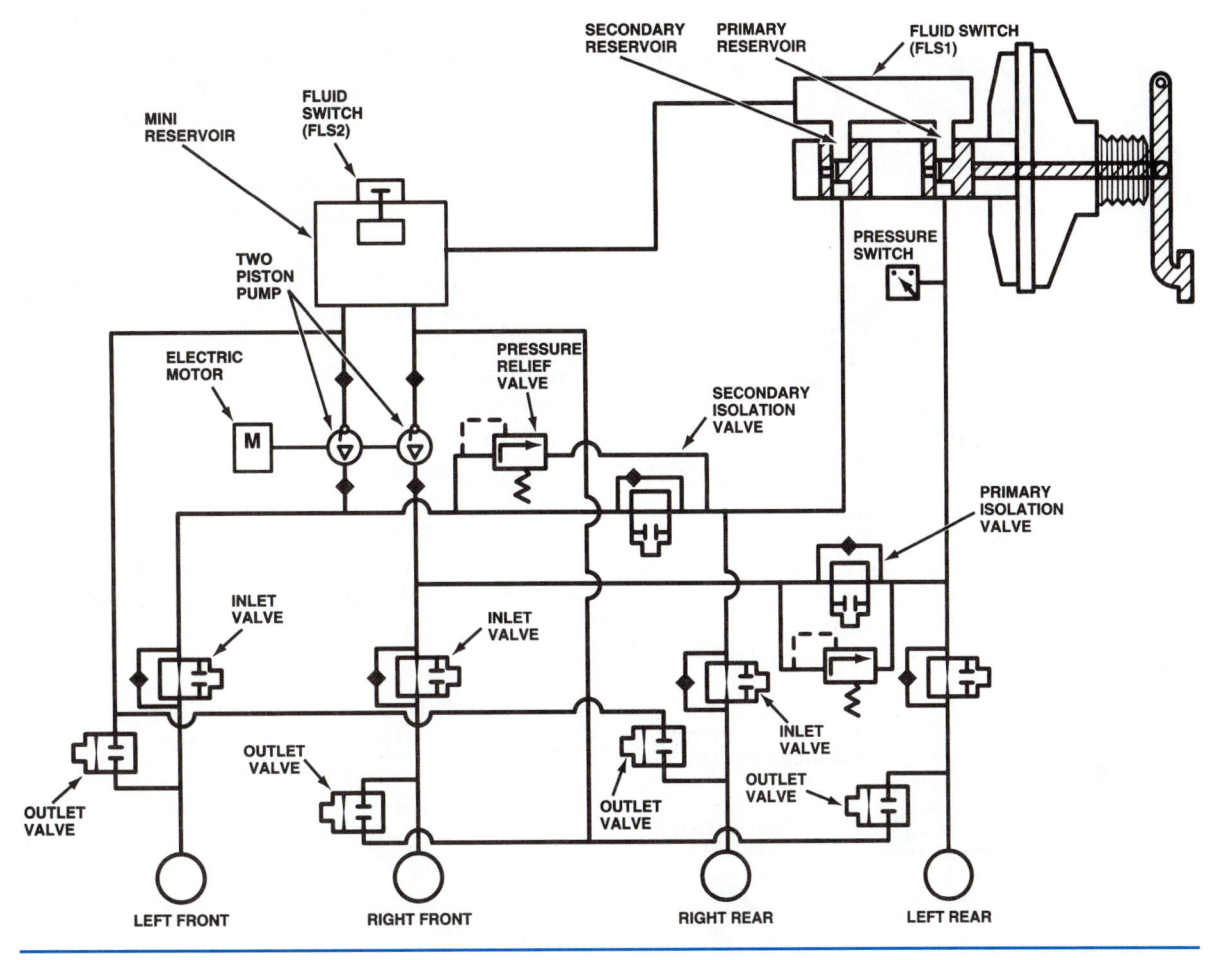

Figure 15-70. A Teves Mark IV hydraulic circuit.

valves, a pump and motor, and a "mini" fluid reservoir with a fluid level indicator. An indicator is needed because the reservoir is sealed and cannot be checked manually. If the fluid level is low for any reason, the amber ABS warning light will come on and the ABS system will be deactivated—but no fault code will set.

On applications with traction control, the modulator is larger and contains two additional isolation valves, one for each front wheel in rear-wheel drive applications. The isolation valves, close when the traction control function prevents application of the front brakes.

Control Module

The control module, which GM refers to as the electronic brake control module (EBCM) and Chrysler calls the controller for antilock brakes (CAB), contains two microprocessors as a redundant failsafe. Each microprocessor receives identical inputs from the wheel speed sensors and makes its own calculations. The results are then compared. If there is any disagreement, the controller disables the ABS/traction control system and illuminates the ABS and traction control warning lights. The controller is located in the front of the engine compartment on the Lincoln Town Car, in the trunk of Thunderbirds and Cougar XR7s, and on the hydraulic modulator assembly of the Chrysler LH cars. Other versions place it elsewhere in the vehicle.

The EBCM monitors wheel speed continuously through its four wheel speed sensors, which are mounted in the front knuckles and rear brake rotor backing plates. Ford applications use two types of front wheel speed sensors. Type 1 has an adjustment screw and sleeve for setting the air gap. Type 2 has an adjustable mounting bracket. There are also two different types in the rear. Type 1 has an

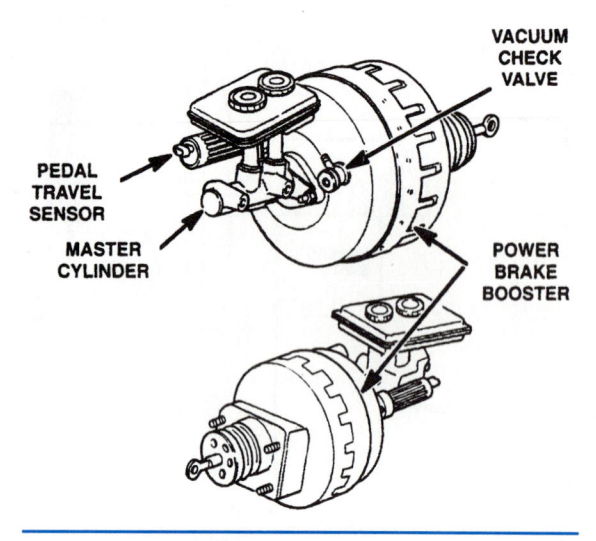

Figure 15-71. Teves Mark IV ABS may use a pedal position sensor on some models. On Jeep vehicles it is mounted on the vacuum booster. (Courtesy of DaimlerChrysler Corporation. Used with permission)

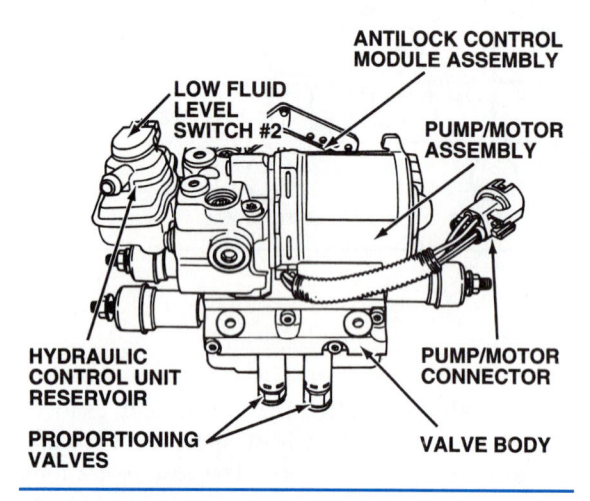

Figure 15-72. A Teves Mark IV hydraulic control unit used on a Chrysler.

adjustable mounting bracket, whereas Type 2 is not adjustable. Leave as is wheel speed sensors on GM and Chrysler applications are non-adjustable.

The control module may also receive additional input from a switch mounted on the brake pedal, which provides pedal travel information. The switch is normally closed but opens when the driver presses the brake pedal beyond 40 percent of its normal travel. This signals the control module to energize the pump in the modulator assembly, which then pumps brake fluid back to the master cylinder. The returning fluid causes the pedal to rise until the switch closes, at which point the pump is turned off. This maintains proper pedal feel during ABS stops and prevents the pedal from sinking too far toward the floor. The driver may hear some noise while this is occuring and feel the pedal pulsate and rise, which is all normal operation.

Proper adjustment and operation of the brake pedal travel switch is very important. If it is not correct, the driver may notice objectionable pedal feel during the entire ABS stop. The pedal will become very firm, pushing the driver's foot back unusually far.

A pump motor sensor in the modulator assembly signals the control module when the pump is running. If the pump runs excessively and the brake pedal switch remains open, the control module recognizes that something is wrong and sets a fault code.

ABS Operation

During normal braking, fluid from the master cylinder enters the modulator assembly through two inlet ports located at the back of the modulator. The fluid then passes through four normally open isolation solenoid valves, one for each wheel, and passes into each of the four separate brake circuits.

If the control module senses that a wheel is about to lock, the control module energizes the isolation solenoid valve to close off the brake circuit and hold pressure in the line. If the wheel is still decelerating too rapidly, the control module opens the normally closed dump solenoid valve to release pressure in the circuit. The controller can the de-energize both solenoids, allowing pressure to be reapplied in the circuit. At the same time, the pump in the modulator is energized to build pressure in the system and to route fluid back to the master cylinder. The Mark IV system can cycle through this scenario up to 15 times a second for as long as the system requires antilock braking.

Traction Control

Mark IV traction control can be found on rear-wheel drive Fords and front-wheel drive GM and Chrysler applications. Traction control involves braking the drive wheels only and does not include reducing engine torque (no throttle relaxer, spark retard, or injector disabling functions).

Antilock Brake Systems

When one or both drive wheels begin to spin while accelerating, the difference in front-to-rear wheel speeds signals the control module to engage traction control. The control module does this by first closing the appropriate traction control isolation valve so pressure does not enter the wrong brake circuit when braking pressure is applied to the drive wheel. At the same time, the pump generates brake pressure, which the control module routes to the brake. A TRAC CNTL or TRACTION ASSIST indicator light illuminates to inform the driver that traction control is functioning.

If the driver steps on the brake pedal while the system is in the tractin control mode, either the stop lamp or brake pedal travel switch will signal the control module to immediately discontinue traction control. On most Ford rear-wheel drive and GM front-wheel drive applications, traction control only functions at speeds of less than 25 mph.

On Chryslers, the control module contains a thermal limiter program to discontinue traction control braking after a predetermined lengh of time so the brakes do not overheat. When this occurs, the TRAC OFF warning light illumintes to alert the driver that traction control is temporarily disabled. After the pads have cooled down a predetermined length of time, the traction control system will reset itself and the warming light will go off.

The control module used with vehicles equipped with traction control is different from the one used on applications with ABS only and is not interchangeable.

Teves Mark 20 and Mark IV G

The Mark 20 ABS first appeared on 1995 Volkswagens and 1997 Chrysler minivans, Jeep Cherokees and Grand Cherokees, and Dodge and Plymouth Neons. It was added to the Chrysler Sebring and LH-series passenger cars in 1998. Minivans, the Chrysler Sebring, and LH-series passenger cars can be optionally equipped with the low speed traction control system (LTCS).

A number of variations of the Mark 20 ABS are available. Of these, the Mark 20i and the Mark 20e are the most common. The main difference between the two is whether or not the system includes electronic variable brake proportioning (EVBP).

- Teves Mark 20i: ABS, with or without traction control
- Teves Mark 20e: ABS, EVBP, with and without traction control.

The Teves Mark 20 is a nonintegral, four-wheel ABS. The hydraulic modulator and pump assembly, which Chrysler calls an integrated control unit (ICU) is separate from the master cylinder. A dual compensating port master cylinder is used on vehicles without traction control. A center valve master cylinder is used on vehicles with traction control.

The Mark IV G is used on 1997 and later Jeep Wranglers. It is essentially the same as the Mark 20, but the controller (CAB) is separate from the HCU, rather than being combined as a single unit. For this reason, Chrysler also refers to the Mark IV G as the "Mark 20 nonintegrated" system.

System Description

The Mark 20 has two different hydraulic circuits, depending on application. Rear-wheel drive vehicles separate the base brake system into front and rear hydraulic circuits. The ABS uses three channels and four wheel sensors, figure 15-73. The front wheels are controlled independently; the rear wheels are controlled together using the select-low principle. Front-wheel drive vehicles separate the base brake system into diagonally split circuits (right rear–left front and left rear–right front). The ABS uses four channels and four wheel sensors, figure 15-74. Like the RWD version, it controls the rear wheels together using the select-low principle. On applications with traction control, the system applies braking force to the spinning drive wheel to slow it to approximately the same speed as the nonspinning drive wheel, figure 15-75.

Master cylinder designs differ depending on whether the vehicle is equipped with traction control. On applications without traction control, a standard compensating port design is used. The compensating port allows residual brake pressure to flow from the master cylinder bore to the reservoir when the pedal is released. The lip seals of the piston cups close off the compensating ports between the reservoir chambers and the master cylinder bore.

On applications with traction control, each master cylinder piston is equipped with a center valve, figure 15-76. The center valves open

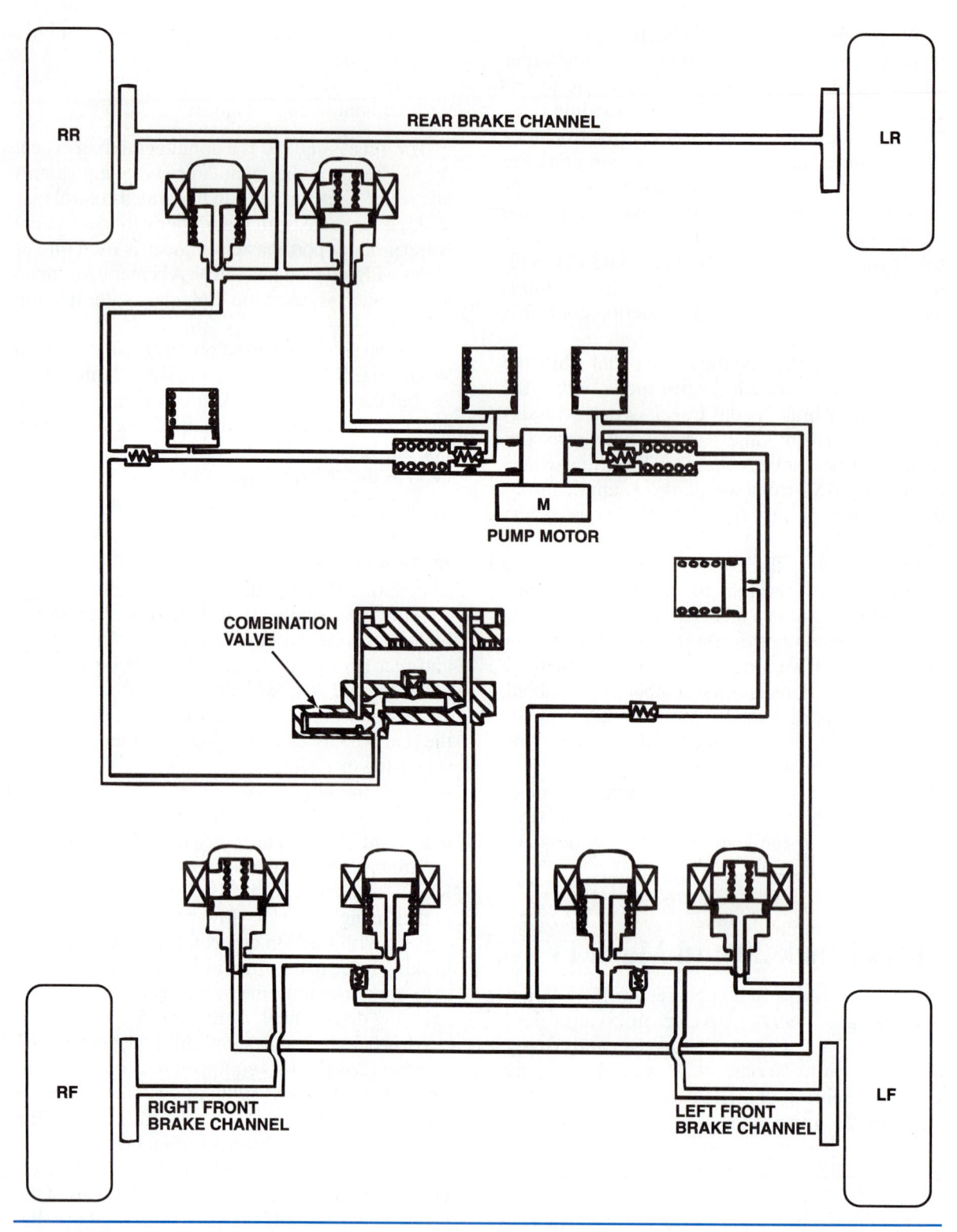

Figure 15-73. The Teves Mark 20 ABS for rear-wheel drive applications.

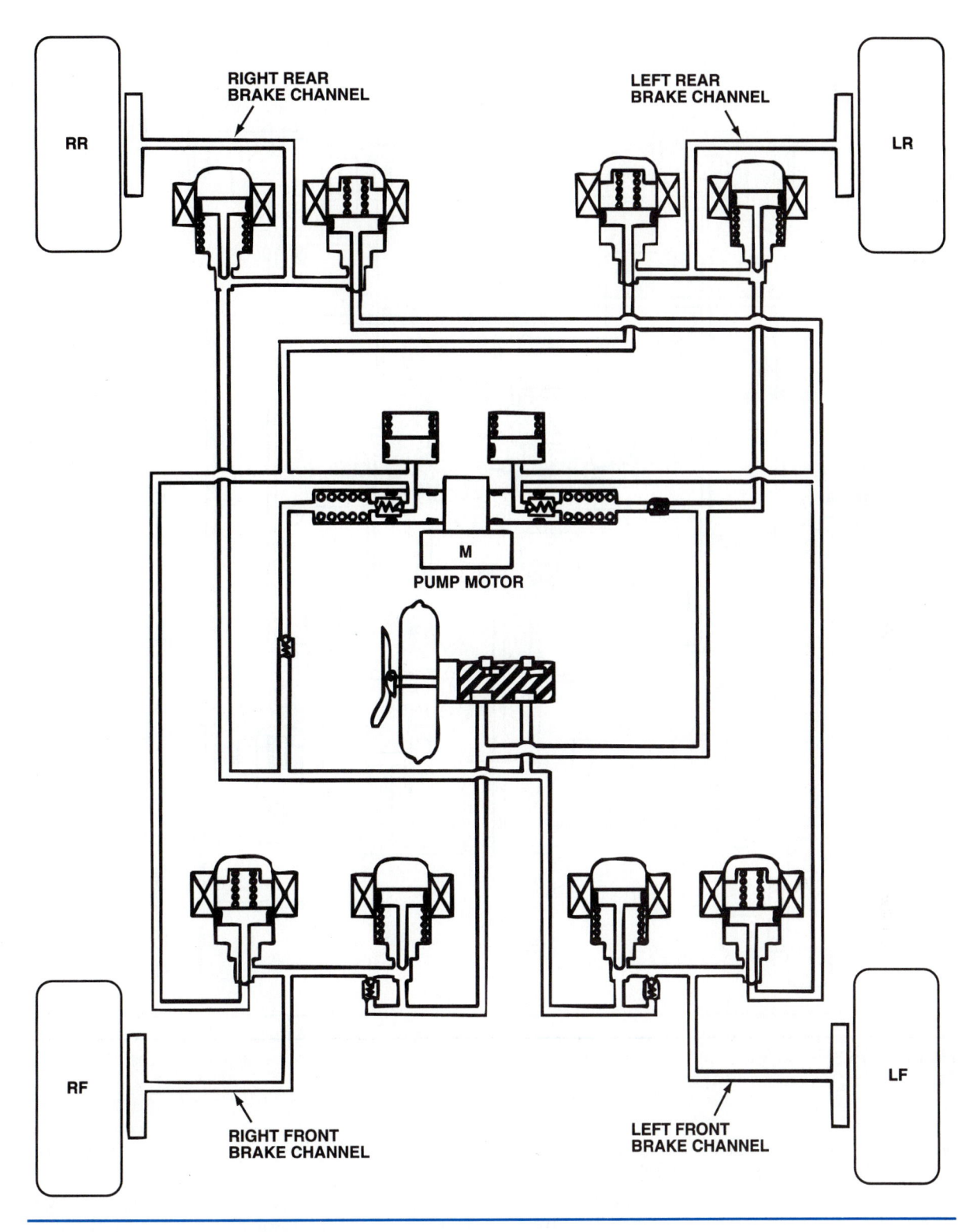

Figure 15-74. The Teves Mark 20 ABS for front-wheel drive applications.

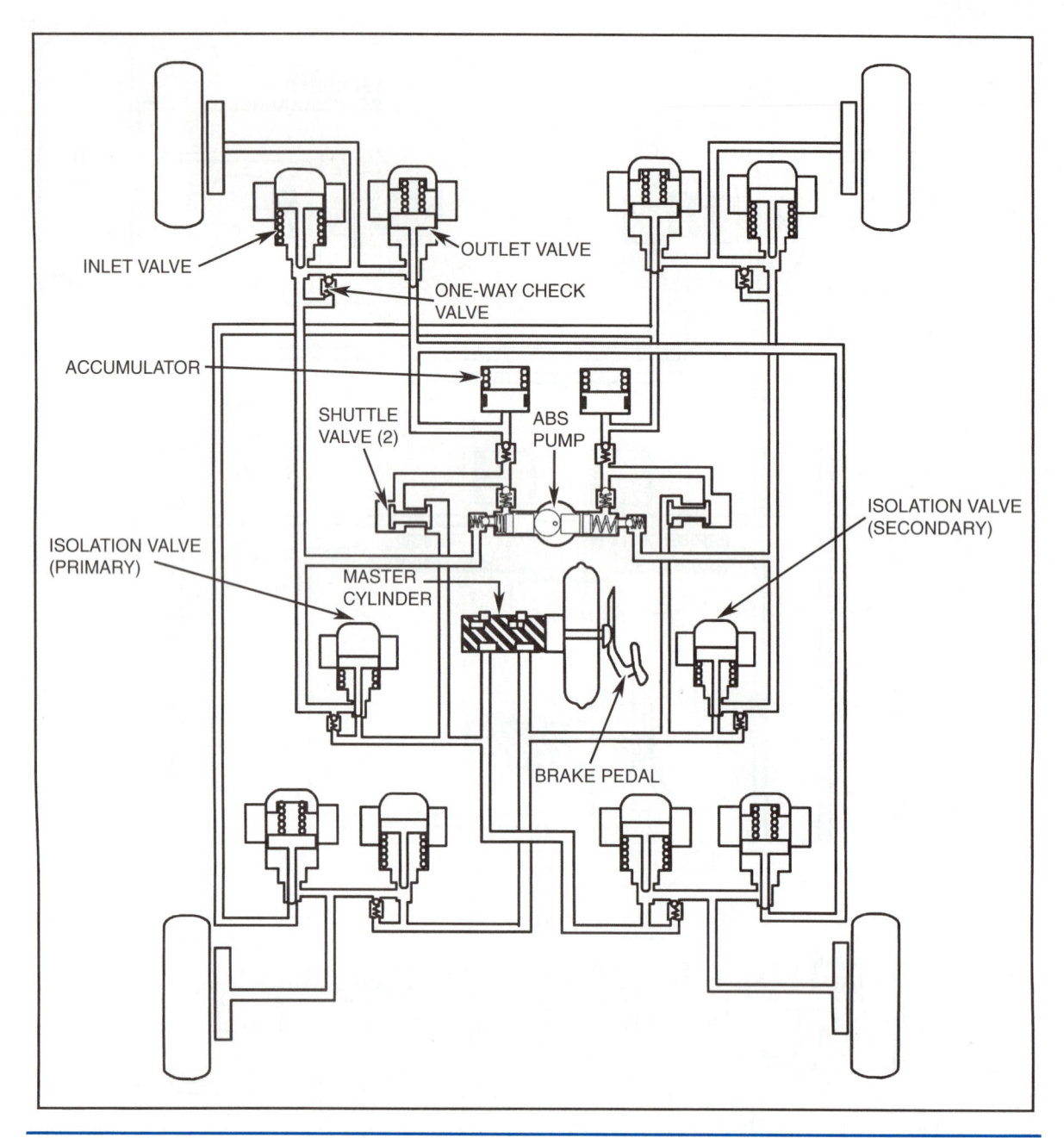

Figure 15-75. Teves Mark 20 ABS with traction control, showing the two additional isolation valves. (Courtesy of DaimlerChrysler Corporation. Used with permission)

when the traction control system is in operation, allowing brake fluid to be drawn from the master cylinder reservoir. This protects the lip seals.

The solenoid valves are contained in a unit that incorporates the CAB on all Chryslers except Jeep Wranglers, figure 15-77. Chrysler refers to this as an integrated control unit (ICU). On Wrangler applications, the CAB is a separate unit, whereas the solenoids and pump are contained in HCU. As with the Teves Mark IV, an extra pair of isolation solenoids is used for the nondriving wheels of vehicles with traction control.

Jeep vehicles equipped with the Mark 20 system use a deceleration sensor, also called a G-switch, to monitor deceleration in both forward and rearward braking.

Control Module

The control module on all Mark 20-equipped vehicles except the Jeep Wrangler includes the CAB,

Antilock Brake Systems

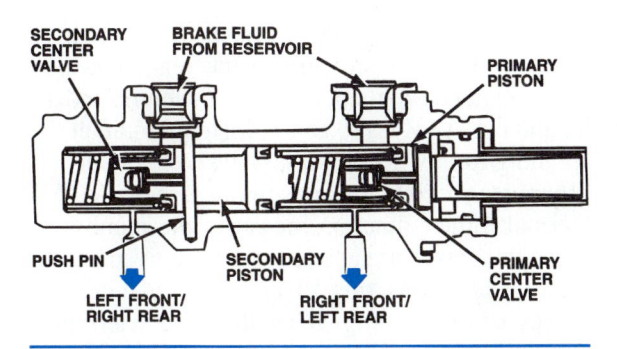

Figure 15-76. On Teves 20 systems with traction control, the master cylinder pistons have center valves.

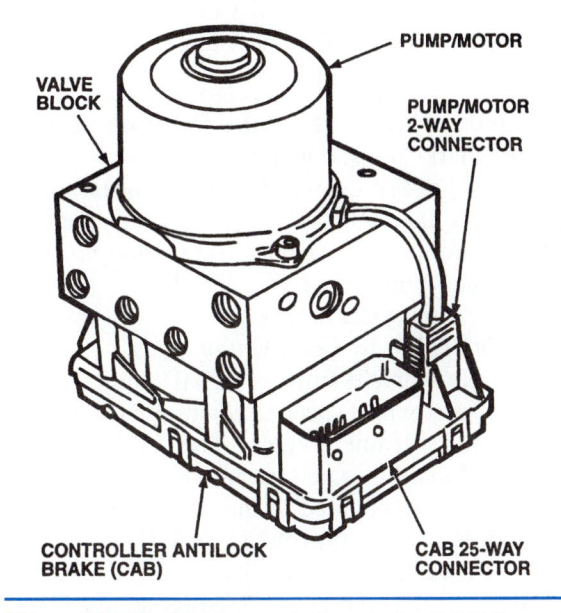

Figure 15-77. The integrated control unit (ICU) used on Chrysler vehicles except Jeep Wranglers.

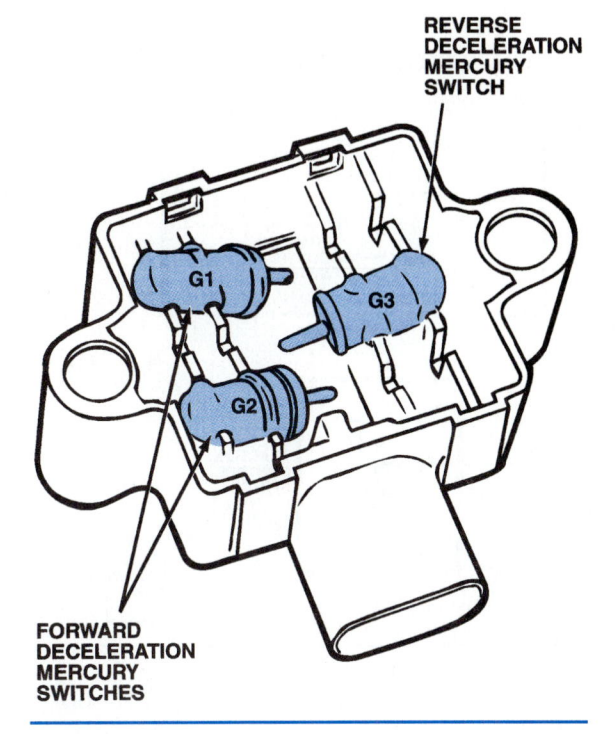

Figure 15-78. The deceleration sensor in Mark 20 ABS for Jeep applications uses three mercury switches that open in response to forward and reverse deceleration G-force.

the hydraulic pump, its motor, and a valve block that contains the solenoid valves. On front-wheel drive vehicles, which have a diagonally split base braking system, the valve block contains an inlet valve and an outlet valve block contains an inlet valve and an outlet valve for each wheel. The CAB is mounted in various locations in the engine compartment, depending on model. The letters "TCS" are stamped on CABs used with traction control. The number of valves in the modulator varies according to application: six for rear-wheel drive (three channels), eight for front-wheel drive (four channels), and ten for traction control.

On Wranglers, the CAB is a separate unit from the hydraulic control module, and is mounted under the instrument panel to the right of the steering column.

The CAB receives two constant power inputs, one for the solenoids and one for the pump motor.

When the ignition key is turned to run or start, a third power input causes the CAB to run a self-diagnostic check. If a problem is found, the CAB sets a code and disables the ABS. Two separate grounds are also provided.

The CAB monitors signals from the four wheel speed sensors, as well as a brake switch. The brake switch enables the CAB to switch to ABS operation sooner than it otherwise would, but the ABS will continue to function if it fails.

On Jeep vehicles, the CAB also monitors signals from a deceleration sensor, figure 15-78. Three normally closed mercury switches contained in the switch housing open under the G-force of deceleration. There are two switches to measure forward deceleration and one for rearward deceleration. The forward switches are mounted at different angles in the housing, so they open at different G-forces. Under hard braking on a high-friction surface, such as dry pavement, the G-force from deceleration is enough to open both of the forward switches. Under hard braking on a medium-friction surface, such as gravel, only forward switch G2 will open. On a slippery surface, neither of the switches will open. Switch G3 opens under hard braking when the vehicle is moving rearward.

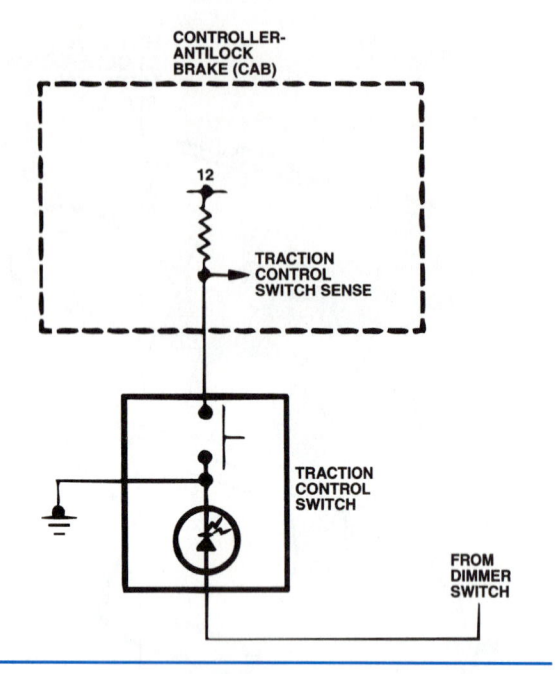

Figure 15-79. The CAB in the Mark 20 traction control system uses this circuit to sense the position of the traction control switch.

The CAB also monitors the traction control switch when the vehicle is so equipped. It sends a constant 12-volt signal to the switch and uses changes in the voltage to determine whether the switch is open or closed, figure 15-79. When the driver pushes the switch, the circuit is grounded, the voltage signal drops, and the CAB disables traction control.

ABS Operation
The MARK 20 ABS has one inlet valve and one outlet valve for each hydraulic channel (three channels with rear-wheel drive and four channels with front-wheel drive). During normal braking and during the pressure-build cycle of an ABS stop, the inlet and outlet valves are open and the outlet valves are closed. During the hold cycle of an ABS stop, both the inlet and outlet valves are closed, maintaining constant hydraulic pressure at the brake. During the decay cycle of an ABS stop, the outlet valve opens, releasing hydraulic pressure to a pair of accumulators for storage.

Traction Control Operation
The TCS operates at vehicle speeds up to 35 mph. The system is automatically switched on whenever the ignition is turned on. On front wheel drive vehicles, the traction control system uses a pair of isolators to block pressure to the rear brakes so they will not be applied, then uses the pump and inlet and outlet valves to apply brake pressure at the spinning front wheel until it slows to the same approximate speed as the nonspinning wheel. A CAB thermal limiter function measures the number of times the brakes are applied within a given period of time and uses this to calculate whether the brakes are overheating. When the CAB determines that the brakes are overheating, it disengages the traction control system for a predetermined period of time to facilitate cooling. The CAB will also interrupt traction control if the driver steps on the brake pedal during TCS operation.

Electronic Variable Brake Proportioning
On the Mark 20e system, the proportioning valves that limit brake pressure to the rear wheels are replaced by the ABS, which limits and controls pressure to the rear wheels. The feature is called electronic variable brake proportioning (EVBP) by DaimlerChrysler or electronic brake force distribution (EBD) by Continental/Teves.

The electronic controller (CAB) compares individual rear wheel speed deceleration rates to the rates of the front wheels. The CAB is programmed to allow a certain amount of rear wheel deceleration, compared to the front wheel deceleration, depending on vehicle model.

When the CAB senses that EVBP cycles are needed, it enters a pressure hold mode to prevent any more pressure increase in the affected rear hydraulic channel. If additional braking balance is required, the CAB pulse modulates (rapidly switches on and off) the outlet valve solenoid. This allows the CAB to vent the pressure in a controlled manner, rather than simply dumping the pressure, as is done during ABS braking.

ABS Plus
Another function of the Mark 20e EBD software is the ability to sense when the vehicle is braking in a turn. Called ABS plus by DaimlerChrysler, the software balances braking forces at the wheels (side to side) to counteract a yaw movement (the tendency for the vehicle to rotate around its center of gravity) and improve vehicle stability while braking in cornering maneuvers. ABS plus is active during all braking, not only during ABS events.

The CAB monitors the wheel speed sensors to determine whether ABS plus cycles are needed. The CAB then uses the ABS modulator valves to

Antilock Brake Systems

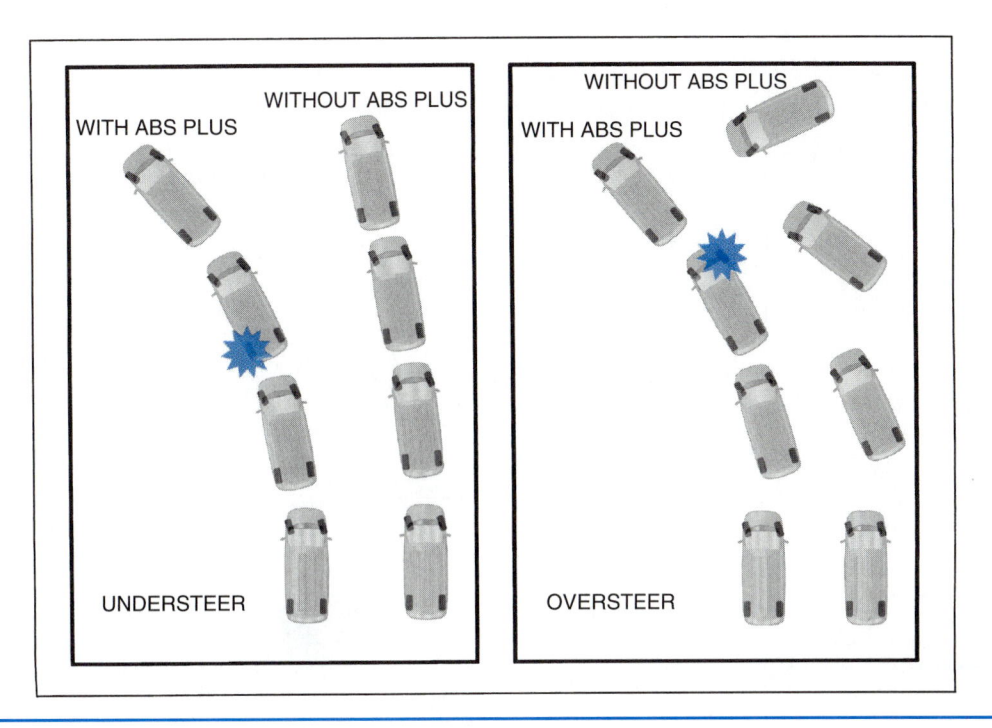

Figure 15-80. Teves Mark 20 ABS with ABS plus helps maintain vehicle control. (Courtesy of DaimlerChrysler Corporation. Used with permission)

help balance braking from side to side, thus reducing the tendency of the driver to understeer or oversteer, figure 15-80.

Teves Mark 25 and Mark 60

Used on some 2001 and later European imports, the Teves Mark 25 and Mark 60 ABS designs are lighter and more compact than the Mark 20 but operate in the same way as described previously, figure 15-81. The Mark 60 is found on high-volume models in the European market, whereas the Mark 25 may be found on luxury-class cars, vans, and light trucks. Some improvements include a new pump design, an electronic controller with a "two-chip" processor redundancy concept, and the option of traction control/stability control without an increase in the controller size.

The Mark 25e and Mark 60e, introduced in 2003 on some European models, include additional features as follows:

- A deflation detection system senses a tire that is losing air pressure and warns the driver of low tire pressure
- A hill start assist on vehicles with manual transmissions can sense when the vehicle is

Figure 15-81. The Teves Mark 60e modulator assembly. (Courtesy of Continental Teves)

stopped on a sloping surface and maintain the brakes applied as the driver moves from the brake to the accelerator pedal, thus keeping the vehicle from rolling; the brakes remain on until the accelerator is pressed

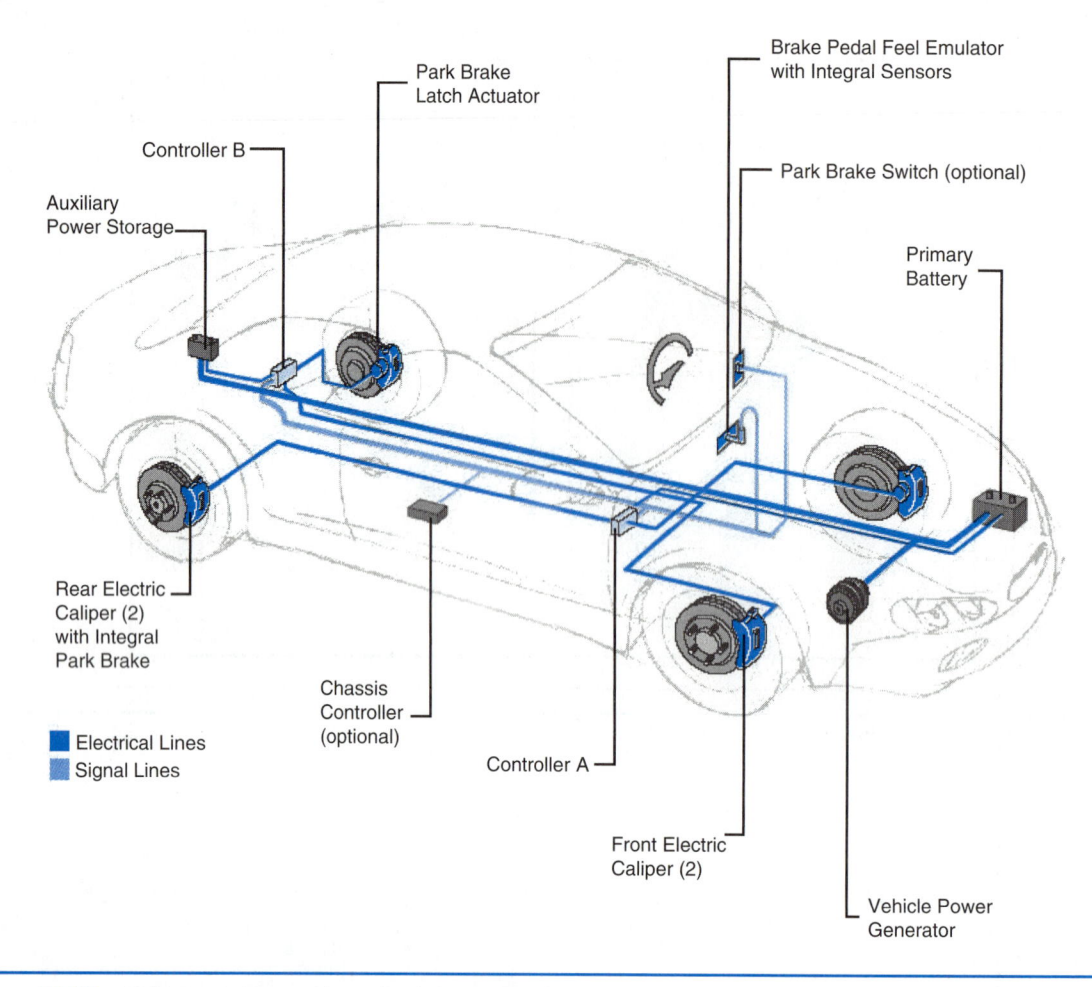

Figure 15-82. Vehicles may soon have brake systems that are entirely operated by electronics, using the vehicle electrical system and no hydraulics. (Courtesy of Delphi Corporation)

- An adaptive cruise control integrates the ABS into the cruise control system to maintain a safe distance from the vehicle ahead; if that vehicle brakes unexpectedly, cruise control is released and the brakes applied, if necessary, to maintain the set distance
- Additional features as needed, such as CAN communications and brake-by-wire capability, figure 15-82.

TOYOTA ABS

Toyota has used four basic ABS designs in its vehicles. These include a rear-wheel system used in some pickup trucks, and three different four-wheel systems used in passenger cars and four-wheel drive vehicles.

Rear-Wheel ABS

In 1990, Toyota introduced a new rear-wheel antilock brake system on the SR5 Truck and 4Runner. The same system was added to the T-100 intermediate-sized truck in 1993.

Toyota's rear-wheel ABS system, figure 15-83, is different from other rear-wheel ABS systems. It uses hydraulic pressure generated by the power steering pump to apply additional brake pressure to the rear wheels in the ABS braking mode, figure 15-84. All other rear-wheel ABSs (Kelsey-Hayes) rely solely on brake pressure generated by the driver's foot through the master cylinder and vacuum power booster. Because the other rear-wheel ABSs use no pump. Brake pressure can bleed off rather quickly during an ABS stop. This, in turn, requires increased pedal effort by the driver and causes a drop in pedal height.

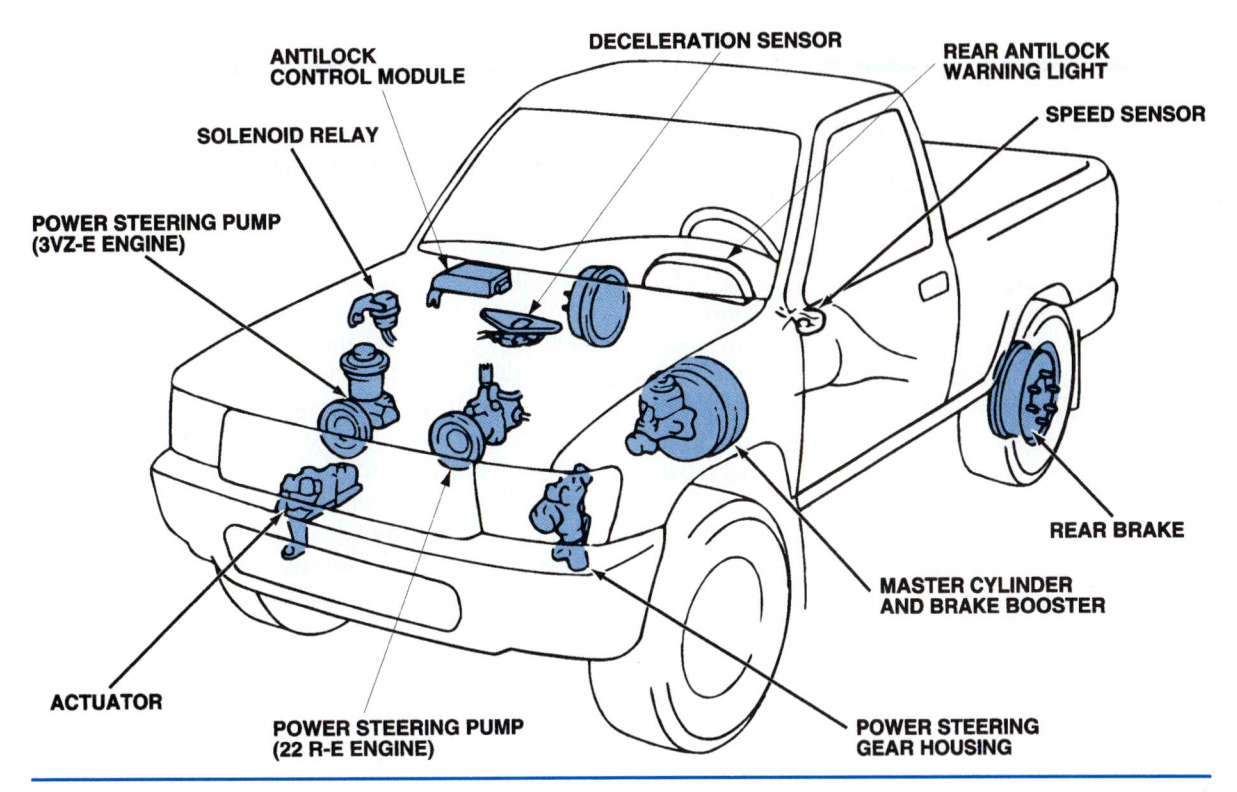

Figure 15-83. The Toyota rear-wheel ABS.

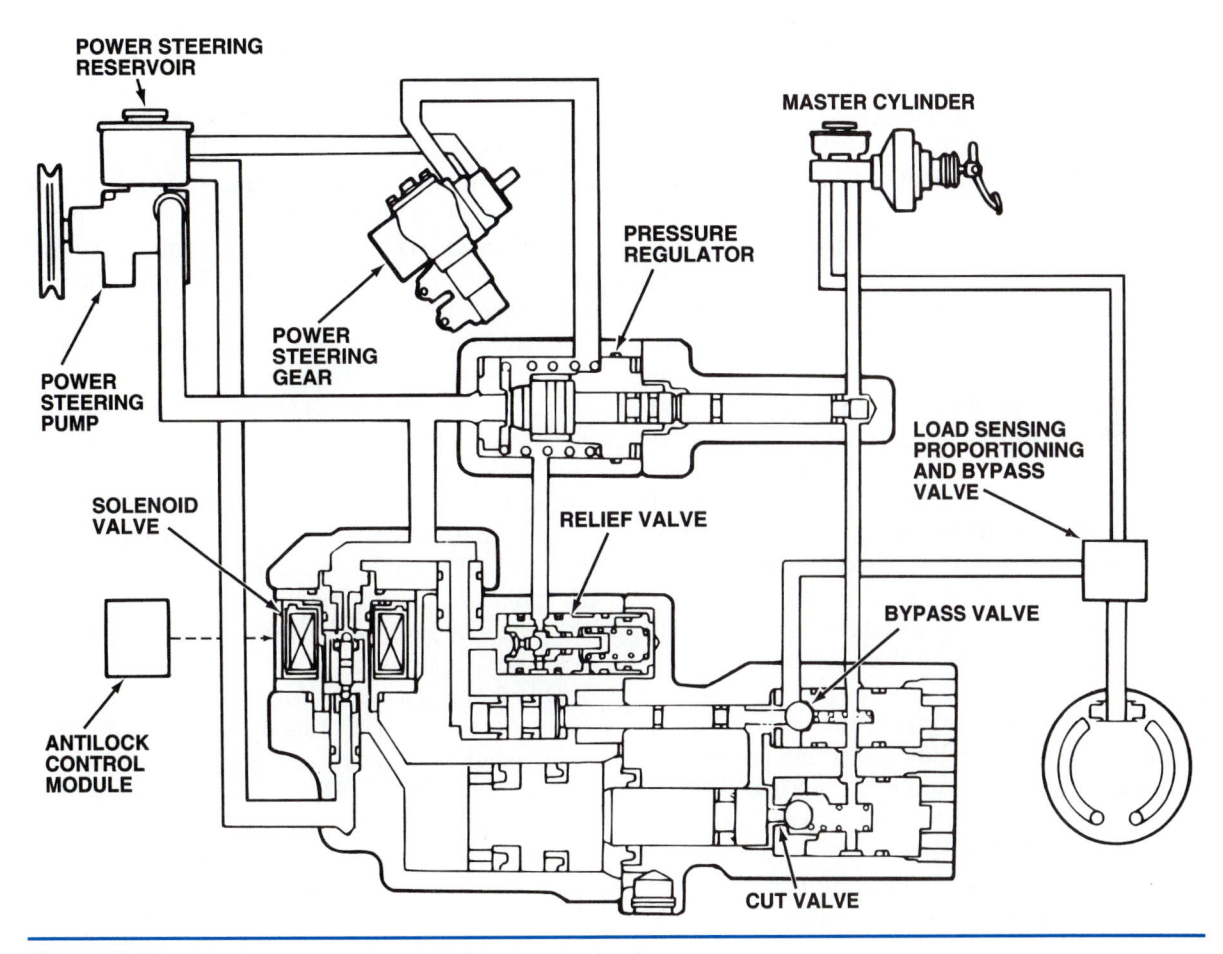

Figure 15-84. The Toyota rear-wheel ABS hydraulic circuit.

361

When Toyota designed this system for their trucks, they wanted the simplicity and low cost of a rear-wheel-only ABS without the increase in pedal effort and drop in pedal height that other systems allow during ABS braking. The result was a unique system that uses a standard master cylinder and vacuum booster but functions much like an integral ABS during the ABS mode by shunting hydraulic pressure from the power steering pump to generate additional brake pressure. This allows sustained ABS braking without increased pedal effort or a loss of pedal height.

The Toyota rear-wheel ABS works only two-wheel drive on trucks with 4WD. When driving in 4WD mode, the ABS is disengaged and brake balance is controlled by a load-sensing proportioning valve that prevents rear brake lockup.

System Components

Toyota's rear-wheel ABS uses a single speed sensor. It is a magnetic induction sensor that monitors rear wheel speed by reading the ring gear in the rear differential. A deceleration sensor detects vehicle deceleration with an inertia slit plate and LED/photo transistor circuit. The sensor is located inside the cab under the front center console.

The stoplight switch, located on the brake pedal, signals the ABS control module when the brakes are applied. A REAR ANTILOCK warning light warns the driver if a problem occurs in the system.

An electronic control unit (ECU) monitors inputs from the sensors and brake switch and decides when ABS braking is needed. The ECU is also located inside the cab behind the glove box.

The ABS solenoid relay, located on the right inner fender near the bulkhead, routes voltage to the ABS solenoid in the hydraulic actuator. The actuator is mounted on the frame under the battery. It contains a pressure regulator valve to control power steering pressure in relation to brake pressure, a bypass piston (opens and closes a bypass valve according to power steering pressure), a relief valve (relieves power steering pressure if the pressure in the actuator is too high), and a single ABS solenoid valve. The ABS solenoid valve is controlled by the ECU via the ABS solenoid relay and performs the hold-release-reapply function that modulates pressure to the rear brakes. The actuator has no replaceable parts.

System Operation

During normal braking, the rear-wheel ABS does nothing. But if the rear wheels start to slip, the control module energizes the ABS solenoid relay, which passes current to the ABS solenoid in the actuator. The solenoid moves upward, isolating the rear brake circuit, and releasing pressure at the same time to prevent wheel lockup. The control module then cycles the solenoid on and off to maintain pressure in the rear brake lines within a narrow range. If pressure needs to be increased, the control module alters the duty cycle or on–off ratio of the ABS solenoid. Increasing the percentage of off time to on time allows pressure from the power steering pump to increase pressure to the rear brakes.

Four-Wheel ABS

Toyota has used three types of four-wheel ABS in its vehicles. The Bosch 2 and a Nippondenso system, discussed earlier in this chapter, were used in earlier models. The Nippondenso system is still used in the Land Cruiser. A three-channel system was first offered in the 1993 Corolla and later offered in all models through 1997 except the Land Cruiser. We discuss the Toyota three-channel system in detail.

The system uses three channels and four wheel sensors in all models except the Supra, which has four channels. Three-channel applications control the rear wheels simultaneously; the four-channel system controls them independently, figure 15-85. The system is designed to maintain a wheel lockup rate of 10 to 30 percent.

There is a pair of solenoids in each channel, one for pressure hold and one for decay (six solenoids for three-channel versions and eight solenoids for four-channel applications).

The hold solenoids are normally open. During normal braking, hydraulic pressure from the master cylinder passes through the hold solenoids to the wheel brakes, figure 15-86. When the pedal is released, a spring-loaded check valve in the hold solenoid opens to let hydraulic pressure return to the master cylinder. During as ABS stop, the hold solenoid closes, blocking hydraulic pressure from the master cylinder. The decay solenoid is also closed, hydraulic pressure constant. If the wheel still is approaching lockup, the decay solenoid opens, releasing hydraulic pressure to the reservoir. During the build cycle, the hold valve opens again and the pump supplies hydraulic pressure to the

Antilock Brake Systems

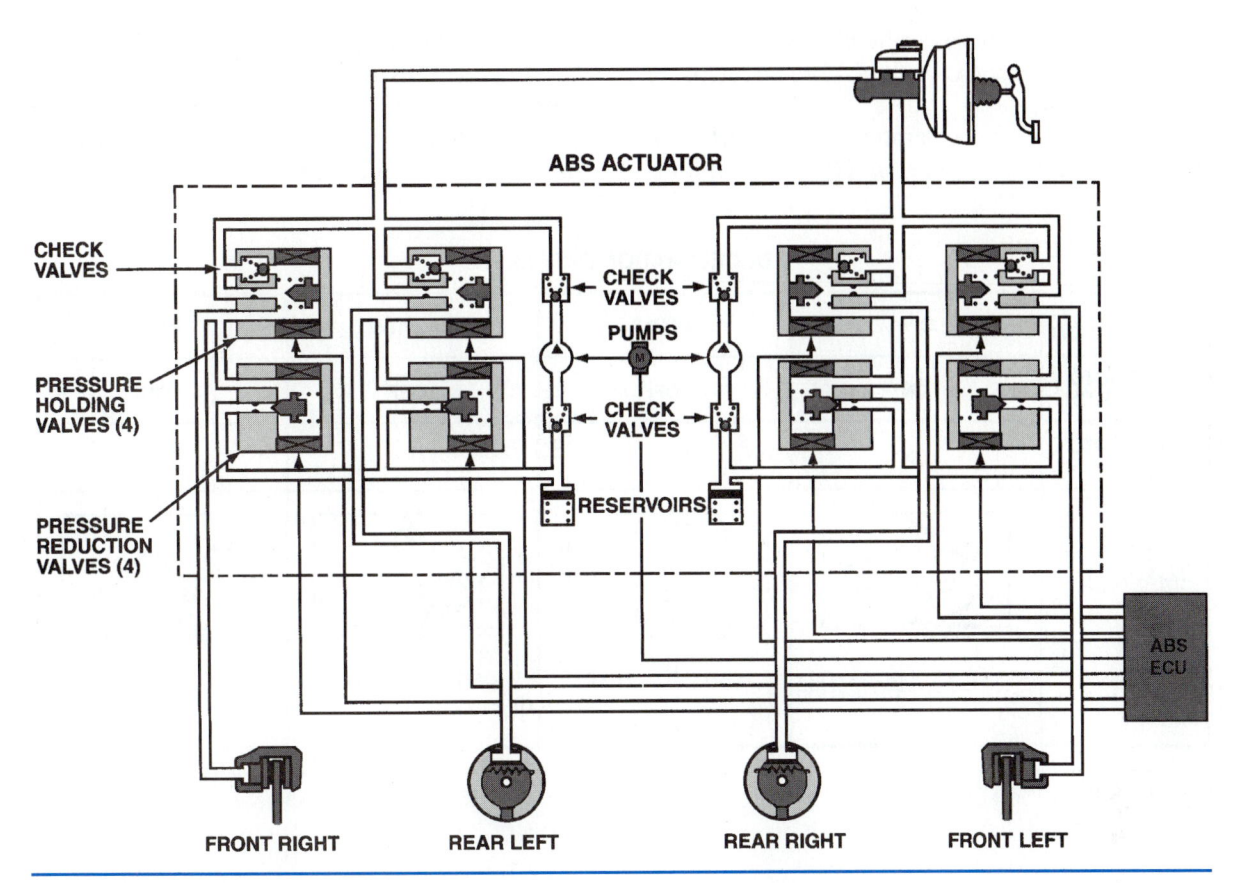

Figure 15-85. The Toyota four-wheel ABS uses a hold solenoid and decay in each hydraulic channel. (Courtesy of Toyota Motor Sales U.S.A., Inc)

Figure 15-86. The normally open hold solenoid closes at the start of an ABS stop; the check valve allows fluid to flow to the master cylinder when the brake pedal is released. (Courtesy of Toyota Motor Sales U.S.A., Inc)

wheel brake. It also supplies pressure to the master cylinder, alerting the driver to ABS operation.

Deceleration and Lateral Acceleration Sensors

A deceleration sensor, also called a G-switch, is used on full-time four-wheel drive models equipped with Toyota ABS. Two different designs are used. The most widely used type has a plate that swings pendulum fashion in the front-to-rear direction of the vehicle in response to the G-force of deceleration, figure 15-87. A pair of LEDs emit light, which passes through slits in the moving plate to a pair of phototransistors. The ECU's signal-conversion curcuit uses the on–off pattern of the light hitting the phototransistors to calculate the rate of deceleration and adjust ABS operation accordingly.

Supras, starting in mid-1993, are equipped with a lateral acceleration sensor. This is the same design as the LED-type deceleration sensor but is mounted crosswise in the vehicle so the plate swings from side to side in response to the G-force developed during cornering.

A second type of deceleration sensor is used in RAV4 vehicles, beginning in 1996. It employs two semiconductor pressure sensors, each mounted at 45 degrees to the centerline of the vehicle, figure 15-88. The sensors convert pressure, which results from the G-force of deceleration, into electrical signals that are sent to the ECU.

DECELERATION RATE LEVEL

RATE OF DECELERATION	LOW-1	LOW-2	MEDIUM	HIGH
NO. 1 PHOTO TRANSISTOR	ON	OFF	OFF	ON
NO. 2 PHOTO TRANSISTOR	ON	ON	OFF	OFF
POSITION OF SLIT PLATE	NO. 1 PHOTO TRANSISTOR (ON) NO. 2 PHOTO TRANSISTOR (ON)	(OFF) (ON)	(OFF) (OFF)	(ON) (OFF)

Figure 15-87. The LED-type deceleration sensor used with Toyota ABS in four-wheel drive vehicles. (Courtesy of Toyota Motor Sales U.S.A., Inc)

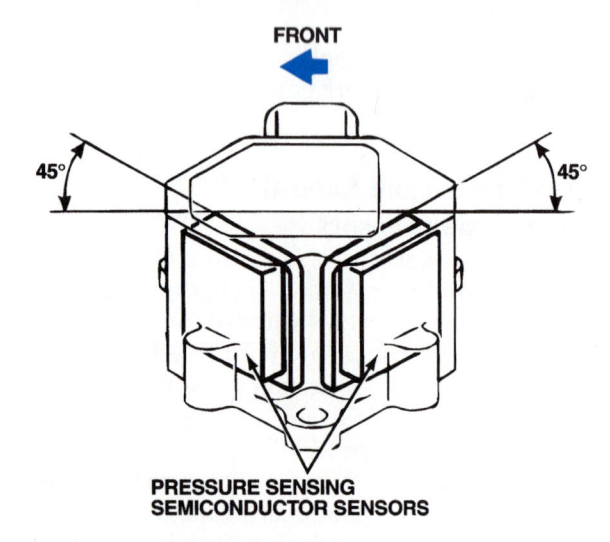

Figure 15-88. The semiconductor-type deceleration sensor used with the 1996 and later RAV4. (Courtesy of Toyota Motor Sales U.S.A., Inc)

SUMMARY
ABS APPLICATION GUIDE

Year Started	Year Ended	Vehicle Application
ADVICS		
2002	& up	Toyota Sienna
2003	& up	Toyota Matrix
2003	& up	Pontiac Vibe
2003	& up	Mazda (some models)
2003	& up	Isuzu (some models)
2004	& up	Scion
2004	& up	Chevrolet Colorado
2004	& up	GMC Canyon
BENDIX LC4		
1994	1996	Chrysler Town and Country Minivan
1994	1995	Chrysler LeBaron
1994	1995	Dodge Caravan, Spirit, Shadow
1994	1996	Plymouth Acclaim, Sundance, Voyager
BENDIX ABX-4		
1995	1997	Dodge/Plymouth Neon
1995	1997	Chrysler Cirrus, Dodge Stratus
BENDIX 6		
1991	1993	Dodge Daytona, Spirit
1991	1993	Chrysler LeBaron
1991	1993	Plymouth Acclaim and Laser
BENDIX 9 (JEEP)		
1989	1991	Jeep Cherokee and Wagonee
BENDIX 10		
1990	1993	Chrysler New Yorker, Imperial, Fifth Ave
1990	1993	Dodge Dynasty
1991	1993	Dodge Caravan/Plymouth Voyager
1991	1993	Eagle Premier
BENDIX MECATRONIC II		
1995	1998	Ford Contour
1995	1998	Mercury Mystique
BOSCH 2 and 2S		
1985	1996	Audi
1987	1991	Bentley
1985	1993	BMW
1986	1989	Chevrolet Corvette
1989	1993	Chrysler Imports
1990	1993	Infiniti
1991	& up	Isuzu Rodeo
1992	& up	Isuzu Trooper
1987	1989	Jaguar XJ-6
1990	1996	Lexus
1987	1993	Mazda Miata, RX7, 929
1991	& up	Mitsubishi Diamante, Expo
1993	& up	Mitsubishi Mirage
1989	& up	Mitsubishi Sigma
1989	& up	Nissan 240SX, Maxima
1990	& up	Nissan 300ZX, Stanza
1993	1998	Nissan Altima
1991	1998	Nissan NX, Sentra
1985	1994	Porsche
1987	1991	Rolls Royce
1987	1988	Sterling
1990	1992	Subaru Legacy
1991	& up	Suzuki (some models)
1987	1996	Toyota Camry
1987	& up	Toyota Celica
1987	& up	Toyota Cressida
1991	& up	Toyota MR2
1991	& up	Toyota Previa
1987	& up	Toyota Supra
1987	1993	Volvo

Year Started	Year Ended	Vehicle Application
BOSCH 2E		
1991	& up	Dodge Stealth
1991	& up	Eagle Talon
1991	& up	Mitsubishi 3000 GT
1991	1998	Mitsubishi Eclipse
1991	& up	Mitsubishi Galant
1991	& up	Plymouth Laser
BOSCH 2S MICRO		
1990	1993	Corvette
BOSCH 2U		
1991	1994	Buick Estate Wagon, Roadmaster
1991	1992	Buick Reatta, Riviera
1991	1994	Cadillac Brougham, Eldorado, Seville, Ttouring Coupe
1991	1994	Chevrolet Caprice, Caprice Wagon
1994	1998	Ford Mustang
1993	1998	Mercury Villager
1993	1998	Nissan Quest
1992	1994	Oldsmobile Custom Cruiser Wagon
1991	1993	Oldsmobile Toronado, Trofeo
BOSCH 2U ASR		
1992	1995	Corvette
1993	1994	Cadillac (some models)
BOSCH 3		
1987	1993	Cadillac Allante
1988	1990	Dodge Dynasty
1988	1990	Chrysler Imperial, New Yorker, Fifth Avenue
BOSCH ABS/ASR		
1993	1994	Cadillac, All
1992	1994	Chevrolet Corvette
BOSCH 5.0		
1995	1996	Corvette
1995	1996	Cadillac (All)
1995	1998	Porsche 911
1996	2001	Ford Taurus
1996	2001	Mercury Sable
1995	1996	Chevrolet Impala
BOSCH 5.0 (Bosch/Delco Hybrid)		
1996	1999	Buick Estate Wagon, Park Avenue, Roadmaster
1996	2002	Cadillac Eldorado, Fleetwood, Seville
1997	2000	Chevrolet Corvette
BOSCH VEHICLE DYNAMICS CONTROL (VDC)		
1996	1997	Mercedes (some models)
BOSCH 5.3		
1997	2001	Audi (All)
2002	& up	Buick Rendezvous
1998	& up	Cadillac Catera
2003	& up	Cadillac DeVille
1998	& up	Chevrolet Camaro
2001	& up	Chevrolet Corvette
1999	& up	Ford Contour, Mustang
2003	& up	Hummer H2
1999	& up	Mazda 626, MPV
1999	& up	Mercury Cougar, Mystique, Villager
1999	& up	Mitsubishi Eclipse
1999	& up	Nissan Altima, Frontier, Quest, Sentra
1998	2002	Oldsmobile Intrigue
1998	& up	Pontiac Firebird, Grand Prix
2004	& up	Pontiac GTO
2001	& up	Pontiac Aztec

Year Started	Year Ended	Vehicle Application
BOSCH 5.3—*continued*		
1999	2003	Porsche 911
1998	2003	Porsche Boxter
2000	2003	Saturn
1997	& up	Subaru Legacy
1997	& up	Toyota (depending on model)
BOSCH 5.7		
2004	& up	Porsche 911
2001	& up	BMW (some models)
1998	& up	Mercedes (some models)
BOSCH 8.0		
2003	& up	Saturn Vue, Ion
DELPHI CHASSIS (DELCO MORAINE) POWERMASTER III		
1989	1991	Buick Regal 1989 1991 Oldsmobile Cutlass
1989	1991	Pontiac Grand Prix 1990 1991 Pontiac Grand Prix GTU
DELPHI CHASSIS (DELCO) ABS VI		
1992	1999	Buick Regal 1991 1998 Buick Skylark GS
1991	1996	Chevrolet Beretta, Corsica
1993	1997	Chevrolet Camaro
1992	1999	Chevrolet Cavalier
1992	2001	Chevrolet Lumina
1992	1999	Chevrolet Lumina APV/Venture Van
1999	2000	Daewoo (All)
1992	1997	Geo Prism
1992	1998	Oldsmobile Achieva, Cutlass, Silhouette
1999	2000	Oldsmobile Alero
1993	1997	Pontiac Firebird
1991	2000	Pontiac Grand Am SE
1992	1997	Pontiac Grand Prix
1992	1993	Pontiac LeMans
1992	1994	Pontiac Sunbird
1992	1999	Pontiac Trans Sport
1991	1999	Saturn (All)
DELPHI CHASSIS DBC 7		
1999	& up	Buick Century, Regal
2000	& up	Chevrolet Cavalier, Impala, Monte Carlo
2000	2003	Chevrolet Malibu
1999	& up	Chevrolet Tracker
2001	& up	Oldsmobile Cutlass, Alero
2000	2004	Oldsmobile Silhouette
2001	& up	Pontiac Grand Am
2000	& up	Pontiac Montana, Sunfire
DELPHI DBC 7.2		
2003	& up	Cadillac CTS
2004	& up	Cadillac SRX, XLR
2004	& up	Chevrolet Malibu
KELSEY-HAYES EBC2 RWAL and RABS/RABS II		
1989	1990	Chevrolet Astro
1989	1991	Chevrolet Blazer (S-series)
1988	1993	Chevrolet C/K Pickup
1990	1992	Chevrolet G-van
1989	1993	Chevrolet S/T-series Trucks
1990	1991	Chevrolet Suburban
1989	& up	Dodge Dakota, Ram, Ramcharger
1990	1991	Ford Aerostar
1992	& up	Ford Aerostar (RABS II)
1987	1992	Ford Bronco
1987	1990	Ford Bronco II
1990	1994	Ford Econoline
1991	1992	Ford Explorer
1987	1994	Ford F-Series
1987	1998	Ford F250, 350, 450 (depending on model)
1989	1992	Ford Ranger (RABS)
1993	& up	Ford Ranger (RABS II)
1991	1995	Geo Tracker

Year Started	Year Ended	Vehicle Application
1988	1993	GMC C/K Pickup
1989	& up	GMC Jimmy
1988	1991	GMC R/V
1989	1993	GMC S/T Series
1989	1993	GMC Safari
1989	1992	GMC Safari Cargo Van
1992	1995	GMC Sierra
1990	1992	GMC Suburban
1990	& up	Isuzu Amigo
1990	& up	Isuzu Pickup
1990	& up	Isuzu Rodeo
1990	& up	Isuzu Trooper
1990	1998	Mazda MPV
1990	& up	Mazda Navajo
1990	& up	Mazda Pickup
1991	& up	Nissan Pathfinder
1991	& up	Nissan Pickup
1990	& up	Suzuki Sidekick
KELSEY-HAYES EBC4 4WAL		
1990	1993	Chevrolet Astro
1991	1993	Chevrolet Blazer (S/T)
1994	1995	Chevrolet S/T-series Trucks
1992	1994	Chevrolet Suburban
1991	1994	GMC Jimmy (4-door)
1992	1994	GMC Jimmy (2-door)
1991	1994	GMC S/T Series
1990	1994	GMC Safari
1992	1994	GMC Suburban
1992	1994	GMC Yukon
1991	1994	Oldsmobile Bravada
KELSEY-HAYES EBC4-VCM		
1994	1995	Chevrolet S/T (certain engines)
KELSEY-HAYES EBC5U/5H		
1993	1997	Dodge Dakota
1994	1997	Dodge Ram and Ram van
1994	1997	Ford Aerostar, Econoline
KELSEY-HAYES (TRW) EBC310		
1994	2000	GM C/K trucks
1995	2000	GM Vans
1995	2000	Chevrolet/GMC Suburban
1995	2000	Chevrolet Blazer
1995	2000	GM S/T 4 door trucks
KELSEY-HAYES (TRW) EBC 325		
1998	& up	Dodge Dakota, Ram
1998	& up	Isuzu Rodeo
1998	& up	Kia Sportage
2000	& up	GM C/K trucks
2000	& up	GM Vans
2000	& up	Chevrolet/GMC Suburban
2000	& up	Chevrolet Blazer
2000	& up	GM S/T 4 door trucks
2004	& up	Chevrolet SSR
KELSEY-HAYES (TRW) EBC 410		
1997	& up	Ford Expedition, Navigator
1997	& up	Ford F150, F250
1994	& up	Ford Windstar
2000	& up	Ford Excursion
KELSEY-HAYES (TRW) EBC 430		
2003	& up	Chevrolet Avalanche
MANDO		
1992	1993	Hyundai (some models)
NIPPONDENSO 2L		
1990	1996	Infiniti
1990	1996	Lexus
1988	1996	Toyota (some models)
1990	1997	Subaru (depending on model)

Antilock Brake Systems

Year Started	Year Ended	Vehicle Application
NIPPONDENSO 5.3		
1996	& up	Subaru (some models)
1996	& up	Toyota (some models)
1998	& up	Mazda (some models)
2000	& up	Daewoo (some models)
SUMITOMO 1		
1989	Mar-91	Ford Probe
1991	& up	Honda
1998	Mar-91	Mazda 626
1989	Mar-91	Mazda MX-6
1987	1991	Mazda RX-7
SUMITOMO 2		
1994	1996	Ford Escort GT Mar-91
1997		Ford Probe
1994	1997	Ford Aspire
1994	1999	Hyundi (some models)
Mar-91	1998	Mazda 626
1990	1998	Mazda Miata
1992	1997	Mazda MX-3
1995	1997	Mazda Protégé
Mar-91	1997	Mazda MX-6
1994	1996	Mercury Tracer LTS
TEVES MK II		
1986	1989	Buick LeSabre, Electra, Park Avenue
1986	1990	Cadillac Fleetwood, Deville
1987	1992	Ford Thunderbird
1990	1994	Jaguar XJ6, XJS
1985	1989	Lincoln Continental
1985	1992	Lincoln Mark VI
1988	1989	Merkur Scorpio
1989	1992	Mercury Cougar
1988	1992	Peugeot (some models)
1987	1993	Saab (some models)
1990	1993	Volkswagen Corrado, Eurovan, Golf, Jetta, Passat
TEVES MARK IV		
1990	1996	BMW (some models)
1991	1995	Buick LeSabre, Park Avenue

Year Started	Year Ended	Vehicle Application
1991	1998	Buick Riviera
1991	1993	Cadillac DeVille
1993	1998	Chrysler Concorde, LHS, New Yorker
1993	1998	Dodge Intrepid
1993	1998	Eagle Vision
1993	1994	Ford Bronco, Explorer
1992	& up	Ford Crown Victoria
1991	1995	Ford Taurus
1990	1995	Ford Taurus SHO
1993	1996	Ford Thunderbird
1992	1996	Jeep Cherokee, Grand Cherokee, Wagoneer
1990	& up	Lincoln Continental 1994 1996 Lincoln Mark VIII
1990	1996	Lincoln Town Car
1993	1999	Mercury Cougar
1992	& up	Mercury Grand Marquis
1991	1995	Mercury Sable
1991	1995	Oldsmobile Delta 88, 98
1995	1999	Oldsmobile Aurora
1991	1995	Pontiac Bonneville, SSE, SSEi
1994	1996	Volvo (some models)
1994	& up	Volkswagen Corrado, Jetta TEVES MARK 20
1997	& up	BMW (depending on model)
1998	& up	Chrysler 300M, Cirrus, Concorde, LHS
1999	2000	Chrysler Sebring 1997 2000 Chrysler Town & Country
1998	& up	Dodge Neon
1998	2000	Dodge Stratus
1997	& up	Honda
1997	& up	Jeep Grand Cherokee, Wrangler
1998	& up	Kia (some models)
1997	1998	Lincoln Mark VIII
1998	2000	Plymouth Breeze
1998	& up	Plymouth Neon
1996	& up	Volvo (some models)
1995	& up	Volkswagen (some models)
TOYOTA REAR WHEEL ABS		
1990	1995	Toyota 4Runner, Pickup
1993	1994	Toyota T-100 Pickup

Review Questions

Choose the letter that represents the best possible answer to the following questions:

1. Technician A says the Bendix 9 ABS on the Jeep Wagoneer has a speed sensor on each wheel.
   Technician B says the Bendix 9 modulator assembly contains nine solenoid valves. Who is right?
   a. Technician A
   b. Technician B
   c. Both A and B
   d. Neither A nor B

2. Technician A says the Bendix 10 ABS has 10 solenoid valves.
   Technician B says the Bendix 10 ABS has adjustable wheel speed sensors. Who is right?
   a. Technician A
   b. Technician B
   c. Both A and B
   d. Neither A nor B

3. Technician A says the Bendix ABX-4 system uses accumulators just like the other systems.
   Technician B says the Bendix ABX-4 system works like a two-channel system. Who is right?
   a. Technician A
   b. Technician B
   c. Both A and B
   d. Neither A nor B

4. Technician A says the Bendix Mecatronic II ABS uses a "closed-loop" hydraulic system for traction control.
   Technician B says the Bendix Mecatronic II ABS uses a single pump motor with a dual-piston pump. Who is right?
   a. Technician A
   b. Technician B
   c. Both A and B
   d. Neither A nor B

5. Technician A says the Bendix III system uses isolation solenoids to control each brake.
   Technician B says the Bendix III and the LC4 systems are essentially the same. Who is right?
   a. Technician A
   b. Technician B
   c. Both A and B
   d. Neither A nor B

6. Technician A says the Chrysler ABCM and the Cadillac EBCM are the same Bosch control unit.
   Technician B says the 1990–92 Allante uses a different module with a 55-pin connector. Who is right?
   a. Technician A
   b. Technician B
   c. Both A and B
   d. Neither A nor B

7. Technician A says the three-way solenoid valves in a Bosch ABS are normally open.
   Technician B says the three-way valves in a Bosch ABS are normally closed. Who is right?
   a. Technician A
   b. Technician B
   c. Both A and B
   d. Neither A nor B

8. Technician A says the Bosch 3 valve block unit must be replaced if the sensor block is faulty.
   Technician B says the sensor block can be replaced as a separate unit. Who is right?
   a. Technician A
   b. Technician B
   c. Both A and B
   d. Neither A nor B

9. Technician A says the hydraulic assembly of the Bosch 3 ABS can be overhauled.
   Technician B says the plastic fluid reservoir, fluid level sensor, sensor block, control pressure switch, and two rear proportioning valves must be replaced with the hydraulic assembly. Who is right?
   a. Technician A
   b. Technician B
   c. Both A and B
   d. Neither A nor B

10. Technician A says the Bosch 3 ABS has spring-loaded accumulators like the other Bosch systems.
    Technician B says the replenishing valve remains open as long as ABS braking occurs. Who is right?
    a. Technician A
    b. Technician B
    c. Both A and B
    d. Neither A nor B

Antilock Brake Systems

11. Technician A says the Bosch 2 unit is a nonintegral unit with a conventional master cylinder, vacuum power booster and a separate hydraulic modulator assembly. Technician B says the hydraulic modulator on the Bosch 2 unit is part of the assembly. Who is right?
    a. Technician A
    b. Technician B
    c. Both A and B
    d. Neither A nor B

12. Technician A says the brake systems on the Delco Moraine ABS IV system are split front and rear with the primary side of the master cylinder serving the front wheels and the secondary side serving the rear wheels. Technician B says each wheel has its own adjustable wheel speed sensor. Who is right?
    a. Technician A
    b. Technician B
    c. Both A and B
    d. Neither A nor B

13. Technician A says the Delco VI system modulates brake pressure during antilock braking by closing an isolation solenoid valve to hold pressure in a given brake circuit and then opening a decay valve to release pressure.
    Technician B says the Delco ABS-VI system modulates brake pressure during antilock braking by positioning a small piston up and down inside a fluid chamber. Who is right?
    a. Technician A
    b. Technician B
    c. Both A and B
    d. Neither A nor B

14. Technician A says the Kelsey-Hayes ABS EBC2, RABS, RWAL, and ZPRWAL systems on Ford and Dodge applications have the speed sensor in the differential and the sensor ring on the ring gear. Technician B says that on GM applications the speed sensor is located in the transmission tailshaft and the sensor ring is on the transmission output shaft. Who is right?
    a. Technician A
    b. Technician B
    c. Both A and B
    d. Neither A nor B

15. Technician A says the Kelsey-Hayes 4WAL system is a nonintegral, four-wheel, three-channel ABS that functions in four-wheel drive.
    Technician B says the 4WAL ABS is deactivated when switching from two- to four-wheel drive. Who is right?
    a. Technician A
    b. Technician B
    c. Both A and B
    d. Neither A nor B

16. Technician A says the Teves Mark II ABS is an integral four-wheel, three-channel system with a combination master cylinder, hydraulic modulator, and pump motor and accumulator assembly.
    Technician B says the integral hydraulic control unit contains four ABS solenoids. Who is right?
    a. Technician A
    b. Technician B
    c. Both A and B
    d. Neither A nor B

17. Technician A says the Teves Mark IV ABS controls all four wheels independently. Technician B says the Mark IV system is a four-channel ABS with individual wheel speed sensors and separate ABS solenoids for each brake circuit. Who is right?
    a. Technician A
    b. Technician B
    c. Both A and B
    d. Neither A nor B

18. Technician A says the Teves Mark 20 ABS can be equipped with traction control. Technician B says the traction control system in the Mark 20 ABS uses throttle control at speeds above 35 mph. Who is right?
    a. Technician A
    b. Technician B
    c. Both A and B
    d. Neither A nor B

19. Technician A says the Teves Mark 20 ABS uses a diagonally split hydraulic system in all applications.
    Technician B says the Teves Mark 20 ABS has a deceleration sensor when installed in four-wheel drive vehicles. Who is right?
    a. Technician A
    b. Technician B
    c. Both A and B
    d. Neither A nor B

20. Technician A says Toyota's rear-wheel ABS uses hydraulic pressure generated by the power steering pump to apply additional brake pressure to the rear wheels in the ABS braking mode.
    Technician B says the Toyota rear-wheel ABS allows sustained ABS braking without increased pedal effort or a loss of pedal height. Who is right?
    a. Technician A
    b. Technician B
    c. Both A and B
    d. Neither A nor B

21. Technician A says the Toyota rear-wheel ABS uses speed sensors on each rear wheel. Technician B says the Toyota rear-wheel ABS functions only in two-wheel drive mode on four-wheel drive trucks. Who is right?
    a. Technician A
    b. Technician B
    c. Both A and B
    d. Neither A nor B

22. Technician A says the Toyota four-wheel ABS system does not have isolation valves. Technician B says the Toyota four-wheel ABS has two valves per hydraulic channel. Who is right?
    a. Technician A
    b. Technician B
    c. Both A and B
    d. Neither A nor B

23. Technician A says Toyota uses more than one type of deceleration sensor. Technician B says the Supra lateral acceleration sensor is of the same design as the LED deceleration sensor but mounted crosswise in the vehicle. Who is right?
    a. Technician A
    b. Technician B
    c. Both A and B
    d. Neither A nor B

16

The Brake System and Vehicle Suspension

OBJECTIVES

Upon completion and review of this chapter, you will be able to:

- List and describe different types of vehicle tires and how each relates to braking.
- Explain how tires are sized and how tire size affects braking.
- Explain P-metric tire designations.
- Define the term "aspect ratio."
- Explain the term "tire contact patch."
- Describe how tire sizes can affect the operation of the antilock brake system.
- Explain how tread design relates to traction.
- Describe how tire inflation affects traction.
- Describe the condition and effects of tire and wheel runout.
- Explain wheel offset and its effect on braking.
- Define the term "scrub radius."
- Explain the effects of suspension and wheel alignment on braking.
- List and describe the different types of wheel bearings and the effects of bearing problems on braking.

KEY TERMS

aspect ratio
caster
hydroplane
lateral runout
loaded runout
radial runout
scrub radius
section width
tire contact patch
wheel offset

INTRODUCTION

Modern automobile brakes generate more than enough friction to lock the wheels and, under all but extreme conditions, have enough mass to resist fade. As a result, the primary limiting factor on the braking performance of most cars is tire traction. Because the tires mount on the wheels and the wheels bolt to the suspension, any problems in these areas that adversely affect tire traction will also affect braking power or balance.

This chapter covers the contribution the tires, wheels, and suspension make to braking. It also focuses on the problems that can be created by defects in, or damage to, any of these parts. The effect of mismatched tires, wheels, and suspension settings on braking is also examined.

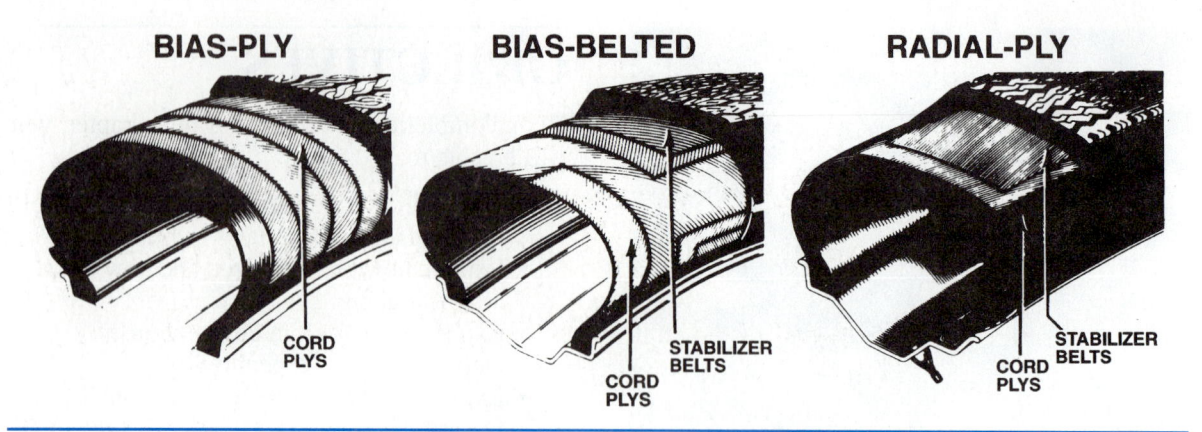

Figure 16-1. There are three types of tire construction.

TIRES AND BRAKING

Tires are important to braking because they form the friction link between the vehicle and the road. In every brake system, the size of the wheel friction assemblies and the range of force used to apply them are engineered based on the amount of traction available between the tires and the road. For predictable and powerful braking performance, it is essential the tires at all four wheels be similar in the following respects:

- Construction
- Size
- Tread design.

TIRE CONSTRUCTION

Three types of tire construction have been used on passenger cars, figure 16-1:

- Radial-ply
- Bias-ply
- Bias-belted.

Radial-ply tires are the standard type used on today's cars and light trucks. In a radial-ply tire, the cord layers run at a 90-degree angle from bead to bead. The rubber that makes up the sidewalls and tread is molded and cured onto the casing. This makes the sidewalls very flexible, which increases the size of the **tire contact patch** on the road, figure 16-2. Because of their flexibility, radial-ply tires must have belts around their circumferences to stabilize the tread. When compared to the older style bias-belted tires, radial tires have a larger "footprint" for better braking and handling.

In a bias-ply tire, the cord layers, or plies, run from bead to bead at an angle of approximately 26

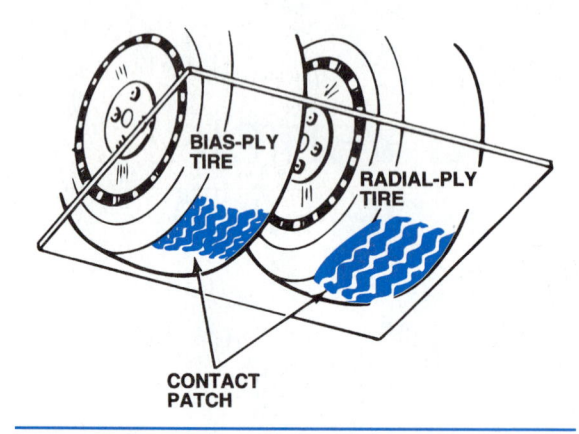

Figure 16-2. The size of the tire contact patch has a direct effect on braking traction.

to 38 degrees. The ply layers are made of nylon or polyester fabric and form the tire casing. Bias-ply tires are rarely used in automotive applications today. Bias-belted tires use the same casing construction as bias-ply tires, however, they have additional fabric belts around the circumference of the tire under the tread.

Tire Construction and Braking

When tires of similar size but different construction are compared, radial tires provide the most traction, followed by bias-belted and bias-ply tires in that order. Radial tires also respond to turning inputs much more rapidly and with greater force than either bias-ply or bias-belted tires. These differences can cause braking or handling problems if tires of differing construction are used on the same vehicle.

For the most predictable performance, tires with the same construction should be used on all four wheels. Tire manufacturers recommend that

The Brake System and Vehicle Suspension

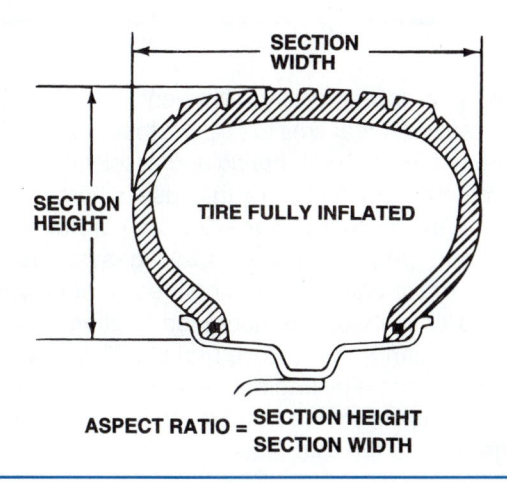

Figure 16-3. Section width and section height are important tire dimensions.

radial-ply tires not be mixed with other types, but if they must be, the radial-ply tires should be mounted on the rear wheels.

TIRE SIZE

Passenger car tire sizes are based on three measurements: inside diameter, section width, and aspect ratio. A tire's inside diameter corresponds to the wheel size the tire fits, for example a 13-, 14- or 15-″ wheel; a few tires have metric wheel diameters, such as 360 millimeters. The **section width,** figure 16-3, is the width of the tire at its widest point. A wider section width provides a wider tire tread, although the two measurements are not the same. The **aspect ratio** of a tire is the ratio of its section height to its section width expressed as a percentage, figure 16-4. The higher the aspect ratio, the taller and narrower the tire. Most modern tires have aspect ratios between 45 and 80 percent.

The values in figure 16-4 are expressed in a size designation molded into the sidewall of the tire. The designation may also include additional information about the performance of the tire. The markings on the tire sidewall give information on the following:

- Tire size
- Speed rating
- Load index
- Temperature, traction, and tread wear ratings
- Mud and snow designation
- Maximum pressure
- Type of construction
- Manufacture date.

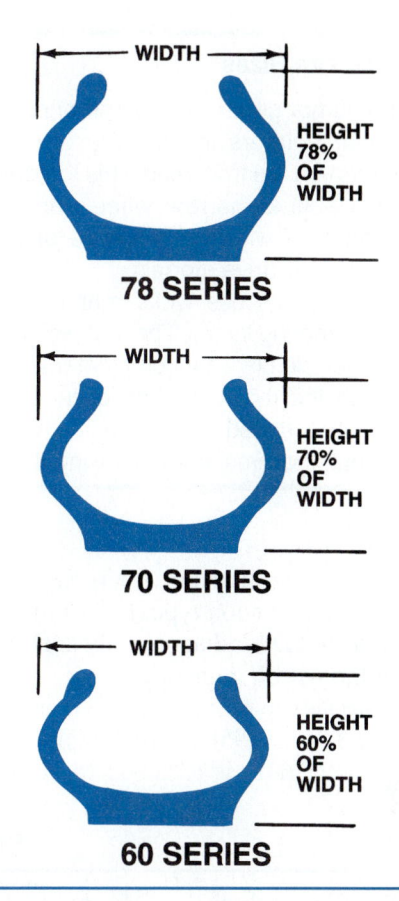

Figure 16-4. The aspect ratio describes the relationship between tire section width and height.

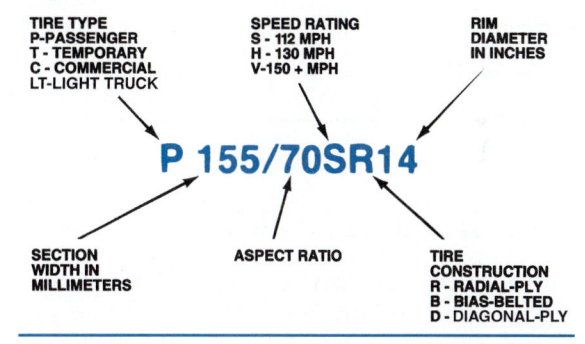

Figure 16-5. The P-metric tire size designation system.

When considering the tire–brake relationship, the size, speed rating, and load index should be considered.

P-Metric Size Designations

In the P-metric system the tire size is written in the format "PWWW/AATDD," for example, P155/70SR14, figure 16-5. P is the tire type: P for passenger cars, LT for light truck, C for commercial, and T for temporary spare. WWW is the tire

Unequal Tire Sizes

Although it is not generally recommended, some vehicle manufacturers install wider tires on the rear axle than on the front axle. This increases the traction available at the rear wheels and is done primarily on cars with rear engines or powerful front-engine, rear-drive models.

Rear-engine cars use wider rear tires to help counteract cornering forces. The rear weight bias of this type of vehicle places much greater lateral traction loads on the rear tires than occur in a front-engine car. The added traction of the wider tires helps prevent the tires from breaking loose and causing a spin. Powerful front-engine, rear-drive cars use wider rear tires to help put their horsepower to the ground while cornering and accelerating.

Vehicle manufacturers that use different size tires on the front and rear axles are careful to design and proportion their braking systems so proper brake balance is maintained. In both types of cars described, the increased traction of wide rear tires can help improve braking. This is especially true in rear-engine cars, where the rear weight bias allows the rear tires to provide a much larger part of the total braking force.

section width in millimeters. AA is the aspect ratio. T is the construction type: R for radial-ply, B for bias-ply, and D for diagonal-ply. DD is the diameter of the wheel, in inches.

Older tire size markings may be marked without an aspect ratio (195R13, for example). In this case the aspect ratio is assumed to be 80. This older marking system may include the aspect ratio and speed rating as part of the tire size, 195/60VR15, where 60 is the aspect ratio and V is the speed rating. In some cases the load rating and the speed rating may be combined, 195/55R15 84V, where 84 is the load rating and V is the speed rating.

Speed Rating Table

Rating	N	Q	S	T	U	H	V	Z	W	Y
Maximum Speed	88	100	112	118	124	130	150	Over 150	169	188

Load Rating Table

Rating	75	82	84	85	87	88	91	92	93	105
Capacity (lbs.)	853	1047	1102	1135	1201	1235	1356	1389	1433	2039

Tire Size and Braking

There are two factors to consider when examining the effects of tire size on braking: the area of the tire contact patch and the overall diameter of the tire. Separately or in combination, differences in these values can create unusual braking behavior.

Contact Patch Area

Chapter 3 explained how the coefficient of friction between two surfaces is not affected by the size of the area where they come into contact, except where material is transferred from one surface to the other. This is what happens between a tire and the road. All other things being equal, a tire with a large contact patch has a higher coefficient of friction, and, therefore, greater traction than a tire with a small contact patch. This means it can provide more braking power and accept greater brake application force before it begins to skid.

If tires with different contact patch areas are mounted on the same axle, the tire with the smaller contact patch will lock first. The car will then try to pivot around the tire with the larger contact patch, causing a pull to that side. When tires with different contact patch areas are mixed front to rear, the tires that have smaller contact patches will again lock first, resulting in a problem with brake balance. Front-brake lockup will cause a loss of steering; rear-brake lockup will affect vehicle stability and may cause a spin.

There is no practical way to determine the contact patch area of a tire, so the best way to avoid

The Brake System and Vehicle Suspension 375

problems is to mount tires of the same size and type on all four wheels unless otherwise specified by the vehicle manufacturer. If this is not possible, the same size tires must be used on each axle. However, keep in mind that tires with the same size designation from two different manufacturers may not have exactly the same contact patch areas because of differences in construction.

Tire Diameter

The overall diameter of a tire also affects braking action because a larger tire has a longer radius that exerts greater leverage, or torque, on the wheel friction assembly. As a result, a larger tire requires higher brake application force to create the same amount of friction at the road. If tires with different diameters are mounted on the same axle, all other factors being equal, the smaller tire will lock first, causing a pull toward the side with the larger tire. When tires with different diameters are mixed front to rear, the result is usually a problem with brake balance. The brakes on the axle with the smaller-diameter tires will lock first.

Tire Diameter and ABS

It is especially important to maintain the proper tire and wheel sizes on vehicles with ABS. Because the wheel speed sensor information is determined by the rolling diameter of the tires, any change in tire size will change the information from the wheel sensor. Even on high-performance vehicles that may use 16″ wheels on the front and 18″ wheels on the rear, the ABS software is programmed for this difference.

TIRE TREAD DESIGN

The very best dry-pavement traction, for both braking and cornering, is achieved with racing "slicks," tires whose tread is perfectly smooth. Although this type of tire puts all of its tread rubber in contact with the road, it is impractical for street use because it causes the car to **hydroplane** on wet pavement.

To maintain traction in poor weather, all street tires have a tread pattern, figure 16-6, that consists of ribs and grooves that allow water to be displaced from under the tire. Generally, a more open tread pattern displaces water better than a closed tread pattern.

In addition to the openness of the tread pattern, braking traction is also affected by the stability of the individual tread elements that contact the road. Small tread elements squirm a great deal under braking and tend to reduce traction. Larger tread elements are more stable and do not move around as much, which increases the traction they can provide.

For each type of tire, manufacturers try to strike the best possible compromise between dry and wet road traction. Generally, open tread patterns with small individual tread elements (such as those on all-weather or snow tires) work well on wet roads but have less traction on dry pavement. Tread patterns with less open areas and large individual tread elements (such as those on high-performance tires) do not work as well in the wet but have much greater traction on dry pavement.

Because tread life is less of a concern with performance tires, they often use softer tread rubber compounds that help restore some of the wet-road traction that their tread patterns sacrifice for dry-road handling. In addition, some performance tires have unidirectional tread patterns designed to rotate in a specific direction to help channel water out from under the tire.

Tire Tread Design and Braking

Just as with tires of different construction or size, uneven braking will result if tires with different

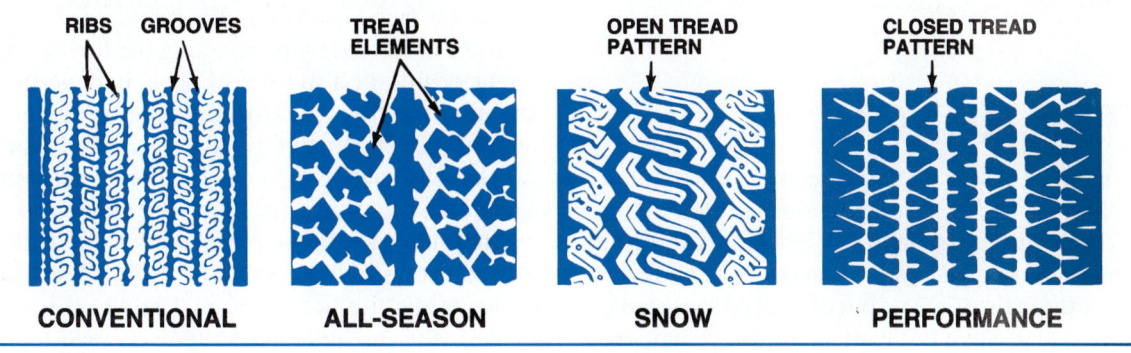

Figure 16-6. Tires come in a variety of tread patterns, depending on their intended use.

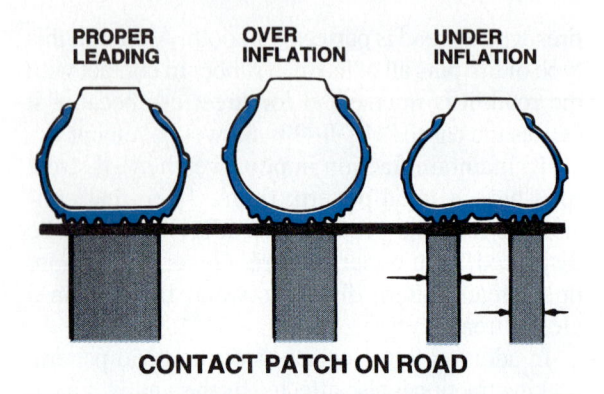

Figure 16-7. Improper tire inflation pressures affect the size of the contact patch.

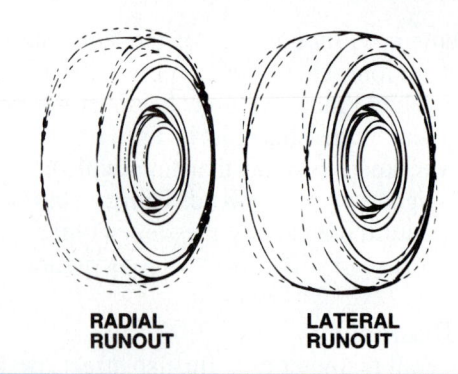

Figure 16-8. Tires suffer from both radial and lateral runout.

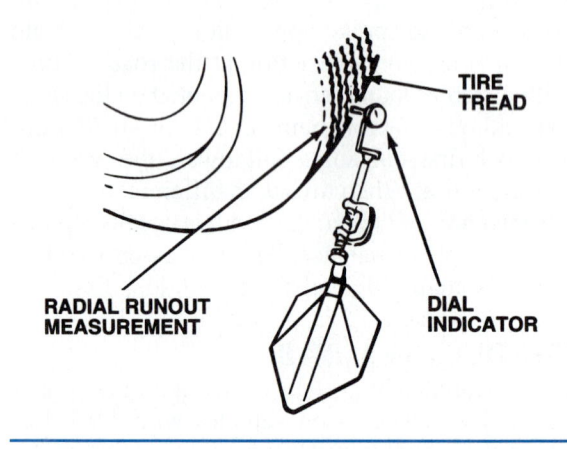

Figure 16-9. Measuring tire radial runout.

tread patterns are mixed. A pull to one side during braking results from side-to-side mixing, and brake balance is upset by a front-to-rear mix. The car will usually pull toward tires that have more closed and stable tread patterns because those types of tires have higher coefficients of friction and are the last to lock.

TIRE INFLATION

Although it is not actually a part of the tire, the air inside is an important consideration in both the design and braking performance of a tire. In fact, improper inflation is one of the most common causes of brake pull. Every tire is designed to operate within a narrow range of pressures, and improper inflation, whether lower or higher than the recommended value, affects braking because it reduces the size of the tire contact patch, figure 16-7, and, therefore, traction. Like most other tire-related braking problems, improper inflation can cause either a pull or an imbalance, depending on whether the traction difference is side to side or front to rear.

TIRE RUNOUT

Even if the tires on all four wheels are identical, it is still possible to have tire problems that affect braking. Other than improper inflation, the most common of these is runout. Runout is a condition that causes a tire to rotate out of true. All tires have some runout, but a tire with an excessive amount cannot be accurately balanced and will not maintain proper contact with the road, thus reducing traction. There are two basic forms of tire runout:

- Radial
- Lateral.

Radial Runout

Radial runout, figure 16-8, is when the tire tread surface is out of round. Because wheels suffer runout the same as tires, radial runout, can sometimes be corrected by remounting the tire in a different position on the wheel. This is done in an attempt to get tire and wheel runout to counteract one another. In more extreme cases, radial runout can be eliminated by shaving the tread round again on a tire truing machine. The amount of radial runout is determined by placing a dial indicator against a smooth portion of the tread near the center of the tire, figure 16-9. The wheel is then rotated, and the amount of runout is read on the

The Brake System and Vehicle Suspension

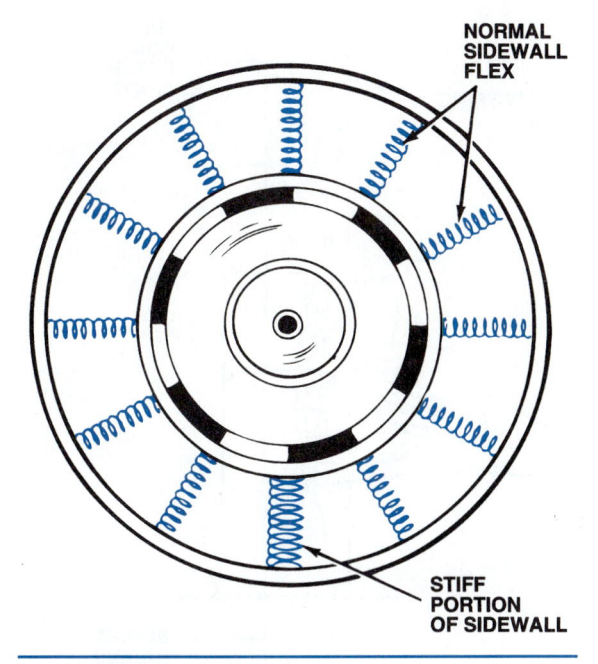

Figure 16-10. Loaded runout causes the tire sidewall to act like a variable spring.

measuring device. Total radial runout should not exceed .060″ (1.5 mm).

Loaded Runout

Loaded runout is a form of radial runout caused by variations in the stiffness of the tire sidewall. These variations are created in manufacturing, and prevent the tire from flexing normally and rolling concentrically, figure 16-10. A tire with loaded runout appears round when checked in an unloaded state. The latest model computer-driven wheel balancing equipment can measure loaded runout and help the technician determine whether the tire should be replaced.

Lateral Runout

Lateral runout, figure 16-8, is when the tire sidewall does not run true. As with radial runout, it may be possible to change the tire position on the rim to correct for small amounts of lateral runout. In extreme cases, however, the tire must be replaced.

Lateral runout is measured in a manner similar to radial runout, except that the dial indicator is placed against a smooth area of the tire sidewall near its widest point, figure 16-11. Lateral runout should not exceed .080″ (2.0 mm).

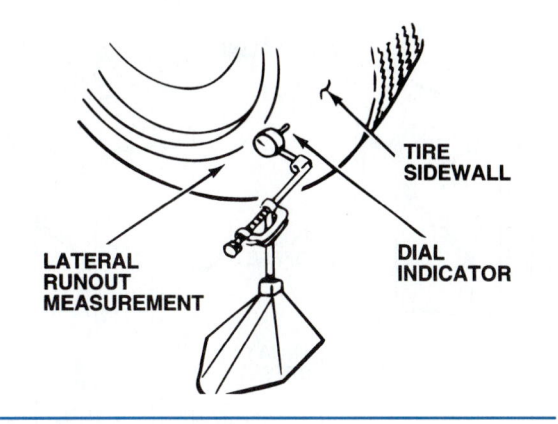

Figure 16-11. Measuring tire lateral runout.

WHEELS AND BRAKING

The wheels on a car affect the suspension geometry, as well as the size and shape of the tire contact patches. Although these affects take place at all four wheels, they are far more significant at the front suspension, which must provide both steering and braking control. There are two areas of wheel design that affect brake performance:

- Offset
- Rim width.

The wheels, hubs, suspension, and brakes are designed by engineers to work together as a package when installed on the vehicle. Often, the manufacturer that supplies the braking system will supply the whole brake/suspension unit as an assembly. Any modification to wheel width and offset can change braking performance, sometimes for the better, sometimes for the worse.

Wheel Offset

Wheel offset is the distance between the centerline of the rim and the mounting plane of the wheel, figure 16-12. When the wheel mounting plane is outboard of the rim centerline, the wheel has positive offset; most cars have wheels with positive offset. When the wheel mounting plane is inboard of the rim centerline, the wheel has negative offset. Negative offset is found primarily in "deep-dish" aftermarket wheels sold to widen the vehicle track and enhance appearance. The

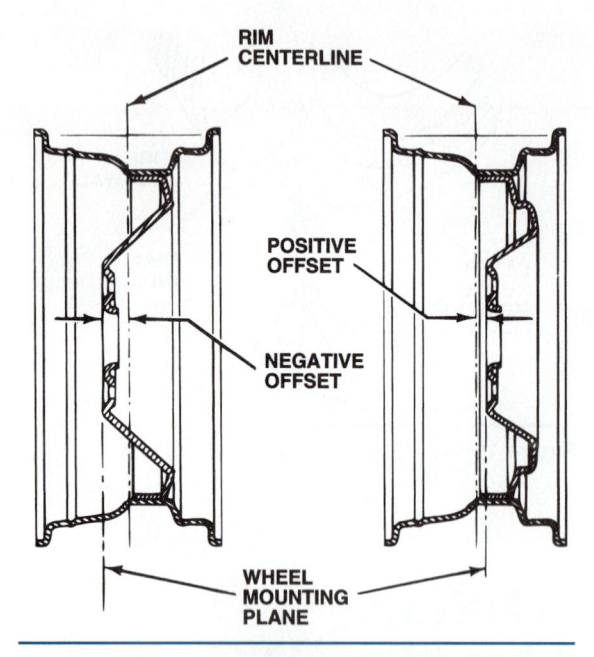

Figure 16-12. Wheel offset is an important factor in vehicle handling and braking.

amount and direction of any offset are important because they affect the load-carrying ability of the hubs, spindles, axles, and wheel bearings. The wheel and suspension are designed to work together so that vehicle weight is evenly distributed on the wheel bearings and other parts. If a wheel with nonstock offset is fitted, particularly one with more negative offset, increased bearing wear will result. In cases of extreme offset, axle or spindle failure may occur.

Scrub Radius

Wheel offset also affects the stability of the vehicle because it helps determine the **scrub radius** of the suspension. Scrub radius is the distance from the centerline of the tire contact patch to the point where a line drawn through the steering axis intersects the road, figure 16-13. Like wheel offset, scrub radius is either positive or negative. If the tire contact patch centerline is outboard of the point where the steering axis intersects the road, the car has positive scrub radius. If the contact patch centerline is inboard of the point where the steering axis intersects the road, the car has a negative scrub radius.

The type and amount of scrub radius for each particular vehicle are designed at the factory. Positive scrub radius pulls the tires outward to remove slack from the suspension; negative scrub radius pushes the tires inward for the same purpose. Generally, negative scrub radius provides more stable

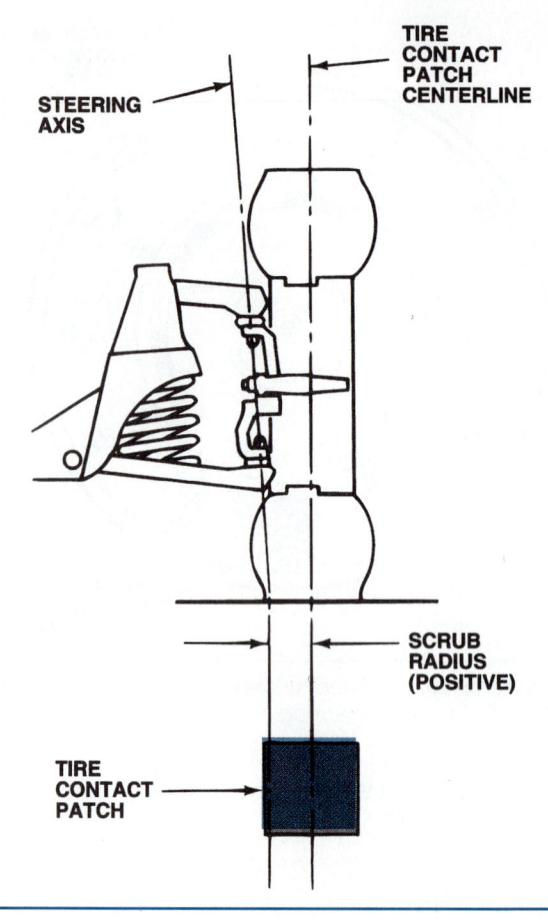

Figure 16-13. The scrub radius of the suspension can help maintain vehicle stability in an emergency.

handling in the event of a tire blowout or failure of one circuit in a diagonal-split brake hydraulic system. Because they must transmit braking, cornering, and acceleration forces through the front wheels, many FWD cars have negative scrub radius for more stable performance.

To provide a wider track and a better appearance, some aftermarket wheels have increased negative offset. This results in increased positive scrub radius, which can destabilize the vehicle, particularly in FWD cars. If the difference in offset is great enough to change the suspension geometry from negative to positive scrub radius, it may be very difficult to retain steering control if one circuit of a diagonal-split brake system fails.

Wheel Rim Width

As discussed earlier, the size of the tire contact patch affects traction. If tires with contact patches of different sizes are mounted on a car, the result can

The Brake System and Vehicle Suspension

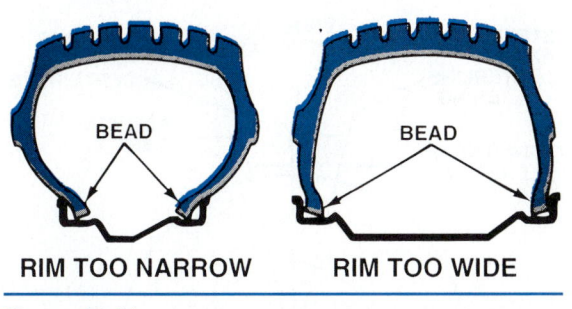

Figure 16-14. A tire on a rim of the incorrect width will neither seat nor seal properly.

be a brake pull to one side or problems with front-to-rear brake balance. Wheel rim width is important to braking because it affects the size of the tire contact patch and the stability of the tire under braking.

If a tire is mounted on a rim that is too narrow or too wide, the beads are distorted and the size of the contact patch is reduced. In addition, the tire will not seat properly on the rim, figure 16-14, and may flex excessively from side to side, making the car harder to control during a stop. In extreme cases, the tire can lose air or come off the rim entirely.

WHEEL RUNOUT

Wheels can be damaged in an accident, or even in a light brush with a curb. A damaged wheel, or one that has been incorrectly manufactured, will have radial or lateral runout similar to the tire problems already discussed. In fact, whenever tire runout is encountered, the wheel should be checked for runout as well.

Measured at the points shown in figure 16-15, radial runout should not exceed .040″ (1.0 mm) for steel wheels and .030″ (.75 mm) for aluminum wheels. Lateral runout should not exceed .045″ (1.1 mm) for steel wheels and .030″ (.75 mm) for aluminum wheels. Because hubs suffer runout as well, minor amounts of wheel runout can sometimes be compensated for by repositioning the wheel on the hub two or three bolt holes from the original location.

Wheel Runout and Braking

Slightly bent or distorted wheels can be hard to spot unless they are specifically checked for runout. Such wheels are usually difficult to balance or cannot be balanced at all. This results in vibrations and erratic tire wire—both of which reduce traction and

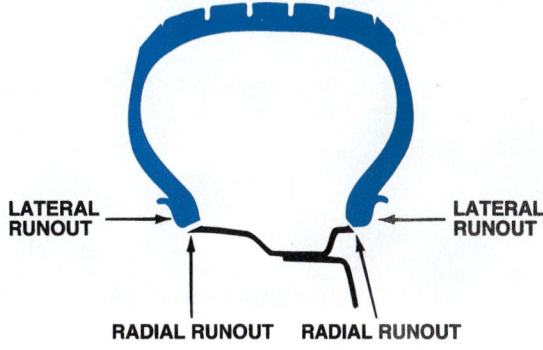

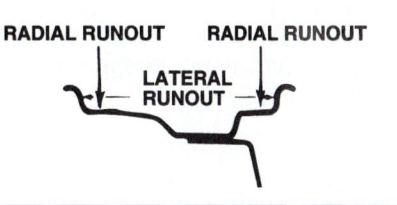

Figure 16-15. Checking locations used to measure wheel runout.

braking performance. Damaged wheels also prevent the front suspension from being properly aligned, which can contribute to pull under braking.

BRAKE SHROUDING

Brake shrouding occurs when the wheel prevents adequate cooling airflow over the friction assembly. Some shrouding is needed to protect the brake from rain, mud, and dust; however, too much can cause the brakes to overheat and fade, thus reducing stopping power.

In some cases, brake shrouding is a design problem; the wheels or wheel covers simply do not provide enough room or venting for adequate airflow over the drum or caliper. Some manufacturers address this problem with wheels and wheel covers that have vent holes or slots to improve brake cooling. Certain new wheel designs go one step further and work as fans to draw cooling air over the brakes, figure 16-16.

Brake shrouding may also result from aftermarket dust shields that install between the wheel and the brake friction assembly to keep the wheels free of brake dust. Although these shields do keep the wheels clean, they sometimes do so at the expense of brake cooling. Actually, the

Figure 16-16. A wheel designed to draw cooling air over the brake friction assembly.

brake dust that covers the wheels of some cars is an indication that ample cooling air is flowing over the brakes.

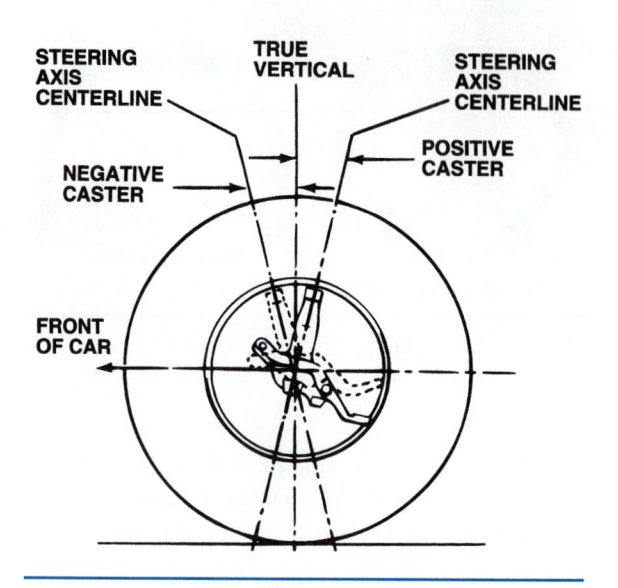

Figure 16-17. Side-to-side variation in front-wheel caster can cause a braking pull.

SUSPENSION AND BRAKING

A suspension that is in good condition and properly adjusted is essential for maximum traction and vehicle stability. When the suspension is in good condition, it damps the bumps in the road, equalizes tire wear, enables the tires to follow the contours of the road more closely, keeps the tire contact patches in better contact with the pavement, and holds the vehicle stable so it is easy to stop and steer in a controlled manner. Any problem with the suspension, such as component wear or damage or incorrect wheel alignment, prevents the automobile from operating as it was designed and reduces braking performance.

Suspension Wear and Damage

Wear in the suspension affects braking in a number of ways. Sagging springs or uneven torsion bar adjustment can cause the vehicle to pull to one side during braking. If the shock absorbers are worn, the car will move around on its suspension too much, making it difficult to modulate the brake pedal for smooth stops. Worn ball joints, tie-rod ends, or suspension bushings allow the suspension components to shift position when loaded under braking, and this can cause the car to dart or steer to one side or another. Finally, a worn suspension cannot be aligned accurately.

Wheel Alignment

Even though the suspension may be in perfect condition, a car still may not stop well if the wheel alignment is out of specification. Although incorrect camber and toe affect tire wear, it is primarily large **caster** variations, on the order of several degrees, that cause braking problems. This is particularly true if the wheel on one side of the car has positive caster while the wheel on the other side has negative caster, figure 16-17. Problems with caster variation may cause the car to dart or pull toward the side with positive caster during braking.

WHEEL BEARINGS AND BRAKING

The final links between the suspension and the wheel and tire are the wheel bearings. If these bearings are worn and out of adjustment, braking performance will be affected. There are actually three types of wheel bearings on cars today:

- Straight roller bearings
- Ball bearings
- Tapered roller bearings.

Straight cylindrical roller bearings, figure 16-18, consist of rollers, a cage to space the rollers apart, and an outer race. Straight roller wheel bearings are found only on rear-driven axles and use the axle shaft itself as the inner race. Once straight

The Brake System and Vehicle Suspension

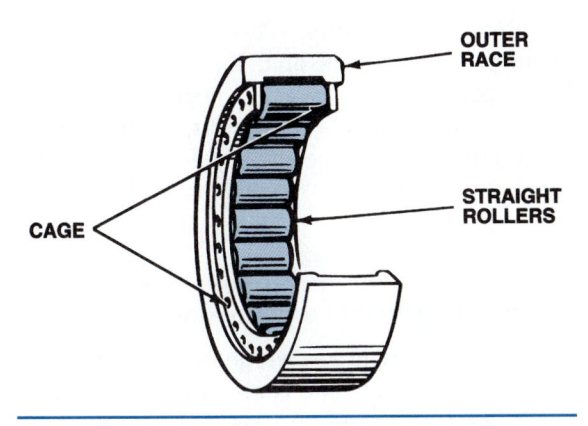

Figure 16-18. Straight roller bearings are used only on rear-drive axles.

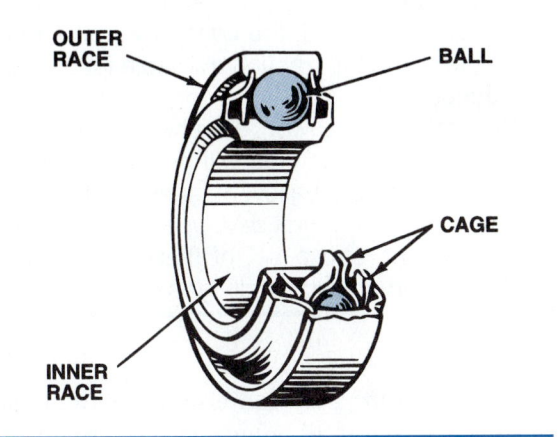

Figure 16-19. Ball bearings are used on both front and rear axles.

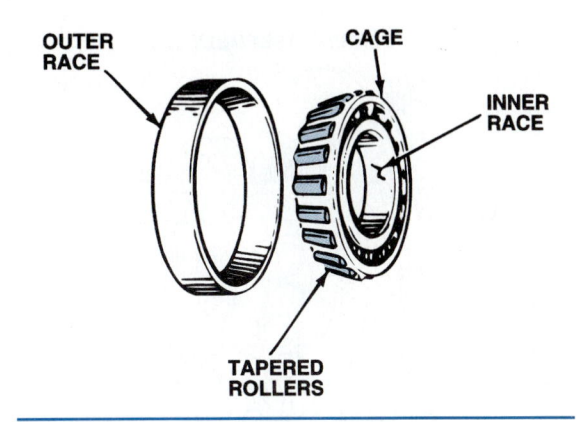

Figure 16-20. The tapered roller bearing is a common type of wheel bearing.

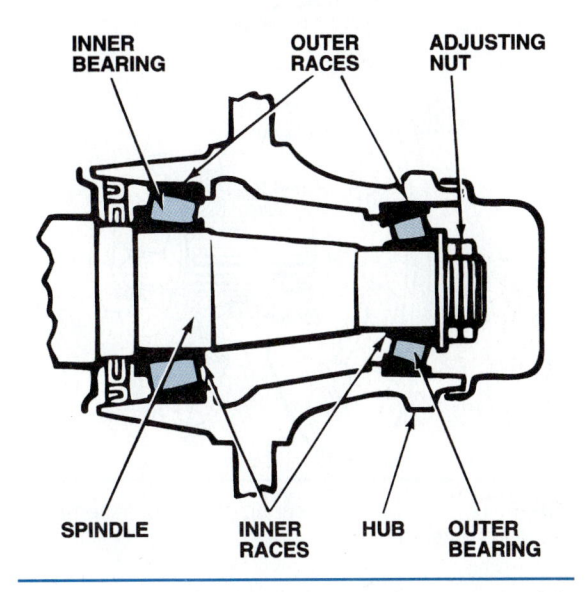

Figure 16-21. Tapered roller bearings are used in pairs for wheel-bearing service.

roller bearings are pressed into place, they need no adjustment. And because they are lubricated by rear axle oil, they also do not require periodic lubrication.

Ball bearings, figure 16-19, consist of an inner race, spherical steel balls, a cage to space the balls apart, and an outer race. Ball bearings are used on both front and rear axles, and double-row ball bearings are used on the front wheels of many FWD cars. Most modern ball bearings are permanently sealed and do not require periodic lubrication. Ball bearings are also nonadjustable; if they exhibit play outside the specified limits, they must be replaced.

Tapered roller bearings, figure 16-20, consist of an inner race, tapered cylindrical rollers, a cage to space the rollers apart, and an outer race. Pairs of tapered roller bearings (one inner bearing and one outer bearing) are commonly used on non-driven axles at both the front and rear ends of the car, figure 16-21. Most tapered roller bearings must be periodically cleaned, repacked with grease, and adjusted for end play.

Some cars use sealed, double-row ball or tapered roller bearing assemblies, figures 16-22 and 16-23, that integrate the wheel hub, bearings, and races. This type of bearing is adjusted, lubricated, and sealed during manufacture. If it wears out, the entire assembly is replaced as a unit.

Wheel Bearing Adjustment and Braking

Wheel bearings are important to braking because loose wheel bearings allow excessive runout of the disc brake rotors. This causes increased brake pad knockback and, therefore, increased brake pedal travel. If the wheel bearings are loose on only one side of the car, brake application on that side will be slightly delayed, causing a pull to the opposite side when the brakes are first applied.

■ Unsprung Weight and Braking

Unsprung weight is any weight not supported by the vehicle springs. The wheels, tires, brake friction assemblies, and springs make up the bulk of a car's unsprung weight. In functional terms, unsprung weight is mass the suspension cannot fully control. When wheel and tire movement is not fully controlled, tire contact with the road is affected, and the traction available for braking is reduced.

Generally, the lower a vehicle's unsprung weight, the better its tires will remain in contact with the road. Light-alloy magnesium and aluminum wheels were first developed to reduce unsprung weight, and a set of these wheels can contribute to improved handling as well as braking.

Wheel bearing play can also affect front-end alignment. Even if all alignment settings are adjusted correctly, loose wheel bearings will alter the settings as soon as the car rolls down the road. An overtightened wheel bearing can also lead to premature bearing failure and increased drag on one side of the car. This will cause a pull during braking.

Wheel Bearing Seals and Braking

All wheel bearings are lubricated with either grease or the oil used in the rear axle. These lubricants are contained in the bearings by grease seals. If a worn or damaged seal allows the bearing lubricant to escape, the lubricant may contaminate the brake linings. Contaminated brake linings have greatly reduced stopping power, which usually makes the car pull toward the opposite side under braking. In some cases, contaminated linings can make the brakes grab or cause premature lockup.

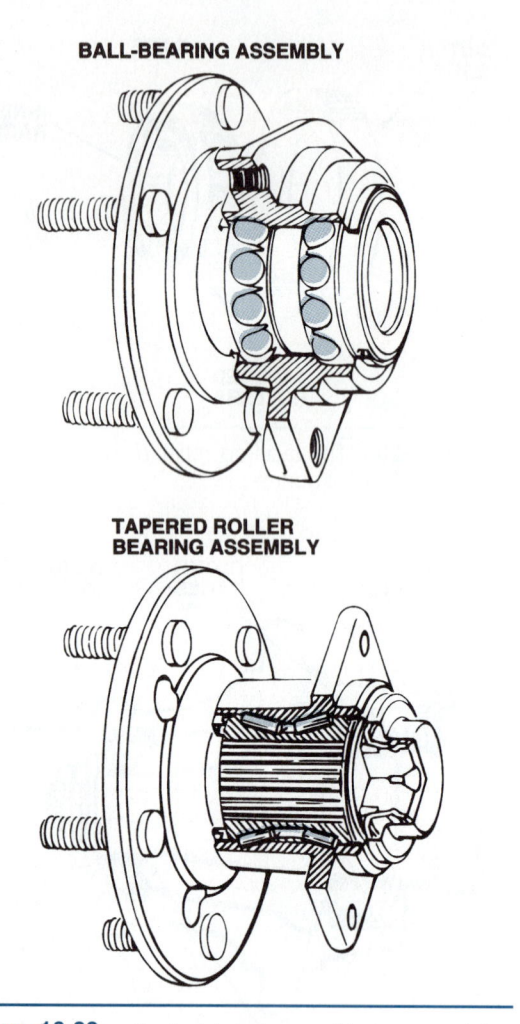

Figure 16-22. Sealed bearing-and-hub assemblies are used on some newer cars.

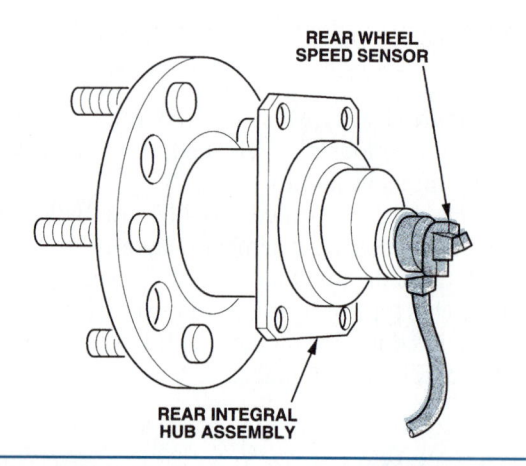

Figure 16-23. This sealed hub assembly also contains the antilock brake system wheel sensor. (Courtesy of General Motors Corporation, Service and Parts Operations)

The Brake System and Vehicle Suspension

SUMMARY

Tire traction is the limiting factor in modern brake performance. Traction is affected by tire construction, size, and tread design. The three types of passenger car tire construction are radial-ply, bias-ply, and bias-belted. Tire construction affects braking because each type provides a different amount of traction and responds with different speed and force to braking loads.

Tires are sized by inside diameter, sectional width, and aspect ratio. These values are molded into the sidewall of the tire in a size designation. They use the P-metric system. Tire size affects braking in two ways: It changes the size of the contact patch with the road, and a larger diameter tire exerts greater leverage on the friction assembly.

All tires used on passenger cars have a tread pattern so that water can be channeled out from under the tire in wet weather. Tread patterns affect braking because they are a compromise between wet- and dry-road traction; open patterns that work well in the wet reduce the amount of rubber in contact with the road, and thus dry-weather traction.

In general, tires should not be mixed by construction, size, or tread design. Side-to-side variations can cause a pull during braking. Front-to-rear variations can cause an imbalance that will lock the brakes on one axle before those on the other.

Other than mounting mismatched tires, braking problems can also be caused by improper inflation or tire runout. Improper inflation reduces the size of the tire contact patch, and thus traction. Tire runout, whether radial, loaded, or lateral, affects the true rotation of the tire, and reduces contact between the tire and road, and thus traction.

Wheels can cause erratic braking if they have the incorrect offset or width. Incorrect offset increases wear on the wheel bearings and, in severe cases, may cause suspension failure. Incorrect offset also affects the scrub radius of the suspension, which can create major instability problems in the event of a tire blowout or brake failure.

Wheels with the incorrect rim width create braking problems because the tire does not fit properly. This reduces the size of the contact patch and allows the tire to flex excessively, making the car difficult to control. Wheels also suffer from runout, and, just as with runout in tires, this creates problems in keeping the tire contact patch in firm contact with the road. A wheel that shrouds the friction assembly may limit the amount of cooling airflow and lead to overheating and brake fade.

A suspension that is worn or damaged, misaligned, or has wheel bearings that are out of adjustment will cause braking performance to suffer. Suspension wear allows the car to move around excessively when the brakes are applied; this wear affects wheel alignment as well. The main wheel-alignment-related braking problem is a pull caused by extreme caster variation from one side of the car to the other.

Three types of wheel bearings are used on today's cars: straight roller bearings, ball bearings, and tapered roller bearings. Tight wheel bearings can cause a brake pull. Loose wheel bearings throw off alignment settings and may allow excessive brake rotor runout. Usually, only certain tapered roller bearings can be adjusted; all other wheel bearings must be replaced if they become worn. Leaking wheel bearing grease seals can cause the brakes to pull on application or lock unexpectedly.

Review Questions

Choose the letter that represents the best possible answer to the following questions:

1. Which of the following tire factors does not affect dry-road braking performance?
   a. Size
   b. Rotation direction
   c. Ply arrangement
   d. Tread shape

2. Radial tires have better traction because they:
   a. Have more flexible sidewalls
   b. Are belted tires
   c. Both a and b
   d. Neither a nor b

3. Which of the following is an allowable front-to-rear mix of tires with different construction?
   a. Bias-belted front, bias-ply rear
   b. Radial-ply front, bias-belted rear
   c. Bias-belted front, radial-ply rear
   d. None of the above

4. A tire's aspect ratio is:
   a. Expressed as a percentage
   b. The ratio of section width to tire diameter
   c. Both a and b
   d. Neither a nor b

5. In the tire size designation "P195/50VR15," the number "195" is the:
   a. Speed rating
   b. Tread width
   c. Metric rim diameter
   d. None of the above

6. A tire with a larger contact patch:
   a. Provides more traction than a tire with a smaller contact patch
   b. Allows the brakes to be applied harder without locking
   c. Both a and b
   d. Neither a nor b

7. Which of the following is an allowable front-to-rear mix of tires with different diameters?
   a. 78 series front, 80 series rear
   b. 80 series front, 60 series rear
   c. 50 series front, 60 series rear
   d. None of the above

8. An open tread pattern with many small individual tread segments:
   a. Will provide the best dry-road stopping power
   b. Offers superior tread stability under braking
   c. Displaces water quickly from under the tire
   d. Is made with a very soft rubber compound

9. Improperly inflated tires:
   a. Can create front-to-rear brake imbalance
   b. Reduce tire contact patch area
   c. May cause a brake pull to one side
   d. All of the above

10. Loaded tire runout:
    a. Is caused by tire wear
    b. Is a form of lateral runout
    c. Both a and b
    d. Neither a nor b

11. Wheel offset:
    a. Is positive when the rim center line is outboard of the wheel mounting plane
    b. Affects the distribution of load on the wheel bearings
    c. Affects the suspension scrub radius
    d. All of the above

12. Wheel rim width is important to braking because it affects the:
    a. Size of the tire contact patch
    b. Stability of the tire on the rim
    c. Both a and b
    d. Neither a nor b

13. Which of the following is not a problem caused by wheel runout?
    a. Brake shrouding
    b. Difficulty balancing tires
    c. Reduced traction
    d. Inaccurate wheel alignment

The Brake System and Vehicle Suspension

14. The primary wheel alignment problem that can contribute to brake pull is side-to-side:
    a. Camber variation
    b. Toe change
    c. Positive scrub radius
    d. None of the above

15. Incorrect wheel bearing adjustment can cause:
    a. Brake rotor runout
    b. An intermittent pull during braking
    c. Increased brake pedal travel
    d. All of the above

Glossary of Technical Terms

Accumulator: A container that stores hydraulic fluid under pressure. A heavy metal spring or compressed gas behind a diaphragm provides the opposing force.

Active Digital or Magneto-Resistive Wheel Speed Sensor: A type of wheel speed sensor that must be powered by the brake control module before it can produce a wheel speed signal.

Actuating System: The parts of a brake system that transmit the braking force applied at the brake pedal to the wheel friction assemblies and increase it to a usable level.

Air Chambers: The parts of an air brake system that convert air pressure into the mechanical force used to apply the wheel friction assemblies.

Air Gap: The distance or clearance between the end of a wheel speed sensor and its toothed ring.

Air-Hydraulic System: An air brake system that uses a single air chamber to power a hydraulic master cylinder that applies the wheel friction assemblies through conventional brake calipers and wheel cylinders.

Anchor Eyes: The semicircular notches at the ends of some shoe webs where they contact the shoe anchor on the backing plate.

Anchor Plate: A bracket, solidly attached to the vehicle suspension, on which a floating or sliding caliper mounts.

Annular Electric Brake: An electric brake design in which the electromagnet is shaped like a circle or ring (annulus) inside the brake shoes.

Anodize: A method of electrolytically coating a metal surface with a protective oxide.

Aramid Fiber: A synthetic product used as a reinforcing agent in brake linings. Aramid fiber has impressive wear properties.

Arcing: A grinding process that machines drum brake shoe linings to the proper curvature for a given drum size and brake design.

Asbestos: The generic name for a group of minerals used in brake friction materials and made up of millions of individual fibers. The fibers pose a serious health hazard if inhaled or ingested.

Asbestosis: A progressive and disabling lung disease caused by inhalation of asbestos fibers over an extended period of time.

Aspect Ratio: The ratio of tire section height to section width expressed as a percentage.

Atmospheric Pressure: The pressure on the Earth's surface caused by the weight of air in the atmosphere. At sea level, this pressure is 14.7 psi at 32°F (0°C).

Atmospheric-Suspended Power Chamber: A booster power chamber that has atmospheric pressure on both sides of its diaphragm when the brakes are not applied.

Automatic Adjusters: Brake adjusters that use shoe movement, or parking brake application to continually reset the lining-to-drum clearance.

Axial Cooling Fins: Brake drum cooling fins that are perpendicular to the centerline of the axle.

Backing Plate: The fixed part of a drum brake that is attached to the vehicle suspension. The wheel cylinder and brake shoes are mounted on the backing plate.

Banjo Fitting: A banjo-shaped connector with a hollow bolt through its center that enables brake lines to exit hydraulic components at a right angle.

Barrel Wear: A type of brake drum wear in which the center of the friction surface is worn more than the edges.

Bellmouth: A form of brake drum distortion in which the open edge of the drum has a larger diameter than the closed edge.

Bimetallic Brake Drum: A drum with an aluminum outer drum cast around a preformed iron liner.

Binders: Glues used to hold the various elements of a friction material together.

Bleeder Screw: The hollow screw that is tightened to close a bleeder valve and loosened to open the valve, allowing air and contaminated fluid to be flushed from the brake hydraulic system.

Bleeder Valve: A valve on the wheel cylinders and brake calipers that is opened when purging the hydraulic system of air.

Body Control Module: An onboard computer system that controls various vehicle functions, such as remote door locks, power windows, seat heaters, parking brake release, theft deterrent, interior and exterior lighting, and many other functions.

Bonded Linings: Linings that attach to the lining table or backing plate with a high-temperature adhesive cured under heat and pressure.

Booster Holding Position: The point at which a booster maintains a constant level of power assist.

Booster Vacuum Runout Point: The point at which a vacuum booster is supplying the maximum amount of power assist possible. Any additional braking force must be applied at the brake pedal.

Brake Adjuster: A device or mechanism used to set and maintain the proper clearance between the brake lining and drum.

Brake Balance: The split of braking power between the front and rear axles. Proper brake balance allows the driver to use the full braking ability at all four wheels when stopping the car.

Brake Band: A flexible circular metal band that surrounds a brake drum and contracts around it to create the friction necessary to stop the vehicle. The rubbing surface of a brake band is faced with some type of brake lining material.

Brake Block: A molded piece of friction material. Brake block is attached to shoes and pads to serve as the brake lining.

Brake Caliper: The device in a disc brake that converts hydraulic pressure back into the mechanical force used to apply the brake pads against the rotor.

Brake Drum: The rotating part of a drum brake assembly that is contacted by the brake shoes to create the friction necessary to stop the vehicle.

Brake Dust: The dust created when brake friction materials wear during use. This dust contains hazardous levels of asbestos fibers and is the most common source of asbestos exposure.

Brake Fade: The partial or total loss of braking power that occurs when excessive heat reduces friction between the brake linings and the rotor or drum.

Brake Lathe: A machine that rotates drums or rotors and uses cutting bits to remove metal from the friction surfaces.

Brake Lines: The network of steel tubing and rubber hoses used to transmit brake system hydraulic pressure from the master cylinder to the wheel friction assemblies.

Brake Lining: A material that is good at generating friction and resisting heat. The lining is molded, bonded, or riveted to the rubbing surfaces of brake shoes and pads.

Brake Pads: The parts of a disc brake assembly that are faced with the lining material that rubs against the rotor to create the friction necessary to stop the vehicle.

Brake Shoes: The curved metal parts of a brake assembly that are faced with the lining material that rubs against the rotor to create the friction necessary to stop the vehicle.

Bridge Bolts: Special high-strength bolts used to join the halves of split brake calipers.

Burnish: The initial seating process in which the brake shoes or pads wear to conform to the exact contours of the brake drum or disc.

Cam: A stepped or curved eccentric wheel mounted on a rotating shaft. As a cam is turned, objects in contact with it are raised or lowered.

Cam Grind: A type of brake shoe arcing that produces a lining thinner at its ends than at its center.

Caster: The angle at which the steering axis leans forward (negative) or rearward (positive) from a perpendicular position.

Centrifugally Cast Brake Drum: A drum with a pressed-steel outer drum and a cast-in iron liner.

Clevis: A U-shaped bracket with holes in each end through which a retaining pin or bolt is inserted. A clevis is often used to attach the brake pedal pushrod to the pedal arm.

Coefficient of Friction: A numerical value expressing the amount of friction between two objects. It is obtained by dividing tensile force by weight force.

Compensating Port: The opening between the fluid reservoir and high-pressure chamber that allows fluid to enter or exit the hydraulic system to adjust for changes in volume.

Composite Brake Drum: A drum made from two different metals. All composite drums have cast-iron friction surfaces.

Controller: A device that uses a variable resistor to regulate current flow to an electric brake friction assembly based on hand, foot, hydraulic, or air pressure.

Control Module: The electronic microprocessor that monitors and regulates the operation of the ABS. Its primary inputs are from the wheel speed sensors and brake

Glossary of Technical Terms

pedal switch. Outputs control the operation of the ABS solenoids in the modulator, and the pump and motor assembly on systems that have a pump.

Core Charge: A surcharge added to the cost of a rebuilt component if the original cannot be rebuilt.

Corrosion: The chemical dissolving or etching away of metals. Also, the name given the residue left by the process.

Cup Seals: Circular rubber seals with a depressed center surrounded by a raised sealing lip. Cup seals can contain high pressure in one direction but do not seal in the other.

Curing Agents: Substances used in friction materials to ensure the various elements bond together properly into brake block.

Diaphragm: A flexible membrane used to isolate two substances or areas from one another. A rubber diaphragm is often used to isolate brake fluid in the reservoir from moisture in the air.

Discard Diameter: The largest inside diameter at which a brake drum can safely operate.

Discard Dimension: The thinnest width at which a rotor can safely operate.

Disc Brake: A type of brake in which friction is generated by brake pads rubbing against the sides of a brake rotor.

Dished Rotor: A brake rotor that is thinner at the inner edges of its friction surfaces. Dishing is a form of taper variation.

Double-Leading Brake: A nonservo brake in which both shoes are energized.

Double-Trailing Brake: A nonservo brake in which neither shoe is energized.

Drum Brake: A type of brake in which friction is generated by brake shoes rubbing against the inside of a brake drum.

Drum-in-Hat: A type of parking brake used with fixed and sliding calipers that uses the inside of the rotor hub (hat) as a brake drum, fitted with a pair of brake shoes, for the mechanical parking brake function.

Drum Web: The metal plate or structure that fills the closed edge of the drum.

Dual-Circuit Brake System: A brake system that actuates the wheel friction assemblies using two separate hydraulic circuits.

Dual Master Cylinder: A master cylinder that contains two pistons that supply the hydraulic pressure for a dual-circuit braking system.

Dual-Power Brake System: A system that uses both a vacuum booster and a hydraulic booster to increase brake application force.

Dual-Servo Brake: A drum brake that has servo action in both the forward and reverse directions.

Eccentric: Off center or out of round. A bolt whose center section is offset from its primary axis.

Eccentric Distortion: A form of brake drum distortion in which the geometric center of the circle described by the friction surface is offset from that of the axle.

Economic Commission for Europe (ECE): The committee established by the European Union for establishing governing economic policies and regulations for the European Union. These policies also govern the safety characteristics of vehicles manufactured for use in the European Union.

Edgebrand: A series of codes on the side of a brake lining that identify the manufacturer, the specific lining material, and the friction coefficient of the lining. The edgebrand is only for identification and comparison; it does not indicate lining quality.

Effective Pedal Travel: The portion of brake pedal travel converted to piston movement in the master cylinder.

Elastic Limit: The point beyond which a deformed piece of metal will no longer return to its original shape.

Electro-Hydraulic (E-H) Booster: A power booster that uses an electric motor and pump to create hydraulic pressure that is then used to increase brake application force.

Emergency Brakes: Another name sometimes given to the parking brakes because of their limited ability to stop the car in the event of total failure of the service brakes.

Energy: The ability to do work. Energy available to do work but not actually being used is called potential energy.

Equalizer: A bracket or cable guide used in a parking brake linkage to ensure both wheel friction assemblies receive equal application force.

Equilibrium Reflux Boiling Point (ERBP): The boiling point of a brake fluid as determined by a special test procedure. Both dry and wet ERBPs are used in evaluating brake fluids.

European Union (EU): An economic and political union of more than 15 countries in Europe established for developing common laws and regulations.

Fillers: Substances added to friction materials to obtain specific performance characteristics.

First Law of Thermodynamics: The natural law that states energy cannot be created or destroyed; it can only be converted into another form.

Fixed-Anchor Grind: A variation of undersize grinding that compensates for the size and location of a fixed shoe anchor.

Fixed Brake Caliper: A brake caliper whose body is solidly attached to the vehicle suspension and does not move when the brakes are applied.

Flash Codes: Diagnostic trouble codes that are read by counting flashes of the ABS warning lamp. The numeric value of the code corresponds to a specific problem or failure.

Floating Caliper: A caliper whose body moves on the anchor plate but does not make metal-to-metal contact with it. Floating calipers are supported by bushings and O-rings that slide on guide pins and sleeves.

Forsterite: A crystalline chemical component of brake dust created by the action of heat on asbestos.

Foundation Brakes/Base Brakes: The primary vehicle brake system (same as service brakes). This does not include the components of the antilock or traction control systems.

Freeplay: The portion of brake pedal travel not converted into piston movement in the master cylinder.

Friction: The resistance to motion between two surfaces in contact.

Friction Material: A blend of substances with a relatively consistent friction coefficient over a wide range of conditions. The friction materials used in car brakes are organic, metallic, semimetallic, and synthetic.

Friction Modifiers: Additives used to alter the friction coefficient of a brake lining material.

Gas Fade: Brake fade caused by hot gases and dust particles that reduce friction between the brake linings and drum or rotor under hard, prolonged braking.

Glazed Lining: An overheated brake lining with a smooth, shiny appearance. Glazed linings have reduced stopping power and cause noise.

Grinding: The process of using a brake lathe and a power-driven abrasive stone to remove metal from drums to refinish their friction surfaces.

Grommets: Ring-shaped parts made of a third material that prevent problems where two other dissimilar materials come into contact. Plastic brake fluid reservoirs attach to metal master cylinders with rubber grommets.

Gross Vehicle Weight Rating (GVWR): The manufacturer's specified maximum allowable weight for a vehicle, including passengers and cargo.

Ground: The wiring and connections that return electrical current to the battery. The ground is common to all circuits in a vehicle's electrical system.

Hard Spots: Circular, bluish/gold, glassy areas on drum and rotor friction surfaces where extreme heat has altered the structure of the metal. Hard spots are also called chill spots.

Heat Checking: Small cracks on the friction surface of the drum or rotor. Heat checks do not penetrate through the friction surface and can usually be machined out of it.

Homogeneous: Being of a similar nature. Homogeneous liquids blend together completely; no part of either liquid remains separate.

Hybrid Pad Sets: Brake pad sets that contain an organic pad for one side of the rotor and a semimetallic pad for the other.

Hydroplane: When a tire rolls on a layer of water instead of staying in contact with the pavement. Hydroplaning occurs when all of the water on the pavement cannot be displaced from under the tire tread.

Hygroscopic: An affinity or attraction for water. Polyglycol brake fluids are hygroscopic.

Inboard Disc Brakes: Brakes mounted near the differential rather than out at the wheels.

Inert: A substance that exhibits no chemical activity, or does so only under extreme conditions.

Inertia: The property of a body at rest to remain at rest, and a body in motion to remain in motion in a straight line unless acted on by an outside force.

Insulator: A material that prevents or slows the transfer of heat from one area to another.

Integral ABS: An antilock system that combines the functions of the master cylinder, power booster, and antilock unit. Integral antilock systems mount in place of the conventional master cylinder and power booster.

Integral Vacuum Booster: A vacuum booster that installs between the brake pedal and the master cylinder. Integral boosters are actuated by brake pedal movement.

Intercable Adjuster: An adjuster built into some parking brake cables that allows the outer housing to be made longer or shorter to adjust the parking brake.

Intermediate Lever: A parking brake linkage lever (other than the parking brake pedal, lever, or handle) used to increase parking brake application force.

Isolation Solenoid: A solenoid in an ABS modulator assembly that

Glossary of Technical Terms

isolates or holds hydraulic pressure in a given brake circuit during antilock braking.

Jam Nuts: Two nuts that are tightened against each other to lock them in position.

Kinetic Energy: The energy of mass in motion. All moving objects possess kinetic energy.

Kinetic Friction: The amount of friction that exists between two surfaces in motion.

Knockback: Brake caliper piston retraction caused by rotor runout driving a piston back into its bore when the brakes are released.

Lateral Runout: Side-to-side movement of the friction surfaces of a brake rotor, or a side-to-side variation in the true rotation of the tire sidewall.

Lathe-Cut: A process of cutting rubber seals to a precise shape on a rotating drum.

Leading Shoe: Any shoe in a non-servo brake that is energized by drum rotation.

Leading-Trailing Brake: A non-servo brake with one energized and one de-energized shoe.

Leverage: The use of a lever and fulcrum to create a mechanical advantage. The brake pedal is an automotive part that employs leverage.

Lightly Loaded Vehicle Weight (LLVW): The unloaded weight of the vehicle, plus a load of 386 pounds (180 kg), which includes the weight of both the test instrumentation and the driver.

Lining Fade: Brake fade caused by a drop in the brake lining coefficient of friction as a result of excessive heat. Lining fade can affect either disc or drum brakes.

Lining Table: The outermost part of the brake shoe that supports the brake lining friction material.

Loaded Runout: Radial runout that appears only when the tire is supporting the weight of the car. Loaded runout is caused by stiff sections accidentally built into the tire sidewall.

Manual Adjusters: Brake adjusters that require periodic resetting with a hand tool.

Mass: The property of an object that leads to the concept of inertia.

Master Cylinder: The device that converts mechanical pressure from the brake pedal into hydraulic pressure that is routed to the wheels to operate the friction assemblies.

Mechanical Advantage: The ratio of the force exerted to the force applied. A manual brake pedal may have a mechanical advantage, or pedal ratio, of 5 to 1.

Mechanical Fade: Brake fade caused by heat expansion of the brake drum away from the brake linings. Mechanical fade is not a problem with disc brakes.

Mechanical-Hydraulic Booster: A power booster that uses hydraulic pressure from the power steering pump to increase brake application force.

Metallic Linings: All-metal sintered brake linings once used in heavy-duty drum brake applications.

Micro-Inches: A measurement system used to express the roughness of a machined or ground surface.

Mold Bonded Linings: Brake pad linings with the friction material cured in place on a backing plate drilled to provide physical engagement. A bonding adhesive is also used between the backing plate and lining.

Multiplier Vacuum Booster: A vacuum brake power booster that installs between the master cylinder and the wheel friction assemblies. Multiplier boosters are actuated by hydraulic pressure from the master cylinder.

Natural Frequency: The frequency at which an object is most prone to vibration. In general, large heavy objects have low natural frequencies, whereas small light objects have high natural frequencies.

Nibs: Small indentations on the edge of the brake shoe lining table that contact the shoe support pads on the backing plate.

Nondirectional Brake: A non-servo brake in which both shoes are energized in either direction.

Nonintegral ABS: An ABS that uses a conventional master cylinder and vacuum booster with a separate hydraulic modulator assembly.

Non-Servo Brake: A drum brake in which each shoe is applied individually; the operation of one shoe has no effect on the other.

Occupational Safety and Health Administration (OSHA): A division of the U.S. Department of Labor that oversees and regulates matters affecting safety in the workplace.

Organic Linings: Linings made from friction materials whose main reinforcing agent is asbestos fibers.

Out-of-Round: A form of brake drum distortion in which the friction surface no longer forms an exact circle.

Oversize Brake Shoes: Brake shoes with thicker linings designed for use in drums that have been machined oversize.

Over-Travel Spring: A special assembly on some cable-actuated star-wheel automatic adjusters that prevents overadjustment or damage.

Oxidation: A form of corrosion caused by combining a substance with oxygen. Rust, more precisely called iron oxide, is a common type of oxidation.

Pad Wear Indicators: Devices that alert the driver when the brake linings have worn to the point where they need replacement. Wear indicators may be mechanical or electrical.

Parallelism: The state that exists when two surfaces are an equal distance apart at every point.

Parking Brakes: The secondary vehicle brake system that is controlled by a hand lever or foot pedal, operates on only two wheels, and holds a parked car in position.

Particulates: Small pieces (particles) of matter; dust is a common particulate.

Pawl: A hinged or pivoted part that fits into a toothed wheel to provide rotation in one direction while preventing it in the other.

Pedal Ratio: On a brake pedal, the ratio of foot pedal travel to pedal pushrod travel.

Power Booster: A vacuum-, hydraulic-, or electro-hydraulic-powered device that multiplies the mechanical pressure applied to the brake pedal by the driver and relays the increased force to the master cylinder.

Power Chamber: The main housing of a vacuum booster that is internally divided in half by a flexible diaphragm. Pressure differentials between the halves move the diaphragm and create application force.

Pressure: A measurement of the load placed on an object based on the amount of force applied to a specific area.

Pressure Differential: The difference in pressure between two areas. Vacuum boosters use a pressure differential to create brake application force.

Primary Shoe: The shoe in a servo brake that transfers a portion of its stopping power to the secondary shoe. The primary shoe provides approximately 30 percent of the total stopping power.

Quick-Take-Up Master Cylinder: A type of dual master cylinder that supplies a large volume of fluid on initial brake application to take up the clearance designed into low-drag brake calipers.

Quick-Take-Up Valve: The part of a quick-take-up master cylinder that controls fluid flow between the reservoir and the primary low-pressure chamber.

Radial Cooling Fins: Brake drum cooling fins that are parallel to the centerline of the axle.

Radial Runout: A variation in true rotation of the tire tread; an out-of-round condition.

Ratchet: A mechanism with interlocking teeth that allow movement in one direction but not the other.

Red Brake Warning Lamp (RBWL): The lamp on the dash or instrument cluster that warns of a failure in part of the base brake system. It is so-named to differentiate it from the ABS warning lamp which is usually orange.

Regenerative Braking System (RBS): An electrical energy system that is installed in an electric vehicle (EV) for recovering or dissipating kinetic energy and that uses the propulsion motor(s) as a retarder for partial braking of the EV while returning electrical energy to the propulsion batteries or dissipating electrical energy.

Release Solenoid: An ABS solenoid in the hydraulic modulator assembly that vents or releases pressure from a brake circuit during antilock braking.

Replenishing Port: The opening between the fluid reservoir and low-pressure chamber that keeps the chamber filled with fluid.

Residual Pressure: A constant pressure held in the brake hydraulic system when the brakes are not applied.

Resurfacing: The process of using a brake lathe and a spinning abrasive disc to remove minor damage and contaminants from the rotor friction surface. Resurfacing is also done after a rotor is turned to give the friction surfaces a nondirectional finish.

Riveted Linings: Brake linings that attach to the lining table or backing plate with copper or aluminum rivets.

Rotor: The rotating part of a disc brake assembly that is contacted by the brake pads to create the friction necessary to stop the vehicle; also called a brake disc.

Rotor Hat: The raised center section of a brake rotor designed to be installed on a separate hub assembly.

Runout: Side-to-side deviation in the movement of a spinning brake rotor.

Scan Tool: A handheld electronic tool that plugs into a vehicle's diagnostic connector or the ABS control module harness (with adapters) to access ABS diagnostic information. A scan tool may also be required on some applications to bleed the ABS modulator.

Scoring: Extreme wear on a drum or rotor friction surface. Scoring is often caused by metal-to-metal contact or foreign materials between the linings and friction surface.

Scrub Radius: The distance between the centerline of the tire con-

Glossary of Technical Terms

tact patch and the point where a line drawn through the steering axis intersects the road.

Secondary Shoe: The shoe in a servo brake that receives extra application force from the primary shoe. The secondary shoe provides approximately 70 percent of the total stopping power.

Section Width: The dimension obtained when a tire is measured from one sidewall to the other across its widest point.

Select-Low Principle: The controlling principle for rear-wheel antilock systems that states that pressure to both wheels shall be limited to the level required by the wheel with the least traction.

Self-Energizing Action: A characteristic of drum brakes in which the rotation of the drum increases the application force of a brake shoe by wedging it tighter against the drum.

Semimetallic Lining: A heavy-duty lining molded from iron powder, steel fibers, and organic materials. Semimetallic linings work especially well at high temperatures.

Series: A method of connecting several parts in a row so that one feeds into the next.

Service Brakes: The primary vehicle brake system that is controlled by the brake pedal, operates the friction assemblies at all four wheels, and slows or stops the car in normal driving.

Servo Brake: A drum brake that uses the stopping power of one shoe (primary) to help increase the application force of the other shoe (secondary).

Shoe Web: The portion of the brake shoe below the lining table that receives the application force from the wheel cylinder.

Sintering: The process of fusing a metal-powder mixture together under high heat and pressure.

Sliding Caliper: A caliper whose body moves on the anchor plate and makes metal-to-metal contact with it. Sliding caliper bodies move on machined ways and are retained by special keys.

Slope: The percentage of full hydraulic system pressure supplied to the rear brakes by the proportioning valve. Slope is expressed as the ratio of rear pressure to front pressure.

Solenoid: An electromagnetic device that moves a plunger up or down when electric current is sent through a coil that surrounds the plunger. Solenoids are used to operate the hydraulic control valves in ABS.

Solid Brake Drum: A drum made entirely of cast iron.

Solid Brake Rotor: A rotor with solid metal between its friction surfaces.

Solvent: A liquid capable of dissolving another substance.

Specific Gravity: The ratio comparing the mass of a solid or liquid to that of an equal volume of water.

Speed Nuts: Spring-steel clips used to hold floating drums and rotors in place during vehicle assembly.

Split Point: The pressure at which a brake proportioning valve begins to limit hydraulic pressure to the rear brakes.

Spool Valve: A sliding valve that uses lands and valleys machined on its surface to control the flow of hydraulic pressure through ports in its bore.

Spot Magnetic Electric Brake: An electric brake design in which the electromagnet is a small disc (spot) attached to an actuating lever.

Static Friction: The amount of friction that exists between two surfaces at rest.

Straight-Air System: An air brake system that uses an air chamber at each wheel friction assembly to apply the brakes through a mechanical lever and cam arrangement.

Swaged: A method of locking a part in place by permanently deforming a portion of it or the surrounding material.

Swept Area: The area of a brake drum or rotor that contacts the brake linings.

Synthetic Lining: Nonasbestos, nonmetallic brake linings reinforced with fiberglass or aramid fiber.

Tandem Booster: A vacuum power booster that uses two diaphragms to increase brake application force. Tandem boosters are smaller in diameter than single-diaphragm boosters.

Taper Variation: A difference in the thickness of a brake rotor measured at the inner and outer edges of its friction surfaces. Most rotors with taper variation are thinner at the outer edge.

Taper Wear: A type of brake drum wear in which the closed edge of the friction surface is worn more than at the open edge.

Thermodynamics: The area of the physical sciences that deals with the interactions of heat energy and mechanical energy.

Throttle Relaxer: A device used with some traction control systems. It decreases the engine's throttle opening when power output needs to be reduced to control wheel spin when accelerating.

Tire Contact Patch: The area of tire rubber that actually touches the road at any one time; also called the tire footprint.

Tire Slip: The difference between vehicle speed and the speed at which the tire tread moves along the pavement. Tire slip is commonly expressed as a percentage.

Torque: The turning or twisting force applied at the end of a rotating shaft.

Traction: The amount of grip between the tire tread and the road surface. Higher traction allows greater braking and cornering force to be generated.

Trailing Shoe: Any shoe in a non-servo brake that is deenergized by drum rotation.

Turning: The process of using a brake lathe to cut metal from drums and rotors to refinish their friction surfaces.

Undersize Grind: A type of brake shoe arcing that produces a lining with a constant thickness ground to a radius slightly smaller than that of the brake drum.

Uni-Servo Brake: A drum brake that has servo action in the forward direction only.

Unsprung Weight: The weight of any suspension and brake components not supported by the vehicle's springs. High unsprung weight makes suspension movement more difficult to control.

Vacuum: Technically, a complete absence of pressure (0 psi), although the term is commonly used to describe any pressure less than atmospheric pressure.

Vacuum Servo: A flexible diaphragm with a linkage attached to it installed in a sealed housing. When vacuum is applied to one side of the diaphragm, atmospheric pressure on the other side moves the diaphragm and linkage to perform work.

Vacuum-Suspended Power Chamber: A booster power chamber that has vacuum on both sides of its diaphragm when the brakes are not applied.

Vapor Lock: The condition in which brake fluid boils into a compressible vapor causing a loss in braking performance.

Vented Brake Rotor: A rotor with cooling passages between its friction surfaces.

Viscosity: A liquid's resistance to flow. Viscosity usually varies with temperature.

Water Fade: A delay in brake application caused by water contamination that reduces friction between the brake linings and drum or rotor.

Ways: Special sliding surfaces machined into the anchor plate and caliper body where these parts of a sliding caliper make contact and move against one another.

Weight Bias: An element of vehicle design that results in either the front or rear suspension having to support more than half of the vehicle's weight. Most cars have a forward weight bias.

Weight Transfer: The shift of weight toward the front of a vehicle that occurs when the brakes are applied while driving forward.

Wheel Cylinder: The device in a drum brake that converts hydraulic pressure back into the mechanical force used to apply the brake shoes against the drum.

Wheel Friction Assemblies: The axle-mounted components of a brake system that create the friction necessary to stop a vehicle.

Wheel Offset: The distance between the rim centerline and the mounting plane of the wheel.

Wheel Speed Sensor: A magnetic or electronic sensor that generates a wheel speed signal for the ABS. The frequency of the signal is proportional to the speed of the wheel.

Work: The transfer of energy from one system to another, particularly through the application of force.

Yaw Movement: The tendency for the vehicle to rotate around its center of gravity.

Index

ABS control module, 279–281
Accumulator, 260, 285–286
Active digital wheel speed sensors, 277, 278
Actuating system, 11
Adjustable brake pedals, 79–80
Adjusting link, 147–148
Adjusting mechanism, 139–140
Air gap, 278
Air-hydraulic systems, 16
Air pressure, 245–248
Alfa Romeo, 147, 164
Aluminum pistons, 127
Anchor eyes, 180
Anchor plate, 123, 170
Annular electric brake, 16
Antilock brakes (ABS), 17–20
    ABS components, 17–19
    ABS functions, 19–20
    advanced functions, 286–289
    characteristics of, 269–272
    components, 276–286
    operation, 272–273
    origins, 272
    system configurations, 273–276
    and tire diameter, 375
Aramid brake linings, 188
Aramid fiber, 188
Archimedes, 43
Arcing, 193–194
Asbestos exposure, 27–32
    asbestosis, 28, 29
    cancer, 29
    effects, 28
    precautions, 30–31
    sources, 28
    standards, 27–28
    waste disposal, 31–32
Asbestosis, 28, 29
Aspect ratio, 373
Atmospheric pressure, 245

Atmospheric-suspended power chamber, 249
Automatic adjusters, 150–156
Axial cooling fins, 200

Backing plate, 10
    of drum brakes, 140–141
Banjo fittings, 70–71
Barrel wear, 208
Base brakes. *See* Service brakes
Bellmouth, 211
Bendix, 274, 275
    ABS, 293–304
Bendix Hydro-Boost, 256, 260
Bimetallic brake drum, 202
Binders, 184
Body control module (BCM), 225
Bonded linings, 191
Booster holding position, 253
Booster vacuum runout point, 253
Bosch. *See also* Robert Bosch Company
    ABS, 304–319
Brake adjusters, 150
Brake balance, 102
Brake band, 4
Brake bleeding, 64–66
Brake block, 184
Brake-by-wire, and ABS brakes, 20
Brake calipers, 10, 122–133
    body of, 123–125
    brake fluid routing, 125
    dust boots, 131–132
    operation of, 132–133
    pistons, 126–128
    piston seals, 128–131
Brake disc, 167
Brake drum, 144
Brake drum and rotor damage, 208–210

Brake drum and rotor distortion, 210–214
Brake drum and rotor wear, 207–208
Brake drum construction, 200–202
Brake dust, 28
Brake fade, 52–54, 139
Brake fluid, 58
    brake bleeding, 64–66
    changes, 66–67
    specifications, 58–61
    storage and handling, 63–64
    types, 61–63
Brake fluid reservoir, 82–84
Brake fluid routing, 125
Brake hoses, 68–70
Brake lathe, 214
Brake light switch operation, 114
Brake line fittings, 70–73
Brake lines, 8, 67–73
    brake hoses, 68–70
    brake tubing, 67–68
    fittings, 70–73
    routing, 73
Brake linings, 2–3
    assembly methods, 190–192
    material coefficient of friction, 188–190
Brake operation
    brake fade, 52–54
    energy principles of, 38
    friction principles, 48–51
    hydraulic principles, 42–48
    inertia, 40–41
    kinetic energy and, 38–40
    mechanical principles, 41–42
Brake pads, 5, 167
    construction, 182–184
    friction materials, 184–188
Brake pedal freeplay, 81–82
Brake pedal ratio, 80

395

Brake repair
  and asbestos exposure, 27–32
  and chemical poisoning, 32–33
  and health, 26–27
  and health care rights, 33–34
  and the law, 26
Brake rotor, 167–168
Brake rotor construction, 202–205
Brake shoes, 2, 142–143
  construction of, 180–181
  friction materials, 184–188
Brake shoe-to-drum fit, 192–194
  proper, 193
Brake shrouding, 379–380
Brake standards, federal, 24–26
Brake systems
  antilock brakes, 17–20
  and asbestos exposure, 27–32
  designs, 3–6
  early brake designs, 2–3
  and health, 26–27
  methods of brake actuation, 11–16
  operation of, 6
  parking brakes, 10–11
  service brakes, 6–8
  wheel friction assemblies, 8–10
Brake system switches, 108–109
Brake tubing, 67–68
Braking
  and suspension, 380
  and tires, 372
  and wheel bearings, 380–382
  and wheel runout, 379
  and wheels, 377–379
Brazing, 192
Bridge bolts, 123

Caliper-actuated disc parking brakes, 234–237
Cam, 107
Cam manual adjusters, 156
Caster, 380
Cast-iron pistons, 127–128
Center port master cylinder, 95
Centrifugally cast brake drum, 202
Chemical poisoning, 32–33
  effects of, 33
Chrysler Corporation, 14, 128, 164, 272, 274. *See also* DaimlerChrysler
Citroen, 80
Clevis, 81

Coefficient of friction, 48–50
Combination master cylinder, 95
Combination valves, 108
Compensating port, 90
Composite brake drum, 201–202
Compression fittings, 70–71
Controller, 16
Control module, 273
Conventional dual-piston master cylinder, 89–92
Core charge, 181
Corrosion, 60
Cup seals, 86
Curing agents, 185

DaimlerChrysler, 108, 277. *See also* Chrysler Corporation
Dayton Engineering Laboratories Company. *See* Delco
Delco, 273, 274, 275
Delphi, 275
Delphi Chassis (Delco Moraine), ABS, 319–333
Diaphragm, 83
Discard diameter, 216–217
Discard dimension, 217
Disc brakes, 5–6, 10
  advantages, 162–165, 165–168
  in aircraft, 174
  design, 168–172
  inboard, 174
  rear, 173–174
Dished rotor, 208
Double-leading brake, 146
Double-trailing brake, 145–146
Driveline auxiliary parking brakes, 237–239
Drum and rotor metal removal limits, 216–217
Drum and rotor mounting methods, 205–207
Drum and rotor refinishing, 214–216
  special considerations, 217–219
Drum brakes, 9–10
  advantages of, 137–138
  automatic brake adjusters, 150–156
  brake adjusters, 150
  construction of, 140–144
  design of, 144
  disadvantages of, 138–140
  dual-servo brakes, 147–150

  manual brake adjusters, 156–157
  non-servo brakes, 144–146, 147
  uni-servo brakes, 150
Drum-in-hat, 173, 233
Drum parking brakes, 231–233
  integral, 232
  rear-disc auxiliary, 232–234
Dry boiling point, 59
Dual-circuit brake systems, 87–89
Dual master cylinder, 7–8, 87
Dual-power brake systems, 265
Dual-servo brakes, 147–150
  construction of, 147–148
  operation of, 148–150
Duesenburg, A., 14
Duesenburg, F., 14
Dunlop Tire, 164, 272
DuPont, 188
Dynamic proportioning, and ABS brakes, 19

Eccentric, 82
Eccentric distortion, 212
Economic Commission for Europe (ECE), 24
Edgebrand, 189
Effective pedal travel, 80
Elastic limit, 211
Electric actuation, of brakes, 16, 17
Electro-hydraulic (E-H) boosters, 261
Electronic brake control module (EBCM), 112, 277, 278
Emergency brakes, 11
Energy, 38
Environmental Protection Agency (EPA), 27
Equalizer, 228
Equilibrium Reflux Boiling Point (ERBP), 59
European Union (EU), 24
External contracting-band brakes, 3–4

Fade resistance, in brake design, 162–165
Federal Motor Vehicle Safety Standard. *See* FMVSS
Ferodo, 30
Ferrari, 164
Fiberglass brake linings, 188
Fillers, 184–185

Index

Fixed caliper design, 168–169
Fixed-piston seals, 129–131
Flare fittings, 71–73
Flash codes, 281
Floating and sliding caliper design, 169–171
    advantages, 170–171
    disadvantages, 171
Floating calipers, 171
Fluid compatibility, of brake fluid, 60–61
Fluid level switch, 111–112
    operation of, 112
FMVSS 105, 161
FMVSS 106, 68–70
FMVSS 116, 60
FMVSS 135 brake test, 24–26
Ford Motor Company, 61, 123, 164, 216–217, 272
Foundation brakes. *See* Service brakes
Four-channel ABS, 273
Frame-mounted brake pedals, 78–79
Freeplay, 81–82
Friction material, 184–188
Friction modifiers, 184
Friction principles
    coefficient of friction, 48–50
    friction and heat, 51–52
Front-wheel-drive (FWD), 87, 103, 107, 205
Front-wheel parking brakes, 236
Frood, H., 30
Fusing, 192

Gas fade, 54, 139, 164
General Motors (GM), 108, 235, 274, 277
    Powermaster brake booster, 261–264
Glazed lining, 192
Gravity bleeding, 65
Grinding, 215
Grommets, 82
Gross vehicle weight rating (GVWR), 24
Ground, 112

Hard spots, 210
Health care rights, 33–34
Heat checking, 209
High Efficiency Particulate Air (HEPA), 31
Hispano-Suiza, 9
Homogeneous, 61
Honda, 276
Hybrid pad sets, 186
Hydraulic actuation, of brakes, 13–14
Hydraulic boosters, 256
Hydraulic principles, 42–48
    and brake design, 48
    constancy of pressure, 44
    hydraulic pressure and piston size, 44–47
    noncompressibility of liquids, 43–44
Hydraulic system mineral oil (HSMO), 62–63
Hydraulic valves, 100
Hydroplane, 375
Hygroscopic, 59

Inboard disc brakes, 174
Inertia, 40–41
Insulator, 126
Integral, 274–276
Integral vacuum boosters, 250–256
Intercable adjuster, 231
Intermediate lever, 228
Internal expanding-band brakes, 4–5
Internal expanding-shoe brakes, 5
International Organization for Standardization (ISO), 70
    flare fittings, 72–73
Isolation solenoid, 282
Isotta-Fraschini (I-F), 9

Jaguar, 164
Jamnuts, 231

Kelsey-Hayes, 275
    ABS, 333–344
Kevlar, 188
Kinetic energy, 38–40
    and brake design, 40
    weight and speed effects, 39–40
Kinetic friction, 50–51
Knockback, 129

Lateral runout, 212–213, 377
Lathe-cut, 129
Leading shoe, 144
Leading-trailing brake, 146
Leverage, 41
Lever-latch automatic adjuster, 154
Lightly loaded vehicle weight (LLVW), 24
Lining fade, 53–54, 139, 163
Lining table, 180
Linkage cables, 227–228
Linkage design, 229–231
Linkage rods, 226–227
Liquids, noncompressibility of, 43–44
Loaded runout, 377

Magneto-resistive wheel speed sensors, 277, 278–279
Mag wheels, 190
Manual adjusters, 156–157
Manual bleeding, 64–65
Master cylinder, 7–8
Master cylinder body, 84–85
Master cylinder construction, 82–87
    brake fluid reservoir, 82–84
    master cylinder body, 84–85
    master cylinder pistons, 85
    piston seals, 85–87
Master cylinder operation, 89–95
    center port master cylinder, 95
    combination master cylinder, 95
    conventional dual-piston master cylinder, 89–92
    portless master cylinders, 94–95
    quick-take-up master cylinder, 92–94
Master cylinder pistons, 85
Mechanical actuation, of brakes, 12–13
Mechanical advantage, 42
Mechanical compatibility, of brake fluid, 60
Mechanical fade, 53, 139, 163
Mechanical-hydraulic boosters, 256–261
Metallic brake linings, 186
Metering valves, 101–104
    and brake bleeding, 103–104
    operation of, 102–103
    systems without, 103
Micro-inches, 219
Model A Duesenburg, 14
Mold bonded linings, 192
Multiplier vacuum boosters, 256

Nibs, 180
Nippondenso, 275
  ABS, 344–345
Nondirectional brake, 146
Non integral, 274–276
Non-servo brakes, 144–146, 147
  specific, 145–146

Occupational Safety and Health Administration (OSHA), 27, 30, 31, 33
One-channel ABS, 274
Organic linings, 184–186
Out-of-round, 211
Oversize brake shoes, 194
Over-travel spring, 152
Oxidation, 60

Pad wear indicators, 183–184
Parallelism, 213–214
Parking brake automatic adjusters, 153
Parking brake linkage, 143–144
Parking brakes, 10–11
  caliper-actuated disc, 234–237
  driveline, 237–239
  drum, 231–233
  front-wheel, 236
  linkages, 226–231
  pedals, levers, and handles, 224–226
  warning lights and switches, 226
Particulates, 63
Pawl, 151
Pedal assemblies, 7, 78–82
  adjustable brake pedals, 79–80
  brake pedal freeplay, 81–82
  brake pedal ratio, 80
  frame-mounted brake pedals, 78–79
  suspended brake pedals, 79
Pedal ratio, 80
Phenolic pistons, 128
Piston bores, 123–125
Piston seals, 85–87
Piston size, and hydraulic pressure, 44–47
Piston stops, 141
Pneumatic actuation, of brakes, 14–16
Polyglycol brake fluid, 61
Porsche, 164
Portless master cylinders, 94–95

Power booster, 7
Power brakes
  air pressure, 245–248
  dual-power brake systems, 265
  electro-hydraulic (E-H) boosters, 261
  hydraulic boosters, 256
  integral vacuum boosters, 250–256
  mechanical-hydraulic boosters, 256–261
  multiplier vacuum boosters, 256
  need for, 244
  Powermaster brake booster, 261–264
  vacuum booster theory, 248–250
  ways to increase braking power, 244–245
Power chamber, 248
Powermaster brake booster, 261–264
Pressure, constancy of, 44
Pressure bleeding, 65
Pressure differential, 245
Pressure differential switch, 109–111
  operation of, 110–111
  resetting, 111
Primary shoe, 148
Proportioning valve, 104–108
  adjustable, 106
  and brake bleeding, 108
  electronic brake proportioning, 108
  height-sensing, 107–108
  operation of, 105–107

Quick-take-up master cylinder, 92–94

Racing calipers, 127
Radial cooling fins, 200
Radial runout, 376
Ratchet automatic adjusters, 154–156
Rear-wheel-drive (RWD), 87, 103, 205
Red brake warning light (RBWL), 109
Release solenoid, 284
Replenishing port, 90
Reservoir covers, 83–84
Residual pressure, 100

Residual pressure check valve, 100–101
Resurfacing, 215–216
Riveted linings, 190–191
Robert Bosch Company, 274, 275. *See also* Bosch
Rotor, 5
Rotor distortion, 212–214
Rotor hat, 202
Runout, 129, 376–377

SAE, 58, 66, 70
  flare fittings, 71–72
Scan tool, 282
Scoring, 209
Scrub radius, 378
Secondary shoe, 148
Section width, 373
Select-low principle, 274
Self-energizing action, 144–145
Semimetallic linings, 186–187
Series, 120
Service brakes, 6–8
Servo action, 138
Servo brake, 147
Shoe anchors, 140–141
Shoe arcing. *See* Arcing
Shoe support pads, 141
Shoe web, 180
Silicone brake fluid, 61–62, 67
Sintering, 186
Sliding caliper, 172
Slope, 104
Society of Automotive Engineers (SAE). *See* SAE
Solenoid, 273
Solid brake drum, 201
Solid brake rotor, 203
Specific gravity, 62–63
Speed nuts, 206
Splash shield, 167
Split point, 104
Spool valve, 258
Spot magnetic electric brake, 16
Stability control, and ABS brakes, 19
Starwheel automatic adjusters, 150–153
Starwheel manual adjusters, 156
Static friction, 50–51
Steel pistons, 127–128
Stoplight switch, 112–114
Straight-air system, 15

Index

Stroking piston seals, 128–129
Strut-quadrant automatic adjuster, 155–156
Strut-rod automatic adjuster, 154–155
Sumitomo, 275
    ABS, 345–347
Supplemental brake assist (SBA), 247–248
Suspended brake pedals, 79
Suspension, and braking, 380
Swaged, 69
Swept area, 162–163
Synthetic lining, 187–188

Tandem booster, 251
Taper variation, 208
Taper wear, 207
Teves, 274, 276
    ABS, 347–360
Three-channel ABS, 273–274
Throttle relaxer, 288
Tire contact patch, 372
Tires
    and braking, 372
    construction, 372–373
    inflation, 376
    runout, 376–377
    size, 373–375
    tread design, 375–376
Tire slip, 270
Torque, 103
Toyota, 276
    ABS, 360–364
Traction, 270
Traction control, 286–288
    and ABS brakes, 19
Trailing shoe, 144
Turning, 214

Uni-servo brakes, 150
Unsprung weight, 126, 174

Vacuum, 245
    booster supply, 246–248
    measuring, 245–246
Vacuum bleeding, 65
Vacuum booster theory, 248–250
Vacuum servo, 225
Vacuum-suspended power chamber, 249
Vapor lock, 58
Vented brake rotor, 203

Water fade, 139, 164
Ways, 172
Wedge manual adjusters, 156, 157
Weight bias, 40–41
Weight transfer, 40
Wet boiling point, 59
Wheel bearings, and braking, 380–382
Wheel cylinders, 9, 119–122, 123, 141–142
    body of, 120
    cup expanders, 120–121
    designs, 121–122
    dust boots, 121
    operation of, 122
    pistons and seals, 120
Wheel friction assemblies, 8–10
Wheel offset, 377–378
Wheel runout, 379
Wheels
    and braking, 377–379
    rim width, 378–379
Wheel speed sensors, 272, 276–279

Yaw movement, 288